Echinoderm Research and Diversity in Latin America

Juan José Alvarado
Francisco Alonso Solís-Marín
Editors

Echinoderm Research and Diversity in Latin America

Springer

Editors
Juan José Alvarado
Centro de Investigaciónes en Ciencias del Mar y Limnologia
Universidad de Costa Rica
San José
Costa Rica

Francisco Alonso Solís-Marín
Instituto de Ciencias del Mar, y Limnologia
Universidad Nacional Autónoma de México
México City
Mexico

ISBN 978-3-642-20050-2 ISBN 978-3-642-20051-9 (eBook)
DOI 10.1007/978-3-642-20051-9
Springer Heidelberg New York Dordrecht London

Library of Congress Control Number: 2012941234

Printed on acid-free paper

Springer is part of Springer Science+Business Media (www.springer.com)

We would like to dedicate this book to the pioneers of echinoderms research in Latin American. To María Elena Caso, Irene Bernasconi, Luis Roberto Tomassi and Evelyn Zoppi de Roa.

Foreword

This is a remarkable book. First and foremost, it compiles an immense amount of literature on all aspects of echinoderm biology and ecology over an extremely broad geographical range. This will be extremely useful because much of the literature is relatively unknown to the world at large. The description of the oceanography and marine environments of each country in a comprehensive way will be of considerable interest to marine biologists in general.

One of the most interesting aspects of this book is the section devoted to the history of echinoderm studies in each country. The first reports about echinoderms were isolated and sporadic, not systematic studies. These generally did not begin until the 1850s with various foreign expeditions along the coasts of Latin America. Studies of echinoderms by workers in Latin American countries themselves did not begin to develop appreciably until the 1950s. This has changed in a dramatic way. In the latter part of the twentieth century, interest in marine biology and recognition of its importance in terms of basic science, the environment and conservation, and economics increased greatly. This led to an increase in the number of marine biologists in general and of echinoderm biologists in particular. The amount of research activity that has developed since then is impressive. Another thing that impressed me, apparent in the book, is the passion of Latin American biologists for echinoderms, their commitment to conservation, their desire to assist their countries through their efforts, and their collegiality.

Florida, USA, March 2012

John M. Lawrence
Department of Integrative Biology,
University of South Florida,
Tampa

Foreword

Juan José Alvarado is one of those students who, from very early on, denoted a clear interest in Marine Science and at some point defined his area of interest. When he started looking at the echinoderms of Costa Rica and later, at those reported and collected from Coco Island National Park, he found what he was looking for. Since then, he has been actively carrying out research on echinoderms, and most important, he has drawn attention to the activities regarding this group that are taking place in Latin America. First, with the publication of *Research on Echinoderms in Latin America* in 2005 (Revista de Biología Tropical, Volume 53, Supplement 3: 387 p.). Then, with the creation of the Iberoamerican Echinoderm Network Red Iberoamericana de Equinodermos (RIE) in 2006. And in 2008, with the publication of another Special Issue on echinoderm research in Latin America: Revista de Biología Tropical, 56 (Suppl. 3): 360 p. Now, together with Francisco Solís-Marín from the Universidad Nacional Autónoma de México (UNAM), they have put together an outstanding compilation of echinoderm research in Ibero-america. The dedication of this book to the pioneering Latin American echinoderm researchers is a tribute to their great work and an acknowledgment of the fruit their early and many times solitary work has borne. It is a great honor for me to write this Foreword, for I consider their work a significant step for biological research in the region..

San José, Costa Rica, March 2012

Jorge Cortés
Centro de Investigación en Ciencias
del Mar y Limnología (CIMAR),
Universidad de Costa Rica

Preface

The Iberoamerican Echinoderms Network (Red Iberoamericana de Equinodermos) (RIE) (http://zicatela.umar.mx/ ~ redequinos/2/) was created in August 2006 at the 12th International Echinoderms Congress in New Hampshire, United States, as an initiative to strengthen academic connections between Iberoamerican echinoderm researchers. The RIE seeks to fortify the links of researchers in the different countries of the region, being a bridge of communication and exchange of ideas. It seeks to be the platform for discussion and communication of priority projects for Latin America echinoderms. It seeks to offer the opportunity of collaborative work, to place the region among the principal research and conservation regions of the world. Likewise, the RIE intends to support students interested in echinoderm research. To enrich their formation, it will offer information about researchers who could help in their projects, courses, seminars, or symposia, as well as providing literature. The goal of the RIE is to develop workshops, congresses, special journals supplements and books, and to support the development of the knowledge of the echinoderms in Latin America.

The first effort of the RIE was the preparation and publication of a Special Issue of the Revista de Biología Tropical (International Journal of Tropical Biology and Conservation) (Alvarado, J. J. and Cortés, J. (Editors). 2005. Research on Echinoderms in Latin America, Volume 53, Supplement 3: 387 p.) (http://www.ots.ac.cr/tropiweb/intpages/suppl/sup53-3.html). This initiative was born during the 11th International Echinoderms Congress in Münich, Germany, in 2003. This Special Issue compiled 28 papers from 12 countries by 62 authors, touching topics as diverse as paleontology in the Uruguayan region and fisheries models of sea cucumbers in Baja California. The second output of the RIE was a second supplement of the Revista de Biología Tropical (Alvarado, J.J. and Cortés, J. (Editors). 2008. Research on Echinoderms in Latin America II, Volume 56, Supplement 3: 360 p.) (http://www.ots.ac.cr/tropiweb/intpages/suppl/sup56-3.html). This issue compiled 21 papers from eight countries by 46 authors in Latin America (including an invited paper from the Canary Islands). Those papers demonstrated the high degree of knowledge of some countries in topics such as diversity, ecology, aquaculture, and fisheries. It also identified countries where more efforts are needed.

The third effort was the establishment of the first Latin American Echinoderm Congress (Congreso Latinoamericano de Equinodermos-CLE) held in Puerto Madryn, Argentina, between 13 and 18 November 2011. This congress was organized by the Centro Nacional Patagónico (CENPAT), the Museo Argentino de Ciencias Naturales "Bernardino Rivadavia" (MACN), and the Universidad Nacional de la Patagonia San Juan Bosco. A total of 112 people participated in this event with 76 oral presentations and 54 Poster, covering the following themes: biogeography, biochemistry, ecology, physiology, paleontology, fisheries and aquaculture, reproduction and development, and taxonomy, systematics, and evolution.

This book represents the fourth output and the confirmation of the commitment of the RIE with the international scientific community. The book represents an extensive recompilation of information about all the research done in the region from theses, reports, books, and scientific journals. We were able to identify the strengths and weaknesses of each country and also allow the international scientific community to be able to access information that was held in reports, theses, and local papers that were written in Spanish, allowing a greater diffusion of the results and conclusions obtained. Our goal is to improve the echinoderm research on the region and with this road map establish future research in collaboration with the different laboratories and researchers that participated in this book.

Juan José Alvarado
Francisco Solís-Marín

Red
Iberoamericana
de Equinodermos

Acknowledgments

First of all, we thank each of the authors of the chapters that have made this compilation about Echinoderm Research and Diversity in Latin America possible, to share their knowledge and passion for such a star group. Muchas gracias!

We also would like to acknowledge the help and support of all the reviewers who with their comments and suggestions improved this book: Chris Pomory, Cynthia Lara de Castro, David Pawson, Harilaos Lessios, John Pearse, Manuel Rey Méndez, Marc Eléaume, Martin Brogger, Nieves Elvira, Pablo González, Philip Lambert, Rosa del Valle, Tim O'hara, Tom Hopkins. We wish to grant special credit to John Lawrence and Jorge Cortés who with all the patience, interest, and care reviewed each of the chapters, leaving their wisdom on every page.

We wish to grant special credit to all the students from the Laboratorio de Sistemática y Ecología de Equinodermos from the Instituto de Ciencias del Mar y Limnología, Universidad Nacional Autónoma de Mexico, for their help with the final elaboration of the appendix: Mauricio Valdés De Anda, Julio Adrián Arriaga Ochoa, Andrea Alejandra Caballero Ochoa, Tania Pineda Enríquez, Carolina Martín Cao-Romero, Yoalli Quetzalli Hernández Díaz, Alejandra Martínez Melo, Viridiana Tapia Ramírez, Guadalupe Bribiesca Contreras, Pedro Josué Garcés Solchaga, Nancy Escandón Flores, Lucia Alejandra Hernández Herrejón and Carlos García Linares.

We appreciate the help of Monica Chavez in preparing all the figures in this book and Cindy Fernandez who always believed in this project and was a keystone for the final elaboration. JJ Alvarado is grateful to CONACYT (México) and CONICIT-MICIT (Costa Rica). Finally, we thank Marion Schnider and SPRINGER for their interest in this book.

Contents

Chapter 1
Echinoderm Research and Diversity in Latin America

Juan José Alvarado and Francisco Alonso Solís-Marín

1.1 Introduction

The phylum Echinodermata consists of approximately 7,000 living species and 13,000 extinct species (Pawson 2007). Studies on echinoderms have focused on experimental ecology due to their important role in marine ecosystems, with several species identified as "keystone species" (Uthicke et al. 2009). These species can have effects on community structure and function through several trophic levels and on a wide geographic scale (Birkeland 1989). However, growing global pressures resulting from fishing echinoderms have put them under threat (Micael et al. 2009).

Latin America (LA) comprehends all the countries in America that speak Spanish or Portuguese; including the French speaking country of Haiti, that is located in Hispaniola Island, shared with Dominican Republic. LA includes Mexico, Central America, and the islands of the Caribbean and South America. However, the term has been used with different meanings that have created some confusion (Colburn 2002). In total it includes 22 countries with an area of

J. J. Alvarado (✉)
Centro de Investigación en Ciencias del Mar y Limnología (CIMAR),
Universidad de Costa Rica, San Pedro, 11501-2060, San José, Costa Rica
e-mail: juanalva76@yahoo.comjuan.alvarado@ucr.ac.cr

J. J. Alvarado
Posgrado en Ciencias Marinas y Costeras, Universidad Autónoma
de Baja California Sur, La Paz, México

F. A. Solís-Marín
Colección Nacional de Equinodermos "Ma. E. Caso Muñoz",
Laboratorio de Sistemática y Ecología de Equinodermos, Instituto de Ciencias
del Mar y Limnología (ICML), Universidad Nacional Autónoma de México (UNAM),
Apdo. Post 70-305, 04510, México, DF, Mexico
e-mail: fasolis@cmarl.unam.mx

J. J. Alvarado and F. A. Solís-Marín (eds.), *Echinoderm Research and Diversity in Latin America*, DOI: 10.1007/978-3-642-20051-9_1,

21,069,501 km^2, and according to the Sea around Us project. It possesses 17,010,693 km^2 of exclusive economic zones and 2,706,433 km^2 of shelf area (www.seaaroundus.org). LA is a megadiverse region, including six of the 17 megadiverse countries around the world: Brazil, Colombia, Ecuador, Mexico, Peru and Venezuela (Michán and Llorente-Bousquets 2010). However, scientific production in terms of science and engineering is considerably lower (3.2 %) in 2000 than in the United States and Canada (36.7 %) (Hermes-Lima et al. 2007).

This book presents a change in the research paradigm of echinoderms in this region. It gives an exhaustive review of all studies that has been done. All chapters have been written by local scientists. It compiles reports, thesis, and papers in regional and international journals. John Pearse (personal communication) noted that the common opinion that the echinoderms knowledge in LA was "scarce with information often fragmented, but quite the contrary, showing that there is an impressive amount of work". Until today the greatest regional reviews about echinoderms were the compilation by Linda Yvonne Maluf (1988) on composition and distribution along the Eastern Pacific and the masterpiece presented by Clark and Downey (1992) about taxonomy and distribution of the starfishes of the Atlantic. The contribution by Hendler et al. (1995) about the echinoderms in the Caribbean Sea not only compiled taxonomic and distribution information, but also provided ecological data and high quality pictures of the species and their identifiable features.

This book is divided in 16 chapters (including this one) and two taxonomic lists (Appendix) (one from the Pacific Ocean and one from the Caribbean Sea and Atlantic Ocean including the Canary Islands) with information about presence/absence, bathymetric range and the type of habitat or substrate where the species have been found. These taxonomic lists were built following the criteria of Clark (1931, 1947) for Crinoidea; Clark (1989, 1993, 1996), Clark and Downey (1992), Clark and Mah (1999) for Asteroidea; Matsumoto (1915), Stöhr and O'Hara (2007) for Ophiuroidea; Smith (1984) for Echinoidea and Panning (1929, 1934a,b, 1935), Pawson and Fell (1965) for Holothuroidea. The World Register of Marine Species (Appeltans et al. 2011) was also consulted for all species. From those taxonomic lists we present only the species that are fully identified, avoiding the presence of "sp." reports. However, in each chapter the authors included "sp." reports in their species numbers. Fourteen chapters deal with studies done in countries, regions or islands (Fig. 1.1): Mexico, Central America (Belize, Guatemala, Honduras, El Salvador, Nicaragua, Costa Rica) and Panama, Colombia, Ecuador, Venezuela, Peru, Brazil, Uruguay, Argentina, Cuba, Hispaniola Island (Haiti and Dominican Republic), Puerto Rico and the Canary Islands. The final chapter has a biogeographical and diversity analysis of the two taxonomic lists. Unfortunately a chapter about Chilean Echinoderms could not be included in this book, but the Pacific taxonomic list includes all the species present in that country plus Easter Island.

Each country chapter contains six sections. 1. An introduction with a brief description of: the marine environments, length of the coast, continental shelf area and oceanographic conditions. 2. A historic account of the investigations that have

Fig. 1.1 The American continent with the countries participating on the book indicated in colors

been done on echinoderms with information on the first expeditions and collections, as well as ecology, reproduction, paleontology, bioerosion, competition, among others studies. 3. Distributional patterns and diversity. 4. If specific fishery or aquaculture activity exists, this is explained in a section, giving the name of the species extracted, numbers, and seasons of extraction, regulations, aquaculture activities, marketing and the benefits to local communities from these activities. 5. A section on threats to echinoderms, explaining the main environmental and anthropogenic threats that are affecting the echinoderm populations and how. It also examines whether some echinoderms are affecting the environment (overpopulations, extreme grazing, bioerosion). 6. The final section contains general conclusions from the review and recommendations for future research to meet the needs of the country.

This book covers a large part of America, excluding only Canada, the United States, the Lesser Antilles, Guyana, Surinam and French Guiana. For the latter two countries and the French overseas territory, information on echinoderms is available from the publications by Cherbonier (1959a, b, c, d, e), Walenkamp (1976, 1979), Jangoux (1978) and Le Loeuff and Von Cosel (2000).

1.2 Echinoderm Research in LA: An Overview

Latin American echinoderm studies started in the 13th century. Bernardino de Sahagún (1499–1590) a Spanish Franciscan friar who wrote about the Aztec culture in Mexico (*The General History of the Things of New Spain*) before the Spanish conquerors arrived in Central America (Florentine Codex). He describes and explains ancient native religion, beliefs, practices, deities of the Aztecs. His magnificent document was prepared in 12 books. Book number 11 describes the "Earthly Things" (properties of animals, birds, fish, trees, herbs, flowers, metals, stones, and colors). In this volume he draws, describes and explains some invertebrate forms used by the Aztecs during their daily life, including some sea urchins, starfish and probably sea cucumbers. This is probably the earliest record of echinoderms in the Americas.

Louis Éconches Feuillée (1660–1732), a French member of the Order of the Minims, was an explorer, geographer, and botanist who travelled to Chile, Argentina and Peru on different expeditions from 1709 to 1711. He gave the first echinoderm references for these Latin American countries. The first formal taxonomic description of a Latin American echinoderm was made by the famous naturalist Carl Nilsson Linnæus (1707–1778) in his 1758 publication of Systema Naturae. In this classification of all living organisms, he formally described several echinoderm species from Latin American coasts (i.e. *Oreaster reticulatus*, *Echinometra lucunter* and *Clypeaster rosaceus*). Jean-Baptiste Pierre Antoine de Monet de Lamarck (1734–1829), a French naturalist and an early proponent of the idea that evolution occurred and proceeded in accordance with natural laws, published in 1816 one of his masterpieces, "Histoire Naturelle des Animaux sans Vertébres". He reported new echinoderm species collected in LA. A contemporary of Lamarck was the Chilean Juan Ignacio Molina (1740–1829) better known as "Abate Molina". In 1782 he published his work "Saggio sulla Storia Naturale del Chile" where he described two new species of sea urchins from Chilean waters, Echinus albus (=*Loxechinus albus*), and *Tetrapygus niger*.

Jean-Louis-Rodolphe Agassiz (1807–1873), a Swiss paleontologist, geologist and a prominent zoologist, was a professor at Harvard University in the United States. In several papers he described echinoid fauna from the LA. Rudolf Amandus Philippi Krumwiede (1808–1904) was a German-Chilean paleontologist and zoologist who described and published new and already known echinoderm species from Chile. His most "famous" species is probably the sea urchin *Diadema antillarum*, from the Caribbean, which he described in 1845. Theodore

Lyman (1833–1897), an U.S. zoologist who studied under the supervision of Professor Louis Agassiz. He worked many years at the Museum of Comparative Zoology of Harvard University and described many species of Latin American echinoderms, mainly ophiurans. Alexander Emanuel Rodolphe Agassiz (1835–1910), a Swiss-born U.S. marine zoologist and oceanographer, was the son of Louis Agassiz. He emigrated in 1849 to the United States where he conducted significant systematic zoological work on Latin American echinoderms (e.g. starfish and echinoids).

Rafael Arango y Molina (1837–1893), a Cuban zoologist who wrote the first echinoderm records of the archipelago, worked together with L. F. Pourtalés (1824–1880), a Swiss who also worked on LA echinoderms and visited Cuba around 1867.

The transition from 19th to 20th centuries was characterized in general by large expeditions that reported new discoveries on the echinoderm fauna of LA. These included studies by several of the most prominent echinoderm researchers such as Addison Emery Verrill (1839–1926), Theodoro Mortensen (1868–1952), Hubert Lyman Clark (1870–1947), Austin Hobart Clark (1880–1954), Elizabeth Deichmann (1896–1975), María Codocedo Rojas (1909–1989), Irene Bernasconi (1896–1989), Gustave Cherbonier (1909–1995), María Elena Caso Muñoz (1915–1991), Maureen E. Downey (1920–2000), Ailsa McGown Clark (1926), Ignácio M. Brito (1927–2001), Patricio Sánchez Reyes (1928–1999), Cinthya Anne Gust Ahearn (1952–2008) and Isabel Pérez Farfante de Canet (1916–2009).

In this century, many active echinoderm specialists still work on the Latin American fauna: Luiz Roberto Tommasi (Universidade de São Paulo, Brasil), David L. Pawson (Smithsonian Institution, Washington, D.C, U.S.A.), Claude Massin (Institut Royal des Sciences naturelles de Belgique, Brussels, Belgium), John M. Lawrence (University of South Florida, U.S.A.), Alfredo Laguarda Figueras (Instituto de Ciencias del Mar y Limnología, Universidad Nacional Autónoma de México), Gordon Hendler (Los Angeles County Museum, Los Angeles California, U.S.A), Rich Mooi (California Academy of Sciencies, California, U.S.A.), Alberto Larraín Prat (Universidad de Concepción, Chile), Harilaos A. Lessios (Smithsonian Tropical Research Institute, Panama) and many others who are still actively increasing our knowledge of this fascinating invertebrate group.

1.3 General Results

Most Latin American countries have a common beginning in echinoderm research, with the presence of European and United States expeditions, followed by a period of research by regional scientists in the middle of the twentieth century. Researchers like María Elena Caso, Irene Bernasconi and Luiz Roberto Tomassi are prominent. Most of the specimens collected by these expeditions are in foreign museums in the United States (Los Angeles County Museum of Natural History,

Museum of Comparative Zoology at Harvard University, California Academy of Sciences Natural History Museum, Smithsonian Institution National Museum of Natural History, The Scripps Institution of Oceanography Benthic Invertebrate Collection), France (Museum National d'Histoire Naturelle), Germany (Museum für Naturkunde) and Denmark (Zoologisk Museum-Staten Naturistorike Museum). However the Coleccción Nacional de Equinodermos Ma. Elena Caso Muñoz at the Instituto de Ciencias del Mar y Limnología, Universidad Nacional Autonoma of Mexico, holds one of the largest collection of Latin American specimens with more than 81,000.

During the last three decades each country has set its own path, being specialized in one or two areas. The common line of research is basic ecology (distribution, population sizes, and diversity) and taxonomy. However some countries stand out in particular lines of research: physiology (Puerto Rico and Argentina), larval development (Venezuela and Mexico), aquaculture (Venezuela, Mexico, Brazil), fisheries (Mexico), evolution (Panama), biogeography (Panama, Mexico), paleontology (Mexico, Uruguay), reproduction (Panama, Mexico, Brazil) and secondary metabolites (Argentina). Most of the research has focused on the Echinoidea and Asteroidea, with very few works on Crinoidea. This last class needs an increased research effort. Regular echinoids have taken much of the attention. Among them is the genus *Diadema*, due to the massive die-offs of *D. antillarum* in the Caribbean, the barrens grounds produce by over populations of this species on the Canary Islands, and the predominance along the coral reefs of the Eastern Tropical Pacific of *D. mexicanum* with differential bioerosives impacts depending on its abundance. There is a need to increase the research on echinoderms in Uruguay, Central America, Haiti, Dominican Republic, Cuba, and Peru. It is also necessary to develop taxonomical, ecological (i.e. behavior, predation), physiological and development studies. It is essential to improve research in the Ophiuroidea and Holothuroidea.

On the other hand, most of the threats to echinoderms and their needs are shared. The most common threats are bad coastal planning, sedimentation, coastal pollution, illegal extraction, inefficient fisheries management, overfishing of predators, lack of protection, lack of baselines studies before the establishment of a sea cucumber fishery, lack of resources and lack of taxonomists. It is extremely urgent to support fishery practices with basic ecological studies, to understand the biology and ecology of each species before the establishment of an extraction activity or quota.

A great part of the research that has been done in LA can be found in BSc., "Licenciatura", MSc. and PhD thesis. Some are regional and national reports or abstracts of projects that have not had international distribution. The majority of those publications are in Spanish, and published in regional journals like Acta Científica Venezolana, Boletín de Investigaciones Marinas y Costeras, Ciencias Marinas, Gayana Zoológica, Hidrobiológica, Océanides, Revista de Biología Marina y Oceanografía, Revista de Biología Tropical and Revista Ciencia y Mar. Many publications are in journals of museums or research centers (e.g. Revista del Museo Argentino de Ciencias Naturales "Bernardino Rivadavia, Anales del

Table 1.1 Number of species in each Echinodermata class from the taxonomic list (TL) in the Appendix

	Total	Pacific TL	Caribbean and Atlantic TL	Common species between TLs
Crinoidea	82	21	63	2
Asteroidea	392	193	228	29
Ophiuroidea	521	211	340	30
Echinoidea	242	118	137	13
Holothuroidea	302	174	158	30
Total	1539	717	926	104
spp. with bathymetric information	–	620 (86 %)	836 (92 %)	–
spp. with habitat information	–	473 (66 %)	630 (70 %)	–

Insituto de Ciencias del Mar y Limnología de la Universidad Nacional Autónoma de México). These have a good level of national and local distribution, but not regionally. There is a need to promote the publication of thesis in local and regional journals, as well to encourage publication in English to have a broader audience and distribution.

A total of 1,539 species (Table 1.1) (21.8 % of the worldwide species accordingly to Pawson 2007) can be found in the taxonomic lists (see Appendix), with 717 species in the Pacific and 926 in the Caribbean and Atlantic, and 104 species shared in both ocean basins. The richest class is Ophiuroidea with 521 species, followed by the class Asteroidea with 392 species. All classes have the richest number of species in the Caribbean and Atlantic in contrast with the Pacific, except for the class Holothuroidea, which is richest in the latter.

From the Pacific list, 86 % of the species give bathymetric information and 66 % give information on the habitat. From the Caribbean and Atlantic list, 92 % and 70 % of the species give bathymetric and habitat information, respectively. This indicates there is a greater understanding of the environment conditions were this organisms inhabits in the Caribbean and Atlantic than in the Pacific. This is probably due to a greater and continuous research effort in this region (i.e. Pawson et al. 2009).

References

Appeltans W, Bouchet P, Boxshall GA, Fauchald K, Gordon DP, Hoeksema BW, Poore GCB, van Soest RWM, Stöhr S, Walter TC, Costello MJ (eds) (2011) World Register of Marine Species. http://www.marinespecies.org

Birkeland C (1989) The influence of echinoderms on coral-reef communities. In: Jangoux M, Lawrence J (eds) Echinoderm studies, vol 3. AA. Balkema, Rotterdam, pp 1–79

Cherbonier G (1959a) Echinodermes de la Guyane Francaise (Crinoides, Astérides, Ophiurides, Échinides, Holothurides). Bull Mus 31:105–111
Cherbonier G (1959b) Echinodermes de la Guyane Francaise (Crinoides, Astérides, Ophiurides, Échinides, Holothurides) (2ª note). Bull Mus 31:168–172
Cherbonier G (1959c) Echinodermes de la Guyane Francaise (Crinoides, Astérides, Ophiurides, Échinides, Holothurides) (3ª note). Bull Mus 31:261–265
Cherbonier G (1959d) Echinodermes de la Guyane Francaise (Crinoides, Astérides, Ophiurides, Échinides, Holothurides) (4ª note). Bull Mus 31:367–372
Cherbonier G (1959e) Echinodermes de la Guyane Francaise (Crinoides, Astérides, Ophiurides, Échinides, Holothurides) (5ª et dernière note). Bull Mus 31:440–447
Clark AH (1931) A monograph of the existing crinoids. Vol. 1. The comatulids. Part 3. Superfamily Comasterida. Smith Inst US Nat Mus Bull 82:1–816
Clark AH (1947) A monograph of the existing crinoids. Vol. 1. The comatulids. Part 4b. Superfamily Mariametrida (Concluded -the family Colobometridae) and Superfamily Tropiometrida (Excep the families Thalassometridae and Charitometridae). Smith Inst US Nat Mus Bull 82:1–473
Clark AM (1989) An index of names of recent Asteroidea. Part 1. Paxillosida and Notomyotida. In: Jangoux M, Lawrence JM (eds) Echinoderms studies. AA. Balkema, Rotterdam, pp 225–347
Clark AM (1993) An index of names of recent Asteroidea -Part 2: Valvatida. In: Jangoux M, Lawrence JM (eds) Echinoderms studies. Rotterdam, AA Balkema, pp 187–366
Clark AM (1996) An index of names of recent Asteroidea -Part 3: Velatida and Spinulosida. In: Jangoux M, Lawrence JM (eds) Echinoderms studies. Rotterdam, AA Balkema, pp 183–250
Clark AM, Downey ME (1992) Starfishes of the Atlantic. Chapman & Hall, London
Clark AM, Mah C (1999) An index of names of recent Asteroidea. Part 4: Forcipulatida and Brisingida. In: Jangoux M, Lawrence JM (eds) Echinoderms Studies. AA Balkema, Rotterdam, pp 229–347
Colburn FD (2002) Latin America at the end of politics. Princeton University Press, New Jersey
Hendler G, Miller JE, Pawson DL, Kier PM (1995) Sea stars, sea urchins, and allies: echinoderms of Florida and the Caribbean. Smithsonian Institution Press, Washington
Hermes-Lima M, Santos NCF, Alencastro ACR, Ferreira ST (2007) Whither Latin America? Trends and challenges of science in Latin America. IUBMB Life 59:199–210
Jangoux M (1978) Les étoiles de mer (Echinodermata: Asteroidea) récoltées par le M/V Calamar au large des Guyanes et du Venezuela. Bull Zool Mus Univ Amsterdam 6:93–101
Le Loeuff P, Von Cosel R (2000) Aperçus sur la mcrofaune benthique du plateau continentl de la Guyane Française (résultats de la champagne GREEN 0, 16 au 20 avril 1999). Doc Sei Tech Cent RD Bretagne 86
Maluf LY (1988) Composition and distribution of the Central eastern Pacific echinoderms. Nat Hist Mus LA County Tech Rep 2:1–242
Matsumoto H (1915) A new classification of Ophiuroidea with descriptions of new genera and species. Nat Sci Phil 67:43–93
Micael J, Alves MJ, Costa AC, Jones MB (2009) Exploitation and conservation of echinoderms. Oceanogr Mar Biol Annu Rev 47:191–208
Michán L, Llorente-Bousquets J (2010) Bibliometría de la sistemática biológica sobre América Latina durante el siglo XX en tres bases de datos mundiales. Rev Biol Trop 58:531–545
Panning A (1929) Die gattung Holothuria. (1. Teil). Mitteil Zool Staatsinst Zool Mus Hamburg 44:91–138
Panning A (1934a) Die gattung Holothuria. (2. Teil). Mitteil Zool Staatsinst Zool Mus Hamburg 45:25–50
Panning A (1934b) Die gattung Holothuria. (3. Teil). Mitteil Zool Staatsinst Zool Mus Hamburg 45:65–107
Panning A (1935) Die gattung Holothuria. 5. Teil, Schlub. Mitteil Zool Staatsinst Zool Mus Hamburg 46:1–18
Pawson DL (2007) Phylum Echinodermata. Zootaxa 1668:749–764

Pawson DL, Fell HB (1965) A revised classification of the Dendrochirote holothurians. Brevioria 214:1–7

Pawson DL, Vance DJ, Messing CG, Solis-Marin FA, Mah C (2009) Echinodermata of the Gulf of Mexico. In: Felder DL, Camp DK (eds) Gulf of Mexico: Origin, waters, and biota, Vol 1. Biodiversity. Texas A&M University Press, Texas, pp 1177–1204

Smith AB (1984) Classification of the Echinodermata. Palaeontol 27:431–459

Stöhr S, O'Hara T (2007) World Ophiuroidea database. Available online at http://www.marinespecies.org/ophiuroidea.

Uthicke S, Schaffelke B, Byrne M (2009) A boom-bust phylum? Ecological and evolutionary consequences of density variations in echinoderms. Ecol Monogr 79:3–24

Walenkamp JHC (1976) The Asteroids of the coastal waters of Surinam. Zool Verhandelingen 147:3–102

Walenkamp JHC (1979) Asteroidea (Echinodermata) from the Guyana shelf. Zool Verhandelingen 170:3–112

Chapter 2
The Echinoderms of Mexico: Biodiversity, Distribution and Current State of Knowledge

Francisco A. Solís-Marín, Magali B. I. Honey-Escandón, M. D. Herrero-Perezrul, Francisco Benitez-Villalobos, Julia P. Díaz-Martínez, Blanca E. Buitrón-Sánchez, Julio S. Palleiro-Nayar and Alicia Durán-González

F. A. Solís-Marín (✉) · M. B. I. Honey-Escandón · A. Durán-González
Laboratorio de Sistemática y Ecología de Equinodermos, Instituto de Ciencias del Mar y Limnología (ICML), Colección Nacional de Equinodermos "Ma. E. Caso Muñoz", Universidad Nacional Autónoma de México (UNAM), Apdo. Post. 70-305, 04510, México, D.F., México
e-mail: fasolis@cmarl.unam.mx

A. Durán-González
e-mail: aliciad@cmarl.unam.mx

M. B. I. Honey-Escandón
Posgrado en Ciencias del Mar y Limnología, Instituto de Ciencias del Mar y Limnología (ICML), UNAM, Apdo. Post. 70-305, 04510, México, D.F., México
e-mail: mbihoney@hotmail.com

M. D. Herrero-Perezrul
Centro Interdisciplinario de Ciencias Marinas, Instituto Politécnico Nacional, Ave. IPN S/n Col., Playa Palo de Santa Rita, CP. 23096, La Paz, BCS, México
e-mail: dherrero@ipn.mx

F. Benitez-Villalobos · J. P. Díaz-Martínez
Instituto de Recursos, Universidad del Mar, Carretera a Zipolite km 1.5, Puerto Ángel, Oaxaca, México
e-mail: fvb@angel.umar.mx

J. P. Díaz-Martínez
e-mail: lobi2@hotmail.com

B. E. Buitrón-Sánchez
Departamento de Paleontología, Instituto de Geología UNAM, 04510, Mexico, D.F., México
e-mail: blancab@servidor.unam.mx

J. S. Palleiro-Nayar
Centro Regional de Investigación Pesquera, Instituto Nacional de Pesca, Apartado postal 1305, Ensenada, BC, México
e-mail: juliopalleiro@yahoo.com.mx

J. J. Alvarado and F. A. Solís-Marín (eds.), *Echinoderm Research and Diversity in Latin America*, DOI: 10.1007/978-3-642-20051-9_2,

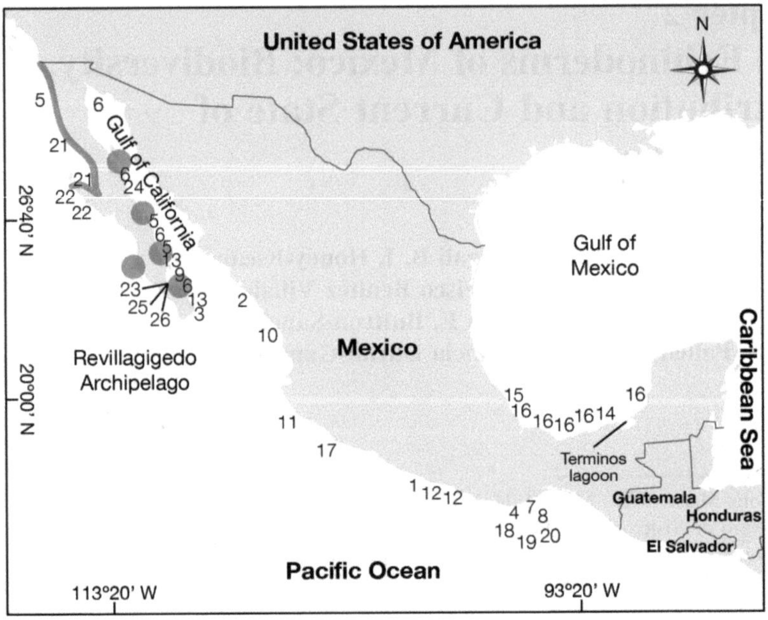

Fig. 2.1 Mexico with its principal islands, archipelago, marine lagoons and coasts. Fishing area of sea urchins: *Strongylocentrotus franciscanus* (*grey areas*), fishing areas of *Isostichopus fuscus* in the Gulf of California and Bahia Magdalena (*grey circles*), and the geographic location where ecological studies of echinoderms have been carried out. (*1*) Salcedo-Martínez et al. (1988); (*2*) Caso et al. (1994); (*3*) Reyes-Bonilla and Calderon-Aguilera (1999); (*4*) Benítez-Villalobos (2001); (*5*) Reyes-Bonilla et al. (2005); (*6*) Cintra-Buenrostro et al. (2005); (*7*) Zamorano and Leyte-Morales (2005a); (*8*) Zamorano and Leyte-Morales (2005b); (*9*) Holguín-Quiñones et al. (2008); (*10*) Rios-Jara et al. (2008); (*11*) López-Uriarte et al. (2009); (*12*) Zamorano and Leyte-Morales (2009); (*13*) Herrero-Perezrul et al. (2010); (*14*) Caso et al. (1996); (*15*) Celaya-Hernández et al. (2008); (*16*) Vazquez-Bader et al. (2005); (*17*) Espino-Barr et al. (1996); (*18*) Herrera-Escalante et al. (2005); (*19*) Benitez-Villalobos et al. (2008); (*20*) Benitez-Villalobos et al. (2009); (*21*) Tapia-Vasquez et al. (1996); (*22*) Fajardo-Leon et al. (2008); (*23*) Herrero-Perezrul et al. (1999); (*24*) Reyes-Bonilla et al. (2008); (*25*) Herrero-Perezrul and Reyes-Bonilla (2008); (*26*) Skarbnik-López et al. (2010)

2.1 Oceanography and Marine Environments in Mexico

Mexico is the twelfth largest country in the world in terms of extent of coastline and marine surface area. It has an exclusive economic zone of 314,992,000 ha, a continental platform of 39,460,300 ha and a coastline of 11,500 km. 1,600,000 ha are estuarine areas and about 1,250,000 ha are coastal lagoons. Approximately 6,000 km^2 of islands are found in the Pacific Ocean (including the remote Guadalupe Island and the Revillagigedo Islands), Gulf of Mexico, Caribbean, and Gulf of California (Fig. 2.1), which explains the country's extraordinary biodiversity in terms of coastal and marine resources and ecosystems (wetlands, mangrove forests, barrier islands, dunes, coral reefs, sea grass meadows and nearshore islands) that are distributed along these four seas (INEGI 2007).

The Pacific coast of Mexico extends from 14° to 33°N, and from 92° to 117°W. This includes the Baja California Peninsula, the second largest in the world (about 1,600 km in length), numerous islands and the eastern coast of the Gulf of California; the Gulf of Tehuantepec, making a total coast line well above 8,000 km in length (Fig. 2.1).

The Gulf of California (GC) has several oceanographic features that make it unique among semi enclosed seas of similar latitude and dimensions, the most important being strong tidal mixing, some of it close to the deep stratified oxygen concentration zone. The GC is 1,400 km long and its width in the inner region is 150–200 km, with an area of 177,000 km^2, a mean depth of 818 m, and a volume of 145,000 km^3. The GC is an evaporative basin, but in general it gains heat through the surface. Lacking a sill at the point of connection with the Pacific Ocean, the GC is constantly affected by a wide spectrum of signals coming from the Pacific Ocean, including tides and El Niño, which is the most important interannual anomaly (Lavin and Marinone 2003) (Fig. 2.1).

The Revillagigedo Archipelago or Revillagigedo Islands are a group of four volcanic islands off the Mexican Pacific coast, known for their unique ecosystem. They have been part of the Mexican State of Colima since 1861, but are under Mexican federal jurisdiction. They lie 386 km southwest of Cabo San Lucas, the southern tip of Baja California and between 720 and 970 km west of Manzanillo. They are located around 18°49′N and 112°46′W. The Revillagigedo Islands are home to many endemic plant and animal species and often are called Mexico's "little Galapagos". They are recognized as a distinct terrestrial ecoregion, part of the Neotropic ecozone. Socorro Island is the largest and has the most diverse flora, fauna, and topography (Fig. 2.1).

The Gulf of Mexico (GM) consists of several ecological and geological provinces, such as the continental shelf, the continental slope, and the abyssal plain. The coastal zone consists of tidal marshes, sandy beaches, mangrove-covered areas, and many bays, estuaries, and lagoons. The continental shelf forms an almost continuous terrace around the margin of the GM. Its width varies from a maximum of more than 320 km to a minimum of about 40 km. Off the coast of Campeche and the Yucatan Peninsula, the continental shelf consists of a broad area composed primarily of carbonate material. Off the states of Campeche and Yucatan, the bank of Campeche extends from 19° and 23° N, and 89° and 93°W in the southern GM (Gore 1992). It is delimited in the north by the Campeche slope; in the south by the shoreline of the states of Campeche, Yucatan and Quintana Roo; in the west by the Campeche canyon; and in the east by the Yucatan strait. This area is an almost flat carbonate plateau that has an average depth of about 51 m (Fig. 2.1).

The remainder of the GM shelf consists of sand, silt, and clay sediments. On the shelf and the slope that dips downward to the abyssal plain, buried salt domes occur at various depths. The abyssal plain, which forms the floor of the gulf, consists of a large triangular area near the centre, bounded by abrupt fault scarps toward Florida and the Yucatan Peninsula and by more gentle slopes to the north and west. The basin is very flat, having a gradient of only about 0.3 m per 2,440 m. The deepest point in the Mexico Basin (Sigsbee Deep), is 5,203 m.

The Terminos Lagoon, the largest lagoon-estuarine ecosystem in Mexico, is located in the southern GM. Terminos Lagoon is a large ($\sim$2,500 km^2), shallow (mean depth 3.5 m) coastal lagoon bordering the southern Gulf of Mexico in Campeche, with 200,108 ha of open water including associated lagoons and channels, surrounded by about 259,000 ha of mangroves and cattail marshes. The lagoon borders two geological provinces: to the east the Yucatan Peninsula (low rainfall, calcareous soils, and no significant surface drainage); to the west and south, the lowlands of Tabasco and the highlands of Chiapas and Guatemala. Offshore is the Bay of Campeche, a region supporting one of the largest marine fisheries in Mexico. The Usumacinta-Grijalva river systems discharge into the Gulf by three main rivers: the Candelaria, the Chumpan, and the Palizada. The lagoon is surrounded by shores and is fully incorporated into a National Flora and Fauna Reserve that comprises 705,016 ha of open water and associated wetlands and upland (Yañez-Arancibia et al. 1983) (Fig. 2.1).

The Mexican Caribbean and the Yucatan Peninsula is a northeastern projection of Central America, lying between the GM to the west and north and the Caribbean Sea to the east, encompassing some 197,600 km^2. It includes the Mexican states of Campeche, Quintana Roo, and Yucatan and, to the south, large parts of Belize and Guatemala. The peninsula has a mean width of about 320 km and a coastline of about 1,100 km (De la Lanza-Espino 1991). The Yucatan Peninsula subsurface Caribbean water (16–20 °C, 36.1–36.5 ‰) upwells along the eastern slope of the Yucatan shelf. This water, which originates at depths of about 220–250 m in the Yucatan Channel, rises at about 10 cm s^{-1} into the euphotic zone and only occasionally breaks the ocean surface. This upwelling follows a seasonal cycle. During spring and summer the upwelled water intrudes over the Yucatan shelf to create a two-layered water column. Yucatan is one of the most important upwelling regions on a western oceanic margin (Merino 1997) (Fig. 2.1).

2.2 Echinoderm Fauna of Mexico: An Inventory and Analysis

2.2.1 Historical Review

Studies on Mexican echinoderms started in the nineteenth century, when some specimens were collected and analyzed by Louis Agassiz in 1841. He reported the presence of the sand dollar *Mellita quinquiesperforata* (as *Mellita hexapora*) in the Gulf of Mexico (Veracruz). Later, Verrill (1870, 1871) published on some specimens collected by Yale University in Mexican waters. The *Blake* expeditions to the Gulf of Mexico and Caribbean (1878 and 1884) produced several papers, such as Perrier (1881) on Asteroidea, Lyman (1883) on Ophiuroidea, Théel (1886) on Holothuroidea and Agassiz (1878–1879, 1888) on Echinodea. The *Challenger* expeditions in the Mexican Pacific produced papers by Lyman (1879, 1882), Théel (1879), Agassiz (1881a, b), Carpenter (1884) and Sladen (1889). The *Albatross* expedition in 1884 to the Mexican Caribbean resulted in the reports of Rathbun (1885) for some sea urchins. The Mexican Pacific expedition in 1891 resulted in reports by Ludwig (1893) and Hartlaub (1895).

Few expeditions have been made during the twentieth century. Several non-Mexican authors wrote different reports that included echinoderms from Mexican waters. The reports of Ludwig (1905) and Clark (1917, 1920a, b) were based on material collected by the last *Albatross* expedition from 1899 to 1905. The expeditions to the East Pacific made by the *Velero III* under the direction of Captain Allan Hancock during 1931–1941 resulted in the description of Mexican echinoderms by Clark (1948), Ziesenhenne (1940, 1942), Deichmann (1941, 1958) and Domantay (1953, 1961). The Templeton Crocker Expedition on board of the *Zaca* took place in the Gulf of California in 1936, Deichmann (1937) and Ziesenhenne (1937) described the holothuroids and echinoids of that area. In 1937 and 1938, again on board of the *Zaca*, the Eastern Pacific Expeditions of the New York Zoological Society resulted in publications by Deichmann (1938) and Clark (1940). In 1940, Steinbeck and Ricketts made an expedition to the Gulf of California on board the *Western Flyer* and cataloged more than 550 species of animals, including echinoderms (Steinbeck and Ricketts 1941). Fisher (1906), Clark (1913, 1923a, b, 1933, 1916, 1918, 1954), Boone (1926) and Deichmann (1930) published on Mexican echinoderms from different collections.

During the second half and the end of the twentieth century, several authors continued to describe new species or published checklists of the echinoderms of Mexico, such as Clark (1940), Deichmann (1941, 1954, 1963), Fell (1962), Thomas (1962), Parker (1963), Schroeder (1964), Phelan (1970), Downey (1972, 1973), Brusca (1973, 1980), Litvinova (1975), Lessios et al. (1984a, b), Hendler and Turner (1987), Hendler and Peck (1988), Maluf (1988), Kerstitch (1989), Solís-Marín et al. 1993), Hendler et al. (1995), Cutress (1996), Hendler (1996), Roux and Pawson (1999), Sagarin et al. (2008) and Pawson et al. (2009).

Massin and Hendrickx (2010) described a new species of the genus *Synallactes* (*S. virgulasolida*) collected during a deep-water benthic fauna survey off the Pacific coast of Mexico in the East Pacific, on board the R/V *El Puma.* This was the first record of a *Synallactes* in the Gulf of California.

Massin and Hendrickx (2011) studied the deep-water holothurians (377–2,200 m) collected during the research cruises aboard the R/V "El Puma" from the Mexican Pacific and Gulf of California, recording 13 species, two of which were new to science (*Ypsilocucumis californiae* and *Mitsukuriella unusordo*). The small number of specimens collected during this survey compared to other areas of the world could be linked to the limiting effect of the Pacific Mexico Oxygen Minimum Zone.

Honey-Escandón et al. (2011) described a new of species of holothuroid (*Holothuria carere*) from the shallow waters of the Mexican Pacific.

Mexican expeditions did not take place until 1939 and added more species to the list of echinoderms. The Universidad Nacional Autónoma de México (UNAM)

Fig. 2.2 Maria Elena Caso Muñoz (December 1915–October 1991)

started the first attempt to organize and classify mainly the Mexican echinoderm fauna into a comprehensive scientific collection. Dr. Maria Elena Caso began a series of taxonomic studies that lasted for more than 50 years and that resulted in more than 50 papers.

Maria Elena Caso Muñoz (Fig. 2.2) was born in Mexico City (December 18, 1915). She was the daughter of Antonio Caso (a noted Mexican philosopher and rector of the former Universidad Nacional de Mexico, now the National Autonomous University of Mexico) and Josefina Muñoz. Maria Elena childhood was in a warm and loving family environment. The end of the Spanish Civil War, with Franco's victory in 1939, sparked the start of one of the most dramatic exoduses in contemporary history. Exiled scientists carried out tremendously important work in the countries offering them asylum, helping to promote scientific and technological progress. Enrique Rioja Lo-Bianco arrived to Mexico in 1939 to work at the Instituto de Biologia, Universidad Nacional de Mexico (UNAM). There, Ma. Elena Caso fell under the tutelage of Enrique Rioja and immediately started the systematic study of the Mexican echinoderms by collecting specimens from Mexican waters, initiating the Mexican Echinoderm National Collection (MENC).

She studied in the Faculty of Sciences at the UNAM (Mexico City) (1937–1940) where she obtained her Masters degree in 1943. In 1961 she received the degree of Doctor of Philosophy at the same University, presenting a thesis on the present knowledge of the Mexican echinoderm fauna. In the course of her career she published 56 papers on echinoderms and one book honoring Rioja's scientific endeavor. She described eleven new taxa (five species, one subfamily, one subgenus and two varieties). It was Ma. Elena Caso's intention to provide for future researches reference works that would eliminate the need to study the older literature. Consequently, her work (all exclusively written in Spanish) contains a staggering amount of details pictured in highly illustrated volumes. After her death (October 23, 1991) some of her papers were still in press. They were published *postmortem* (Caso 1992, 1994a, b, 1996a, b, c, d). Her ideas on the taxonomy of echinoderms were often controversial but were always respected. Maria Elena Caso will always be regarded as the pivotal figure in the systematic study of echinoderms in Mexico.

Currently, the Mexican Echinoderm National Collection (MENC) that Dr. Caso started holds more than 74,000 specimens belonging to ~620 species distributed along Mexican coasts. The MENC also holds specimens from different parts of the world. Recently, the collection has entered a rapid growth phase both in terms of numbers and of geographic and taxonomic coverage. Since 1984, yearly expeditions of Mexican research vessels "Puma" in the Pacific Ocean and "Justo Sierra" in the Atlantic, and since 1992, twice-yearly expeditions to both shallow and deep sea areas in Mexican waters, have added numerous taxa previously unrepresented in the collection. Important collections previously minor at MENC, such as frozen tissues, have grown quickly, establishing MENC as an important national repository. The MENC also possesses an excellent echinoderm library, with more than 5,100 reprints, books, proceedings, dissertations, etc., covering a great variety of topics from Aristotle to the present. Such records have been much more complete with generous requests from people around the world since 1939.

Knowledge of Mexican echinoderms continued to increase in the MENC at the end of the twentieth century and the beginning of the twenty-first century, with new curators and several students. The papers published include description of new species, checklists, taxonomic and biogeographic studies as well as morphological and molecular phylogenies. Some examples are Buitron and Solis-Marin (1993), Reyes-Bonilla (1995), Solís-Marín et al. (1997a, b, 2003, 2005, 2007a, b, 2009), Cintra-Buenrostro et al. (1998), Bravo-Tzompantzi et al. (1999), Godinez-Dominguez and Gonzalez-Sanson (1999), Solis-Marin and Laguarda-Figueras (1999, 2008, 2010a, b), Barbosa-Ledesma et al. (2000), Laguarda-Figueras et al. (2001, 2002, 2004, 2005a, b, 2009), Duran-Gonzalez et al. (2005), Fuentes-Farias et al. (2005), Luna and Gonzalez-Vallejo (2006), Hernandez-Herrejon et al. (2008, 2010), Honey-Escandón et al. (2008), Solis-Marin (2008), Torres-Martinez et al. (2008), Laguarda-Figueras and Solis-Marin (2009) and Martinez-Melo and Solis-Marin (2010).

The number of thesis or dissertations on the taxonomy of echinoderms increased during this period, with the highest number between 2001–2010: Orbe-Mendoza (1971), Gamboa-Contreras (1978), Worbis-Torres (1986), Olivares-Gonzalez (1986), San Juan-Ruiz (1988), Herrero-Perezrul (1990), (Sánchez-Domínguez 1993), Anzo-Martinez (1994), Bravo-Tzompantzi (1996), Castañeda-Sarabia (1996), Hernandez-Pliego (1998), Solis-Marin (1998), Cortes-Fernandez (1999), Benitez-Villalobos (2000), Gomez-Carriedo (2001), Frontana-Uribe (2002, 2005), Celaya-Hernandez (2006), Arriaga-Ochoa (2007, 2010), Dominguez-Castanedo (2007), Hernandez-Herrejon (2007, 2010),

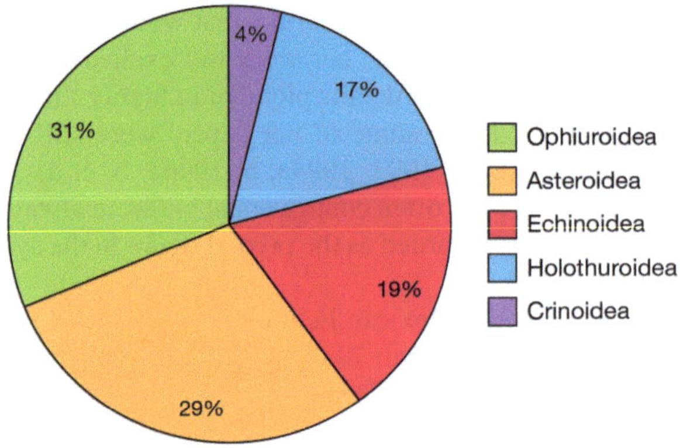

Fig. 2.3 Echinoderm composition (number of species per class) in Mexican waters

Gonzalez-Azcarraga (2009), Bribiesca-Contreras (2011), Estrada-Rodriguez (2011), Hernandez-Diaz (2011), Pineda-Enriquez (2011) and Zarate-Montes (2011).

2.2.2 Biodiversity and Distribution of the Mexican Echinoderms

Mexico's extensive littoral zone and great diversity of habitats supports 643 species of echinoderms (see Appendix), approximately 10 % of the world total. Class Crinoidea is less well represented with 29 species (4 % of the world total), while the Class Ophiuroidea is the richest of all with 197 species (31 %); Class Asteroidea is second with 185 species (29 %); Class Echinoidea is the third most diverse group with 119 species (19 %); Class Holothuroidea is represented by 113 species (17 %) (Fig. 2.3). Diversity and distribution of echinoderms in Mexico is described for the four coastal regions of this country: Gulf of California, Pacific Ocean, Gulf of Mexico and Caribbean Sea (Fig. 2.4).

2.2.3 Gulf of California

Although the Gulf of California is a diverse zone for some classes of echinoderms, there are no records of crinoids in this area. The asteroids in Mexico are well represented in all the coasts. In the Gulf of California there is a vast variety and abundance of specific generic types of asteroids higher than the Caribbean Sea. There are 58 species. Twenty nine occur in deep waters (>200 m depth). The most conspicuous genera are: *Astropecten, Luidia, Nidorellia, Oreaster, Pharia, Phataria* and *Heliaster* (Solis-Marin et al. 1993) (Fig. 2.5).

Seventy one ophiuroid species occur in the Gulf of California in two orders, 14 families and 34 genera. *Ophiactis savignyi* has the widest distribution in shallow

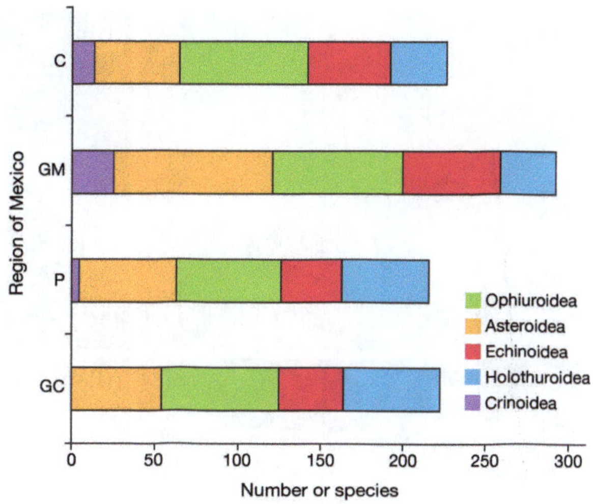

Fig. 2.4 Number of echinoderm species per class in each region of Mexico. *C* Caribbean, *GM* Gulf of Mexico, *P* Pacific Ocean and *GC* Gulf of California

waters. *Ophiothrix spiculata* is the most abundant and frequent species in the deep waters (Fig. 2.6).

The Gulf of California is the second area in terms of diversity of echinoids, with a total of 39 species (Solis-Marin et al. 2005), 12 of them in deep water. The most characteristic are: *Hesperocidaris asteriscus*, *H. perplexa*, *Arbacia incisa*, *Astropyga pulvinata*, *Centrostephanus coronatus*, *Echinometra vanbrunti*, *Lovenia cordiformis*, *Encope wetmorei*, *E. micropora*, *Clypeaster europacificus*, *Agassizia scrobiculata* and *Mellita grantii* (Figs. 2.6 and 2.7).

This region holds the greatest diversity of the sea cucumbers in Mexico, with 64 species. Most (48 species) have wider distributions; they occur from the Gulf of California to Central America or Galapagos Islands. Only two have a northern distribution, from Alaska to the Gulf of California. The majority of the species have shallow bathymetric distributions. Only 20 of the 58 species were collected from depths greater than 200 m. The most characteristic species of sea cucumbers are *Holothuria* (*Selenkothuria*) *lubrica*, *H.* (*Thymiosycia*) *arenicola*, *H.* (*T.*) *impatiens*, *H.* (*Halodeima*) *inornata*, *H.* (*Stauropora*) *fuscocinerea*, *H.* (*Platyperona*) *difficilis*, *Isostichopus fuscus*, *Neothyone gibbosa* and *N. gibber* (Solis-Marin et al. 2005, 2009) (Fig. 2.8).

2.2.4 Mexican Pacific Ocean

On the Pacific coast, the crinoids are represented by four species (13.7 % of the total for Mexico) off the western coast of the Baja California peninsula, and in the polymetallic nodule fields near the Revillagigedo archipelago and the Clarion

Fig. 2.5 **a** *Davidaster* sp., Mexican Caribbean. **b** *Nidorellia armata* (Gray, 1840), Gulf of California. **c** *Heliaster kubiniji* Xantus, 1860, Gulf of California. **d** *Amphiaster insignis* Verrill, 1868, Gulf of California. **e** *Linckia columbiae* Gray, 1840, Gulf of California. **f** *Astropecten armatus* Gray, 1840, Gulf of California. Images **a** by Valeria Mas, **b** and **e** by Carlos Sanchez-Ortiz, **c**, **d** and **f** by Israel Sanchez

fracture zone. These species, one stalked crinoid and three comatulids, are well adapted to thrive in extreme depths. The stalked crinoid is known from 3,000 to 4,500 m, while the comatulids occur from 12 to 3,234 m. The hyocrinid species *Hyocrinus foelli* Roux and Pawson, 1999, has been reported only near the Revillagigedo archipelago.

Fig. 2.6 **a** *Ophiothrix galapagensis* Lütken and Mortensen, 1899, central Gulf of California. **b** *Astrodictyum panamense* (Verrill, 1867), central Gulf of California. **c** *Ophiothela mirabilis* Verrill, 1867, Guerrero, Mexican Pacific. **d** *Ophiothrix rudis* Lyman, 1874, Guerrero, Mexican Pacific. **e** *Echinometra viridis* A. Agassiz, 1863, Veracruz, Gulf of Mexico. **f** *Echinometra vanbrunti* A. Agassiz, 1863, Jalisco, Mexican Pacific. Images **a** and **b** by Carlos Sanchez-Ortiz, **c** and **d** by Valeria Mas, **e** by Fred Boulag and **f** by Magali Honey-Escandón

The Class Asteroidea off the Mexican Pacific coast is of great interest, not only because of the high number of genera characteristic of this area, but also for the close relationships that the endemic species have with species in the Indo-Pacific, the Mediterranean and the Caribbean (Laguarda-Figueras et al. 2004). There are 59 species in total; 68 % (41 species) are shared with the Gulf of Baja California, 47

Fig. 2.7 **a** *Tripneustes depressus* A. Agassiz, 1863, Jalisco, Mexican Pacific. **b** *Toxopneustes roseus* (A. Agassiz, 1863), Jalisco, Mexican Pacific. **c** *Astropyga pulvinata* (Lamarck, 1816), Guerrero, Mexican Pacific. **d** *Holothuria pluricuriosa* Jaeger, 1833, Gulf of California. **e** *Isostichopus fuscus* Ludwig, 1874, central Gulf of California. **f** *Cucumaria flamma* Solis-Marin and Laguarda-Figueras, 1999, Jalisco, Mexican Pacific. Images **a** by Magali Honey-Escandón, **b** by Francisco Solis, **c** by Valeria Mas, **d** and **e** by Carlos Sanchez-Ortiz, **f** by Mauricio Valdes

are distributed in shallow waters and 12 occur only in the West coast of Baja California Peninsula (Fig. 2.4). Only 10 species are found also in the Revillagigedo Islands (Honey-Honey-Escandón et al. 2008). Conspicuous genera of asteroids include: *Astropecten, Luidia, Nidorellia, Pharia, Phataria, Heliaster,* and *Henricia* (Solis-Marin et al. 1993) (Fig. 2.5).

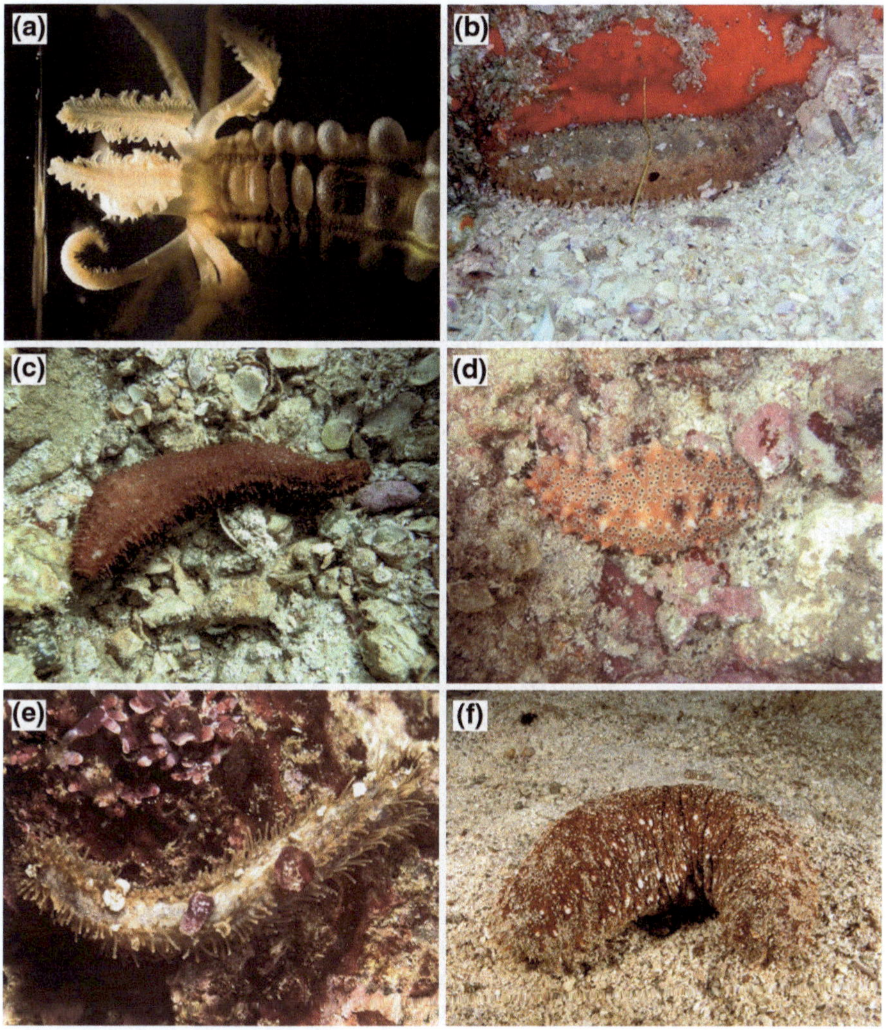

Fig. 2.8 **a** *Euapta godeffroyi* (Semper, 1868), central Gulf of California. **b** *Holothuria arenicola* Semper, 1868, Jalisco, Mexican Pacific. **c** *Labidodemas cf. macullochi* (Deichmann, 1958), Jalisco, Mexican Pacific. **d** *Holothuria kefersteini* Selenka, 1867, south of Gulf of California. **e** *Pentamera chierchia* (Ludwig, 1887), Guerrero, Mexican Pacific. **f** *Holothuria cf. inhabilis* Selenka, 1867, central Gulf of California. Images **a** by Offer Keter, **b** and **c** by Francisco Solis, **d** by Luis Carballo, **e** by Valeria Mas and **f** by Carlos Sanchez-Ortiz

The Pacific Ocean has the lowest diversity of ophiuroids of all Mexican waters, with only 63 species distributed in two orders, 12 families and 28 genera. Only 23 species are restricted to the Mexican Pacific Coast and the rest (40 species) are common to the Pacific Ocean and the Gulf of California. The order Ophiurida is the richest from the class Ophiuroidea, with nine families. The most speciose genus is *Amphiodia*, with seven species. *Ophiactis savignyi* is the ophiuroid species with the widest distribution in the shallow waters of the Pacific Ocean; it is distributed along the Gulf of California, the Gulf of Mexico, and also in the Mexican Caribbean. It has been recorded for eight States, the Revillagigedo Islands, and for the Maria Islands. Also, *Ophiocoma aethiops* is widely distributed in the shallow waters of the Pacific Ocean and it has been recorded for seven States, as well as for the Revillagigedo Islands.

Thirty seven species of echinoids are found in the Mexican Pacific Ocean, including the west coast of Baja California. These are classified in 25 genera, 14 families and seven orders. Twenty seven species are shared with the Gulf of California. The most common species are: *E. thouarsii*, *D. mexicanum*, *C. coronatus*, *A. incisa*, *Echinometra vanbrunti*, *Clypeaster europacificus*, *Encope micropora*, *E. perspectiva* and *E. wetmorei* (Honey-Escandón et al. 2008) (Figs. 2.7 and 2.8).

The Pacific coast of Mexico is the second in biodiversity for holothuroids, with 53 species (Fig. 2.4). Ten occur only on the west coast of Baja California. Seven occur also in the Pacific Coast. Of these 53 species, 46 also occur in the Gulf of California and the range of distribution can extend south to the Galapagos Islands. The characteristic species are almost the same for the Gulf of California: *Afrocucumis ovulum*, *Holothuria (Selenkothuria) lubrica*, *H. (Semperothuria) imitans*, *H. (Thymiosycia) arenicola*, *H. (T.) impatiens*, *H. (Halodeima) inornata*, *H. (Mertensiothuria) hilla*, *H. (Stauropora) fuscocinerea*, *Isostichopus fuscus*, *Neothyone gibbosa* and *N. gibber* (Honey-Escandón et al. 2008) (Figs. 2.7, 2.8).

2.2.5 Gulf of Mexico

The Gulf of Mexico is the most diverse area for the Classes Crinoidea, Asteroidea, Ophiuroidea and Echinoidea. Twenty five species (86 % of the total for Mexico) of crinoids live in the Gulf of Mexico and most of them occur in Campeche Bank. This Bank holds 22 of the 25 species present in the entire Gulf; about 75.8 % of the total species of crinoids in Mexican waters. Most of those crinoids can be found in both shallow (<200 m), and deep (>200 m) water environments; 17 species (68 % of the species in the Gulf of Mexico) are found in waters of less than 200 m depth, while 23 species inhabit the deepest part of the gulf. The unstalked or comatulid crinoids represent 76 % (19 species) of the total number of species present in the Gulf of Mexico, the stalked crinoids comprise 21 % (six species) and there is one species (4 %) of cyrtocrinid crinoid, *Holopus rangii,* present on Campeche Bank (Fig. 2.4).

There are 96 species of sea stars in the Gulf of Mexico. Half (48 species) are distributed in deep waters (>200 m depth), and 29 species have a wide bathymetric range. The most common genera are *Luidia*, *Astropecten*, *Cheiraster*, *Linckia*, *Pteraster* and *Echinaster* (Duran-Gonzalez et al. 2005).

Seventy nine species of ophiuroids occur in the Gulf of Mexico. Yucatan has the greatest number of species (46). *Ophiactis savignyi* is the species most frequently found in the Gulf of Mexico, especially in the State of Veracruz.

The number of species of echinoids in the Gulf of Mexico is 59. Of these, 30 (52 %) are present in deep waters (>200 m depth). The most common shallow water species with wide bathymetric (0–80 m) distribution are *Eucidaris tribuloides, Astropyga magnifica, Arbacia punctulata, Lytechinus variegatus carolinus* and *Encope michelini*. On the outer shelf *Clypeaster ravenelii* and *Brissopsis* are present in 80–190 m. Finally, *Plesiodiadema antillarum* and *Brissopsis atlantica* occur on the slope to 200 m. Geographically, *Encope aberrans, Clypeaster subdepressus* and *Echinolampas depressa* are restricted to Campeche Bank, Yucatan and Cape Catoche, Quintana Roo (Laguarda-Figueras et al. 2005a).

The Gulf of Mexico has the lowest diversity of sea cucumbers with only 33 species. Two species also occur off the Pacific coast and 20 are distributed also in the Mexican Caribbean in shallow waters. The rest (15 species) are present in deep waters of the Gulf of Mexico (Fig. 2.4), more than 200 m deep. The most common species are: *Holothuria (Halodeima) grisea, H. (H.) floridana, H. (Selenkothuria) glaberrima, H. (Semperothuria) surinamensis* and *Isostichopus badionotus* (Duran-Gonzalez et al. 2005).

Mexican Caribbean

In the Mexican Caribbean Sea, the crinoid fauna is not as diverse as it is in the Gulf of Mexico; nonetheless, the diversity of ecosystems (coral reefs, sandy bottom plains, deep slopes) provide habitats for 13 species, that represent 45 % of the sea lilies and feather stars reported in Mexican waters. In the Caribbean, the stalked crinoids represent 23 % with three species reported, inhabiting the Arrowsmith Bank and the Yucatan Channel area; while the comatulids comprise the major percentage (76 %) with 10 species in the same area. In this area, the isocrinid stalked crinoid *Endoxocrinus parrae* is the most frequently found species, along with the unstalked *Davidaster rubiginosus* (Fig. 2.5), *Comactinia meridionalis* and *Crinometra brevipinna*. A total of 13 species (84 %) are shared with the Gulf of Mexico (Fig. 2.4). Stalked crinoids are present in the Bank of Campeche and the Caribbean at depths that range from 86 to 143 m in the shallow areas; and from 747 to 1,245 m as the deepest limit. In the same region, the comatulid crinoids can be found from 2 to 200 m in the shallow range; and from 200 to more than 3,500 m in the deepest areas.

For the Asteroidea, the Mexican Caribbean is least diverse, with only 51 species. Forty two (79 %) of these are shared with the Gulf of Mexico. Of these 51 species, 14 occur in deep waters (>200 m depth) and 29 have a batimetric range at depths from 0 to more than 200 m. The most common genera are almost the same as in the Gulf of Mexico: *Luidia, Astropecten, Cheiraster* and *Echinaster* (Laguarda-Figueras et al. 2005b) (Fig. 2.5).

The Mexican Caribbean is the second most species-rich area of ophiuroids with 78 species. The distribution patterns of the species are diverse. The ones with the widest distribution in shallow waters of the Gulf of Mexico and Mexican Caribbean Sea are *Ophiolepis elegans, O. impressa, Ophiocoma echinata, Ophioderma cinereum, Ophiactis savignyi,* and *Ophiothrix angulata.* The deep-sea species with the widest distribution for both areas are *Ophiolepis elegans, O. impressa, Ophioderma cinereum,* and *Ophiothrix angulata.* Forty nine species are shared with the Gulf of Mexico.

Echinoids of the Mexican Caribbean coast are less studied than the other areas. Nevertheless, it is the second area in diversity with 50 species; 39 of them shared with the Gulf of Mexico. The most characteristic species occur in shallow waters. These are *Encope aberrans, E. michelini, Clypeaster subdepressus, Cassidulus caribaearum, Diadema antillarum, Echinometra lucunter, E. viridis, Eucidaris tribuloides* and *Tripneustes ventricosus* (Duran-Gonzalez et al. 2005).

The Mexican Caribbean has 33 species of holothuroids. The great majority of them (27 species) also occur throughout the Caribbean Sea and the Florida Keys. The most characteristic species are *Holothuria (Halodeima) floridana, H. (H.) grisea, H. (H.) mexicana, H. (Semperothuria) surinamensis, H. (Thymiosycia) arenicola, H. (T.) impatiens, H. (T.) thomasi, Isostichopus badionotus, I. macroparentheses* and *Euapta lappa* (Laguarda-Figueras et al. 2005b).

Endemism

The delimitation of endemism in marine species is difficult due to dispersion of larvae through oceanic currents. Nevertheless, Mexico possesses a few species that are endemic to a biogeographic province.

The Gulf of California, due to its unique oceanographic and geologic characteristics, has the majority of endemismic species (5). The echinoid *Encope grandis* is reported only for the Gulf of California and is present both along the continental coast and the islands within the Gulf. Three species of ophiuroids are endemic. These are *Ophiacantha hirta, Amphiophiura oligopora* and *Amphiura seminuda.* One species of shallow-water holothuroid *Athyone glasselli* is endemic.

For the North Pacific Ocean of Mexico, two species of crinoids, *Fariometra parvula* and *Florometra taneri* appear as endemic to the Californian province. In the southern part of Mexico, the sea star *Luidia latiradiata* (formely *Platasterias latiradiata*) seems to be endemic to the region from Chiapas, Mexico to Colombia. This species possesses archaic characters that can describe it as a "living fossil" (Caso 1945, 1970, 1972a; Fell 1962; Solis-Marin et al. 1993).

The only reported endemic species of ophiuroid in the Gulf of Mexico is *Amphiodia guillermosoberoni* from Terminos Lagoon, in Campeche State. This species was discovered and described by Caso (1979b). The species is more abundant inhabiting sites were the salinity ranges from 16 to 21 ppm. *Amphiodia guillermosoberoni* is the only Mexican ophiuroid species known to inhabit low salinity sites.

Only one species of sea star is endemic to the Mexican Caribbean, *Copidaster cavernicola*. This species, described by Solis-Marin and Laguarda-Figueras (2010a) is the only known asteroid that inhabits an anchialine-water cave ("Aerolito") in Cozumel Island (Solis-Marin et al. 2007a, Solis-Marin and Laguarda-Figueras 2008).

2.2.6 Mexican Fossil Echinoderms

The records of fossil Mexican echinoderms have been the subject of numerous publications since Nyst and Galeotti (1839). Buitron (1968) produced the first catalog on fossil echinoids of Mexico. Buitron (1990a) published an illustrated catalog of the echinoids from the Upper Jurassic and Lower Cretaceous of Mexico. Nieto-Lopez and Garcia-Barrera (2006) reviewed the Cretaceous echinoids of Mexico. Research on this matter has been carried out in extended areas such as the northern region (Boese 1910; Boese and Cavins 1927; Cooke 1955; Buitron 1970a, 1974a), the eastern region (Dickerson and Kew 1917; Lambert 1928; Maldonado-Koerdell 1953) and the northeastern region (Caso 1951b). Studies on the Baja California fossil echinoderms are mainly by Jordan and Hertlein (1926), Jackson (1937), Durham (1950, 1961), Wilson and Rocha (1955), and Chase (1956).

The fossil crinoids of Sonora State have been reported by Vachard et al. (2004) and Buitron et al. (2007a, 2007b, 2008). Strimple (1971) mentioned the crinoids from Coahuila State. The fossil echinoderms from Tamaulipas State have been reported by Dumble (1918), Israelsky (1924), Muir (1936) and Caso (1956). The fossil echinoderms of Veracruz State were studied by Boese (1906), Dickerson and Kew (1917), Caso (1956), Buitron (1978a) and Buitron and Silva (1979). Buitron et al. (1987) studied the fossil crinoids of Hidalgo State. Buitron (Buitrón 1973a, b) and Garcia-Barrera and Pantoja (1991) described the fossil echinoid fauna from Michoacan, Guerrero and Colima States. Fossil echinoderms from Puebla State were reported by Alencaster and Buitron (1965), Buitron (1970b), Esquivel-Macias et al. (2004), Martin-Medrano and Garcia-Barrera (2006), Applegate et al. (2009), and Martin-Medrano et al. (2009). Buitron (1976) and Garcia-Barrera and Pantoja (1991) studied the fossil echinoids from the State of Guerrero, and Buitron (1990b) from Oaxaca State. Lambert (1936), Jackson (1937), Muellerried (1951) and Buitrón (1974b, 1974c, 1978b) have recorded the fossil echinoids from Chiapas State.

The fossil echinoderm fauna in Mexico is represented by approximately 248 species. The Eocrinoidea (an extinct class of echinoderms that lived between the Early Cambrian and Late Silurian periods) is represented by one species (*Gogia granulosa* Robison 1965) (Nardin et al. 2009). Seventeen species of fossil Crinoidea have been recorded for the Mexican Paleozoic and Mesozoic; Class Asteroidea with only one species reported for the Cretaceous (*Pentasteria* sp.).

Six species of Ophiuroidea have been recorded from the Mexican Cretaceous and Paleogene-Neogene (Pliocene) from different localities in the mainland territory (*Ophiactis applegatei*, *Ophiura* sp., aff. *Amphiura*, aff. *Ophiomusium*; aff. *Stegophiura* and aff. *Ophiura*). The Echinoidea is the most diverse and widely distributed echinoderm fossil group in Mexico, with 221 species. One hundred of these are from the Mesozoic and 121 species are from the Cenozoic. Both groups are represented in 17 States of Mexico (Baja California, Sonora, Chihuahua, Coahuila, Nuevo Leon, Tamaulipas, San Luis Potosi, Hidalgo, Veracruz, Jalisco, Michoacan, Colima, Puebla, Guerrero, Oaxaca, Chiapas and Yucatan States). The Holothuroidea is represented by two Cretaceous species (Applegate et al. 2009).

The distribution of Mexican echinoderms in the Mesozoic and Cenozoic indicates that there was a large marine faunal province in eastern North America, which differs at the species level from the marine faunal province on the west coast of North America. This implies the existence of some kind of barrier during the Tertiary that prevented the faunal interchange between these two provinces. Echinoderms and other invertebrate groups in the region of the Atlantic and Pacific show the mark of a moderate exchange, superposed on the old pattern of Tertiary provincial aforementioned.

2.3 Research

2.3.1 Ecology and Reproduction Studies on Mexican Echinoderms

Ecological studies of the Echinodermata in Mexico have contributed to the knowledge of species richness, diversity and communities of echinoderms in several zones, mainly in the Gulf of California, southern Pacific and the Gulf of Mexico (Fig. 2.1).

Gulf of California

Salcedo-Martinez et al. (1988) generated a benthic inventory of echinoderms in Zihuatanejo, Guerrero. Data for each species were provided on bathymetric distribution, substrate, and some characteristics of the habitat.

Morgan and Cowles (1996) studied the effects of temperature on the behaviour and physiology of *Phataria unifascialis* and its implications for species distribution in the Gulf of California. They determined whether cool temperature was an important factor limiting this species distribution in the northern regions of the Gulf. Pyloric caecae and gonads showed that food limitation or lack of reproductive potential was a principal limiting factor. They also suggested that low temperatures of the northern gulf may well be an important factor limiting the northern distribution of this species in the Gulf.

Caso et al. (1996) performed qualitative and quantitative analyses of the echinoderm fauna from the littoral of Mazatlan Bay, Sinaloa and identified 25 species. The results showed that *P. unifascialis, Ophiocoma alexandri* and *Echinometra vanbrunti* were conspicuous because of their density and abundance. Cluster analysis showed the existence of three groups: the first included species with a limited distribution and low abundance, the second and the third had species widely distributed on sandy substrates in the bay. The highest diversity and specific richness index values corresponded to the species located in protected areas of the Bay, whereas the lowest diversity and specific richness index values corresponded to the species found in the exposed areas.

Reyes-Bonilla and Calderon-Aguilera (1999) analyzed the abundance, spatial distribution and consumption rate of *Eucidaris thouarsii* and *Acanthaster planci* at Cabo Pulmo, Gulf of California. The results showed that the action of corallivore organisms in terms of their population density and consumption rates was not a key factor determining scleractinian abundance in the reef of Cabo Pulmo.

Holguin-Quiñones et al. (2000) analyzed the spatial and temporal variation of Asteroidea, Echinoidea and Holothuroidea from shallow water at Loreto Bay, Gulf of California. The greatest mean abundance by transect and mean densities in order of importance for Echinoidea were *E. vanbrunti, Centrostephanus coronatus, Tripneustes depressus, E. thouarsii* and *Diadema mexicanum*; and for Asteroidea, *P. unifascialis, Mithrodia bradleyi* and *A. planci*. Asteroidea was the dominant class with 12 species, followed by Echinoidea.

Reyes-Bonilla et al. (2005) examined and compared the community structure of asteroids in four regions of the Gulf of California, characterized by the presence of rocky reefs. They found Loreto Bay had the highest richness and abundance of asteroids. However, there were no statistically significant interregional differences among ecological indices and no groups of locations singled out in terms of species composition. *Phataria unifascialis* was the dominant species followed by *Pharia pyramidatus*. There were statistically significant positive associations between three pairs of species, indicating competition between these asteroids in the Gulf of California is not particularly relevant.

Cintra-Buenrostro et al. (2005) determined the correlation between species richness of shallow water (depth < 200 m) asteroids and key oceanographic factors in nine sections of one degree latitude (from 23 to 31°N) of the Gulf of California. They reported continental shelf area (at 100 and 200 m depth), mean temperature and range at three depth levels (0, 60 and 120 m), thermocline depth, surface nutrient concentrations (nitrates, phosphates and silicates), surface photosynthetic pigment concentration, and productivity. Sea star species were assigned to different feeding guilds: predators of small mobile invertebrates, predators of colonial organisms, generalist carnivores, detritivores, and planktivores. The results showed that total species richness and guild species richness had strong latitudinal attenuation patterns and were higher in the southernmost Gulf. Nutrients and surface pigments always presented negative relationships with species richness, indicating that productive environments set limits to the diversity of sea stars in the study area.

The number of echinoderm species varies along the Gulf of California, increasing towards tropical latitudes (Cintra-Buenrostro 2001). However it seems the community structure of echinoderms in the Gulf is relatively homogeneous and dominated by asteroids (Reyes-Bonilla et al. 2005; Herrero-Perezrul et al. 2008). However, echinoderm distribution is probably related to habitat type. For instance, shallow bays favor coral communities and species associated with them.

Holguin-Quiñones et al. (2008) analyzed the spatial and temporal variation of Echinoidea and Holothuroidea in shallow water at Isla San Jose, Gulf of California. The most abundant species were *T. depresus*, *E. vanbrunti*, *P. unifascialis*, *C. coronatus*, *D. mexicanum* and *E. thouarsii*. There were no statistically significant spatial or temporal differences in diversity, evenness and species richness. The analysis identified two groups from locations of the eastern and western coasts.

Herrero-Perezrul et al. (2010) analyzed the community structure of conspicuous species of echinoderms of three islands in the southern Gulf of California: San Jose, Espiritu Santo and Cerralvo. They identified 23 species. The ecological indices showed similar values between the islands. However, the species composition was different. The community structure was stable in the study area. It seems that habitat type has a strong influence on the distribution and abundance of echinoderms.

Luna-Salguero and Reyes-Bonilla (2010) compared the community and trophic structure of rocky bottom asteroids in two regions of the Gulf of California. The starfish assemblages in both zones were dominated by a single species, *Phataria unifascialis*, and the Loreto area had higher values of richness, abundance and diversity of asteroids than the Ligüi area. In both areas, herbivores predominated above detritivores and carnivores. Abundance at each trophic level was statistically significantly higher in Loreto with more trophic groups per transect. These higher values of richness, abundance and diversity in Loreto are probably a consequence of a higher number of habitats and food resources.

South Pacific

Benitez-Villalobos (2001) characterized the community of echinoderms at two coral reef communities (La Entrega and Casa Mixteca) in Bahias de Huatulco region, Oaxaca State. The La Entrega community showed the lowest values of diversity and evenness and the highest values of dominance, associated with a more intense human activity. At the same locality Zamorano and Leyte-Morales (2005a) repeated the analysis of Benitez-Villalobos (2001) and compared the results obtained by both studies. Species richness and abundances they recorded were higher than those obtained in 2001. However there were no statistically significant differences in the diversity calculated by both studies. Ophiuroidea was the dominant class of echinoderms in 2001, whereas Holothuroidea was dominant in 2005. The authors attributed this change in dominance to the intense dredging that had occurred in Santa Cruz Bay. This probably produced an increase in suspended matter in the water column, supporting the holothuroids, whereas ophiuroids had been affected by the loss of natural refuges.

Zamorano and Leyte-Morales (2005b) worked in La Entrega and assessed population densities at two depth ranges (0–6 m and 6–12 m) of four echinoid species: *E. thouarsii, E. vanbrunti*, *T. roseus* and *D. mexicanum*. Density of the four populations varied considerably with the highest densities in the deep zone. Total abundance of sea urchins, including shallow and deep samples, did not vary significantly over time. However, the abundance per depth range was different over time for all species except for *E. thouarsii*. The authors established that the reef at La Entrega was undergoing an advanced state of erosion that could favor an increase in sea urchin abundance and therefore an increase in bioerosion activity.

Rios-Jara et al. (2008) described the specific composition of echinoderms from Isla Isabel National Park, in the central Mexican Pacific, on rocky and sandy substrates in subtidal areas to 19 m depth. The most important species according to the biological value index were *D. mexicanum, T. roseus*, *P. unifascialis, Pharia pyramidatus*, *Ophicoma alexandri*, *Holothuria lubrica*, *Isostichopus fuscus* and *E. thouarsii*. They established these species showed more affinity to the Gulf of California than to the central-south Pacific region.

Lopez-Uriarte et al. (2009) described the specific composition and community structure of the macrobenthic invertebrates from Bahia de Chamela, Jalisco, using quadrants along transects of 50 m length. Echinoderms showed the highest species richness, and the abundance and number of species were related to the heterogeneity of the substrate, which was characterized by stable mixed substrates including rocks, medium sized sand and scattered coral colonies. These kinds of substrates provide shelter in their cavities and crevices, and surfaces for the attachment of sessile species.

Zamorano and Leyte-Morales (2009) described and characterized the community of echinoderms associated with coral reefs in Zihuatanejo and Acapulco, Guerrero during the rainy and dry seasons. The authors proposed the mean values obtained for diversity in the 13 study sites are low compared to the maximum possible diversity, due to the dominance of the sea urchin *D. mexicanum* and the sea star *P. unifascialis*. These species, *T. roseus* and *Hesperocidaris asteriscus* were the most frequent and predominant species. Compared to other sites of the Eastern Tropical Pacific, densities were considered low. However, they did not find statistically significant differences in the densities of *D. mexicanum* between localities, depths, and seasons, although in some localities, the sea urchin densities increased considerably during the rainy season. The results also showed that, in localities close to Acapulco, Guerrero, a heavily visited touristic place, the diversity of echinoderms was lower and the abundance of *D. mexicanum* was higher.

Gulf of Mexico

In terms of ecology of Echinodermata, the Gulf of Mexico is one of the less studied zones. Only four studies have been published presenting ecological analyses of the communities of echinoderms in the region (Fig. 2.1).

Caso et al. (1994) carried out qualitative and quantitative studies on the echinoderm community from the Laguna de Terminos, Campeche. The relative abundance, geographic and local distribution, type of nutrition and habitat of 12 species of Echinodermata were analyzed. *Lytechinus variegatus*, *Luidia clathrata*, *Echinaster serpentarius*, *Ophiophragmus würdermanii* and *Amphiodia guillermosoberoni* stood out because of their relative importance and abundance. The higher diversity and specific richness index values corresponded to the species located in the areas with major oceanic influence, and the low diversity and specific richness index values corresponded to species found in the lagoons.

Celaya-Hernandez et al. (2008) analyzed the diversity, abundance, distribution and substrate preferences of the regular sea urchins found at the southern region of the Isla Verde lagoon reef, Veracruz. The substrate types considered in the analysis were: coral-rocks, rocks, rocks-sand, and sand and *Thalassia testudinum* beds. *Diadema antillarum, Echinometra lucunter lucunter* and *E. viridis* were mainly associated with coral-rock, rock and sand substrates, whereas *Lytechinus variegatus* and *Tripneustes ventricosus* were mainly associated with sea grass. The authors established that the distribution of the sea urchins is dependent upon substrate type and feeding habits which are intimately linked. The seven species reached their largest abundances during the rainy season. Jorgensen et al. (2008) recorded the high population density survival of *D. antillarum* after a category five hurricane in the southern Mexican Caribbean.

Vazquez-Bader et al. (2008) described the distribution and relative abundance of epifaunal echinoderms in the southwestern region of the Gulf of Mexico, based on 181 collections by the use of an Otter Trawl during different years. This study recorded 25 species of Asteroidea, 15 species of Ophiuroidea, 14 species of Echinoidea, four species of Crinoidea, and one species of Holothuroidea. The most frequently found species was *L. clathrata,* reported at 68 % of all stations. The highest seasonal density of echinoderms was recorded in autumn on carbonate substrate, whereas the highest species richness was found in summer.

Gonzalez-Azcarraga (2009) compared the community structure and morphological diversity of regular sea urchins in the Gulf of Mexico "Veracruz Reef System" and the Cozumel reefs in the Mexican Caribbean. They found statistically significative differences for all indexes among localities and at all depths. Veracruz Reef System showed the highest values due to their greater substrate heterogeneity and a high organic material deposition.

2.3.2 Population Ecology Studies

Few population ecology studies focused on particular species of echinoderms have been carried out in Mexico. This is regrettable, especially in the case of echinoids and holothurians, considering their importance as key organisms in the structure and dynamics of benthic communities, and their potential for commercial exploitation. Important species include the sea urchin *D. mexicanum* and the sea cucumbers *Isostichopus fuscus* and *Parastichopus parvimensis.*

Echinoids

Espino-Barr et al. (1996) analyzed the density and population structure of *D. mexicanum* on the rocky shore of the state of Colima in the Mexican Pacific as part of a technical assessment for requests of permission for sea urchin fishing along Colima and Jalisco shores. The results indicated an average density of 17,000 ind ha^{-1}. There was an increase in gonad weight with respect to body weight from May to August, followed by a decrease, indicating spawning during those months. They propose the commercial exploitation of the species between May and September when the gonads reach the greatest weight. Nevertheless they proposed it be restricted to protect the species.

Herrera-Escalante et al. (2005) evaluated the importance of *D. mexicanum* as a bioerosive agent of coral carbonate at Bahias de Huatulco and the relative magnitude of coral accretion and bioerosion. The results showed that, in general, *D. mexicanum* did not exert a significant role on coral reef community structure at the study area, probably because of their relatively small size and low numbers. Regarding the balance between bioerosion and carbonate production, they established that coral accretion exceeded sea urchin erosion at all sites examined, indicating that at Bahias de Huatulco coral reef communities were actively growing.

Benitez-Villalobos et al. (2008) evaluated the temporal variation of population densities and bioerosion by *D. mexicanum* at Isla Montosa, La Entrega, Isla Cacaluta and San Agustin, Bahias de Huatulco. Mean density of the sea urchins was relatively constant during the study at all localities and there were significant differences between La Entrega and the other localities. The calculated amount of $CaCO_3$ removed by *D. mexicanum* showed variations between localities. The lowest rate of bioerosion occurred at San Agustin and the highest at La Entrega. They compared their results with those obtained by Herrera-Escalante et al. (2005) and established that the recent urchin population decrease means a reduction in their removal of coral $CaCO_3$, and this reduction could be beneficial for coral reef growing and recovery from stress. Nevertheless it could also signify an increase in algal biomass and competition with corals for light and space.

Benitez-Villalobos et al. (2009) reported that until May 2009, the density of *D. mexicanum* at the different reef communities of Bahias de Huatulco remained relatively constant over time, with the highest density at La Entrega, where up to aggregations of up to 100 ind m^{-2} were observed. However, in May 2009, not a single individual was observed in the entire reef area (4–12 m) between 2.30 and 3.20 m and surrounding rocky bottom. Instead there were a high number of spines and hundreds of bare tests scattered over the reef, indicating a mass mortality event in the sea urchin population. Observations made by the authors during further monitoring suggested disease as the cause (Benitez-Villalobos et al. 2009). The observed signs were similar to those described during the mass mortality event of *D. antillarum* that occurred in the 1980s throughout the geographical range of this species in the Western Atlantic.

Fodrie et al. (2007) studied the mechanisms that regulate the feedings position and feeding mode selection of the sand dollar *Dendraster excentricus* in some Mexican localities. Zamorano and Leyte-Morales (2009) worked on the association of the echinoderms with reef formations on the central Mexican Pacific (Guerrero). Torres-Martinez and Solis-Marin (2010) analyzed sediment utilization and feeding niche breadth of the echinoid *Meoma grandis* in the central Mexican Pacific (Guerrero). Ebert (2010) studied the demographic patterns of *Strongylocentrotus purpuratus* along a latitudinal gradient and included one site on the Mexican Pacific territory.

Francisco and Herzka (2010) measured the role of physical and biological factors regulating the mode of feeding of the sand dollar *D. excentricus* in a shallow estuary in Baja California. The percentage of inclined sand dollars was strongly and positively correlated with tidal level but no relationship was found with current velocity, density and organic matter content of the water. The prone position, indicative of deposit feeding, was largely limited to low tidal levels. Sand dollars were only oriented parallel to the prevailing currents during the strongest currents of spring tides (>20 cm s^{-1}). The regulation of the feeding mode of sand dollars in shadow and hydrodynamically complex estuarine systems differs from the feeding mode found in exposed coastal environments.

Sonnenholzner et al. (2010) evaluated the effect of three tagging methods on the growth and survival of *Strongylocentrotus purpuratus*. The tags tetracycline and two internal tags, Decimal Coded Wire Tags (CWT) and Passive Integrated Transponder tags (PIT) were used alone or in combination to evaluate growth rates, survival, tag retention and post-tagging stress. The results suggested that the tags used did not affect growth or survival of the urchins. Post-tagging stress was significantly affected by urchin size but not by tagging methods.

Palleiro-Nayar et al. (2011) analyzed spatial variation in *Strongylocentrotus franciscanus* recruitment and assessed the impact of substrate availability and adult sea urchin density on recruitment along the west coast of Baja California. There were significant differences in recruit density among sites. Substrate structure was the main factor that explained these differences. Adult densities did not explain recruitment differences among sites. Temporal analysis showed that both substrate structure and adult densities were important in explaining recruit densities at both sites. The absence of a clear spatial pattern suggests other factors may explain the differences observed in density and recruitment among sites.

Holothuroids

Reproduction is maybe the most studied trait of sea cucumbers in Mexico, especially for commercial species (*Isostichopus fuscus* and *Parastichopus parvimensis*) (Herrero-Perezrul 1994, 2004; Fajardo-León et al. 1995; Tapia-Vasquez et al. 1996; Herrero-Perezrul et al. 1999). However, some other sea cucumbers (*Holothuria pluricuriosa* and *H. lubrica*) have been studied (Herrero-Perezrul 1994; Skarbnik-Lopez 2006; Skarbnik-Lopez et al. 2010).

Tapia-Vasquez et al. (1996) analyzed the gonadal cycle of *P. parvimensis* at the west coast of Baja California. They showed that the gonad develops during spring and spawning occurs during summer (July and August). They estimated first maturity at lengths between 25 and 29 cm. The reproduction in this species was similar at Isla Natividad and Bahia Tortugas, Baja California Sur (Fajardo-Leon et al. 2008). One interesting trait they described is evisceration prior to gonad development from August to October. Therefore they proposed a "reproductive ban" from February to May and another from August to October because the evisceration process.

Herrero-Perezrul et al. (1999) analyzed the reproduction and growth of *I. fuscus* at Isla Espiritu Santo, Baja California Sur. They described an annual reproductive pattern, with a single spawning period during summer, which was also influenced by sea surface temperature. The reproductive cycle was monitored histologically. Five gonadal stages were described according to cell types present in the gonad. Reyes-Bonilla et al. (2008) evaluated the abundance of *I. fuscus* in the National Park Bahia de Loreto, Gulf of California, during the fishing season 2005–2006 at 29 sites or "banks". The average abundance of *I. fuscus* in the park was 1.41 ± 0.02 individuals per transect, with an average density of 0.0280 ± 0.0004 ind m^{-2}. The density was slightly higher in the southern banks. The lowest density was at Isla Montserrat, which is close to the coastline of the Baja California Peninsula where most of the fishery occurs. They concluded that the low levels of density do not represent a critical situation for the populations.

Herrero-Perezrul and Reyes-Bonilla (2008) estimated the relative condition of an exploited population of *I. fuscus* at Isla Espiritu Santo, Baja California Sur. Length and weight were monitored each month and the weight-length relationship and the index of relative condition were calculated. The weight-length relationship showed that *I. fuscus* grew allometrically at this site. The index of relative condition exhibited a parabolic relationship with total length, and peaked at 21 cm length, which is the size of first maturity. They concluded this pattern may indicate the condition of individuals improves slowly with age up to a certain point and then decreases gradually. This finding is important because provides possible evidence of aging. It is the first indication of senescence in holothurians.

Skarbnik-Lopez et al. (2010) described the reproductive biology and growth of *Holothuria* (*Selenkothuria*) *lubrica* at Bahia de la Paz, Baja California Sur. According to data of the length-weight relationship they determined this species grows allometrically at the site. Five gonadal stages were described according to cell types present in the gonad. The results showed this holothurian spawns annually during summer. Variations of the gonad index were directly related to gonad development and sea-surface temperature, with highest values in summer corresponding with ripe gonads and the highest sea-surface temperature.

Some other studies have been directed to chemical analysis. Bakus (1974) demonstrated the toxicity to fishes of some holothurian species (*Isostichopus fuscus*, *Holothuria impatiens*, *H. imitans*, *Neothyne gibbosa* and *Parastichopus parvimensis*) occurring in the Mexican Pacific (Gulf of California). The evidence supports the hypothesis that toxicity in this species has evolved as a chemical defense mechanism against predation by fishes.

The small number of publications about ecology of echinoderms in Mexico provides evidence of unequal and limited development of ecological studies related to this group (Fig. 2.4). There are few papers analyzing echinoderm communities in Mexico, and only four species have been studied in terms of population ecology. Furthermore, basically all the studies have been focused mainly on the Gulf of California and the Mexican Pacific, leaving the rest of the shoreline practically unstudied ecologically, except for two projects performed in the Gulf of Mexico (Fig. 2.1). This is an indicator of the geographical location and field of influence of the institutions that hold specialists who are dedicated to the study of ecology of echinoderms; therefore it is important and essential to become involved with other institutions in order to develop collaborations that make possible an increase of knowledge of the current status and the dynamics of populations of echinoderms, in terms of their relationships with their biotic and abiotic environments, in order to obtain a more comprehensive knowledge of echinoderm communities in the entire country.

2.3.3 Chemical Studies of Echinoderms in Mexico

There are few studies about the chemistry on Mexican echinoderms. The first was by Lara-Guadarrame (1985), who determinated the basic biochemical compounds in the gonads of the sea urchin *Strongylocentrotus franciscanus*. Encarnacion et al. (1989) isolated the first marine natural product from a Mexican echinoderm, "Neothyside A" from the sea cucumber *Neothyone gibbosa*. This tryterpenoid tetraglycoside was isolated with cromatographic techniques. Another natural product isolated from *N. gibbosa* is Neothyoside B, a triterpenoid diglycoside (Encarnacion et al. 1996).

Encarnacion et al. (1996) wrote a book about the traditional southern California medicine. They mentioned that Indians used some marine organism as traditional drugs, including the sea urchin *Echinometra vanbrunti* and the sea star *Astropecten armatus*. Both are still used as traditional medicines as anti-abortion agents, as agents to remove scars and as a preventive for vaginal bleeding.

Table 2.1 Mexican echinoderms used in the assays of biological activity by Bryan et al. (1996, 1997)

	Antibacterial	Antifouling		Feeding deterrence		
		B. amphirite	*B. neritina*	*L. rhomboidies*	*C. variegatus*	*L. emarginata*
Class Asteroidea						
Anthenoides piercei		√			√	
Echinaster sp.			√	nt	nt	nt
Astropecten articulatus		√	√	√	√	
Goniaster tesselatus	√	√	√	√	√	
Linckia nodosa			√	√		
Luidia clathrata	√				√	
Narcissia trigonaria		√		√	√	
Oreaster reticulatus		√		√	√	
Tamaria halperni				√	√	√
Tethyaster grandis					√	
Tosia parva (=*Pawsonaster parvus*)				√	√	
Class Holothuroidea						
Holothuria lentiginosa				√	√	
Isostichopus badionotus				√	√	
Class Ophiuroidea						
Astrocyclis caecilia				√	√	
Asteroporpa annulata		√	√	√	√	
Astrophyton muricatum	√			√	√	
Class Crinoidea						
Comactinia merdionalis			√	√	√	

√ Activity significative, *nt* No tested

Bryan et al. (1996) studied the antibacterial and antifouling activity of the ethanolic extracts of 20 species of echinoderms (11 asteroids, three holothuroids, three ophiuroids and one crinoid) collected in the Gulf of Mexico. They used cultures of *Deleya marina* and *Alteromonas luteo-violacea* for antibacterial assays, and larvae of the barnacle *Balanus amphirite* and the bryozoan *Bugula neritina* in the antifouling assays. They found that only three species had significant antibacterial activity and 11 species antifouling activity (Table 2.1). Bryan et al. (1997) determinated the chemical defense of 19 species of echinoderms (10 asteroids, three holothuroids, three ophiuroids and crinoid). For this assay they made pellets with a tissue of every species and used them in feeding assays with two marine fish *Lagodon rhombodies* and *Cyprinodon variegatus* and a crustacean *Libinia emarginata*. They found that 15 species deterred feeding by *L. rhombodies* and 17 species deterred feeding by *C. variegatus*. Only two species deterred feeding *L. emarginata* (Table 2.1). Juárez-Espinoza (2010) screened antibacterial, antioxidant and hemolitic activity from ethanolic extracts of *D. antillarlum*.

2.3.4 Molecular Systematics Studies on Mexican Echinoderms

McCartney et al. (2000) sequenced 630 bp of the cytochrome oxidase I (COI) mitochondrial gene of seven species of the genus *Echinometra* (including *E. vanbrunti* from the Mexican Pacific) to provide comparable information on the eastern Pacific and Atlantic species, to estimate dates for cladogenic events. They found that an earlier split, assumed to have been coincident with the completion of the Isthmus of Panama, separated the eastern Pacific *E. vanbrunti* from the Atlantic common ancestor and concluded gene flow within species is generally high.

Lessios et al. (2001) reconstructed the phylogeny of the sea urching genus *Diadema* using mitochondrial DNA sequences from individuals around the world to determine the efficacy of barriers to gene flow and to ascertain the history of dispersal and vicariance events that led to speciation. They used *D. mexicanum* from the Gulf of California in the analysis. The *D. mexicanum* specimens grouped with species from the West Pacific and western and central Atlantic (*D. antillarum*, *D. savignyi* and *D. paucispinun*). They found that Indo-Pacific populations apparently maintained genetic contact with Atlantic populations around the southern tip of Africa for some time after the Isthmus of Panama was completed.

2.3.5 Biogeography Studies

Studies on echinoderm distribution patterns in Mexico are relatively recent. In addition, they are poorly documented. The first approaches have been made to understand the affinities between faunal provinces or regions in the Mexican territory. One of the earliest works in Mexico was made by Parker (1963). His study analyzed the zoogeography and ecology of benthic invertebrates in the Gulf of California, recognizing 11 faunal assemblages based on the characterization of its environments. He found that, compared with benthic communities elsewhere in the world, the diversity of shallow-water species in the Gulf of California is striking. No single species is dominant. Comparisons between the macro-invertebrate assemblages of the Gulf of California and those of the Gulf of Mexico and other parts of the world demonstrate that great similarities, generally at the subgeneric level, exist in similar environments throughout the subtropical and tropical regions of the world.

Laguarda-Figueras et al. (2002) analyzed the geographical and bathymetric distribution of Mexican Caribbean asteroids, evaluating the distributional patters and their affinities. They found that the "extended-Western Atlantic" and "warm

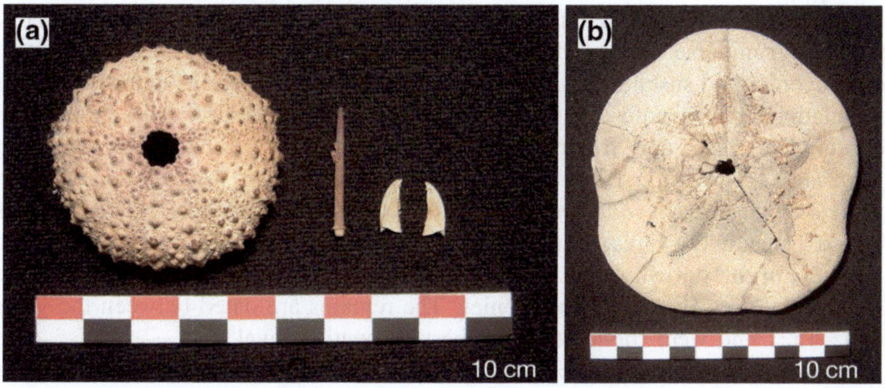

Fig. 2.9 **a** *Echinometra vanbrunti* A. Agassiz, 1863. Test, primary spine and teeth from the offerings at the great temple of the Aztecs; **b** *Clypeaster speciosus* Verrill, 1870. Test from the offerings at the great temple of the Aztecs

water Eastern Atlantic" were the patterns best represented, suggesting that the asteroid fauna in the Mexican Caribbean includes an important group of tropical species that have extended towards the north in zones of colder waters like the northern Gulf of Mexico and east coast of USA. Work by Caballero-Ochoa and Laguarda-Figueras (2010) used Parsimony Analysis of Endemism (PAE) to analyze the holothurian faunal affinities between the provinces of the Mexican tropical Pacific Ocean.

They found that the Cortes Province was the most diverse, including 62 holothurian species. The Panamic Province was not closely related with the Mexican, Cortes, Californian, Clipperton, Revillagigedo and Islas Marianas Provinces. Another important work is that of Hernandez-Diaz (2011) who also used PAE method to solve the relationship between echinoderms of the reefs of Yucatan. She found a nested pattern of distribution, which suggests they all belong to the same zoogeographic region with faunal elements from the Gulf of Mexico and Caribbean Sea.

2.3.6 Ethnobiology of Echinoderms in Mexico

Animal remains have been part of the archaeological discoveries around the world and are present in different cultures. Ethnobiology is defined as the study of the past and present interrelationships between human cultures and the animals in their environment. It includes classification and naming of zoological forms, cultural knowledge and use of wild and domestic animals (Seixas and Begossi 2001).

In 1978, the Aztec Great Temple ruins were discovered beneath the central plaza of Mexico City. Over the next few decades, this sacred structure was excavated and studied, revealing a wealth of information about Aztec religious life. Five-hundred years ago, this multi-tiered pyramid marked the literal center of the Aztec universe, where elaborated ceremonies were performed to maintain cosmic order and sustain the gods. Among the abundant offerings entombed in the foundations, the seventh constructive stage of the Great Temple held the remains of complete tests and fragments of three species of echinoids: *Echinometra vanbrunti* (Fig. 2.9a), *Clypeaster speciosus* (Fig. 2.9b), *Mellita* sp., *Meoma grandis*. *Echinometra vanbrunti* represents the most abundant echinoderm remains buried beneath the Great Temple (Solis-Marin et al. 2010). Only one arm tip fragment of the starfish *Astropecten* sp. has been found, showing that starfish were also used during the burial ceremonies. The remains are dated between the years 1502–1521 and were found at the deepest excavation level.

Because the Aztecs received tribute of other towns that they had dominated, we believe these echinoderms came from the tribute payment of some group dominated by the Mexicas. For example, *C. speciosus* presently occurs along the northwestern coast of Mexico. We believe it was transported from that zone. The Mexica Empire at that time (1502–1521) had conquered part of the national territory (Guerrero, Oaxaca, Chiapas States) and other coastal areas of the Mexican Pacific. This record is very important because it indicates they received materials far more to the north than normally considered (limits of the Michoacan and Guerrero States). Some specimens found in the offerings probably have they origin from the west coast of North Mexico, perhaps Mazatlan, Sinaloa. Although the Aztecs did not conquer the groups of these areas, we believe that the settler groups of "Zihuatlan" province gave tributes to the Aztecs. Although they never knew the exact limits of their empire, they knew that it was in the North of Mexico. The Mexicas never defeated the Tarascos, but they dominated the province of Zihuatlan. There is reason to believe they received tribute of some group or groups that were more to the north of the country.

Animal offerings in the burials were made by the highest priests. Their rituals were closed events. In the massive events they were not celebrated with this type of offerings. The offerings were made to inaugurate and to consecrate new building structures and possibly to commemorate festivities of the ritual calendar. The marine animals, including the sea urchins and starfish (and very probably other classes of echinoderms) were brought alive to the site. *Echinmetra vanbrunti* found in the offerings have remains of pedicellariae on the top of the test and some fragments of the Aristotle's lantern. The transport was made using emissaries who ran very long distances and for a time. In average they ran 30 km daily.

2.4 Aquaculture and Fisheries

Echinoderm fisheries in Mexico are based mainly on the sea urchins *Strongylocentrotus franciscanus* and *S. purpuratus*, the sea cucumbers *I. fuscus* and *Parastichopus parvimensis,* on the Pacific coast of Mexico and in the Gulf of California. Except for *P. parvimensis*, the other species were heavily exploited during the 1990s. It is important to mention that all echinoderm catch is exported to other countries, mostly Eastern countries. In the same way, other echinoderms are also captured for aquarium trade (Micael et al. 2009). Mexico is considered one of the main exporters together with Indonesia, Singapore, Fiji, Sri Lanka, Philippines and Vanuatu. For instance, the populations of the Atlantic sea star *Oreaster reticulatus*, used as souvenir, have declined in the Mexican Caribbean (Lunn et al. 2008). Also in the Gulf of California there are other sea stars (*P. unifascialis, P. pyramidatus* and *Pentaceraster cumingi*) that are currently extracted for ornamentation purposes. However, artisanal fishing has no regulations and the limited information available comes from the results of a survey of the Mexican echinoderm trade. Currently, an estimated 62 stakeholders in Mexico collect an average of 12,000 sea stars annually for the souvenir industry.

Some studies have focused on the effect of chemicals related to metamorphosis of commercially important sea urchins. Carpizo-Ituarte et al. (2002) tested the response of three species of sea urchin (*L. pictus*, *S. purpuratus*, and *S. fransicanus*) to excess K^+ in sea water as artificial inducers of metamorphosis. The results showed that K^+ was an effective metamorphic inducer in all three species. The response depended on concentration and length of exposure. Castellanos-Kotkoff et al. (2004) showed the effect of thyroxine on metamorphosis and the presence of the nuclear receptor COUP-TF in the larvae of three species of sea urchins (*L. pictus*, *S. purpuratus*, and *S. fransicanus*) from the Mexican Pacific.

Salas-Garza et al. (2005) summarize some of the basic studies and main achievements in the larval development of the red sea urchin *S. franciscanus* in Baja California. Spawning of red sea urchins was routinely induced with KCl; eggs were fertilized using a 100,000 sperm ml^{-1} solution. KCl proved the most consistent metamorphic inducer, regularly yielding metamorphosis percentages higher than 90 %. Metamorphosis was considered complete when the functional jaw that juveniles use for first benthic feeding appeared (as soon as 20 days after induction). With this method several thousands of red sea urchin juveniles were produced.

Amador-Cano et al. (2006) used artificial inducers to study signal-transduction pathways involved in metamorphosis of *S. purpuratus* postlarvae. Participation of protein kinase C (PKC), G-protein-coupled receptors (GPCRs), and calcium were investigated during its metamorphosis, showing that the GPCRs may be shared between the artificial (KCl) and natural (biofilm) inducers.

2.4.1 The Fishery of Sea Urchins in the Baja California Peninsula, Mexico

The Baja California fishery is based on two species, the red urchin *S. franciscanus* and the purple urchin *S. purpuratus*. Both species are found in the Pacific coast of North America, including the Baja California peninsula (Ebert and Southon 2003). The former is one of the largest species in the world, growing to about 200 mm test diameter, weighing more than 1 kg and can live for more than 100 years. The purple sea urchin is small, growing to about 70 mm test diameter.

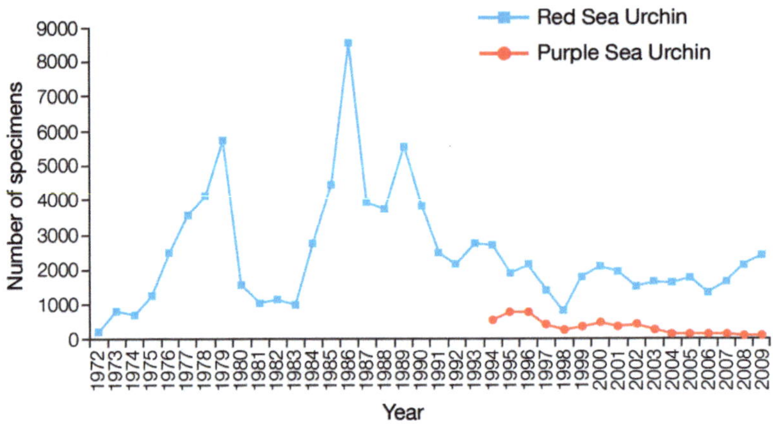

Fig. 2.10 Red and purple sea urchin catch in Baja California, Mexico

The fishery is limited from Coronado Islands to Isla Natividad in Baja California Sur (Fig. 2.10). The fishery for *S. franciscanus* began in 1972 to supply the Japanese market. The fishery grew quickly and is characterized by three peaks in catch, in 1979 (5,707 tons), 1986 (8,493 tons) and 1989 (5,536 tons). On the other hand, *S. purpuratus* has been harvested only since 1993. Although catches have remained relatively small (Fig. 2.10). Densities of this echinoid have risen in the last ten years, but inconsistent roe quality has retarded development of the fishery (Palleiro-Nayar et al. 2008).

Sea urchins are collected by "hooka" or air compressor to depths up to 30 m. The divers use short-handled rakes to dislodge the sea urchin from the reef. Because other fishermen are paid on the basis of roe recovery rather than the whole animal, divers test roe quality as they work. There is no recreational fishery and illegal harvest has not been estimated. However, is believed to be relatively small (Palleiro-Nayar 2009). To date there are some management measures, including reduction in fishing effort (300 divers in 1987, 229 divers in 2010). Catch rates for *S. franciscanus* have declined significantly from 309 kg day^{-1} in 1988 to 87 kg day^{-1} in 2002 and to 176 kg day^{-1} in 2009. The sea urchin roe produced over 1.3 million US dollars in 2009. However, the maximum profits were obtained in 1994 with 10.4 million US dollars (Fig. 2.10).

*2.4.2 The Brown Sea Cucumber Fishery (*Isostichopus fuscus*) in the Gulf of California, Mexico*

The brown sea cucumber *I. fuscus* is a common inhabitant of coral and rocky bottoms from the Gulf of California, Mexico to Ecuador (Maluf 1988; Hearn et al. 2005; Solis-Marin et al. 2009). The fishery of this sea cucumber is without a doubt a most peculiar story. From the beginning, there were no regulations for fishing activities, and the only information that existed was related to taxonomy (Caso 1961b, 1967a). The fishery started in the late 1980s, mostly in the Gulf of California, although some trade was done in Jalisco and Oaxaca (Sierra-Rodriguez 1994; Nuño-Hermosillo 2003) (Fig. 2.1). The maximum recorded catch occurred in 1991, with 1,800 tons but mean annual catch was around 300 tons (Herrero-Perezrul and Chavez-Ortiz 2005). In 1994, due to overfishing, authorities closed the fishery and included *I. fuscus* in the list of protected species (Anonymous 1994). The fishery was closed until 2000. With biological data that had been generated by then, some regulations were established and the fishery reopened in Baja California Sur. Stakeholders were required to record biological data (size, fishing logs, reproduction, etc.) in exchange for the license. Currently the fishery in Baja California Sur represents a sustainable activity which is been going now for 10 years with no evidence of overfishing (Herrero-Perezrul 2010). The fishery management plan in Baja California Sur includes a minimum size of 20 cm and 400 g, total allowable quotas, estimations of biomass and population density, and a reproductive ban from June 1st to October 31.

*2.4.3 The warty Sea Cucumber Fishery (***Parastichopus parvimensis***) in the Baja California Peninsula, Mexico*

The warty sea cucumber *P. parvimensis* can grow to a length of 60 cm and a width of 5 cm. Mobility is limited, though individuals can move up to 4 m day^{-1} while feeding (Morris et al. 1980). It is commonly found in rocky bottoms from 0 to 60 m (Woodby et al. 2000). However, exploitation takes place mostly at depths no deeper than 30 m. The distribution along the west coast of Baja California Peninsula is from Coronado Islands to Punta Abreojos (Fajardo-Leon and Turrubiates-Morales 2009).The fishery started in 1989. In general, fishing takes place when the sea urchin season closes, from March to June. The stock is exploited by the same fishermen exploiting sea urchins. Until 2004, the fishery was only in Baja California, but expanded to Baja California Sur in 2004 (Fajardo-Leon and Turrubiates-Morales 2009). The maximum catch of warty sea cucumber in Baja California was 1992 with 723 tons. However, landings have decreased to around 230 tons in recent years with around 100 divers (Salgado-Rogel et al. 2009). In Baja California Sur the maximum catch was in 2008 with 381 tons. The sea cucumber is boiled, dried and salted before export, while lesser quantities are frozen. In Asia, sea cucumber are claimed to have a variety of beneficial health enhancing properties like lowering high blood pressure, aiding proper digestive function, and others (Rogers-Bennett and Ono 2006).

2.5 Echinoderm Threats in Mexico

The marine biodiversity of Mexico is threatened by various human activities. The main factors are pollution and fishing activities.

2.5.1 *Over Exploitation*

Of the five classes of echinoderms, sea urchins and sea cucumbers are most heavily exploited. The sea cucumbers (bêche-de-mer or trèpang) that are fished commercially include 42 species from around the world. They are consumed for their high-protein content of the body wall, which is boiled and dried (Purcell et al. 2010). During the 1990s, the number of producing countries increased worldwide with established fisheries in Canada, USA, Mexico and Ecuador. Despite efforts to regulate the sea cucumber fisheries in Mexico, illegal fishing continues. One of the greatest dangers to unregulated fishing of this resource is potential population collapse and therefore the risk of species extinction (Pech 2010). This situation is not new. In the late 80's began in the Gulf of California: *I. fuscus* catches exceeded one thousand tons in 1991 (Reyes-Bonilla et al. 2008). However, only 5 years later the fishery authorities declared it endangered and prohibited its capture. This species is currently on the list of protected species (NOM-059-ecol-2001).

So far, there have been four species of sea cucumbers exploited in the Gulf of Mexico and the Mexican Caribbean: *Holothuria mexicana*, *H. floridana*, *Astichopus multifidus*, *Isostichopus badionotus* (Zetina-Moguel et al. 2003) and two species in the Mexican Pacific, *I. fuscus* and *P. parvimensis*.

Fisheries are also a threat to sea urchins in Mexico. Two species of sea urchins from the Mexican Pacific have been exploited: *S. purpuratus* and *S. franciscanus*. One of these fisheries occurs in Baja California Mexico, with the exploitation of *S. purpuratus*, the purple sea urchin. Gonads are exported to the Northeast Pacific (Micael et al. 2009). The organisms are collected by hand or with dredges. Since the quantity and quality of roe cannot be assessed externally, the individuals are cracked open, causing a high level of mortality and discard of animals.

2.5.2 *Trawling and Accidental Catch*

The catch depends on the gear used in particular fishery and the amount of effort applied. One of the major problems in trawl fisheries is the large numbers of other species that are caught and discarded. Northwestern Mexico, and in particular the Gulf of California, is the most important fishing area in the country (Robadue 2002). Seventy-seven percent of the volume of fish production in Mexico is obtained from the Pacific Ocean, and discharged in Baja California, Baja

California Sur, Sonora, Sinaloa and Nayarit States. Most of the catch (80 %) is from the Gulf of California, where coastal fishing generates 114,000 tons y^{-1}: 31 % in Baja California Sur, 28 % in Sonora, 22 % in Sinaloa and 19 % in Baja California. In general, 10 % of coastal catches are sharks and dogfish, 17 % bony fish, 15 % mollusks, 6 % crustaceans and 52 % unidentified (Rodriguez-Valencia and Cisneros-Mata 2006).

Echinoderms are affected directly by incidental catch when fishing with trawls because of the significant bycatch effect. The number of species collected this way is difficult to estimate because they are not recorded in catches. Discards from trawling are dumped at sea and sink to the bottom where their decomposition reduces the levels of oxygen and make the habitat less suitable for benthic organisms. Invertebrates account for up to 90 of the numbers of animals discarded and up to 73% of these are echinoderms (Pranovi et al. 2001).

2.5.3 Ornamental Species

Many species of echinoderms are popular in aquarium trade. Mexico is included in the main exporters together with Indonesia, Singapore, Fiji, Sri Lanka, Philippines and Vanuatu. One Caribbean sea star captured for souvenir industry is *Oreaster reticulatus*. Other countries in the Caribbean, including Mexico, allow the extraction of this species without apparent restrictions.

Little is known about the use of echinoderms as souvenirs or for aquariums. The limited information available comes from the results of a survey of the Mexican echinoderm trade. Currently an estimated 62 fisheries in Mexico collect an average of 12,000 sea stars annually for the souvenir industry (Lunn et al. 2008).

2.5.4 Management and Conservation

Among the possible actions for conservation and management of marine resources, one of the most important is the recognition of priority sites. Protected areas have been advocated as one of the most important and effective tools for safeguarding the world's biodiversity. A major reason for this is that they protect species from their greatest threat: habitat loss. In March 2007, Mexico had 61 Marine Protected Areas with valid establishment decrees, occupying 13,336,390 ha or 58.5 % of its total federal protected area ($\sim$42 % of its total marine exclusive economic zone) (Ortiz-Lozano et al. 2009). It is noteworthy that most of the known species of echinoderms are distributed in shallow water areas, and many of them are located within protected sites.

Until now, no comprehensive work on the population dynamics of the different exploited species have been made. There is no current information on the number of possible species that might suggest its economic importance and its distribution along the Mexican littoral (for example the genus sea star *Pisaster*). Likewise, we need to control and regulate the sea cucumbers fisheries in Mexico. For the sea cucumber *I. fuscus*, the permit holders have to fish in specific areas called management unit for wild life or UMA (*Unidad de Manejo Ambiental*). The exploitation permits are given to organized fishermen who must submit reports on this activity. UMA controls a harvest quota for each area. It has restricted collection to sea cucumbers with a minimum 370 g of total weight and has established a closing season from June to September. On the other hand, there is no official management plan for *P. parvimensis*, although the permit holders must collect the sea cucumber in specific areas and observe a minimum size limit of 200 g body wall (without gut) (Perez-Plascencia 1995). Assessments of the sea cucumber fishery are based on annual analyses of fishery-derived information such as catch rate and mean size of individuals in the landed catch as well as fishery-independent surveys. These are used to determinate biomass and set a quota of no more 10 % of the total biomass estimation.

The main life history traits of holothurians suggest that they constitute fragile stocks. For instance, they are big, slow growing organisms which hardly move, and therefore easily detected by divers. Population density is less than 1 ind m^{-2}, similar to values for many of the most important species around the world (Herrero-Perezrul et al. 1999; Uthicke et al. 2009). Consequently overfishing is likely to occur due to their high vulnerability. Conservation efforts have been made, especially for the fished populations. The permits are based on management regulations. They limit the number of boats and divers in order to control fishing effort, and assess the population before and after the fishing season. A private company has been successful in culturing larvae to produce juvenile *I. fuscus* (F.A. Solis-Marin pers. obs.).

The red sea urchin fishery was unregulated until 1987 when a suite of management reforms transformed the fishery from an open-access competitive model to one in which individual permit holders (with one or more divers) had exclusive access to one area. Prior to 1987, divers were able to fish everywhere, with the result that many areas became severely overfished. In 1994, the Federal Government strengthened the holders permit rights by giving them long term concessions (20 years) for exclusive areas access. The permits are tradable. These changes were designed to provide investment security and promote greater commitment to long-term sustainable use. Other management measures introduced in 1987 included the introduction of a minimum legal size of 80 mm test diameter, a closed season between April and June (later extended to March–June), and a catch and effort reporting scheme (Palleiro-Nayar 2004).

Assessments of the red sea urchin fishery are based on annual analyses of fishery-derived information such as catch rate and mean size of individuals in the catch as well as fishery-independent surveys. Video surveys of reefs also indicate declines in densities of red sea urchins and increased in density of purple sea urchins that have replaced red urchins in deeper water (Salgado-Rogel and Palleiro-Nayar 2008).

In addition, FAO recognizes that some countries have established measures to reduce bycatch in certain fisheries, through which the ships find it much easier to avoid capture of unwanted species (Sommer 2005). The pervasive trend of overfishing, and increasing examples of local extinctions requires immediate action for conserving stocks, biodiversity, ecosystem functioning and resilience from other stressors than overfishing (e.g. global warming and ocean acidification) to ensure the ecological, social and economic benefits of these natural resources (Toral-Granda et al. 2008).

2.6 Concluding Remarks and Recommendations

The geographic position of Mexico surrounded by two oceans, two gulfs and its tropical location, explains its extraordinary biodiversity in terms of coastal and marine resources and ecosystems, making this country one of the most biodiverse on the planet.

The earliest published article that included work on echinoderms from Mexico dates back to Louis Agassiz in 1841, who reported the existence of the sand dollar *Mellita hexapora* (as *M. quinquiesperforata*) for the Gulf of Mexico (Veracruz). Thus began a period of exploration in which most of the species from Mexican waters were described (Verrill 1870, 1871; Perrier 1881; Lyman 1883; Théel 1886; Agassiz 1878–1879, 1888). During the twentieth century, few expeditions took place but several non-Mexican authors wrote reports that included echinoderms from Mexican marine waters (Ludwig 1905; Clark 1913, 1917, 1920a, b, 1923a, b, 1933, 1948, 1916, 1918, 1954; Boone 1926; Ziesenhenne 1937, 1940, 1942; Deichmann 1930, 1937, 1951, 1958; Steinbeck and Ricketts 1941; Domantay 1953, 1961). It was not until 1939 when Mexican expeditions began and more species were added to the developing list of species (Caso 1941, 1943, 1944, 1945, 1946, 1947, 1948a, b, 1949, 1951a, 1953, 1954, 1955, 1958, 1961a, b, 1962a, b, 1963, 1964, 1965, 1967a, b, 1968a, b, c, 1970, 1971a, b, c, 1972a, b, c, 1974a, b, c, 1975, 1976, 1977, 1978a, b, 1979a, b, c). The UNAM began the first attempt to organize and classify Mexican echinoderm fauna into a comprehensive scientific collection. The knowledge of the Mexican echinoderms continued to increase in the Mexican Echinoderm National Collection at the end of the twentieth century and the beginning of the twenty-first century, with new curators and the incorporation of several students. The papers published include description of new species, checklists, taxonomic and biogeographic studies as well as morphological and molecular phylogenies (Caso 1979a, b, c, 1980a, b, 1983a, b, 1984, 1986a, b, 1990, 1992; Buitron and Solis-Marin 1993; Caso et al. 1994a, b, 1996; Caso 1996a, b, c, d; Solis-Marin et al. 1997a, b, 2003, 2005, 2007a, 2007, 2009; Cintra-Buenrostro et al. 1998; Bravo-Tzompantzi et al. 1999, Solis-Marin and Laguarda-Figueras 1999, 2010a, b; Godinez-Dominguez and Gonzalez-Sanson 1999; Barbosa-Ledesma et al. 2000; Laguarda-Figueras et al. 2001, 2002, 2004, 2005a, b, 2009; Duran-Gonzalez et al. 2005; Trujillo-Luna and Gonzalez-Vallejo 2006;

Solis-Marin 2008; Honey-Escandón et al. 2008; Torres-Martinez et al. 2008; Hernandez-Herrejon et al. 2008, 2010; Laguarda-Figueras and Solis-Marin 2009; Martinez-Melo and Solis-Marin 2010; Honey-Escandón et al. 2011).

Mexico has 643 species of echinoderms in its territorial seas. The Class Crinoidea is the least represented with 29 species (4 %) while the Class Ophiuroidea is the richest class with 197 species (31 %). The Class Asteroidea is the second richest class with 185 species (29 %) and the Class Echinoidea is the third most diverse group with 119 species (19 %) reported in both Pacific and Atlantic oceans. The Class Holothuroidea has 113 species (17 %).

It is important to make some recommendations: It is necessary to carry out more ecological studies. The low number and limited scope of publications about the ecology of echinoderms in Mexico provides evidence of an unequal and limited development of ecological studies of echinoderms. Few papers exist concerning analyses of echinoderm communities in Mexico. It is important to generate more information about their density and other population parameters, including growth, reproduction, feeding and their relation with the biotic and abiotic environment. Evaluation must be made of environmental factors that could influence their presence in a certain location, including studies about their vulnerability to climatic change. This information will be of great importance, not only for the scientific knowledge but for verification of ecological theories and a better description of the autoecologic (individual) and synecologic (community) relations. It could also establish strategies leading towards the sustainable production and administration of economically important echinoderms as a resource and search for alternatives to increase their production through aquaculture.

The study of sea cucumber aquaculture in Mexico has not begun despite their high market value and their detritivorous feeding which makes them attractive for cultivation. Since the body wall is the commercial product from sea cucumbers, it is necessary to perform studies on the optimization of its production. Considering that the techniques for holothurian aquaculture are being optimized, particularly in Asiatic regions, we should take advantage of this technological expansion to set up culture programs for our own species (i. e. *I. fuscus* and *I. badionotus*). This could also bring environmental profit, since it would diminish the illegal fishery that is currently observed extensively in Mexican waters.

It is very possible that the list of echinoderms present in Mexican waters is larger than the one presented here. In particular, the deep sea of the country has been little studied. Most expeditions covering this area have been made by institutions of other countries where a great amount of information that is difficult to access is deposited. Because of this, it is necessary to have available catalogues of echinoderms with information about their biology, ecology, distribution and fisheries, in order to improve the access to information and with it the promotion of studies on echinoderms of Mexico.

As we prepare and train more specialists in echinoderm studies in our country, it will be possible to amply advance the range of studies that are now required.

Acknowledgments To Dr. Alfredo Laguarda-Figueras for his critical review during the earliest drafts of this chapter. Students of the Laboratory of Systematics and Ecology of Echinoderms at UNAM made possible the creation of this chapter, together with the realization of the date extensive review, figures and plates design, and data base reports: J.I. Alvarez-Rojas, J.A. Arriaga-Ochoa, G. Bribiesca-Contreras, A.A. Caballero-Ochoa, N. Escandón-Flores, K. Evangelista-García, P. Garcés-Solchaga, L.A. Hernández-Herrejón, Y.Q. Hernánez-Díaz, P.A. Juárez Espinoza, C. Martín Cao-Romero, A. Martínez-Melo, T. Pineda-Enríquez, V. Tapia-Ramírez and M. Valdés-DeAnda. BEBS thanks DGAPA-PAPIIT (UNAM, Mexico) project IN105012, and CONACyT 165826 for funding field trips. FASM thanks DGAPA-PAPIIT IN207011 for funding field trips.

References

Agassiz A (1878–1879) Reports of the results of dredging, under the supervision of Alexander Agassiz, in the Gulf of Mexico, by the United States Coast Survey Steamer "Blake". II. Report on the Echini. Bull Mus Comp Zool 5:181–195

Agassiz A (1881a) Report of the echinoidea dredged by the H.M. *Challenger* during the year 1873–76. Report of the Scientific Results of the Voyage of the H. M. S. Challenger during the years 1873–1876. Zoology 3:1–321

Agassiz A (1881b) Monographies d'Echinodermes vivants et fossiles. Monographies d'Echinodermes, Scutelles. Anatomie. Neuchàtel, Fr. Nos. 1–4:155

Agassiz A (1888) Characteristic deep-sea types—Echinoderms. Three cruises of the United States Coast and Geodetic Survey Steamer "Blake" in the Gulf of Mexico, in the Caribbean Sea, and along the Atlantic Coast of the United States, from 1877 to 1880, vol 2. The Riverside Press, Cambrige, pp 84–127

Alencáster G, Buitron BE (1965) Fauna del Jurásico Superior de la región de Petlalcingo, Estado de Puebla. Paleontol Mex 21:1–53

Amador-Cano G, Carpizo-Ituarte E, Cristino J (2006) Role of protein kinase C, G-protein coupled receptors and calcium flux during metamorphosis of the sea urchin *Strongylocentrotus purpuratus*. Biol Bull 210:121–131

Anonymous (1994) Diario Oficial de la Federación. NOM-059-ECOL-1994. Determinación de las especies y subespecies de flora y fauna silvestre terrestres y acuáticas en peligro de extinción, amenazadas, raras y las sujetas a protección especial, y que establece especificaciones para su protección. Diario Oficial de la Federación, México, vol 16. Mayo, pp 2–59

Anzo-Martínez AD (1994) Ofiuroideos (Echinodermata: Ophiuroidea) de las playas Pedregosa y Manzanillo, Municipio de Zihuatanejo, Guerrero, México. Licenciatura thesis, Fac Biol, Univ Michoacana de San Nicolás de Hidalgo, Morelia, Michoacán, México

Applegate SP, Buitron BE, Solis-Marin FA, Laguarda-Figueras A (2009) Two Lower Cretaceous (Albian) fossil holothurians (Echinodermata) from Tepexi de Rodriguez, Puebla, Mexico. Proc Biol Soc Wash 122:91–102

Arriaga-Ochoa JA (2007) Holoturoideos (Echinodermata: Holothuroidea) del Golfo de California. Licenciatura thesis, Fac Biol, Univ Michoacana de San Nicolás de Hidalgo, Morelia, Michoacán, México

Arriaga-Ochoa JA (2010) Análisis Filogenético del género *Lissothuria* Verrill, 1867 (Dendrochirotida: Psolidae). Master thesis, Univ Nac Aut México, Mexico DF, Mexico

Bakus GJ (1974) Toxicity in holothurians: a geographical pattern. Biotropica 6:229–236

Barbosa-Ledesma IF, Solis-Marin FA, Laguarda-Figueras A (2000) New records for cidaroid echinoids (Echinodermata: Echinoidea) of the Gulf of Mexico, Mexico. Rev Biol Trop 48:721

Benítez-Villalobos F (2000) Comunidad de equinodermos asociados a la zona arrecifal Puerto Escondido Bahías de Huatulco, Oaxaca, México. Licenciatura thesis, Univ del Mar, Oaxaca, México

Benítez-Villalobos, F. 2001. Comparación de la comunidad de equinodermos asociada a arrecifes en dos localidades de las Bahías de Huatulco, Oaxaca, México. Cien Mar 5:19–24
Benitez-Villalobos F, Diaz Martinez JP, Martinez-Garcia M (2009) Mass mortality of the sea urchin *Diadema mexicanum* in La Entrega at Bahias de Huatulco, Western Mexico. Coral Reefs 28:1017
Benítez-Villalobos F, Domínguez MT, Lopez-Perez RA (2008) Temporal variation of the sea urchin *Diadema mexicanum* population density at Bahias de Huatulco Western México. Rev Biol Trop 56(Suppl 3):255–263
Boese E (1906) Nota preliminar sobre la fauna pliocénica de Santa Maria Tatetla. Bol Soc Geol Mex 2:1–51
Boese E (1910) Monografía geológica y paleontológica del cerro de Muleros cerca de Ciudad Juárez, Estado de Chihuahua. Bol Inst Geol 25:1–189
Boese E, Cavins A (1927) The Cretaceous and Tertiary of Southern Texas and Northern Mexico. Univ Texas Bull 2748:7–142
Boone L (1926) Echinoderms from the Gulf of California and the Perlas Islands. Scientific Results of the Second Oceanographic Expedition of the "Pawnee" 1926. Bull Bingham Oceanogr Coll 2:1–14
Bravo-Tzompantzi D (1996) Contribución al conocimiento de los equinodermos (Echinodermata: Echinoidea) del Caribe Mexicano: Puerto Morelos, Quintana Roo, México. Licenciatura thesis, Esc Biol, Ben Univ Aut Puebla, Puebla, México
Bravo-Tzompantzi D, Solís-Marín FA, Laguarda-Figueras A, Abreu-Pérez M, Durán-González A (1999) Equinoideos (Echinodermata: Echinoidea) del Caribe Mexicano: Puerto Morelos, Quintana Roo, México. Avicennia 10/11:43–72
Bribiesca-Contreras G (2011) Biodiversidad de ofiuroideos (Echinodermata: Ophiuroidea) del arrecife Isla Verde y análisis de registros históricos para el Parque Nacional Sistema Arrecifal Veracruzano (PNSAV). Licenciatura thesis, Fac Cien, Univ Nac Aut México, México DF, México
Brusca RC (1973) A handbook to the common intertidal invertebrates of the Gulf of California. University of Arizona Press, Tucson
Brusca RC (1980) Echinodermata. In: Brusca RC (ed) Common intertidal invertebrates of the Gulf of California. University of Arizona Press, Tucson, pp 398–422
Bryan PJ, Rittschof D, McClintock JB (1996) Bioactivity of echinoderm ethanolic body-wall extract: an assessment of marine bacterial attachment and marcoinvertebrate larval settlement. J Exp Mar Biol Ecol 196:79–96
Bryan PJ, McClintock JB, Hopkins TS (1997) Structural and chemical defenses of echinoderms from the nortern Gulf of Mexico. J Exp Mar Biol Ecol 210:173–186
Buitrón BE (1968) Catálogo de equinoides fósiles de México. Paleontol Mex 26:1–50
Buitrón BE (1970a) Equinoides del Cretácico del norte de México. Paleontol Mex 31:1–70
Buitrón BE (1970b) Equinoides del Cretácico Inferior de la región de San Juan Raya-Zapotitlán, estado de Puebla. Paleontol Mex 30:1–64
Buitrón BE (1973a) *Tetragramma gloriae* n. sp., equinoide de la Formación San Lucas (Aptiano) en al región de Huetamo, Michoacán. Paleontol Mex 36:33–41
Buitrón BE (1973b) Equinoides del Cretácico Inferior de Colima. Paleontol Mex 36:21–29
Buitrón BE (1974a) *Caenocidaris* cf. *C. cucumifera* (Agassiz) equinoide del Jurásico Superior de Chihuahua. Paleontol Mex 40:48–51
Buitrón BE (1974b) Algunas especies de la Familia Conulidae Lambert, 1911 (Echinoidea) del Cretácico Superior de Chiapas. Paleontol Mex 39:1–24
Buitrón BE (1974c) Algunos equinoides del Oligoceno de Chiapas. Paleontol Mex 39:24–38
Buitrón BE (1976) Tres especies de equinoides (Echinodermata-Echinoidea) del Cretácico Superior de Atenango del Río, Guerrero. Bol Soc Geol Mex 37:18–24
Buitrón BE (1978a) Distribución de los equinoides terciarios en la Planicie Costera del Golfo de México en América Central, en el norte de América del sur y en Las Antillas. Inst Geol 101:66–113

Buitrón BE (1978b) Hemicidáridos (Echinodermata-Echinoidea) del Jurásico Superior de Oaxaca y Chiapas. Inst Geol 2:65–68
Buitrón BE (1990a) Catálogo de equinoides del Jurásico Superior y del Cretácico Inferior de México. Inst Mex del Petróleo, Subdirección de Exploración, Mexico DF, México
Buitrón BE (1990b) Presencia de *Angulocrinus polyclonus* (Felix) en el Oxfordiano del sur de México. Rev Subdir Explor 3:19–25
Buitrón BE, Silva SB (1979) Dos especies de equinoides (Echinodermata: Echinoidea) del Eoceno Tardío de Tantoyuca, Veracruz. Inst Geol 3:122–125
Buitrón BE, Solis-Marin FA (1993) La biodiversidad en los equinodermos fósiles y recientes de México. Rev Soc Mex Hist Nat 44:209–231
Buitrón BE, Patiño J, Moreno A (1987) Crinoides del Paleozoico Tardío (Pensilvánico) de Calnali, Hidalgo. Soc Mex Paleontol 1:125–136
Buitrón BE, Almazán VE, Vachard D (2007a) Middle Permian crinoids (Echinodermata, Crinoidea) from Cerros Los Monos, Caborca, Sonora, México and paleogeographic considerations. Rev Mex Geol 24:344–353
Buitrón BE, Almazán VE, Gómez EC, Vachard D (2007b) A late Atokan regional encrinite (early late Moscovian, Middle Pennsylvanian) in the Sierra Agua Verde, Sonora State, NW Mexico. In: Alvaro JJ, Aretz M, Boulvain F, Munnecke A, Vachard D, Vennin E (eds) Palaeozoic Reefs and Bioaccumulations: Climatic and Evolutionary Controls. Geol Soc, London Special Publications, London, pp 201–209
Buitrón BE, Almazán VE, Vachard D, Laguarda-Figueras A, Solis-Marin FA (2008) A review of a crinoid columnals (Echinodermata-Crinoidea) from the Carboniferous of Mexico. Rev Biol Trop 56(Suppl 3):1–12
Caballero-Ochoa AA, Laguarda-Figueras A (2010) Zoogeography of holothurians (Echinodermata: Holothuroidea) of the Mexican Pacific Ocean. In: Reich M, Reitner J, Roden V, Thuy B (eds) Echinoderm research 2010, 7th European Con Echinoderms, Göttingen: abstract volume and field guide to excursions. University of Göttingen, Germany, pp 26–27
Carpenter PH (1884) Report of the crinoidea collected by H.M.S. Challenger during the years 1873-76. Report of the Scientific Results of the Voyage of the H. M. S. Challenger during the years 1873–1876. Zoologica 3:1–21
Carpizo-Ituarte E, Salas-Garza A, Pares-Sierra G (2002) Induction of metamorphosis with KCl in three species of sea urchins and its implications in the production of juveniles. Cien Mar 28:157–166
Caso ME (1941) Contribución al conocimiento de los asteridos de México. I. La existencia de *Linckia guildinguii* Gray, en la costa pacífica. An Inst Biol UNAM 12:155–160
Caso ME (1943) Contribución al conocimiento de los asteridos de México. Master thesis, Fac Cien, Univ Nac Aut México, México DF, México
Caso ME (1944) Estudio sobre astéridos de México. Algunas especies interesantes de astéridos litorales. An Inst Biol UNAM 15:237–257
Caso ME (1945) Modificación de la Familia Luidiidae Verrill. Las subfamilias nuevas de la Familia Luidiidae y observaciones de *Platasterias latiradiata*. An Inst Biol UNAM 16:459–473
Caso ME (1946) Contribución al conocimiento de los Equinodermos de México. I. Distribución y morfología de *Mellita quinquiesperforata* (Leske), *M. lata* Clark y *M. longifissa* Michelin. An Inst Biol UNAM17:247–259
Caso ME (1947) Estudios sobre Astéridos de México. Descripción de una nueva especie del género *Moiraster* de Santa Rosalía, Golfo de California. An Inst Biol UNAM 18:225–231
Caso ME (1948a) Contribución al conocimiento de los equinodermos de México. II. Algunas especies de equinoideos litorales. An Inst Biol UNAM 19:183–231
Caso ME (1948b) Datos históricos y estado actual de la fauna de astéridos de México. Soc Mex Hist Nat 1:21–32
Caso ME (1949) Contribución al conocimiento de los equinodermos litorales de México. An Inst Biol UNAM 20:341–355

Caso ME (1951a) Contribución al conocimiento de los ofiuroideos de México. 1. Algunas especies de ofiuroideos litorales. An Inst Biol UNAM 22:219–312

Caso ME (1951b) Los equinoideos fósiles del Cenozoico de México. Bol Asoc Mex Geól Petroleros 3:57–96

Caso ME (1953) Estado actual de los conocimientos acerca de la fauna de los equinodermos de México. Mem Congr Cient Mex UNAM 7:209–222

Caso ME (1954) Contribución al conocimiento de los holoturoideos de México. Algunas especies de Holoturoideos litorales y descripción de una nueva especie *Holothuria portovallartensis*. An Inst Biol UNAM 25:417–422

Caso ME (1955) Contribución al conocimiento de los holoturoideos de México. II. Algunas especies de los holoturoideos litorales de la costa Atlántica Mexicana. An Inst Biol UNAM 26:501–525

Caso ME (1956) El género *Clypeaster* Lamarck 1801, en el Terciario de México. An Inst Biol UNAM 27:487–528

Caso ME (1958) Contribución al conocimiento de los holoturoideos de México. III. Algunas especies de Holoturoideos litorales de la costa pacífica de México. An Inst Biol UNAM 28:309–338

Caso ME (1961a) Estudios sobre Astéridos de México. Observaciones sobre especies de *Tethyaster* de las costas de México. An Inst Biol UNAM 31:449–461

Caso ME (1961b) Los Equinodermos de México. PhD thesis, Fac Cien, Univ Nac Aut México, México DF, México

Caso ME (1962a) Estudios sobre Astéridos de México. Observaciones sobre especies pacíficas del género *Acanthaster* y descripción de una subespecie nueva, *Acanthaster ellisi pseudoplanci*. An Inst Biol UNAM 32:313–331

Caso ME (1962b) Estudios sobre equinodermos de México. Contribución al conocimiento de los equinodermos de la islas Revillagigedo. An Inst Biol UNAM 33:293–330

Caso ME (1963) Contribución al conocimiento de los holoturoideos de México. Descripción de una n. sp. de *Holothuria* de un nuevo subgénero (*Paraholothuria* n. sg.). An Inst Biol UNAM 33:367–380

Caso ME (1964) Contribución al conocimiento de los Holoturoideos de México. Descripción de un nuevo subgénero del Género *Microthele* y una nueva especie *Microthele* (*Paramicrothele*) *zihuatanensis*. An Inst Biol UNAM 35:105–114

Caso ME (1965) Estudio sobre Equinodermos de México. Contribución al conocimiento de los Holoturoideos de Zihuatanejo y de la Isla de Ixtapa (primera parte). An Inst Biol UNAM 36:253–291

Caso ME (1967a) Contribución al estudio de los holoturoideos de México. Morfología interna y ecología de *Stichopus fuscus* Ludwig. An Inst Biol UNAM 37:175–181

Caso ME (1967b) Contribución al estudio de los holoturoideos de México. Morfología y sistemática de *Euthyonidium ovulum* (Selenka). An Inst Biol UNAM Ser Cienc Mar Limnol 38:1–10

Caso ME (1968a) Contribución al estudio de los holoturoideos de México. La Familia Psolidae. Descripción de una nueve especie del género *Psolus*. *Psolus conchae* sp. nov. An Inst Biol UNAM Ser Cienc Mar Limnol 39:1–19

Caso ME (1968b) Contribución al conocimiento de los holoturoideos de México. Ecología y morfología de *Holothuria glaberrima* Selenka. An Inst Biol UNAM Ser Cienc Mar Limnol 39:21–30

Caso ME (1968c) Contribución al estudio de los holoturoideos de México. Un caso de parasitismo de *Balcis intermedia* (Cantraine) sobre *Holothuria glaberrima* Selenka. An Inst Biol UNAM Ser Cienc Mar Limnol 39:31–40

Caso ME (1970) Contribución al conocimiento de los Asterozoa de México. Situación taxonómica actual, morfología externa y datos ecológicos de *Platasterias latiradiata* Gray. An Inst Biol UNAM Ser Cienc Mar Limnol 41:1–62

Caso ME (1971a) Contribución al estudio de los Holoturoideos de México. Morfología externa e interna y ecología de *Holothuria grisea* Selenka. An Inst Biol UNAM Ser Cienc Mar Limnol 42:31–40

Caso ME (1971b) Contribución al conocimiento de los Equinoideos de México. Estudio morfológico de *Brissopsis alta* Mortensen, erizo de profundidad. An Inst Biol UNAM Ser Cienc Mar Limnol 42:41–56

Caso ME (1971c) Los pepinos de mar. Nat 2:19–22

Caso ME (1972a) La importancia de *Platasterias latiradiata* Gray en el estudio de los Equinodermos. Mem IV Congr Nac Oceanogr México 217–221

Caso ME (1972b) El aspecto económico de las holoturias o pepinos de mar en la alimentación humana. El Trépang. Rev Soc Mex Hist Nat 33:85–98

Caso ME (1972c) El género *Acanthaster*. Su biología, ecología y su efecto destructor de los arrecifes. Rev Soc Mex Hist Nat 33:51–83

Caso ME (1974a) Morfología externa de *Acanthaster planci* (Linnaeus). J Mar Biol Assoc India 16:83–93

Caso ME (1974b) Contribución al estudio de los Equinoideos de México. El género *Tripneustes* Agassiz. Morfología y ecología de *Tripneustes ventricosus* (Lamarck). An Cent Cien Mar Limnol UNAM 1:1–24

Caso ME (1974c) Contribución al estudio de los Equinoideos de México, Morfología de *Tripneustes depressus* Agassiz y estudio comparativo entre *T. ventricosus* y *T. depressus*. An Cent Cien Mar Limnol UNAM 1:25–40

Caso ME (1975) Contribución al conocimiento de los Asterozoa de México. La Familia Mithrodiidae. Descripción de una nueva especie del género *Mithrodia*. *Mithrodia enriquecasoi* sp. nov. An Cent Cien Mar Limnol UNAM 2:1–28

Caso ME (1976) El estado actual del estudio de los equinodermos de México. An Cent Cien Mar Limnol UNAM 3:1–56

Caso ME (1977) Especies de la Familia Asterinidae en la costa Pacífica de México. Descripción de una nueva especie del género *Asterina*, *Asterina agustincasoi* sp. nov. An Cent Cien Mar Limnol UNAM 4:209–232

Caso ME (1978a) Ciencia y Técnica de los Equinodermos en relación con el hombre. Primera parte aspecto científico. An Cent Cien Mar Limnol UNAM 5:255–286

Caso ME (1978b) Los Equinoideos del Pacífico de México. Parte 1. Ordenes Cidaroidea y Aulodonta; Parte 2. Ordenes Stiridonta y Camarodonta. An Cent Cien Mar Limnol UNAM Publ Esp 1:1–244

Caso ME (1979a) Los Equinodermos de la Bahía de Mazatlán, Sinaloa. An Cent Cien Mar Limnol UNAM 6:197–368

Caso ME (1979b) Descripción de una nueva especie de ofiuroideo de la laguna de Términos, *Amphiodia guillermosoberoni* sp. nov. An Cent Cien Mar Limnol UNAM 6:161–184

Caso ME (1979c) Los equinodermos (Asteroidea, Ophiuroidea y Echinoidea) de la Laguna de Términos, Campeche. An Cent Cien Mar Limnol UNAM Publ Esp 3:1–186

Caso ME (1980a) Contribución al estudio de los Echinozoa de México. La Familia Mellitidae Stefanini. Descripción de una nueva especie del género *Mellita*, *Mellita eduardobarrosoi* sp. nov. An Cent Cien Mar Limnol UNAM 7:141–180

Caso ME (1980b) Los equinoideos del Pacífico de México. Parte Tercera. Orden Clypeasteroida. An Cent Cien Mar Limnol UNAM Publ Esp 4:1–252

Caso ME (1983a) Los equinoideos del Pacífico de México. Parte Cuarta. Ordenes Cassiduloida y Spatangoida. An Cent Cien Mar Limnol UNAM Publ Esp 6:1–200

Caso ME (1983b) Especies del género *Amphiochondrius* de la costa del Pacífico Americano. Descripción de una nueva especie del género *Amphichondrius*. *Amphichondrius unamexici* sp. nov. (Ophiuroidea). An Cent Cien Mar Limnol UNAM 10:209–230

Caso ME (1984) Descripción de un género nuevo y una especie nueva de Holoturoideo. *Parathyonacta* gen. nov. y *Parathyonacta bonifaznuñoi* sp. nov. colectada en la campaña oceanográfica SIPCO III a bordo del Buque Oceanográfico "El Puma". An Cent Cien Mar Limnol UNAM 11:181–210

Caso ME (1986a) Los Equinodermos del Golfo de California colectados en las campañas SIPCO I-II-III a bordo del B/O "El Puma". An Cent Cien Mar Limnol UNAM 13:91–184

Caso ME (1986b) Descripción de una nueva especie de ofiuroideo de la Bahía de Mazatlán, Sin. *Ophioderma sodipallaresi* sp. nov. y comparación con *Ophioderma variegatum* Lütken. An Cent Cien Mar Limnol UNAM 13:223–248

Caso ME (1990) Un nuevo asteroideo del Caribe mexicano. *Astropecten caribemexicanensis* sp. nov. y comparación con la especie afín *Astropecten nitidus* Verrill. An Cent Cien Mar Limnol UNAM 17:107–130

Caso ME (1992) Los equinodermos (asteroideos, ofiuroideos y equinoideos) de la Bahía de Mazatlán, Sinaloa. An Cent Cien Mar Limnol UNAM Publ Esp 11:1–214

Caso ME (1994a) Los pepinos del mar. InforMAR. Gaceta Informativa de la Unidad de Educación en Ciencia y tecnología del Mar, SEP, SELP 1:21–23

Caso ME (1994b) Estudio morfológico, taxonómico, ecológico y distribución geográfica de los asteroideos colectados durante las campañas oceanográficas Cortés 1, 2, 3. An Cent Cien Mar Limnol UNAM Publ Esp 12:1–111

Caso ME (1996a) III Asteroideos del Caribe mexicano colectados en las campañas oceanográficas PROIBE II-III-IV-V a bordo del B/O "Justo Sierra". An Cent Cien Mar Limnol UNAM 22:27–57

Caso ME (1996b) IV Asteroideos del Caribe mexicano colectados en las campañas oceanográficas PROIBE I-II-III-IV-V a bordo del B/O "Justo Sierra". An Cent Cien Mar Limnol UNAM 22:59–82

Caso ME (1996c) Las especies del género *Luidia* Forbes y *Astropecten* Gray del Caribe Mexicano, colectadas en las campañas oceanográficas PROIBE II-III-IV-V a bordo del B/O "Justo Sierra". An Cent Cien Mar Limnol UNAM 22:1–25

Caso ME (1996d) I. Las especies del género *Hesperocidaris* Mortensen, *Astropyga* Gray, *Arbacia* Gray y *Lytechinus* Agassiz, colectadas en las campañas oceanográficas Cortes 1, 2, 3 a bordo del B/O "El Puma". An Cent Cien Mar Limnol UNAM 22:83–100

Caso ME, Laguarda-Figueras A, Solís-Marín FA, Ortega-Salas A, Durán-González AL (1994) Contribución al conocimiento de la ecología de las comunidades de equinodermos de la Laguna de Términos, Campeche, México. An Cent Cien Mar Limnol UNAM 21:67–85

Caso ME, Laguarda-Figueras A, Solís-Marín FA, Ortega-Salas A, Durán-González AL (1996) Contribución al conocimiento de la ecología de las comunidades de equinodermos de la bahía de Mazatlán, Sinaloa, México. An Cent Cien Mar Limnol UNAM 22:101–119

Castañeda-Sarabia O (1996) Contribución al conocimiento del macrobentos de la infauna del margen continental del Golfo de México. Licenciatura thesis, Fac Cien, Univ Nac Aut México, México DF, México

Castellanos-Kotkoff M, Carpizo-Ituarte E, De-La-Rosa-Vélez J (2004) Thyroxine effect in metamorphosis and presence of the nuclear receptor COUP-TF in the larvae of the sea urchin *Lytechinus pictus*. Integr Comp Biol 44:681

Celaya-Hernández EV (2006) Erizos regulares (Echinodermata: Echinoidea) de la parte sur de la laguna arrecifal de Isla Verde, Veracruz, Ver., México. Licenciatura thesis, Fac Cien, Univ Nac Aut de México, México DF, México

Celaya-Hernández EV, Solis-Marin FA, Laguarda-Figueras A, Durán-González A (2008) Asociación a sustratos de los erizos regulares (Echinodermata: Echinoidea) en la laguna arrecifal de Isla Verde, Veracruz, México. Rev Biol Trop 56(Suppl 3):281–295

Chase ER (1956) Additional notes on the Pliocene and Pleistocene fauna of Turtle Bay area, Baja California, Mexico. Trans San Diego Soc Nat Hist 12:177–180

Cintra-Buenrostro CE (2001) Los Asteroideos (Echinodermata: Asteroidea) de aguas someras del Golfo de California, México. Oceánides 16:49–90

Cintra-Buenrostro CE, Reyes-Bonilla H, Arizpe-Covarrubias O (1998) Los equinodermos (Echinodermata) del arrecife de Cabo Pulmo, Pacífico de México. Rev Biol Trop 46:341–344

Cintra-Buenrostro CE, Reyes-Bonilla H, Herrero-Pérezrul MD (2005) Oceanographic conditions and diversity of sea stars (Echinodermata: Asteroidea) in the Gulf of California, Mexico. Rev Biol Trop 53(Suppl 3):245–261

Clark AH (1916) Six new starfishes from the Gulf of California and adjacent waters. Proc Biol Soc Wash 29:51–62

Clark AH (1918) A new genus and species of multibrachiate ophiuran of the family Gorgonocephalidae from the Caribbean Sea. Proc U S Nat Mus 54:637–640

Clark AH (1954) Echinoderms (other than holothurians) of the Gulf of Mexico. U S Fish Wildl Serv Fish Bull 55:373–379

Clark HL (1913) Echinoderms from Lower California, with descriptions of new species. Bull Am Mus Nat Hist 32:185–236

Clark HL (1917) Ophiuroidea. Report XVIII and XXX on the scientific results of the tropical Pacific in charge of Alexander Agassiz, on the U. S. Fish Commission steamer "*Albatross*" from August 1899 to March 1900 and from October 1904 to March 1905. Bull Mus Comp Zool Harv Coll 61:429–453

Clark HL (1920a) Asteroidea. XXXII Report on the scientific results of the expedition to the eastern tropical pacific, in charge of Alexander Agassiz, by the U. S. Fish. Comission Steamer "*Albatross*", from October 1904 to March 1905, Lieut. Commander L. M. Garrett, U.S.N., Commanding. Bull Mus Comp Zool Harv Coll 39:73–113

Clark HL (1920b) Holothuroidea. XXXIII. Reports on the scientific results of the expedition to the eastern tropical pacific, in charge of Alexander Agassiz, by the U. S. Fish. Comission Steamer "Albatross", from October 1904 to March 1905, Lieut. Commander L. M. Garrett, U.S.N., Commanding. Mem Mus Comp Zool Harv Coll 39:119–154

Clark HL (1923) Echinoderms from Lower California, with descriptions of new species: supplementtary report. Scientific results of the Expedition to the Gulf of California in charge of C. H. Towsend, by the U. S. Fisheries Steamship "*Albatross*" in 1911. Commander G. H. Burrage, U. S. N. Commanding. Bull Mem Am Mus Nat Hist 48:147–163

Clark HL (1933) Scientific survey of Porto Rico and Virgin Islands. Part. 1. A handbook of the littoral echinoderms of Puerto Rico and the other West Indian Islands. NY Acad Sci 16:1–147

Clark HL (1940) Notes on Echinoderms from the West Coast of Central America. Eastern Pacific Expeditions of the New York Zoological Society. Zoologica 25:331–352

Clark HL (1948) A report of the Echini of the warmer Eastern Pacific, based on the collections of the "*Velero*" III. Allan Hancock Pac Exp 8:225–351

Cooke CW (1955) Some Cretaceous echinoids from the Americas. US Geol Surv Prof Pap 254E:87–112

Cortés-Fernández H (1999) Los ofiuroideos (Echinodermata: Ophiuroidea) de Isla los Islotes La Bahía de la Paz, B.C.S., México. Licenciatura thesis, Univ Aut Baja California Sur, La Paz, México

Cutress BM (1996) Changes in dermal ossicles during somatic growth in Caribbean littoral sea cucumbers (Echinodermata: Holothuroidea: Aspidochirota). Bull Mar Sci 58:44–116

De la Lanza-Espino G (1991) Oceanografía de mares mexicanos. AGT Editor SA, México

Deichmann E (1930) The holothurians of the western part of the Atlantic Ocean. Bull Mus Comp Zool Harv Coll 71:41–226

Deichmann E (1937) The Templeton Crocker Expedition. IX. Holothurians from the Gulf of California, the West Coast of Lower California and Clarion Island. Zoologica 22:161–176

Deichmann E (1938) Eastern Pacific Expeditions of the New York Zoological Society. XVI. Holothurians from the western coasts of Lower California and Central America, and from Galápagos Islands. Zoologica 23:361–387

Deichmann E (1941) The holothuroidea collected by the *Velero III* during the years 1932 to 1938. Part I. Dendrochirota. Allan Hancock Pac Exp 8:61–195

Deichmann E (1954) The holothurians of the Gulf of Mexico. Fish Bull Fish Wildl Serv 55:381–410

Deichmann E (1958) The Holothuroidea collected by the *Velero III* and *IV* during the years 1932 to 1954. Part. II Aspidochirota. Allan Hancock Pac Exp 11:253–348

Deichmann E (1963) Shallow water holothurians known from the Caribbean waters. Stud fauna Curacao Other Carib Isl 14:100–118

Dickerson RE, Kew WS (1917) The fauna of a Middle Tertiary formation and the associated horizons of Northeastern Mexico. Proc Calif Acad Sci 4:125–156

Domantay JS (1953) A brief summary of the Pacific and Atlantic Holothuroidea of the Allan Hancock Foundation Collections. Philip J Sci 82:133–140

Domantay JS (1961) New forms of Holothuroidea from the vicinity of Southern California and Mexico in the collection of the Allan Hancock Foundation, University of Southern California in Los Angeles, California. Philip J Sci 90:333–346

Domínguez-Castanedo NC (2007) Estudio de la macrofauna bentónica de la Laguna Arrecifal de Isla Sacrificios, Veracruz. Master thesis, Univ Nac Aut México, México DF, México

Downey ME (1972) *Midgardia xandaros* new genus, new species, a large brisingid starfish from the Gulf of Mexico. Proc Biol Soc Wash 84:421–426

Downey ME (1973) Starfishes from the Caribbean and the Gulf of Mexico. Smith Control Zool 126:1–158

Dumble ET (1918) Geology of the northern and of Tampico embayment area. Proc Calif Acad Sci 4:113–156

Durán-González A, Laguarda-Figueras A, Solis-Marin FA, Buitrón Sánchez BE, Ahearn CG, Torres-Vega J (2005) Equinodermos (Echinodermata) de las aguas mexicanas del Golfo de México. Rev Biol Trop 53(Suppl 3):53–68

Durham JW (1950) E.W Scripps 1940 cruises to the Gulf of California. Part. 2: Megascopic paleontology and marine stratigraphy. Mem Geol Soc Am 43:1–216

Durham JW (1961) The echinoid *Mellita* in the Pacific coast Cenozoic. LA Coun Mus Cont Sci 48:1–12

Ebert TA (2010) Demographic patterns of the purple sea urchin *Strongylocentrotus purpuratus* along a latitudinal gradient, 1985–1987. Mar Ecol Prog Ser 406:105–120

Ebert TA, Southon JR (2003) Red sea urchins (*Strongylocentrotus franciscanus*) can live over 100 years: confirmation with A-bomb ^{14}carbon. Fish Bull 101:915–922

Encarnación R, Carrasco G, Espinoza M (1989) Neothyside A, proponed structure of a triterpenoid tetraglycoside from the pacific sea cucumber, *Neothyone gibbosa*. J Nat Prod 52:248–251

Encarnación R, Murillo JI, Nielsen J, Christophersen C (1996) Neothyoside B, a triterpenoid diglycoside from the pacific sea cucumber *Neothyone gibbosa*. Acta Chem Scand 50:848–849

Espino-Barr E, Cibrián RR, García-Boa A (1996) Estructura y densidad de la población del erizo tinta *Diadema mexicanum* en el litoral rocoso del estado de Colima. Cienc Pesq 2:60–67

Esquivel-Macías C, Solis-Marin FA, Buitrón-Sánchez BE (2004) Nuevos registros de placas columnares de crinoides (Echinodermata, Crinoidea) del Paleozoico Superior de México, algunas implicaciones paleogeográficas y paleoambientales. Coloq Paleontol 54:15–23

Estrada-Rodríguez P (2011) Estrellas de mar (Echinodermata: Asteroidea) del Golfo de California. Licenciatura thesis, Univ Nac Aut de México, México DF, México

Fajardo-León MC, Turrubiates-Morales JR (2009) Distribución y densidad de *Parastichopus parvimensis* de Puerto Escondido a Punta Abreojos en la costa noroeste de Baja California Sur. Cienc Pesq 17:13–20

Fajardo-León MC, Michel E, Singh J, Vélez JA, Masso A (1995) Estructura poblacional y ciclo reproductor del pepino de mar *Isostichopus fuscus* en Santa Rosalía, BCS México. Cienc Pesq 11:45–53

Fajardo-León MC, Suárez-Higuera MCL, Valle-Manríquez A, Hernández-López A (2008) Reproductive biology of the sea cucumber *Parastichopus parvimensis* (Echinodermata: Holothuroidea) at Isla Natividad and Bahía Tortugas, Baja California Sur, México. Cienc Mar 34:165–177

Fell HB (1962) A surviving somasteroid from the eastern Pacific Ocean. Science 136:633–636

Fisher WK (1906) New starfishes from the Pacific Coast of North America. Proc Acad Sci Wash 8:111–139

Fodrie FJ, Herzka AJ, Lucas AJ, Ramos FV (2007) Intraspecific density regulates positioning and feeding mode selection of the sand dollar *Dendraster excentricus*. J Exp Mar Biol Ecol 340:169–183

Francisco V, Herzka SZ (2010) Regulation of feeding mode by the sand dollar *Dendraster excentricus* in a shallow estuarine habitat. J Exp Mar Biol Ecol 383:146–155

Frontana-Uribe SC (2002) Contribución al conocimiento de la composición faunistica de los anélidos poliquetos y equinodermos asociados a sustratos duros de la Isla Socorro, Archipiélago de Revillagigedo, México. Licenciatura thesis, Univ Nac Aut México, México DF, México

Frontana-Uribe SC (2005) Filogeografía de *Pisaster ochraceus* Brandt, 1835 (Echinodermata: Asteroidea) a lo largo de su distribución. Master thesis, Cen Inves Cien Ed Sup Ensenada, Ensenada, México

Fuentes-Farias AM, Villarroel-Melo MD, Solis-Marin FA (2005) Los invertebrados marinos. In: Villaseñor-Gómez LE (ed) La Biodiversidad en Michoacán. Estudio de Estado, Suma, CONABIO, UMSNH y Gobierno del Estado de Michoacán, pp 85-88

Gamboa-Contreras JA (1978) Estudio preliminar acerca de la fauna de equinodermos de la costa norte del estado de Veracruz. Licenciatura thesis, Univ Aut Nuevo León, Monterrey, México

García-Barrera P, Pantoja J (1991) Equinoides del Albiano Tardío de la Formación Mal Paso de la región de Chumbítaro, estados de Guerrero y Michoacán, México. Soc Mex Paleontol 4:23–41

Godínez-Domínguez E, González-Sansón G (1999) Diversidad de macroinvertebrados de fondos blandos de la plataforma continental de Jalisco y Colima, México. Cienc Mar 25:609–627

Gómez-Carriedo BS (2001) Relación del Phylum Echinodermata con su hábitat en la costa Michoacana. Licenciatura thesis, Univ Michoacana de San Nicolás de Hidalgo, Morelia, México

González-Azcárraga A (2009) Estructura de las Asociaciones y Diversidad morfológica de erizos de mar (Echinoidea) en los Parques Nacionales Sistema Arrecifal Veracruzano y Arrecifes de Cozumel, México. Masters thesis, Univ Nac Aut México, Mazatlán, Sinaloa, México

Gore RH (1992) The Gulf of Mexico. Pineapple, Sarasota

Hartlaub C (1895) Reports on the dredging operations off the west coast of Central America to the Galapagos, to the west coast of Mexico, and in the Gulf of California, in charge of Alexander Agassiz, carried on by the U. S. Fish Commission steamer "*Albatross*", during 1891, Lieut. Commander Z. L. Tanner, U. S. N., commanding. 13. Die Comatulien. Bull Mus Comp Zool Harv Coll 27:129–152

Hearn A, Martinez P, Toral-Martinez MV, Murillo JC, Polovina J (2005) Population dynamics of the exploited sea cucumber *Isostichopus fuscus* in the western Galapagos Islands Ecuador. Fish Oceanogr 14:377–385

Hendler G (1996) Echinodermata collected at Rocas Alijos. In: Schmieder RW (ed) Rocas Alijos. Scientific results from the cordell expeditions. Monogr Biol. Kluwer Acad Publishers, London, pp 319–338

Hendler G, Peck RW (1988) Ophiuroids off the deep end: Fauna of the Belizean fore-reef slope. In: Burke RD, Randewov PV, Lambert P, Parsley RL (eds) Echinoderm biology: Proceedings of the 6th int echinoderm Conference, Victoria. AA Balkema, Rotterdam, Brookfield, pp 411–419

Hendler G, Turner RL (1987) Two new species of *Ophiolepis* (Echinodermata: Ophiuroidea) from the Caribbean Sea and Gulf of Mexico: With notes on ecology, reproduction and morphology. Contrib Sci 395:1–14

Hendler G, Miller JE, Pawson DL, Kier PM (1995) Sea stars, sea urchins and allies: echinoderms of Florida and the Caribbean. Smith Institute Press, Washington DC

Hernández-Díaz YQ (2011) Zoogeografía de equinodermos (Echinodermata) de los Bajos de Sisal y Arrecife Alacranes Yucatan, México. Master thesis, Univ Nac Aut México, México DF, México

Hernández-Herrejón LA (2007) Ofiuroideos (Echinodermata: Ophiuroidea) del Golfo de México. Licenciatura thesis, Univ Michoacana de San Nicolás de Hidalgo, Morelia, México

Hernández-Herrejón LA (2010) Filogenia del género *Ophioderma* Müller and Troschel, 1840 (Ophiuroidea: Echinodermata). Master thesis, Univ Nac Aut México, México DF, México

Hernández-Herrejón LA, Laguarda-Figueras A, Solis-Marin FA (2008) Ofiuroideos (Echinodermata: Ophiuroidea) de las aguas mexicanas del golfo de México. Rev Biol Trop 56(Suppl 3):83–167

Hernández-Herrejón LA, Solis-Marin FA, Laguarda-Figueras A, Pineda-Enríquez T (2010) First record of *Ophioderma vansyoci* (Echinodermata: Ophiuroidea) in the Gulf of California. Mar Biod Rec 3:DOI 10.1017/S1755267210001028

Hernández-Pliego P (1998) Pepinos de Mar (Echinodermata: Holothuroidea) del estado de Quintana Roo, México, de la Colección Nacional de Equinodermos del Instituto de Ciencias del Mar y Limnología de la UNAM. Licenciatura thesis, Univ Aut Estado de México, México DF, México

Herrera-Escalante T, López-Pérez RA, Leyte-Morales GE (2005) Bioerosion caused by the sea urchin *Diadema mexicanum* (Echinodermata: Echinoidea) at Bahías de Huatulco, Western Mexico. Rev Biol Trop 53(Suppl 3):263–273

Herrero-Pérezrul MD (1990) Pepinos de mar (Echinodermata: Holothuroidea) de "El Faro de Bucerías", Michoacán, México. Licenciatura thesis, Univ Michoacana de San Nicolás de Hidalgo, Morelia, México

Herrero-Pérezrul MD (1994) Estudio comparativo de la reproducción de *Isostichopus fuscus* (Ludwig, 1875) y *Neothyone gibosa* Deichmann, 1941 (Echinodermata: Holothuroidea), en la Bahía de La Paz, Baja California Sur, México. Master Thesis, Cent Interdis Cien Mar, Inst Politec Nac, La Paz, México

Herrero-Pérezrul MD (2004) Análisis de la pesquería de pepino de mar en la Península de Baja California, México. Doctoral thesis, Cent Interdis Cien Mar, Inst Politec Nac, La Paz, México

Herrero-Pérezrul MD (2010) Informe final de actividades. Recurso pepino de mar *Isostichopus fuscus* dentro del Predio Federal de Aprovechamiento LA PAZ en Baja California Sur, México (registro ZF-DGVS-0030-BCS). Temporada 2008-2010. SEMARNAT-DGVS

Herrero-Pérezrul MD, Chávez-Ortíz EA (2005) Optimum fishing strategies of *Isostichopus fuscus* (Echinodermata: Holothuroidea) in the Souther Gulf of California, Mexico. Rev Biol Trop 53(Suppl 3):357–366

Herrero-Pérezrul MD, Reyes-Bonilla H (2008) Weight-Length relationship and relative condition of the holothurian *Isostichopus fuscus* at Espiritu Santo Island, Gulf of California, Mexico. Rev Biol Trop 56(Suppl 3):273–280

Herrero-Pérezrul MD, Reyes-Bonilla H, García-Domínguez F, Cintra-Buenrostro CE (1999) Reproduction and growth of *Isostichopus fuscus* (Echinodermata: Holothuroidea) in the southern Gulf of California, Mexico. Mar Biol 135:521–532

Herrero-Pérezrul MD, Reyes-Bonilla H, González-Azcárraga A, Cintra-Buenrostro CE, Rojas-Sierra A (2008) Equinodermos. In: Daneman GD, Ezcurra E (eds) Bahía de los Angeles: Recursos Naturales y Comunidad, línea Base 2007. Secretaría de Medio Ambiente y Recursos Naturales, Instituto Nacional de Ecología, Pronatura Noroeste AC, México DF, pp 339–357

Herrero-Pérezrul MD, Rojero-León S, Reyes-Bonilla H (2010) Community structure of conspicuous echinoderms of three islands from the Gulf of California, Mexico. In: Harris LG, Böttger SA, Walker CW, Lesser MP (eds) Echinoderms: Durham. Proceedings of the 12th Internatinal Echinoderm Conference. CRC Press, Taylor and Francis group, Balkema, Leiden, pp 615–620

Holguín-Quiñones O, Wrigth LH, Solis-Marin FA (2000) Asteroidea, Echinoidea y Holothuroidea en fondos someros de la Bahía de Loreto, B.C.S., México. Rev BiolTrop 48:749–757

Holguín-Quiñones O, González-Medina FJ, Solis-Marin FA, Félix-Pico EF (2008) Variación espacio temporal de Scleractina, Gorgonacea, Gastropoda, Bivalvia, Asteroidea, Echinoidea y Holothuroidea de fondos someros de la Isla San José, Golfo de California. Rev Biol Trop 56:1189–1199

Honey-Escandón M, Solis-Marin FA, Laguarda-Figueras A (2008) Equinodermos (Echinodermata) del Pacífico Mexicano. Rev Biol Trop 56(Suppl 3):57–73

Honey-Escandón M, Solis-Marin FA, Laguarda-Figueras A (2011) *Holothuria* (*Selenkothuria*) *carere*, a new species of sea cucumber (Echinodermata: Holothuroidea) from the Mexican Pacific. Zootaxa 2922:27–33

INEGI (2007) Información geográfica: aspectos generales del territorio Mexicano. http://mapserver.inegi.gob.mx/geografia/espanol/datosgeogra/extterri/frontera.cfm?s=geoandc=920 Accessed on 15 Jan

Israelsky MC (1924) Notes on some echinoids from the San Rafael and Tuxpan beds of the Tampico Region, Mexico. Proc Calif Acad Sci 4:137–145

Jackson R (1937) Mexican fossil Echini. Proc US Nat Mus 84:227–237

Jordan EK, Hertlein LG (1926) Contribution to the Geology and Paleontology of the Tertiary of Cedros Island and adjacent parts of Lower California. Proc Calif Acad Sci 15:409–464

Jorgensen P, Espinoza-Ávalos J, Bahena-Basave H (2008) High population density survival of the sea urchin *Diadema antillarum* (Philippi, 1845) to a category 5 hurricane in the southern Mexican Caribbean. Hidrobiology 18:257–260

Juárez-Espinoza PA (2010) Determinación de la actividad biológica de los extractos orgánicos del erizo de mar *Diadema mexicanum* A. Agaziss, 1863. Licenciatura Thesis, Univ Aut Baja California Sur, La Paz, México

Kerstitch A (1989) Sea of Cortez Marine Invertebrates. A Guide for the Pacific Coast Mexico to Ecuador. Sea Challengers Publications, Monterey

Laguarda-Figueras A, Solis-Marin FA (2009) *Holothuria* (*Cystipus*) *casoae* a new species of sea cucumber (Echinodermata: Holothuroidea) from the Central Eastern Pacific. Sci Mar 73:573–578

Laguarda-Figueras A, Solis-Marin FA, Durán-González A, Hernández-Pliego P, Del Valle-García R (2001) Holoturoideos (Echinodermata: Holothuroidea) del Caribe Mexicano: Puerto Morelos. Avicennia 14:7–46

Laguarda-Figueras A, Torres-Vega J, Solis-Marin FA, Mata-Pérez E, Durán-González A, Abreu M (2002) Los asteroideos (Echinodermata: Asteroidea) del Caribe Mexicano: Incluyendo comentarios sobre su zoogeografía. Avicennia 15:1–8

Laguarda-Figueras A, Abreu M, Torres-Vega J, Solis-Marin FA, Durán-González A (2004) Ofiuroideos (Echinodermata: Ophiuroidea) del Caribe Mexicano: Puerto Morelos. Avicennia 17:13–34

Laguarda-Figueras A, Gutiérrez-Castro AI, Solis-Marin FA, Durán-González A, Torres-Vega J (2005a) Equinoideos (Echinodermata: Echinoidea) del Golfo de México. Rev Biol Trop 53(Suppl 3):69–108

Laguarda-Figueras A, Solis-Marin FA, Durán-González A, Ahearn CG, Buitrón-Sánchez BE, Torres-Vega J (2005b) Equinodermos (Echinodermata) del Caribe Mexicano. Rev Biol Trop 53(Suppl 3):109–122

Laguarda-Figueras A, Hernández-Herrejón A, Solis-Marin FA, Durán-González A (2009) Los ofiuroideos del Caribe Mexicano y Golfo de México. Conabio, ICMyL-UNAM, México DF, México

Lambert J (1928) Notes sur quelques echinides recuellis par Mr. Walter Staub dans le Neogene de l'est du Mexique. Eclogae Geol Helv 21:272–283

Lambert J (1936) Quelques nouveaux Echinides fossiles du Crétacé de Mexique. Bull Soc Geol France 6:1–6

Lara-Guadarrame JC (1985) Extracción e identificación de compuestos químicos de las gónadas del erizo de mar *Strongylocentrotus franciscanus*. Licenciatura thesis, Univ Aut Baja California, Ensenada, México

Lavin MF, Marinone SG (2003) An overview of the physical oceanography of the Central Gulf of California. In: Velasco-Fuentes OU, Sheinbaum J, Ochoa de la Torre JL (eds) Nonlinear Processes in Geophysical Fluid Dynamics. Kluwer Acad Pub, Dordrecht, The Netherlands, pp 173-204

Lessios HA, Robertson DR, Cubit JD (1984a) Spread of *Diadema* mass mortality through the Caribbean. Science 226:335–337

Lessios HA, Cubit JD, Robertson DR, Shulman MJ, Parker MR, Garrity SD, Levings SC (1984b) Mass Mortality of *Diadema antillarum* on the Caribbean Coast of Panama. Coral Reefs 3:173–182

Lessios HA, Kessing BD, Pearse JS (2001) Population structure and speciation in tropical seas: global phylogeography of the sea urchin *Diadema*. Evolution 55:955–975
Litvinova NM (1975) Ophiuroids of the Caribbean and Gulf of Mexico collected during 14th Cruise of the R/V "*Akademic Kurchatov*". Acad Sci USSR 100:196–204
López-Uriarte E, Ríos-Jara E, Galván-Villa CM, Juárez-Carrillo E, Enciso-Padilla I, Robles-Jarero EG, Pérez-Peña M (2009) Macroinvertebrados bénticos del litoral somero de Punta La Rosada, Bahía Chamela, Jalisco. Scient CUCBA 11:57–68
Ludwig HL (1893) IV. Reports on the Dredging Operations off the West Coast of Central America to the Galapagos, to the West Coast of Mexico, and to the Gulf of California, in charge of Alexander Agassiz, carried on by the U.S. Fish Commission Steamer "*Albatross*" during 1891, Lieut Commander Z. L. Tanner, U.S.N., Comanding. Bull Mus Comp Zool Harv Coll 24:105–114
Ludwig HL (1905) VII. Asteroidea. Reports on an exploration off the West Coast of Mexico, Central and South America, and off the Galapagos islands, in charge of Alexander Agassiz, by the U. S. Fish Commision steamer "Albatross", during 1891. XXXV. Reports on the Scientific results of the expedition to the Tropical Pacific in Charge of A. Agassiz on the Fish Commision Steamer "Albatross", from August, 1899, to March, 1900. Mem Mus Comp Zool Harv Coll 32:1–292
Luna-Salguero BM, Reyes Bonilla H (2010) Estructura comunitaria y trófica de las estrellas de mar (Echinodermata: Asteroidea) en arrecifes rocosos de Loreto, Golfo de California, México. Hidrobiology 20:127–134
Lunn KE, Villanueva-Noriega MJ, Vincent ACJ (2008) Souvenirs from the sea. An investigation into the curi trade in echinoderms from Mexico. Traffic Bull 22:19–32
Lyman T (1879) Ophiuridae and Astrophytidae of the exploring voyage of H.M.S. "*Challenger*", under Prof. Sir W. Thomson, F.R.S. Part II. Ophiuridae and Astrophytidae of the "*Challenger*" expedition. Bull Mus Comp Zool Harv Coll 6:17–83
Lyman T (1882) Report on the Ophiuroidea dredged by H.M.S. "*Challenger*" during the years 1873–1876. Report of the Scientific Results of the Voyage of H.M.S. "*Challenger*" 1873–1876, vol 5. pp 1–386
Lyman T (1883) Lyman T (1883) Report on the Ophiuroidea. Reports on the results of dredging, under the supervision of Alexander Agassiz, in the Caribbean Sea (1878-79), and on the east coast of the United States, during the summer of 1880, by the U.S. Coast survey Steamer "*Blake*", commander J.R. Bartlett, U.S.N. Commanding. Bull Mus Comp Zool Harv Coll 10: 227–287
Maldonado-Koerdell M (1953) Los equinoides regulares del Mesozoico de México. An Esc Nac Cienc Biol 7:15–44
Maluf LY (1988) Composition and Distribution of the Central Eastern Pacific Echinoderms. Nat Hist Mus LA County Tech Rep 2:1–242
Martinez-Melo A, Solis-Marin FA (2010) Revision of the genera *Cassidulus* and *Rhyncholampas*. In: Harris LG, Böttger SA, Walker CW, Lesser MP (eds) Echinoderms: Durham. Proceedings of the 12th International Echinoderm Conference. CRC Press, Taylor and Francis group, Balkema, Leiden, pp 229–232
Martín-Medrano L, García-Barrera P (2006) Fossil ophiuroids of Mexico. In: Vega F, Nyborg TG, Perrilliat MC, Montellano-Ballesteros M, Cevallos-Ferriz SR, Quiroz-Barroso SA (eds) Studies of Mexico paleontology. Springer, The Netherlands, pp 115–131
Martín-Medrano L, Ben T, García-Barrera P (2009) New Albian (Early Cretaceous) ophiuroids from the Tlayúa Quarry, Puebla, Mexico. Palaeontol 52:83–94
Massin C, Hendrickx ME (2010) A new species of deep-water Holothuroidea (Echinodermata) of the genus *Synallactes* from off Wester Mexico. Sci Mar 74:599–603
Massin C, Hendrickx ME (2011) Deep-water Holothuroidea (Echinodermata) collected during the TALUD cruises off the Pacific coast of Mexico, with the description of two new species. Rev Mex Biodiver 82:413–443
Mcartney MA, Keller G, Lessios HA (2000) Dispersial barriers in tropical oceans and speciation in Atlantic and eastern Pacific sea urchins *Echinometra*. Mol Ecol 9:1391–1400

Merino M (1997) Upwelling on the Yucatan Shelf: hydrographic evidence. J Mar Sys 13:101–121

Micael J, Alves MJ, Costa AC, Jones MB (2009) Exploitation and conservation of echinoderms. Oceanogr Mar Biol An Rev 47:191–208

Morgan MB, Cowles DL (1996) The effects of temperature on the behaviour and physiology of *Phataria unifascialis* (Gray) (Echinodermata, Asteroidea). Implications for the species distribution in the Gulf of California, Mexico. J Exp Mar Biol Ecol 208:13–27

Morris RH, Abbot DP, Haderline EC (1980) Intertidal invertebrates of California. Stanford University Press, California

Mullerried FK (1951) Algunos fósiles marinos del Terciario Inferior y Medio de Palenque, Chiapas. Soc Mex Hist Nat 12:209–227

Muir JM (1936) Geology of the Tampico region, Mexico. Am Assoc Petrol Geol, Tulsa, Oklahoma

Nardin E, Almazán-Vázquez E, Buitrón BE (20099 First report of *Gogia* (Eocrinoidea, Echinodermata) from the early-middle Cambrian of Sonora (Mexico), with biostratigraphical and palaeoecological comments. Geobios 42:233–242

Nieto-López I, García-Barrera P (2006) Fossil Cretaceous echinoids of Mexico. In: Vega F, Nyborg TG, Perrilliat MC, Montellano-Ballesteros M, Cevallos-Ferriz SR, Quiroz-Barroso SA (eds) Studies of Mexico paleontology. Springer, The Netherlands, pp 101–114

Nuño-Hermosillo A (2003) Ecología poblacional, ciclo reproductivo e historia de la pesquería del pepino de mar *Isostichopus fuscus* (Ludwig, 1875) (Echinodermata: Holothuroidea) en Bahía Chamela, Jalisco, México. Master thesis, Univ Guadalajara, Puerto Vallarta, México

Nyst H, Galeotti H (1839) Description de quelques fossiles du calcaire jurassique de Tehuacan, Mexique. Bull Acad R Sci B Let Bruxelles 7:218–222

Olivares-González E (1986) Algunos aspectos sobre la biología del erizo de espina larga *Diadema mexicanum* (Echinoidea: Echinodermata). I. Periodos reproductivos y II. Parasitismo. Licenciatura thesis, Univ Aut Baja California Sur, La Paz, México

Orbe-Mendoza A (1971) Contribución al conocimiento de los cidaroideos de México. Licenciatura thesis, Univ Nac Aut México, México DF, México

Ortiz-Lozano L, Gutiérrez-Velázquez AL, Granados-Barba A (2009) Marine and terrestrial protected areas in Mexico: Importance of their functional connectivity in conservation management. Ocean Coast Manag 52:620–627

Palleiro-Nayar JS (2004) Dinámica de la población de erizo rojo *Strongylocentrotus franciscanus* sujeta a extracción comercial en Baja California. Master Thesis, Cent Interdis Cien Mar, Ensenada, México

Palleiro-Nayar JS (2009) Análisis poblacional del erizo rojo *Strongylocentrotus franciscanus* en la costa occidental de la Península de Baja California. Puede considerarse un metapoblación. Doctoral Thesis, Cent Interdis Cien Mar, Ensenada, México

Palleiro-Nayar JS, Salgado-Rogel ML, Aguilar-Montero D (2008) La pesquería de erizo morado *Strongylocentrotus purpuratus* y su incremento poblacional en Baja California. Cienc Pesq 16:29–35

Palleiro-Nayar JS, Sosa-Nishizaki O, Montaño-Moctezuma G (2011) Impact of substrate and the presence of adults on recruitment of the red sea urchin *Strongylocentrotus franciscanus* in Baja California. Cienc Mar 37:59–70

Parker RH (1963) Zoogeography and ecology of macroinvertebrates of the Gulf of California and continental slope of Western Mexico. In: Van Andel TH, Shor GGJr (eds) Marine geology of the Gulf of California. Am Ass Petrol Geol, Tulsa, Oklahoma, pp 331–376

Pawson DL, Vance D, Messing CG, Solis-Marin FA, Mah CL (2009) Echinodermata of the Gulf of Mexico. In: Felder DL, Camp DK (eds) Gulf of Mexico, Origin, Waters, and Biota. 1. Biodiversity. Harte Research Institute for Gulf of Mexico Studies Series. Texas A&M Univesity Press, Texas, pp 1177–1204

Pech D (2010) Un breve panorama del pepino de mar en el mercado mundial y en el litoral de Campeche. Jaina. Bol Inf Centro EPOMEX Univ Aut Campeche 21:13–19

Pérez-Plascencia G (1995) Crecimiento y reproducción del pepino de mar en la Bahía de Todos Santos, Baja California, México. Master Thesis, Univ Baja California, Ensenada, Mexico
Perrier E (1881) Description sommaire des espéces nouvelles d'astéries. Reports on the results of dredging under the supervision on Alexander Agassiz, in the Gulf of Mexico, 1877–78, by the United States coast survey steamer "Blake", Lieut-Commander C. D. Sigsbee, U. S. N., commanding and in the Caribbean sea, 1878–79, by the U. S. C. S. S. "Blake", commander J. R. Bartlett, U. S. N., commanding. Bull Mus Comp Zool Harv Coll 9:1–31
Phelan T (1970) A field guide to the cidaroid echinoids of the Nortwestern Atlantic Ocean, Gulf of Mexico, and the Caribbean Sea. Smith Control Zool 1:1–22
Pineda-Enríquez T (2011) Biodiversidad de los equinodermos (Echinodermata) del Golfo de California, de 0 a 300 m de profundidad, utilizando el submarino Deep See como herramienta de muestreo. Licenciatura thesis, Univ Nac Aut México, México DF, México
Pranovi F, Raicevich S, Franceschini G, Torricelli P, Giovanardi O (2001) Discard analysis and damage to non-target species in the "rapido" trawl fishery. Mar Biol 139:863–875
Purcell SW, Lovatelli A, Vasconcellos M, Ye Y (2010) Managing sea cucumber fisheries with an ecosystem approach. FAO Tech paper 520, Rome, Italy
Rathbun R (1885) Report upon the echini collected by the U.S. Fish Commission Steamer "*Albatross*" in the Caribbean Sea and Gulf of Mexico, January to May, 1884. Proc US Nat Hist Mus 3:1–83
Reyes-Bonilla H (1995) Asteroidea and Echinoidea (Echinodermata) of Isla San Benedicto, Revillagigedo Archipielago, México. Rev Inv Cien Ser Mar UABCS 6:29–38
Reyes-Bonilla H, Calderon-Aguilera LE (1999) Population density, distribution and consumption rates of three corallivores at Cabo Pulmo reef, Gulf of California, Mexico. PZNSI Mar Ecol 20:347–357
Reyes-Bonilla H, González-Azcárraga A, Rojas-Sierra A (2005) Estructura de las asociaciones de estrellas de mar (Asteroidea) en arrecifes rocosos del Golfo de California, México. Rev Biol Trop 53(Suppl 3):233–244
Reyes-Bonilla H, Herrero-Pérezrul MD, González-Romero S, González-Peralta A, Ramírez-Hernández Y (2008) Abundance of the brown sea cucumber *Isostichopus fuscus* at the National Park Bahia de Loreto, Mexico. Rev Biol Trop 56(Suppl 3):265–271
Ríos-Jara E, Galván-Villa CM, Solis-Marin FA (2008) Equinodermos del parque nacional Isla Isabel, Nayarit, México. Rev Mex Biodiver 79:131–141
Robadue D (2002) An Overview of governance of the Gulf of California. University of Rhode Island, Coastal Resources Center Narragansett, Rhode Island
Rodríguez-Valencia JA, Cisneros-Mata MA (2006) Captura incidental de las flotas pesqueras ribereñas del Pacífico Mexicano. Reporte técnico del Programa Golfo de California de WWF-México a la Comisión de Cooperación Ambiental de América del Norte, México DF, Mexico
Rogers-Bennett L, Ono DS (2006) Status report 2006 Sea Cucumber. Depart Fish and Game California. http://www.dfg.ca.gov/marine/status/report2006/seacucumber.pdf
Roux M, Pawson DL (1999) Two new Pacific Ocean species of Hyocrinid Crinoids (Echinodermata) with comments on presumed giant-dwarf gradients related to seamounts and abyssal plains. Pac Sci 53:289–298
Sagarin RD, Gilly WF, Baxter CH, Burnett N, Christensen J (2008) Remembering the Gulf: changes to the marine communities of the Sea of Cortez since the Steinbeck and Ricketts expedition of 1940. Front Ecol Environ 6:372–379
Salas-Garza A, Carpizo-Ituarte E, Pares-Sierra G, Martinez-López R, Quintana-Rodríguez R (2005) Juvenile production of the red sea urchin *Strongylocentrotus franciscanus* (Echinodermata: Echinoidea) in Baja California, Mexico. Rev Biol Trop 53(Suppl 3):345–355
Salcedo-Martinez S, Green G, Gamboa-Contreras A, Gómez P (1988) Inventario de macroalgas y macroinvertebrados bénticos, presentes en áreas rocosas de la región de Zihuatanejo, Guerrero, México. An Inst Cienc Mar Limnol UNAM 15:73–96
Salgado-Rogel ML, Palleiro-Nayar JS (2008) Disminución de la abundancia de erizo rojo y propuesta para su manejo en Baja California, México. Cienc Pesq 16:37–45

Salgado-Rogel ML, Palleiro-Nayar JS, Rivera-Ulloa JL, Aguilar-Montero D, Vázquez-Solórzano E, Jiménez-Quiroz MC (2009) La pesquería y propuestas de manejo del pepino de mar *Parastichopus parvimensis* en Baja California, México. Cienc Pesq 17:17–26

San Juan-Ruiz S (1988) Estudio descriptivo de la morfología externa de *Tripneustes ventricosus* (Lamarck) y *Lytechinus variegatus* (Lamarck) (Echinoidea: Echinodermata) colectados en el arrecife Lobos, Veracruz, Ver. Licenciatura thesis, Univ Veracruzana, Veracruz, México

Sánchez-Domínguez C (1993) Taxocenosis y estructura de la comunidad de equinodermos del arrecife de Isla de En medio, Ver. (Stelleroidea, Crinoidea, Equinoidea: Echinodermata). Licenciatura thesis, Univ Veracruzana, Veracruz, México

Schroeder EE (1964) Notes on the biology of the echinoid *Diadema antillarum*. Carib J Sci 4:421–433

Seixas CS, Begossi A (2001) Ethnozoology of fishing communities from Ilha Grande (Atlantic Forest coast, Brazil). J Ethnobiol 21:107–135

Sierra-Rodríguez P (1994) Prospección de pepino de mar *Stichopus fuscus* en la costa de Oaxaca, México. Resúmenes del Primer taller de evaluación sobre las pesquerías ribereñas del Pacífico mexicano. CRIP Manzanillo, Mexico

Skarbnik-López J (2006) Ciclo reproductivo de *Holothuria* (*Selenkothuria*) *lubrica* Selenka, 1867 (Echinodermata: Holothuroidea) en el sur de la Bahía de La Paz, B.C.S., México. Licenciatura thesis, Univ Aut Baja California Sur, La Paz, Baja California Sur, México

Skarbnik-López J, Herrero-Pérezrul MD, Reyes-Bonilla H, Domínguez FG, Morales JT (2010) Reproductive cycle of *Holothuria (Selenkothuria) lubrica* Selenka, 1867 (Echinodermata: Holothuroidea) in Bahia de la Paz, Mexico. In: Harris LG, Böttger SA, Walker CW, Lesser MP (eds) Echinoderms: Durham. Proceedings of the 12th International Echinoderm Conference. CRC Press, Taylor and Francis group, Balkema, Leiden, pp 487–494

Sladen WP (1889) Report on the Asteroidea. Report on the Scientific Results of the Voyage of *H.M.S. Challenger* during the years 1873–1876. Zoology 30:1–893

Solis-Marin FA (1998) Sistemática, distribución y morfología del género *Mellita* L. Agassiz, 1841 (Echinodermata, Echinoidea, Clypeasteroidea). Master thesis, Univ Nac Aut México, México DF, México

Solis-Marin FA (2008) Echinodermata de México. Catálogo taxonómico de especies de México. In: Sarukhán J, Soberón M, Halffter G, Llorente-Bousquets J (eds) Capital Natural de México, vol I. Conocimiento actual de la biodiversidad. Comisión Nacional para el Conocimiento y Uso de la Biodiversidad, Mexico DF, Mexico, CD

Solis-Marin FA, Laguarda-Figueras A (1999) *Cucumaria flamma*, a new species of sea cucumber from the central eastern Pacific (Echinodermata: Holothuroidea). Proc Biol Soc Wash 112: 778–786

Solís-Marín FA, Laguarda-Figueras A (2008) Equinodermos. In: Mejía-Ortíz LM (ed) Biodiversidad Acuática de la Isla de Cozumel. CONABIO/University of Quintana Roo, México DF, pp 187–214

Solis-Marin FA, Laguarda-Figueras A (2010a) A new species of starfish (Echinodermata: Asteroidea) from an anchialine cave in the Mexican Caribbean. Rev Mex Biodiver 81: 663–668

Solís-Marín FA, Laguarda-Figueras A (2010b) Los equinodermos del estado de Yucatan. In: Durán R, Méndez M (eds) Biodiversidad y Desarrollo Humano en Yucatán. CICY, PPD-FMAM, CONABIO, SEDUMA, México DF, pp 248–249

Solis-Marin FA, Laguarda-Figueras A, Leija-Tristán A (1997) Morphology, systematics and distribution of *Meoma ventricosa grandis* and *M. ventricosa ventricosa* (Echinodermata: Echinoidea: Brissidae) along Mexican Coasts. Proc Biol Soc Wash 110:301–309

Solis-Marin FA, Laguarda-Figueras A, Vázquez Gutierrez F, Mejia L, Yáñez G (2007a) Echinoderm fauna of Anchialine caves in Cozumel Island, Mexico. In: Harris LG, Böttger SA, Walker CW, Lesser MP (eds) Echinoderms: Durham. Proceedings of the 12th International Echinoderm Conference. CRC Press, Taylor and Francis group, Balkema, Leiden, pp 259–261

Solis-Marin FA, Laguarda-Figueras A, Martinez-Melo A, Celaya Hernandez EV, Velasquez-Castro N, Barrera-Rivera JA, Melgar T, Valentin-Maldonado N (2010) Echinoderm remains in the offerings of the Great Temple of the Aztecs. In: Harris LG, Böttger SA, Walker CW, Lesser MP (eds) Echinoderms: Durham. Proceedings of the 12th International Echinoderm Conference. CRC Press, Taylor and Francis group, Balkema, Leiden, pp 77–79

Solís-Marín FA, Herrero-Pérezrul MD, Laguarda-Figueras A, Torres-Vega J (1993) Asteroideos y equinoideos de México (Echinodermata). In: Salazar-Vallejo SI, González NE (eds) Biodiversidad Marina y Costera de México. CONABIO, Cent Invest Quintana Roo AC, México DF, pp 91–105

Solís-Marín FA, Reyes Bonilla H, Herrero-Pérezrul MD, Arizpe-Cobarrubias O, Laguarda-Figueras A (1997a) Sistemática y distribución de los equinodermos de la Bahía de La Paz. Rev Cienc Mar 23:249–263

Solís-Marín FA, Laguarda-Figueras A, Durán-González A, Mata-Pérez E, Buitrón-Sánchez BE (2003) El género *Mellita* (Echinodermata: Echinoidea: Clypeasteroidea) de las costas de México. Avicennia 16:55–70

Solís-Marín FA, Laguarda-Figueras A, Durán-González A, Ahearn CG, J Torres-Vega (2005) Equinodermos (Echinodermata) del Golfo de California, México. Rev Biol Trop 53(Suppl 3):123–137

Solís-Marín FA, Laguarda-Figueras A, Gordillo-Hernández MA (2007b) Estudio Taxonómico de los equinodermos del Parque Nacional Sistema Arrecifal Veracruzano. In: Granados-Barba A, Abarca-Arenas L, Vargas-Hernández JM (eds) Investigaciones Científicas en el Sistema Arrecifal Veracruzano. Univ Aut Campeche, Campeche, pp 73–100

Solís-Marín FA, Arriaga-Ochoa JA, Laguarda-Figueras A, Frontana-Uribe CS, Durán-González A (2009) Holoturoideos del Golfo de California. CONABIO, ICMyL-UNAM, México DF, México

Sommer M (2005) Pesca de arrastre. Aniquilación silencionsa. Rev Electron Vet 6:1695–7504

Sonnenholzner JI, Montaño-Moctezuma G, Searcy-Bernal R (2010) Effect of three tagging methods on the growth and survival of the purple sea urchin *Strongylocentrotus purpuratus*. Pan-Am J Aquat Sci 5:414–420

Steinbeck J, Ricketts E (1941) Sea of Cortez A Leisurely journal of travel and research. The Viking Press, New York

Strimple HL (1971) A Permian crinoid from Coahuila, Mexico. J Paleontol 45:1040–1042

Tapia-Vázquez OM, Castro-González JJ, Valles-Ríos H (1996) Madurez gonádica del pepino de mar *Parastichopus parvimensis* en la costa occidental de Baja California, México, en 1994. Cienc Pesq 12:5–11

Théel H (1879) Preliminary report on the Holothurioidea, of the exploring voyage of H.M.S. "*Challenger*". Part I. Bihang Till Kongl Svenska vetenskapsakademien Akade-miens Handlingar 5:1–20

Théel H (1886) Report on the Holothurioidea dredged by the HMS "*Challenger*" during the years 1873–1876 Part II. Report of the Scientific Results of the Voyage of H.M.S. Challenger 1873–1876. Zoology 14:1–290

Thomas LP (1962) The shallow water amphiurid brittle stars (Echinodermata, Ophiuroidea) of Florida. Bull Mar Sci 12:623–694

Toral-Granda V, Lovatelli A, Vasconcellos M (2008) Sea Cucumbers. A global review of fisheries and trade. FAO Tech Paper 516, Rome, Italy

Torres-Martinez MA, Solis-Marin FA (2010) Sediment utilization and feeding-niche breadth of *Meoma ventricosa grandis* Lamarck, 1816 (Echinodermata: Brissidae) at Boca Chica Channel, Acapulco, Guerrero, Mexico. In: Harris LG, Böttger SA, Walker CW, Lesser MP (eds) Echinoderms: Durham. Proceedings of the 12th International Echinoderm Conference, CRC Press, Taylor and Francis group, Balkema, Leiden, pp 643–645

Torres-Martínez MA, Solís-Marín FA, Laguarda-Figueras A (2008) Poros filodiales en la identificación de dos subespecies de erizos de mar: *Meoma ventricosa grandis* (Pacifico) y *Meoma ventricosa ventricosa* (Atlántico). Rev Biol Trop 56(Suppl 3):13–17

Trujillo-Luna BR, González-Vallejo NE (2006) Equinodermos (Echinodermata) de la colección de referencia de bentos costero de ECOSUR. Univ Cienc 22:83–88

Uthicke S, Schaffelke B, Byrne M (2009) A boom–bust phylum? Ecological and evolutionary consequences of density variations in echinoderms. Ecol Monogr 79:3–24

Vachard D, Flores de Dios A, Buitrón BE (2004) Guadalupian and Lopigian (middle and late permian) deposits from Mexico and Guatemala, a review data. Geobios 37:99–115

Vázquez-Bader AR, Laguarda-Figueras A, Gracia-Gasca A, Solis-Marin FA, Celaya-Hernández EV, Durán-González A (2008) Seasonal changes in the density and species composition of the epifaunal echinoderms recorded from the Southwester Gulf of Mexico. Rev Biol Trop 56(Suppl 3):297–310

Verrill AE (1870) Descriptions of echinoderms and corals from the Gulf of California. Trans Connec Acad Art Sci 61:93–98

Verrill AE (1871) Notes on the radiata in the museum of Yale College, with descriptions of new genera and species. Additional observations on echinoderms, chiefly from the Pacific Coast of America. Trans Connect Acad Art Sci 61:568–593

Wilson IF, Rocha V (1955) Geology and mineral deposits of the Boleo Copper district, Baja California, México. US Geol Surv Prof Pap 273:1–134

Woodby D, Smiley S, Larson R (2000) Depth and habitat distribution of *Parastichopus californicus* near Sitka, Alaska. Alaska Fish Res Bull 7:22

Worbis-Torres H (1986) Introducción al conocimiento de los equinoideos (Echinodermata: Echinoidea) de la costa de Tamaulipas y Norte del estado de Veracruz, México. Licenciatura thesis, Univ Noreste, Tampico, Tamaulipas, México

Yáñez-Arancibia A, Lara-Domínguez AL, Chavance P, Hernández DF (1983) Environmental Behavior of Terminos Lagoon Ecological System, Campeche, Mexico. An Inst Cienc Mar Limnol UNAM 10:137–176

Zamorano P, Leyte-Morales GE (2005a) Densidad poblacional de cuatro especies de erizo (Echinodermata: Echinoidea) en el arrecife de La Entrega, Oaxaca. Oceánides 20:65–72

Zamorano P, Leyte-Morales GE (2005b) Cambios en la diversidad de equinodermos asociados al arrecife coralino de La Entrega, Oaxaca, México. Cienc Mar 9:19–28

Zamorano P, Leyte-Morales GE (2009) Equinodermos asociados a formaciones arrecifales en Zihuatanejo y Acapulco, Guerrero, México. Bol Invest Mar Cost 38:7–28

Zárate-Montes CM (2011) Biocenosis y distribución de los Asteroidea en el talud continental del Golfo de California. Master thesis, Univ Nac Aut México, México DF, México

Zetina-Moguel C, Ríos-Lara V, Koyoc M, Cruz Hernández-Herrera I, Cervera–Cervera K, Anda-Fuentes D, Arceo-Briseo P, Ortí EE, Guevara-Ortíz M (2003) Sea cucumber (*Astichopus multifidus*, *Isostichopus badionotus* and *Holothuria floridana*) biomass estimation in two areas of Yucatan Coast between October 2000 to March 2001. Proc Gulf Carib Fish Inst 54:297–306

Ziesenhenne FC (1937) The Templeton Crocker Expedition. X. Echinoderms from the West Coast of Lower California, the Gulf of California and Clarion Island. Zoology 22:209–239

Ziesenhenne FC (1940) New ophiurans of the Allan Hancock Pacific Expeditions. Allan Hancock Pac Exp 8:9–58

Ziesenhenne FC (1942) New Eastern Pacific sea stars. Allan Hancock Pac Exp 8:197–223

Chapter 3
Central America Echinoderms: Diversity, Ecology and Future Perspectives

Juan José Alvarado, Enrique Barraza and Tito Isaac Sancho-Mejías

3.1 Introduction

Central America (CA) is a narrow strip of land that divides the two largest oceans of the world, which are the Atlantic and the Pacific (Cortés 2007). It include seven countries: Belize, Guatemala, El Salvador, Honduras, Nicaragua, Costa Rica and Panama. According to the "Sea Around US Project" (www.seaaroundus.org) CA posseses an Exclusive Economic Zone of 1,518,705 km^2 (1,005,922 km^2 on the Pacific and 512,783 km^2 on the Caribbean), which is three time larger than its terrestrial area (521,876 km^2). CA also posseses 156,251 km^2 of continental shelf (56,217 km^2 on the Pacific and 100,034 km^2 on the Caribbean), with 3.7 % of the coral reefs area of the world (0.4 % on the Pacific and 3.3 % on the Caribbean) (www.seaaroundus.org), and 6,603 km of coastline (2,856 km on the Caribbean Sea and 3,747 km on the Pacific Ocean) (www.cia.gov/library/publications/the-world-factbook).

J. J. Alvarado (✉)
Centro de Investigación en Ciencias del Mar y Limnología (CIMAR),
Universidad de Costa Rica San Pedro, San José, 11501-2060, Costa Rica
e-mail: juanalva76@yahoo.comjuan.alvarado@ucr.ac.cr

J. J. Alvarado
Posgrado en Ciencias Marinas y Costeras, Universidad Autónoma de Baja California Sur,
La Paz, Mexico

E. Barraza
Gerencia de Vida Silvestre, Ministerio de Medio Ambiente y Recursos Naturales,
San Salvador, El Salvador
e-mail: ebarraza@marn.gob.sveulaliabarr@yahoo.com

T. I. Sancho-Mejías
Museo de Zoología, Escuela de Biología, Universidad de Costa Rica,
San José, Costa Rica
e-mail: sancho.ekin02@gmail.com

J. J. Alvarado and F. A. Solís-Marín (eds.), *Echinoderm Research and Diversity in Latin America*, DOI: 10.1007/978-3-642-20051-9_3,

The formation of the Isthmus of Panama united South and North America, and gave rise to the great biological interchange of terrestrial faunae and floras, but cutting the seaway between the Atlantic and the Pacific (O'Dea et al. 2007). This process was the result of two geologic processes: plate tectonics and volcanisms (Coates and Obando 1996). Twenty million years ago (m.y.a.; Miocene), Central and South America were separated by a wide and deep seaway. This seaway became narrower due to the movement of the South American plate to the Northwest until the arc of Panama collided with South America 10 m.y.a. Subsequently, it began to rise and produced the emergence of land around 5 m.y.a., causing a final closure of the isthmus between 3.5 and 2.5 m.y.a. (Coates and Obando 1996; O'Dea et al. 2007).

The Pacific coast is characterized by extensive mangrove systems; sandy beaches, rocky points, cobbles beaches, deltas, and isolated coral reefs of limited development (Cortés 2007). This coast experiences seasonal upwelling of cold, nutrient-rich waters from December to April, and produces an increase of the plankton productivity (D'Croz and Robertson 1997; Fiedler 2002; Amador et al. 2006). This elevated productivity caused by upwelling adds to the high production from river run-off caused by high precipitation, and increases the nutrients contribution of terrestrial origin (O'Dea et al. 2007). This coast is under the influence of the Intertropical Convergence Zone (ITCZ), characterized by high irradiance, variable winds and high precipitation (Amador et al. 2006). The depth of the mixing layer is shallow (30–40 m) with a permanent thermocline (Longhurst 1998), and localized near the surface (25 m depth) (Brenes et al. 1990). Productivity is considered as moderate to high, due to the presence of upwelling areas such as the Costa Rica Thermal Dome (Fiedler 2002) and seasonal upwelling in the Gulfs of Tehuantepec, Papagayo and Panama (Bakun et al. 1999). The Pacific coast has six mayor gulfs: (1) The Gulf of Fonseca (shared by El Salvador, Honduras and Nicaragua), (2) the Gulf of Papagayo, (3) the Gulf of Nicoya and (4) the Golfo Dulce (2, 3 and 4 in Costa Rica), and the Gulfs of (5) Chiriquí and (6) Panama (in Panama) (Cortés 2007). Golfo Dulce is one of four fjord-like gulfs in the tropics with anoxic conditions at the bottom (Quesada-Alpízar and Cortés 2006).

The Pacific side of CA also has an oceanic island, Cocos Island (Isla del Coco), of volcanic origin, and located approximately 500 km Southwest of Costa Rica and 630 km Northeast of the Galápagos Islands (Fig. 3.1). Cocos Island is the only subaereally exposed portion of the aseismic Cocos Ridge. Its volcanic origin is associated with the Galápagos hot spot (Castillo et al. 1988). This oceanic ridge is more than 1,000 km long and approximately 200 km wide with depths of 2,000 m (Harpp et al. 2005; Marcaillou et al. 2006). Moreover, the middle America Trench, in front of the Pacific coast of CA extends from Mexico to Costa Rica with a maximum depth of 6,669 km (Astiz et al. 1987). Hydrothermal and cold seeps and oxygen minimum zones have been reported in the continental shelf (Bohrmann et al. 2002; Helly and Levin 2004, respectively).

The currents on the Pacific side of CA (Fig. 3.1) are under the influence of the North Equatorial Countercurrent (NECC), and are responsible for the heat transport from the Tropical Pacific to CA. A portion of the water is first carried out to

Fig. 3.1 Central America main currents and distribution of reefs (R), mangroves (M) and seagrass (S) ecosystems (modified from Lessios et al. 1984; Glynn et al. 1996; Cortés 2007)

the Gulf of Panama and then moved northward by the Costa Rica Coastal Current along the coast of CA, and finally joins the North Equatorial Current (Wyrtki 1967; Brenes and Gutiérrez 1998; Kessler 2006).

The Caribbean coast of CA is characterized by relatively stable water temperatures, extensive coral reefs and seagrass beds, and large areas of carbonate platforms (Cortés 2007). This side of CA is considered a passive margin with few elevated outcrops (Coates 1997). This coast is characterized by lower levels of nutrients than the Pacific and is less productive, although there are some embayments that have nutrient enrichment due to river run-off (D'Croz and Robertson 1997; O'Dea et al. 2007). The principal currents of the Caribbean Sea go from east to west. The isthmus deviates some water to the south along Nicaragua, Costa Rica and Panama (Fig. 3.1). The rest of the water continues to the west towards Honduras and then north into Belize (Jackson and D'Croz 1997). This coast has two main Gulfs: (1) the Gulf of Honduras shared by Belize, Guatemala and Honduras (Harborne et al. 2001; Heyman and Kjerfve 2001) and (2) the Gulf de los Mosquitos in Panama which includes the semi-enclosed bay of Bocas del Toro (Cortés 2007). The Caribbean of CA has four carbonate platforms in Belize, Honduras and Nicaragua (Miskitos), and Panama (Bocas del Toro and San Blas) (Cortés 2007). Belize has the largest atolls of the Caribbean and one of the largest barrier reefs of the world (Gibson and Carter 2003). Long stretches of this coast are low-lying and may be associated with coastal lagoons or canals and swamps (Cortés 2007).

The aim of this chapter is to review and detail echinoderm research in CA, from Belize to Costa Rica on the Caribbean side, and from Guatemala to Costa Rica on the Pacific side. The echinoderm research from Panama is reviewed by Coppard and Alvarado (Chap. 4) due to the greater amount of information from that country. In order to highlight future needs of new research lines, we present a review of the main expeditions in the region, and describe ecological and other studies from both coasts as well as the other research efforts carried out.

3.2 Echinoderm Research

The knowledge of the CA echinoderm fauna is disperse and scarce for some of its groups. Alvarado and Cortés (2004, 2009) divided the research of echinoderms in the region into two periods. The first period, from mid-nineteenth century to the mid-twentieth century, was characterized by expeditions from North America and Europe, in a period when most of the local species were described. The second period, from mid-twentieth century to the present, has been characterized by widespread interest in ecology done by local scientists at research stations in CA.

Echinoderm research in CA started on the Caribbean side in 1884 (although the first report was from Panama by Verrill (1867) when the US Fish Commission Steamer *Albatross* expedition collected the asteroid *Sclerasterias tanneri* from Honduran waters. Between 1888 and 1891, the *Albatross* visited the Pacific side of CA. Those expeditions, led by Alexander Agassiz, collected mostly from Cocos Island and Cocos Ridge (Ludwig 1894, 1905; Lütken and Mortensen 1899; Alvarado 2010). Ludwig (1894, 1905), Agassiz (1898) and Lütken and Mortensen (1899) described many new species of asteroids, ophiuroids, echinoids and holothurians from the eastern Pacific. The asteroid *Pauliella aenigma*, the echinoids *Hesperocidaris panamensis* and *Centrocidaris doederleini*, and the holothurian *Psolus diomedeae* were species described from specimens collected at Cocos Island, Costa Rica (Alvarado 2010).

Between 1930s and 1940s, as a result of the expeditions to the eastern Pacific by the Allan Hancock Foundation and the New York Zoological Society, there was a significant increase of research on echinoderms (Ziesenhenne 1937, 1940, 1942, 1955; Clark 1939, 1940, 1948; Deichmann 1938, 1941, 1958). The *Velero III* from the Allan Hancock Foundation collected 37 species of echinoderms from the Pacific of CA (35 species from Costa Rica, one species from Nicaragua and one species from Guatemala), from eight type of species localities (seven from Costa Rica and one from Nicaragua). The species described from Costa Rica were the ophuiroid *Ophionereis dictyota*, the asteroid *Tamaria obstipa*, the echinioids *Encope cocosi* (endemic to Cocos Island) and the holothurians *Labidodemas americanum*, *Euthynodium veleronis*, *Psolus diomedae* and *Pentamera beebei*, and the irregular echinoid *Encope laevis* from Corinto Port, Nicaragua.

The Zoological Society of New York expeditions aboard the SY *Arcturus* and SY *Zaca* were led by William Beebe. They reported the presence 58 species of echinoderms (Fisher 1928; Clark 1940). However, *Zaca* collected a total of 822 specimens: 73 asteroids, 645 ophiuroids and 104 echinoids. This included 57 species: 13 species of asteroids, 29 species of ophiuroids and 16 species of echinoids. From that material Clark (1940) described two new species of ophiuroids: *Amphiodia vicina* from the Gulf of Nicoya, Costa Rica, and *Ophiolepis grisea* from the Gulf of Fonseca (Monypenny Point) and the presence of *Echinaster* (*Othilia*) *aculeata,* a species that was thought to be extinct.

Between 1979 and 1981, as part of a study of the megabenthic invertebrates communities of Gulf of Nicoya on board the R/V *Skimmer,* nine species of echinoderms were found (two species of asteroids: *Luidia tessellata* and *Astropecten armatus*; five species of ophiuroids: *Ophiocoma* sp., *Ophiothrix* sp., *Ophiopallus* sp., one species each of Ophiocanthidae, Ophiodermatidae and Ophiuridae; and two species of echinoids: *Astropyga pulvinata* and *Arbacia incisa*). All the echinoderms together accounted for only 0.5 % of the total number of individuals and only 1.6 % of the total biomass. They were more abundant in the inner parts of the gulf (Maurer et al. 1984). Between 1980 and 1981, Maurer and Vargas (1983) continued with the study of the benthic communities, but reported only four species of ophiuroids (*Amphipholis* sp., *Amphipholis squamata, Ophiactis savignyi* and *Ophiothrix spiculata*), which accounted for 0.5 % of the biomass.

Later expeditions at the end of the twentieth century and beginning of the twenty first centuries were carried out on board the RV *Urracá* from the Smithsonian Tropical Research Institution in collaboration with authorities in Costa Rica (University of Costa Rica) (Cortés 2009) and El Salvador (Ministerio de Medio Ambiente y Recursos Naturales) (Barraza and Hasbún 2005). The El Salvadorian vessels, the RV *Fengur,* collected many new records of echinoderms (Barraza 1995), and the RV *Miguel Olivier* collected new deep-sea asteroids and ophiuroids at 900 m in December 2009. The Costa Rican expeditions from the Centro de Investigaciones en Ciencias del Mar y Limnología collected specimens at Cocos Island at depths between the years 2005 and 2009. Other Pacific expeditions that visited and collect echinoderms were the RV *Boric* (1927), RV *Stranger* (1939, 1969), and the RV *Te Vega* (1968), last one from Stanford University

(Cortés 2009). Most of these expeditions did not publish accounts of their expeditions, but the species collected are recorded in the data base of the National Museum of Natural History, Smithsonian Institution, Washington, DC. The species are recently reported by Alvarado et al. (2010).

On the Caribbean side, Devaney (1974) reviewed the first expeditions that visited Belize. He indicated that the first research ship was the *Pawnee I* of the Harry Payne Bingham Institute in April 1925, from which Boone (1928) described two species of echinoids, three species of asteroids and eight species of crinoids from Glover's Reef. From the 1937–1938 *Rosaura* expedition, John and Clark (1954) reported four species of asteroids, five species of ophiuroids, two species of echinoids, and one species of holothuroids. Some of them from depths of more than 914.4 m (500 fathoms). During the 1959–1960 Cambridge Expedition, Stoddard (1962) made brief remarks on echinoderms in his comparisons of ecological zonation (Devaney 1974). These expeditions were complemented with the extensive taxonomic work carried out at Carrie Bow Cay (Kier 1975; Pawson 1976; Hotchkiss 1982; Macurda 1982).

According to Miller (1984), the asteroid family Ophidiasteridae is represented by 13 species in the tropical western Atlantic, Caribbean and Gulf of Mexico. For four species, no new material has been reported from the western Atlantic since the description of the holotypes. Miller (1984) collected specimens of *Copidaster lymani* at Carrie Bow Cay, Belize. Thus, these are the first specimens reported for western Atlantic since the holotype had been described from Florida. Miller (1984) also included a key to the western Atlantic species of Ophidiasteridae.

During the 1980s, Gordon Hendler and a number of colleagues made several studies on Belizean echinoderms diversity and distribution, and included the descriptions, of at least, five new species of ophiuroids species. Hendler and Miller (1984) described two new ophiuroid species from the western Atlantic: *Ophioderma devaneyi* collected from the outer continental shelf and upper slope off Florida, and *Ophioderma ensiferum* collected the reef slope of the Belize Barrier Reef, off Carrie Bow Cay. Hendler and Littman (1986) examined the size distribution of 33 ophiuroid species from distinct substrata and reef zones located in four zones on the Belize Barrier Reef. They related size-distribution and population density data to the mode of reproduction and the biological interactions between the ophiuroids, and their habitat in order to understand the biology of juvenile individuals and other reef cryptofauna.

Hendler and Turner (1987) described two new *Ophiolepis* species from the Caribbean Sea and the Gulf of Mexico, *O. gemma* from deep-reef habitats and *O. ailsae* from bathyal depths. *Ophiolepis gemma* was collected at Carrie Bow Cay located on the seaward side of the Belize Barrier Reef, Barbados, and the Gulf of Mexico. They found two brooding specimens from Belize that held embryos in their bursae. They concluded that the occurrence of embryos at the same stage of development in one specimen suggests synchronous brooding in the individual. *Ophiolepis ailsae* was collected at several locations through the Caribbean Sea. Bursae examination did not reveal indication of brooding. Thus the reproductive mode of this species is not known. They described specialized hooked plates on the distal arms in both new species.Moreover, they suggested the hooks are used in feeding.

Hendler (1988a) described *Ophiolepis pawsoni,* bringing to seven the number of *Ophiolepis* species reported from the western Atlantic. The specimens were collected on the seaward slope of the Belize Barrier Reef where it occurs simpatrically with *O. gemma*. Hendler (1988a) also included a key to the species of the genus *Ophiolepis* that comprises three groups based on different morphology of the disk, arms, and central rosette.

Hendler and Pawson (2000) made an exhaustive description of the echinoderms present at several localities in Belize and reported the presence of more than 90 species, mostly from barrier reefs localities (including Carrie Bow, Rhomboidal and Pelican Cays). They described the environments and gave some information about the natural history of the species. Most of this information is also present in the compilation of echinoderms of the Caribbean by Hendler et al. (1995).

Between 1935 and 1978, research vessels visited the Caribbean coast: RV *Leonidas* (1935), RV *Oregon* (1962, 1967), RV *Oregon II* (1970), RV *Alaminos* (1970), RV *Pillsbury* (1971), RV *Pescanica 25* (1974) and RV *Alpa Helix* (1977). These expeditions had several collection stations in Honduras, Nicaragua, Costa Rica and Panama. No scientific accounts of the echinoderms collected were published. Most of the specimens are deposited at the National Museum of Natural History, Smithsonian Institution, at Washington, DC. More recently those species are reported by Alvarado (2011).

From 1867 to 2011, 230 papers related to echinoderms in CA (including Panama) have been published (Fig. 3.2). Between 1860 and 1970, there were less than ten publications per decade (Fig. 3.2a). Most of those publications are related to expedition's reports and monographs of specific groups. Their contribution was enormous, setting the foundations for future research, describing the majority of species that nowadays we know for CA. The last four decades (1970–2011) have been the most productive in terms of numbers of publications. This period was not only dedicated to describing new species, but also their relationship with their habitat, biological processes, reproduction, evolution, and their relationship with disturbances like the El Niño-Southern Oscillation. In these four decades echinoderm research in the region was greatly advanced by (1) the Smithsonian Tropical Research Institute (STRI) in Panama with their laboratories since 1960s in Naos, Punta Galeta and Bocas del Toro (Robertson et al. 2009), (2) the Carrie Bow Cay Marine Field Station that was established in 1972 by the Smithsonian National Museum of Natural History (Rützler 2009), and (3) the Marine Science and Limnology Research Center (Centro de Investigaciones en Ciencias del Mar y Limnología-CIMAR) at the University of Costa Rica that was established in 1979 (Cortés 2009). The research was carried out mostly by Gordon Hendler, Peter W. Glynn and Harilaos Lessios, and they without any doubt were the pioneers in the study of echinoderm ecology and evolution in CA and established the base line for future research. The effect of the presence of these institutions is evident in the number of publications per country in CA. Panama (92), Belize (49) and Costa Rica (39) published most of these, with only 13 publications for the remaining

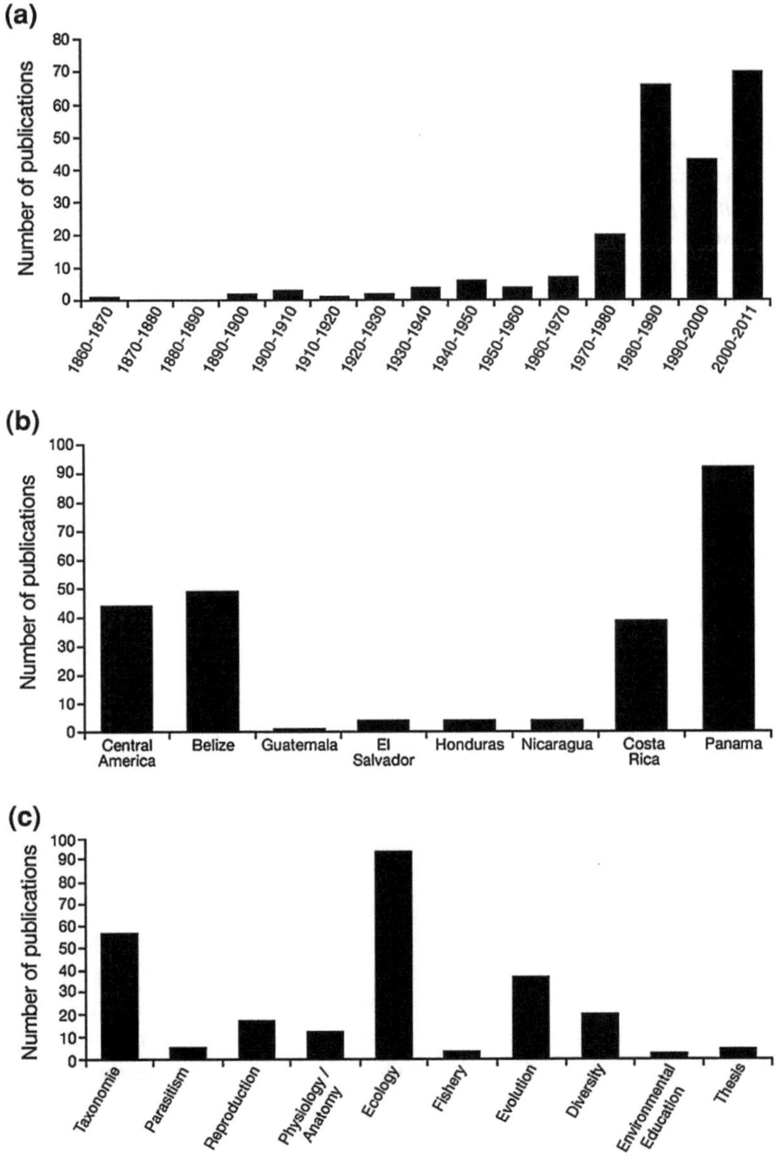

Fig. 3.2 **a** Number of echinoderm publications per decade since 1860 until 2011 in Central America, **b** per country and by **c** research area

countries (Fig. 3.2b). Forty-four publications are common to all CA countries resulted from those major early expeditions to the region or reviews. The main focuses of research have been ecology (94 publications), taxonomy (57 publications), evolutionary processes (37 publications), diversity (20 publications) and reproduction (17 publications). Very little research has been done on physiology, anatomy or fisheries, and none in either conservation issues or aquaculture (Fig. 3.2c). These indicate future areas of research.

3.2.1 Ecology

Crinoid Ecology

Crinoid research has being carried out only in Belize. There are no reports on the presence of crinoids on the Pacific side, with the exception of *Thalassometra agassizii* and *Fariometra parvula* found in deeper waters at Cocos Ridge (Alvarado and Cortés 2009; Alvarado 2010).

Macurda (1982) described the presence of four shallow water crinoids at Carrie Bow Cay, Belize: *Davidaster discoidea*, *Davidaster rubiginosa*, *Ctenantedon kinziei* and *Analcidometra armata*. Both *Davidaster* species were the most abundant crinoids on the forereef slope, and located on the top of the outer ridge, and in the spur and groove zone of the inner fore reef. Macurda (1982) argued that this distribution reflects their need of moderate water motion without direct wave surge. He described three color varieties for *D. discoidea* and seven for *D. rubiginosa*. The three forms of *D. discoidea* have black-tipped pinnules, but differ in the color of the arms (green, yellow and dark green). For *D. rubiginosa,* he describes only three forms that are all orange, but with different pinnules colors (black, yellow and white). The black form is the most common. For these six color forms, he provided the depth range, and the average arm numbers with their length. *Ctenantedon kinziei* was a rare species, occurring only on the top of the outer ridge and down the fore-reef slope, from depths of 13–25 m, and completely hidden within coral colonies (mostly *Montastraea*). *Analcidometra armata* was associated with the gorgonians (*Pseudopterogorgia* spp., *Muriceopsis* sp. and *Gorgonia mariae*) (Macurda 1982).

Humes (2000) described a new parasitic species of siphonostomatoid copepod (*Asterocheres crinoidicola*) on *Nemaster grandis* and *D. rubiginosa*. This was the first report of a copepod parasitic on a crinoid in the Caribbean.

Asteroid Ecology

Caribbean

The asteroid *Oreaster reticulatus* (Fig. 3.3) has been recorded as common on shallow waters throughout the Caribbean Sea. Increased collection for ornamental shell and aquarium trade has resulted in a striking decline in Caribbean populations (See Scheibling and Metaxas 2010 for references). In Belize, Scheibling and Metaxas (2010) examined the potential role of mangroves and fringing patch reefs

Fig. 3.3 Common Central American Caribbean echinoderms. **a** *Oreaster reticulatus*; **b** *Lytechinus variegatus*; **c** *Diadema antillarum*; **d** *Tripnesustes ventricosus*; **e** *Echinometra lucunter*; **f** *Echinometra viridis*; **g** *Actinopyga agassizi*; **h** *Holothuria mexicana*. (Photographs by I.S. Wehrtmann, C. Fernández and J.J. Alvarado)

as recruitment sites for *O. reticulatus*. They sampled populations at sites within these habitat types. They included a population in a seagrass bed and another on an open sand flat for comparison. The radius of each individual was measured. They also recorded the coloration pattern to distinguish juveniles from adults. Their results showed that *O. reticulatus* density varied among sites. It was lowest at the seagrass bed and highest at sand bottom, while intermediate densities were recorded at mangrove and fringing reef sites. They noted that asteroid populations in mangrove and coral habitats were mainly composed of juveniles, but lowest in the seagrass beds and intermediate on the sand flat. Based on the abundance of juveniles on mangrove banks and in fringing reefs, they concluded that these are important habitats for recruitment and serve as nurseries for *O. reticulatus*. Furthermore, they proposed that species such as *O. reticulatus* may serve as sentinels of ecological change, and they may have broad-ranging effects in the health of coastal ecosystems.

Pacific

Guzman and Cortés (1989), as part of a study of the coral reefs at Caño Island, indicated the presence of the sea star *Acanthaster planci*. This species was found in densities between 0.4 and 1.8 ind ha^{-1}, and feeding mainly on the corals *Pavona*, *Psammocora* and *Pocillopora*. *Acanthaster planci* was also observed at Cocos Island, where it had densities of 10.4 $\pm$ 3.1 ind ha^{-1}in 1987. It fed on *Porites lobata*, a coral that was not part of its diet (Guzman and Cortés 1992). Guzman and Cortés (2007) reported densities of 7.4 $\pm$ 2.3 ind ha^{-1} in 2002. Alvarado and Chiriboga (2008) reported a density of 0.02 ind m^{-2} on only two of 19 study sites around the island. In other studies the seastar was present in four of five reefs studied (Guzman and Cortés 1992, 2007).

Ophiuroid Ecology

Caribbean

During the 1990s, Richard B. Aronson evaluated the impact of hurricanes and storms on predator–prey dynamics in Caribbean coral reefs. Aronson (1991) assessed predation, physical disturbance and sublethal arm damage in ophiuroids (*Ophioderma* spp.) in Belize and Jamaica in order to compare to Jurassic-Present trends. This study supported the hypothesis that predation on ophiuroids increased when durophagous teleostean fishes and decapods crustaceans diversified in the Late Mesozoic.

Aronson (1992) investigated predation intensity experienced by *Ophiothrix oerstedi* within and among back reef sites in Jamaica, Barbados and Belize, and provided assessment of the impact of severe storms on predator–prey interactions in back reef habitats. Aronson (1998) evaluated the ecology of predation on ophiuroids in the Caribbean on two temporal scales: seasonally over a 3 years period and annually over a decade. His research suggests that some ecological relationships are stable at multiple temporal scales. Aronson's investigations concluded that strong hurricanes did not cause detectable damage to the ophiuroids subpopulations and did not affect predator–prey interactions (Aronson 1991, 1992, 1998).

Hendler (1982) studied the feeding behavior of the basketstar *Astrophyton muricatum* at Carrie Bow Cay, Belize. He described the sequence of events in nocturnal behavior from 18:16 to 05:40 h, by filming an indivdual and measuring the direction and intensity of currents. *Astrophyton muricatum* fed during the periods of moderate water speed associated with ebb and flood currents. It changed feeding position in response to tidal flow by directing its arms into the current.

Hendler (1984a) examined the mutualistic association between the sponge *Callyspongia vaginalis* and its obligate commensal *Ophiothrix lineata* on Carrie Bow Cay. The diet of *O.lineata* consisted of detrital particles adhering to the sponge which were too large for it to use as food. Thus, the brittlestar cleans the inhalant surface of the sponge as it feeds, increasing filtration capability of the sponge. This association allows *O. lineata* to hide on the sponge from predatory fishes. Hendler (1984a) reports the presence of six species of ophiuroids (*Ophiothrix lineata, O. suensoni, O. angulata, O. oerstedi, Ophiactis savignyi* and *O. quinqueradia*) associated with the sponges *C. vaginalis* and *Neofibularia nolitangere*, and provides their size frequency (disk diameter) and their location inside or outside of the sponges. *Ophiothrix lineata* and *O. oerstedi* were restricted to *C. vaginalis*, while only *O. quinqueradia* was restricted to *N. nolitangere*. *Ophiothrix lineata* activity outside the sponges begins at dusk and diminishes in the morning (Hendler 1984a).

Hendler and Peck (1988) compared Belize Barrier Reef ophiuroids collected at a depth of about 24 m on the fore-reef slope with those collected in shallow waters (0–15 m depth) by Hendler and Littman (1986). They found 44 species, eight species were restricted to the shallow reef, and nine species were found only on the fore-reef. They measured the disk diameter of each individual and grouped the species according to four modes of reproduction: (1) planktotrophic (with pluteus larvae), (2) lecitotrophic (with vitellaria larvae), (3) brooders, and (4) fissiparous. They noted that species richness of brooding species increased with depth, but the number of fissiparous, planktotrophic and lecitotrophic species did not vary with depth. Their results showed that species richness increases with depth. They found differences in the species richness composition, relative species abundance, and reproductive modes between the ophiuroids occurring on the shallow-reef and the fore-reef slope.

Hendler et al. (1999) described a previously unreported phenomenon of planktonic dispersal of juvenile ophiuroids. This observation suggests that juveniles reenter the plankton and drift, or perhaps raft on algal fragments, after having been recruited to the benthos. The water-borne juvenile ophiuroids are compared to plantigrade and planktonic postlarva stages of other marine invertebrates.

Parasitic crustaceans in ophiruorids from Belize were studied by Maddocks (1987), Humes and Hendler (1999) and Hendler and Kim (2010). Maddocks (1987) described the ostracod *Pontocypria hendleri* as a commensal of the ophiuroid *Ophioblenna antillensis* from Carrie Bow and Wee Wee Cays. Humes and Hendler (1999) described the presence of three copepods (*Ophiosyllus latus*, *O. reductus* and *Pseudanthessius deficiens*) which are parasitic on four species of Belizian ophiuroids (*Ophioderma cinereum*, *O. brevispinum*, *Ophiocoma echinata* and *Ophiocomella ophiactoides*). *Pseudanthessius deficiens* live in *O. cinereum* and *O. brevispinum.* The other two species of copepods are host-specific: *O. latus*–*Ophiocomella ophiactoides* and *O. reductus*–*O. echinata*. One to four *O. latus* occurred on 17–57 % of *O. ophiactoides* examined. These copepods generally clasp the first or second arm joint near the edge of the disk. Hendler and Kim (2010) described the presence of a parasitic metanauplius larva of thaumatipsyllid copepods (*Caribeopsyllus chawayi* and *Caribeopsyllus* sp.) in *Ophiothrix angulata* and *Ophiactis savignyii* in Carrie Bow. This parasitic larva used mandibular chelae to grasp the host's stomach wall. The larvae of *C. chawayi* were found only in *O. savignyi*, and the larvae of *Caribeopsyllus* sp. was found only in *O. angulata* living on *Halimeda* spp. Between 18–27 % of *O. angulata* were parasitized. Most of the parasites were small compared to their hosts. Infected *O. angulata* tended to be small and were both sexually mature and immature (Hendler and Kim 2010).

Hendler (2005) described two new species of ophiuroids from the Southern Caribbean of Costa Rica and Panama, *Ophiothrix cimar* and *O. stri* were named in honor of the Centro de Investigaciones en Ciencias del Mar y Limnología (CIMAR) and the Smithsonian Tropical Research Institute (STRI) respectively. Individuals or both species are cryptic and were found beneath coral rubble in shallow water fringing coral reefs. *Ophiothrtix stri* was found also in the bore holes of the sea urchin *Echinometra lucunter*, but the association was not obligatory. *Ophiothrix cimar* was relatively more common than *O. stri* in back reef micro-habitats, but neither species occurred in nearby lagoonal and mangrove habitats. Both species are gonochoric and do not brood. Their eggs are of a size indicative of abbreviated larval development. The gametes of *O. stri* developed into demersal embryos that developed within a large, adhesive fertilization envelope. Hendler (2005) also discussed their systematic affinities, and indicated that *O. stri* and *O. synoecina* could be sister species with a greater systematic affinity to *O. lineata* than to *O. cimar*. He also states that the reefs where the species were previously observed have deteriorated because of environmental degradation and produced a reduction of the populations.

Most recent research on Belizean ophiuroids involves genetic connectivity and phylum diversity on a regional scale of the Caribbean. Richards et al. (2007) assessed gene flow between Florida and Belize of the broadcast-spawing brittle star *Ophiothrix lineata.* They concluded this species showed significant genetic structure between Florida and Belize.

Pacific

Between 1987 and 1996, a series of studies of different benthic communities were conducted on the mud flats at Punta Morales, Gulf of Nicoya, Costa Rica (Vargas

1987, 1989). In those studies the sand dollar *Mellitella stokesii* (as *Encope stokessi*) and the ophiuroid *Microphiopholis geminate* (as *Amphipholis germinata*) appeared as regular members of those communities. This research continued until 1996, including 76 collecting dates from February 1984 until September 1996 (Vargas and Solano 2011). Using sediment cores (265.5 cm^3; 14 cores/date), Vargas and Solano (2011) studied the presence of both species in an area of 400 m^2. They found a total of 149 individuals of *M. geminata* and 63 of *M. stokesii*. Both species had an irregular temporal pattern of abundance, with slight increases at the end of the rainy season or during the dry seasons.

Echinoid Ecology

Caribbean

Strathmann (1981) studied the role of spines in preventing structural damage to echinoids tests at Carrie Bow Cay and Twin Cay, Belize. He analyzed echinoid tests by breaking them by way of dropping lead fishing sinkers on living sea urchins submerged in sea water. He observed specimens from five species representing four echinoid orders: *Eucidaris tribuloides, Diadema antillarum, Tripneustes, Lytechinus variegatus* and *Echinometra viridis* (Fig. 3.3). The results suggest that spines and tissue apparently protect the test apparently by both spreading the load and absorbing energy. He also noted breaks on podial pore tracks are common in all the species. The breaks at sutures were common in living specimens of *E. tribuloides*. In contrast, the breaks on sutures in *D. antillarum, L. variegatus, T. ventricosus*, and *E. viridis* were often absent and comprised a small portion when present. Thus, he suggests that tracts of podial pores are sites of relative weakness. In addition, he evaluated the strength of the test of the different species under impact. He found that, within species, strength appears to increase with size. Strathmann's (1981) observations on test strength and thickness led him to set up a ranking of strength under impact. *Eucidaris tribuloides* was the strongest, followed by *E. viridis* and *D. antillarum*, then *T. ventricosus. Lytechinus variegatus* was the weakest species. Thus, it appears that, among the species examined, the most primitive design is the strongest. He extrapolated the results from modern echinoids to extinct classes and he argued that is speculative, because it is difficult to determine the causes of failure of skeletons in most fossil forms.

Information about the decline and recuperation of populations of *D. antillarum* in Central America is scarce because there are few records prior to the mass mortality and only for several of localities in Belize (Hay 1984; Lewis and Wainwright 1985); Honduras (Hay 1984), and Costa Rica (Valdez and Villalobos 1978; Cortés 1981). Post mortality information comes from Costa Rica (Cortés 1994; Fonseca 2003; Alvarado et al. 2004; Bolaños et al. 2005; Fonseca et al. 2006; Myhre and Acevedo-Gutiérrez 2007; Cortés et al. 2010), Nicaragua (Ryan et al. 1998), Honduras (Lessios 1998), Guatemala (Fonseca and Arrivillaga 2003) and Belize (Aronson and Precht 1997; Carpenter and Edmunds 2006).

Valdez and Villalobos (1978) studied the spatial distribution, and the correlation with the substrate and the degree of aggregation of *D. antillarum* on the coral reef at Cahuita National Park, Costa Rica in 1977. They reported densities between 3.83 and 8.8 ind m^{-2} and a higher degree of aggregation during morning hours that could be a protection mechanism. In the afternoon, *D. antillarum* started to disperse to the reef substrate to feed. Cortés (1981) reported *D. antillarum* densities in this same location in 1980 between 4 and 33 ind m^{-2}. In June 1983, a high mortality of this sea urchin was observed in Cahuita decreasing the densities to 0.25–2 ind m^{-2} (Murillo and Cortés 1984). By 1992, the densities were low (0.01 ind m^{-2}) and *D. antillarum* had almost disappeared from this reef (Cortés 1994). These low densities remained from 1999 until 2003 (Fonseca 2003; Alvarado et al. 2004; Bolaños et al. 2005). Bolaños et al. (2005) indicated a low density of *D. antillarum* (0.03 ind m^{-2}) on the reef lagoon of Cahuita National Park, in October 2003. In this same area, they reported the presence of 15 echinoderms species (one asteroid, seven ophiuroids, five echinoids and two holothurians) that were more diverse at sites with coral substrate and algae than seagrass or sand sites.

Between January and November, 1984, Morales and Murillo (1996) performed the first study on the distribution, abundance and composition of the zooplankton present at the coral reef of Cahuita National Park. As part of the merozooplankton, they described the echinopluteus larvae of the sea urchins *D. antillarum* and *L. variegatus*. In general, the larvae were scarce, with a single peak in abundance in November (36 ind m^{-2}). This peak probably corresponded to spawning during the new moon as observed in Panama (Lessios 1981). The low numbers of *D. antillarum* larvae are presumed to be related to the high adult mortality in 1983 (Morales and Murillo 1996).

At the CARICOMP monitoring site in Cahuita (Meager Shoal), between 1999 and 2008, the density of *D. antillarum* oscillated between 2 and 7 ind m^{-2}, while densities of *E. viridis* decreased significantly from 20 to 0.6 ind m^{-2} from 2000 to 2008 (Fonseca et al. 2006; Cortés et al. 2010). At this same site, densities of *E. tribuloides* were always low (<1 ind m^{-2}; Cortés et al. 2010). Thus there has been a recovery of urchin grazers at Cahuita, and probably with a replacement of the main grazers.

For the Gandoca–Manzanillo Wildlife Refuge, Costa Rica, to the south of Cahuita National Park, Myhre and Acevedo-Gutiérrez (2007) reported a slight increase on *D. antillarum* from 2000 to 2004. In 2000, Fonseca (2003) reported a mean density of 0.10 ± 0.22 ind m^{-2}, that increased to 0.2 ± 0.2 in adult urchins (40–70 mm test diameter) m^{-2} in 2004 (Myhre and Acevedo-Gutiérrez 2007). An increased abundance of live coral within the refuge seems to be associated with this small recovery in density of urchins, because in zones with high densities of *D. antillarum*, the cover of non-calcareous macroalgae was low, presumably as a result of herbivory (Myhre and Acevedo-Gutiérrez 2007).

At Corn Island, Nicaragua, Ryan et al. (1998), reported *D. antillarun* densities being below 4 ind m^{-2} in some shallow (< 5 m) nearshore reefs but not at the CARICOMP sites. On those sites the mean density of *E. viridis* was 29.2 ± 25.2 ind m^{-2}.

Carpenter and Edmunds (2006) reported intermediate densities (0.5–4 ind m^{-2}) of *D. antillarum* between 2003 and 2004 at Turneffe Atoll, Belize compared to other Caribbean localities. The size frequency of the urchins ranged between < 20 and 100–110 mm, with a modal size of 40–49 mm, a normal distribution, and with high recruitment (ind < 30 mm). According to their data, *D. antillarum* in Belize, as well as other five sampling sites along the Caribbean, is recovering and helping with the recovery of corals reefs by facilitating coral recruitment.

Lessios et al. (2001), using mtDNA from *Diadema* populations from the Caribbean (*D. antillarum*), eastern Pacific (*D. mexicanum*) and eastern Atlantic (*D. aff. antillarum*), and including specimens from Belize, Honduras, Costa Rica and Panama, showed that *D. antillarum* was abundant in the Caribbean long before (100,000 years before present) humans could have affected ecological processes. This conclusion agreed with Jackson's (1997) hypothesis in that *D. antillarum* was abundant before humans began to remove fish from the Caribbean and contradicting the idea of Hay (1984) that past high abundances, before the 1983 mass mortality were a recent phenomenon due to the removal of its fish predators and competitors by humans. However, it is possible that their populations were not affected by human pressure, but their grazing impact is, reason why they became known as one of the most important carbonate removers form the Caribbean reefs (Ogden 1977; Scoffin et al. 1980; Bak 1994).

Vadas et al. (1982) studied the presence of *Lytechinus variagetus* in seagrass beds at the Miskitos Cays, Nicaragua, reporting densities between 0.8 and 10.8 ind m^{-2}, test diameters and weights between 43.2 and 51.6 mm and 49.4–63.4 g respectively, and a gonad volume between 0.7 and 2.9 cm^3. In 2001, Fonseca (2008) visited this area to describe the coral reef localities using the AGGRA protocol. She reported the presence of nine species of echinoderms at the sampling sites (*O. reticulatus, Linckia guildingii, Asterina folium, E. viridis, E. lucunter, D. antillarum, Meoma ventricosa, Isosticopus badionotus* and *Holothuria mexicana;* Fig. 3.3). No density or distribution was reported.

In Cayos Cochinos Biological Reserve, Honduras, Lessios (1998) reported the presence of 13 echinoids species (*E. tribuloides, D. antillarum, Arbacia punctulata, L. variegatus, Tripneustes ventricosus, E. lucunter, E. viridis, Clypeaster rosaceus, Paraster doederleini, P. jloridiensis, Brissopsis elongata, Brissus unicolor* and *Meoma ventricosa*). He provided information about field identification characteristics, habitat and density. Lessios (1998) indicated that the shallow water echinoid fauna of this Reserve appears to be typical of the Caribbean. However, the presence of *A. punctulata* (0.010 ind m^{-2}) is noteworthy, because this species is often absent or very rare elsewhere in the tropical West Atlantic. Moreover, *L. variegatus,* that is normally abundant in *Thalassia tetudinum* beds everywhere in the Caribbean, was rare (0.005 ind m^{-2}) at Cayos Cochinos in 1995, as was *D. antillarum* (0.190 ind m^{-2}). *Echinometra lucunter* had the highest density (3.805 ind m^{-2}), followed by *E. tribuloides* (0.467 ind m^{-2}).

Fig. 3.4 *Astropyga pulvinata* aggregation in Punta Ulloa, Cocos Island, August 2007 (Photograph by J. Garay)

Pacific

On the Pacific side, Alvarado (2008) studied the seasonal occurrence of *Astropyga pulvinata* in Bahía Culebra, located on the north Pacific side of Costa Rica. During upwelling months (December to April), this sea urchin forms aggregations as a protection mechanism from predators while feeding. During the rainy season when non-upwelling occurs, the populations of *A. pulvinata* were reduced in the bay, and probably migrating into deeper waters. Alvarado (2008) calculated an average of 679 ind month^{-1} for the upwelling season and 159 ind month^{-1} for the non-upwelling season. The average number of aggregations per month was 21.6, with an average of 27 ind per aggregation. Most individuals have a test diameter between 70 and 120 mm. These aggregations act as refuges for other animals like fish juveniles and crustaceans, thus having an important conservation value. This aggregation behavior was also observed in Cocos Island in August 2007 (J. Garay pers. comm.; Fig. 3.4).

Alvarado and Fernández (2005) reported the presence of 25 species of echinoderms (four species of asteroids, six species of ophiuroids, five species of echinoids and ten species of holothuroids) from Marino Ballena National Park, on the south Pacific of Costa Rica. Echinoderms densities were low (0.01–1.92 ind m^{-2}). *Diadema mexicanum* (Fig. 3.5) had densities between 0.05 and 1.03 ind

m^{-2}, being highest on the farthest site from the coast. They argued that the lower densities were probably the result of the high sedimentation, producing a negative effect on the diversity of echinoderms and on the development of coral reefs (Alvarado and Fernández 2005; Alvarado et al. 2009).

In Cocos Island, Costa Rica, Alvarado and Chiriboga (2008) reported the presence of 28 species of echinoderms (seven species of asteroids, six species of ophiuroids, nine species of echinoids and six species of holothuroids) located between 3 and 18 m deep at 18 sites around the Island. The sea urchin *D. mexicanum* was the only echinoderm present at all sites and depths studied, and provided a density between 0.05 and 6.53 ind m^{-2}.

For the Pacific side of Nicaragua, on the Rivas Department, Alvarado et al. (2011) reported the presence of 11 species of echinoderms (one species of asteroid, six species of echinoids, and four species of holothuroids). The sea urchin *D. mexicanum* was present in nine of the ten study sites, with a density between 0.04 and 1.10 ind m^{-2}, while *Echinometra vanbrunti* had the highest density (4.09 ind m^{-2}) of all echinoderms studied.

Carballo-Bolaños and Pocasangre-Orellana (2007) describe the echinoderm intertidal fauna to a depth of 2 m in the Los Cobanos rocky reefs system, a natural protected area in El Salvador. They reported the presence of 13 echinoderms species (five species of ophiuroids, two species of echinoids and six species of holothuroids). The ophiuroids were the most abundant group with 1,771 individuals at the study sites, and followed by the sea cucumbers with 877 individuals, while there were only 22 individuals of *Echinometra vanbrunti* and *Astropyga pulvinata*. The ophiuroids *Ophiocoma aethiops* and *O. alexandri* (Fig. 3.5), and the sea cucumber *Holothuria kefersteini* had the highest densities (0.15, 0.14 and 0.13 ind m^{-2}, respectively). These species were more common in the upper intertidal zone than in the lower intertidal zone. They also reported, for the first time in El Salvador, the presence of the ophiuroid *Ophiothrix rudis*. Moreover, underwater surveys in the same protected area found subtidal echinoids: *A. pulvinata, Toxopneustes roseus*, *D. mexicanum* and *Eucidaris thouarsii*. *Astropyga pulvinata* was absent on subtidal boulders at Pirigallo and Menaguera Islands, Gulf of Fonseca (2001–2008) (E. Baraza pers. observ). In November 2009 some scattered individuals were found and in 2011 *A. pulvinata* were widespread at both places (E. Baraza pers. observ).

Holothuroid Ecology

Caribbean

Hasbún and Lawrence (2002) describe the presence of five holothurians (*Isostichopus badionotus*, *Holothuria mexicana, H. thomasi, H. arenicola* and *Actinopyga agassizi*) in Cayos Cochinos Biological Reserve, Honduras Caribbean. They provide detailed information about the spicule width, length and height from

Fig. 3.5 Common Central America Pacific echinoderm species. **a** *Phataria unifacialis*; **b** *Nidorellia armata*; **c** *Pharia pyramidata*; **d** *Pentaceraster cumingi*; **e** *Echinothrix calamaris*; **f** *Diadema mexicanum*; **g** *Ophiocoma aethiops*; **h** *Ophiocoma alexandri*. (Photographs by G. Edgar, S. Sunagawa, B. Bezy and J.J. Alvarado)

the tegument and tentacles. They also reported the presence of the commensal pearlfish, *Carapus bermudensis*, living in *H. mexicana* and *A. agassizi* (Fig. 3.3).

Pawson and Miller (1988) describe, with specimens of *Thyone tanyspeira* (Phyllophoridae), collected in Honduras, Colombia and Trinidad. The specimens were collected during the expedition of the RV *Pillsbury* in 1969 and 1971 into the Caribbean.

Pacific

Bakus (1974) compared the toxicity of holothurianss in order to test the hypothesis that toxicity in holothurians is inversely related with latitude and that this phenomenon is in part a reflection of fish predation. He conducted the study at Cocos Island (Costa Rica), Bahía de Bacochibampo (México), Santa Catalina Island (California), and Friday Harbor (Washington). Six of the seven species of holothurians at Cocos Island were toxic to fishes, three of six holothurians were toxic to fish in Mexico, while one of two were toxic to fish in California and three of twelve species in Washington were mildly toxic to fish. At Cocos Island the toxic species were *Holothuria theeli*, *H. atra*, *H. impatiens*, *H.* sp. 1, *H.* sp. 2 and *Bohadschia* sp. The body wall, viscera, and Cuverian tubules (in *H. theeli* and *H.* sp.2) were the most toxic to fishes.

3.2.2 Grazing and Bioerosion

Caribbean

In the Miskitos Cays, Nicaragua, Vadas et al. (1982) studied the abundances, feeding and reproductive activities of the sea urchin *Lytechinus variegatus* on a seagrass bed (*Thalassia testudinum*). Intermediate (2–4 m) and deeper (4–8 m) depths contain fewer urchins and have higher plant diversities. Although highest plant diversities were correlated with intermediate grazer densities ($\sim$5 ind m^{-2}). The diet of these urchins appears to be governed primarily by availability and the palatability of food items. Detrital *T. testudinum* and sediment were the two major food items in the diet (40 and 25 %, respectively). This feeding strategy minimizes the impact of urchin grazing on the structure and the viability of *T. testudinum* and the algal-seagrass community. Over 68.5 % of the volume of the gut contents of *L. variegatus* consisted of particle sizes in the 0.5–3.0 mm category. Intestinal tract pH ranged from 6.0 in the esophagus and 6.5–7.0 in mid-stomach. The average dry weight of the gut contents was 113 mg cm^{-3}. Of this 68 % was $CaCO_3$. Food passage in the gut averaged 0.22 cm^3 ind^{-1} h^{-1} for the first 14.5 h, then values decreased rapidly. The consumption rates of *L. variegatus* that was taken from *T. testudinum* was 0.6 g dry wt day^{-1}.

Hay (1984) compared the pattern of fish and urchin grazing on overfished and less fished reefs in Haiti, Saint Croix, Saint Thomas, Belize, Honduras and Panama. In Belize he worked at Carrie Bow Cay and Lighthouse reefs and in

Honduras on Media Luna and Becerro reefs. On the overfished reefs, the rate of *T. testudinum* removal increased significantly with depth, urchin densities were high, and urchin grazing was equal to, or greater than, fish grazing in shallow habitats. At the overfished reefs (Haiti, Saint Croix and Saint Thomas) the percent of *T. testdinum* eaten per hour by sea urchins was between 0 and 10.8 % with a *D. antillarum* density range between 5–20 ind m^{-2}. At the less fished reefs (Honduras and Belize) it was between 0 and 1 % with *D. antillarum* densities of < 1 ind m^{-2} for the Honduras sites and around 0.2 and 8 ind m^{-2} for the Belize sites. He concluded that on the reef unaffected by humans, the overall impact of urchins on macrophytes was minor relative to the impact of grazing fishes.

The combined effects of a reduction in herbivory due to *D. antillarum* mass mortality in 1983–1984 (Lessios et al. 1984) and the increase in vacant substratum because of coral death (Aronson et al. 2000) have caused dramatic changes in the Caribbean. This has led to several attempts to understand the ecosystem dynamics after these events. McClanahan (1999) examined the ecology of a grazer living sympatrically with *D. antillarum* and *Echinometra viridis*. He assessed the role that finfish and invertebrate predators play in controlling the distribution of *E. viridis* as well the ability of *E. viridis* to control exposed fleshy algae on patch reefs on the Glovers Reef Atoll lagoon. McClanahan's (1999) results showed that the abundance of *E. viridis* is constrained by predation, which restricts *E. viridis* to cryptic locations. He proposed that this sea urchin is unable to control exposed erect fleshy algae as well as *D. antillarum* (McClanahan 1999).

In despite of McClanahan's (1999) results, Aronson et al. (2002a) reported that high densities of the sea urchin *E. viridis* kept the cover of fleshy and filamentous macroalgae to low levels. They concluded that herbivorous activity of *E. viridis* could help re-establishment of the coral *Agaricia tenuifolia*. This trend has been observed by Aronson and his colleagues throughout the Caribbean as well in Belizean reefs (Aronson 2002b; Aronson et al. 2005). Carpenter and Edmunds (2006) reported that dense populations of *Diadema* now occur over a multi-kilometer-wide scale across the entire Caribbean. They conclude that population recovery of *D. antillarum* is occurring at both local and regional scales and that grazing by this echinoid is creating conditions that favor the recruitment of corals (Carpenter and Edmunds 2006).

Brown-Saracino et al. (2007) examined urchin population characteristics, bioerosion rates, their fish predators, and potential competitors on three unprotected reefs and one reef within Hol Chan Marine Reserve in the lagoonal regions of the reef system off Belize. Their report presents densities of Balistidae and Labridae (as predator fish of urchins) and Labridae (potential competitors of urchins). Population data on coral reefs and algae were collected. They recorded urchin wet weight and test diameter. They also analyzed gut contents to estimate yearly bioerosion rates per square meter of reefs. Six species of sea urchin were found at the study sites, *D. antillarum*, *E. lucunter*, *E. viridis*, *E. tribuloides*, *L. variegatus* and *L. williamsi*. *Echinometra viridis* was the dominant species at all but one of the patch reef sites surveyed. Furthermore, *D. antillarum* was found in low abundances and restricted to northern sites. Balistids were rare at all sites, although labrids were relatively abundant. A significant inverse relationship between the abundance of Labridae and urchin densities was found. Parrotfish and urchin densities were not related. Calcium carbonate comprised approximately 90 % of the urchin gut content, and followed by organic material. Urchin bioerosion rates ranged from lowest at Hol Chan Reserve ($\sim$0.2 kg $CaCO_3$ m^{-2} yr^{-1}) to highest at Coral Gardens ($\sim$1.4 kg $CaCO_3$ m^{-2} yr^{-1}), an unprotected area close to Hol Chan. Small urchins and low proportion of organic material in gut contents were found at low algae covered sites, suggesting that the herbivors were food limited. Taken together their results indicate that both top-down and bottom-up factors contribute to the control of sea urchin populations in the lagoons off Belize. They also suggested that Marine Protected Areas (MPA) may provide the means for limiting sea urchin densities and bioerosion of coral reefs and result in reef recovery, returning the calcium carbonate cycle to a more balanced state.

Pacific

Fischer (1990) described the importance of *D. mexicanum* as a bioeroder in Costa Rica. This species is found in very shallow waters (down to 8 m), and produces stronger bioerosion on the limestone. At Caño Island, Guzman (1988) found low densities of this sea urchin, but nevertheless it caused bioerosion at the base of massive corals. Also, at Cocos Island grazing activities of this sea urchin caused high levels of bioerosion and reduced coral recruitment (Guzman and Cortés 1992). After the 1982–83 El Niño event, Guzman and Cortés (1992) stated that coral recovery would take centuries due the high densities of bioeroders and low coral reproduction rates. Guzman and Cortés (1992, 2007) indicated that part of the deterioration of the reef structures at Cocos Island was due to the bioerosive action of *D. mexicanum.*

In 2002, there was a five-fold increase of coral cover and a notable sea urchin reduction (Guzman and Cortés 2007). Guzman and Cortés (2007) determined that *D. mexicanum* was not playing a important role on the reef bioerosion. However, the urchin could still fulfill its key herbivorous role in assisting the recruitment of corals, as has been observed in other reefs on the Caribbean (Sammarco et al. 1974; Sammarco 1980, 1982a, b; Mumby et al. 2006). Lessios et al. (1996) reported the presence of an Indo-Pacific echinoid (*Echinothrix diadema* and *Echinothrix calamaris*; Fig. 3.5) at Cocos Island, and that probably arrived to the island during the 1982–1983 El Niño (Lessios et al. 1996, 1998). These two species are well known bioeroders in the Indo-Pacific (Bak 1990; Peyrot-Clausade et al. 2000; Carreiro-Silva and McClanahan 2001). However, their impact in Cocos Island is unknown. But due to their current densities (0.03–0.02 ind m^{-2}; Alvarado and Chiriboga 2008) it could be predicted that they would be less compared to that of *D. mexicanum*.

3.2.3 Reproduction and Ontogenetic Development

Caribbean

Hendler (1988b) analyzed the ontogenesis of the oral papillae in representatives of the four major groups of Amphiuridae in order to ascertain if comparisons of growth series during postlarval ontogenesis can be used to distinguish homologous structures in closely related taxa. Based on the development of the oral papillae, Hendler (1988b) concluded that the post-larval ontogenesis of the amphiurids is more reliable than larval morphology as an indicator of phylogenetic affinity. Byrne (1991) studied reproduction and development of the ophiuroid *Ophionereis olivacea*, a protandric hermaphrodite species that brood its young. She established that sex reversal occurs at disk diameters ranging between 2.2 and 4.0 mm.

Byrne (1988) examined the oogenesis of *Ophiolepis paucispina* to determine the vitellogenic mechanism involved with the production of the large oocytes and the role of the genital haemal sinus. Anatomical examination of the gonads showed that the ovary wall consists of two parts, the inner and outer sac, which is separated by the genital coelom. The inner sac is composed of the coelomic epithelium, which is a connective tissue layer that contains the haemal sinus, and the germinal epithelium. Experimental evidence revealed that each oocyte is surrounded by the haemal sinus during oogenesis. It appears that that the oocytes bulge into the haemal sinus as they enlarge, and thereby, develop an evagination in the genital sinus. The Golgi complex and endoplasmic reticulum are probably involved in the elaboration of yolk bodies. Endocytotic activity is ubiquitous during oogenesis, and indicated the uptake of exogenous material.

Byrne (1988) concluded that the large oocytes of *O. paucispina* are produced through encytotic incorporation of yolk precursors derived from a somatic source, and that the haemal sinus functions as the proximate source and store of these precursors. Byrne's (1988) results together with the results of other studies (see Byrne 1988 for references) suggests that extraovarian vitellogenesis may be common in echinoderms. Additionally, Byrne (1989) described the ultrastructure of the ovary and oogenesis of *O. paucispina*. She documented the developmental of oocyte growth from the early proliferative stage through the previtellogenic and vitellogenic stages. Byrne (1988, 1989) contributed evidence that the echinoderm genital haemal sinus is an intragonadal nutrient store. Byrne (1989) concluded that her experimental evidence, taken together with the biochemical evidence for the somatic and ovarian synthesis of vitellogenin, indicate that entirely autosynthetic yolk formation may be rare, whereas mixed auto- and heterosyntheticy yolk formation may be common in Echinodermata.

Pacific

The only reference to reproduction in echinoderms in the Pacific is that of Pearse (1968) with *D. mexicanum* at Cocos Island. He reported that all the individuals studied showed all gametogenic stages in the gonads, and suggested that the reproduction occurs year-round. Furthermore, there is no research presently available on crinoid, asteroid or holothuroid reproduction from the Caribbean or Pacific coasts.

3.2.4 Pollution

Rojas et al. (1998) studied the effect of coastal pollution at Cahuita coral reef on the sea cucumber *Holothuria mexicana*. They determined metal concentrations in the sea cucumbers and suggested their use as bio-indicators in the future. The respiratory tree was the structure that showed the highest levels of metals, with the exception of Cu and Pb which were respectively higher in the muscles and the body wall, respectively (Rojas-Guitart 1990; Rojas et al. 1998).

3.2.5 Physiology and Anatomy

Lawrence (1967) studied the gut's lipid reserves of three sea urchins (*E. lucunter, T. ventricosus,* and *L. variegatus*) from Portete (Costa Rica). He calculated the ratio of the amount of total and neutral lipid (in mg) in the gut with respect to size of the individuals. *Echinometra lucunter* had the greatest amounts of total lipid and neutral lipid. The ratio of total lipid to body size was 1.00 for *E. lucunter,* while it was only 0.68 for *T. ventricosus* and 0.40 for *L. variegatus.* Also, *E lucunter* had the highest gonad index, test index, and lantern index. *Tripneustes ventricosus* appears to be the most fragile of the three urchins because its test and lantern indices were the lowest. The high levels of neutral lipid found are comparable to that in the gut of temperate urichins, such as *Strongylocentrotus purpuratus purpuratus*, and higher than that in the gut of *Stomopneustes variolaris.*

Hendler (1984b) made the first report of color change in brittlestars. He studied individuals of all the Caribbean species of the genus *Ophiocoma*: *O. echinata, O. paucigranulata, O. pumila* and *O. wentii*. All four species were found to change color in natural light regimes. Color change was observed in situ at Panamá, while experimental data was recorded from individuals at Carrie Bow Cay, Belize. He observed five individuals of each *Ophicoma* species in order to document diel color-change. He also analyzed severed arms of *O. wendtii* and other individuals with broken arms in order to find out if isolated arms would maintain the same pattern of color-change as intact individuals. Additionally, portions of dorsal arm plates chipped from *O. wendtii* arms were examined to find whether the chips change color at the same time as whole brittlestars and severed arms. His results showed that the day-night color change involves a replacement of brown shades by gray and black. Individuals of all four species change color in the late afternoon. Experimental evidence demonstrated that transformation occurs over a period of three to four hours and it is effected by chromatophores. Furthermore, as seen with the separated body parts, they appear to respond to light independently of the nervous central system.

During the day *O. paucigranulata* was brownish-black. At night the overall color fades to dark gray. In *O. echinata* regions that are brown during the day turn into gray at night. *Ophiocoma pumila* undergoes a comparatively subtle color-change during the day from yellowish-brown to brownish-gray. At night, the brown shades turn to gray and the gray ones turn black. The color-change of *O. wendtii* is striking. During the day, specimens had a homogeneous brownish-gray color-phase, including the spines that sometimes have brownish-orange tips. At night, the disk turns grayish–brown and black and the dorsal side of the arm develops conspicuous gray or white and black bands while the ventral side of the arm remains unbanded. At night the arm spines often turn white basally, gray along the shaft, and orange.

Hendler (1984b) demonstrated that the four species are not equally sensible to light. Three species withdrew their arms into crevices in response to illumination, while *O. pumila* showed little reaction to illumination. Hendler (1984b) tested whether *O. paucigranulata* was more sensitive than *O. echinata* by measuring the speed of reaction. *Ophiocoma wendtii* was the most sensitive species, reacting rapidly to the lights even when the light was attenuated with two layers of red cellophane. The other three species did not react to dim red light. *Ophiocoma wendtii* reacts to lower levels of illumination at night more than during the day, and exhibits negative phototaxis in moonlight as well as in sunlight. He argued that color-change is either an endogenous rhythm, or a response to the light regime and that the color-change may expose photosensitive tissues that control the ophiuroid's detection of shaded places in the reef. Finally, he concluded that negative phototaxis, coordinated with color-change, may facilitate the defensive shadow-seeking response under varying illumination levels. Color change also confers pigmentation patterns which may camouflage ophiuroids from predatory fish.

Hendler and Byrne (1987) described for the first time a presumed photoreceptor system in the body wall of the ophiocomid brittlestar *O. wendtii*, and it was based on examination of the fine structure of the dorsal arm plate (DAP). This led Cobb and Hendler (1990) to test the photoreceptor hypothesis, that microscopic, transparent structures of the skeletal plates, and that their associated chromatophores and nerve bundles comprise a photoreceptor system in *O. wendtii* as porposed by Hendler and Byrne (1987).

In order to examine the functional, systematic, and paleontological importance of their distinctive integument, Byrne and Hendler (1988) undertook a comparative examination of the arm structure of shallow water and bathyal ophiomyxid brittlestars. Additionally, they used the behavior and arm morphology to evaluate the specialization of Paleozoic ophiuroids. Specimens of *Ophiomyxa flaccida*,

Ophiomyxa tumida and *Ophiogeron supinus,* from the subfamily Ophiomyxinae; and *Ophiobyrsa serpens,* from the subfamily Ophiobyrsidae, were collected at several Caribbean locations. *Ophiomyxa flaccida* was collected using SCUBA in Belize, Panamá and Bahamas Islands at depths between 5–25 m. *Ophiobyrsa serpens* was collected at a depth of 24 m on the Belize Barrier Reef. *Ophiomyxa tumida* and *O. supinus* were collected with the Johnson-Sea-Link submersible in the Bahamas Islands at depths between 300–600 m. Their analyses showed that the arms of *O. supinus* were largely feltwork connective tissue fibers with the predominance of interfibrillar matrix, while the arm integument of *O. flaccida*, *O. tumida* and *O. serpens* was composed of a thin layer of feltwork connective tissue and a thick layer of fibrillar connective tissue. The authors' observations on the locomotion behavior and morphology of the ophiomyxid brittlestar argued against the sedentary nature of Paleozoic ophiuroids. They noted that limited arm mobility, rather than integumentary structure, is evidence of functional similarity between *O.supinus* and some Paleozoic ophiuroids. They suggestsed Paleozoic ophiuroids may have been epifaunal forms with a pattern of locomotion more like some recent astropectinid asteroids than amphiurid ophiuroids.

3.3 Diversity and Distribution

Maluf (1988a, b) synthesized echinoderm distribution and composition from California to Peru. She reported 630 species, and peaks of richness in the Gulf of California, Panama and the Galapagos Archipelago. The last two are places involved a high sampling effort. Then Maluf (1988b) compared species similarity and found that CA, from Guatemala to Costa Rica, is a separate unit. She also indicated that the zone between Panama and Costa Rica possess a high degree of endemism and is a transition point, while Panama, Colombia and Ecuador had a greater affinity.

Alvarado and Cortés (2004) considered both the Caribbean and Pacific echinoderms of Central American. They found 315 species, with Panama and Costa Rica having the greatest number of species. However, their calculations of similarity were done without any coast differentiation or discrimination of the presence of the oceanic island, Cocos Island that belongs to Costa Rica. This fact prevents a detailed evaluation of the richness of the Central American Pacific coast. In later investigations, Alvarado et al. (2008), Miloslavich et al. (2010) and Alvarado (2011) reported 253 species of echinoderm fauna for the Caribbean of CA, including the Archipelago of San Andres and Providencia (Colombia). In comparisons with the other countries of the Caribbean, such as Costa Rica and Guatemala, they are the less diverse countries, but probably the CA coast is the least studied (with the exceptions of Panama and Belize) (Alvarado 2011). For the Pacific side of CA, Alvarado et al. (2010) reported 287 echinoderm species.

In this chapter, excluding Panama (Table 3.1, Appendix) we report a total of 416 species of echinoderms for both the Pacific and the Caribbean coast of CA.

Table 3.1 Number of echinoderm species per class in each country by coast (from Alvarado et al. 2008, 2010, Alvarado and Cortés 2010, Alvarado 2011, and Appendix)

Coast	Country	Crinoidea	Asteroidea	Ophuiroidea	Echinoidea	Holothuroidea	Total
Pacific	Guatemala	0	0	7	5	4	16
	El Salvador	0	10	17	13	18	58
	Honduras	0	0	3	4	4	11
	Nicaragua	0	6	12	14	7	39
	Costa Rica	0	12	38	28	28	106
	Cocos Island	2	33	30	31	31	127
Caribbean	Belize	5	18	58	27	26	134
	Guatemala	2	5	4	8	4	23
	Honduras	11	30	14	29	11	95
	Nicaragua	9	30	6	19	2	66
	Costa Rica	4	7	16	13	4	44
Caribbean coasts		16	54	66	47	31	214
Pacific coasts		2	46	60	48	53	207
Total		18	100	123	94	80	416

According to Pawson (2007) 7,000 echinoderm species have been reported globaly. CA is a very diverse area, with 5.9 % of the world species. The Caribbean and Pacific coasts are very similar in the number of total species (214 and 207 species, respectively, with seven in common; Table 3.1), and differing more notably in the numbers of holothuroids and crinoids. The first group is more diverse in the Pacific and the second group is more diverse in the Caribbean. It is important to note that the number of species from the Pacific coast without considering Cocos Island and the Cocos Ridge, CA possesses only 126 species. Cocos Island and the Cocos Ridge contributed with 39 % of the echinoderm species of the Pacific of CA. Even without Cocos Island, the Pacific coast of Costa Rica is the richest in CA with 106 species (Table 3.1). Honduras is the least diverse (11 species). However, Cocos Island is the richest location on the Pacific with 127 species. Belize (134 species) is the richest country on the Caribbean, while Guatemala (23 species) is the least diverse. Ophiuroidea is the most diverse class on both coasts, while Crinoidea is the least diverse (Table 3.1).

Seven species of echinoderms are found on both coasts: three ophiuroids (*Amphiodia violacea*, *Amphipholis squamata* and *Ophiactis savignyi*), one echinoid (*Clypeaster speciosus*) and three holothuroids (*Holothuria arenicola*, *Holothuria impatiens* and *Benthodytes sanguinolenta*). *Amphiodia violacea* is present on the Pacific coast of Costa Rica and Cocos Island, and in the Caribbean coast of Belize. *Amphipholis squamata* is present on the Pacific coasts of Costa Rica and Nicaragua, and on the Caribbean coast of Belize, Guatemala and Honduras. *Ophiactis savignyi* is present on the Pacific coast of El Salvador, Nicaragua, Costa Rica and at Cocos Island, and on the Caribbean coast of Belize, Honduras and Costa Rica. *Clypeaster speciosus* is present on the Pacific side at Cocos Island, and on the Caribbean coast of Costa Rica. *Holothuria arenicola* is present on the Pacific coast of Costa Rica and at Cocos Island, and on the Caribbean coasts of Belize, Guatemala and Honduras. *Holothuria impatiens* is present on the Pacific coast of El Salvador, Costa Rica and at Cocos Island, and on the Caribbean coast of Belize. *Benthodytes sanguinolenta* is present on the Pacific side at Cocos Island and on the Caribbean coast of Honduras (see Appendix).

Six species are distributed on all the countries coasts of the Pacific side and Cocos Island: *Ophiocoma aethiops*, *Ophiocoma alexandri*, *Echinometra vanbrunti*, *Clypeaster rotundus*, *Holothuria difficilis* and *Isostichopus fuscus*. Ten species are found on the Caribbean coasts of all the CA countries: *Davidaster discoidea*, *Astropecten articulatus*, *Oreaster reticulatus*, *Astrophyton muricatum*, *Ophiocoma echinata*, *Eucidaris tribuloides*, *Diadema antillarum*, *Echinometra lucunter*, *Meoma ventricosa* and *Holothuria mexicana*. One hundred and twenty-eight species are reported for the Pacific coast of only one country, 43 species for two countries, 14 species for three countries, three species for five countries and six species for six countries. On the Caribbean coasts of CA 130 species are reported for only one country, 43 species for two countries, 25 species for three countries, five species for four countries and ten species for five countries (see Appendix).

3.4 Paleontology

Donovan et al. (2005) indicated that the knowledge of echinoid fossils from Belize is poorly documented. In their paper they reported the presence of two echinoids: *Echinolampas* sp. cf. *aldrichi* from the Mid Miocene to Quaternary, and *Echinocorys* sp. cf. *scutata* from the Mid Eocene. Gischler (2010) analyzed five Holocene sediment cores from Glovers and Lighthouse reefs, and Turneffe islands, Belize. He picked out echinoid spines, plates, and teeth. These were identified to genus based on comparison with these pieces of echinoids collected live and treated with dilute sodium hypochlorite and dried. A few fragments of the massive spines of *E. tribuloides* and the burrowing echinoid *Clypeaster* were found in cores from Turneffe. *Diadema* and *Echinometra* pieces were the most commonly encountered. Gischler (2010) reported that three significant echinoid mass mortality events occurred during 6.3–6.1 kyr BP (Glovers reef) as well as 4 and 1.3 kyr BP (Turneffe Islands). The first mass mortality event apparently only affected *Diadema.* The second and third events affected both *Diadema* and *Echinometra.*

Alvarado et al. (2006) reviewed echinoderm paleontological research that occurred from 1880 until 2001 in Costa Rica, and listed 46 species. The fauna is composed of Crinoidea with one genus and Echinoidea with seven orders, 18 families, 28 genera and 45 species. Of the echinoids, the order Spatangoida was the most diverse, with seven families, 15 genera and 26 species. Within this order, the most diverse family was Brissidae, with eight genera. The genus *Schizaster* had seven species. They indicated that the majority of species are Cretaceous and Miocene, from 11 geological formations. The fossil echinoderm fauna of Costa Rica is similar to other areas of CA and México, with a good record from the Miocene and a notable decline to the Pliocene (Buitrón 1978). Alvarado et al. (2006) pointed out that a greater effort is needed in the investigation of fossil echinoderms to improve its understanding, and to have a real appreciation of its composition. Specimen identification, due to lack of specialists and literature, is one of the greatest problems. There is also a need to explore more areas in CA.

3.5 Threats to Echinoderms

Echinoderms are vulnerable to a number of threats from natural and anthropogenic origin. The threats can also be classified as directly or indirectly related to echinoderms. Any deterioration of the environment where the echinoderms live can be classified as an indirect threat. Because the majority of echinoderms in CA live in shallow coastal waters, the most important habitats are coral reefs, rocky reefs and seagrass meadows. All threated due to human activity.

Cortés (2007) indicated the main natural disturbances to coral reefs in CA are the following: (1) extensive mortality of corals due to bleaching associated with extremes in water temperature or salinity, sediment or bacterial infection;

Fig. 3.6 Portete reef platforms uplifted after a 7.5 earthquake on the Southern Caribbean of Costa Rica in 1991 (Photograph by J. Cortés)

(2) hurricanes; and (3) tectonic activity. On April 22, 1991, a 7.5 magnitude earthquake in Southern Caribbean of Costa Rica resulted in an uplift of the coast of between 50 and 190 cm (Fig. 3.6), and caused mass mortalities among the intertidal and reef communities (Cortés et al. 1993, 1994). Among those populations most affected were the corals, algae, sea urchins and chitons. Dense populations found in a narrow band just below sea level before the uplift of the sea urchins *E. lucunter, L. variegatus* and *M. ventricosa* were killed by the exposure in Portete reef platform; the most impacted was *E. lucunter* (Cortés et al. 1993; Alvarado et al. in prep.).

Among the main anthropogenic impacts to coral reefs in CA, Cortés (2007) emphasized pollution, including garbage, oil, heavy metals from farming and industry, and sedimentation. Cortés (2007) also mentions tourism as an important source of disturbance induced by curio collection, construction of tourism facilities, anchor and boat-related damage and damage caused by divers and snorkelers. In this sense, tourism can also be viewed as a direct threat to echinoderms, and especially to regular and irregular intertidal sea urchins that are collected as a personal souvenir or for aquaria. Even if there is no official register of these collections, anecdotal comments from local residents indicate that the populations of some sand dollars have diminished over time. Inside Marine Protected Areas (MPA) this activity is banned, but due to lack of patrolling and conscience this practice continues.

The most important direct threat to echinoderms in CA is illegal fisheries and by-catch of longline and trawling fisheries. Sea cucumbers fisheries is probably one of the main illegal fisheries in CA. In El Salvador, this activity is strictly regulated since holothurians are considered threatened species due to unauthorized fisheries for overseas smuggling (Decree No. 36 Ministry of the Environment and Natural Resources; Diario Oficial No. 383 (103) 5 June 2009, p. 75). The activity stopped after legal prosecutions and holothurians are recolonizing depleted areas. In Costa Rica sea cucumber fisheries are forbidden from any MPA, but an official decree (N°23.138/MAG; La Gaceta N°80, 27th April 1994, p. 4) of the Ministry of Agriculture establishes a maximum of ten permits for artisanal fishers for fisheries of sea cucumbers of the species *Holothuria inornata* and *Isostichopus fuscus*. The decree established a maximum of 1,200 sea cucumbers per fisher per month per defined area. The permit is for two months between October and April and the sea cucumbers can be disembarked only at official ports. They are subject to inspection and fishermen have to present a monthly report. We have contacted the Costa Rican authority (Instituto Costarricense de Pesca y Acuicultura-INCOPESCA) to get more information about the decree and its implementation, because there was no baseline investigation to establish the quotas or the target species, or any research about the status of the populations, or reproduction seasons for the establishment of that decree. The answer from the national authority is contradictory. Some officials tell us that there is no sea cucumbers fishery at the present. Other officials said that there are Japanese and Mexican carrying out this activity on the North Pacific coast of the country. There are no records of any permit granted. According to Toral-Granda (2008), in a Global FAO report of sea cucumber fishery, Hong Kong imported from Costa Rica 1,268 kg dry weight of sea cucumbers from 1999 to 2004 (1999: 108 kg; 2000: 664 kg; 2001: 325 kg; 2003: 7 kg; 2004: 164 kg). One threat to sea cucumbers in Costa Rica, and probably for the rest of CA, is the lack of coordination, register, control and patrolling by the national authorities, INCOPESCA and the ministry of Agriculture, the Environment and Foreign Trade.

Toral-Granda (2008) indicated that sea cucumber fishing activities in Nicaragua dated back to 1994 where the first exports are recorded. In 2005, oriental traders requested a permit to exploit this resource. A one-year permit was granted while scientific research was undertaken on the Pacific coast. No taxonomic information exists on the species harvested during that year. Currently three species are known to be commercially harvested. At least another seven unidentified sea cucumbers are harvested. In the Pacific coast, the most important species is *I. fuscus*. In the Caribbean, the most important species are *H. mexicana*, *H. arenicola* and *I. badionotus*. All products are exported to China and Taiwan as there is no consumption in Nicaragua. Toral-Granda (2008) indicated that Hong Kong reported imports from Nicaragua involving 6,078 kg dry weight from 2003 to 2005 (2003: 252 kg; 2005: 5,826 kg). Currently, no sea cucumber fishing activities are regulated. However, there is a plan to start applying total allowable catches and fishing seasons to both Pacific and Caribbean species in Nicaragua (Toral-Granda 2008).

By-catch is one of the most significant issues affecting fisheries management today, due to the incidental mortality of non-target species. This affects biodiversity and trophic food webs (Hall et al. 2000). According to Allan Bolaños

(Programa de Restauración de Tortugas Marinas-PRETOMA, Costa Rica, pers. comm. 2011), a recent study of artisanal fishery on the Pacific coast of Costa Rica showed that every 1,000 hooks captured 15–20 sea stars (*Astropecten aramatus*, *Luidia tessella*, *L. superba*, and *Tethyaster canaliculatus,* this last two new reports for Costa Rica). Ingo Wehrtmann (Universidad de Costa Rica, pers.comm 2011) pointed out that deep shrimp trawling (depth of 150–200 m) did not capture any echinoderms. However, in shallow waters (depth of 20–50 m) there was a by-catch of echinoderms (Allan Bolaños, pers. comm. 2011). In El Salvador asteroids are captured during shrimp trawling activities (10–50 m). There are no studies on the possible disturbances associated with this activity.

The natural direct threats to echinoderm populations are less numerous or common in CA. The only well documented one has been over the mass mortality of *D. antillarum* by a waterborne pathogen in the Caribbean (see ecology section). For the Pacific side, there are no reports of disease or mass mortalities, or research focused on test abnormalities, or parasites.

3.6 Concluding Remarks and Recommendations

CA is a region with high echinoderm diversity, a products of its rich evolutionary and geological history and its particular geographical location. This fact has attracted the attention of researchers for the last 150 years, as indicated in the number of publications that we have reported. Nevertheless, these 150 years of investigations have not been equal for all countries of the region. The scientific history of the region cannot be separated from the political history. The various armed conflicts and the political instability were factors that resulted in a low priority given scientific research. Panama, Belize and Costa Rica have been relatively more stable compared to Guatemala, Honduras, El Salvador and Nicaragua and they lead in research and consequently in greater numbers of publications, and hence, more species reported. The case of the Caribbean of Honduras and Nicaragua is interesting. For example, the results of the RV *John Elliot Pillsbury* expedition reported a high diversity of asteroids, however, the sampling was limited to a few stations. Comparing the richness of species in the Caribbean, Alvarado (2011) indicates that the number of species from Central America would increase considerably if research increases. Moreover, Alvarado (2011) explains that the richness of species in Belize, Costa Rica and Panama is a reflection of the presence of important research centers like CIMAR, STRI and the station at Carrie Bow. Nevertheless, the richness of environments that these countries possess also contributes to the high numbers of species. Guatemala is a special case, since its Caribbean coastline is very short and, its Pacific coast consists of sandy beaches. This results in a smaller number of species.

This chapter has provided evidence that the great majority of investigations have been focused on the Ophuiroidea and Echinoidea, but few efforts on Asteroidea and Holothuroidea and practically none on Crinoidea. Most of the research has been on the Pacific, and indicates that strong efforts are needed on the Caribbean coast. There are few studies focused directly on Asteroidea and Holothuroidea since they generally only mentioned as part of the echinoderm fauna of a particular region.

Ecology, taxonomy and evolution have been the main lines of investigations in CA. These research efforts should continue, and at the same time, studies focused on reproduction and physiology should be encouraged. Moreover, it is necessary to start research in fisheries, aquaculture and management that help stakeholders and authorities make decisions for more sustainable extractions. To be able to conduct research in CA, it is necessary to support and to promote the exchange of information among the countries, as well as to fund the formation of new researchers for this group.

Acknowledgments We would like to thank A. Bolaños, I. Wehrtmann, C. Fernández, M. Springer, R. Vargas, F.A. Solís-Marín and the late C.G. Ahearn for their help with literature, comments and museum access. M. Chavez for the elaboration of the figures. JJAB thanks CONICIT, MICIT and CONACYT. We appreciate the comments made by J. Cortés, J. Lawrence and T. Hopkins that improved this paper.

References

Agassiz A (1898) Preliminary report on the echini. Dredging operations of the Albatross. Bull Mus Comp Zool 32:71–86

Alvarado JJ (2008) Aggregation behavior of the sea urchin *Astropyga pulvinata* (Lamarck, 1816) (Echinodermata: Echinoidea) in Punta Flor, Bahía Culebra, Costa Rica. Pac Sci 62:579–592

Alvarado JJ (2010) Isla del Coco (Costa Rica) Echinoderms: State of knowledge. In: Harris LG, Böttger SA, Walker CW, Lesser MP (eds) Echinoderms: Durham. Proceedings of the 12th international Echinoderm conference. CRC Press, Taylor and Francis, Balkema, Leiden, pp103–113

Alvarado JJ (2011) Echinoderm diversity from the Caribbean province. Mar Biodiv 41:261–285

Alvarado JJ, Cortés J (2004) The state of knowledge on echinoderms of Costa Rica and Central America. In: Heinzeiller T, Nebelsick JH (eds) Echinoderms: München Proceedings of the 11th international Echinoderm conference. Taylor and Francis, London, pp 149–155

Alvarado JJ, Fernández C (2005) Equinodermos del Parque Nacional Marino Ballena, Pacífico,Costa Rica. Rev Biol Trop 53(3):275–284

Alvarado JJ, Chiriboga A (2008) Distribución y composición de los equinodermos de las aguas someras en la Isla del Coco, Pacifico Oriental, Costa Rica. Rev Biol Trop 56(2):99–111

Alvarado JJ, Cortés J (2009) Echinoderms. In: Wehrtmann IS, Cortés J (eds) Marine biodiversity of Costa Rica, Central America. Monographiae Biologicae, Vol 86. Springer, Berlin, pp 421–433

Alvarado JJ, Cortés J, Salas E (2004) Population densities of *Diadema antillarum* Philippi (Echinodermata: Echinoidea) at Cahuita National Park (1977–2003), Costa Rica. Carib J Sci 40:257–259

Alvarado JJ, Zeledón LA, Boyd R (2006) Equinodermos fósiles de Costa Rica: notas preliminares. Rev Biol Trop 54(1):287–299

Alvarado JJ, Fernández C, Cortés J (2009) Water quality conditions on coral reefs at the Marino Ballena National Park, Pacific Costa Rica. Bull Mar Sci 84:137–152

Alvarado JJ, Solis-Marin FA, Ahearn C (2008) Equinodermos (Echinodermata) del Caribe Centroamericano. Rev Biol Trop 56(3):37–55

Alvarado JJ, Solis-Marin FA, Ahearn C (2010) Echinoderms (Echinodermata) diversity off Central America Pacific. Mar Biodiv 40:45–56
Alvarado JJ, Aguirre-Rubí J, Ayala A, Buitrago F, Fernández C, Reyes-Bonilla H, Sánchez A (2011) Characterization of the coral communities off San Juan del Sur, Pacific Nicaragua. Bull Mar Sci 87:129–146
Amador J, Alfaro E, Lizano O, Magaña V (2006) Atmospheric forcing in the Eastern Tropical Pacific: a review. Progr Oceanogry 69:101–142
Aronson RB (1991) Predation, physical disturbance, and sublethal arm damage in ophiuroids: a Jurassic–Recent comparison. Mar Ecol Prog Ser 74:91–97
Aronson RB (1992) The effects of geography and hurricane disturbance on a tropical predator-prey interaction. J Mar Biol Ecol 162:15–33
Aronson RB (1998) Decadal-scale persistence of predation potential in coral reef communities. Mar Ecol Prog Ser 172:53–60
Aronson RB, Precht WF (1997) Stasis, biological disturbance, and community structure of a Holocene coral reef. Paleobiology 23:326–346
Aronson RB, Pretch WF, Macintyre IG, Murdoch TJT (2000) Coral bleach-out in Belize. Nature 405:36
Aronson RB, Precht WF, Toscano MA, Koltes KH (2002a) The 1998 bleaching event and its aftermath on a coral reef in Belize. Mar Biol 141:435–447
Aronson RB, Macintyre IG, Precht WF, Murdoch TJT, Wapnick CM (2002b) The expanding scale of species turnover events on coral reefs in Belize. Ecol Monogr 72:233–249
Aronson RB, Macintyre IG, Lewis SA, Hilbun NL (2005) Emergent zonation and geographic convergence of coral reefs. Ecology 86:2586–2600
Astiz L, Kanamori H, Eissler H (1987) Source characteristics of earthquakes in the Michoacan seismic gap in Mexico. Bull Seismol Soc Am 77:1326–1346
Bak RPM (1990) Patterns of echinoid bioerosion in two Pacific coral reef lagoons. Mar Ecol Prog Ser 66:267–272
Bak RPM (1994) Sea urchin bioerosion on coral reefs: place in the carbonate budget and relevant variables. Coral Reefs 13:99–103
Bakun A, Csirke J, Lluch-Belda D, Steer-Ruiz R (1999) The Pacific Central American Coastal LME. In: Sherman K, Tang Q (eds) The large marine ecosystems of the Pacific rim: assessment, sustainability, and management. Blackwell Science, Cambridge, pp 268–281
Bakus GJ (1974) Toxicity in holothurians: a geographical pattern. Biotropica 6:229–236
Barraza, JE (1995). Equinodermos de El Salvador, II. In: Serrano F (ed.) Historia Natural y Ecología de El Salvador. Ministerio de Educación, pp 87–99
Barraza JE, Hasbún CR (2005) Los equinodermos (Echinodermata) de El Salvador. Rev Biol Trop 53(3):139–146
Bohrmann G, Heeschen K, Jung C, Weinrebe W, Baranov B, Cailleau B, Heath R, Huhnerbach V, Hort M, Masson D, Trummer I (2002) Widespread fluid expulsion along the seafloor of the Costa Rica convergent margin. Terra Nova 14:69–79
Bolaños N, Bourg A, Gómez J, Alvarado JJ (2005) Diversidad y abundancia de equinodermos en la laguna arrecifal del Parque Nacional Cahuita, Caribe, Costa Rica. Rev Biol Trop 53(3):285–290
Boone L (1928) Scientific results of the first oceanographic expedition of the Pawnee 1925. Echinodermata from tropical east American seas. Bull Bingham Oceanogr Coil 1:1–22
Brenes CL, Gutierrez A (1998) Oceanographic aspects of the Central American Pacific. Top Meteor Oceanogr 5:3–15
Brenes CL, León S, Gutiérrez A, Arroyo G (1990) Condiciones hidrográficas en la región de los "Papagayos". Rev Geofis 33:5–19
Brown-Saracino J, Peckol P, Curran HA, Robbart ML (2007) Spatial variation in sea urchins, fish predators, and bioerosion rates on coral reefs of Belize. Coral Reefs 26:71–78
Buitrón BE (1978) Distribución de los equinoideos terciarios en la planicie costera del Golfo de México, en América Central, en el norte de América del Sur y en la Antillas. Univ Nac Aut Mex Inst Geol 101:66–113

Byrne M (1988) Evidence for endocytotic incorporation of nutrients from haemal sinus by the oocytes of the brittlestar *Ophiolepis paucispina*. In: Burke RD, Randewov PV, Lambert P, Parsley RL (eds) Echinoderm Biology Proceedings of the 6th International Echinoderm Conference. Victoria AA Balkema, Rotterdam, Brookfield, pp 557–563

Byrne M (1989) Ultrastructure of the ovary and oogenesis in the ovoviviparous ophiuroid *Ophiolepis paucispina* (Echinodermata). Biol Bull 176:79–95

Byrne M (1991) Reproduction, development and population biology of the Caribbean ophiuroids *Ophionereis olivacea*, a protandric hermaphrodite that broos its young. Mar Biol 111:387–399

Byrne M, Hendler G (1988) Arm structures of the ophiomyxid brittlestars (Echinodermata: Ophiuroidea: Ophiomyxidae). In: Burke RD, Randewov PV, Lambert P, Parsley RL (eds) Echinoderm Biology Proceedings of the 6th International Echinoderm Conference. Victoria, AA Balkema, Rotterdam, Brookfield, pp 687–695

Carballo-Bolaños RE, Pocasangre-Orellana XM (2007) Composición y estructura de la fauna intermareal de Equinodermos en el sistema arrecifal rocoso Los Cóbanos, Departamento de Sonsonate, El Salvador. Licenciatura Thesis, Escuela de Biología, Facultad de Ciencias Naturales y Matématica, Univ El Salvador, San Salvador, El Salvador

Carpenter RC, Edmunds PJ (2006) Local and regional scale recovery of *Diadema* promotes recruitment of scleractinian corals. Ecol Lett 9:271–280

Carreiro-Silva M, McClanahan TR (2001) Echinoid bioerosion and hervibory on Kenyan coral reefs: the role of protection from fishing. J Exp Mar Biol Ecol 262:133–153

Castillo P, Batisa R, Vanko D, Malavassi E, Barqueo J, Fernández E (1988) Anomalously young volcanoes on hot-spot traces: I. Geology and petrology of Cocos Island. Geol Soc Amer Bull 100:1400–1414

Clark AH (1939) Echinoderms (other than Holothurians) collected on the Presidential Cruise of 1938. Smithson Miscell Collec 98:1–22

Clark HL (1940) Eastern Pacific expeditions of the New York zoological society. XXI notes on Echinoderms from the west coast of Central America. Zoologica 25:331–352

Clark HL (1948) A report on the Echini of the warmer eastern Pacific, based on the collections of the Velero III. Allan Hancock Pac Exp 8:225–352

Coates AG, Obando JA (1996) The Geologic evolution of the Central American Isthmus. In: Jackson JBC, Budd AF, Coates AG (eds) Evolution and Environment in Tropical America. The University of Chicago Press, Chicago, pp 21–58

Coates AG (1997) Central America: a Natural and Cultural History. Yale University Press, New Haven

Cobb JLS, Hendler G (1990) Neurophysiological characterization of the photoreceptor system in a brittlestar, *Ophiocoma wendtii* (Echinodermata: Ophiuroidea). Comp Biochem Physiol 97:329–333

Cortés J (1981) The coral reef at Cahuita, Costa Rica: a reef under stress. MSc thesis, McMaster Univ, Ontario, Canada

Cortés J (1994) A reef under siltation stress: a decade of degradation. In: Ginsburg RN (ed) Proceedings of the colloquium on the global aspects of coral reefs: health, hazards and history, miami 1993. RSMAS, University of Miami, Florida, pp 240–246

Cortés J (2007) Coastal morphology and coral reefs. In: Bundschuh J, Alvarado GE (eds) Central America: geology, resources and hazards, vol 1. Taylor and Francis, London, pp 185–200

Cortés J (2009) A history of marine biodiversity scientific research in Costa Rica. In: Wehrtmann IS, Cortés J (eds) Marine biodiversity of Costa Rica, Central America. Monogr Biol 86 Springer + Business Media BV, Berlin, pp 47–80

Cortés J, Soto R, Jiménez C, Astorga A (1993) Earthquake associated mortality of intertidal and coral reef organisms (Caribbean of Costa Rica). Proceedings of the 7th International Coral Reef Symposium. Guam 1:235–240

Cortés J, Soto R, Jiménez C (1994) Efectos ecológicos del terremoto de Limón. Rev Geol Amér Central Vol especial, Terremoto de Limón:187–192

Cortés J, Jiménez CE, Fonseca AC, Alvarado JJ (2010) Status and conservation of coral reefs in Costa Rica. Rev Biol Trop 58(Suppl 1):33–49

D'Croz L, Robertson R (1997) Coastal oceanographic conditions affecting coral reefs on both sides of the Isthmus of Panama. Proceedings of the 8th International Coral Reef Symposium 2:2053–2058

Deichmann E (1938) Holothurians from the western coast of Lower California and Central America, and from the Galápagos Islands; Eastern Pacific Expeditions of the New York Zoological Society XVI. Zoologica 23:361–387

Deichmann E (1941) The Holothurioidea collected by the Velero III during the years 1932 to 1938. Part I: Dendrochirota. Allan Hancock Pac Exp 8:61–195

Deichmann E (1958) The Holothurioidea collected by theVelero III and IV during the years 1932 to 1954. Part II: Aspidochirota. Allan Hancock Pac Exp 11:253–349

Devaney DM (1974) Shallow water echinoderms from Brithish Honduras with a description of a new species of *Ophiocoma* (Ophiuroidea). Bull Mar Sci 24:122–164

Donovan SK, Lewis DN, Davis P (2005) Fossil echinoids from Belize. Carib J Sci 41:323–328

Fiedler PC (2002) The annual cycle and biological effects of the Costa Rica Dome. Deep-Sea Res I 49:321–338

Fisher WK (1928) Sea stars from the Arcturus oceanographic expedition. Zoologica 8:487–493

Fischer R (1990) Biogenetic and nonbiogenetically determined morphologies of the Costa Rican Pacific coast. Z Geomorph N F 34:313–321

Fonseca AC (2003) A rapid assessment at Cahuita National Park, Costa Rica, 1999 (Part I: Stony corals and algae). Atoll Res Bull 496:249–257

Fonseca AC (2008) Coral reefs of Miskitus Cays, Nicaragua. Gulf Carib Res 20:1–10

Fonseca AC, Arrivillaga A (2003) Coral reefs of Guatemala. In: Cortés J (ed) Latin American Coral Reefs. Elsevier, Amsterdam, pp 159–169

Fonseca AC, Salas E, Cortés J (2006) Monitoreo del arrecife coralino Meager Shoal, Parque Nacional Cahuita (sitio CARICOMP). Rev Biol Trop 54:755–763

Gibson J, Carter J (2003) The Reefs of Belize. In: Cortés Jorge (ed) Latin American coral reefs. Elsevier, Amsterdam, pp 171–202

Gischler E (2010) Possible fossil echinoid mass mortality detected in Holocee lagoons, Belize. Palaios 25:260–268

Glynn PW, Veron J, Wellington GM (1996) Clipperton Atoll (eastern Pacific): oceanography, geomorphology, reef-building coral ecology and biogeography. Coral Reefs 15:71–99

Guzman HM (1988) Distribución y abundancia de organismos coralívoros en los arrecifes coralinos de la Isla del Caño, Costa Rica. Rev Biol Trop 36:191–207

Guzman HM, Cortés J (1989) Coral reef community structure at Caño Island, Pacific Costa Rica. PSZNI: Mar Ecol 10:23–41

Guzman HM, Cortés J (1992) Cocos Island (Pacific of Costa Rica) coral reefs after the 1982-83 El Niño disturbance. Rev Biol Trop 40:309–324

Guzman HM, Cortés J (2007) Reef recovery 20 years after the 1982–1983 El Niño massive mortality. Mar Biol 151:401–411

Hall MA, Alverson DL, Metuzals KI (2000) By-Catch: problems and solutions. Mar Poll Bull 41:204–219

Harborne AR, Afzal DC, Andrews MJ (2001) Honduras: caribbean coast. Mar Poll Bull 42: 1221–1235

Harpp KS, Wanless VD, Otto RH, Hoernle K, Werne R (2005) The Cocos and Carnegie Aseismic Ridges: a trace element record of long-term plume–spreading center interaction. J Petrol 46:109–133

Hasbún CR, Lawrence AJ (2002) An annotated description of shallow water holothurians (Echinodermata: Holothuroidea) from Cayos Cochinos, Honduras. Rev Biol Trop 50:669–678

Hay ME (1984) Patterns of fish and urchin grazing on Caribbean coral reefs: are previous results typical? Ecology 65:446–454

Helly JJ, Levin LA (2004) Global distribution of naturally occurring marine hypoxia on continental margins. Deep-Sea Res I 51:1159–1168

Hendler G (1982) Slow flicks show star tricks elapsed-time analysis of basketstar (*Astrophyton muricatum*) feeding behavior. Bull Mar Sci 32:909–918

Hendler G (1984a) The association of *Ophiothrix lineata* and *Callyspongia vaginalis*: a brittlestar-sponge cleaning symbiosis. Mar Ecol 5:9–27

Hendler G (1984b) Brittlestar color-change and phototaxis (Echinodermata: Ophiuroidea: Ophiocomidae). Mar Ecol 5:379–401

Hendler G (1988a) Western Atlantic *Ophiolepis* (Echinodermata: Ophiuroidea): a description of *O. pawsoni* new species, and a key to the species. Bull Mar Sci 42:265–272

Hendler G (1988b) Ophiuroid skeleton ontogeny reveals homologies among skeletal plates of adults: a study of *Amphiura filiformis*, *Amphiura stimpsonii* and *Ophiophragmus filograneus* (Echinodermata). Biol Bull 174:20–29

Hendler G (2005) Two new brittle star species of the genus *Ophiothrix* (Echinodermata: Ophiuroidea: Ophiotrichidae) from coral reefs in the southern Caribbean Sea, with notes on their biology. Carib J Sci 41:583–599

Hendler G, Miller JE (1984) *Ophioderma devaneyi* and *Ophioderma ensiferum* new brittlestar species from the western Atlantic (Echinodermata: Ophiuroidea). Proc Biol Soc Wash 97:462–491

Hendler G, Littman BS (1986) The ploys of sex: relationships among the mode of reproduction, body size and habitats of coral-reef brittlestars. Coral Reefs 5:31–42

Hendler G, Byrne M (1987) Fine structure of the dorsal arm plate of *Ophiocoma wendtii*: evidence for a photoreceptor system (Echinodermata, Ophiuroidea). Zoomorphology 107:261–272

Hendler G, Turner RL (1987) Two new species of *Ophiolepis* (Echinodermata: Ophiuroidea) from the Caribbean Sea and the Gulf of Mexico: with notes on ecology, reproduction and morphology. Nat Hist Mus LA County Mus Contrib Sci 395:1–14

Hendler G, Peck RW (1988) Ophiuroids off the deep end: fauna of the Belizean fore-reef slope. In: Burke RD, Randewov PV, Lambert P, Parsley RL (eds) Echinoderm Biology Proceedings of the 6th International Echinoderm Conference. Victoria, AA Balkema, Rotterdam, Brookfield, pp 411–419

Hendler G, Pawson J (2000) Echinoderms of the Rhomboidal Cays: biodiversity, distribution and ecology. Atoll Res Bull 49:273–299

Hendler G, Kim IH (2010) Larval biology of Thaumatopsyllid copepods endoparasitic in Caribbean ophiuroids. J Crust Biol 30:206–224

Hendler G, Baldwin CC, Smith DG, Thacker CE (1999) Planktonic dispersal of juvenile brittle stars (Echinodermata: Ophiuroidea) on a Caribbean reef. Bull Mar Sci 65:283–288

Hendler G, Miller JE, Pawson DL, Kier PM (1995) Sea Stars, Sea Urchins, and Allies: Echinoderms of Florida and the Caribbean. Smith Inst Press, Washington

Heyman WD, Kjerfve B (2001) The Gulf of Honduras. In: Seeliger U, Kjerfve B (eds) Coastla marine ecosystems of Latin America. Springer, Berlin, pp 17–32

Hotchkiss FHC (1982) Ophiuroidea (Echinodermata) from Carrie Bow Cay, Belize, I. Structure and Communities. In: Rützler K, Macintyre IG (eds) The Atlantic Barrier Reef Ecosystem of Carrie Bow Cay, Belize. Smith Inst Press, Washington, pp 387–411

Humes AG (2000) *Asterocheres crinoidicola* n. sp., a copepod (Siphonostomatoida: Asterocheridae) parasitic on crinoids in Belize. System Parasitol 47:103–110

Humes AG, Hendler G (1999) Biology and taxonomy of species of *Ophiopsyllus* and *Pseudanthessius* (Copepoda) associated with brittle stars (Ophiuroidea) in Belize. Bull Mar Sci 65:699–713

Jackon JBC (1997) Reefs since Columbus. Coral Reefs 16:S23–S32

Jackson JBC, D'Croz L (1997) The ocean divided. In: Coates AG (ed) Central America: a natural and cultural history. Yale University Press, New Haven, pp 38–70

John DD, Clark AM (1954) The "Rosaura" Expedition 1937–1938 3. The Echinodermata. Bull Hr Mus Nat Hist (Zool) 2:139–162

Kessler WS (2006) The circulation of the eastern tropical Pacific: a review. Prog Oceanogr 69:181–217

Kier PM (1975) The echinoids of Carrie Bow Cay, Belize. Smith Contr Zool 206:1–45

Lawrence JM (1967) Lipid reserves in the gut of three species of tropical sea urchins. Carib J Sci 7:65–68

Lessios HA (1998) Shallow water echinoids of Cayos Cochinos. Honduras. Rev Biol Trop 46(Suppl 4):95–101

Lessios HA, Robertson DR, Cubit JD (1984) Spread of *Diadema* mass mortality through the Caribbean. Science 226:335–337

Lessios HA (1981) Reporductive periodicity of the echinoid Diadema and Echinometra on the two coasts of Panama. J Exp Mar Biol Ecol 50:47–61

Lessios HA, Kessing BD, Wellington GM, Graybeal A (1996) Indo-Pacific echinoids in the tropical eastern Pacific. Coral Reefs 15:133–142

Lessios HA, Kessing BD, Robertson DR (1998) Massive gene flow across the world's most potent marine biogeographic barrier. Proc R Soc Lob B 265:583–588

Lessios HA, Garrido MJ, Kessing BD (2001) Demographic history of *Diadema antillarum*, a keystone herbivore on Caribbean reefs. Proc R Soc London B 268:1–7

Lewis SM, Wainwright PC (1985) Herbivore abundance and grazing intensity on a Caribbean coral reef. J Exp Mar Biol Ecol 87:215–228

Longhurst A (1998) Ecological geography of the sea. Academic Press, San Diego

Ludwig H (1894) Reports on an exploration off the west coast of Mexico, Central America and South America, and off the Galápagos Islands, in charge of Alexander Agassiz, by the U.S. Fish Commission steamer "Albatross", during 1891, XII. The Holothurioidea. Mem Mus Comp Zool XVII:1–183

Ludwig H (1905) Asteroidea. In: Explorations of "Albatross" in Tropical Pacific, 1891 and 1899–1900. Mem Mus Comp Zool XXII:1–290

Lütken CF, Mortensen T (1899) Reports on an exploration off the west coast of Mexico, Central America and South America, and off the Galápagos Islands, in charge of A. Agassiz, during 1891 on the "Albatross", XXV. The Ophiuridae. Mem Mus Comp Zool XXIII:93–208

Macurda DB Jr (1982) Shallow-water Crinoidea (Echinodermta) from Carrie Bow, Belize, I. Structure and Communities. In: Rützler K, Macintyre IG (eds) The Atlantic barrier reef ecosystem at Carrie Bow Cay, Belize. Smith Inst Press, Washington, pp 413–416

Maddocks RF (1987) An ostracod commensal of an ophiuroid and other new species of *Pontocypria* (Podocopida: Cypridacea). J Crust Biol 7:727–737

Maluf LY (1988a) Composition and distribution of the Central eastern Pacific echinoderms. Nat Hist Mus LA County Tech Rep 2:1–242

Maluf LY (1988b) Biogeography of the central eastern shelf echinoderms. In: Burke RD, Mlademov PV, Lambert P, Parsley RL (eds) Echinoderm biology. AA Balkema, Rotterdam, pp 389–398

Marcaillou B, Charvis P, Collot JY (2006) Structure of the Malpelo Ridge (Colombia) from seismic and gravity modelling. Mar Geophys Res 27:289–300

Maurer D, Vargas JA (1983) Benthic studies in the Gulf of Nicoya, Costa Rica. University of Delaware, College of Marine Studies, pp 1–105

Maurer D, Epifanio C, Dean H, Howe S, Vargas JA, Dittel A, Murillo MM (1984) Benthic invertebrate of a tropical estuary: Gulf of Nicoya, Costa Rica. J Nat Hist 18:47–61

McClanahan TR (1999) Predation and the control of the sea urchin *Echinometra viridis* and fleshy algae in the patch reefs of Glovers Reef, Belize. Ecosystems 2:511–523

Miller JE (1984) Sytematics of the ophidiasterid sea stars *Copidaster lymani* A. H. Clark, and *Hacelia superba* H. L. Clark (Echinodermata: Asteroidea) with a key to species of Ophidiasteridae from the West Atlantic. Proc Biol Soc Wash 97:194–208

Miloslavich P, Díaz JM, Klein PE, Alvarado JJ, Díaz C, Gobin J, Escobar-Briones E, Cruz-Motta JJ, Weil E, Cortés J, Bastidas AC, Robertson R, Zapata F, Martín A, Castillo J, Kazandjan A, Ortiz M (2010) Marine biodiversity in the Caribbean: regional estimates and distribution patterns. PLoS ONE 5:e11916. doi:10.1371/journal.pone.0011916

Morales A, Murillo MM (1996) Distribution, abundance and composition of coral reef zooplankton, Cahuita National Park, Limon, Costa Rica. Rev Biol Trop 44:619–630

Mumby PJ, Hedley JD, Zychaluk K, Harbone AR, Blackwell PG (2006) Revisiting the catastrophic die-off of the urchin *Diadema antillarum* on Caribbean coral reefs: fresh insights on resilience from a simulation model. Ecol Model 196:131–148

Murillo MM, Cortés J (1984) Alta mortalidad en la población del erizo de mar *Diadema antillarum* Philippi (Echinodermata: Echinoidea), en el Parque Nacional de Cahuita, Limón, Costa Rica. Rev Biol Trop 32:167–169

Myhre S, Acevedo-Gutiérrez A (2007) Recovery of sea urchin *Diadema antillarum* populations is correlated to increased coral and reduced macroalgal cover. Mar Ecol Prog Ser 329:205–210

O'Dea A, Rodriguez F, De Gracia C, Coates AG (2007) La paleontología marina en el Istmo de Panamá. Canto Rodado 2:149–179

Ogden JC (1977) Carbonate-sediment production by parrot fish and sea urchins on Caribbean reefs. Stud Geol 4:281–288

Pawson DL (1976) Shallow-water sea cucumbers (Echinodermata: Holothuroidea) from Carrie Bow Cay, Belize. Proc Biol Soc Wash 89:369–382

Pawson DL (2007) Phylum Echinodermata. Zootaxa 1668:749–764

Pawson DL, Miller JE (1988) *Thyone tanyspeira*, a new species of sea cucumber from the Southern Caribbean sea. Bull Mar Sci 42:310–316

Pearse JS (1968) Gametogenesis and reproduction in several abyssal and shallow water echinoderms of the Eastern Tropical Pacific. Stanford Oceanogr Exp Cruise 20 Te Vega Rep:225–234

Peyrot-Clausade M, Chabanet P, Conand C, Fontaine MF, Letourner Y, Harmelin-Vivien H (2000) Sea urchin and fish bioerosion on La Réunion and Moorea reefs. Bull Mar Sci 66:477–485

Quesada-Alpízar MA, Cortés J (2006) Los ecosistemas marinos del Pacífico sur de Costa Rica: estado del conocimiento y perspectivas del manejo. Rev Biol Trop 54(Suppl 1):101–145

Richards VP, Thomas JD, Stanhope MJ, Shivji MS (2007) Genetic connectivity in the Florida reef system: comparative phylogeographym of commensal invertebrate with contrasting reproductive strategies. Mol Ecol 16:139–157

Robertson R, Christy JH, Collin R, Cooke RG, D'Croz L, Kaufmann KW, Heckadon-Moreno S, Mate JL, O'Dea A, Torchin ME (2009) The Smithsonian tropical research institute: marine research, education, and conservation in Panama. Smith Contr Mar Sci 38:73–93

Rojas-Guitart MT (1990) Determinación de Cadmio, Cromo, Cobre, Hieroo, Manganeso, Plomo y Zinc en el pepino de mar *Holothuria* sp (Echinodermata) del arrecife coralino del Parque Nacional Cahuita, Costa Rica. Licenciatura thesis, Fac Cienc, Esc Quím, Univ Costa Rica

Rojas MT, Acuña JA, Rodríguez OM (1998) Metales traza en el pepino de mar *Holothuria* (*Halodeima*) *mexicana* del Caribe de Costa Rica. Rev Biol Trop 46(6):215–220

Rützler K (2009) Caribbean coral reef ecosystems: thirty-five years of smithsonian marine science in belize. Smith Contr Mar Sci 38:43–71

Ryan JD, Miller LJ, Zapata Y, Downs O, Chan R (1998) Great Corn Island, Nicaragua. In: Kjerfve B (ed) CARICOMP: Caribbean Coral Reef, Seagrass and mangroves sites. UNESCO, Paris, pp 95–105

Sammarco PW (1980) *Diadema* and its relationship to coral spat mortality: grazing, competition and biological disturbance. J Exp Mar Biol Ecol 45:245–472

Sammarco PW (1982a) Echinoid grazing as a structure force in coral communities: whole reef manipulations. J Exp Mar Biol Ecol 61:31–55

Sammarco PW (1982b) Effects of grazing by *Diadema antillarum* Phillipi (Echinodermata: Echinoidea) on algal diversity and community structure. J Exp Mar Biol Ecol 65:83–105

Sammarco PW, Levington JS, Ogden JC (1974) Grazing and control of coral reef community structure by *Diadema antillarum* Phillipi (Echinodermata: Echinoidea): a preliminary study. J Mar Res 32:47–53

Scheibling RE, Metaxas A (2010) Mangroves and fringing reefs as nursery habitats for the endangered Caribbean sea star Oreaster reticulates. Bull Mar Sci 86:133–148

Scoffin TP, Stearn CW, Boucher D, Frydl P, Hawkins CM, Hunter IG, MacGeachy JK (1980) Calcium carbonate budget of a fringing reef on the west coast of Barbados. Part II: Erosion, sediments and internal structure. Bull Mar Sci 30:475–508

Stoddard DR (1962) Three Caribbean Atolls: Turneffe Islands, Lighthouse Reef, and Glover's Reef, British Honduras. Atoll Res Bull 87:1–151

Strathmann RR (1981) The role of spines in preventing structural damage to echinoids tests. Paleobiology 7:400–406

Toral-Granda V (2008) Population status, fisheries and trade of sea cucumbers in Latin America and the Caribbean. In: Toral-Granda V, Lovatelli A, Vasconcellos M (eds) Sea cucumbers. A global review of fisheries and trade. FAO Fisheries and Aquaculture Technical Paper. No. 516. Rome, pp 213–229

Vadas RL, Fenchel T, Ogden JC (1982) Ecological studies on the sea urchin, *Lytechinus variegatus*, and the algal-seagrass communities of the Miskito Cays, Nicaragua. Aquat Bot 14:109–125

Valdez MF, Villalobos CR (1978) Distribución espacial, correlación con el sustrato y grado de agregación en *Diadema antillarum* Philippi (Echinodermata: Echinoidea). Rev Biol Trop 26:237–245

Vargas JA (1987) The benthic community of an intertidal mud flat in the Gulf of Nicoya, Costa Rica. Description of the community. Rev Biol Trop 35:299–316

Vargas JA (1989) A three year survey of the macrofauna of an intertidal mud flat in the Gulf of Nicoya. In: Magoon O, Converse M, Miner D, Tobin LT, Clark D (eds) Coastal Zone 89. American Society of Civil Engineers, New York, pp 1905–1919

Vargas JA, Solano S (2011) On *Mellitella stokesii* and *Amphipholis geminata* (Echinodermata), from an intertidal flat in the upper Gulf of Nicoya estuary, Pacific, Costa Rica. Rev Biol Trop 59:193–198

Verrill AE (1867) Notes on the echinoderms of Panama and west coast of America, with description of new genera and species. Trans Connect Acad Arts Sci 1:251–322

Wyrtki K (1967) Circulation and water masses in the Eastern Equatorial Pacific Ocean. Int J Oceanol Limnol 1:117–147

Ziesenhenne FC (1937) The Templeton Crocker Expedition. X. Echinoderms from the West coast of Lower California, the Gulf of California and Clarion Island. Zoologica 22:209–239

Ziesenhenne FC (1940) New ophiurans of the Allan Hancock Pacific Expeditions. Allan Hancock Pac Exp 8:9–59

Ziesenhenne FC (1942) New eastern Pacific sea stars. Allan Hancock Pac Exp 8:197–223

Ziesenhenne FC (1955) A review of the genus *Ophioderma*. Essays in the natural sciences in honor of Captain Allan Hancock on the occasion of his birthday, 26 July 1955. University of Southern California Press, Los Angeles

Chapter 4
Echinoderm Diversity in Panama: 144 Years of Research Across the Isthmus

Simon E. Coppard and Juan José Alvarado

4.1 Introduction

The Republic of Panama is the southernmost country in Central America between 07°00′–10°00′ north and 77°00′–83°00′ west and is located between Costa Rica to the north and Colombia to the south. The country runs from east to west with a length of 772 km and a width that varies from 60 to 177 km (Meditz and Hanratty 1987). Panama has both Caribbean and Pacific coastlines that total some 2,857 km and which incorporate a vast diversity of marine habitats and ecosystems. The 80 km Panama Canal bisects the country and although this is theoretically a freshwater barrier, it provides a link between the Atlantic and Pacific Oceans. Approximately one hundred and fifty rivers flow into the Caribbean, while over three hundred rivers flow into the eastern Pacific.

Panama's weather is governed by the Inter-Tropical Convergence Zone which dictates the seasonal patterns of wind and rainfall in the region. The dry season (mid-December to mid-April) is characterised by low precipitation but strong winds as a result of the northeast trade winds (Forsbergh 1969; Amador et al. 2006). The rainy season (May to November) has light winds with heavy rainfall

S. E. Coppard (✉)
Smithsonian Tropical Research Institute, PO BOX 0843-03092Balboa, Ancón, Panamá
e-mail: CoppardS@si.edu

J. J. Alvarado
Centro de Investigación en Ciencias del Mar y Limnología (CIMAR), Universidad de Costa Rica, San Pedro, 11501-2060, San José, Costa Rica
e-mail: juanalva76@yahoo.com

J. J. Alvarado
Posgrado en Ciencias Marinas y Costeras, Universidad Autónoma de Baja California Sur, La Paz, México

J. J. Alvarado and F. A. Solís-Marín (eds.), *Echinoderm Research and Diversity in Latin America*, DOI: 10.1007/978-3-642-20051-9_4,

which varies regionally from less than 1.3 m to more than 3 m per year, with rainfall on the Caribbean side typically being heavier than on the Pacific side of the continental divide (Meditz and Hanratty 1987).

The formation of the Isthmus of Panama is considered to be one of the most important geologic events of the past 60 million years as it had an enormous impact on Earth's climate and environment. It occurred due to the gradual collision of the Pacific, Cocos, Nazca and Caribbean plates which led to uplift of the sea floor and the formation of volcanic islands. Over the next 12 million years vast quantities of sediment from North and South America were deposited in between the newly emerging islands, with final closure of the straits occurring in the late Pliocene, 2.8 million years ago (mya) (Coates et al. 2005). This formation blocked the North Equatorial Current which had previously flowed westward through the Central American Seaway and forced the current northwards forming the Gulf Stream of today (Berggren and Hollister 1974; Burton et al. 1997). With the Pacific and Atlantic no longer mixing, there were significant changes in temperature, salinity and primary productivity. This resulted in both species extinctions and species radiations as environments changed (Jackson et al. 1993). Sister species that were isolated on either side of the isthmus and that subsequently diverged are referred to as geminates (Jordan 1908). These represent initially similar genomes that were isolated and diverged over the last 3 million years (Lessios 2008) and provide insights into evolutionary adaptations in response to the changing environments.

Today the Caribbean coast of Panama incorporates coral reefs, seagrass beds, rocky shores, mud-flats and mangroves and has a narrow continental shelf. It continues to be influenced by the Atlantic North Equatorial Current which flows through the Lesser Antilles (Kinder et al. 1985). The coast of the south-western Caribbean including Panama is also affected by a cyclonic gyre that develops from this current (Gordon 1967; Kinder et al. 1985; Gómez et al. 2005). For an in-depth review of the Caribbean habitats of Panama, see Gómez et al. (2005). The Caribbean experiences very small tidal ranges (less than 0.5 m) and has relatively stable conditions. Its waters are clear and nutrient poor (D'Croz and Robertson 1997; D'Croz et al. 2005; Collin et al. 2009) and unlike the north coast of Colombia there is no upwelling. These conditions are suitable for growth of extensive coral reefs which date back to the early Pleistocene (Glynn 1982; Cortés 1993). Surface waters of the Caribbean have an upper 200 m layer of warm water heated by solar radiation, which has a correspondingly high salinity. Surface water temperatures vary from 27 °C in the dry season to 30 °C in the rainy season, while salinity varies from 36 psu in the dry season to 32 psu in the rainy season (D'Croz and Robertson 1997). Two archipelagos are present off the Caribbean coast of Panama, the San Blas (Kuna Yala) Archipelago which has a very narrow continental shelf and is subject to strong ocean influence with an oligotrophic coastal area (D'Croz and Robertson 1997) and the Bocas del Toro Archipelago, with a larger continental shelf, and greater influence from continental run-off (D'Croz et al. 2005). Five national parks have been created along the Caribbean coast of

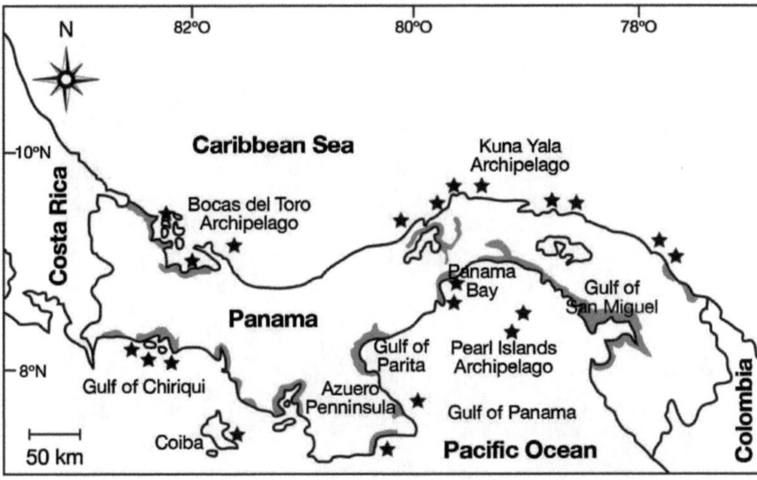

Fig. 4.1 Distribution of coral reefs and mangrove forests along the coasts of Panama (adapted from Gómez et al. 2005). *Black stars*: coral reefs; *grey shading*: mangroves

Panama: Bastimentos Island National Marine Park, Portobelo, Soberania, Galeta, Chagres, and the wetlands of San San Pond Sank (Gómez et al. 2005).

The Pacific coast of Panama is comprised of two gulfs; the Gulf of Chiriquí and the Gulf of Panama which are separated by the Azuero Peninsula (Fig. 4.1). These regions experience broad semi-diurnal tides with a range of up to 6 m (Glynn 1972). Despite their close proximity, they experience very different hydrological conditions. The Gulf of Panama includes the Gulf of San Miguel, the Gulf of Parita and Panama Bay, as well as the Pearl Islands Archipelago which is a Marine Special Management Zone (Guzman et al. 2008). The gulf is broad and shallow, with much of it being less than 40 m deep due to the extension of the continental shelf. The coastal environments consist of muddy and sandy beaches with extensive mangrove forests, which receive large amounts of freshwater runoff from rivers such as the Rio Chepo and Rio Chico as well as the discharge from the Panama Canal. Coral reefs are only moderately developed and are associated with the Pearl Islands Archipelago and the inshore island of Taboga. Throughout the rainy season the surface water temperature ranges from 26–28 °C, while salinity is typically below 30 psu (D'Croz et al. 1991). During the dry season (December to April) the Gulf of Panama experiences wind-driven, seasonal upwelling of colder, more saline, nutrient-rich water (Forsberg 1963; Forsbergh 1969; D'Croz et al. 1991; D'Croz and Robertson 1997; Gómez et al. 2005; Pennington et al. 2006), which results in significantly increased primary productivity. This occurs due to the displacement of the surface water by northern winds and can result in a surface water temperature decrease of 12 °C within 24 h (Glynn 1972). The Gulf of Panama is also subjected to the effects of the El Niño-Southern Oscillation (ENSO) every two to seven years (Guzman and Cortes 2007) resulting in the warming of the water up to 31 °C (Glynn 1985a).

Fig. 4.2 *Pocillopora* spp. framework on Uva Reef, Gulf of Chiriquí, 2010

In contrast, the Gulf of Chiriquí has a narrower continental shelf and less freshwater input from continental runoff. It contains Chiriquí National Park and Coiba Nacional Park which has World Heritage Site status. Several island complexes are present, including the islands of Boca Brava, Los Ladrones, Islas Secas, Parilla, Isla Palenque, Montuosa, and Isla Jicaron, with Isla Jicarita being the island furthest south of the Panama mainland. In the northwest of the gulf, water depth gradually increases down the continental shelf. However, in the east, water depth sharply drops to 160 m before plunging down the continental slope and reaching a depth of 1,000 m less than a 1 km south of the Azuero Peninsula. The Gulf of Chiriquí is shielded from the trade winds out of the North by the Central Cordillera and therefore experiences no seasonal upwelling. The Gulf of Chiriquí contains some of the most developed coral reefs on the continental shelf in the eastern Pacific. These are predominantly formed of a loose *Pocillopora* spp. framework (Fig. 4.2) that originated in the Holocene (Glynn 1976; Cortés 1993).

4.2 Research

The first records of the echinoderm fauna of Panama date back to the late 19th and early 20th century and are the result of the expeditions of the U.S. Fish Commission Steamer *Albatross*. Addison Emery Verrill (Fig. 4.3a) wrote the first paper in 1867 that included echinoderms from Panama. It contained descriptions of many new genera and species. In 1891 Alexander Agassiz (Fig. 4.3b), onboard the *Albatross* collected specimens from the intermediate depths of the eastern Pacific and the Caribbean Sea to study the relationship between forms of marine life on either side

Fig. 4.3 **a** Addison Emery Verrill, **b** Alexander Agassiz, **c** Christian Frederik Lütken, **d** Theodore Mortensen (photos courtesy of the Smithsonian National Museum of Natural History)

of the Isthmus of Panama. The collections from this expedition were used by many authors to describe new echinoderm species (e.g. Agassiz 1892, 1898, 1904; Ludwig 1894, 1905; Lütken and Mortensen 1899, (Fig. 4.3c, d), Clark 1917).

In the 1930's and 1940's expeditions to the eastern Pacific were undertaken by the New York Zoological Society onboard the Arcturus 1925 and Zaca 1937–1938, and by the Allan Hancock Foundation onboard the Velero III 1937–1941. As a result of these expeditions, a significant amount of echinoderm research was published (e.g. Clark 1939; Clark 1940, 1948; Ziesenhenne 1940,

1942, 1955; Deichmann 1941, 1958). In 1915 Mortensen spent several months collecting echinoderms and other invertebrates in the Gulf of Panama, including from St. Elmo in the Pearl Islands Archipelago and the islands of Taboga and Taboguilla in Panama Bay. His findings were later included in a broader work on larval development (Mortensen 1921) and as part of his monumental monograph on echinoids (Mortensen 1928, 1935, 1940, 1943a, b, 1948a, b, 1950, 1951). Other researchers in the 1920's, 30's and 40's made collection trips to specific regions in Panama (e.g. Boone 1928, Clark 1946) focusing more on specific taxa. In later years echinoderms were collected from the Caribbean Sea off Panama as part of general benthic sampling expeditions undertaken by the RV *Oregon* in 1962, the RV *Pelican* in 1963, the RV *Oregon II* in 1970, the RV *Pillsbury* in 1971 and the RV *Alpha Helix* in 1977.

The Smithsonian Tropical Research Institute (STRI) established its first field station on Barro Colorado Island in the Panama Canal Zone in 1923. The marine program at STRI began fifty-one years ago, in 1961. Today STRI runs two field stations on the Caribbean Coast (Galeta Marine Laboratory in Colón and Bocas del Toro Research Station) and two in the Pacific (Naos Island Laboratories complex and a Field Station on Ranchería Island, Coibita). The first research vessel was acquired by STRI in 1970, the 65-ft RV *Tethys* (1970–1972), followed by the 45-foot RV *Stenella* in 1972 (1972–1978) and the 63-foot RV *Benjamin* in 1978 (1978–1994). These research vessels allowed STRI scientists to collect and document the shallow water echinoderm fauna. In 1994 STRI purchased the 96-foot RV *Urraca* (1994–2008) that had an A-frame and oceanographic winch which finally provided the means to trawl and dredge to depths of 250 m. In 1995 the submersible DSRV *Johnson Sea Link* was used to collected samples from the Gulf of Chiriquí at depths over 1,000 m, providing a further insight into the deep water species present off Panama's Pacific coast. STRI researchers who worked on echinoderms in the 70's and 80's included Peter Glynn, Gordon Hendler and Harilaos Lessios. For just over 30 years Lessios has worked at STRI on the life histories and evolution of echinoderms in Panama. His research has focused primarily on echinoids, including work on speciation, molecular biogeography and the evolution of reproductive isolation. Lessios (2005a) published a review of the echinoids of the Pacific waters of Panama, providing information on the species and new records of their distributions.

4.3 Ecology

Ecological research in Panama has included work on all classes of echinoderms, but with fewer studies on crinoids. Eleven species of echinoderms were recorded in the seagrass meadows of *Thalassia testudinum* Banks and Sol. ex K.D. Koenig (1805) in the Canal Zone off the Caribbean coast of Panama (Heck 1977). Species richness of echinoderms correlated with proximity to well developed reefs and was found to be stable throughout the study period (July 1974 to May 1975), without any major decrease in population numbers during periods of reduced salinity.

Fig. 4.4 **a** *Echinometra lucunter*, **b** *Echinometra viridis* at Galeta Point Panama

Mortality of sea urchins on the fringing reef-flat off Galeta Island has been reported to occur twice yearly as the result of physical stress generated by extended aerial exposure of the reef platform (Glynn 1968; Hendler 1977; Cubit et al. 1986). *Tripneustes ventricosus* (Lamarck, 1816), *Lytechinus variegatus* (Lamarck, 1816) and *Diadema antillarum* Philippi, 1845 were found to be less tolerant to high temperature than either *Echinometra lucunter* (Linnaeus, 1758) (Fig. 4.4a) or *Echinometra viridis* Agassiz, 1863 (Fig. 4.4b), and are therefore more likely to die during the periods of exposure.

In the Caribbean *D. antillarum* has been shown to be the dominant herbivore on coral reefs that have been subjected to heavy human fishing pressure, while herbivorous fishes dominate in more pristine conditions where urchin populations are kept in check by fish predators (Hay et al. 1983; Harborne et al. 2009). Off Galeta Island, where human fishing pressure is reported to be low (Hay 1984), almost all grazing on bioassays of *T. testudinum* was attributed to herbivorous fishes (Hay 1984). Off the San Blas Islands of Panama, the grazing activity of *D. antillarum* has been shown to be the primary determinant of fleshy algal biomass on shallow reefs (<3 m) (Foster 1987a). At the time of Foster's study (pre-mass mortality of *D. antillarum*, see Sect. 4.6) this was particularly true in the back reef environment, where *D. antillarum* occurred at densities three times higher ($\sim$6 ind m^{-2}) than in Elkhorn coral habitats (<2 ind m^{-2}) (Foster 1987a).

Chemical defence by coral reef macroalgae has been shown to have an import role in deterring grazers (Hay et al. 1987). Extracted terpenoid compounds (naturally occurring in macroalgal species of *Dictyota*, *Dilophus*, *Laurencia* and *Stypopodium*) were applied to assays of *T. testudinum* and shown to deter grazing by herbivorous fishes and *D. antillarum* at Galeta (Hay et al. 1987). However, Cymopol, a terpenoid bromohydroquinone from the green alga *Cymopolia barbata* (Linnaeus) Lamouroux, 1816, significantly reduced feeding by reef fishes but significantly stimulated feeding in *D. antillarum*.

The ability of the damselfish *Stegastes dorsopunicans* (Poey, 1868) to exclude and therefore reduce the grazing impacts of *D. antillarum* from their territories was studied by Foster (1987b) in San Blas. Despite the major negative impact of

D. antillarum on the algal biomass in damselfish territories, this damselfish species was found to rarely defend its feeding areas against *D. antillarum* and therefore had little effect on its distribution.

On the Pacific coast of Panama grazing by *D. mexicanum* results in the bioerosion of the predominantly *Pocillopora* spp. coral reefs (Glynn 1988). This species contributes 78 % of the overall bioerosion 13 kg m^{-2} y^{-1} produced by benthic eroders (Glynn 1988; Eakin 1991). Analysis of faecal pellets has revealed that the diet of *D. mexicanum* consists of 75.2 ± 11.03 % coral and only 23.0 ± 11.4 % of crustose coralline algae (CCA). In contrast the diet of *Toxopneustes roseus* (A. Agassiz, 1863) is composed of approximately 20 % coral and 80 % of CCA (Glynn 1988). This highlights the potential damage *D. mexicanum* poses to reef structures, particularly when present in large numbers. Such a threat was realised following the 1982–83 El Niño Southern Oscillation event (ENSO). This resulted in mass coral bleaching and a 50 % reduction of live coral cover on Uva Reef (Glynn 1985a), which was followed by a dramatic increase in the population size of *D. mexicanum* from 2–5 ind m^{-2} pre 1983 to 60–100 ind m^{-2} post 1982-83 ENSO (Glynn 1988, 1990; Eakin 1991, 1992). Such large numbers of *D. mexicanum* resulted in high rates of reef framework erosion, shifting the reef environment into a state of near accretionary stasis (Eakin 1992; Glynn 1997). Low predation pressure (with the removal of potentially important fish predators as a result of over-fishing), increased habitat availability, and the increased abundance of benthic algae as a result of the newly dead coral substratum allowed *D. mexicanum* to attain high population densities (Glynn 1988). However, such a large population of *D. mexicanum* could not be sustained and in the year 2000 the population of *D. mexicanum* dropped to near 1974 levels (>10 ind m^{-2}) (Eakin 2001).

The ability of damselfish in the eastern Pacific to exclude *Diadema* from their territories appears to be a very different situation to that reported in the Caribbean. The territorial action of *Stegastes acapulcoensis* (Fowler, 1944) on Uva Reef results in few sea urchins being present in their territories (Glynn 1990), with consequently lower levels of erosion (6.3 mm y^{-1}) than elsewhere on the reef (21.8 mm y^{-1}) (Eakin 1992).

Ecological studies on irregular echinoids in Panama have focused on sand dollars. Seilacher (1979) studied the distribution of *Mellita quinquiesperforata* (Leske, 1778) (as *Mellita lata* Clark, 1940), a subjective junior synonym of *M. quinquiesperforata*) at Maria Chiquita on the Caribbean coast of Panama. This species lives partially buried in the sand (with some of the apical surface exposed) and is restricted to a 4 m wide region of the surf-zone that runs parallel to the beach (Seilacher 1979). This species was observed feeding along this zone, but migrated to deeper water when the intensity of the wave action increased.

The behavioural ecology of a population of *Mellitella stokesii* (L. Agassiz, 1841) (as *Encope stokesi* L. Agassiz, 1841 [sic]) was studied over a year at Venado Beach, on the eastern Pacific coast of Panama by Dexter (1977). Densities of this species were reported to vary from a low of 19 ind m^{-2} in the rainy season to a

Fig. 4.5 *Mellitella stokesii* at Venado beach, **a** aboral view, **b** feeding during low tide

high of 65 ind m^{-2} in the dry season, with numbers concentrated in the mid-intertidal zone where the median sediment grain size was between 170 and 200 μm. No sand dollars were found in the muddy sand bar which had a high silt and clay fraction. Migrations towards the substrate surface were observed at low tide which correlated with the appearance of an organic film on the substrate, composed of a combination of diatoms and fine detrital material. Observations by one of the authors (SEC) at Venado Beach (see Fig. 4.5a, b) agree with Dexter's findings. During high tide this species typically remains buried in the substrate, perhaps to avoid predation, and only migrates to the surface at low tide, where it moves through the upper sediment layer in the shallow tide pools ingesting small particulate mater (Fig. 4.3b). Water temperatures in these pools regularly gets in excess of 40 °C (Coppard unpublished data), demonstrating that *M. stokesii* has a tolerance for high water temperatures. Dexter (1977) reports that this species has a high growth rate, in conjunction with a small maximum size (she recorded a maximum horizontal diameter of 57 mm) and a short life expectancy of typically

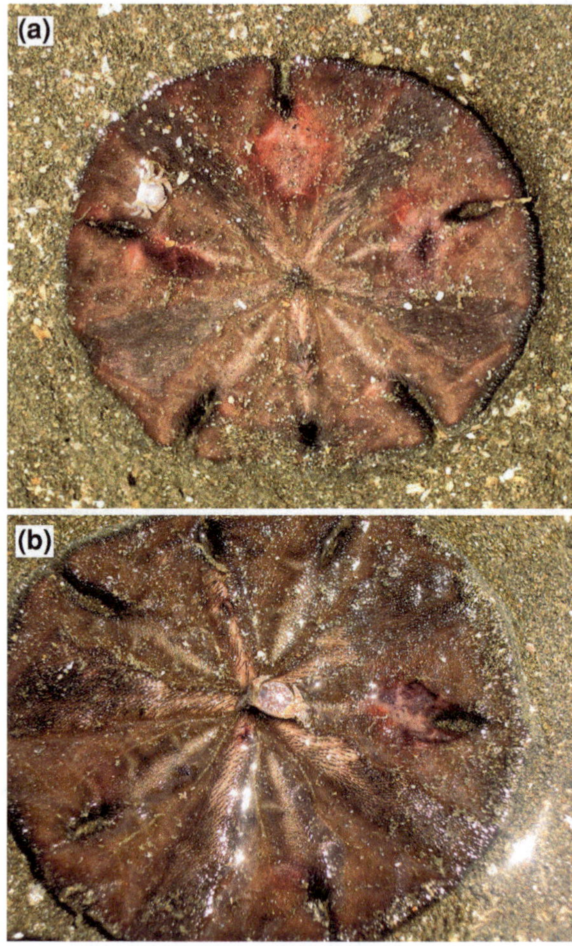

Fig. 4.6 *Mellitella stokesii* (oral views) on Venado beach with *Dissodactylus* spp., **a** multiple wounds, **b** a single wound

less than 1 year. Recruitment of the population was reported to occur throughout the year, but with a peak in settlement during February, in the middle of the Panamanian dry season. This is the season of increased food availability for planktotrophic larvae, due to the upwelling of nutrients and subsequent increase in phytoplankton (Martin et al. 1970).

Two species of pinnotherid crab (pea crab), *Dissodactylus nitidis* Smith, 1870 and *Dissodactylus xantusi* Glassell, 1936 were found associated with *M. stokesii* (Dexter 1977). These species remove the spines near the lunules or marginal slits on the oral surface, creating a wound in which they live (see Fig. 4.6a, b). Dexter questioned whether this relationship is really commensal, as she found that these crabs in aquaria removed larger numbers of spines on the oral surface, which decreased the ability of the sand dollar to feed and move, ultimately resulting in the death of the host.

Telford (1982) described the feeding habits of four species of *Dissodactylus* and found that all of them feed extensively upon their sand dollar hosts, with 50–100 % of their food intake coming from their host's tissues. This is supported by the fact that some *M. stokesii* often have multiple wounds with only a single crab present (Fig. 4.6a).

Ecological work on ophiuroids in Panama is primarily the result of research by Gordon Hendler. Hendler and Meyer (1982) reported the presence of the polychaete *Branchiosyllis exilis* (Gravier, 1900) associated with *Ophiocoma echinata* (Lamarck, 1816) on the Caribbean coast of Panama. This association has only been reported from Panama, despite the fact that both species' distributions are overlapping and widespread. The polychaete displays active recognition of its host and seeks out *O. echinata* in preference to other species of the same genus. They consistently found one polychaete per host, indicating that *B. exilis* is aggressive to member of its own species, and concluded that *B. exilis* is parasitic on *O. echinata*.

An interesting association between juvenile and adult *Ophiocoma aethiops* Lütken, 1859, from Punta Paitilla in Panama was documented by Hendler et al. (1999). Juveniles were found in the bursae (respiratory structures that in brooding and viviparous ophiuroids also serve as brood chambers) of adults that live intertidally at Punta Paitilla but not those that live subtidally at Isla Taboguilla. They proposed this association helped protect juveniles against desiccation and predators. No similar association was found in other species of *Ophiocoma* form either the Pacific (*O. alexandri* Lyman, 1860) or from the Caribbean (*O. echinata, O. wendtii* Müller and Troschel, 1842) of Panama.

Much of the ecological work on starfish in Panama has focused on *Acanthaster planci* (Linnaeus, 1758) (e.g. Glynn 1973, 1974, 1976, 1977, 1981– 1985a,b, 1990). Glynn (1973) assessed the possible effect of *A. planci* on the coral species on Uva Reef in the Gulf of Chiriquí. He reported an average density of 25–36 ind ha^{-1}, with a disc diameter of 17–19 cm and a size frequency that had a unimodal distribution. Population densities of *A. planci* remained fairly stable on Uva Reef from 1970 to 1980, ranging from 7 to 30 ind ha^{-1} (Glynn 1981). At that time these densities were comparable to population sizes in the Indo-Pacific that were not considered to have a serious impact on coral communities (Glynn 1974). However, prey preference data (Glynn 1974) indicate that *A. planci* selectively eats rarer non-branching corals (e.g. species of *Pavona, Gardineroseris, Porites, Millepora,* Fig. 4.7), which are replaced by faster-growing species (e.g. *Pocillopora* spp.), resulting in *A. planci* having a negative effect on species diversity (Glynn 1976). Such a preference can partly be explained by the presence of crustacean symbionts (*Trapezia* sp. and *Alpheus* sp.) in large branching pocilloporid colonies which repulse *A. planci* by nipping at its vulnerable oral surface (particularly the sensory tube feet) as it attempts to mount the coral to feed (Glynn 1976).

Between 1982 and 1983 a strong El Niño-Southern Oscillation had a dramatic impact on the coral reefs of the Eastern Tropical Pacific, causing a die off of between 70–95 % of coral on Pacific Panamanian reefs (Glynn 1985a, b). However, population densities of *A. planci* before and after the 1983 event were not statistically different (Glynn 1990), with *A. planci* continuing to feed on the

Fig. 4.7 *Acanthaster planci* feeding on *Millepora intricata* Milne-Edwards and Haime, 1860, Coiba National Park, Gulf of Chiriquí (photo courtesy of Angel Chiriboga)

dispersed, surviving coral. As the reefs recovered the relative effect of predation by *A. planci* intensified due to the loss of the crustacean guards (*Trapezia* and *Alpheus*), therefore reducing local coral species diversity (Glynn 1985b).

Fong and Glynn (1998) modelled the impacts of predation by *A. planci* and that of the El Niño-Southern Oscillation (ENSO) on the structure of a population of the massive coral *Gardineroseris planulata* (Dana, 1846) on Uva Reef. The results of the simulations suggested that predation was far more important than the ENSO in controlling the size structure of the coral *G. planulata* in the eastern Pacific.

Although *A. planci* has few predators, it has been observed in Panama being attacked by the Harlequin shrimp *Hymenocera picta* Dana, 1852 and by the polychaete *Pherecardia striata* (Kingberg, 1857) (Glynn 1981). The shrimp attacks the aboral surface of the starfish, eventually amputating the arms and using its chelae to expose and feed on the internal soft parts. However, the harlequin shrimp typically prefers other starfish species such as *Phataria unifacialis* (Gray, 1840), *Pharia piramidatus* (Gray, 1840) and *Nidorellia armata* (Gray, 1840) (Glynn 1977). The wounds created by the harlequin shrimp are used also by the polychaete *P. striata* to enter *A. planci* and feed on the internal tissues. Glynn (1981) calculated that 5–6 % of the *A. planci* population is under attack by *H. picta* and about half of these starfish are killed over a 3-week period.

Guzman and Guevara (2002a) studied the spatial distribution, abundance, and size structure of *Oreaster reticulatus* (Linnaeus, 1758) (Fig. 4.8) in the Bocas del Toro Archipelago. They found a mean density of 149.7 ind ha^{-1} and estimated the population size to be over seven million. Approximately 45 % of the population was found at a density of 308.3 ± 50.9 ind ha^{-1}, in areas dominated by the

Fig. 4.8 *Oreaster reticulatus* in the *Thalassia testudinum* seagrass bed, Bocas del Toro (photo courtesy of Edgardo Ochoa)

seagrass *T. testudinum*, with a substratum that consisted of coarse, calcareous sand. The lowest density (5.6 ind ha^{-1}) was observed in coral reef habitats.

Feeding patterns of *O. reticulatus* off the San Blas Islands were studied by Wulff (1995). She found that 61.4 % of the starfish were eating sponges; however, this depended on their availability. *Oreaster reticulatus* fed upon sixteen of twenty species that normally grow only on the reefs, but surprisingly only one of fourteen species that live in the seagrass meadows and rubble flats surrounding the reefs. Therefore *O. reticulatus* actively searches for and consumes edible sponges that wash into the seagrass beds from reefs, or off mangrove roots (Wulff 1995, 2006).

Meyer (1973) described the feeding behaviour, microgeographic distribution, population size, bathymetric distribution and adaptive skeletal morphology of eight species of Caribbean crinoids from four main study areas, including San Blas and Galeta, in Panama. Based on his findings he divided the species into two categories: 1) those forming a filtration fan (*Nemaster grandis* Clark, 1909, *Tropiometra carinata* (Lamarck, 1816), *Comactinia echinoptera* (Müller, 1840), *Analcidometra caribbea* (A. H. Clark, 1908) and *Neocomatella pulchella* (Pourtalès, 1878)) and 2) those with a radian posture (*Davidaster rubiginosus* (Pourtalès, 1869), *Davidaster discoidea* (Carpenter, 1888) and *Ctenantedon kinziei* Meyer, 1972). Species diversity and population size of shallow-water comatulids was found to be greatest in Panama (seven species). Meyer (1973) proposed that the diversity and abundance of comatulid crinoids can be correlated to increased primary productivity which occurs close to continental land masses.

Ecological studies on holothuroids in Panama have been carried out by Guzman and Guevara (2002b). They studied the population structure, distribution and abundance of the sea cucumbers *Isostichopus badionotus* (Selenka, 1867), *Holothuria* (*Halodeima*) *mexicana* Ludwig, 1875 and *Astichopus multifidus* (Sluter, 1910) in the Bocas del Toro Archipelago. *Holothuria* (*Halodeima*) *mexicana* was found to be more abundant in seagrass, *I. baniodotus* occurred at higher densities in interspersed areas, while *A. multifudus* showed no specific habitat preference. For the Bocas del Toro Archipelago Guzman and Guevara

(2002b) estimated populations of 7,630,164 individuals for *H. (H.) mexicana*, 5,536,349 for *I. badionotus*, and 231,074 for *A. multifidus*. However, the densities of these sea cucumbers in Panama were 50 times lower than those measured at other Caribbean sites (Guzman and Guevara 2002b).

4.3.1 Physiology

Lawrence and Glynn (1984) studied the diet of the sea urchin *Eucidaris thouarsii* (Valenciennes, 1846) through the absorption of nutrients from the predation on the coral *Pocillopora damicornis* (Linnaeus, 1758) in the Gulf of Panama. The absorption efficiency was determined to be 44.7 % for carbohydrates, 85.3 % for protein and 69.3 % for lipids. The urchin absorbs 0.402 mg carbohydrate, 2.814 mg protein and 1.108 mg lipid for each gram of *P. damicornis* ingested; 0.201 mg carbohydrate, 1.407 mg protein and 0.554 mg lipid per individual per day; and absorbs 0.824×10^{-3} kcal carbohydrate, 7.950×10^{-3} kcal protein and 235×10^{-3} kcal lipid per individual per day. The authors concluded that despite the high efficiency of digestive absorption of organic constituents of the coral ingested, *E. thouarsii* needs to consume large quantities to meet its nutritional requirements.

Work by Hendler and Byrne (1987) on *Ophiocoma wendtii* collected in Galeta, Panama Caribbean, Belize and Florida, resulted in the identification and description of three structures in the dorsal arm plate (DAP) that comprise a photoreceptor system. The upper surface of the DAP bears transparent microscopic structures which are expanded peripheral trabeculae of the calcite stereom. These structure are part of the photoreceptor system and can facilitate light transmission through the DAP by decreasing light refraction, reflection and absorption that occur at stereom/stroma interfaces. Bundles of nerve fibres located below the expanded peripheral trabeculae function as primary photoreceptors. The third elements of the system are the chromatophores that regulate the intensity of light impinging on the sensory tissue are regulated by the diurnal activity cycle. During the day the chromatophores cover the expanded peripheral trabeculae and thereby shade the nerve fibres. At night they retract into inter-trabecular channels, uncovering the expanded peripheral trabeculae and thereby exposing the nerve fibres to transmitted light. Thus, the transparent stereom may play a role in photoreception, in addition to its generally recognized skeletal function. Although ciliated cells that may be sensory are present in the epidermis of *O. wendti*, they do not appear to be photoreceptors (Hendler and Byrne 1987).

4.3.2 Reproduction

Reproductive cycles of echinoids on both sides of the Panama Isthmus have been studied by Lessios (1981a, 1984, 1985a, 1991). Populations of *D. mexicanum* and *E. vanbrunti* from Punta Paitilla, Isla Urabá and Culebra Island in the seasonal Gulf of Panama have well-defined, synchronous, reproductive cycles (Lessios 1981a). The majority of individuals reached a peak gonad size in September, with spawning predicted to occur from September to October. This would allow the newly metamorphosed sea urchins, rather than the larvae, to benefit from the increased food production as a result of the dry season upwelling (Lessios 1981a). Populations of *D. antillarum* and *E. lucunter* from Maria Chiquita and Fort Randolph in the less seasonal Caribbean demonstrated only slight, if any synchrony between individuals. However, *E. viridis* which is rarer at these locations had a well defined, population wide cycle. Based on these data Lessios (1981a) proposed that in a constant environment the intensity of selection for synchrony between individual gametogenic cycles may be inversely proportional to population density. Later work by Lessios (1985a) on Caribbean echinoids did not directly support this hypothesis. However, Lessios (1985a) suggested that population density may be one of the factors determining the degree of synchronicity. This study (1985a) revealed that *E. viridis*, *L. williamsi*, *Clypeaster rosaceus* (Linnaeus, 1758) and *Leodia sexiesperforata* (Leske, 1778) are reproductively active during Panama's rainy season from May to November, but quiescent during the dry season (Lessios 1985a), while *L. variegatus*, *T. ventricosus*, and *Clypeaster subdepressus* (Gray, 1825) have ripe gonads throughout the year. Only *E. lucunter* showed non-periodic fluctuations in readiness to spawn.

Lessios (1984, 1991) also worked on lunar spawning periodicity in Panamanian echinoids. *Diadema mexicanum* was found to spawn on the third quarter of the lunar cycle, with a peak three to four days after full moon, while its allopatric congener *D. antillarum*, spawned during the first lunar quarter, one to two days after new moon. Lessios (1984) wrote that it is therefore possible that geminate species of *Diadema* have acquired a prezygotic isolating mechanism while in allopatry and therefore reproductive isolation need not be the product of natural selection against hybridization, thus casting doubt on the speciation by reinforcement theory.

Lunar spawning rhythms were not found in *T. ventricosus*, *E. viridis*, *L. sexiesperforata*, *C. rosaceus* or *L. williamsi* (Lessios 1991). *Eucidaris tribuloides* and *D. antillarum* were found to have a distinct lunar rhythm spawning at different lunar phases, while *L. variegatus* has a semilunar rhythm spawning every new and full moon (Lessios 1991). Such spawning rhythms are unlikely to be cued by tides as the Atlantic coast experiences only small tidal ranges (<0.5 m) where water level is determined as much by meteorological conditions as by tides. Lunar periodicity was therefore shown to not be a lineage specific trait inherited from a common ancestor, but has evolved independently in different echinoid taxa (Lessios 1991).

Temporal and spatial variation in egg size of thirteen Panamanian echinoids was studied by Lessios (1987). His results showed that mean egg size from different females collected from the same locality on the same day were significantly different in all species studied. However, daily (within-month) variation in egg size was only significant in *E. tribuloides*, *D. antillarum*, and *L. variegatus*, while no significant differences were found in the daily means of egg volume in *D. mexicanum* and *E. viridis*. Monthly means of egg volume were different in *L. variegatus*, *L. williamsi*, *T. ventricosus*, *E. viridis*, *E. vanbrunti*, *C. rosaceus* and *C. subdepressus*. Between-years variation was also significant in all of these species except *L. variegatus*. All Caribbean species showed a decline in egg size after September, however, this could not be explained by any obvious environmental fluctuation. *Echinometra lucunter* showed no significant variation between months of the same year but exhibited differences between years. No significant monthly or annual variation was observed in *E. tribuloides*, *E. thousarii* (Valenciennes, 1846), *D. antillarum*, *D. mexicanum* or *Leodia sexiesperforata*. Correlations between size of eggs collected at a particular time and the intensity of spawning by the population at that time was not significant, suggesting that size of mature eggs is not determined by the reproductive state of the parental population (Lessios 1987).

Lessios (1990) looked at adaptation and phylogeny as determinants of egg size in twenty-two species of echinoids and two species of asteroids from the Caribbean and Pacific coasts of Panama. Lessios (1990) identified four trends: (1) species have eggs that are significantly different from those of other congeneric species; (2) species evolving for three million years in separate environments have accumulated differences in their egg sizes and the direction of these differences is not random; (3) egg size of congeneric species is similar where developmental modes are similar (n.b. the volume of *Clypeaster rosaceus* eggs is eight times greater than that of its congener *C. subdepressus*, an adaptation that reflects developmental mode as the larvae of *C. rosaceus* are facultative planktotrophs, while those of *C. subdepressus* are obligate planktotrophs); (4) there is no similarity between the egg sizes of species belonging to the same family or higher taxonomic category, even if these species inhabit the same ocean. He concluded that because the phenomena of smaller eggs in the Pacific (a pattern attributed to higher levels of primary productivity in the eastern Pacific) holds for lineages that have not shared a common ancestor for a long time, it is likely that independent adaptation by the members of each pair has occurred due to the environments of the two oceans during the last three million years.

McAlister (2008) studied the plasticity of larval feeding arms for geminate species pairs to determine how food availability can influence growth and the arm length of the larvae. No plasticity in arm length was found in either Caribbean or Pacific species, however, arm length was significantly longer in Caribbbean species (including in *D. antillarum*). McAlister (2008) suggests that historical changes in food levels led to the development of longer feeding arms in Caribbean species, while plasticity may be limited by egg size or the timing of reproduction.

The evolution of gametic incompatibility in neotropical *Echinometra* was investigated by Lessios and Cunningham (1993). They used fertilization experiments to determine the degree of reproductive isolation both between the sympatric *E. lucunter* and *E. viridis*, in the Caribbean, and with the allopatric *E. vanbrunti* from the eastern Pacific. They reported that, contrary to the predictions of the 'speciation by reinforcement hypothesis', the degree of incompatibility between the allopatric *E. lucunter* and *E. vanbrunti* is higher than between the sympatric *E. lucunter* and *E. viridis*. Crosses between *E. viridis* and *E. vanbrunti* produced rates of fertilization almost equal to those of homogamic crosses, while sperm of *E. lucunter* were found to be able to fertilize eggs of the other two species. However, few *E. lucunter* eggs are fertilized by heterospecific sperm (Lessios and Cunningham 1993). In the wild *E. lucunter* and *E. viridis* maintain their genetic integrities despite having this unidirectional gamete isolation. Allozyme data placed the speciation event of the sympatric species after the rise of the isthmus. Thus there is no correlation between genetic divergence and strength of reproductive isolation (Lessios and Cunningham 1993).

Later work by McCartney and Lessios (2002) confirmed these findings and demonstrated a much stronger block to cross-species fertilization of *E. lucunter* eggs than was previously shown. They reported that gamete incompatibility in these species is weaker in sympatry than in allopatry, and based on the mtDNA data in McCartney et al. (2000), such gamete incompatibility in these species arose in the last 1.5 million years.

Zigler et al. (2008) looked at egg energetics, fertilization kinetics, and population structure in echinoids with facultatively feeding larvae. This work included *C. rosaceus*, one of only two echinoid species known to have facultatively planktotrophic larvae. The eggs of *C. rosaceus* resemble those of echinoid species with obligately feeding larvae in egg energetic density, egg buoyancy and fertilization kinetics, but resemble non feeding larvae in time to metamorphosis, the early formation of the coelom and the reduced allocation to larval feeding structures (Zigler et al. 2008).

Cross-fertilization experiments between *C. rosaceus* and *C. subdepressus* by Zigler et al. (2008) revealed that bidirectional gametic incompatibility has evolved between the two species. These species are commonly found together or within close proximity to one another in the shallow waters of the Caribbean in Panama (both species live in the sand, but *C. rosaceus* is typically more common in turtle grass beds). These species have no temporal reproductive isolation, as both have ripe gonads for much of the year, but with peaks from July to October (Lessios 1984). The results of Zigler et al. (2008) indicate that the small eggs of *C. subdepressus* are particularly resistant to fertilization by *C. rosaceus* sperm, while the larger *C. rosaceus* eggs are slightly more vulnerable to fertilization by *C. subdepressus* sperm, but at levels that are unlikely to results in conspecific fertilizations in nature.

Research on the sea urchin sperm protein bindin in sea urchin species found off the coasts of Panama has been conducted by Zigler and Lessios (2003a, b, 2004), McCartney and Lessios (2004), Zigler et al. (2005), and Geyer and Lessios (2009). Bindin is a sea urchin sperm protein that mediates sperm-egg attachment and membrane fusion. Across echinoid genera, its divergence has been shown to be correlated with heterospecific incompatibility in fertilization (Zigler et al. 2005).

In sympatric species, a region of the bindin molecule reportedly evolves at high rates under strong positive selection (Metz and Palumbi 1996), whereas in allopatric species (e.g. *T. ventricosus* in the Caribbean and *T. depressus* in the eastern Pacific), bindin variation was reported to be low, with no evidence of positive selection. However, this does not appear to be the case for all sea urchin species. *Lytechinus williamsi* and *L. variegatus*, which have overlapping distributions in the Caribbean, show no clear evidence of selection (Zigler and Lessios 2004). While in spatially overlapping *E. lucunter* and *E. viridis*, positive selection was only found to have accumulated along the bindin lineage of *E. lucunter* (McCartney and Lessios 2004). Further, Geyer and Lessios (2009) found no evidence to suggest that the source of selection on bindin in *E. lucunter* is reinforcement, as bindin showed no evidence of stronger selection in areas of sympatry relative to areas of allopatry. Geyer and Lessios (2009) propose that the evolution of this molecule is a result of the processes acting within species, such as sexual selection and sperm competition.

The reproductive cycle of the starfish *O. reticulatus* in Isla Solarte, Bocas del Toro Archipelago, was studied from February 2000 to February 2001 by Guzman and Guevara (2002a). Males and females with ripe gonads were observed almost every month, however, the gonad index revealed peaks in reproductive activity, during July, November, and January. Gonad maturation was not triggered by changes in water temperature (these ranged from 27 °C to 30 °C), but may have responded to chemical cues in the water (e.g. nutrients, salinity) (Guzman and Guevara 2002a).

Guzman et al. (2003) assessed the reproductive status of two commercial species of sea cucumber, *I. badionotus* and *H. (H.) mexicana* in the Caribbean of Panama. They reported that for both species the minimum reproductive length was 13–20 cm with a weight of 150 g. Both species were gonochoric with a 1:1 sex ratio. Gametogenesis and spawning occurred throughout the year; however there were periods of peak reproductive activity, between July and November for *I. badionotus* and from February to July for *H. (H.) mexicana* (Guzman et al. 2003).

4.3.3 Evolution

Molecular work on echinoids in Panama started with Lessios (1979), who used geminate species of Panama (*E. tribuloides* and *E. thouarsii*, *D. antillarum* and *D. mexicanum*, *E. viridis*, *E. lucunter* and *E. vanbrunti*) to test the molecular clock hypothesis. This hypothesis predicts that each protein changes at a constant rate, so that the degree of divergence between two species is linearly related to the time for which their lineages have remained separate (Wilson et al. 1977). Lessios (1979) found that this does not hold for Panamanian echinoids, as very different rates of divergence were found in the enzymatic proteins between geminates in the different genera. *Eucidaris* and *Echinometra* exhibited transisthmian distances sixteen and thirty-seven times greater than intraspecific ones respectively, while the trans-isthmian distance in *Diadema* was twenty times smaller than for the *Echinometra*. Lessios (1979) proposed that the most plausible explanation was that the molecules used in this investigation were evolving under the influence of natural selection.

In 1981(b) Lessios combined this molecular work with morphometric data using twenty morphological measurements. The morphometric data showed that differentiation between members of each geminate pair are not significantly different from local variation within each species. However, the ratio of inter to intraspecific Mahalanobis distance was congruent with the molecular data, being lowest in *Diadema*, intermediate in *Eucidaris*, and highest in *Echinometra*. Morphological rates of divergence were found to be more rapid in sympatric species, possibly due to habitat separation between congeners (Lessios 1981b).

Bermingham and Lessios (1993) used mitochondrial DNA (mtDNA) and isozymes in these sea urchin species to measure the degree of divergence. Trans-isthmian isozyme divergence in the geminate pairs differed by an order of magnitute, while mtDNA divergence was equivalent in all pairs demonstrating that mtDNA can provide accurate estimates of time since separation (Bermingham and Lessios 1993).

After the mass mortality of *D. antillarum* in the Caribbean (see Sect. 4.6. Threats to echinoderms), Lessios (1985b) looked at the genetic consequences of the restriction in the genetic structure of populations. His results showed no loss of genetic variability to that recorded in the same populations of *D. antillarum* in 1977. Lessios suggested that organisms with planktonic larvae are likely to recover from local extinction events without significant losses of their genetic variability, as larval exchange allows re-seeding of affected areas by recruits not exposed to the bottleneck.

A phylogenetic survey of sea urchin retroviral-like (SURL) retrotransposable elements in 33 species of echinoid (regular sea urchins, sand dollars, and heart urchins) was conducted by Gonzalez and Lessios (1999). High ratios of synonymous to nonsynonymous substitutions suggest that the reverse transcriptase of the elements is under strong purifying selection (Gonzalez and Lessios 1999). They report that despite the predominance of vertical transmission, sequence similarity of 83–94 % for SURL elements from hosts that have been separated for 200 Myr suggests four cases of apparent horizontal transfer between the ancestors of the extant echinoid species. In three additional cases, elements with identical RT sequences were found in sea urchin species separated for a minimum of 3 Myr. Thus, horizontal transfer plays a role in the evolution of this retrotransposon family (Gonzalez and Lessios 1999).

Cytochrome oxidase I (COI) divergence was assessed in Atlantic and eastern Pacific *Echinometra* by McCartney et al. (2000), using the closure of the Isthmus of Panama to date cladogenic events. They reported that the Atlantic species *E. lucunter* and *E. viridis* diverged 1.27–1.62 mya at a time in the Pleistocene when sea levels fell and Caribbean coral speciation and extinction rates were high. The fact that these species split so recently, yet do not hybridize, demonstrates that reproductive barriers between marine species can occur in less than 1.6 million years (McCartney et al. 2000).

A global molecular phylogeny of *Diadema* (including *D. antillarum* and *D. mexicanum* from Panama) was published by Lessios (2001) and Lessios et al. (2001). This molecular study showed that *Diadema* is composed of reciprocally monophyletic mtDNA clades, and that the eastern Atlantic *D. antillarum* (*D. antillarum*-b) from Madeira, Canary Islands, Cape Verde and Sao Tome is genetically distinct from the western Atlantic *D. antillarum*. These species are separated by the long stretch of deep water separating the eastern from the western Atlantic. *Diadema antillarum ascensionis* Mortensen, 1909 from the central Atlantic islands of Ascension and St. Helena was found to be genetically isolated, but nested within the Brazilian clade of the western Atlantic *D. antillarum*. Lessios suggests that the biogeographic barrier between the Caribbean and Brazil may be caused by the outflow from the Amazon and Orinoco Rivers. High levels of gene flow were reported among populations of *D. mexicanum* in the eastern Pacific, with *D. mexicanum* from the Galapagos and Cocos Island belonging to the same population as sea urchins from Panama and Mexico (Lessios et al. 2001).

A study of the phylogeography of the pantropical sea urchin *Tripneustes* by Lessios et al. (2003) showed that based on morphology, COI, and bindin data, *T. depressus* A. Agassiz, 1863 from the eastern Pacific is in fact the same species as *T. gratilla* (Linnaeus, 1758) from the western Pacific. The formation of the Isthmus of Panama, the deep water separating the eastern and western Atlantic, and the freshwater plume of the Orinoco and Amazon rivers between the Carribbean and the coast of Brazil were shown to be important barriers to the evolution of *Tripneustes*. However, the Eastern Pacific Barrier (5,000 km of deep water) between the central and eastern Pacific was unimportant in the subdivision in the genus.

A phylogeographic study of the genus *Lytechinus* by Zigler and Lessios (2004) using mitochondrial COI, and the entire molecule of nuclear bindin shows that the genus *Lytechinus* is paraphyletic (using *Toxopneustes* and *Tripneustes* as out-groups), with *Sphaerechinus granularis* (Lamarck, 1816) coming out as the sister species to *L. euerces* Clark, 1912 . The authors propose that *L. euerces* should be moved into the monotypic *Sphaerechinus*, rather than placing *Sphaerechinus* in *Lytechinus* which would increase the range of the genus to the temperate eastern Atlantic and Mediterranean (Mortensen 1943a). *Lytechinus semituberculatus* (Valenciennes in L. Agassiz, 1846) and *L. panamensis* formed a clade with no distinction between the two species, while *L. anamesus* and *L. pictus* were also phylogenetically indistinguishable. A well supported Atlantic clade was formed of *L. williamsi* and the three subspecies of *L. variegatus* (*L. variegatus variegatus*, *L. variegatus atlanticus*, *L. variegatus carolinus*). *Lytechinus williamsi* from the Caribbean was found to share mtDNA haplotypes with *L. variegatus variegatus*, however, their bindin was shown to be distinct and to coalesce within each morphospecies. A small private clade of mtDNA was found in some *L. williamsi* from Panama and Belize. Zigler and Lessios (2004) suggest that this may be indicative of former differentiation in the process of being swamped by introgression, or of recent speciation.

Molecular work on the starfish *A. planci*, including specimens from Panama, revealed that it is not in fact a single species, but a pan-Indo-Pacific species complex consisting of four deeply diverged clades (Pacific, Red Sea, Northern Indian Ocean, and Southern Indian Ocean) (Vogler et al. 2008). These clades reportedly diverged between 1.95 and 3.65 mya (Pliocene to early Pleistocene) and have genetic distances (8.8–10.6 %) equivalent to the distances between other sibling species of starfish (Waters et al. 2004). Vogler et al. (2008) propose that this speciation process was driven by sea level changes (Pillans et al. 1998) that isolated populations.

Zulliger and Lessios (2010) published a global phylogeny of the genus *Astropecten* based on 117 specimens belonging to 40 species. This genus is one of the most species-rich genera among starfish containing over 150 described species with six species recorded from Panama: *A. armatus* Gray, 1840, *A. articulatus* (Say, 1825), *A. cingulatus* Sladen, 1883, *A. exiguus* Ludwig, 1905, *A. fragilis* Verrill, 1870 and *A. regalis* Gray, 1840. Such diversity is remarkable, because most species of *Astropecten* have a long-lived planktotrophic larval stage. Consequently one would expect this to lead to a low speciation rate (Zulliger and Lessios 2010). Zulliger and Lessios (2010) compared their molecular phylogeny to Döderlein's (1917) morphological phylogeny. They found high levels of congruence on the whole, but many discrepancies on a local scale. Phylogentic inference from their data reveals that that morphological and ecological convergence has taken place in *Astropecten*, resulting in allopatric non-sister taxa with similar morphologies and habitat preferences. Varieties of several species were also shown to exhibit genetic distances large enough to justify recognizing them as separate species.

Reviews on speciation in organisms separated by the Isthmus of Panama have been written by Lessios (1998, 2008). Lessios (1998) wrote a chapter in "Endless Forms: Species and Speciation" on the first stage of speciation as seen in organisms separated by the Isthmus of Panama. Lessios points out that although geminate species have had a long history of contributing evidence relevant to speciation, they remain an underutilised tool for understanding vicariant speciation. Lessios's (2008) review showed that a total of 38 regions of DNA have been sequenced in nine clades of echinoids, 38 clades of crustaceans, 42 clades of fishes, and 26 clades of molluscs with amphi-isthmian subclades. Of these, 34 clades are likely to have been separated at the final stages of Isthmus completion, 73 clades split earlier and eight clades maintained post-closure genetic contact (e.g. via a circumglobal route or through the Panama Canal). As no vicariant event is better dated than that of the Isthmus of Panama, molecular divergence between species pairs remains to be of great interest, particularly as adaptive divergence can be seen in species life history parameters.

4.4 Diversity and Distribution of Echinoderms

Chesher (1972) published a list of Pacific and Caribbean echinoids and identified potential geminate species pairs. Maluf (1988) synthesised and tabulated the composition and distribution of central eastern Pacific echinoderms from the literature, which has proved to be a valuable resource for researchers. Lessios (2005a) updated the records of echinoids found in the Pacific waters of Panama, and distinguished species that have been documented for Panama from those that have collection records that straddle Panama, but with no published records of actual presence in Panama. In the latter category Lessios (2005a) lists *Kamptosoma asterias* (Agassiz, 1881), *Plesiodiadema horridum* (Agassiz, 1898), *Cystocrepis setigera* (Agassiz, 1898), *Pourtalesia tanneri* Agassiz, 1898, *Homolampas fulva* Agassiz, 1879, *Echinometra oblonga* (Blainville, 1825) and *Encope perspectiva* Agassiz, 1841. The inclusion of *E. perspectiva* in this category is a mistake as it was collected in Panama from Playa Grande, on San Jose Island, Pearl Islands Archipelago by Clark (1946) and more recently by Coppard (2010) from the same location. *Plesiodiadema horridum* is listed in the Smithsonian, National Museum of Natural History collections as collected off Panama (USNM 21050). However, the coordinates of the collection site (3°09.00′N, 82°08.00′W) indicate that it was collected off Colombia. Lessios (2005a) gives details of why the subtropical *Centrostephanus coronatus* (Verrill, 1867) and central and western Pacific species *E. oblonga* are unlikely to occur in Panama, and that the record of *Caenocentrotus gibbosus* (Agassiz, in Agassiz and Desor, 1846) in Panama is a mistake in the literature.

Alvarado et al. (2008) assessed the diversity of echinoderms in the Caribbean of Central America and gave faunal lists for each country. Panama had the greatest species richness, with 154 species listed, comprised of 15 species of crinoids, 23 species of asteroids, 56 species of ophiuroids, 30 species of echinoids and 30 species of holothuroids. Among the crinoids they included both *Davidaster insolitus* Clark 1917 and *Davidaster discoideus* (Carpenter, 1888). The number of species of crinoids should be reduced to 14 because *D. insolitus* is currently considered a junior synonym of *D. discoideus* (Messing 2010). Among the asteroids, two species are missing from their list for Panama. These are *Echinaster (Othilia) echinophorus* (Lamarck, 1816) collected from Limon Bay (Smithsonian National Museum of Natural History, USNM 38235) and Fort Randolph (USNM E 26417, USNM E 26418), and *Persephonaster patagiatus* (Sladen, 1889) collected from the Gulf of Mosquitos (USNM E 31501). The number of ophiuroid species is incorrect as *Amphiodia riisei* (Lütken, 1869) [sic] is a junior synonym of *Ophiophragmus riisei* (Lütken in: Lyman 1860) (Stöhr 2010). Among the holothuroids they list *Thyone deichmannae* Madsen, 1941 and *Thyone inermis* Heller, 1868 which are junior synonyms of *Havelockia inermis* (Heller, 1868) (Hansson 2011). Although Alvarado et al. (2008) state that there are 30 echinoid species recorded from off the Caribbean coast of Panama they only list 29 species which agrees with more recent work by Coppard (2010).

Table 4.1 Number of echinoderm species recorded from Panamanian waters

	Pacific	Caribbean	Both pacific and caribbean	Total
Crinoidea	7	14	0	21
Ophiuroidea	75	51	4	130
Asteroidea	56	25	2	83
Echinoidea	49	27	2	78
Holothuroidea	73	27	2	102
Total	260	144	10	414

Alvarado et al. (2010) documented the echinoderm diversity in the Pacific coast of Central America, again providing faunal lists for each country. Although Alvarado et al. (2010) included the paper by Lessios (2005a) in their bibliography they unfortunately ignored Lessios's comments and tables. These indicate echinoid species that have distributions in the eastern Pacific that straddle Panama (but have not been recorded from Panama), those unlikely to occur in Panama, and errors in the literature. Alvarado et al. (2010) also incorrectly listed *Encope grandis* Agassiz, 1841 in Guatemala, El Salvador, Honduras, Nicaragua and Panama. This species is endemic to the Gulf of Mexico; *Encope michelini* Agassiz, 1841 in Costa Rica (this is an Atlantic species that occurs off Florida), *Dendraster excentricus* (Eschscholtz, 1829) in Panama (this temperate species has a distribution range from Alaska to Baja California), as well as including many junior and senior synonyms as separate taxa (e.g. *Ophionereis annulata* (Le Conte, 1851) and *Ophionereis dictyota* Ziesenhenne 1940; *Sclerasterias alexandri* (Ludwig 1905) and *Hydrasterias diomediae* Ludwig 1905; *Lissothuria ornata* Verrill 1867 and *Thyonepsolus beebei* Deichmann, 1937). Alvarado et al. (2010) list 253 species of echinoderms from Panama, the country with the highest species richness in Central America. This is explained by the long coastline (1,690 km) of Panama which incorporates high levels of coastal heterogeneity and reflects the greater echinoderm research effort than in other Central America countries (Alvarado et al. 2010).

A web-based project by Coppard (2010) ("The Echinoderms of Panama") has documented 414 species of echinoderms that occur off the Caribbean and Pacific coasts of Panama (Table 4.1 and Appendix). This work was carried out in conjunction with the Encyclopaedia of Life and contains images (including type material), diagnostic descriptions and references with links to nomenclators, GenBank, Barcode of Life, and depth, distribution and specimen data. All species names were checked with the World Register of Marine Species and lists of junior synonyms are given. Where physical specimens were not found in museum collections, records from the literature were used.

Since the final closure of the Panama Isthmus unequal rates of speciation, extinction and migration have occurred in the Caribbean and Pacific, resulting in very different levels of biodiversity seen on either side of the isthmus today (Vermeij 1978; Jackson et al. 1996; Budd 2000). Among invertebrates, greater levels of diversity have been reported in the Caribbean than in the Pacific for

Fig. 4.9 *Nemaster grandis* with arms spread for feeding, Bocas del Toro, Caribbean Sea (photo courtesy of Edgardo Ochoa)

corals (Glynn 1982), sponges (Van Soest 1994), benthic foraminiferans (Collins 1999), cheilostome bryozoans (Cheetham et al. 2001) and cupuladriid bryozoans (O'Dea et al. 2004). With the exception of crinoids, the diversity of echinoderms off Panama shows a very different trend, with a far greater diversity in the Pacific (270 species) than in the Caribbean (154 species). The high diversity of crinoids in the Caribbean is most likely a consequence of greater coral reef development, which provides a broad range of suitable habitats for species such as *Nemaster grandis* (Fig. 4.9).

For the other classes of echinoderms, the greater number of species in the Pacific of Panama can partially be explained by a number of species having crossed the Eastern Pacific Barrier (e.g. *Toxopneustes roseus*, *Tripneustes depressus*, *Acanthaster planci*) and thus have a West Pacific origin, with no sister species in the Caribbean. However, in many genera, particularly those that contain mud and sand dwelling species, a genus in the Caribbean often contain one or two species, while in the Pacific the same genus will often have more than double the number of species (e.g. in Panama the genus *Luidia* has three species recorded from the Caribbean, but nine species from the eastern Pacific; *Astropecten*, two species from the Caribbean, four species from the Pacific; *Encope*, one species from the Caribbean, five species from the Pacific). The different hydrological conditions in the Gulf of Chiriquí and the Gulf of Panama have created high levels of coastal heterogeneity providing a diverse range of habitats. However, such morphological variation that has given rise to authors erecting many species names in the Pacific may simply reflect ecophenotypic varriation within a species.

Sampling the sharp drop down the continental slope to a depth over 1,000 m just south of the Azuero Peninsula has resulted in a number of deep water species being recorded from the Pacific of Panama, such as sea lilies (e.g. *Calamocrinus diomedae* Agassiz, 1890, depth 1,430 m), brittle stars (e.g. *Astrodia plana* (Lütken and Mortensen 1899), depth 3,058 m; *Asteronyx loveni* Müller and Troschel, 1842,

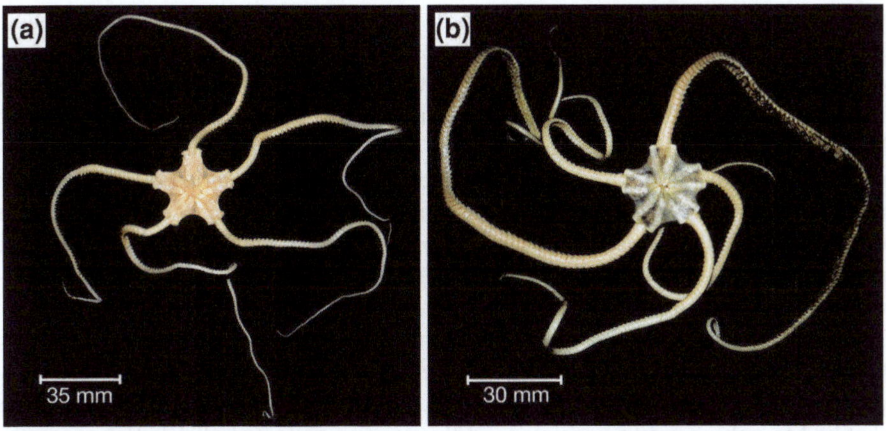

Fig. 4.10 Deep water ophiuroids, **a** *Astrodia plana* and **b** *Asteronyx loveni*

depth 1,866 m, Fig. 4.10), starfish (e.g. *Dytaster gilberti* Fisher, 1905, depth 2,690 m; *Pectinaster agassizi* Ludwig 1905, depth 2,323 m), sea urchins (e.g. *Salenocidaris miliaris* (Agassiz 1898), depth 3,058 m; *Caenopedina diomedeae* Mortensen, 1939, depth 850 m; *Araeosoma leptaleum* Agassiz and Clark, 1909, 1,062 m) and sea cucumbers (e.g. *Molpadia spinosa* (Ludwig, 1893), depth 3,279 m; *Protankyra brychia* (Verrill, 1885), depth 4,000 m).

Ten species have collection records from both sides of the Panama Isthmus. These are the ophiuroid species *Hemipholis elongata* (Say, 1825), *Ophiactis savignyi* (Müller and Troschel, 1842), *Ophioderma panamensis* Lütken, 1859 and *Amphipholis squamata* (Delle Chiaje, 1828); the asteroid species *Linckia guildingii* Gray, 1840, *Ctenodiscus crispatus* (Retzius, 1805); the echinoid species *Moira atropos* (Lamarck, 1816) (the subspecies *Moira atropos clotho* Michelin, 1855 is recorded from the Pacific), *Meoma ventricosa* (Lamarck, 1816) (the subspecies *Meoma ventricosa grandis* Gray, 1851 is recorded from the Pacific); and the holothuroids *Holothuria (Thymiosycia) arenicola* Semper, 1868 and *Holothuria (Thymiosycia) impatiens* (Forskål, 1775). With the final closure of the isthmus (2.8 mya) there has been no gene flow between populations in the Pacific and Caribbean. However, the opening of the freshwater Panama Canal in 1914 provided a means for echinoderms to cross the isthmus, either as larvae in the ballast tanks of ships or associated with fouling organisms on the hulls of ships. Roy and Sponer (2002) found identical mtDNA haplotypes of *Ophiactis savignyi* in the Caribbean and Pacific. This led them to suggest that these sponge-dwelling brittle stars have been recently transported between oceans on the hulls of ships. Global phylogenies have also shown that genes of certain species have travelled around the world during the last 3 million years providing a connection between tropical species on the two sides of tropical America (Bowen et al. 2001, 2006; Lessios et al. 2001; Rocha et al. 2005).

Some echinoderm species have records in Panama that may represent transient populations. *Arbacia spatuligera* (Valenciennes, 1846) is recorded (specimen not examined) from the reef off French Fort (USNM 1017377), in Panama Bay. This may be a misidentification or represent a transient population of this species, with larvae carried up from Ecuador into the Gulf of Panama but with insufficient recruitment to establish a viable population. This species has therefore not been included in the Pacific data for Panama, since to the best of our knowledge it is the only record of this species in Panama. Records of *Lytechinus pictus* in Panama are probably erroneous. Mortensen (1921) described *L. panamensis* based on specimens he obtained from Taboga Island. Clark (1948) collected one specimen from Bahia Honda as well as one specimen from Gorgona Island, Colombia. However, Clark (1948) states that these may be specimens of *L. panamensis*, a view supported by Zigler and Lessios (2004) and Lessios (2005a). Lessios (2005a) reports that he has only seen two specimens of this species collected off Isla Saboga in the Perlas Archipelago (8°38′N, 79°03′W). Clark (1925) and Mortensen (1928–1951) speculated that *Lytechinus* from Peru and Ecuador may not be *L. semituberculatus* but *L. panamensis*, with the larvae of this species occasionally settling in the Gulf of Panama.

Panama is the type locality of 82 species of echinoderms (nine Caribbean, 73 eastern Pacific, one of these species occurs in both the Caribbean and Pacific). However, four of these species (starfish) are now considered subjective junior synonyms of other species. The majority of the Pacific species were established from Panamanian specimens in the late 1800s and early 1900s by authors such as Lütken and Mortensen (1899) and Ludwig (1894, 1905), as a result of the material collected on expeditions by the *Albatross*. The majority of new Caribbean species were discovered much later in the 1960s, 1970s and 1980s by authors such as Chesher (1968, 1970, 1972) and Hendler (1995 2005), as a result of the laboratories of the Smithsonian Tropical Research Institute in Panama. In recognition of this, Hendler (2005) named the ophiuroid species *Ophiothrix stri* Hendler 2005.

4.5 Aquaculture and Fisheries

The sustainable aquaculture of echinoderms in Panama has not been developed. In 1997 the Panamanian government granted a permit for the harvesting and processing of bêche-de-mer (sea cucumbers) in Bocas del Toro. However, this exploitation was short lived, with the permit being revoked 30 days later. Nevertheless, in that short time period Guzman and Guevara (2002b) estimate that 750,000 sea cucumbers were harvested (based on 25 fishermen extracting an average of 1,000 and a maximum of up to 1,500 cucumbers day^{-1}). This consisted primarily of three species *Holothuria* (*Halodeima*) *mexicana*, *Isostichopus badionotus* and *Astichopus multifidus*. This quantity represents a biomass of ca. 180 tons (based on average body wet weight of 240 g). If such fishing pressure was permitted, Guzman and Guevara (2002b) estimate that stocks of the three

Fig. 4.11 *Astichopus multifidus* photographed in Bocas del Toro (photo courtesy of Juan Sanchez)

species of holothuroids would collapse within a very short period of time (305 days for *H. (H.) mexicana*, 221 days *I. badionotus* and only 9 days for *A. multifidus*), based on a catch effort of 25,000 ind day^{-1} (Fig. 4.11)

Commercial bêche-de-mer fishing currently is banned in all Panamanian waters under Decree 157–2003. However, there are reports of illegal fishing activities, with five species being commercially exploited in Panama: *Actinopyga agassizi* (Selenka, 1867) *H. (H.) mexicana*, *A. multifidus*, *I. badionatus* (Caribbean) and *I. fuscus* (eastern Pacific) (Toral-Granda 2008). In Parque Nacional Marino Isla Bastimentos, *H. (H.) mexicana* showed a maximum density of 956 ind ha^{-1} in 480 ha representing only 10 % of the shallow water protected area, while *I. badionotus* and *A. multifidus* were strikingly absent in 95 % of the protected insular area around Cayos Zapatillas. This indicates that that illegal fishing has taken place inside the park (Guzman and Guevara 2002b). Data from the statistics department in Hong Kong, revealed that 281 kg (dry weight) of bêche-de-mer were imported from Panama in 2004 and 408 kg (dry weight) in 2005 (Toral-Granda 2008).

4.6 Threats to Echinoderms

Between 1983 and 1984, *D. antillarum* suffered one of the most extensive and severe mass mortality events ever recorded for a marine animal. This was probably caused by a waterborne pathogen transported by ocean currents, since plots of surface currents in the Caribbean Sea coincided with the spread of the die-off (Lessios 1988; Miller et al. 2003). The mass mortality of *D. antillarum* has been a major factor leading to a shift from coral-dominated to algae-dominated communities on many Caribbean reefs during the past 20 years (Lessios 1988). The die-off was first observed in January 1983 at Galeta Point, near the entrance of the Panama Canal. In approximately three months it had extended to the San Blas Archipelago, before reaching the Panama-Colombia border by the end of June. By

Fig. 4.12 *Diadema antillarum* on a reef in San Blas, depth 2 m, June 2010

January 1984 the die-off had spread throughout the rest of the Caribbean with population densities in Panama reduced to 5.8 % of their previous levels (Lessios et al. 1984a). The first symptom of the presumed disease was accumulation of sediment on the spines and sloughing off of the spine ectoderm. This was followed by loss of pigmentation in the skin covering the spine muscles, the peristome, and in the periproctal cone. Spines became brittle and broke, while others fell off completely, exposing areoles and tubercules. Tube feet became flaccid, no longer retracted fully upon stimulation, and were unable to cling to the substrate. Many individuals at this stage were found on their sides or being moved to and fro by the surge. In advanced stages large patches of skin pealed off the test, the peristome and the anal cone lost practically all pigmentation, and the spines could be pulled off with little force (Lessios et al. 1984b).

Before the mass mortality, mean densities of *D. antillarum* on reefs at Punta Galeta and San Blas where 1.82 and 1.91 ind m^{-2}, respectively (Lessios et al. 1984b). After the mass mortality the densities dropped to an average of 0.05 ind m^{-2} in San Blas, and to 0.004 ind m^{-2} in Punta Galeta. This density reduction was not observed in other species of echinoids (*E. lucunter*, *E. viridis*, *Echinoneus cyclostomus* (Leske, 1758), *E. tribuloides*, *L. variegatus* and *T. ventricosus* (Lamarck 1816)) on the same reefs. In San Blas, the mean sizes were reduced from 48.6 mm to 25.0 mm (Lessios et al. 1984b). However, this mass reduction on the population did not have any consequence on its genetic variability (Lessios 1985b).

Five years after the event, Lessios (1988) did not observe any recovery of the populations on the San Blas reefs, with little influx of juveniles and no increase in population densities of other sea urchins species (*E. viridis*, *E. lucunter*, *E. tribuloides*, *T. ventricosus*, or *L. variegatus*). He reported no alteration in the reproductive behaviour of *D. antillarum*, proposing that the low recruitment success could be an effect of low populations densities and low adult numbers.

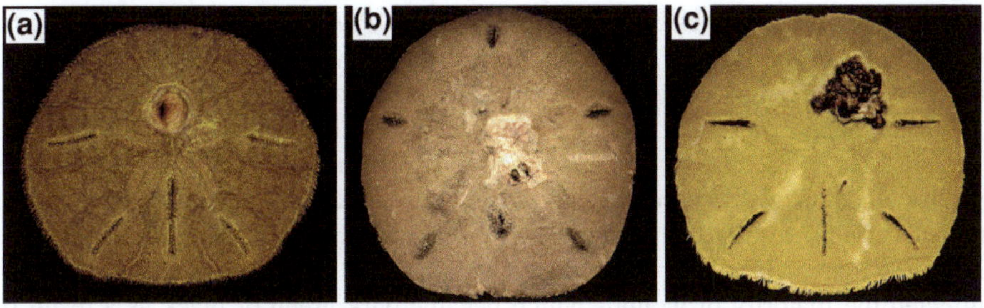

Fig. 4.13 Mellitid sand dollars showing varying degrees of predation and secondary infection, **a** *M. longifissa*, collected alive (sublethal predation), **b** *E. laevis* (lethal predation, no secondary infection), **c** *M. kanakoffi*, found dead (probable death from secondary infection due to the large amount of necrotised tissue)

Ten years after the mass mortality densities remained at less than 3.5 % of their pre-mortality levels, with *D. antillarum* being absent in its preferred habits in San Blas (Lessios 1995). Moreover, the mean size of individuals remained low (17.3–35.2 mm). This was thought to be a direct result of the disease, as food availability was not a limiting factor (Lessios 1995).

Twenty years after the mass mortality populations remained at less than 6.5 % of their pre-mortality levels in San Blas (Lessios 2005b). Some successful recruitment events have been observed on a few reefs. However, these have not been sustained over time (Lessios 2005b). The lack of apparent recruitment could have been caused by too few adults in upstream areas to reproduce at a rate that would overcome the normal sources of mortality (the Allee effect) or by the continuing presence of the pathogen that devastated the populations in 1983–1984. Recent observations (2010) on some San Blas reefs indicate a slight recovery (Fig. 4.12), however, it remains to be seen if these populations will be sustained.

On 27 April 1986 more than 8 million liters of crude oil spilled into a complex region of mangroves, seagrassses, and coral reefs east of the Caribbean entrance to the Panama Canal (Jackson et al. 1989). At the seaward edge, numbers of the most abundant sea urchin *E. lucunter* was reduced by 80 % within a few days of the spill, and the reef flat was littered with its skeletons (Jackson et al. 1989). However, further inshore no such mortality of *E. lucunter* was reported (Jackson et al. 1989).

During 2008–2009 Coppard (unpub. obs.) found large numbers of dead and dying sand dollars of the genera *Encope* and *Mellita* washed up on beaches from Punta Abreojos, Baja California Sur, Mexico, to Santa Elena Bay, La Libertad in Ecuador in the eastern Pacific and off Costa Rica and Belize in the Caribbean. In Panama dead mellitid sand dollars were found at Playa Las Lajas in the Gulf of Chiriqui, but not on beaches in the Gulf of Panama. These dead sand dollars were unusual in that they appeared to have an infection that originated at one or more circular denuded regions on the aboral surface of the test (Fig. 4.13a). The cause of such scars is most likely the result of attacks by gastropods from species such as

Northia pristis (Deshayes in Lamarck, 1844) and *Polinicies* sp. in the Pacific and *Cassis tuberosa* (Linneaus, 1758) in the Caribbean which have been reported to attack, and bore into the tests of mellitid sand dollars (McClintock and Marion 1993; Sonnenholzner and Lawrence 1998). Attacks always occurred on the aboral surface, and in the majority of cases centred over one or more petals. When the gastropods managed to bore through the test into the coelom (Fig. 4.13b), death of the sand dollar resulted. However, the presence of multiple scars of varying size indicates that such attacks are not always lethal. Following such attacks bacteria infected the majority of wounds which became dark green or black as a result of necrosis (Fig. 4.13c).

In many instances where the coelom had not been penetrated, the bacteria in the wound spread and caused the death of the sand dollar. Such pathogenic bacteria may be similar to that reported in *Meoma ventricosa* in Curaçao by Nagelkerken et al. (1999). In this instance a tetradotoxin-producing bacterial strain of the genus *Pseudoalteromonas* was found in the sediment and was ingested by *M. ventricosa*. The density of such strains of bacteria may have increased as a result of the outflow of polluted water from the harbour.

In Bocas del Toro, only 2.8 % of the shallow habitats in the archipelago are within the protected area (Guzman and Guevara 2002a). Since 1997, coastal environments in Bocas del Toro have changed, as infrastructure such as roads, marinas, and hotels have developed. Destruction of these seagrass habitats is resulting in an increase in runoff and sedimentation in the coastal zone (Guzman and Guevara 1998). This has already started to affect the coral reefs (Guzman and Guevara 2001), and could potentially modify the function of the remaining coastal ecosystems.

The illegal harvesting of sea cucumbers continues to put pressure on the remaining populations particularly on the Caribbean coast of Panama, along the coastal perimeter of Laguna de Chiriquí, from the point of Península Valiente through the southern section of Isla Popa and Cayo Agua (Cruz 2000; Guzman and Guevara 2002a).

4.7 Conclusions and Recommendations

Echinoderm diversity in Panama is reported to be greater on both the Caribbean (Alvarado et al. 2008) and eastern Pacific coasts (Alvarado et al. 2010) than in other Central American countries. The length of coastline, with high levels of coastal heterogeneity may explain high levels of species richness. However, this diversity relative to other Central American countries is probably more of a reflection of the research effort that has taken place in Panama. The first phase of echinoderm research in Panama began in the late 1800's with the collections made by the U.S. Fish Commission Steamer *Albatross* in the eastern Pacific. This period of exploration resulted in many new species being described. The second phase of echinoderm research in Panama in the 1930's and 1940's continued to explore the diversity of species and was the result of expedition by the New York Zoological Society and the Allan Hancock Foundation. These expeditions resulted in the collection of a vast number of specimens (some 28,835 echinoids alone from eastern Pacific of the Americas) with the description of many new species. The third phase of echinoderm research in Panama began with the inception of the marine program at STRI in 1961. This has resulted in extensive ecological and evolutionary echinoderm research, which has increased knowledge of echinoderms not only in Panama, but in many countries in tropical America, and other regions in the Atlantic and Pacific. Most of this knowledge concerns the ecology, reproduction and evolution in echinoids and certain species of asteroids (e.g. *A. planci* and *O. reticulatus*). However, less is known about holothuroids, ophiuroids and crinoids, particularly on a molecular and ecological level, highlighting a need for more research in these classes of echinoderm.

The continued monitoring of populations of *Diadema antillarum* needs to be maintained to determine whether the recently observed population increase is sustained, while the monitoring of sea cucumber populations, particularly in Bocas del Toro is needed to protect against illegal fishing. Guzman and Guevara (2002b) have proposed that protected areas should be established allowing the monitoring of natural populations. These could be part of the Parque Nacional Marino Isla Bastimentos (PNMIB) or may be located close to fishing communities willing to protect the resource.

Acknowledgments Simon Coppard thanks SENACYT (La Secretaría Nacional de Educación Superior Ciencia Tecnología e Innovación) (Panama) who helped fund this project as part of project COL08-002 and David M. Rubenstein who funded Simon through a Smithsonian Rubenstein fellowship. Juan José Alvarado thanks Consejo Nacional de Ciencia y Tecnología (CONACYT; Mexico) and Consejo Nacional para Investigaciones Científicas y Tecnológicas (CONICIT; Costa Rica). We are grateful to Edgardo Ochoa, Juan Sánchez, Angel Chiriboga and Dave Pawson for providing us with photographs. We appreciate the comments made by J. Cortés, J. Lawrence and H.A. Lessios that improved this paper.

References

Agassiz A (1892) Reports on an exploration off the west Coast of Mexico, Central and South America, and off Galápagos Island, in Charge of Alexander Agassiz by the U.S. Fish Commission Steamer "*Albatross*" during 1891, Lieut. Commander Tanner ZL, U. S. N. Commanding. I. *Calamocrinus diomedae* Agassiz A new stalked crinoid. With notes on the apical system and the homologies of echinoderms.Mem Mus Comp Zool Harvard Coll 17:1–95

Agassiz A (1898) Preliminary report on the echini. Dredging operations of the Albatross. Bull Mus Comp Zool 32:71–86

Agassiz A (1904) The Panamic deep sea Echini. Report XXXII on an exploration off the west coasts of Mexico, Central and South America, and off the Galapagos Islands. Mem Mus Comp Zool Harvard Coll 31:1–243

Alvarado JJ, Solis-Marin FA, Ahearn C (2008) Equinodermos (Echinodermata) del Caribe Centroamericano. Rev Biol Trop 56(Suppl 3):37–55

Alvarado JJ, Solis-Marin FA, Ahearn C (2010) Echinoderms (Echinodermata) diversity off Central America Pacific. Mar Biodiv 40:45–56
Amador JA, Alfaro EJ, Lizano OG, Magaña VO (2006) Atmospheric forcing of the eastern tropical Pacific: a review. Prog Oceanogr 69(2–4):101–142
Berggren WA, Hollister CD (1974) Paleogeography, paleobiogeograph and the history of circulation in the Atlantic ocean, in Studies in paleo-oceanography. Soc Econ Paleontol Mineral Spec Publ 20:126–186
Bermingham E, Lessios HA (1993) Rate variation of protein and mitochondrial DNA evolution as revealed by sea urchins separated by the Isthmus of Panama. Proc Nat Acad Sci USA 90:2734–2738
Boone L (1928) Echinoderms from the Gulf of California and the Perlas Islands. Bull Bingham Oceanogr Coll Yale Univ 2:1–14
Bowen BW, Bass AL, Rocha LA, Grant WS, Robertson DR (2001) Phylogeography of the trumpetfishes (*Aulostomus*): ring species complex on a global scale. Evolution 55:1029–1039
Bowen BW, Muss A, Rocha LA, Grant WS (2006) Shallow mtDNA coalescence in Atlantic pygmy angelfishes (Genus *Centropyge*) indicates a recent invasion from the Indian Ocean. J Hered 97:1–12
Budd AF (2000) Diversity and extinction in the cenozoic history of Caribbean reefs. Coral Reefs 19:25–35
Burton KW, Ling HF, O'Nions RK (1997) Closure of the Central American Isthmus and its effect on deepwater formation in the North Atlantic. Nature 386:382–385
Cheetham AH, Jackson JBC, Sanner J (2001) Evolutionary significance of sexual and asexual modes of propagation in neogene species of the bryozoan Metrarabdotos in tropical America. J Paleontol 75:564–577
Chesher RH (1972) The status of knowledge of Panamanian echinoids, 1971, with comments on other echinoderms. Bull Biol Soc Wash 2:139–157
Clark AH (1939) Echinoderms (other than *Holothurian*s) collected on the presidential cruise of 1938. Smithson Miscell Collec 98:1–22
Clark AH (1946) Echinoderms from the Pearl Islands, Bay of Panama with a revision of the Pacific species of the genus *Encope*. Smithson Miscell Collec 106:1–11
Clark HL (1917) *Ophiuroidea*. Report XVIII and XXX on the scientific results of the expedition of the "*Albatross*" to the tropical Pacific 1899–1900 and 1904–1905. Bull Mus Comp Zool Harvard Univ 61:429–453
Clark HL (1940) Eastern Pacific expeditions of the New York Zoological Society. XXI. Notes on echinoderms from the West Coast of Central America. Zoology 25:331–352
Clark HL (1948) A report on the *Echini* of the warmer eastern Pacific, based on the collections of the Velero III. Allan Hancock Pac Exp 8:225–352
Coates AG, McNeill DF, Aubry MP, Berggren WA, Collins LS (2005) An introduction to the geology of the Bocas del Toro archipelago, Panama. Carib J Sci 41:374–391
Chesher RH (1968) The systematics of sympatric species in West Indian spatangoids. Studies Tropic Oceanogr 7:1−168, pls 1–35
Chesher RH (1970) Evolution in the genus *Meoma* (Echinoidea: Spatangoida) and a description of a new species from Panama. Bull Marine Sci 20:731–761
Clark HL (1925) A catalogue of the recent sea urchins (Echinoidea) in the collection of the British Museum (Natural History). Oxford Univ. Press, London. 250 p
Collin R, D'Croz L, Gondola P, Del Rosario JB (2009) Climate and hydrological factors affecting variation in chlorophyll concentration and water clarity in the Bahia Almirante, Panama. Smith Contr Mar Sci 38:323–334
Collins LS (1999) The miocene to recent diversity of Caribbean benthic foraminifera from the Central American Isthmus. Bull Am Paleontol 357:91–107
Coppard SE (2010) The Echinoderms of Panama. Accessed online at http://echinoderms.lifedesks.org/ on 30 May 2011
Cortés J (1993) Comparison between Caribbean and eastern Pacific coral reefs. Rev Biol Trop 41:19–21

Cruz LG (2000) Campaña de educación ambiental para la conservación y protección del pepino de mar (Echinodermata: Holothuroidea) en 15 comunidades del Archipiélago de Bocas del Toro y Península Valiente. MSc Thesis, Universidad Santa María La Antigüa, Panama

Cubit JD, Windsor DM, Thompson RC, Burgett JM (1986) Water-level fluctuations, emersion regimes, and variations of echinoid populations on a Caribbean reef flat. Estuar Coast Shelf Sci 22:719–737

D'Croz L, Robertson R (1997) Coastal oceanographic conditions affecting coral reefs on both sides of the Isthmus of Panama. Proceedings of 8th international coral reef symposium, Panama 2:2053–2058

D'Croz L, Del Rosario JB, Gómez JA (1991) Upwelling and phytoplankton in the Bay of Panama. Rev Biol Trop 39:233–241

D'Croz L, Del Rosario JB, Góndola P (2005) The effect of fresh water runoff on the distribution and dissolved inorganic nutrients and plankton in the Bocas del Toro Archipelago, Caribbean Panama. Carib J Sci 41:414–429

Deichmann E (1941) The *Holothurioidea* collected by the Velero III during the years 1932–1938. Part I. Dendrochirota. Allan Hancock Pac Exp 8:61–195

Deichmann E (1958) The *Holothurioidea* collected by theVelero III and IV during the years 1932 to 1954. Part II Aspidochirota. Allan Hancock Pac Exp 11:253–349

Dexter D (1977) A natural history of the sand dollar *Encope stokesi* L Agassiz in Panama. Bull Mar Sci 27:544–551

Döderlein L (1917) Die Asteriden der Siboga-Expedition. I. Die Gattung Astropecten und ihre Stammesgeschichte. In:Brill EJ (ed) *Siboga-Expeditie. Uitkomsten op zoölogisch, botanisch, ozeanographisch en geologisch gebied verzameld in Nederlandsch Oost-Indie 1899 – 1900 aan boord H.M. "Siboga"*, 46 (a), Leiden, pp 191

Eakin CM (1991)The damselfish-algal lawn symbiosis and its influenced on the bioerosion of an El Niño impacted coral reef, Uva Island, pacific Panama. PhD dissertation, University of Miami, Florida

Eakin CM (1992) Post-El Niño Panamanian reefs: less accretion, more erosion and damselfish protection. Proceedings of 7th International Coral Reef Symposium, Guam 1:387–396

Eakin CM (2001) A tale of two ENSO events: carbonate budgets and the influence of two warming disturbances and intervening variability, Uva Island, Panama. Bull Mar Sci 69:171–186

Fong P, Glynn PW (1998) A dynamic size-structured population model: does disturbance control size structure of a population of the massive coral *Gardineroseris planulata* in the Eastern Pacific? Mar Biol 130:663–674

Forsberg ED (1963) Some relationships of meteorological, hydrographic and biological variables in the Gulf of Panama. Bull Inter Amer Trop Tuna Comm 7:1109

Forsbergh ED (1969) On the climatology, oceanography and fisheries of the Panama Bight. Bull Inter Amer Trop Tuna Comm 14:49–385

Foster SA (1987a) The relative impacts of grazing by Caribbean coral reef fishes and *Diadema*: effects of habitat and surge. J Exp Mar Biol Eco 105:1–20

Foster SA (1987b) Territoriality of the dusky damselfish: influence on algal biomass and on the relative impacts of grazing by fishes and *Diadema*. Oikos 50:153–160

Geyer LB, Lessios HA (2009) Lack of character displacement on the male recognition molecule, bindin, in Atlantic sea urchins of the genus *Echinometra*. Mol BiolEvol 26:2135–2146

Glynn PW (1968) Mass mortalities of echinoids and other reef flat organisms coincident with midday, low water exposures in Puerto Rico. Mar Biol 1:226–243

Glynn PW (1972) Observations on the ecology of the Caribbean and Pacific coasts of Panama. Bull Biol Soc Wash 2:13–30

Glynn PW (1973) *Acanthaster*: effect on coral reef growth in Panama. Science 180:504–506

Glynn PW (1974) The impact of *Acanthaster* on corals and coral reefs in the eastern Pacific. Environ Conser 1:295–304

Glynn PW (1976) Some physical and biological determinants of coral community structure in the Eastern Pacific. Ecol Monogr 46:431–456

Glynn PW (1977) Interaccions between *Acanthaster* and *Himenoceras* in the field and laboraroty. Proceedings of 3rd international coral reef symposium, Florida 2: 209–215

Glynn PW (1981) *Acanthaster* population regulation by a shrimp and a worm. Proceedings of 5th international coral reef symposium, Manila 2:607–612

Glynn PW (1982) Indiviual recognition and phenotypic variability in *Acanthaster planci* (Echinodermata: *Asteroidea*). Coral Reefs 1:89–94

Glynn PW (1983) Crustacean symbionts and the defense of corals: coevolution on the reef. In: Nitecki MH (ed) Coevolution. University Chicago Press, Chicago, pp 111–178

Glynn PW (1984) An amphinomid worm predator of the crown-of-thorns sea star and general predation on asteroids in Eastern and Western Pacific corals. Bull Mar Sci 35:54–71

Glynn PW (1985a) El Niño associated disturbance to coral reefs and post disturbance mortality by *Acanthaster planci*. Mar Ecol Prog Ser 26:295–300

Glynn PW (1985b) Corallivore population size and feeding effects following El Niño (1982–83) associated coral mortality in Panama. Proceedings of 5th international coral reef symposium, Tahiti 4:183–188

Glynn PW (1988) El Niño warming, coral mortality and reef framework destruction by echinoid bioerosion in the eastern Pacific. Galaxea 7:129–160

Glynn PW (1990) Coral mortality and disturbances to coral reef in the Tropical Eastern Pacific. In: Glynn PW (ed) Global Ecological Consequences of the 1982–83 El Niño-Southern Oscillation. Elsevier, Amsterdam, pp 55–126

Glynn PW (1997) Assessment of the present health of coral reefs in the eastern Pacific. In: Grigg RW, Birkeland C (eds) Status of coral reefs in the Pacific. Univ Hawaii Sea Grant Coll Progr, Hawaii, pp 31–40

Gómez JA, Villalaz JR, D'Croz L (2005) Panama. In: Miloslavich P, Klein E (eds) Caribbean Marine Biodiversity: the Known and the Unknown. DEStech Publications, Lancaster, pp 157–168

Gonzalez P, Lessios HA (1999) Molecular evolution of the sea urchin retroviral-like (SURL) family of transposable elements. Mol Biol Evol 16:938–952

Gordon AL (1967) Circulation of the Caribbean sea. J Geophys Res 72:6207–6223

Guzman HM, Cortes J (2007) Reef recovery 20 years after the 1982–1983 El Niño massive mortality. Mar Biol 151:401–411

Guzman HM, Guevara CA (1998) Arrecifes coralinos de Bocas del Toro, Panama. I. Distribución, estructura y estado de conservación de los arrecifes continentales de la Laguna de Chiriquí y la Bahía Almirante. Rev Biol Trop 46:601–622

Guzman HM, Guevara CA (2001) Arrecifes coralinos de Bocas del Toro, Panamá: IV. Distribución, estructura y estado de conservación de los arrecifes continentales de Península Valiente. Rev Biol Trop 49:53–66

Guzman HM, Guevara CA (2002a) Annual reproductive cycle, spatial distribution, abundance, and size structure of *Oreaster reticulatus* (Echinodermata: *Asteroidea*) in Bocas del Toro, Panama. Mar Biol 141:1077–1084

Guzman HM, Guevara CA (2002b) Population structure, distribution and abundance of three commercial species of sea cucumber (Echinodermata) in Panama. Carib J Sci 38:230–238

Guzman HM, Guevara CA, Hernandez IC (2003) Reproductive cycle of two commercial species of sea cucumber (Echinodermata: *Holothuroidea*) from Caribbean Panama. Mar Biol 142: 271–279

Guzman HM, Benfield S, Breedy O, Mair JM (2008) Broadening reef protection acroos the marine conservation corridor of the Eastern Tropical Pacific: distribution and diversity of reefs in Las Perlas Archipelago, Panama. Environ Conser 35:46–54

Hansson H (2011) *Havelockia inermis* (Heller, 1868). Accessed. World Register of Marine Species at http://www.marinespecies.org/aphia.php?p=taxdetails&id=124634 on 2011-06-04

Harborne AR, Renaud PG, Tyler EHM, Mumby PJ (2009) Reduced density of the herbivorous urchin *Diadema antillarum* inside a Caribbean marine reserve linked to increased predation pressure by fishes. Coral Reefs 28:783–791

Hay ME, Colbum T, Downing D (1983) Spatial and temporal patterns in herbivory on a Caribbean fringing reef: the effects on plant distribution. oecologia 58:299–308
Hay ME (1984) Patterns of fish and urchin grazing on Caribbean coral reefs: are previous results typical? Ecology 65: 446–454
Hay ME, Fenical W, Gustafsson K (1987) Chemical defense against diverse coral reef herbivores. Ecology 68:1581–1591
Heck KLJr (1977) Comparative species richness, composition, and abundance of invertebrates in Caribbean seagraa (*Thalassia testudinum*) meadows (Panamá). Mar Biol 41:335–348
Hendler G (1977) The differential effects of seasonal stress and predation on the stability of reef-flat echinoid populations. Proceedings of 3rd international Coral Reef Symposium, Miami 1:217–223
Hendler G (1995) New species of brittle stars from the western Atlantic, Ophionereis vittata, Amphioplus sepultus, and Ophiostigma siva, and the designation of a neotype for Ophiostigma isocanthum (Say) (Echinodermata: Ophiuroidea). Contrib Sci, Natural History Museum of Los Angeles County 458:1–19
Hendler G (2005) Two new brittle star species of the genus *Ophiothrix* (Echinodermata: *Ophiuroidea*: *Ophiotrichidae*) from coral reefs in the Southern Caribbean Sea, with notes on their biology. Carib J Sci 41:583–599
Hendler G, Meyer DL (1982) An association of a polychaete, *Branchiosyllis exilis* with an ophiuroid, *Ophiocoma echinata*, in Panama. Bull Mar Sci 32:736–744
Hendler G, Byrne M (1987) Fine structure of the dorsal arm plate of *Ophiocoma wendti*: Evidence for a photoreceptor system (Echinodermata, *Ophiuroidea*). Zoomorphology 107:261–272
Hendler G, Grygier MJ, Maldonado E, Denton J (1999) Babysitting brittle stars: heterospecific symbiosis between ophiuroids (*Echinodermata*). Invert Biol 118:190–201
Jackson JBC, Cubit JD, Keller BD, Batista V, Burns K, Caffey HM, Caldwell RL, Garrity SD, Getter CD, Gonzalez C, Guzman HM, Kaufmann KW, Knap AH, Levings SC, Marshall MJ, Sterger R, Thompson RC, Weil E (1989) Ecological effects of a major oil spill on Panamanian coastal marine communities. Sci 243:37–44
Jackson JBC, Jung P, Coates AG, Collins LS (1993) Diversity and extinction of Tropical American mollusks and emergence of the Isthmus of Panama. Sci 260:1624–1626
Jackson JBC, Jung P, Fortunato H (1996) Paciphillia revisited: transisthmian evolution of the Strombina Group (Gastropoda: columbellidae). In: Jackson JBC, Budd AF, Coates AG (eds) Evolution and environment in tropical America. Univ Chicago Press, Chicago, pp 234–270
Jordan DS (1908) The law of the geminate species. Am Nat 42:73–80
Kinder TH, Heburn GW, Green AW (1985) Some aspects of the Caribbean circulation. Mar Geol 68:25–52
Lawrence JM, Glynn PW (1984) Absorption of nutrients from the coral *Pocillopora damicornis* (L.) by the echinoid *Eucidaris thouarsii* (VAL.). Comp Biochem Physiol A 77:111–112
Lessios HA (1979) Use of Panamanian sea urchins to test the molecular clock. Nature 280:599–601
Lessios HA (1981a) Reproductive periodicity of the echinoid *Diadema* and *Echinometra* on the two coasts of Panama. J Exp Mar Biol Ecol 50:47–61
Lessios HA (1981b) Divergence in allopatry: molecular and morphological differentiation between sea urchins separated by the Isthmus of Panama. Evolution 35:618–634
Lessios HA (1984) Possible prezygotic reproductive isolation in sea urchins separated by the Isthmus of Panama. Evol 38:1144–1148
Lessios HA (1985a) Annual reproductive periodicity in eight echinoid species on the Caribbean coast of Panama. In: Keegan B,F O'Connor BDS (eds) Echinodermata. Proceedings of 5th international Echinoderm Conference, Balkema AA, Rotterdam, pp 303–312
Lessios HA (1985) Genetic consequences of mass mortality in the Caribbean sea urchin *Diadema antillarum*. Proceedings of 5th International Coral Reef Symposium, Tahiti, 4:119–126
Lessios HA (1987) Temporal and spatial variation in egg size of 13 Panamanian echinoids. J Exp Mar Biol Ecol 114:217–239

Lessios HA (1988) Population dynamics of *Diadema antillarum* (Echinodermata: *Echinoidea*) following mass mortality in Panama. Mar Biol 99:515–526

Lessios HA (1990) Adaptation and phylogeny as determinants of egg size in echinoderms from the two sides of the Isthmus of Panama. Am Nat 135:1–13

Lessios HA (1991) Presence and absence of monthly reproductive rhythms among eight Caribbean echinoids off the coast of Panama. J Exp Mar Biol Ecol 153:27–47

Lessios HA (1995) *Diadema antillarum* 10 years after mass mortality: still rare, despite help from a competitor. Proc R Soc Lond B 259:331–337

Lessios HA (1998) The first stage of speciation as seen in organisms separated by the Isthmus of Panama. In: Howard D, Berlocher S (eds) Endless forms: Species and Speciation. Oxford University Press, Oxford, pp 186–201

Lessios HA (2001) Molecular phylogeny of *Diadema*: systematic implications.In: Barker M (ed)Echinoderms 2000: Proceedings of 10th international Echinoderm Conference Swets and Zeitinger, Lisse 487–495

Lessios HA (2005a) Echinoids of the Pacific waters of Panama: status of knowledge and new records. Rev Biol Trop 53(Suppl 3):147–170

Lessios HA (2005b) *Diadema antillarum* populations in Panama twenty years following mass mortality. Coral Reefs 24:125–127

Lessios HA (2007) Reproductive isolation between species of sea urchins. Bull Mar Sci 81:191–208

Lessios HA (2008) The Great American Schism: divergence of marine organisms after the rise of the Central American Isthmus. An Rev Ecol Evol Syst 39:63–91

Lessios HA (2010) Speciation in sea urchins. In: Harris LG, Böttger SA, Walker CW, Lesser MP (eds) Echinoderms: Durham. Proceedimngs of 12th international echinoderm conference. CRC Press, Taylor and Francis group, Balkema, pp 91–101

Lessios HA, Robertson DR, Cubit JD (1984a) Spread of *Diadema* mass mortality through the Caribbean. Science 226:335–337

Lessios HA, Cubit JD, Robertson DR, Shulman MJ, Parker MR, Garrity SD, Levings SC (1984b) Mass mortality of *Diadema antillarum* on the Caribbean coast of Panama. Coral Reefs 3:173–182

Lessios HA, Cunningham CW (1993) The evolution of gametic incompatibility in neotropical *Echinometra*: a reply to McClary. Evol 47:1883–1885

Lessios HA, Kessing BD, Pearse JS (2001) Population structure and speciation in tropical seas: global phylogeography of the sea urchin *Diadema*. Evolution 55:955–975

Lessios HA, Kane J, Robertson DR (2003) Phylogeography of the pantropical sea urchin *Tripneustes*: contrasting patterns of population structure between oceans. Evolution 57:2026–2036

Ludwig H (1894) Reports on an exploration off the west coast of Mexico, Central America and South America, and off the Galápagos Islands, in charge of Alexander Agassiz, by the U.S. Fish Commission steamer "*Albatross*", during 1891. XII. The Holothurioidea. Mem Mus Comp Zool XVII:1–183

Ludwig H (1905) Asteroidea in: explorations of "*Albatross*" in Tropical Pacific, 1891 and 1899–1900. Mem Mus Comp Zool XXII:1–290

Lütken CF, Mortensen T (1899) Reports on an exploration off the west coast of Mexico, Central America and South America, and off the Galápagos Islands, in charge of A. Agassiz, during 1891 on the "*Albatross*". XXV. The Ophiuridae. Mem Mus Comp Zool XXIII:93–208

Maluf LY (1988) Composition and distribution of the Central eastern Pacific echinoderms. Nat Hist Mus LA County Tech Rep 2:1–242

Martin WE, Duke A, Bloom SG, McGinnis JT (1970) Possible effects of a sealevel canal on the marine ecology of the American Isthmian Region. Bioenvironmental and radiological-safety feasibility studies, Atlantic-Pacific interoceanic canal. Battelle Mem Inst, Columbus, Ohio

McAlister JS (2008) Evolutionary responses to environmental heterogeneity in Central American echinoid larvae: plastic versus constant phenotypes. Evolution 62:1358–1372

McCartney MA, Lessios HA (2002) A quantitative analysis of gamete incompatibility between closely related species of neotropical sea urchins. Biol Bull 202:166–181

McCartney MA, Lessios HA (2004) Adaptive evolution of sperm bindin tracks egg incompatibility in neotropical sea urchins of the genus *Echinometra*. Mol Biol Evol 21: 732–745

McCartney MA, Keller G, Lessios HA (2000) Dispersal barriers in tropical oceans and speciation in Atlantic and eastern Pacific sea urchins of the genus *Echinometra*. Mol Ecol 9:1391–1400

McClintock JB, Marion KR (1993) Predation by the king helmet (Cassis tuberose) on six-holed sand dollars (*Leodia sexiesperforata*) at San Salvador, Bahamas. Bull Mar Sci 52:1013–1017

Meditz SW, Hanratty DM (1987) Panama: a country study. GPO for the Library of Congress, Washington

Messing C (2010) *Davidaster discoideus* (Carpenter, 1888). In: Messing C (ed) World List of the Crinoidea. Accessed. World Register of Marine Species at http://www.marinespecies.org/aphia.php?p=taxdetails&id=246783 on 2011-06-04

Metz EC, Palumbi SR (1996) Positive selection and sequence rearrangements generate extensive polymorphism in the gamete recognition protein bindin. Mol Biol Evol 13:397–406

Meyer DL (1973) Feeding behavior and ecology of shallow-water unstalked crinoids (Echinodermata) in the Caribbean Sea. Marine Biology, 22:105–130

Miller RJ, Adams AJ, Ogden NB, Ebersole JP (2003) *Diadema antillarum* 17 years after mass mortality: is recovery beginning on St. Croix? Coral Reefs 22:181–187

Mortensen T (1921) Studies of the development and larval forms of echinoderms. CA Reitzel, Copenhagen

Mortensen T (1928) A Monograph of the *Echinoidea I. Cidaroidea*. CA Reitzel, Copenhagen

Mortensen T (1935) A Monograph of the *Echinoidea. II. Bothriocidaroidea, Melonechinoidea, Lepidocentrotida* and *Stirodonta*. CA Rietzel, Copenhagen

Mortensen T (1940) A Monograph of the *Echinoidea*. III. 1. Aulodonta. CA Rietzel, Copenhagen

Mortensen T (1943a) A Monograph of the *Echinoidea*. III. 2. Camarodonta. I. Orthopsidæ,Glyphocyphidæ, Temnopleuridæ and Toxopneustidæ. CA Rietzel, Copenhagen

Mortensen T (1943b) A Monograph of the *Echinoidea*. III. 3. Camarodonta. II. Echinidæ, Strongylocentrotidæ, Parasaleniidæ, Echinometridæ. CA Rietzel, Copenhagen

Mortensen T (1948a) A Monograph of the *Echinoidea*. IV. 1. Holectypoida, Cassiduloida. CA Rietzel, Copenhagen

Mortensen T (1948b) A Monograph of the *Echinoidea*. IV. 1. Holectypoida, Cassiduloida. CA Rietzel, Copenhagen

Mortensen T (1950) A Monograph of the *Echinoidea*, vol 1. Spatangoida, CA Rietzel, Copenhagen

Mortensen T (1951) A Monograph of the *Echinoidea*, vol 2. Spatangoida, CA Rietzel, Copenhagen

Nagelkerken I, Smith GW, Snelders E, Karel M, James S (1999) Sea urchin *Meoma ventricosa* die-off in Curaçao (Netherlands Antilles) associated with a pathogenic bacterium. Dis Aquat Org 38:71–74

O'Dea A, Herrera-Cubilla A, Fortunato H, Jackson JBC (2004) Life history variation in cupuladriid bryozoans from either side of the Isthmus of Panama. Mar Ecol Progr Ser 280:145–161

Pennington JT, Mahoney KL, Kuwahara VS, Kolber DD, Calienes R, Chavez FP (2006) Primary production in the eastern tropical Pacific: a review. Prog Oceanogr 69:285–317

Pillans B, Chappell J, Naish TR (1998) A review of the Milankovitch climatic beat: template for Plio-Peistocene sea-level changes and sequence stratigraphy. Sediment Geol 122:5–21

Rocha LA, Robertson DR, Rocha CR, Van Tassell JL, Craig MT, Bowen BW (2005) Recent invasion of the tropical Atlantic by an Indo-Pacific coral reef fish. Mol Ecol 14:3921–3928

Roy MS, Sponer R (2002) Evidence of a human-mediated invasion of the tropical western Atlantic by the 'world's most common brittlestar'. Proc R Soc London Ser B 269:1017–1023

Seilacher A (1979) Constructional morphology of sand dollars. Paleobiology 5:191–221

Sonnenholzner J, Lawrence JM (1998) Disease and predation in *Encope micropora* (Echinoidea: Clypeasteroida) at Playas, Ecuador. In: Mooi R, Telford M (eds) Echinoderms: San Francisco. Proceedings of9th international echinoderm conference, San Francisco. AA Balkema, Rotterdam, pp 829–833

Stöhr S (2010) *Ophiophragmus riisei* (Lütken in: Lyman 1860). In: Stöhr S, O'Hara T (eds) World Ophiuroidea database. Accessed. World Register of Marine Species at http://www.marinespecies.org/aphia.php?p=taxdetails&id=246617 on 2011-06-04

Telford M (1982) Echinoderm spine structure, feeding and host relationships of four species of *Dissodactylus* (Brachyura: Pinnotheridae). Bull Mar Sci 32:584–594

Toral-GrandaV (2008) Population status, fisheries and trade of sea cucumbers in Latin America and the Caribbean. In: Toral-Granda V, Lovatelli A, Vasconcellos M (eds) Sea cucumbers. A global review of fisheries and trade. FAO Fisheries and Aquaculture Technical Paper. No. 516. Rome, pp 213–229

Van Soest RWM (1994) Demosponge distribution patterns. In: Van Soest RWM, Kempen TMG, Van Braekman JC (eds) Sponges in Space and Time. Biology, Chemistry, Paleontology. Proceedings of 4th international Porifera Congress, Amsterdam. AA Balkema, Rotterdam, pp 213–223

Vermeij GJ (1978) Biogeography And Adaptation: Patterns Of Marine Life. Harvard University Press, Cambridge

Verrill AE (1867) Notes on the echinoderms of Panama and west coast of America, with description of new genera and species. Trans Connect Acad Arts Sci 1:251–322

Vogler C, Benzie J, Lessios HA, Barber P, Wörheide G (2008) A threat to coral reefs multiplied? Four species of crown-of-thorns starfish. Biol Lett 4:696–699

Waters JM, O'Loughlin PM, Roy MS (2004) Cladogenesis in a starfish species complex from southern Australia: evidence for vicariant speciation? Mol Phylogen Evol 32:236–245

Wilson AC, Carlson SS, White TJ (1977) Biochemical evolution. An RevBiochem 46:573–639

Wulff JL (1995) Sponge-feeding by the Caribbean starfish *Oreaster reticulatus*. Mar Biol 123:313–325

Wulff JL (2006) Sponge systematics by starfish: predators distinguish cryptic sympatric species of Caribbean fire sponges, *Tedania ignis* and *Tedania klausi* n. sp. (Demospongiae, Poecilosclerida). Biol Bull 211:83–94

Ziesenhenne FC (1940) New ophiurans of the allan hancock Pacific expeditions. Allan Hancock Pac Exp 8:9–59

Ziesenhenne FC (1942) New eastern Pacific sea stars. Allan Hancock Pac Exp 8:197–223

Ziesenhenne FC (1955) A review of the genus *Ophioderma*. Essays in the Natural Sciences in Honor of Captain Allan Hancock on the Occasion of his Birthday, July 26, University of Southern California Press, Los Angeles, 1955

Zigler KS, Lessios HA (2003a) Evolution of bindin in the pantropical sea urchin Tripneustes: comparisons to bindin of other genera. Mol Biol Evol 20:220–231

Zigler KS, Lessios HA (2003b) 250 million years of bindin evolution. Biol Bull 205:8–15

Zigler KS, Lessios HA (2004) Speciation on the coasts of the new world: phylogeography and the evolution of bindin in the sea urchin genus *Lytechinus*. Evolution 58:1225–1241

Zigler KS, Raff E, Popodi E, Raff R, Lessios HA (2003) Adaptive evolution of bindin in the genus Heliocidaris is correlated with the shift to direct development. Evolution 57:2293–2302

Zigler KS, McCartney MA, Levitan DR, Lessios HA (2005) Sea urchin bindin divergence predicts gamete compatibility. Evolution 59:2399–2404

Zigler KS, Lessios HA, Raff RA (2008) Egg energetics, fertilization kinetics, and population structure in echinoids with facultatively feeding larvae. Biol Bull 215:191–199

Zulliger DE, Lessios HA (2010) Phylogenetic relationships in the genus *Astropecten* (Paxillosida: Asteroidea) on a global scale: molecular evidence for morphological convergence, occurrence of species-complexes and possible cryptic speciation. Zootaxa 2504:1–19

Chapter 5
Echinoderms of Colombia

Milena Benavides-Serrato, Giomar Helena Borrero-Pérez, Jaime R. Cantera K, Mishal Cohen-Rengifo and Raúl Neira

5.1 Introduction

Colombia, together with Indonesia, Brazil, and Mexico, shares a privileged position at the top of the list of countries with the highest species richness in the world, the so called megadiverse countries (IAvH 1998a, b). However, this reputation is based upon the relatively well documented diversity of terrestrial biota such as orchids, amphibians, birds, mammals, and butterflies (Andrade et al. 1992). Biodiversity in Colombia is related to the high diversity of habitats that occur on the mainland. Close to half of the country is covered by the Amazon River system and the Orinoco savannas. The Andes Mountains are separated into three ranges, and drainage from the Magdalena and Cauca River basins forms the wetlands of the Caribbean and feeds the tropical dry forest ecosystems.

M. Benavides-Serrato (✉) · G. H. Borrero-Pérez
Instituto de Investigaciones Marinas y Costeras—INVEMAR, A.A 1016,
Santa Marta, Colombia
e-mail: milena_benavides@invemar.org.co

G. H. Borrero Pérez
e-mail: giomar_borrero@invemar.org.co

J. R. Cantera K · R. Neira
Departamento de Biología, Universidad del Valle, A.A. 25360, Cali, Colombia
e-mail: jaime.cantera@correounivalle.edu.co

R. Neira
e-mail: rneira56@yahoo.com

M. Cohen-Rengifo
Centre d'Oceanologie de Marseille, Université de la Méditerranée,
Marseille Cedex, France
e-mail: ablocsberg@gmail.com

J. J. Alvarado and F. A. Solís-Marín (eds.), *Echinoderm Research and Diversity in Latin America*, DOI: 10.1007/978-3-642-20051-9_5,

Fig. 5.1 Location of the land and sea areas of Colombia (System Information Laboratory LABSIS-INVEMAR)

The watershed of the western Andes and the Pacific lowlands is an extremely rich region, isolated for many years, constituting the Chocó biodiversity hotspot. The Sierra Nevada de Santa Marta, the highest coastal mountain in the world, is an isolated mountain in the Caribbean, rising 5,500 m above the hot deserts of the Guajira Peninsula (Díaz and Acero 2003).

However, with nearly 1,000,000 km^2 of territorial waters, nearly half the territory of Colombia is maritime. It is the only South American country with coasts on both the tropical Pacific Ocean and the Caribbean Sea, with a total shoreline of nearly 3,000 km. Because Colombia possesses oceanic islands located far offshore, wide sections of both the tropical eastern Pacific and the southern-central Caribbean are included within the boundaries of its Economic Exclusive Zone (Fig. 5.1).

There are remarkable differences between the two coasts in terms of climatic, geological, oceanographic, ecological, and biological features (Botero and Alvarez-León 2000; Marrugo-González et al. 2000). In addition, coastal morphology, climate, and physical oceanographic processes vary greatly along spatial gradients as well as seasonally within both the Caribbean and the Pacific. These features allow the great majority of tropical marine habitats to be well represented in Colombia. Considering that each habitat harbors its own particular biota, the total diversity of marine organisms in such a variety of environments is likely to be very high. Their study represents a huge challenge for future biological research.

The information presented in this chapter is derived from a compilation and review of the knowledge of Colombian echinoderms, based on diverse sources of information such as published echinoderm studies, undergraduate thesis, technical reports and extracted information from databases of museums as National Museum of Natural History (NMNH) and other databases available such as Biodiversity Marine Information System (SIBM) of Colombia, Ocean Biogeographic Information System (www.iobis.org) and Inter-American Biodiversity Information Network (www.iabin.net). This chapter represents a baseline study of echinoderms in Colombia, useful for continuing and developing further research. Because of the physical and biological differences for the Caribbean and Pacific coasts of Colombia, as well as historical differences in the research carried out on these two regions, the information is presented separately for the two coasts.

5.1.1 Colombian Caribbean Sea

The Colombian Caribbean coast extends from 8°40′N and 77°21′W at the Panama-Colombia border, northeastwards to 11°55′N and 71°18′W at the Colombia-Venezuela border (Fig. 5.2). San Andrés and Providencia archipelago (between 12°–15° N and 79°–82° W) are two high-standing oceanic islands and a series of atolls and coral banks which are aligned in a NNE direction along the lower Nicaragua Rise. The total coastline of the Colombian Caribbean is 1,937 km, with a terrestrial area of 6,958 km^2, insular area of 79 km^2, and territorial sea of 532,162 km^2. It includes eight departments: Guajira, Magdalena, Atlántico, Bolivar, Sucre, Cordoba, Antioquia and San Andres, Providencia and Santa Catalina (INVEMAR 2010).

There is a remarkable climatic gradient along the Caribbean mainland coast. The northernmost part of the Guajira peninsula is extremely dry and hot and the land vegetation is very sparse. In contrast, annual precipitation in the area around the gulf of Urabá, close to the Panama boundary is continue and is over 3,000 mm $year^{-1}$, and coastal plains and mountains there are covered by dense tropical rain forest. The Urabá region represents a northern outpost of the Chocó diversity hot spot for the terrestrial biota (Gentry 1986).

On the other hand, geomorphologic, physical oceanographic and ecological processes vary greatly along spatial gradients as well as seasonally within the

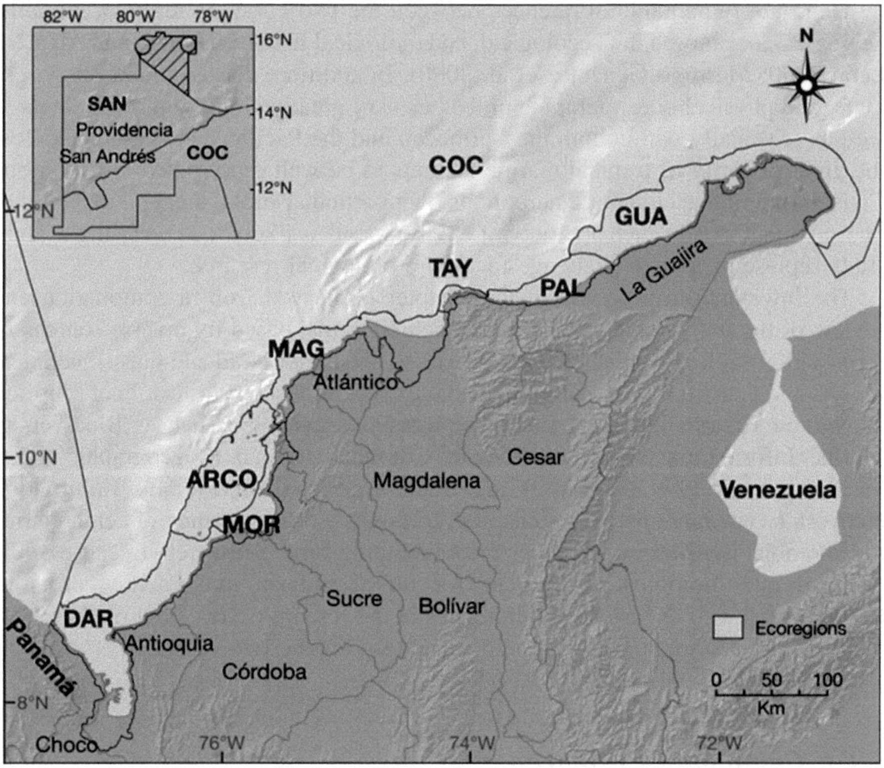

Fig. 5.2 Ecoregions and coastal departments of the Colombian Caribbean Sea: GUA: Guajira, PAL: Palomino, TAY: Tayrona, MAG: Magdalena, ARCO: Coralline Archipelagos, MOR: Morrosquillo, DAR: Darién, SAN: San Andrés and Providencia Archipelago, COC: Oceanic Caribbean. (System Information Laboratory LABSIS-INVEMAR)

Colombian Caribbean. This results in an irregular or heterogeneous distribution of major habitats and ecosystems in the region. Upwelling in La Guajira peninsula promotes the abundance of marine vegetation (seagrass and algal beds) but limits the development of coral reefs. The water conditions of a large section of the middle part of the coast are influenced by heavy discharge of freshwaters and sediments by major rivers (Magdalena, Canal del Dique, Sinú) which form alluvial plains, and deltaic formations with coastal lagoons. Estuarine conditions prevail in these lagoons and mangrove forests are extensive. In contrast, warm waters and high isolation around the offshore islands on the continental shelf and at the oceanic archipelago of San Andrés and Providencia foster the development of coral reefs. Finally, the northern ranges of the Sierra Nevada de Santa Marta sink steeply into the sea. This result in an irregular shoreline with cliffs and pocket beaches creating a diverse mosaic of rocky and sandy shores, fringing reefs, and seagrass meadows (INVEMAR 2000).

Insular Formations of Colombian Caribbean Sea

Two main island groups are present in the Colombian Caribbean Sea, the archipelago of San Andrés, Providencia, Santa Catalina and its cays, and the archipelago of Rosario, San Bernardo and Isla Fuerte. The former was declared a World Biosphere Reserve in the year 2000 and is characterized by volcanic elevations covered by carbonate platforms produced by the reef biota. These emerge to form islands and cays, surrounded by coral reefs and sedimentary plains, vegetated by algae or seagrasses that form a shallow island shelf (down to 40 m deep), surrounded by seafloor exceeding 600 m in depth. The second archipelago is part of the Marine Protected Area Corales del Rosario, San Bernardo and Isla Fuerte, the only marine protected area of Colombia, enclosing the most developed coral reefs of the Colombian north coast and a high biodiversity of other ecosystems (INVEMAR 2010).

Caribbean Ecoregions

The Colombian Caribbean Sea has been divided into nine ecological regions according to natural characteristics like geomorphology, hydrography, sedimentology, and ecosystems. These will be used to present the diversity and distribution of echinoderms. They are: Guajira (GUA), Palomino (PAL), Tayrona (TAY), Magdalena (MAG), coralline archipelagos (ARCO), Morrosquillo (MOR), Darién (DAR), San Andrés and Providencia Archipelago (SAN) and Oceanic Caribbean (COC) (Fig. 5.2). Six of these regions are coastal and limited by the 200 m isobath that defines the continental shelf platform. SAN is an insular sector that comprises the atolls, coral reefs and islands of the San Andres and Providencia archipelago. The Oceanic Caribbean sector (COC) includes the continental slope extending from 200 to 2,700 m in the Colombian Basin (Tabares et al. 1996; INVEMAR 2000). Table 5.1 summarizes the main physical, oceanographic and ecological features of these ecoregions.

5.1.2 Colombian Pacific Ocean

The Pacific coast of Colombia is located in the northwest region of South America, between 7°12′N–77°53′W and 1°26′N–78°49′W, bordering Panama in the north and Ecuador in the south. Four departments are in this area: Chocó, Valle del Cauca, Cauca and Nariño, and includes the islands of Gorgona, Gorgonilla and Malpelo. The coastline extends for 1,576 km. The departments have a terrestrial area of 8,181 km^2, an insular area of 14 km^2, and a territorial sea area of 359,995 km^2 (Steer et al. 1997; INVEMAR 2002) (Fig. 5.3).

The Pacific coast of Colombia can be divided into two areas based on physiographic and hydrographic characteristics. The northern coast, from Panama to

Table 5.1 Characteristics of Colombian Caribbean and Pacific ecoregions (Modified from Díaz and Acero 2003)

Ecoregion	Climate and oceanography	Coastal, shelf and seafloor morphology	Main ecosystems represented
Caribbean			
Guajira (GUA)	Highly influenced by strong winds, desert. Seasonal upwelling causing high primary and secondary marine production	Low coastal plains with scattered rocky cliffs. Wide continental shelf	Seagrass beds, sandy beaches, salt marshes, scattered fringing mangrove forests, sandy bottoms
Palomino (PAL)	Dry to semi humid; heavy coastal swell; several small rivers	Highly exposed coast dominated by coarse sand beaches. Continental shelf narrow	Beaches, coastal lagoons and swamps, sandy bottoms
Tayrona (TAY)	Semi-desert to dry; seasonal upwelling	High coastal mountains form a series of small bays and inlets with pocket beaches and cliffs. Continental shelf very narrow or absent	Rocky shores, coral reefs, seagrass beds, beaches, rocky and sandy bottoms
Magdalena (MAG)	Dry. Coastal waters highly influenced by discharges of the Magdalena drainage system with heavy sediment loads (turbid waters)	Coastal morphology very heterogeneous and dynamic. Continental shelf rather narrow made of terrigenous sand and silts	Coastal lagoons and estuaries, mangrove forests, sandy beaches, muddy bottoms
Coralline-Archipelagos (ARCO)	Dry climate. Transparent, oceanic waters alternate seasonally with turbid waters from continental discharges	Offshore shallow carbonate platforms and coral shoals, scattered coralline island cays	Coral reefs, seagrass beds, bioclastic sandy bottoms
Gulf of Morrosquillo (MOR)	Dry to semi-humid. Coastal waters influenced by continental discharges (mostly turbid waters)	Coastal morphology dominated by alluvial plains and deltaic systems; low exposure to wave action. Continental shelf wide and muddy	Seagrass beds, mangrove forests, estuaries, beaches, muddy bottoms
Darién (DAR)	Semi-humid to very humid. High influence of continental discharges (predominantly turbid waters)	Coastal morphology very variable (alluvial plains, terraces, rocky cliffs). Continental shelf wide	Estuaries, mangrove forests, scattered seagrass and corals, muddy bottoms
San Andrés and Providencia Archipelago (SAN)	Dry to semi-humid, frequently exposed to storms and hurricanes. Oceanic waters	Two major high standing islands, sandy cays, submerged carbonate platforms	Coral reefs, sandy beaches and submarine bottoms, scattered seagrass beds and mangroves

(continued)

Table 5.1 (continued)

Ecoregion	Climate and oceanography	Coastal, shelf and seafloor morphology	Main ecosystems represented
Oceanic Caribbean (COC)	Oceanic waters and seafloor beyond continental and island shelves	Continental and insular shelf slopes and abyssal plains, scattered guyots and sea mounts	Pelagic realm; bathyal and abyssal ecosystems, water depths between 150 to more than 3,000 m
Pacific			
Northern Pacific (PAN)	Very rainy; rather clear waters; seasonally affected by upwelling	Very narrow continental shelf; high tectonic activity; coastal morphology defined by mountains with a dense canopy of rain forest; steep rocky (volcanic) cliffs alternating with pocket beaches	Scattered mangrove forests and coral formations
Baudó (BAU)	Marine waters moderately influenced by terrestrial runoff	Coastal morphology defined by a large alluvial plain with extensive sandy beaches and intertidal sand flats; rather narrow continental shelf	Sandy beaches, intertidal sand flats
Buenaventura (BUE)	Turbid waters	Coastal morphology heterogeneous. There are terraces of soft rocks with steep cliffs subject to marine erosion; narrow continental shelf	Alluvial and deltaic plains with mangroves, intertidal mud plains, sandy beaches and littoral terraces
Naya (NAY)	Influenced by many rivers and littoral barrier islands; turbid waters	Coastal morphology defined by an ample alluvial plain; rather wide continental shelf	Highly dynamic estuarine system; well developed mangrove forests behind the barrier islands
Sanquianga (SAQ)	Large deltaic system formed by several meandering rivers which form a complex system of estuaries, channels and open lagoons; turbid waters	Wide continental shelf	Mangrove forests and intertidal sand plains (nesting habitats for sea turtles). A protected area (Sanquianga National Park)
Tumaco (TUM)	Influenced by rivers such as the Mira River, adjacent to the Ecuadorian border; turbid waters	Narrow continental shelf	Alluvial plains; scattered cliffs; fringing mangrove forests

(continued)

Table 5.1 (continued)

Ecoregion	Climate and oceanography	Coastal, shelf and seafloor morphology	Main ecosystems represented
Gorgona Island (GOR)	Relatively clear waters	Mountain island covered by rain forest	Rocky and sandy shores, fringing coral carpets and reefs, submerged rocky and soft bottoms
Malpelo Island (MAL)	Oceanic island surrounded by waters up to 1,000 m depth	Steep basaltic cliffs, insular shelf absent	Rocky bottoms; scattered coral carpets; breeding area for oceanic birds
Oceanic Pacific (PAO)	Moderately high pelagic productivity enhanced by nutrients from terrestrial runoff	Territorial and Economic Exclusive Oceanic Zone from 200 to 3,000 m depth	It is one of the most important commercial tuna fishing areas in American Pacific

Cabo Corrientes, has uplifted shores dominated by rocky cliffs and rocky shores with intrusions of sand and silt, forming a high volcanic coastline (Galvis and Mojica 1993; INGEOMINAS 1998; INVEMAR 2008). The continental shelf is narrow and the 200 m isobaths is located about 15 km from the coastline (Cantera and Contreras 1993). Despite the high rainfall in the area, the proximity of the coast to the Serranía de Baudó does not permit the formation of rivers. The few rivers that exist have a relatively low discharge.

The southern area, from Cabo Corrientes to the border with Ecuador, has lowline coasts with tidal plains. It is mostly covered by mangroves, with cliffs in Málaga, Buenaventura and Tumaco Bay. This area has several rivers with high flow. The main ones are the Baudó and San Juan (INGEOMINAS 1998; INVEMAR 2008). Sediments deposited by these rivers have formed a wide continental shelf, extending about 55 km to Gorgona Island (Cantera and Contreras 1993) (Fig. 5.3).

Insular Formations of Pacific Colombia

The two main islands or systems of islands in the Colombian Pacific, Gorgona and Malpelo, have been designated as protected areas (Fig. 5.3). The first is located 35 km from the continent in the Department of Cauca, between 2°55′N–78°09′W and 3°00′N–78°14′W. It is the largest island territory on the continental shelf of the Pacific, with a total area of 14.1 km^2 and surrounding an area of 610 km^2 (Fig. 5.3). Sixty-one thousand ha of the waters surrounding Gorgona and Gorgonilla islands were designated as a National Natural Park (PNN) in 1983 (UAESPNN 1998; Giraldo et al. 2008). Although the coral reefs of the PNN Gorgona have a relatively small area, they are one of the most developed of the Eastern Tropical Pacific (ETP) (Zapata and Vargas-Ángel 2003).

Malpelo is an oceanic island 490 km west of the Department of Valle del Cauca between 3°51′N and 81°35′W, with a terrestrial area of 3.5 km^2. It was declared a Fauna and Flora Sanctuary (SFF) in 1995. Because of its especially sensitive marine nature, the marine protected area was expanded from 651 to 8,574 km^2 in 2002. In 2006, UNESCO classified it as a World Heritage Site (UAESPNN 1998; Díaz et al. 2000; UNESCO 2006). The island is separated from the mainland by a depth of more than 3,000 m (Graham 1975). It is the only one that emerges from the Malpelo Ridge, a submarine volcanic cordillera 241.4 km long and 80.5 km wide (Chase 1968). The SFF is surrounded by clear oceanic waters that allow the presence of corals to a depth of 30 m (Graham 1975).

Pacific Ecoregions

The Pacific region is subdivided into nine ecoregions based on environmental, social, cultural and political factors: Pacific North (PAN), Baudó (BAU), Buenaventura (BUE), Naya (NAY), Sanquianga (SAQ), Tumaco (TUM), Gorgona

(GOR), Malpelo (MAL), and Pacific Ocean (PAO) (Fig. 5.5). These ecoregions will be used to present the diversity and distribution of echinoderms. Six of these sectors are coastal and limited by the 200 m isobath, which in turn defines the continental shelf. In the cases of GOR, geographical boundaries are the 40 m isobath, on the eastern side and the 200 m isobath on the western side, while MAL emerges from 500 m deep sea floor isobath, beyond the continental shelf. The ecoregion PAO is represented by all jurisdictional Colombian marine areas on the Pacific, adjacent to the conventional boundary of the continental shelf below a depth of 200 m (INVEMAR 2000). Table 5.1 summarizes the main physical, oceanographic, and ecological features of these ecoregions.

5.2 Echinoderm Research

5.2.1 Historical Account of Scientific Studies About Echinoderms on the Caribbean Coast of Colombia

Studies of the echinoderms in the Colombian Caribbean Sea are recent, dating from the 1970s. They have been focused on shallow waters (0–50 m). The studies, which were mostly taxonomic, were published by Allain (1976), Meyer and Macurda (1976), Caycedo (1978, 1979), Álvarez (1981), Quiñones (1981), Gallo (1985, 1988a, b), and Aguilar-Duarte (2005). These authors published taxonomic lists and ecological and biological notes of echinoderms of different ecosystems along the Colombian Caribbean coast, particularly of Tayrona National Natural Park (PNNT), other areas near Santa Marta and Rosario Corals, and San Bernardo Islands National Natural Park (PNNCRSB). Three new species of echinoderms were described from Colombian material, *Ctenantedon kinziei* Meyer, 1972, *Holothuria (Thymiosycia) thomasi* Pawson and Caycedo, 1980, and *Ophiothrix synoecina* Schoppe, 1996.

Some investigations have been focused on the ecology of the echinoderm community in different kinds of habitats, such as sandy beaches and rocky shores (Arevalo 1978; Schoppe 1993). This kind of research has increased during the last decade including soft bottom, coral reefs, and other ecosystems (Gárces and Romero 2003; Vélez 2003; Bejarano et al. 2004; Aguilar-Duarte 2005; Monroy and Solano 2005; Gómez and Rodriguez 2009; Romero 2009; González-Gaviria 2010). On the other hand, echinoderms have been incorporated into the Coral Reefs Monitoring National System (SIMAC) in Colombia. On the Caribbean Colombian coast, echinoderms have been observed from 1998 to 2011. Studies have focused on seven species of sea urchins: *Diadema antillarum*, *Echinometra viridis*, *E. lucunter*, *Eucidaris tribuloides*, *Lytechinus variegatus*, *L. williamsi*, and *Tripneustes ventricosus* (Navas-Camacho et al. 2010).

Some studies have been conducted, such as Allain (1978), who studied the deformations of tests of *Lytechinus variegatus* collected in Cartagena Bay

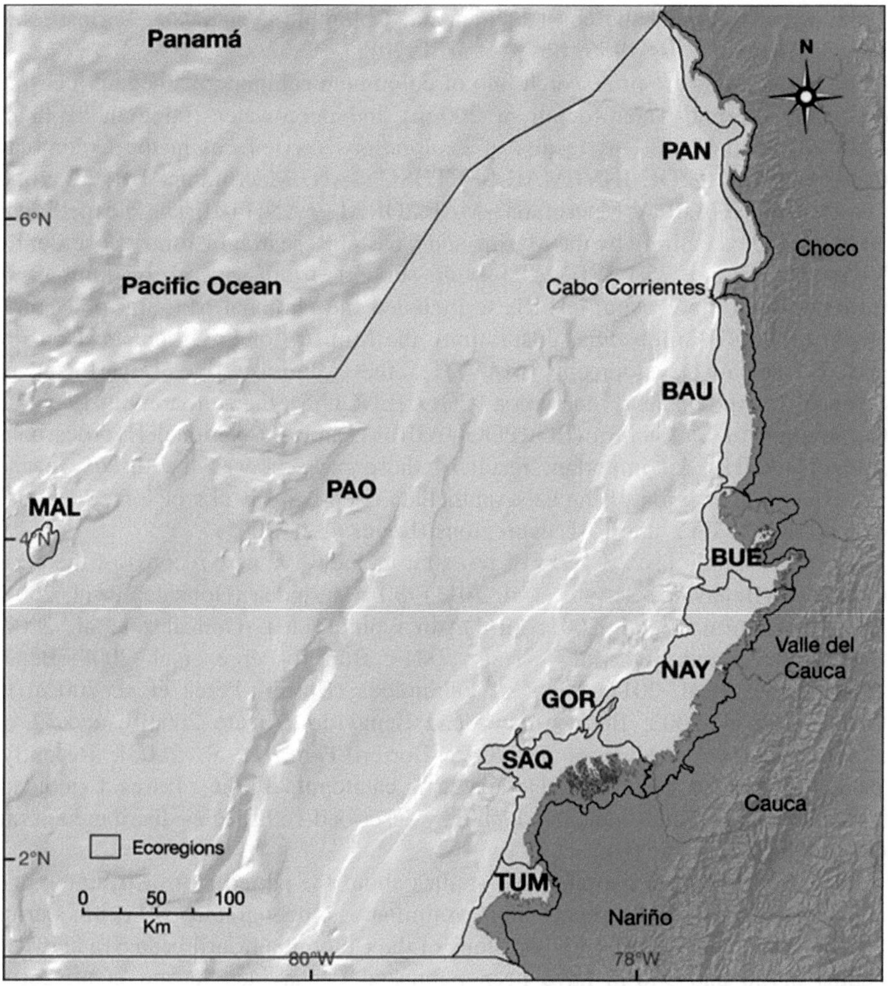

Fig. 5.3 Ecoregions and coastal departments of the Colombian Pacific Sea: PAN: Northern Pacific; BAU: Baudó; BUE: Buenaventura; NAY: Naya; GOR: Gorgona Island; SAQ: Sanquianga; TUM: Tumaco; MAL: Malpelo Island; PAO: Oceanic Pacific (System Information Laboratory LABSIS-INVEMAR)

(MAG ecoregion). Ortiz (2006) studied the change in the ossicle density and shape during growth of some species of the genus *Holothuria* from the Caribbean and Pacific coasts.

With respect to deep waters, there are few records from the decade of the 1960s from samples collected by international research cruises in Colombian Caribbean waters. Among them were five biological voyages by the research vessel *Oregon* in 1963: Oregon I: 87, 92, and 104 and *Oregon* II: 4 and 22 (Palacio 1974); the research vessel of the University of Miami (R/V *John Elliot Pillsbury*) in 1966 and

1968, which sampled the coast of Panama, Colombia, Venezuela, Surinam and French Guyana (Voss 1966; Bayer et al. 1970).

In 1998, a new era of research into of colombian echinoderms began, focusing especially on mid-water (depth of 200 m) and deep-water (up to a depth of 1,000 m) echinoderms, as results of explorations carried out in the Colombian Caribbean Sea: CIOH-INVEMAR-SMITHSONIAN, Macrofauna I-II, Macrofauna-CORPOGUAJIRA, Macrofauna-MARCORAL y ANH I-II. These expeditions were conducted entirely by the Marine and Coastal Research Institute "José Benito Vives de Andreis" (INVEMAR) with support from several institutions in Colombia and other countries. These included the National Museum of Natural History (NMNH-Smithsonian Institution), the Ministry of Environment, Housing and Territorial Development (MAVDT), the Administrative Department of Science, Technology and Innovation (COLCIENCIAS), the Autonomous Regional Corporation of La Guajira (CORPOGUAJIRA), and the National Hydrocarbons Agency (ANH). An important result of these projects was the discovery and subsequent description of the azooxanthellate deep coral banks in some localities of the ARCO, TAY and PAL ecoregions (Reyes et al. 2005).

Result from the last 13 years of research include five new records of the class Crinoidea (Benavides-Serrato et al. 2011), 30 Asteroidea (González et al. 2002; Benavides-Serrato et al. 2005, 2011), 46 Ophiuroidea (González et al. 2002; Borrero-Pérez and Benavides-Serrato 2004a; Borrero-Pérez et al. 2008; Benavides-Serrato et al. 2011), eight of Echinoidea (Borrero-Pérez et al. 2002a, b; González et al. 2002; Borrero-Pérez and Benavides-Serrato 2004b), and 11 of Holothuroidea (González et al. 2002; Borrero-Pérez et al. 2003). Recently, Benavides-Serrato et al. (2011) created a catalogue of the classes Crinoidea, Asteroidea and Ophiuroidea, in which they described 169 species distributed in the Colombian Caribbean Sea.

Based on this information, some studies about the patterns of distribution and abundance, ecological aspects and preliminary biogeography of echinoderms present in the continental shelf and slope of the Colombian Caribbean Sea between depths of 20 and 940 m have been published (Benavides-Serrato et al. 2001; Benavides-Serrato and Borrero-Pérez 2004, 2010; Borrero-Pérez et al. 2004).

Part of this important process was the consolidation of the echinoderm collection of the Colombian Museum of Marine Natural History (MHNMC) into 3,804 lots, principally collected from regions of the Colombian Caribbean Sea. The evolution of the collection coincides with information retrieval by the INVEMAR projects and also with the activities developed in these projects aimed at improving the preservation and management of the collections. At the same time, the Marine Biodiversity Information System (SIBM) was improved and is now available through the INVEMAR web page (www.invemar.org.co), where a special section is designated for the echinoderm collection.

5.2.2 *Historical Account of Scientific Studies About Echinoderms in the Pacific Coast of Colombia*

Echinoderms from the Eastern Tropical Pacific (ETP), particularly on the Pacific coast of Colombia, are not well known. The first records were collections made by European explorers from the eighteenth century, when some descriptions of species of this area were published.

To find works specifically devoted to the understanding and study of the fauna of echinoderms of this area, it is necessary to go to the mid-nineteenth century. At this time, several scientists from the United States and European institutions published new species descriptions and reported new distributional data. The most important contributions to the lists of species of echinoderms at that time in this area are the works of Gray (1840), who described a large number of species of asteroids, and Verrill (1867, 1868, 1869, 1870 and 1871), who published many descriptions and lists of common species of shallow waters and described a large part of the regular echinoids of the ETP. Lütken (1859) and Lyman (1879, 1882) contributed descriptions of new species of ophiuroids. Selenka (1867) contributed to knowledge of the holothurians. The *Challenger* Expedition (1873–1876) took samples near the Galapagos Islands and the Gulf of Panama (Sladen 1889; Théel 1886), including species found in Colombian waters. At the end of the nineteenth and the beginning of the twentieth century, as a result of the *Albatross* expeditions, Agassiz (1891–1905) described a large number of species, mainly sea stars and sea urchins of the Panamic province (Baja California to Peru) (Agassiz 1881, 1888). H.L Clark (1916) published a study of asteroids of this region. Some major expeditions visited the shores of the eastern Pacific Ocean in search of tropical marine fauna, visiting specific places along the Colombian Pacific Coast, and gave some records of echinoderm species. In particular, the ship "*Saint George,*" which visited Gorgona Island in 1924–1925, the "*Askoy Expedition,*" which visited Panama, Colombia and Ecuador in 1941, and *Velero III* (1931–1941) of the Allan Hancock Foundation, which also visited Gorgona and elsewhere, contributed to the knowledge of the marine fauna of the Colombian Pacific. These expeditions resulted in the descriptions of other echinoderm species and important collections of irregular sea urchins (Clark 1940, 1948) and holothurians (Deichmann 1938). In the mid-twentieth century, a series of surveys undertaken by the Zoological Society of New York and led by W. Beebe resulted in publications of echinoderms of shallow waters from northern Baja California to Panama. In 1943, A.G. Fraser made a compilation of the history of exploration, with sampling areas, maps, and information on the cruises of "*Velero III,*" which focused on the echinoderm species of rocky habitats.

Taxonomic knowledge of the species of the ETP was also increased by the work done in Mexico by M.E. Caso between 1950 and 1987 (Caso 1953, 1971, 1974, 1976, 1978a, b, 1979, 1980, 1983, 1987), and by the expedition of the Smithsonian Institution in 1972, which visited Malpelo Island and listed five species of ophiuroids and four of echinoids in a paper about sublittoral communities (Birkeland et al. 1975) and reported the presence of six species of asteroids (Downey 1975). Hendler (1979) contributed to knowledge of ophiuroid biology. The synthetic works of Clark (1989), Clark and Downey (1992), and Maluf (1988) on echinoderms of Central America included published and unpublished reports of the ETP, with new reports and ecological information on this group in the area. More recent works on echinoderms of ETP are those of Pawson (1995), Solis-Marin (1997), and Lessios (2005).

The taxonomic and ecological investigation of Colombian Pacific echinoderms was done almost entirely after the 1980s. In this decade, a research group from the Universidad del Valle, Cali, published papers, mainly species lists or descriptive ecology, for some coastal localities between 1985 and 1999. These include the papers of Neira and Von Prahl (1986), who listed 20 species for the island of Gorgona, including 11 new records; Cantera et al. (1987), who reported two species of ophiuroids (*Ophiactis savingny* and *Ophiothela mirabilis*) associated with the gorgonian *Leptogorgia alba;* and Cantera and Neira (1987) who collected for the first time a parasitic mollusk (genus *Echineulima*) from the carapace and spines of the sea urchin *Centrostephanus coronatus* on Gorgona Island.

Neira and Cantera (1988) reported 17 species of echinoderms of Bahía Malaga: six asteroids, two echinoids, eight ophiuroids, and one holothurian, of which six were new records for the Colombian Pacific. Pardo et al. (1988) recorded the presence in the Colombian Pacific of 70 species, with the first analysis of the descriptive ecology of echinoderms, including distribution and abundance patterns of species of several habitats of the Pacific coast of Colombia. Von Prahl (1990) reported 17 species of echinoderms from Malpelo Island. Neira et al. (1992) reported the presence of 15 species of echinoderms of the PNN Ensenada Utría (five asteroids, five ophiuroids and five echinoids) and a short description of two new records for the Colombian Pacific (*Mithrodia bradleyi* and *Encope ecuadorensis*). Brando et al. (1992) increased the number from Malpelo Island to 18 species. Furthermore, Schoppe and Werding (1996) studied the effect of the activity of the boring urchin *Echinometra vanbrunti* upon rocky shores. Maluf (1991) increased the list of echinoderm species in a paper mentioning 200 species of the Galapagos archipelago, including 50 species shared with Malpelo Island.

Cantera et al. (2001) reported the echinoderms associated with bioerosion of coral reefs of Gorgona Island, finding three species of sea urchins: *Diadema mexicanum* and *Centrostephanus coronatus* that were abundant in the reefs, and *Astropyga pulvinata* that was very occasional. Arboleda (2002) listed 118 species of echinoderms in a paper about the status of knowledge of fishes and some invertebrates of the Pacific coast of Colombia. Toro-Farmer (1998) and Toro-Farmer et al. (2004) studied the patterns of distribution and bioerosion rate of the sea urchin *Centrostephanus coronatus* on Gorgona Island, finding a rate of bioerosion of ca. 0.19 kg m^{-2} $year^{-1}$. Neira and Cantera (2005), in a review of published and unpublished literature and unpublished data of zoological collections on the taxonomy and ecology of Pacific Colombian echinoderms, reported 66 species found in shallow bottoms at 10 localities on different substrates confirmed by presence in collections or by valid scientific data. Ortiz (2006), whose work was based on material in national collections, studied the variations of some ossicles of holothurians of both Caribbean and Pacific coasts of Colombia and contributed to an improvement of the descriptions of species.

In the Management Plan of the Marine Protected Area of Malpelo (2005–2009), 109 species of echinoderms were included. These species had been recorded since 1975 by various institutions and unofficial sources from shallow and deep water (up to a depth of 1000 m) near the island. Cohen-Rengifo (2007) and Cohen-Rengifo et al. (2007) increased the number of species of Malpelo Island to 84. Fifty-six percent of these species occur at depths of more than 200 m, Defined as such, we can consider that the number of species surrounding Malpelo Island are 44, the other are of the Oceanic Pacific Ecoregion (PAO) defined as oceanic zones with depths of more of 200 m.

5.3 Diversity of Echinoderms and Distribution

5.3.1 Geographical Distribution of Species of Echinoderms in the Caribbean Sea

In total, 259 species of echinoderms have been reported from the Colombian Caribbean Sea. These species have been collected on the coasts from depths of 0 to 940 m in all ecoregions that comprise the Colombian Caribbean Sea that were defined earlier. The species that have been studied belong to five classes of echinoderms. Ophiuroidea is the most diverse with 80 species, followed by Asteroidea (67 species), Echinoidea (49 species), Holothuroidea (44 species) and Crinoidea (22 species).

Number of Species Per Class Per Caribbean Ecoregion

In the Colombian Caribbean Sea, ARCO and TAY are the ecoregions with the highest number of recorded echinoderms, 172 and 158 species respectively. This may be the result of the greater diversity of ecosystems present and the greater number of studies carried out in these ecoregions. Furthermore, MAG (134 species), GUA (122 species) and DAR (115 species) are also represented by a high number of species, whereas PAL (82 species), MOR (54 species), and SAN (45 species) have a lower number of recorded species. This is consistent with the smaller number of studies in these ecoregions (Fig. 5.4).

Ophiuroidea has more species in PAL, TAY, MAG, ARCO and DAR than the other classes. In some zones, Asteroidea is better represented, such as GUA and MOR, where a high number of species of this class is present.

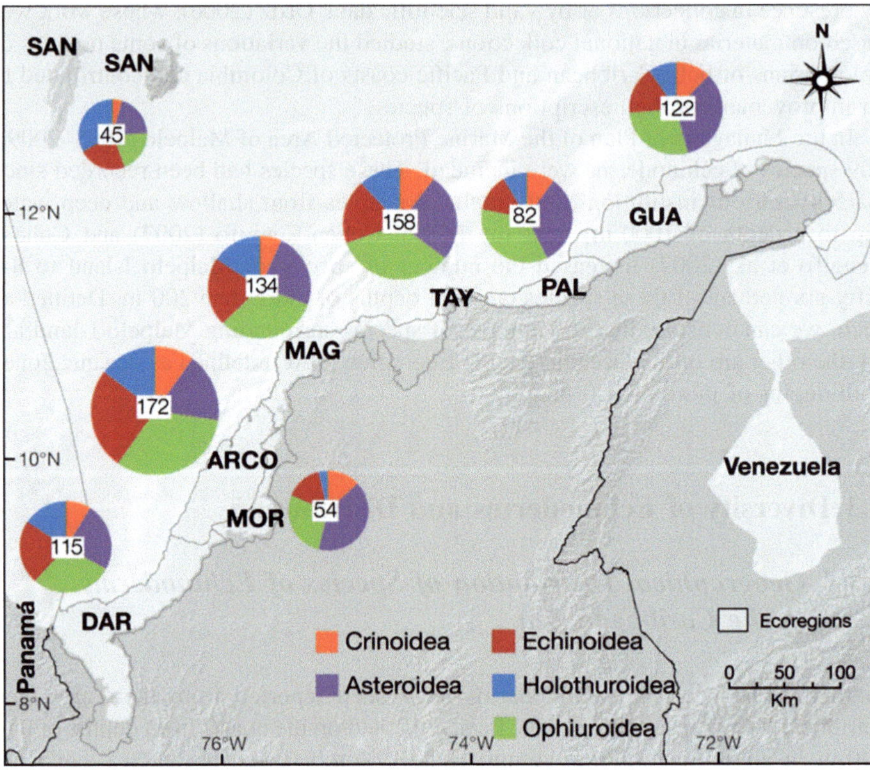

Fig. 5.4 Total number of echinoderms and number of species per class in each ecoregion of the Caribbean Colombian coast. The total number of echinoderms in each ecoregion is presented in the white square and is reflected in the size of the graphic. Details of the number of species per class in each ecorregion are in Table 5.2 (System Information Laboratory LABSIS-INVEMAR)

Table 5.2 Total number of species and species per ecoregion of the Colombian Caribbean Sea

Ecoregion	Crinoidea	Asteroidea	Ophiuroidea	Echinoidea	Holothuroidea	Total
GUA	15	41	34	20	12	122
PAL	8	27	29	11	7	82
TAY	16	40	54	29	19	158
MAG	9	32	43	29	21	134
ARCO	17	30	56	44	25	172
MOR	7	22	15	8	2	54
DAR	10	28	33	26	18	115
SAN	2	9	9	10	15	45
Total	22	67	80	51	44	264

GUA Guajira, *PAL* Palomino, *TAY* Tayrona, *MAG* Magdalena, *ARCO* Coralline MORArchipelagos, DARMorrosquillo, Darién, *SAN* San Andrés and Providencia Archipelago

Distribution of Species by Habitat

The ecoregions of the Colombian Caribbean Sea include different kinds of habitat, such as seagrass beds, sandy beaches, salt marshes, mangroves, coastal lagoons and swamps, sandy bottoms, rocky shores, coral reefs, estuaries, and soft bottoms. Echinoderms could inhabit each of these ecosystems. However, only some of them have been studied and species of echinoderms have been reported in only some of them. Below are descriptions of the habitats and the research that includes information on echinoderms.

Intertidal and Shallow Water Ecosystems

Most studies conducted on shallow water echinoderms of the Colombian Caribbean Sea have been developed in the ecoregion TAY, specifically in Tayrona National Natural Park (PNNT), Punta Betín, and Santa Marta Bay, and in the ecoregion ARCO in Rosario Corals and San Bernardo Islands National Natural Park (PNNCRSB). Among these studies are those by Gallo (1985, 1988a, b), who collected 43 species of Asteroidea, Ophiuroidea and Echinoidea at depths from 0 to 30 m at PNNT; Meyer (1972), who described the new crinoid species *Ctenantedon kinziei* from Punta Betín and Santa Marta Bay; and Caycedo (1978, 1979), who published lists of echinoderms found in PNNT and PNNCRSB from at depths from 0 to 38 m. There are also some lists of echinoderms published by Álvarez (1981), who recorded 70 species of all echinoderm classes and Allain (1976), who recorded 38 species of echinoids of all ecoregions the Colombian Caribbean Sea. It is important to note the study made by Quiñones (1981), who listed 23 species of echinoderms from the SAN ecoregion, specifically in Providencia and Santa Catalina Islands.

Rocky Shores

In the Colombian Caribbean Sea, most of the rocky shores are distributed in a discontinuous line along the continental coast and also around some islands. Generally they are interrupted by beaches, coastal lagoons, and estuaries (Batista-Morales and Diaz-Sanchez 2010). Rocky shores in the Colombian Caribbean Sea have been little studied, so studies focusing on the echinoderms are also scarce. Gárces and Romero (2003) carried out research with the aim of characterizing the structure of the ophiuroid community associated with the mesolittoral rocky shore of Gaira bay (TAY ecoregion), determining spatial-temporary changes of the community between August, 2001, and May, 2002. As a result, 12 species were identified. The most frequent and abundant were *Ophiactis savignyi* and *Ophiothrix angulata*, and the least frequent and abundant was *Ophiocoma wendtii*. Monroy (2003) and Monroy and Solano (2005, 2008) developed an ecological study focused on evaluating the distribution of populations of the Colombian endemic species *Ophiothrix synoecina* and its host *Echinometra lucunter* and other accompanying fauna of this association. They found a broad distribution of *E. lucunter* throughout the Caribbean coast (DAR, ARCO, TAY, GUA), different from the restricted distribution of *O. synoecina,* which is found only on northern Colombian rocky shores (TAY and GUA ecorregions). The results of this research showed a significant reduction in the density of the sea urchin and its associated fauna, including *O. synoecina.* Specifically, the frequency of occurrence of *O. synoecina* has declined about 50 % in recent years in the Santa Marta area (TAY), in contrast to population studies made by Schoppe (1993) in the same area. Recently, Romero (2009) developed a study in Bahía Concha (PNNT) in the ecoregion TAY, where was established the population dynamics of *E. lucunter* in the rocky shore ecosystem and was measured the spatial–temporal variation with respect to size and density. The results indicated that environmental conditions could be affecting population density rocky shores.

Coral Reefs

Coral reefs have been intensively studied in Colombia, focusing on mapping, zonation, health status, and the associated fauna and flora. Twenty-one coral areas of the Caribbean are broadly distributed. 80 % of them are in the SAN ecoregion, where they are also more developed and complex. The other areas are scattered along the continental coast, mostly within ARCO and TAY (Díaz et al. 2000). As mentioned above, the populations of seven species of sea urchins have been monitored in the Colombian Caribbean Sea through the National System of Monitoring of Coral Reefs (SIMAC). It was found that changes in the populations of sea urchins could be determined by the genus *Echinometra,* especially *E. viridis.* Although is not possible to define the causes of the annual variation of the density of sea urchins recorded, the patterns could be the result of factors such as predation, recruitment, inter-specific interactions, and environmental and weather changes. Although there is no specific information about Colombian *Diadema antillarum* populations before their massive mortality (Lessios et al. 1984; Murillo and Cortés 1984; Liddell and Ohlhorst 1986; Hunte et al. 1986; Carpenter 1988; Carpenter 1990; Hendler et al. 1995), González-Gaviria (2010) showed that in the PNNT and Playa Blanca-Rodadero (Santa Marta) there has been a slow and irregular recovery in comparison with other Caribbean populations.

Among echinoderms studies associated with Colombian Caribbean coral reefs it's important to mention Bejarano et al. (2004), who characterized the ophiuroid community in the microhabitats provided by sponges and other reef components at San Bernardo Archipelago (ARCO). The community was composed of 22 morphotypes, including: *Ophiothrix angulata, O. oerstedii* and *O. suensoni.* On the other hand, Gómez and Rodriguez (2009) studied the ophiuroid community associated with sponge populations in Cispata Bay (MOR). They described the different kinds of associations between ophiuroids and sponges and identified 13 and 12 species of echinoderms and poriferans respectively.

Sandy Beaches

Nearly 60 % of the 1,650 km of the Colombian Caribbean coast consists of sandy beaches. Studies of these beaches have focused mainly on the zonation patterns of the faunal communities at a few specific sites. In the Caribbean, beaches with coarse sand exposed to wave action, which occur largely in the northern half of the coast, have a more diverse community dominated by bivalves of the genus *Donax* and "sand dollars" such as *Mellita* sp. A specific echinoderm study was published by Arevalo (1978), who observed the biological and ecological aspects of *Mellita quinquiesperforata* populations on public beaches from Bocagrande to La Boquilla, Cartagena (MAG ecoregion). He noted that there are morphometric and behavioral variations with respect to other coastal populations of the Atlantic, considering the species' distribution, which is from the coast of Massachusetts to southern of Brazil.

Mangrove Areas

Mangroves are among the most biologically productive ecosystems in the world and play an important role in the sustainability of fisheries. They protect the shoreline against erosion, and provide an important resource of wood. Five species of mangrove have been reported in the Caribbean; *Avicennia germinans* and *Rhizophora mangle* are the most common and abundant (Orjuela et al. 2010). Unlike along the Pacific coast, the Caribbean mangroves are limited to narrow sections within lagoons, swamps, estuaries, and river mouths. The major coverage of mangroves is in Ciénaga Grande de Santa Marta (CGSM) in the ecoregion MAG (Orjuela et al. 2010). Echinoderm studies of this habitat are few. Báez (2001) characterized a community associated with roots of *R. mangle* in the Cispatá bay (ecoregion MOR), where two species of ophiuroids were found: *Ophiothrix oerstedii* and *Ophiactis savignyi.*

Seagrass Beds

Seagrass beds are distinct and unique ecosystems with relatively high rates of primary production which support detritus-based food chains. In Colombia, they occur solely in the Caribbean. The seagrass ecosystems provide habitats for a wide variety of marine organisms, promote sediment stabilization, and maintain a dynamic environment for nutrient cycling (Díaz et al. 2003). Seagrass beds and their associated communities were rarely studied in Colombia until very recently. Díaz et al. (2003) developed an integrative study of Colombian seagrass beds, including composition and distribution, community structure, and environmental problems. As a result of this research, more than 550 species of fishes and invertebrates and about 35 species of macroalgae were recorded in the seagrass beds of Colombia (Montoya-Maya 2002). Seventeen species of echinoderms have been found in these beds. *Tripneustes ventricosus* and *Lytechinus variegatus* have been frequently observed, especially in the ecorregions SAN, GUA, TAY and ARCO. In some areas particularly in the ecoregions GUA and ARCO, populations of *Oreaster reticulatus* are frequent. Specific studies focused on echinoderms associated with this ecosystem are few. Moreno-Forero and Hernández-Torres (1992) published an ecological study of size composition of *L. variegatus* associated with sea grass beds of *Thalassia testudinum* in Isla Grande–Islas Rosario (ARCO ecoregion). The size composition and some morphometric relations are described, and some ecological notes on the population are given.

Soft Bottoms

Soft bottoms are extensive marine ecosystems, at depths from 0 to more than 10,000 m. They predominate in the territorial sea of the Colombian Caribbean and Pacific Oceans, with an extension of 889.400 km^2, comprising 99.5 % of underwater ecosystems (Guzmán-Alvis and Solano 2002). Soft bottoms can include sand, clay, silt and mud particles, are not stable substrates, and only form low topographic features (Montoya-Cadavid 2010). Data collected from 1998 to 2011 show increasing of knowledge of this ecosystem in Colombia. Taxonomic studies have concentrated on the Guajira (GUA), Tayrona (TAY), Magdalena (MAG), and Oceanic Caribbean (COC; depth > 200 m) ecosystems (INVEMAR 2010). Most of the specific echinoderm studies on this ecosystem, has been developed at depths between 20 and 1,000 m, which are mentioned in the next section on deep water ecosystems. The only study in shallow soft bottoms was published by Vélez (2003), who studied some structural aspects of the ophiuroid community associated with infaunal bottoms in shallow waters of the San Andrés and Providencia Archipelago (SAN ecoregion). During this research, that identified 15 species of ophiuroids, two climatic periods were included during 1 year (1999–2000).

Deep Water Ecosystems

Research in deep water ecosystems has been conducted throughout the Colombian Caribbean Sea and not confined to sites like most studies in shallow water. Most studies have been carried out in the soft bottom ecosystem, which is dominant at depths greater than 200 m. As a result of exploration in soft bottoms, banks of deep azooxanthellate coral communities have been found, which are mentioned below.

Soft Bottoms

For this habitat in deep waters, it is important to consider the marine explorations made before 1998 by international and national investigations focused especially on depths between 200 and 1,000 m. Among them are the explorations made by the R/V *Pillsbury* in 1966–1968, whose results were presented by Bayer et al. (1970). These authors listed more than 100 species of echinoderms collected between Cartagena and the gulf of Urabá along the ecoregions DAR, MOR, ARCO and MAG from 70 to 3,000 m depth. The results of other explorations, carried out by the INVEMAR-CIOH-SMITHSONIAN in 1995, were published by González et al. (2002), who listed 22 echinoderm species collected in shallow and mid waters (30–380 m), between Cartagena and the gulf of Urabá in the same ecoregions, but including some localities in the PNNCRSB (ARCO). However, the most complete information about echinoderm communities of soft bottoms is from explorations carried out between 1998 and 2009, such as Macrofauna I-II, Macrofauna-CORPOGUAJIRA, Macrofauna-MARCORAL, and ANH I-II. In these explorations, 188 species of echinoderms were collected, of which 81 were recorded for the first time in Colombian Caribbean waters; Ophiuroidea was the most abundant and richest class. In addition, an ecological analysis using multivariate and univariate statistics was undertaken, including all echinoderm records collected during those biological expeditions. The results of this analysis showed four assemblages defined by depth: the first, in shallow waters (depth of 20–70 m), was characterized by *Astropecten antillensis* and *Ophiura acervata.* Intermediate waters (depth of 150 m) were characterized by the clypeasteroid *Clypeaster euclastus.* The third, to a depth of 300 m, had *Caryometra atlantidis* and *Amphiura rathbuni* as characteristic species. The last conglomerate, between depths of 500 and 940 m, with the brittle star *Ophiernus adspersus adspersus* and the sea star *Cheiraster (Chiraster) planus* as the characteristic species. According to the results, the first change in faunal composition of the Colombian Caribbean Sea occurs between depths of 150 and 300 m. An interesting result is that in the conglomerates from depths of 300 and 500 m, a different faunal composition was observed between the samples located in the northeastern Colombian Caribbean Sea and those to the southwest with respect to the Magdalena River (Borrero-Pérez et al. 2004; Benavides-Serrato and Borrero-Pérez 2010).

Deep Azooxanthellate Coral Community Banks

Based on results of the Macrofauna I and II explorations (1998–2002) carried out by INVEMAR, there are deep azooxanthellate coral banks in three localities of the Colombian Caribbean Sea: off La Guajira Peninsula (PAL ecoregion), off Santa Marta (TAY ecoregion), and off San Bernardo Islands (ARCO ecoregion) (Fig. 5.5, Table 5.3). *Cladocora debilis* (thin tube coral) was the main coral matrix builder in the bank formation off La Guajira, at a depth of 70 m. Echinoderms were the most abundant taxon collected in these beds, with 1,367 specimens. The crinoid *Analcidometra armata* was the dominant echinoderm, which has been reported as a common species attached to gorgonians in the Caribbean Sea (Meyer 1973; Hendler et al. 1995). *Ophiothrix angulata* was found in several stations along the sampling area. It has also been reported on *Oculina varicosa* banks off

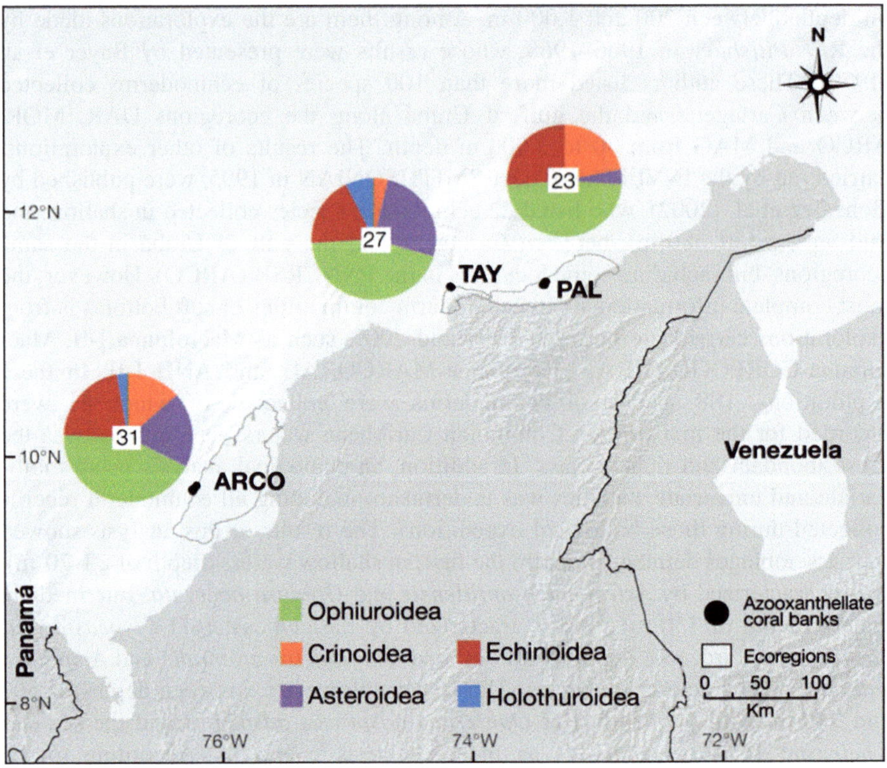

Fig. 5.5 Representation of the total number of echinoderms and number of species per class within each deep azooxanthellate coral bank from north to south along La Guajira Peninsula (PAL), Santa Marta (TAY) and San Bernardo Islands (ARCO) (Reyes et al. 2005). The total number of echinoderms in each bank is presented in the white square and is reflected in the size of the graphic; details of the number of species per class in each bank are in Table 5.3 (System Information Laboratory LABSIS-INVEMAR)

Table 5.3 Number of species per class on each deep coral setting of the Colombian Caribbean Sea

Class	Deep coral bank		
	GUA	SMA	SBE
Crinoidea	5	1	5
Asteroidea	1	7	1
Ophiuroidea	11	12	16
Echinoidea	6	5	8
Holothuroidea	–	2	1
Total	23	27	31

GUAi Guajira, *SMA* Santa Marta, *SBE* San Bernardo

Florida (Reed and Mikkelsen 1987). Species found in the bank formation off Santa Marta, at a depth of 200 m, included *Ophiothrix suensoni*, commonly found on gorgonias (Hendler et al. 1995) and the sea cucumber *Holothuria lentiginosa enodis*, previously recorded off Florida from *Oculina varicosa* banks (Pawson et al. 1982). Another 14 species were found exclusively in this area. Three of them, *Stephanasterias albula*, *Trigonocidaris albida*, and *Palaeobrissus hilgardi*, were the first records for the southern Caribbean. The bank formation off San Bernardo, at a depth between 155 and 160 m, was the most diverse sector for echinoderms, with 38 species. Brittle stars such as *Ophioderma appressa*, *Astrocnida isidis*, *Asteroporpa (Asteroporpa) annulata*, and *Asteroschema oligactes* were found attached to stony corals and gorgonians. Also, some rare species, such as *Ophiosyzygus disacanthus*, previously recorded as dwelling in hard bottoms (Turner and Heyman 1995; Borrero-Pérez and Benavides-Serrato 2004b), the stalked crinoid *Democrinus conifer*, and the reef dwelling comatulid crinoid *Davidaster rubiginosus* were present. These three unique azooxanthellate coral communities sustain one of the highest species diversities found along the Colombian continental shelf (Reyes et al. 2005).

5.3.2 Geographical Distribution of Species of Echinoderms in the Colombian Pacific Ocean

One hundred and twenty-four species of echinoderms have been reported from the Pacific coast of Colombia. These species have been reported from the coast of Chocó (Punta Ardita, Cabo Marzo and Gulf of Cupica, Utria Bight), Valle del Cauca (Malaga bay, Buenaventura bay, Gulf of Tortugas), Cauca (mouth of the Guapi River), Nariño (Mulatos Beach, Tumaco Bight), Gorgona, and Malpelo Island. The species reported are Asteroidea (34 species), Ophiuoroidea (30), Echinoidea (28), and Holothuroidea (32). There are no reports of Crinoidea. Echinoids and asteroids are the classes with highest species richness. Ophiuroids have the highest number of individuals.

Number of Species Per Class Versus Ecoregion in the Colombian Pacific Ocean

The localities with most species are the deeper parts of the oceanic Pacific (52), Malpelo (44), and Gorgona (41 species), and deeper parts of the oceanic Pacific (52) (Neira and Cantera 2005; Cohen-Rengifo 2007). The other area with a high number of species is Buenaventura (31) (Fig. 5.6, Table 5.4). As in the Caribbean, areas with a high number of studies have more species reported. Some studies have shown that deep species can be very abundant.

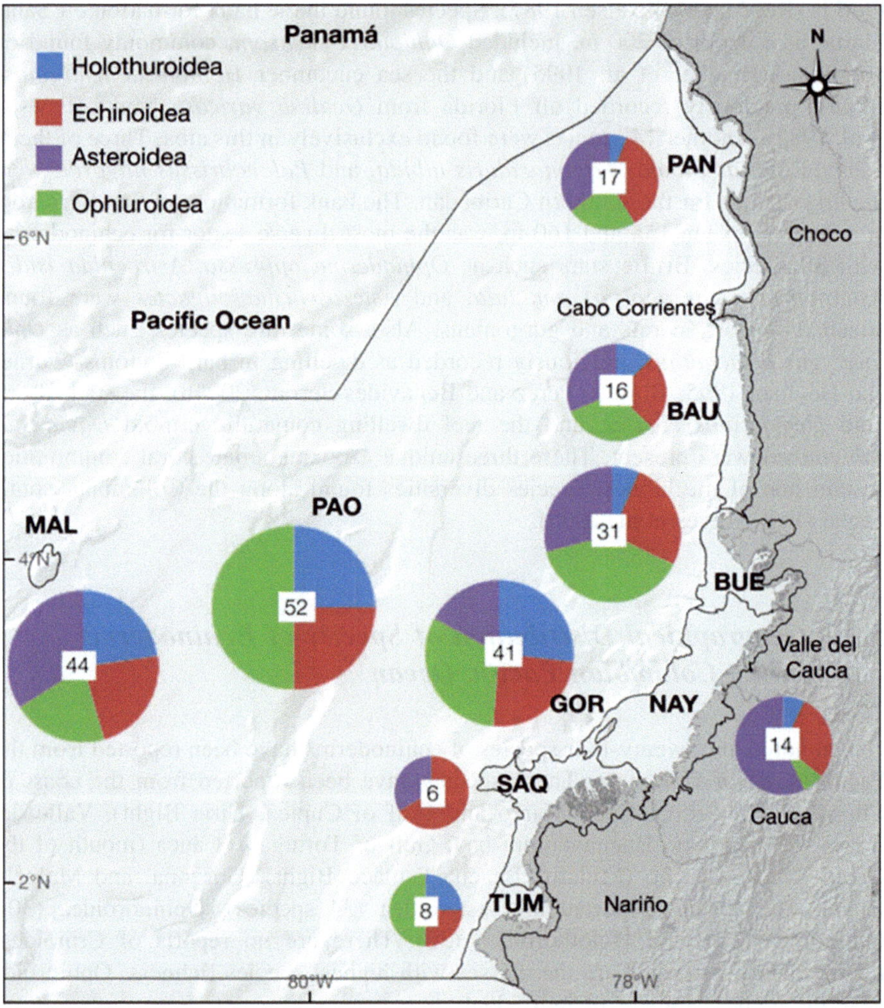

Fig. 5.6 Representation of the number of echinoderms and number of species per class in each ecoregion of the Pacific Colombian coast. The total number of echinoderms in each ecoregion is presented in the white square and is reflected in the size of the graphic; details of the number of species per class in each bank are in Table 5.4 (System Information Laboratory LABSIS-INVEMAR)

Distribution of Species by Habitat

Because of the abundance estuaries and river discharges, echinoderms are not abundant along a large part of the Pacific coast of Colombia coast. Malpelo Island has been studied intensively (Cohen-Rengifo 2007; Cohen-Rengifo et al. 2007). These authors reported 84 species of echinoderms occurring around this island, including species of deeper adjacent areas attributed here to the PAO region.

Table 5.4 Number of species per class in each ecoregion of the Colombian Pacific Ocean

Ecoregion	Asteroidea	Ophiuroidea	Echinoidea	Holothuroidea	Total
PAN	6	4	6	1	17
BAU	5	5	4	2	16
BUE	9	12	8	2	31
NAY	8	1	4	1	14
SAQ	2	0	4	0	6
TUM	0	4	2	2	8
GOR	7	13	10	11	41
PAO	11	11	14	16	52
MAL	15	9	10	10	44
Total	34	30	28	32	124

Pacific North (PAN), Baudó (BAU), Buenaventura (BUE), Naya (NAY), Sanquianga (SAQ), Tumaco (TUM), Gorgona (GOR), Malpelo (MAL) and Pacific Ocean (PAO)

Gorgona island (GOR ecoregion) also has a high number of records (41 species), although it has been relatively less sampled. Among the coastal localities, the tectonic estuary of Málaga Bay (BUE ecoregion) has the highest number of species (25), followed by the localities along the north Pacific coast (e. g. Utría Bight, 19 species). All these localities are characterized by rocky areas as cliffs, shores, and shallow bottoms. All localities also have high salinity (30 PSU or more).

In contrast, estuarine areas of the southern Pacific coast of Buenaventura Bay (BUE), Tumaco Bay (TUM), the mouth of the Guapi River (NAY), and Mulatos Beach (SAN), have a low number of species. But these species are always associated with the external zones of bays or estuaries. The low number of species in these areas is the result of estuarine conditions (low salinity, high turbidity, and sedimentation) and rarity of suitable rocky substrate.

Most species of echinoderms were found in more than one habitat. Rocky substrates and coral reefs share some species of the cryptofauna, because of the similarity of cavity structure of both rocky shores and coral framework. It is also possible that physically contiguous habitats, such as cliffs and rocky shores, or sandy beaches and mud flats also share some species. The moisture that remains in some intertidal substrates during low tide allows the existence of some subtidal species.

The species and the most important characteristics of groups of echinoderms of different habitats are presented below.

Rocky Cliffs

Tectonic processes along the Pacific coast of Colombia have resulted in abundant steep cliffs and to rocky shores with sloping platforms composed of boulders, pebbles, and gravel formed by cliff erosion. In the northern basaltic cliffs the rocks are igneous, very hard, and consolidated. In the southern part most rocks are sedimentary (mudstones and shale), subject to rapid erosion by living organisms (bioerosion). These cliffs have nearly vertical walls that fall directly into the sea or may be separated from it by sandy beaches. Most of these are located in areas exposed to strong waves and high hydrodynamic conditions. The erosion forms caverns and abrasion terraces, which are inhabited by sea urchins of the family Cidariidae (*Eucidaris thouarsii* and *Hesperocidaris asteriscus*), Diadematidae (*Centrostephanus coronatus*), and Echinometridae (*Echinometra vanbrunti*). Some ophiuroids can be found occasionally (*Ophioderma teres*, *O. panamensis*, and *Ophiocoma alexandri*).

Rocky Shores

This habitat is located along the whole Pacific coast. It is characterized by igneous and sedimentary rocks at different stages of fragmentation and by variability of environmental conditions. Rocky platforms are covered with hard blocks of various sizes (20 cm to 1 m in diameter), pebbles (5–10 cm), or gravel (2–5 cm) or are partially covered by sand or mud. In some cases these platforms only emerge during the minimal astronomic tides. This ecosystem is one of the richest in species, with 23 species (two asteroids, seven echinoids, nine ophiuroids, five holothurians). The most common species of rocky shores, living under rocks or in cavities or holes, are: *Phataria unifascialis*, *Pharia pyramidata*, *E. thouarsii*, *H. asteriscus*, *D. mexicanum*, *C. coronatus*, *Echinometra vanbrunti*, *O teres*, *O. panamensis*, *Ophiothrix* (*Ophiothrix*) *spiculata* (associated with clusters of polychaetes and sponges), *Ophiocoma aethiops*, *Ophionereis annulata*, *Ophiactis savignyi*, and *Holothuria (Cystipus) inhabilis*. *O. savignyi* and *Ophiothela mirabilis* are found associated with the gorgonian *Leptogorgia alba* (Neira and Cantera 2005).

Coral Reefs

The coral areas are found mainly in some localities of the Chocó (Cabo Marzo, Cupica Gulf, Utría Bight) and the islands of Gorgona and Malpelo (Zapata and Vargas-Ángel 2003). The major coral reefs are found on the island of Gorgona, where there are true reefs 2 or 3 km long, and some coral patches with species of scleractinian corals of the genera *Pocillopora*, *Pavona*, *Porites, Psammocora.* These reefs have four areas: the backreef, comprising sandy areas and a platform with dead coral and few living colonies (depth between 1 and 3.5 m); a platform-crest region, with the highest coralline coverage (but largely dead) and dominated by *Pocillopora* coral (depth between 0.5 and 3 m); a reef front, with a decreased coverage of *Pocillopora* and an increased number of massive coral colonies (depth between 1 and 4 m at low tide, reaching 8 m at high tide); and finally, a reef slope, depth to 12 m, characterized by low coral coverage, and some dead colonies scattered on the sandy bottom. Patches also occur, with massive and branching corals (Cantera et al. 2001; Zapata 2001). Malpelo Island has coral colonies growing on steep rocky platforms, with *Pocillopora* species having a different external morphology in response to the slope of the substrate and the strength of the currents (Brando et al. 1992).

This ecosystem has 32 species (four asteroids, eight echinoids, 11 ophiuroids, and nine holothurians). In the Gorgona reef, the area with the greatest richness of echinoderms is the reef front with species of both intertidal and subtidal habitats. In this ecosystem occur, from the shallow zones to the outer edge, the sea urchins *A. pulvinata*, *C. coronatus*, *D. mexicanum*, *E. vanbrunti* and *E. thouarsii*, between the coral colonies and in the cavities eroded by them at the base of coral colonies. *Toxopneustes roseus* is found in the sandy patches between corals at different depths. Under the bases of corals and rocks within the reefs are *Ophioderma teres*, *O. panamensis*, *Ophiolepis plateia, O. variegata, O. aethiops, O. alexandri, Ophiocomella schmitti, Ophionereis annulata*, and *Ophiostigma tenue. Ophiactis savigny* is associated with the bivalves *Pinna rugosa* and *Pinctada mazatlanica*), and *O. spiculata* is associated with the bivalve *Pinna rugosa.*

The Asteroidea of the reef areas are *Pentaceraster cumingii* and *Mithrodia bradleyi*, found in the deeper parts of the reef. Cohen-Rengifo (2007) and Narváez and Zapata (2010) have reported *Acanthaster plancii* from the Malpelo reefs, and some unpublished data record it from Gorgona Island. Finally, in the reefs there are holothurians: *Holothuria (Thymiosycia) arenicola, H. (Platyperona) difficilis, H. (Stauropora) fuscocinerea, H. (Mertensiothuria) leucospilota, H. (Thymiosycia) impatiens, H. (Mertensiothuria) hilla*, H. *(Lessonothuria) pardalis, Labidodemas macculochi, H. (Cystipus) inhabilis* and *Pentamera chierchia.* All these species are found at the base of the colonies or out (at night) in the sandy patches.

Sandy Beaches

Sandy beaches of the Pacific coast of Colombia are intertidal zones formed mainly by sediments from the discharge of rivers. These beaches have an upper area, emerged most of the time and covered only at high tides and a lower area that remains submerged much of the time. These beaches may consist of siliceous materials of medium size (between 1 and 2 mm) mixed with silt and clay brought from land or mangrove areas through runoff (Von Prahl et al. 1990). This habitat is associated with external areas of bays which can be subject to strong wave action. The nine species found on these beaches are irregular urchins of the family Mellitidae. Their presence at these beaches is seasonal. Very large numbers of individuals (between 9 and 20 m^{-2}) are found between September and November and are almost absent at other times (Cantera 1991). The species involved are *Encope insularis* and *Mellita longifissa*, which migrate from the sublittoral zone (submerged up to a depth of 16 m) to the upper intertidal zone, in places where tidal changes maintain wet conditions, such as hydraulic dunes characterized by presence of ripple marks.

Mangrove Areas

Echinoderms are virtually absent from the mangroves on the Colombian Pacific coast, except for a few records of the ophiuroids *Amphiodia oerstedii*, *Microphipholis geminata,* and *A. squamata*, which were found under hard objects on mud flats, near the roots of red mangrove (*Rhizophora* sp.), outside of the Bay of Buenaventura.

Submerged Rocky and Sandy Substrates

These bottoms are located within bays or on the continental shelf of the Colombian Pacific area. Most sandy and muddy bottoms are enriched with abundant organic matter from coastal mangroves and rivers, deposited between depths of 5 and 20 m along the coast. The most abundant species found in these submerged muddy and organic substrates are *Astropecten armatus*, *A. verrilli*, *A. regalis* and *A. brasiliensis peruvianus*, *Luidia columbia*,, *L. superba*, and *Luidia latiradiata*. Sandy bottoms are inhabited by the echinoids *Encope laevis* and *E. micropora ecuadoriensis.*

On submerged rocky bottoms located near Gorgona and Malpelo Islands and the northern part of the coast (Chocó) the most abundant species are the asteroids *Nidorellia armata*, *P. unifascalis*, *P. pyramidatus*, the ophiuroid *Ophiothrix spiculata* (on sponges), the echinoid *Echinometra vanbrunti*, and the holothurians *Holothuria portovallartensis* and *Isostichopus fuscus.*

5.4 Aquaculture, Fisheries and Environmental Threats

In the Colombian Caribbean Sea, echinoderms are not included in the diet of native people. However, some marine resources such as species of sea cucumber are caught for export. The over-exploitation of sea cucumbers in areas traditionally fished (Asian countries) has resulted in drastic population declines and even local extinctions. This has led to the expansion of this fishery in the world, including the Caribbean, as well as an increase in the number of exploited species. In Colombia, the first documented record of this fishery is in 2006 (Borrero-Pérez et al. 2010), when there was a massive collection, mainly of *Isostichopus badionotus* and *Holothuria mexicana*, in the area of Portete (Guajira). However, there are records of the export of dried sea cucumber (0.55 t) from Colombia to the Asian market in the year 2001 (Bruckner et al. 2003). In 2005, resolution 2467 of INCODER authorized two exploratory fishing permits in the Guajira. In none of these cases have the harvested species or biomass been reported. Recently, in 2008 and 2009, there were several catches of sea cucumbers in the area of Santa Marta, including the species *Holothuria glaberrima*, *Holothuria grisea*, and *Ocnus suspectus* (Benavides-Serrato 2009). Despite the ecological and economic importance of these invertebrates, there is a lack of information in Colombia and in most Caribbean countries about their biology, population dynamics, and fishery, laws (legislation) for their management are lacking. In Colombia, interest in this resource is growing, and studies on the reproductive biology of some species (e.g. *H. glaberrima*) and on the potential of aquaculture are beginning. However, the current status of natural populations remains unknown, while exploitation continues. At the last workshop organized by the IUCN in Cartagena de Indias, in June 2010, *I. badionotus* and *H. mexicana* were mentioned as vulnerable species. However, due to lack of information, these species were placed in the insufficient data category.

Sea cucumbers were not considered in the Red Book of marine invertebrates of Colombia published in 2002. At this time the only three species evaluated and analyzed were *Diadema antillarum, Ophiothrix synoecina* categorized as Deficient Data (DD), and *Oreaster reticulatus* categorized as Least Concerning (LC). *Diadema antillarum* suffered massive mortality in 1983 due to an unknown pathogen, which spread throughout the Caribbean Sea, causing a mortality of over 97 % of mature individuals. This trend in the Caribbean Sea was also seen in Colombia but was not documented. Even today, there are no detailed data on the population dynamics of this species. However, González-Gaviria (2010) found that abundances appeared to be lower compared to pre-mortality levels, indicating that as in the rest of the Caribbean, recovery in this region has been slow. *Oreaster reticulatus* is especially vulnerable to human exploitation due to its size, its slow locomotion and easy access to its habitat. Individuals of this species, dried and painted in bright colors, have been sold as decorative objects on beaches for many years, including Colombian beaches. Because of overexploitation, *O. reticulatus* is now considered a rare species in some localities of the Caribbean Sea where formerly it was very common. In Colombia this species is still common in some areas. However, there is insufficient information to assess the dynamics of their populations and there are no regulations for their protection. *Ophiothrix synoecina* was included in the Red Book because it has a very limited geographical distribution, being endemic to rocky shores in the area of Santa Marta and Tayrona National Park (Schoppe and Werding 1996). There has been no recent information on its populations. It is now known that the species is only present on the rocky shores of Santa Marta and the Guajira and the density of its populations has declined significantly since 1996. Thus a change in conservation status to vulnerable (VU) has been suggested (Monroy and Solano 2008).

Echinoderms are not used as food by the Pacific coast inhabitants, and for this reason they are not fished or cultivated. Similarly, there are no reports (even oral) of commercial fishing for echinoderms. Some species are threatened by incidental capture (*Astropecten*, *Luidia*) by small trawling nets. Most populations are not impacted by anthropic factors. Some species (*Encope* and *Mellita* in sandy beaches and *D. mexicanum* and probably *C. coronatus* in coral reefs) undergo great variations in number of individuals, as was observed in other echinoderm populations of the world (Uthicke 2009), displaying cyclic population dynamics, probably in relation to specific seasons, or climatic conditions. In some places, populations may decrease because of human activities in coastal areas. These activities can increase sedimentation rates, changing rocky substrates to sand or mud and consequentially, change the relative abundance of different species affect assemblages composition. A conservation status assessment is difficult because of the lack of periodical monitoring and insufficient data from deeper waters.

5.5 Recommendations

The advance in the study of echinoderms has been notable in recent years, mostly at species level in deep habitats, specifically soft bottoms. Interesting results, including the discovery of new habitats, show the need to continue with these investigations, particularly at depths greater than 1,000 m. This requires improved sampling techniques than are currently being used, which are less destructive and more efficient.

Although in 1998 the largest amount of information about echinoderms was on the shallow-water species, during recent years, research has been focused on deeper ecosystems, which has increased our knowledge of these zones, mainly in the Caribbean, while there has been little recent research in shallow ecosystems. It is now clear that more taxonomic and ecological studies are needed of these invertebrates in shallow areas in the Caribbean and of deeper zones in the Pacific.

Although there has been a significant increase in ecological studies of echinoderms in Colombia, there still is little information about this important group of invertebrates. Further studies are necessary especially considering the commercial interest that is being focused on some species such as sea cucumbers; the current status of their natural populations and their dynamics is not known.

Building on all the information that has been collected during recent years and specimens that have been deposited in the echinoderm collections of MHNMC and the Universidad del Valle, future research could utilize new tools such as molecular techniques to investigate cryptic speciation, population connectivity, species adaptation and conservation biology, considering the privileged location of Colombia to answer questions about the evolution of species in the Americas and it high biodiversity and resources to conserve and use sustainably.

Although studies on Colombian Pacific echinoderms have been carried out for many years since the initial international expeditions that found the first species and in recent years have increased, knowledge of this group in this region is still incomplete. Taxonomy and descriptive ecology of the species of coastal ecosystems of the mainland and islands are relatively well known, and there are abundant but not complete references for many species of some deep areas. For these reasons it is necessary to continue to expand the inventory of species, visiting poorly known locations on the coast (i.e. northern Chocó), or increasing the coverage of oceanographic cruises that explore the offshore bottoms of the Pacific Ocean in search of benthic fauna.

Similarly, although there is some biological data for some groups (mainly asteroids and echinoids), it is necessary to increase research on species richness and diversity, ecological studies that include structure of assemblages of species, distribution, population densities, and breeding seasons. These biological characteristics must be related to the environmental and oceanographic conditions of the study areas (e.g. coastal and marine National Natural Parks, such as Ensenada de Utría, Bahia Malaga, and Gorgona and Malpelo Islands).

Bioerosion activity caused by some sea urchins has been studied around Gorgona Island, but there is no information on bioerosion rates and the causes of greater or lesser bioerosion action. Similarly, further studies are necessary to explain the causes of important and sudden variations in the populations that occur in some beach echinoderms (*Encope* and *Mellita*) and coral reefs (*Diadema, Centrostephanus* and *Holothuria*) at certain times of the year. It is also important to study these variations in relation to oceanographic features that affect the Eastern Tropical Pacific, particularly El Niño phenomenon that apparently increased populations of some echinoderms in Gorgona Island in 1982 and 1997.

Finally, although there is no commercial exploitation of echinoderm species in the Colombian Pacific, it is necessary to initiate studies to establish the potential of some species that are exploited in other regions. It may also be important to initiate preliminary assays of culture of the potentially important species.

Acknowledgments The authors wish to express thanks to J.J. Alvarado for the invitation to participate in this synthesis on the Latinoamerican echinoderms. The results presented in this chapter have been produced by different kinds of research projects on both Colombian coasts. We are grateful for the support of the Ministry of Environment, Housing and Territorial Development (MAVDT), the Administrative Department of Science, Technology and Innovation (COLCIENCIAS), the Autonomous Regional Corporation of La Guajira (CORPOGUAJIRA), and the National Hydrocarbons Agency (ANH), among others, all of them with the participation of INVEMAR. For the Pacific coast, the projects "The relationship between bioerosion and growth rates in Colombian Pacific coral reefs of the island of Gorgona (project UNIVALLE 1106-09-069–96), Biomalaga (project INVEMAR-UNIVALLE INCIVA 210509-16821), and Biodiversity of vulnerable stages of benthic fauna of Bahía Malaga (project UNIVALLE-INVEMAR 1106-405-20155) were logistically and financially supported by the Universidad del Valle and COLCIENCIAS with the participation of INVEMAR. We thank G. Hendler, D.Pawson, C. Ahearn, and C. Mah, C. Messing, R. Mooi, J. Lawrenc, and F. Solís-Marín, who have contributed permanently to the echinoderm collection of the Universidad del Valle (CRBMUV), and the Colombian Museum of Marine Natural History MHNMC-INVEMAR through confirmation of species identification, advice, and literature. We also would like thank Colombian researchers dedicated to taxonomy, especially G. Navas and O.D. Solano. Finally, we would like to thank J. Espriella and R.Pinzón Corredor, members of the System Information Laboratory LABSIS-INVEMAR for preparing the maps elaboration in this chapter and P. Silverstone-Sopkin for correcting the English grammar.

References

Agassiz A (1881) Report of the echinoidea dredged by the H.M.S Challenger during the year 1873–76. Zool 3:1–321

Agassiz A (1888) XIX Characteristic deep-sea types-Echinoderms. Three cruises of the United States Coast and Geodetic Survey Steamer "Blake" in the Gulf of Mexico in the Caribbean

Sea, and along the Atlantic Coast of the United States, from 1877 to 1880, vol 2. The Riverside Press, Cambrige, pp 84–127

Aguilar-Duarte C (2005) Caracterización taxonómica de la comunidad de equinodermos presentes en los Arrecifes Coralinos de las Islas de San Bernardo. Parque Nacional Natural Corales del Rosario y San Bernardo, Golfo de Morrosquillo, Bolívar, Colombia. Biología thesis, Univ Antioquia, Medellín, Colombia

Allain J (1976) Erizos de la Costa Norte de Colombia. Informe Museo del Mar 15:1–18

Allain J (1978) Déformations du test chez L'oursin *Lytechinus variegatus* (Lamarck) (Echinoidea) de la Baie de Carthagène. Caldasia 12:363–375

Álvarez LR (1981) Listado preliminar de los equinodermos de la costa Atlántica colombiana. Bol Mus Mar 10:24–39

Andrade GI, Ruiz JP, Gómez R (1992) Biodiversidad, conservación y uso de recursos naturales, Colombia en el contexto internacional. Fescol-Cerec, Bogotá

Arboleda E (2002) Estado actual del conocimiento y riqueza de peces, crustáceos decápodos, moluscos, equinodermos y corales escleractinios del Océano Pacífico colombiano. Biólogo Marino thesis, Inst Invest Mar Cost, Fund Univ Jorge Tadeo Lozano, Santa Marta, Colombia

Arevalo F (1978) Observaciones bioecológicas y biométricas del Dólar de arena Mellita quinquiesperforata Leske 1778 Echinodermata: Echinoidea para las playas de Bocagrande a la Boquilla en la ciudad de Cartagena. Biólogo Marino thesis, Fund Univ Jorge Tadeo Lozano, Bogotá, Colombia

Batista-Morales AM, Diaz-Sanchez CM (2010) Estado de los litorales rocosos. Informe del Estado de los Ambientes Marinos y Recursos Marinos y Costeros en Colombia: Año 2009. Ser Pub Periódicas 8. INVEMAR, Santa Marta

Báez P (2001) Estructura de la biocenosis asociada a raíces sumergidas de mangle rojo (*Rhizophora mangle*) en la isla Tintipán, archipiélago de San Bernardo, Caribe Colombiano. Biology thesis, Univ Javeriana, Bogotá, Colombia

Bayer FM, Voss G, Robins R (1970) Bioenvironmental and radiological safety feasibility studies Atlantic–Pacific interoceanic Canal. Report on the Marine Fauna and Benthic Shelf Slope Communities of the Isthmian Region. Univ Miami, Florida

Bejarano S, Zea S, Diaz JM (2004) Ofiuros (Ophiuroidea: Echinodermata) en ambientes arrecifales del archipiélago de San Bernardo (Caribe colombiano) Bol Invest Mar Cost 33:29–47

Benavides-Serrato M, Borrero-Pérez GH, Solano OD, Navas GR (2001) Equinodermos del talud superior del Caribe colombiano. Resúmenes, IX Colacmar, San Andrés

Benavides-Serrato M, Borrero-Pérez GH (2004) Echinoderms of the continental shelf and upper slope of the Colombian Caribbean Sea: new records and zoogeographical considerations. In: Heinzeiller T, Nebelsick JH (eds) Echinoderms: München. Proceedings of the 11th international Echinoderm conference. Taylor and Francis Group, London, pp 571

Benavides-Serrato M, Borrero-Pérez GH, Navas GR, Solano OD (2005) Listado taxonómico de los asteroideos (Echinodermata: Asteroidea) de la plataforma y el talud superior del Caribe colombiano. Rev Biol Trop 53(Suppl 3):171–194

Benavides-Serrato M (2009) Decomiso de pepinos de mar en la zona costera de Santa Marta. Concepto técnico tipo c. Inst Invest Mar Cost "José Benito Vives de Andréis", Santa Marta

Benavides-Serrato M, Borrero-Pérez GH (2010) Equinodermos de la franja superior de la plataforma y la franja superior del talud continental del Caribe colombiano. In: INVEMAR (eds) Biodiversidad del margen continental del Caribe colombiano. INVEMAR, Ser Publ Esp 20, Santa Marta, pp 255–282

Benavides-Serrato M, Borrero-Pérez GH, Diaz-Sanchez CM (2011) Equinodermos del Caribe colombiano I: Crinoidea, Asteroidea y Ophiuroidea. INVEMAR, Ser Publ Esp 22, Santa Marta

Borrero-Pérez GH, Benavides-Serrato M (2004a) New record of *Ophiosyzygus disacanthus* Clark, 1911 (Echinodermata: Ophiuroidea: Ophiomyxidae) in the Caribbean Sea. Proc Biol Soc Wash 117:541–544

Borrero-Pérez GH, Benavides-Serrato M (2004b) Primer registro de *Clypeaster ravenelli* y *Centrostephanus longispinus longispinus rubricingulus* (Echinodermata: Echinoidea) para el Caribe colombiano. Bol Invest Mar Cost 33:277–282

Borrero-Pérez GH, Solano OD, Benavides-Serrato M (2002a) Lista revisada de los erizos (Echinodermata: Echinoidea) del mar Caribe colombiano. Biota Colombiana 3:141–148

Borrero-Pérez GH, Benavides-Serrato M, Solano OD, Navas GR (2002b) Equinoideos (Echinodermata: Echinoidea) colectados en la franja superior del talud continental del Caribe colombiano. Bol Invest Mar Cost 31:133–166

Borrero-Pérez GH, Benavides-Serrato M, Solano OD, Navas GR (2003) Holothuroideos (Echinodermata: Holothuroidea) recolectados en el talud continental superior del Caribe colombiano. Bol Inst Oceanogr Venezuela 42:65–85

Borrero-Pérez GH, Benavides-Serrato M, Solano OD (2004) Distribution and abundance of Echinoderms on the continental shelf and upper slope of the Colombian Caribbean Sea. In: Heinzeiller T, Nebelsick JH (eds) Echinoderms: München. Proceedings of the 11th international Echinoderm conference. Taylor and Francis Group, London, pp 571

Borrero-Pérez GH, Benavides-Serrato M, Solano OD, Navas GR (2008) Brittle-stars (Echinodermata: Ophiuroidea) from the continental shelf and upper slope of the Colombian Caribbean. Rev Biol Trop 56(Suppl 3):169–204

Borrero-Pérez GH, Santos-Acevedo M, Ortiz-Gómez E (2010) Perspectives and present situation of sea cucumber fisheries in the Colombian Caribbean Sea. In: Harris LG, Böttger SA, Walker CW, Lesser MP (eds) Echinoderms: Durham. Proceedings of the 12th international Echinoderm conference. CRC Press, Taylor and Francis group, Balkema, Leiden, pp 588

Birkeland C, Meyer D, Stames J, Buford C (1975) Subtidal Communities of Malpelo Island, Colombia. In: Graham JB (ed) The biological investigation of Malpelo Island, Colombia. Smith Inst Press, Washington, DC, pp 55–68 (Smith Contr Zool 176)

Botero L, Alvarez-León R (2000) The Caribbean coast of Colombia. In: Sheppard C (ed) Seas at the millenium: an environmental evaluation, vol 1. Pergamon, Amsterdam, pp 663–676

Brando A, Von Prahl H, Cantera J (1992) Malpelo: Isla Oceánica de Colombia. Banco de Occidente: Credencial, Bogota

Bruckner AW, Johnson KA, Field JD (2003) Conservation strategies for sea cucumbers: can a CITES Appendix II listing promote international trade? Beche-de-mer 18:24–33

Cantera JR (1991) Etude structurale des mangroves et des peuplements littoraux des deux baies du pacifique colombien (Málaga et Buenaventura). Rapport avec les conditions du milieu et les perturbations anthropiques. These d'Etat Sciences, Univ d'Aix-Marseille, Marseille

Cantera JR, Contreras R (1993) Ecosistemas Costeros. In: Leyva P (ed) Colombia: Pacífico Tomo I. Fondo para la Protección del Medio Ambiente FEN, Santafé de Bogotá, pp 65–79

Cantera JR, Neira R (1987) Primer registro del género *Echineulima* Luzten y Nilsen, 1975 (Eulimidae) gasterópodo parásito de erizos del mar en la isla Gorgona Colombia. An Inst Inv Mar Punta Betín 17:87–93

Cantera JR, Von Prahl H, Neira R (1987) Moluscos, crustáceos y equinodermos asociados a la Gorgonia *Lophogorgia alba* Duchassaing y Michelloti, 1864 en la isla Gorgona, Colombia. Bol Ecotrop 17:3–24

Cantera JR, Zapata FA, Forero P, Francisco V, Jiménez JM, Londoño E, Narváez K, Neira R, Orozco CA, Toro-Farmer G (2001) Organismos bioerosionadores en arrecifes de la isla Gorgona. In: Barrios LM, López-Victoria M (eds) Gorgona marina: Contribución al conocimiento de una isla única. Inst Invest Mar Cost "José Benito Vives de Andreis". Ser Publ Esp, Santa Marta, pp 79–92

Carpenter RC (1988) Mass mortality of a Caribbean sea urchin: immediate effects on community metabolism and other herbivores. Proc Natl Sci U S A 85:511–514

Carpenter RC (1990) Mass mortality of *Diadema antillarum*. I. Long term effects on sea urchin population-dynamics and coral reef algal communities. Mar Biol 104:67–77

Caso ME (1953) Estado actual de los conocimientos acerca de la fauna de los equinodermos de México. Mem Congr Cient Mex UNAM 7:209–222

Caso ME (1974) Contribución al estudio de los Equinoideos de México, Morfología de *Tripneustes depressus* Agassiz y estudio comparativo entre *T. ventricosus* y *T. depressus*. An Cent Cien Mar Limnol UNAM 1:25–40

Caso ME (1976) El estado actual del estudio de los equinodermos de México. An Cent Cien Mar Limnol UNAM 3:1–56
Caso ME (1978a) Los Equinodermos de la Bahía de Mazatlán, Sinaloa. An Cent Cien Mar Limnol UNAM 6:197–368
Caso ME (1978b) Los Equinoideos del Pacífico de México. Parte 1. Ordenes Cidaroidea y Aulodonta; Parte 2. Ordenes Stiridonta y Camarodonta. An Cent Cien Mar Limnol UNAM Publ Esp 1:1–244
Caso ME (1979) Los equinodermos (Asteroidea, Ophiuroidea y Echinoidea) de la Laguna de Términos, Campeche. An Cent Cien Mar Limnol UNAM Publ Esp 3:1–186
Caso ME (1980) Los equinoideos del Pacífico de México. Parte Tercera. Orden Clypeasteroida. An Cent Cien Mar Limnol UNAM Publ Esp 4:1–252
Caso ME (1983) Los equinoideos del Pacífico de México. Parte Cuarta. Ordenes Cassiduloida y Spatangoida. An Cent Cien Mar Limnol UNAM Publ Esp 6:1–200
Caso ME (1987) Los equinodermos (asteroideos, ofiuroideos y equinoideos de la Bahía de Mazatlán, Sinaloa. An Cent Cien Mar Limnol UNAM Publ Esp 11:1–214
Caycedo IE (1978) Holothuroidea (Echinodermata) de aguas someras en la Costa Norte de Colombia. An Inst Inv Mar Punta de Betín 10:149–198
Caycedo IE (1979) Observaciones de los Equinodermos en las Islas del Rosario. An Inst Inv Mar Punta de Betín 11:39–47
Chase RE (1968) Sea floor topography of the central eastern Pacific Ocean. U S Bur Com Fish Cir 291:1–33
Clark HL (1916) Six new starfishes from the Gulf of California and adjacent waters. Proc Biol Soc Wash 29:51–62
Clark HL (1940) Eastern Pacific Expeditions of the New York Zoological Society. XXI. Notes on Echinoderms from the west coast of Central America. Zool 25:331–352
Clark HL (1948) A report on the Echini of the warmer eastern Pacific, based on the collections of the Velero III. Allan Hancock Pac Exp 8:225–352
Clark AM (1989) An index of names of recent Asteroidea. Part 1. Paxillosida and Notomyotida. In: Jangoux M, Lawrence JM (eds) Echinoderms studies. AA Balkema, Rotterdam, pp 225–347
Clark AM, Downey ME (1992) Starfish of the Atlantic. Chapman and Hall, London
Cohen-Rengifo M (2007) Equinodermos del Santuario de fauna y flora. Biólogo thesis, Fund Univ Jorge Tadeo Lozano, Bogotá, Colombia
Cohen-Rengifo M, Bessudo S, Soler G (2007) Echinoderms, Malpelo Fauna and Flora Sanctuary, Colombian Pacific: new reports and distributional issues. Check List 5:702–711
Deichmann E (1938) Holothurians from the western coast of Lower California and Central America, and from the Galápagos Islands; Eastern Pacific Expeditions of the New York Zoological Society XVI. Zool 23:187–361
Díaz JM, Acero A (2003) Marine biodiversity in Colombia: achievements, status of knowledge, and challenges. Gayana 67:261–274
Díaz JM, Barrios LM, Cendales MH, Garzón-Ferreira J, Geister J, López-Victoria M, Ospina GH, Parra-Velandia F, Pinzón J, Vargas-Angel V, Zapata FA, Zea S (2000) Áreas Coralinas de Colombia. INVEMAR, Ser Publ Esp 5, Santa Marta
Díaz JM, Barrios LM, Gómez-López DI (2003) Las praderas de pastos marinos en Colombia: Estructura y distribución de un ecosistema estratégico. INVEMAR, Ser Publ Esp 10, Santa Marta
Downey ME (1975) Asteroids from Malpelo Island with a description of a new species of the genus *Tamaria*. In: Graham JB (ed) The biological investigation of Malpelo Island, Colombia. Smith Inst Press, Washington, DC, pp 86–90 (Smith Contr Zool 176)
Gárces N, Romero C (2003) Caracterización estructural de la comunidad de ofiuros (Echinodermata: Ophiuroidea) presente en la zona norte de la bahía de Gaira (Caribe colombiano), asociada al mesolitoral rocoso. Biólogo thesis, Fund Univ Jorge Tadeo Lozano, Bogotá, Colombia
Gallo NJ (1985) Contribución al estudio de los equinodermos del Parque Nacional Natural Tayrona: III Asteroidea. Programa de Ecosistemas Marinos, Informes técnicos subproyectos. INDERENA, INVEMAR, Santa Marta

Gallo NJ (1988a) Contribución al conocimiento de los equinodermos del Parque Nacional Natural Tayrona: I Echinoidea. Trianea 1:99–118
Gallo NJ (1988b) Contribución al conocimiento de los equinodermos del Parque Nacional Natural Tayrona: II Ophiuroidea. Trianea 2:383–402
Galvis J, Mojica J (1993) Geología. In: Leiva P (ed) Colombia Pacífico. Tomo 1. Fondo FEN, Santa Fé de Bogotá, Colombia, pp 80–95
Gentry AH (1986) Species richness and floristic composition of Chocó region plant communities. Caldasia 15:71–79
Giraldo A, Rodríguez-Rubio E, Zapata F (2008) Condiciones oceanográficas en isla Gorgona, Pacífico oriental tropical de Colombia. Lat Am J Aquat Res 36:121–128
Gómez PC, Rodriguez ER (2009) Distribución y Ecología de las poblaciones de ofiuroideos (Echinodermata: Ophiuroidea) asociados a esponjas en la bahía de Cispatá. Biologo thesis, Univ Córdoba, Córdoba, Colombia
González DN, Solano OD, Navas GR (2002) Equinodermos colectados por la expedición CIOH-Invemar-Smithsonian desde Cartagena hasta el Golfo de Urabá, Caribe colombiano. Bol Inv Mar Cos 31:85–132
González-Gaviria FT (2010) Estado actual de las poblaciones del erizo negro Diadema antillarum Philippi (Echinoidea: Diadematidae) en el Parque Nacional Natural Tayrona y playa Blanca, Santa Marta, Caribe colombiano. Biologo thesis, Univ Magdalena, Santa Marta, Colombia
Graham J (1975) The biological investigation of Malpelo Island: introduction. In: Graham JB (ed) The biological investigation of Malpelo Island, Colombia. Smith Inst Press, Washington, DC, pp 3–10 (Smith Contr Zool 176)
Gray JE (1840) A synopsis of the genera and species of the class Hypostoma (*Asterias* Linnaeus). An Mag Nat Hist 6(157–184):275–290
Guzmán-Alvis A, Solano OD (2002) Estado de los fondos blandos de la plataforma continental. In: Ospina-Salazar GH, Acero A (eds) Informe del estado de los ambientes marinos y costeros en Colombia: año 2001. INVEMAR, Medellín, pp 71–75
Hendler G (1979) Reproductive periodicity of ophiuroids (Echinodermata: Ophiuroidea) on the Atlantic and Pacific coasts of Panama. In: Stephen ES (ed) Reproductive ecology of marine invertebrates. Univ South Carolina Press, Columbia, pp 145–156
Hendler G, Miller JE, Pawson DL, Kier PM (1995) Sea Stars, Sea Urchins, and Allies: Echinoderms of Florida and the Caribbean. Smith Inst Press, Washington, DC
Hunte W, Coté I, Tomascik T (1986) On the dynamics of the mass mortality of *Diadema antillarum* in Barbados. Coral Reefs 4:135–139
IAvH (Instituto Alexander Von Humboldt) (1998a) Colombia, Biodiversidad Siglo XXI. Propuesta Técnica para la Formulación de un Plan de Acción en Biodiversidad. Instituto Alexander von Humboldt, Ministerio del Medio Ambiente, Bogotá
IAvH (Instituto Alexander Von Humboldt) (1998b) Informe Nacional sobre el Estado de la Biodiversidad 1977—Colombia, vol 1. Instituto Alexander von Humboldt, PNUMA, Ministerio del Medio Ambiente, Bogotá
INGEOMINAS (1998) Geomorfología y aspectos erosivos del litoral Caribe colombiano. Geomorfología y aspectos erosivos del litoral Pacífico colombiano. Publ Geol Esp 21, Bogotá
INVEMAR (2000) Programa Nacional de Investigación en Biodiversidad Marina y Costera PNIBM. Díaz-Merlano JM, Gómez-López DI (eds) Invemar, Fonade, MMA, Santa Marta
INVEMAR (2002) Informe del estado de los ambientes y recursos marinos y costeros en Colombia: 2001. Ser Doc Generales 3, Santa Marta
INVEMAR (2008) Informe del estado de los ambientes y recursos marinos y costeros en Colombia: 2007. Ser Doc Generales 8, Santa Marta
INVEMAR (2010) Informe del estado de los ambientes y recursos marinos y costeros en Colombia: 2009. Ser Doc Generales 8, Santa Marta
Lessios HA, Cubit JD, Robertson DR, Shulman MJ, Parker MR, Garrity SD, Levings SC (1984) Mass mortality of *Diadema* at Jamaica. Exp Mar Biol Ecol 95:271–278
Lessios HA (2005) Echinoids of the Pacific Waters of Panama: status of knowledge and new records. Rev Biol Trop 53(Suppl 3):147–170

Liddell DW, Ohlhorst SL (1986) Changes in benthic community composition following the mass mortality of *Diadema* at Jamaica. J Exp Mar Biol Ecol 95(271):278

Lütken CF (1859) Berskrivelse af de fra Vestindienog Central Amerikas Veskyst hidtil beckjendte Slangestjerner. Additamenta ad historian Ophiuridarum:179-27

Lyman T (1879) Ophiuridae and Astrophytidae of the exploring voyage of H. M. S. "Challenger", under Prof. Sir W. Thomson, F. R. S. Part II. Ophiuridae and Astrophytidae of the "Challenger" expedition. Bull Mus Comp Zool Harvard Coll 6:17–83

Lyman T (1882) Report on the Ophiuroidea dredged by H.M.S. "Challenger" during the years 1873–1876. Report of the Scientific Results of the Voyage of H.M.S. "Challenger" 1873–1876. 5:1–386

Maluf LY (1988) Composition and distribution of the Central eastern Pacific echinoderms. Nat Hist Mus LA County Tech Rep 2:1–242

Maluf LY (1991) Echinoderm Fauna of the Galápagos Islands. In: James MJ (ed) Galápagos marine invertebrates. Taxonomy, biogeography, and evolution in Darwin's Islands. Plenum Press, New York, pp 345–367

Marrugo-González AJ, Fernández-Maestre R, Alm AA (2000) The Pacific coast of Colombia. In: Sheppard C (ed) Seas at the millenium an environmental evaluation, vol 1. Pergamon, Amsterdam, pp 677–686

Meyer DL (1972) *Ctenantedon*, a new Antedonid crinoid convergent with comasterids. Bull Mar Sci 22:53–66

Meyer DL (1973) Feeding behavior and Ecology of shallow-water unstalked crinoids (Echinodermata) in the Caribbean Sea. Mar Biol 2:105–129

Meyer DL, Macurda DB (1976) Distribution of shallow water crinoids near Santa Marta, Colombia. Mitt Inst Colombo-Alemán de Invest Cien 8:141–156

Monroy M (2003) Distribución geográfica y estado actual de Ophiothrix synoecina (Echinodermata: Ophiuroidea) asociado obligado a Echinometra lucunter, en el litoral rocoso del Caribe colombiano. Biólogo thesis, Fund Univ Jorge Tadeo Lozano, Bogotá, Colombia

Monroy M, Solano O (2005) Estado poblacional de *Echinometra lucunter* (Echinoidea: Echinometridae) y su fauna acompañante en el litoral rocoso del Caribe colombiano. Rev Biol Trop 53(Suppl 3):291–297

Monroy M, Solano O (2008) *Ophiothrix synoecina* (Echinodermata: Ophiuroidea:Ophiotrichidae): especie endémica y vulnerable del Caribe colombiano. Bol Invest Mar Cost 37:191–196

Montoya-Cadavid E (2010) Estado del conocimiento de los fondos blandos en Colombia. In: INVEMAR (ed) Informe del estado de los ambientes y recursos marinos y costeros en Colombia: Año 2009. Ser Publ Per 8, Santa Marta, pp 159–172

Montoya-Maya P (2002) Evaluación de la macrofauna epibentónica asociada a praderas de Thalassia testudinum (Banks ex Köning) en el Caribe colombiano. Biólogo thesis, Fund Univ Jorge Tadeo Lozano, Santa Marta, Colombia

Moreno-Forero SK, Hernández-Torres L (1992) Morfometría y composición por tallas de *Lytechinus variegatus* (Equinoidea) en una pradera de *Thalassia* (Isla Grande-Islas del Rosario). VIII Sem Nac Cien Tecnol Mar, Congr Centroamericano y del Caribe de Cien Mar, Santa Marta, pp 619–627

Murillo MM, Cortés J (1984) Alta mortalidad en la población del erizo de mar *Diadema antillarum* Philippi (Echinodermata: Echinoidea), en el Parque Nacional Cahuita, Limón, Costa Rica. Rev Biol Trop 32:167–169

Narváez K, Zapata FA (2010) First record and impact of the crown-of-thorns starfish, *Acanthaster planci* (Spinulosida: Acanthasteridae) on corals of Malpelo Island, Colombian Pacific. Rev Biol Trop 58(Suppl 1):139–143

Navas-Camacho R, Gómez-Campo K, Vega-Sequeda J, López-Londoño T, Duque DL, Abril A, Bolaños N (2010) Estado de los Arrecifes Coralinos. In: INVEMAR (ed) Informe del estado de los ambientes y recursos marinos y costeros en Colombia: Año 2009. Ser Publ Per 8, Santa Marta, pp 75–100

Neira R, Von Prahl H (1986) Notas sobre los equinodermos de la isla Gorgona. In: Von Prahl H, Alberico M (eds) Isla Gorgona. Univ Valle, Cali, pp 29–57

Neira R, Cantera JR (1988) Los Equinodermos de la bahía de Málaga. Mem Sem Nac Cien Tec Mar, Bogotá, pp 121–130

Neira R, Barba P, Pardo R (1992) Equinodermos del Parque Nacional Natural Ensenada de Utría (Pacífico colombiano). Bol Inv Mar Cos 21:77–83

Neira R, Cantera JR (2005) Composición taxonómica y distribución de las asociaciones de equinodermos en los ecosistemas litorales del Pacífico colombiano. Rev Biol Trop 53(Suppl 3): 195–206

Orjuela A, Ricardo R, Villamil C, Perdomo L (2010) Estado del conocimiento de los manglares. In: INVEMAR (ed) Informe del estado de los ambientes y recursos marinos y costeros en Colombia: Año 2009. Ser Publ Per 8, Santa Marta, pp 101–120

Ortiz E (2006) Cambio espicular en algunos pepinos de mar del género Holothuria Linnaeus, 1758 (Echinodermata: Holothurianea) presentes en el Caribe y Pacífico colombianos. Biólogo thesis, Fund Univ Jorge Tadeo Lozano, Bogotá, Colombia

Palacio FJ (1974) Peces colectados en el Caribe colombiano por la Universidad de Miami. Bol Mus Mar 6:3–137

Pardo R, Neira R, Cantera JR (1988) Taxonomía y Ecología de equinodermos litorales del Pacifico Colombiano. Mem Sem Nac Cien Tecnol Mar, Bogotá, pp 87–92

Pawson DL (1995) Echinoderms of the tropical island Pacific: status of their systematics and notes on their ecology and biogeography. In: Maragos JE, Peterson MNA, Eldredge LG, Bardach JE, Takeuchi HE (eds) Marine and coastal biodiversity in the tropical Island Pacific region, vol 1., Species systematics and information management prioritiesEast-West Center, Honolulu, pp 171–192

Pawson DL, Miller JE, Hoskin CM (1982) Distribution of *Holothuria lentiginosa enodis* Miller and Pawson in relation to a deep-water *Oculina* coral reef Fort Pierce, Florida (Echinodermata: Holothuroidea). In: Lawrence JM (ed) Proceedings of the 4th international Echinoderms conference, Tampa Bay, 1981. Balkema Press, Rotterdam, p 321

Quiñones R (1981) Lista preliminar de los equinodermos. In: Werding B, Garzón J, Zea S (eds) Informe sobre los resultados de la expedición a las islas de Providencia y Santa Catalina. INVEMAR, Santa Marta, pp 35–45

Reyes JO, Santodomingo N, Gracia A, Borrero-Pérez GH, Navas GR, Mejía-Ladino LM, Bermúdez A, Benavides M (2005) Southern Caribbean azooxanthellate coral communities off Colombia. In: Freiwald A, Roberts JM (eds) Cold water corals and ecosystems. Springer, Berlin, pp 309–330

Reed JK, Mikkelsen PM (1987) The molluscan community associated with the scleractinian coral *Oculina varicosa*. Bull Mar Sci 40:99–131

Romero JE (2009) Estudio poblacional de Echinometra lucunter (Echinodermata: Echinoidea) en la zona litoral rocosa de la ensenada de Bahía Concha (PNNT) Caribe Colombiano desde Noviembre del 2007 a Mayo del 2008. Biólogo thesis, Univ Atlántico, Barranquilla, Colombia

Schoppe S (1993) Die karpose um den felsbohrenden seeigel *Echinometra lucunter* (L.): Untersuchung der lebensraumbedingungen und der biologie der assoziierten Arten. PhD thesis, Justus-Liebig Univ, Giensen

Schoppe S, Werding B (1996) The boreholes of the sea urchin genus *Echinometra* (Echinodermata: Echinoidea: Echinometridae) as a microhabitat in tropical South America. PSZNI Mar Eco 17:181–186

Selenka E (1867) Beitrage zur Anatomie und Systematik der Holthurien. Zeitschrift fur wissenschaftliche Zoologie 17:291–374

Sladen WP (1889) Asteroidea. Report of the Scientific results Voyage of H.M.S. "Challenger" 1873–1876. 30:1–893

Solis-Marín FA (1997) Catálogo de los Equinodermos recientes de México. Informe Final del Proyecto G010. Univ Nac Aut México, Inst Cien Mar Limnol, Mexico DF, Mexico

Steer R, Arias F, Ramos A, Aguirre P, Sierra P, Alonso D (1997) Documento preliminar de políticas de ordenamiento ambiental de las zonas costeras colombianas. Ministerio del Medio Ambiente, Bogotá

Tabares NG, Soltau JM, Diaz J (1996) Caracterización geomorfológica del sector suroccidental del Mar Caribe. Bol Cien CIOH 17:3–16

Théel H (1886) Report on the Holothurioidea 2. Challenger science results. Zool 14:1–290

Toro-Farmer GA (1998) Estimación de la intensidad bioerosionadora de los erizos (Echinodermata) en un arrecife coralino de la isla Gorgona, Pacífico Colombiano. Biologo thesis, Univ del Valle, Cali, Colombia

Toro-Farmer G, Cantera JR, Londoño-Cruz Orozco C, Neira R (2004) Patrones de distribución y tasa de bioerosión del erizo *Centrostepahnus coronatus* (Diadematoida: Diadematidae), en el arrecife de Playa Blanca, Pacífico colombiano. Rev Biol Trop 52:67–76

Turner RL, Heyman RM (1995) Rediagnosis of the brittlestar genus *Ophiosyzygus* and notes on its type species *O. disacanthus* (Echinodermata: Ophiuroidea: Ophiomyxidae) based on the type specimens from Japanese waters and new material from the Gulf of Mexico. Proc Biol Soc Wash 108:292–297

UAESPNN (Unidad Administrativa Especial del Sistema de Parques Nacionales Naturales) (1998) El sistema de parques nacionales naturales de Colombia. Ministerio del Medio Ambiente, Bogotá

UNESCO (United Nations Educational, Scientific and Cultural Organization) (2006) Extract of the decisions adopted by the 30th sesions of the Word Heritage Comitee. UNESCO, Paris

Uthicke S, Schaffelke B, Byrne M (2009) A boom–bust phylum? Ecological and evolutionary consequences of density variations in echinoderms. Ecol Monogr 79:3–24

Vélez C (2003) Estrellas quebradizas infaunales (Echinodermata: Ophiuroidea) asociadas a fondos blandos someros de las Islas de San Andrés y Providencia. Biólogo thesis, Fund Univ Jorge Tadeo Lozano, Bogotá, Colombia

Verrill AE (1867) Notes on the echinoderms of Panama and west coast of America, with description of new genera and species. Trans Connect Acad Arts Sci 1:251–322

Verrill AE (1869) On new imperfectly know Echinoderms and Corals. Proc Nat Hist Soc Boston :381–396

Verrill AE (1870) Description of Echinoderms and Corals from the Gulf of California. Am J Sci 49:93–100

Von Prahl H (1990) Malpelo la Roca Viviente. Fondo para la Protección del Medio Ambiente "José Celestino Mutis", FEN, Cali, Colombia

Von Prahl H, Cantera JR, Contreras R (1990) Manglares y Hombres del Pacifico colombiano. FEN, Ed Presencia, Bogotá

Voss G (1966) Narrative of the cruise P-6607 to the Southwestern Caribbean, R/V John Elliot Pillsbury. The Rosenstiel Sch Mar Atmos Sci, Univ Miami, Miami, Florida

Zapata FA (2001) Formaciones coralinas de Isla Gorgona. In: Barrios LM, López-Victoria M (eds) Gorgona marina: Contribución al conocimiento de una isla única. Instituto Investigaciones Marinas y Costeras "José Benito Vives de Andreis". Ser Publ Esp 7, pp 27–40

Zapata FA, Vargas-Ángel B (2003) Corals and coral reefs of the Pacific coast of Colombia. In: Cortes J (ed) Coral reefs of Latin America. Elsevier Science, Amsterdam, pp 419–447

Chapter 6
Echinoderms of Ecuador

Jorge Sonnenholzner, Margarita Brandt, Vanessa Francisco, Alex Hearn, María Luzuriaga, Paulina Guarderas and Juan Carlos Navarro

J. Sonnenholzner (✉)
Universidad Laica Eloy Alfaro de Manabí, Departamento Central de Investigación, Manta, Manabí, Ecuador
e-mail: jsonnenholzner@hotmail.com

M. Brandt
Department of Ecology and Evolutionary Biology, Brown University, Providence, RI, USA
e-mail: margarita.brandt@gmail.com

V. Francisco
Laboratorio de Ecología de Ecosistemas de Arrecifes Coralinos, Centro de Investigación y de Estudios Avanzados del Instituto Politécnico Nacional, Mérida, Yucatán, Mexico
e-mail: vanewagen@gmail.com

A. Hearn
Department of Wildlife, Fish and Conservation Biology, University of California, Davis, CA, USA
e-mail: arhearn@ucdavis.edu

M. Luzuriaga
Laboratorio de Plancton e Ictioplancton, Instituto Nacional de Pesca del Ecuador, Guayaquil, Ecuador
e-mail: mluzuriagav@yahoo.com

P. Guarderas
GEOcentro, Universidad San Francisco de Quito, Cumbayá, Ecuador
e-mail: pagigv@yahoo.com

J. C. Navarro
LECROY S.A., Km 2.5, Avenida Carlos Julio Arosemena, Bodega No.5, Guayaquil, Ecuador
e-mail: jcnavarrogo@gmail.com

J. J. Alvarado and F. A. Solís-Marín (eds.), *Echinoderm Research and Diversity in Latin America*, DOI: 10.1007/978-3-642-20051-9_6,

6.1 Introduction

6.1.1 Ecuadorian Continental Shelf

Geography and Climate

The continental coastal territory of Ecuador extends 4,403 km from north to south, including islands, islets, and estuarine borders. The width of the continental shelf exceeds 100 km, particularly off the Gulf of Guayaquil (associated with the most important watershed from the Guayas River basin). The influence of the Guayas River extends over two-thirds of the surface along the continental coast. This causes coastal sedimentation, and a complex estuarine-deltaic system, that extends close to the edge of the continental shelf (Benítez 1995 in Morante-Carballo 2004). Approximately one-third of the coastline in Ecuador consists of stretches of mangroves on mudflats, with major concentrations along the southern half of the northern Ecuadorian coast and in the Gulf of Guayaquil. There are substantial stretches of rocky shores scattered throughout the coast. The longest uninterrupted areas occur in the central coast of Ecuador. Long stretches of sandy beaches are mostly concentrated along the northern-central Ecuadorian shorelines. Coral reef development in this region is prevented by unfavorable conditions that result from freshwater input from river runoff, siltation, nutrient enrichment, and upwelling influences. The majority of reef habitat in this region consists of rocky reefs. Suitable conditions for coral development and persistence are found around islands and rocky promontories located away from the central mainland coast, such as Isla de la Plata, Isla Salango, and Bajo Montañita.

According to the classification of climate and life zones proposed by Cañadas (1983) and Sierra (1999) there is a gradient in rainfall levels from north to south, with a complex spatial and temporal variation. The northern part is a wet tropical region (with $>2{,}000$ mm yr^{-1} of rainfall and without ecologically-dry months throughout the year) and the central and southern regions, range from a dry tropical desert to a sub-tropical desert (with <1.9 mm yr^{-1} of rainfall). The monthly maximum temperatures are higher in the central and northern Ecuador and decrease to the south, at the entrance to the Gulf of Guayaquil. Overall, the average annual temperature is above 22 °C. The maximum temperatures fluctuate between 32 and 38 °C and the minimum is around 15 °C. The relative humidity is above 80 % and with considerable cloudiness. The rainy season runs from January to April, with temperatures averaging 28 °C. The rest of year includes short rainy summers and relatively dry seasons.

Oceanography

Ecuador is located at the confluence of two major marine ecosystems: the Pacific Central American Coastal System and the Humboldt System. The Ecuadorian

Fig. 6.1 The Ecuadorian continental shelf and current system. *Source* Google (Digital Satellite Imagery, TerraMetrics, 2011) modified by L. Molina 2011

marine and coastal area constitutes a remarkable transition zone in the eastern Pacific, characterized by a significant spatial and temporal variability of the physical-oceanographic conditions.

The ocean circulation in the eastern equatorial Pacific area is dominated by the action of the winds and the circulation patterns of several currents, mainly the North and South Equatorial Currents, the Equatorial Counter Current, the California Current, the Panamá Current (PC), the Perú or Humboldt Current, and the Equatorial Undercurrent or Cromwell Current (Allauca 1990). All of these currents carry water that contributes to the formation of the Equatorial Front, which moves from north to south depending on the displacement of the Inter-tropical Convergence Zone (ITCZ, Wyrtki 1965). Stevenson (1970) detected the presence of a circulation system located between the mainland and the Galápagos Islands (Fig. 6.1), which varies in size between the wet and dry seasons. A decrease in the gradient of the prevailing trade winds and a reduction in the intensity of the Perú Current, shifts the balance of forces along the frontal boundary. This results in a southerly movement of a narrow stream of nutrient poor, warm water from the basin of Panamá (Panamá Current) down to the coast of Ecuador and Perú between December and April (Fig. 6.1). On the other hand, during the period of strong NE trade winds, the Panamá Current can extend to 3°S (Enfield 1975).

The Peruvian Currents

Both the oceanic and coastal currents are initiated in at 40–45°S and transport highly saline, cold water (14–23 °C) to the north. The flow continues in a northwesterly direction along the coast of Perú to the Ecuadorian coast where it joins the South Equatorial Current system (Fig. 6.1). This current system prevails from May to November and it represents a "change of season" in Ecuador, as it significantly decreases the sea surface and air temperatures. This system provides favorable conditions for fisheries, because its waters are rich in nutrients. The greatest intensity of this current is between July and September. It gradually weakens and disappears completely in December, when the Panamá Current, warm and low in nutrients, starts to dominate Ecuadorian waters (Fig. 6.1).

The Equatorial Undercurrent

This current runs eastwards along the equatorial Pacific right in front of Ecuador, reaching speeds of approximately 1.5 m s^{-1}. When it reaches the Galápagos Islands it weakens and splits into two branches, one north and one south of the archipelago. The southern branch moves to southeast and generates outcrops that occur along the northern coast of Perú (Fig. 6.1).

The Ecuadorian Coastal Current (ECC)

The Ecuadorian Coastal Current (ECC) moves northwards, paralleling the coast of Ecuador. The northern boundary of this current is linked to the position of the ITCZ, and begins to descend toward the south during December (Moreano 1983). An important aspect to consider is the influence of the Gulf of Guayaquil in the area around its mouth, because it has a well defined circulation regime. Yet it somehow alters the normal course of the circulation of the ECC, mainly through the influence of tidal currents that are stronger into the Gulf (Allauca 1990; Fig. 6.1). Two semidiurnal tides occur both along mainland Ecuador and at the Galápagos. The ECC influences the local currents along the continental coast of Ecuador, and the marine ecosystems and coastal geomorphology.

6.1.2 The Galápagos Islands

Geography and Climate

The Galápagos Islands are situated astride the equator, between 01°40′N and 01°25′S, and between 89°15′W and 92°00′W, about 1,000 km from the Ecuadorian mainland. These oceanic islands have been in existence for at least 3.3 million years. They are made up of 13 major islands, the smallest having an area of 14.1 km^2 (Genovesa Island), and the largest of 4,588 km^2 (130 × 40 km, Isabela Island). There are also over 115 islets and rocks, most of which have vegetation. The most recent islands, Fernandina and Isabela, are moving slowly from west to east, while the oldest islands are San Cristóbal and Española (Cox 1983; Geist 1996). The total land area of the archipelago is about 8,000 km^2. It is surrounded by a Marine Reserve that covers an area of 137,975 km^2, with a total of 1,800 km of coastline (Snell et al. 1995; Toropova et al. 2010) making it the ninth largest marine protected area in the world. Outside the Marine Reserve, national waters extend 200 nautical milles from the island baseline. Climatologically, these islands are not a uniform group, as they extend both north and south of the equator and are influenced by different oceanographic conditions (Abbott 1966). More than 90 % of their surface area is located south of the equator. In spite of their equatorial position, a rather temperate climate (with an annual mean temperature of ~24 °C) prevails because the islands are located in a climatic transition zone between the west coast of South America and the dry zone of the Central Pacific Ocean. These islands are influenced by both warm and cold currents (Figs. 6.1, 6.2) which meet in the Galápagos region. Most climate data available come from the arid coastal zone where the year is divided into a dry season (June–October) and a rainy season (December–April). During the dry season, upwelling cold oceanic water (19 °C) flows along the southern part of the archipelago. The extreme temperatures in the "arid zone" are rarely below 13 °C or above 23 °C, which are cool for this latitude. There is occasional or intermittent drizzle in the coastal zone but almost continuous drizzle in the higher zones of the islands. This kind of rainfall, locally called "garua", is also typical of all South American coasts influenced by the Humboldt Current (Chile, Perú, and southern Ecuador). In the arid zone, the monthly rainfall from May to October never exceeds 25 mm. During the rainy season, the dominant wind changes from south-southeast to east, with light winds in the lower regions. Upwelling of cold oceanic water stops during this period and genuine tropical conditions prevail. The heaviest rainfall is on the southern slopes of the large islands above 250 m altitude. There is almost no rainfall on the leeward side of the coastal region (northern side).

Oceanography

Several marine currents converge at the Galápagos Islands (Fig. 6.2). This determines the great ecological uniqueness of these marine ecosystems. Seven ocean currents have been identified. Three of these are the most decisive currents that determine the seasonal periods of the islands, in terms of intensity, direction and temperatures: (1) the South Equatorial Current (SEC); (2) the warm Panamá Current; and (3) the deep cold Cromwell Current (Chávez and Brusca 1991; Banks 1999).

South Equatorial Current (SEC)

The SEC is the main surface current that flows westward. It is a product of the confluence of two currents with different direction and intensity throughout the

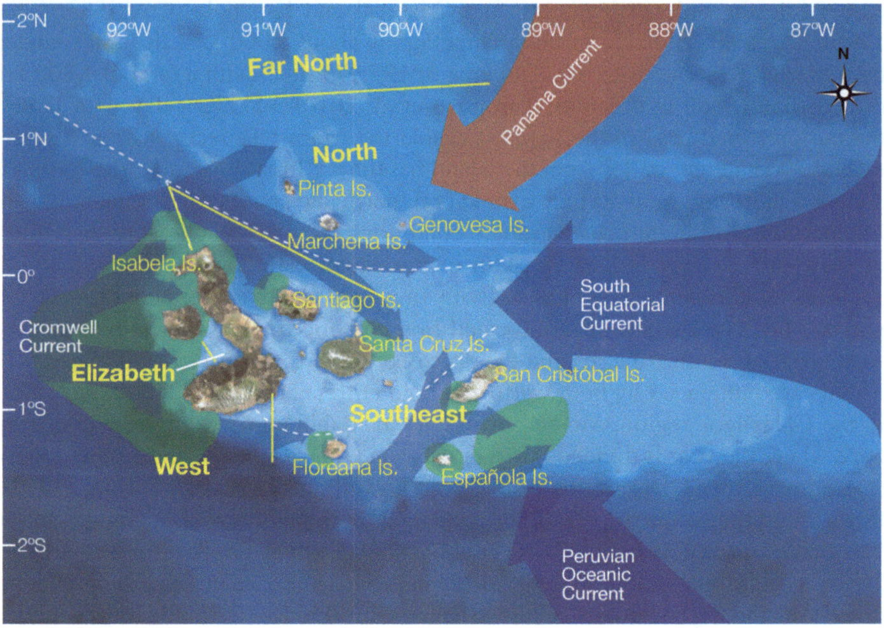

Fig. 6.2 The current system and biogeographic zones of the Galápagos Islands according to Edgar et al. (2002b). *Source* Galápagos National Park Service and Charles Darwin Foundation 2003

year (Fig. 6.2), the tropical Panamá Current (~30–34 ‰, depending on rainfall), and the sub-tropical Perú Current (~35 ‰ close to Ecuador). The dominant flow occurs during the rainy season (from May to November) resulting from the winds blowing from the southeast (Muromtsev 1963). Advection of the cold surface waters from the west (product of the equatorial upwelling, induced by wind at low latitudes) that maintains cold local temperatures in Galápagos (Wyrtki 1966, 1974). The differences in temperature and salinity between the two currents form an oceanic front at the crossroads between the two bodies of waters producing gradients of up to 5 °C at 50 km (Banks 1999). The front descends from north of the Galápagos during the southern summer hemisphere (December to May) reducing the temperature gradient from northwest to southeast, while the sea surface temperature is homogenized in the archipelago (Banks 2002). Both surface currents are relatively impoverished in nutrients after extensive circulation in the open ocean gyres (Thurman 1996). The Perú Current is an extension of cold waters originating in equatorial upwelling zones between Galápagos and the coast of mainland Ecuador and Perú.

The Panamá Current (PC)

Every year a body of surface water (warm and low in salinity and nutrients) that flows from Panamá (Fig. 6.2) reaches Galápagos in January. At this time of year, the SE winds decrease in intensity and the currents flowing westward also weaken. This allows the ITCZ warm water masses to invade from the north. This mass of warm water flows southward along the NW coast of South America. When the PC penetrates to the south and lasts long enough, the event known as El Niño occurs. This has devastating impacts on the marine ecosystems of the region (Jiménez 1987).

The Subsurface Equatorial Current (or Cromwell Current)

The Cromwell Current directly or indirectly generates a series of unusual effects. These include cool surface water temperatures on the equator (17 °C on average) and a large amount of plankton that is responsible for the richness of fauna in the archipelago. It has a thickness of 300 m and a width of 400 km. It is normally positioned at a depth of about 100 m, well below the euphotic zone. Upwelling occurs when it hits the Galápagos platform (Fig. 6.2) bringing nutrient rich, cold waters to the surface (Wyrtki 1985). This current produces persistent upwelling zones enriched with nutrients in the western coasts of most islands and consistently colder water masses. This allows the occurrence and persistence of temperate species in Galápagos, which is an equatorial zone (Feldman 1985). The branches of this water mass come together again in one undercurrent about 100 km east of the archipelago (Banks 1999). Based on analysis of satellite remote sensing data (SeaWiFS), Banks (2002) identified regions where annual primary production is consistently high, in particular to the west of Isabela, known for its high productivity and fishing intensity, but also at some easterly points of Isabela and SE Santa Cruz, which may be due to local patterns of currents and seamounts. It has been suggested that an elevated productivity (10–30 mg Chl *a* m^{-3}) occurs in sporadic but intense events of short duration, while occasional, spatially large but moderately intense (0.5–1.0 mg Chl *a* m^{-3}) productivity occurs throughout the archipelago. This is possibly due to a mixing of points of freshwater runoff and nutrient enrichment from the coast, surface advection from other islands, and retention effects of nutrients from areas with high sediment content and mangrove bays (Wellington 1984; James 1991).

6.2 Diversity of Echinoderms and Distribution

Explorations of echinoderms in Ecuador have been intensive, particularly around the Galápagos Islands, where several expeditions made collections over the past 100 years. These include the *Albatross* in 1894, 1898, 1899, 1905; the *Eagle* and

Table 6.1 Summary of the total number of echinoderms species

	Continental shelf		Galápagos Islands	C	Depth distribution (m)					
	Habitat				0–20	20–100	100–1000	>1000	ND	Total
	Shallow	Open								
Crinoidea	ND	ND	3	0	ND	ND	ND	2	1	3
Asteroidea	19	5	67	15	11	5	12	16	27	71
Ophiuroidea	21	8	82	11	3	3	18	28	42	94
Echinoidea	24	2	43	16	7	4	16	10	14	51
Holothuroidea	19	13	47	11	11	2	2	24	24	63
Total	83	28	242	53	32	14	48	80	108	282

Information is referred for habitats and depth distribution along the continental shelf and the Galápagos Islands, Ecuador. *C* common number of species for both the continental and the Galápagos Islands. *ND* no data

Ara in 1921–1928; the Presidential Cruise in 1938; the *Velero III* and *IV* between 1932–1954; and the New York Zoological Society in the late 1930s. More recently, an expedition was undertaken using the research submersible *Johnson-Sea-Link II* in 1998 (Pawson and Ahearn 2001). These expeditions produced the majority of articles dealing with taxonomy, distribution and diversity of echinoderms from subtropical and tropical waters of the Western Americas. Publications relevant to Ecuador, include Agassiz (1892, 1898, 1904)), Ludwig (1894, 1905), Lütken and Mortensen (1899), Clark (1902, 1917, 1920a, 1920b, 1939, 1940, 1948), Fisher (1911), Boone (1933), Ziesenhenne (1935), Deichmann (1938, 1941, 1958), Caso (1948, 1958, 1961, 1962, 1965, 1976, 1978, 1979), Barham et al. (1973), Brusca (1980), and Maluf (1988a, b).

This chapter includes information about the echinoderms of the continental coast and the Galápagos Islands, based on all known records of the literature and on published data from the catalogues of the Allan Hancock Foundation Collections, Los Angeles County Museum and from holdings at the Scripps Institute of Oceanography and at the National Museum of Natural History of the Smithsonian Institute. Furthermore, species observed in the field, but not necessarily reported in the literature or present in collections (Chiriboga et al. 2011) are also reported here. All species are listed in the Appendix.

6.2.1 Diversity

Ecuador possesses 282 species of echinoderms (Table 6.1). Eighty-three and 242 species have been on the continental shelf of Ecuador, and the Galápagos Islands, respectively (Table 6.1).

6.2.2 Ecuadorian Continental Shelf

In general, the location with the greatest diversity of marine invertebrate species (including echinoderms) seems to be in the Gulf of Guayaquil (which is the main estuary of Ecuador). Maluf (1988a, b) identified this Gulf as a transition area because there are a large number of southern echinoderm species whose ranges are between 1° and 4°S. Transition zones may have a greater number of species because co-occurrence in the ecotone between two similar habitat types is possible. However, this zone may have a higher biodiversity because it has been better studied compared to the other areas along the continental coast (Cruz et al. 2003). Even so, information is still insufficient for echinoderms (Villamar and Cruz 1983; Avilés 1984; Mair et al. 2002; Sonnenholzner and Lawrence 2002). So far, scientific studies have been limited to basic research and the taxonomy of the biota. To date, there is no proper analysis of the distribution of the fauna of echinoderms based on the limited available records (Appendix).

The echinoderm fauna presently known is composed of 26 species: nine species of Asteroidea, two species of Ophiuroidea, ten species of Echinoidea, and four species of Holothuroidea. Of these, 17 species have been reported from the intertidal area and shallow water environments, less than 20 m depth (Mair et al. 2002). Sea urchins, such as *Echinometra vanbrunti* and *Eucidaris thouarsii*; sea cucumbers such as *Holothuria theeli*, and *Holothuria arenicola*; and sea stars such as *Pharia pyramidata* and *Phataria unifascialis* are extremely common along the central and southern coast of mainland Ecuador.

6.2.3 The Galápagos Islands

The Galápagos Islands are the southern distributional limit for many species. Although these islands are separated from the nearest American mainland by 1,000 km of open water, there are several factors influencing the dispersal process that favor affinity between the Panamá Bight (Panamá/Colombia/Ecuador), and the Galápagos, while a smaller affinity is expected between the Galápagos and Perú (Glynn et al. 1983). Only one-third of Perú's echinoderms are shared with the Galápagos Islands (Maluf 1991). The pronounced offshore movement of waters in northern Perú would favor selection for non-planktotrophic species along the mainland coast with reduced larval dispersal to the Galápagos. The islands lie to the south of the normal path of the North Equatorial Countercurrent, the best potential dispersal route for shallow-water tropical forms between the Central and the Eastern Pacific. The greatest influence of this current is between 4° and 10°N latitude, along the coasts of Costa Rica, Panamá, Colombia and the northern part of Ecuador. Trans-Pacific species may have secondarily dispersed to the Galápagos Islands from established mainland populations. The Equatorial Undercurrent could transport (at great depth) central Pacific tropical larvae to the Galápagos.

Two hundred and forty-two species of echinoderms that are known to occur in Galápagos and the Galápagos Ridge, including three crinoids, 67 asteroids, 82 ophiuroids, 43 echinoids, and 47 holothurians. These islands have a high percentage of endemic species relative to other offshore islands of the Central Eastern Pacific (between 34°30′N and 18°00′S), because of the high environmental and spatial heterogeneity of the archipelago, due to its geographic location (Maluf 1988a, 1991). Seventeen percent of echinoderm species in Galápagos (a total of 34 species) are endemic, particularly the asteroids and ophiuroids.

Of these species, 27 species inhabit deep waters between 105 and 925 m (See Appendix), and 107 inhabit depths between 20 and 100 m (30 species of asteroids, 46 species of ophiuroids, and 31 species of echinoids). Deep sea urchins, such as *Plagiobrissus pacificus*, *Brissopsis pacifica* and *Idiobryssus coelus* have been recorded. For intertidal and shallow water environments (<20 m depth), 14 species of asteroids, 28 species of ophiuroids, 25 species of echinoids, and 17 species of holothuroids have been reported. Hickman (1998) and Chiriboga et al. (2011) prepared a list with the most common species of echinoderms along the intertidal and shallow waters around the Galápagos. Among the holothuroids are *Holothuria arenicola*, *H. impatiens*, *H. imitans* and *H. theeli*, and *Isostichopus fuscus*. The latter species is common in the cold western region of Galápagos (Fig. 6.3a). Among the echinoids are *Eucidaris galapagensis* (Fig. 6.3b), *Lytechinus semituberculatus* (Fig. 6.3c), *Tripneustes depressus* (Fig. 6.3d), *Diadema mexicanum* (Fig. 6.3e), *Toxopneustes roseus* (Fig. 6.3f), *Astropyga pulvinata* (Fig. 6.3g), *Agassizia scrobiculata*, and *Clypeaster rotundus*. *Eucidaris galapagensis* was formerly identified as *E. thouarsii*, but new genetic evidence recognized it as a species endemic to the Galápagos Islands, Cocos Islands and Clipperton Atoll (Lessios et al. 1999). Among the asteroids are *Pharia pyramidata* (Fig. 6.3h), *Phataria unifascialis* (Fig. 6.3i–j), *Nidorellia armata* (Fig. 6.3k), *Pentaceraster cumingi* (Fig. 6.3l), and *Mithrodia bradleyi* (Fig. 6.4a) are the most common, while *Astropecten armatus* (Fig. 6.4b), *Leiaster teres* (Fig. 6.4c), *Heliaster cumingi* (Fig. 6.4d), and *Heliaster solaris* are rare. Among the ophiuroids are *Ophiactis savignyi*, *Ophiocoma aethiops*, *O. alexandri*, and *Ophioderma teres*.

The distribution and diversity of echinoderms in the Galápagos Islands differs greatly among biogeographic regions (Maluf 1991). Edgar et al. (2004) proposed a bio-regionalization scheme of five regions within the Galápagos, based on the scheme proposed by Harris (1969) using a data set from communities and species richness of reef fish and macro-invertebrates. These bioregions are the ‘Far-northern’, ‘Northern’, ‘Southeastern’, ‘Western’ and ‘Elizabeth’ (Fig. 6.2). These authors emphasized that both faunal abundance and species richness indicate that Galápagos inshore reef ecosystems are divisible into three major regional components: (1) the far-northern islands of Darwin and Wolf; (2) the central, southern and eastern islands, including the east coast of Isabela; and (3) Fernandina and western Isabela. Concerning the echinoderm fauna, the following patterns were identified:

Fig. 6.3 Common echinoderms off the mainland and at the Galápagos Islands, Ecuador. **a** *Isostichopus fuscus* (Ludwig, 1875), **b** *Eucidaris galapagensis* Döderlein, 1887, **c** *Lytechinus semituberculatus* (Agassiz and Desor, 1846), **d** *Tripneustes depressus* A. Agassiz, 1863, **e** *Diadema mexicanum* A. Agassiz, 1863, **f** *Toxopneustes roseus* (A. Agassiz, 1863), **g** *Astropyga pulvinata* (Lamarck, 1816), **h** *Pharia pyramidata* (Gray, 1840), **i–j** *Phataria unifascialis* (Gray, 1840), **k** *Nidorellia armata* (Gray, 1840), **l** *Pentaceraster cumingi* (Gray, 1840). Photos by: J. Sonnenholzner (**a–b**); Charles Darwin Foundation (**c**); and J.C. Navarro Gómez (**d–l**). Note that photos are not to scale

1. The islands of the Far Northern and the Northern islands include a large component of species from the Indo-Pacific and Panamá, with no endemic species. Species associated with warmer waters and coral environments are

Fig. 6.4 Common echinoderms off the mainland, and at the Galápagos Islands, Ecuador. **a** *Mithrodia bradleyi* Verril, 1870, **b** *Astropecten armatus* Gray, 1840, **c** *Leiaster teres* (Verill, 1871), **d** *Heliaster cumingi* (Gray, 1840), **e** *Asteropsis carinifera* (Lamarck, 1816), **f** *Luidia bellonae* Lütken, 1865, **g** *Linckia columbiae* Gray, 1840, **h** *Caenocentrotus gibbosus* (L. Agassiz and Desor, 1846), **i** *Stichopus horrens* Selenka, 1867, **j** *Paulia horrida* (Gray, 1840), **k** *Acanthaster planci* (Linnaeus, 1758), **l** *Coronaster marchenus* Ziesenhenne, 1942. Photos by: J.C. Navarro Gómez (**a**); X. Romero (**b**); A. Chiriboga (**c–l**). Note that photos are not to scale

Mithrodia bradleyi and *Asteropsis carinifera* (Hawaiian species, see Fig. 6.4e) are found on the northern edge of the Galápagos (Darwin and Wolf Islands).

2. Western islands (west of Isabela and Fernandina) and the Elizabeth bioregion, contain numerous endemic species and those whose distribution includes the cool-temperate Peruvian waters (*Heliaster solaris* and *Luidia superba*). Their

distributions extend to the south of Ecuador. The colder water species (surface temperature can drop to 14 °C in August and December) include *Astrometis sertulifera*, *Luidia bellonae* (Fig. 6.4f), and *Linckia columbiae* (Fig. 6.4g). *Echinometra vanbrunti* is a rare species in this region, despite being common in almost all the islands. It is presumed that the Peruvian species occur here due to the environmental conditions that include colder temperatures and more productive waters, similar to the conditions prevailing along the coast of South America where these species initially evolved. This is also the case of the sea urchin *Caenocentrotus gibbosus* (Fig. 6.4h), an immigrant from the temperate region of the Peruvian-Chilean banks, which is dominant in the lower areas of the western coast of Isabela and Fernandina Islands.

3. The Central/Eastern/Southern bioregions contain a combination of species from various regions. Widely distributed in shallow waters are the sea urchins *Eucidaris galapagensis* (Fig. 6.3b), *Tripneustes depressus* (Fig. 6.3d), *Diadema mexicanum* (Fig. 6.3e), *Toxopneustes roseus* (Fig. 6.3f) and *Centrostephanus coronatus*. *Lytechinus semituberculatus* is native to the Galápagos, and replaces the ubiquitous *L. pictus* of the Eastern Pacific. On sandy and rocky bottoms, the most frequent sand dollars species are *Clypeaster elongatus*, *C. europacificus* and *C. rotundus*.

Edgar et al. (2004) emphasized that the variation in the abundance of echinoderms around individual islands was high, often greater than the variation between islands, thereby significantly lowering the density of certain species of sea stars. For example, *Heliaster cumingi* (Gray, 1840) was not found during baseline surveys despite being previously abundant, occurring in densities greater than 2–3 ind m^{-2} in the intertidal (Wellington 1975). Nevertheless, it has been recently found in Pinzón Island (see Fig. 6.4d). On the other hand, densities of some other species seem to be increasing, for example the sea urchin *Tripneustes depressus*. It appears that its abundance has increased at least an order of magnitude since the 1970s. This species was recorded in just one of ten sites investigated by Strickland (1968), Wellington (1975), and Glynn (Edgar et al. 2002b). However, it is also known that its populations greatly fluctuate through time, absent during some periods, but very abundant during others (Brandt and Guarderas 2002).

Lawrence and Sonnenholzner (2004) documented the occurrence of the most conspicuous echinoderms at seven sites with distinct habitats in central Galápagos (i.e. sites were considered upon spatial gradients, from north to south, on hydrodynamics, wave energy, and sloping). Eighteen species (eight species of echinoids, five species of holothurians, and five species of asteroids) were recorded. Of the echinoids, *Eucidaris galapagensis* occurred in most of the sites (>90 % of the quadrats at every site except one), and reached the highest densities (between 10.1 ± 4.8 and 15.6 ± 6.1 ind m^{-2}). *Lytechinus semituberculatus* was found in >90 % of the quadrats at two sites. Other echinoderms present in more than 35 % of the intervals along transects at a site), were the echinoids *Tripneustes depressus*, *Diadema mexicanum*, *Toxopnesutes roseus* and *Caenocentrotus gibbosus*, the asteroid *Pentaceraster cumingi* and the holothuroid *Holothuria atra*. Diversity was greater at the northern sites than at the southern sites, but there were low densities of asteroids and holothurians. This may be related to a lack of appropriate substrata or food availability at the sites. This is in agreement with Roff et al. (2003) and Edgar et al. (2004) who indicated that, in most cases, abiotic factors such as salinity, temperature, depth, rock type or sediment particle size are important factors for biota that act in synchrony with biological process to affect macro and micro distribution.

Pawson and Ahearn (2001) collected echinoderms from bathyal and abyssal waters and reported four new records for the Galápagos Islands and eight new species. The new species included four holothurians (one representing a new genus), one echinoid, and three asteroids. The cidaroid echinoid *Centrocidaris doederleini* that was formerly considered rare occurs in numerous populations, reaching average densities of 185 ind m^{-2} (Pawson and Ahearn 2001). The ophiuroids *Ophiothix* spp. and *Ophiacantha* spp. often occur together in large aggregations of more than 100 individuals in black sandy areas with negligible current activity (Pawson and Ahearn 2001). The asteroids *Henricia* spp. and *Coronoaster* sp are fairly common on a variety of habitats, but were found mostly on rocky substrates; the stalked crinoids *Thalassometra agassizii* and *Calamocrinus diomedeae* are very rare (Pawson and Ahearn 2001).

6.3 Research

6.3.1 Ecology and Population Biology

Sea Cucumbers

Isostichopus fuscus

Distribution and Density

Isostichopus fuscus is a deposit-feeding aspidochirote species distributed from the Gulf of California to Ecuador, including the Galápagos Islands (Maluf 1991). The cylindrical, tough and flexible body is brownish with darker warts (Fig. 6.3a). The color may change to dark brown depending on the color of the substrate. This species is found mainly on rocky reef areas from 2 to 40 m depth (Hickman 1998) although it has been seen at 60 m. In Galápagos, its distribution is not continuous. Therefore there is a high level of uncertainty regarding its population dynamics. The largest population was concentrated in the western region of the archipelago along the Bolívar Channel (Murillo et al. 2002a). The average size of the individuals was 20 cm in length and the range in wet weight was 100–410 g (average 271 g) (Sonnenholzner 1997). At this location densities fluctuated between 0.27 and 0.40 ind m^{-2} from 1994 to 1997, but decreased to 0.13 ind m^{-2} due to an illegal fishery (Piu 1998, 2000; Martínez 1999; Murillo et al. 2002a, b). In 2003, the lowest adult density was recorded in this area since 1999 (with a reduction of 62 %). Juveniles (<16 cm) were almost completely absent from the survey area from December 1994 until July 1999, after which there was a sudden increase, with juvenile densities increasing by a factor of 20, reaching a peak value of 0.47 ind m^{-2} in April 2001. This maximum density of juveniles implies settlement sometime in early 1998, at the height of El Niño event (Hearn et al. 2005). Following this peak, there was a steady decline, with densities as low as 0.20 ind m^{-2} in the pre-fishery survey of 2003 (Hearn et al. 2005). The Bolívar Channel has been recognized as a sea cucumber nursery ground and has remained closed to fishing activities for the fishing seasons of 2003, 2004 and 2005 (Toral-Granda 2005). Lawrence and Sonnenholzner (2004) reported densities of *I. fuscus* in seven localities at the central region of Galápagos of between 0.02 ± 0.03 and 0.05 ± 0.04 ind m^{-2}.

Growth and Reproduction

There is no knowledge of the recruitment dynamics or growth patterns of the population of *I. fuscus* in Galápagos. Herrero-Pérezrul et al. (1999) and Reyes-Bonilla and Herrero-Pérezrul (2003) obtained preliminary population parameters of *I. fuscus* from México, including estimations of asymptotic length ($L\infty = 32.613$ cm) and growth coefficient ($K = 0.2115$ yr^{-1}). However, Hearn et al. (2005) questioned if those growth parameters were applicable to Galápagos, where individuals might reach 16 cm in length approximately 3.3 yr after settlement and the minimum landing size of 20 cm in length after 4.5–5 yr. *Isostichopus fuscus* reproduces sexually and can spawn throughout the year. Complete information on its reproductive biology is limited for the whole archipelago. Toral-Granda (1996) studied the reproduction of *I. fuscus* in a population at Caamaño Island, Galápagos in 1995. Fifty percent of individuals were sexually mature at 23.6 cm and between 161.0 and 170.0 g dry weight. The smallest individual (less than 16 cm) found with mature gonads had a dry weight between 121.0 and 130.9 g. Four reproductive peaks were found, two reproductive events in each cold and warm water seasons, suggesting a continuous reproduction. The highest gonad index (GI) value was found in December whilst the lowest was recorded in April and September. Mean fecundity (eggs per gonad) for this species was 567,884 ± 95,528. Many animals were regenerating on June 2002 at the western side of Galápagos at the Bolívar Channel (Sonnenholzner, personal observation). These specimens were between 8 and 10 cm in length, and were without a mouth or an anus, indicating an asexual reproduction episode by fission in this population.

Diet and Feeding Habits

Sonnenholzner and Cajas (1993) determined the most dominant food components of *I. fuscus* collected early in the morning at Bolívar Channel at a depth of 12 m. The gut contents contained a high percentage (50.3 %) of polyps of stony corals, followed by benthic diatoms of the genus *Melosira* sp. (12.3 %), *Grammatophora* sp. (11 %) and *Navicula* sp. (9 %), and low percentages of other species of microalgae. A variety of microorganisms, such as foraminifera and tintinnids (0.7 %), as well as abundant detritus was found. *Isostichopus fuscus* is mostly active at night (Hickman 1998; Hearn and Pinillos 2006).

Stichopus horrens

Distribution and Density

Stichopus horrens (Fig. 6.4i) is a large holothurian that can grow to a length of 50 cm (Schoppe 2000). The largest specimen from Galápagos was collected in the western region with a length of 42 cm (Hearn and Pinillos 2006). This species is found in shallow waters at depths between 5 and 20 m, mostly occurs on rocky reef substrates in the central and southern region of Galápagos (Edgar et al. 2002b). *Stichopus horrens* tends to contract to less than half its length during its cryptic behavior during the day. It emerges around dusk to form aggregations with densities exceeding 1–2 ind m^{-2} (Hearn and Pinillos 2006).

Growth and Reproduction

Throughout the Galápagos Islands, the population structure of *Stichopus horrens* has a uni-modal length frequency distribution ranging from 14 to 42 cm. The mean length of individuals was around 28 cm at most islands, although smaller individuals occur at Rábida (22 cm) and Wolf (24 cm). No individuals smaller than 12 cm in length were recorded. Nothing is known about the larval biology and recruitment of this species. Hearn and Pinillos (2006) made preliminary estimates of the Von Bertalanffy population parameters using size frequency distributions ($L\infty = 38.8$ cm, $K = 0.17$), which indicated that an individual of 20 cm in length would correspond to an age of 4 years. *Stichopus horrens* reproduces asexually by fission (Kohtsuka et al. 2005), as well as sexually. Hearn and Pinillos (2006) found individuals greater than 16 cm in length were mature. Fecundity was estimated to vary greatly between 1×10^3 and 1×10^6 eggs per individual.

Holothuria theeli

Distribution and Density

Sonnenholzner and Lawrence (2002) and Sonnenholzner (2003) evaluated the distribution and density by microhabitat preference of *Holothuria theeli* along the continental coast of Ecuador. The maximum density was at the lowest tidal level and there was a progressive decrease in abundance with depth. It was observed on exposed surfaces at low tide, suggesting a resistance to desiccation. The capacity to withstand and exploit air-exposed areas during the tidal cycle has been observed in other species such as *Stichopus badionotus* (Sloan and Von Bodungen1980),

Holothuria atra and *Actinopyga echinites* (Massin and Doumen 1986). *Holothuria theeli* ranged in size from <5 to 20 cm in length. The relative absence of juveniles (individuals less than 5 cm in length) and distinct cohorts upon the midshelf reef suggest different habitat preferences between juveniles and adults (5–11 and 11–20 cm long, respectively). This probably also occurs with the population of *Isostichopus fuscus* in the Galápagos Islands (Martínez et al. 1998), and with *H. atra* and *S. chloronotus* from the Great Barrier Reef (Uthicke 2001). Small and medium juveniles inhabit areas with boulders, cobbles, pebbles and sand, whereas large adults mainly inhabit areas with boulders and cobbles. In order of preference the most used microhabitats were boulders/cobbles > pebbles/sand > thick pavement/bedrock (Sonnenholzner and Lawrence 2002).

Growth and Reproduction

Sonnenholzner (2003) determined abundances of *H. theeli* on transects on a shallow tidal reef flat at the Puntilla Santa Elena (Fig. 6.1) in 1994. Abundances were variable in May but increased by the end of August as a result of asexual reproduction by fission (during the sampling period many individuals were observed without a mouth and none without an anus). The fluctuation in abundance following the period of fission could be related to the survival rates for posterior and anterior body parts (Reichenbach and Holloway 1995). The importance of fission in reproduction is a well-known feature in the biology of this species (Bonham and Held 1963; Uthicke 1997). Many individuals were observed regenerating after asexual reproduction in May. The fission rate of individuals in the intertidal zone to near shore reef was much higher (15–36 %) than at the high-energy midshelf reef zone (3–7 %) (Sonnenholzner 2003).

Diet and Feeding Habits

Sonnenholzner (2003) determined that the distribution pattern of *H. theeli* on a shallow tidal reef flat was apparently not limited by food availability (Yamanouchi 1956). Roberts (1979) and Pomory et al. (1998) suggested differences in feeding time as a means of niche partitioning, where the habitat was partitioned into microhabitats to avoid competition with other holothurians species found in the area. *Holothuria theeli* generally begins burrowing around sunrise and emerges close to sunset, showing a semi-cryptic behavior at night. Sonnenholzner (2003) determined that the feeding activity of *H. theeli* may be low in May compared to March and August, since the intestine length was shorter in May, the capacity of the intestine may fluctuate as a physiological response to food availability (Tanaka 1958). For example, *Stichopus japonicus* ceases feeding after spawning and its intestines degenerate and become smaller in size (Tanaka 1958). The dormant condition in *H. theeli* may result from the change in water temperature from high to low, which may stimulate physiological processes such as feeding inactivity after spawning (Yamanouchi 1942). Under these conditions, *H. theeli* maintains normal feeding activity in March and August but not in May. The dormant condition is evident in juvenile *H. scabra*, where individuals of 4–14 cm in length responded to changes in temperature. Specifically, the maintenance of a constant warm temperature prevents *H. scabra* from burrowing (Mercier et al. 1999). *Holothuria theeli* is undoubtedly a non-selective, generalist feeder throughout the rainy and dry seasons, and appears to be a forager. It feeds at an almost constant daily rate and probably continues to feed through stormy periods. The lowest percentage of prey organisms (such as micro invertebrates and microphytobenthos) occurred in March and the highest in August. The major food sources of *H. theeli* were crustaceans, foraminiferans, diatoms, and detritus. High levels of meiofauna and microphytobenthos in the gut contents of *H. theeli* indicate that both play a significant part in the nutrition of the holothurians without any relationship to the size frequency distribution of the sediment around holothurians (Sonnenholzner 2003).

Sea Urchins

Eucidaris galapagensis

Distribution and Density

Although there were different expeditions since the 1800s to qualitatively study the Galápagos marine diversity, including sea urchins, it was not until the 1950s that information on abundances of sea urchins became available. In 1954, the slate-pencil urchin *Eucidaris galapagensis* Döderlein, 1887 (Fig. 6.3b) occurred at a mean density of 19 ind m^{-2} in the western islands (Malmquist 1991). During the 1970s, some studies indicated that its densities around the archipelago fluctuated between 2–8 and 34–50 ind m^{-2} (Wellington 1975; Glynn et al. 1979; Glynn 1990).

Some years later, Glynn (1988) compared the densities of *E. galapagensis* before and after El Niño 1982–1983 and found that they had increased from 5 to 30 ind m^{-2}. He also found that *E. galapagensis* fed on hermatypic coral polyps and dead corals, which were highly impacted by the warming event. This explained the dramatic increase of the urchin abundances after El Niño (Glynn et al. 1979; Glynn 1988). Similarly, Edgar et al. (2010) examined historical data to document the spatial and temporal scale of recent changes in species distributions and important benthic habitat types. Part of the study included the analysis of underwater images (taken in 1982, 1983 and 1984) to assess changes in the densities of *E. galapagensis* before and after El Niño 1982/1983. They found that densities of this species significantly increased by a factor of about two from 1982 (0.45 ind m^{-2}) to 1984 (1.0 ind m^{-2}).

Brandt and Guarderas (2002) reported *E. galapagensis* was the most common and abundant sea urchin species in the whole archipelago, with an average density of 3.2 ind m^{-2}. They also pointed out that it was abundant in the islands of Rábida and Santiago with average densities of 5.0 ind m^{-2}. The lowest densities occurred in the northern islands of Darwin and Wolf with average densities of 2.1 ind m^{-2}. Irving and Witman (2009) carried out experiments and density surveys of *E. galapagensis* in different habitats in Academy Bay, Santa Cruz Island. They found maximum densities of 28 ind m^{-2} in algal turfs, and only 6 ind m^{-2} in barren habitats. Sonnenholzner et al. (2009) studied the density of *E. galapagensis* in 20 shallow rocky reefs in the southeastern biogeographic region of the Galápagos Marine Reserve (off Santiago, Santa Cruz, Baltra and Seymour Islands), specifically in sites open and closed to fishing. *Eucidaris galapagensis* densities (±SE) were higher in open than in closed sites to fishing (4.5 ± 1.55 vs. 2.2 ± 1.55 ind m^{-2}, respectively). More recently, Paredes (2010) conducted population surveys of *E. galapagensis* at four sites in San Cristóbal Island between August 2009 and March 2010. The highest densities were in Punta Carola and Playa Mann (1.3 ind m^{-2}) followed by La Lobería (1.1 ind m^{-2}) and Tijeretas (0.9 ind m^{-2}).

Size Structure and Gonad Production

Brandt and Guarderas (2002) indicated that the mean test diameter of *E. galapagensis* was 5.12 ± 0.01 cm around the archipelago. The test diameter ranged from 1 to 9 cm and varied between regions/islands ($H = 1462.162$, $p < 0.0001$, $df = 6$), with the largest test diameters occurring in Floreana and the southeastern islands.

During the Base Line survey collections between September 2000 and April 2001 Francisco and Smith (unpublished data) estimated morphometric relationships and gonad index of the most common sea urchins. Twenty individuals of each species (*E. galapagensis*, *Lytechinus semituberculatus* and *Tripneustes depressus*) were collected haphazardly at a depth of approximately 6 m at 12 sites on seven islands. Individuals were dissected and gonad index (GI) was calculated by dividing the drained gonad weight by the total body weight, multiplied by 100. The mean GI (±SE) for *E. galapagensis* was 2.98 ± 0.11 (Francisco and Smith, unpublished data).

Diet and Feeding Habits

Eucidaris galapagensis feeds on encrusting algae (Wellington 1975; Hickman 1998), live hermatypic polyps, dead corals and barnacle plates (Glynn et al. 1979). It contributes to the production of sediment in the subtidal community of the Galápagos (Wellington 1975). It also has the ability to bore deeply into the substrate to create crevices, which reduces its risk of dislodgement from hydrodynamic forces (Witman and Dayton 2001).

Lytechinus semituberculatus

Distribution and Density

The green sea urchin *Lytechinus semituberculatus* (Fig. 6.3c) has been reported as the second most abundant sea urchin species (1.2 ind m^{-2}) of the Galápagos' rocky reefs (Brandt and Guarderas 2002). *L. semituberculatus* was widely distributed in islands with colder and higher productivity water regimes, particularly on Isabela Island (2.7 ind m^{-2}). The lowest densities were in the northern islands (0.1 ind m^{-2}), and on Santa Cruz island (0.2 ind m^{-2}). *L. semituberculatus* shows a significant density pattern associated with depth, suggesting it prefers shallow habitats (<10 m) (Brandt and Guarderas 2002). Edgar et al. (2010) documented changes in its density before and after El Niño 1982/1983. They found a decrease from 0.52 ind m^{-2} in 1982 to 0 individuals in 1984. In another study, *L. semituberculatus* had higher densities in barren grounds (51 ind m^{-2}) than in turf habitats (1.6 ind m^{-2}) (Irving and Witman 2009). Finally, Paredes (2010) conducted population surveys at four sites in San Cristóbal Island between August 2009 and March 2010. Density data showed that Bahía Tijeretas had higher densities of *L. semituberculatus* than the other surveyed sites. The highest density was 4.8 ind m^{-2}, followed by La Lobería (1.6 ind m^{-2}), Playa Mann (0.9 ind m^{-2}) and Punta Carola (0.6 ind m^{-2}).

Size Structure, Allometric Scaling and Gonad Production

Brandt and Guarderas (2002) reported a mean test diameter (±SE) for *L. semituberculatus* in the whole archipelago of 4.3 ± 0.02 cm, with a range of 1–7 cm. There were significant differences in test diameters among islands ($H = 691.363$, $p < 0.0001$, $df = 6$). Populations at Floreana and Santa Cruz islands had the highest mean test diameters (Brandt and Guarderas 2002). In relation to gonad production, Francisco and Smith (unpublished data) found a mean gonad index (±SE) was 7.36 ± 0.43.

In general, body size affects most aspects of an organism's biology (Calder 1984), and scaling of structure and function to body size can be considered in terms of ontogeny and phylogeny. Lawrence et al. (2010) studied the ontogenetic and phylogenetic scaling of lantern components (weight of demipyramid, rotula and epiphysis of the lantern and test diameter) of *L. semituberculatus* from the Galápagos Islands versus three other species of *Lytechinus* (*L. variegatus* from west coast of Florida, U.S.A. and San Blas Islands, Panamá; *L. williamsi* from San Blas Islands, Panamá; and *L. pictus*, northern Baja California, México) and one species of *Tripneustes* (*T. gratilla* from Okinawa, Japan). Results showed that all components were largest in the smallest species (*L. williamsi*), and smallest in the largest species (*T. gratilla*), suggesting a genetic basis and interrelated scaling. All components are larger in *L. variegatus* from Panamá than those from Florida, which could have a genetic and/or environmental basis. These results are very important, since the arguments proposed by Ebert (1980) suggested that large demipyramids of a lantern of sea urchins increase their capacity to feed. Nonetheless, Lawrence et al. (2010) determined that the Aristotle's lantern of sea urchins can show an allometric relation with body size and that it varies intra- and inter-specifically. The reasons for this have not been well established, but these results should be considered for well-focused ecological studies along the Galápagos Islands, concerning growth and reproduction of this abundant species. Because ingestion is the largest term in the balanced growth equation, the feeding apparatus of animals is of particular interest in scaling (Peters 1983).

Diet and Feeding Habits

Hickman (1998) reported *L. semituberculatus* feeds on algae and debris in the Galápagos Islands. It is commonly covered by pieces of seaweed or shells, which would probably be a protection from ultraviolet radiation (UVR). In this sense, Sharp and Gray (1962) provided some of the first evidence that artificial UVR can induce covering in *L. variegatus*.

Tripneustes depressus

Distribution and Density

Idrovo and Sonnenholzner (1994) studied the distribution and abundance of *T. depressus* (Fig. 6.3d) at Punta Los Frailes, Machalilla National Park (Fig. 6.1) along the coast of mainland Ecuador. This species had an aggregated distribution (Morisita index 2.9–1.6), with a density between 1.5 and 1.8 ind m^{-2}.

Luna (2000) conducted density surveys of *T. depressus* at 35 sites around the Galápagos Islands from 1996 to 1997. Density fluctuated from 0 to 1.5 ind m^{-2} with an average value of 0.3 ± 0.4 ind m^{-2}. Higher densities were found in the central region of Galápagos: Rocas Gordon (1.5 ind m^{-2}), Caleta Bucanero (1.2 ind m^{-2}) and Cabo Marshall (0.8 ind m^{-2}), whereas consistent lower densities (mostly no individuals) were found in the western islands (Cabo Berkeley, Bahía Urbina, Bahía Elizabeth, Punta Moreno, Cabo Rosa, Punta Espinoza and Cabo Douglas), with the exception of Caleta Iguana which had densities of 0.2 ind m^{-2}. Luna (2000) also conducted monthly surveys (from October 1996 to April 1997) at one site, Playa Ratonera, Santa Cruz Island. The density of *T. depressus* decreased from 0.5 to 0.1 ind m^{-2} over this period.

To understand spatial and temporal patterns of abundance of *T. depressus,* Guarderas (2000) conducted monthly population surveys (from September 1998 to September 1999) in intertidal and subtidal habitats in Santa Cruz Island. In this study, this species had an aggregated distribution (the ratio variance to average was higher than 1) with high temporal and spatial variation in abundance within sites. Populations in subtidal sites had higher densities and were more stable and more spatially homogenous than in intertidal sites, although there were no statistical differences among habitats. On average, El Barranco had significantly higher average densities (3.0 ind m^{-2}) than Barco Hundido site (0.6 ind m^{-2}), the Charles Darwin Research Station (0.5 ind m^{-2}) and Punta Núñez (0.3 ind m^{-2}). Within all sites, density decreased over time. This result was explained by the movement of sea urchins. It was concluded that *T. depressus* in intertidal sites moved from shallow to deeper areas, probably to find more stable environmental conditions. On the other hand, there was no clear direction or pattern of movement in subtidal sites. Brandt and Guarderas (2002) described abundance of *T. depressus* around the Galápagos Archipelago. The highest densities were found in Floreana (0.9 ind m^{-2}) and in the southeastern islands (Española and San Cristóbal with 0.6 ind m^{-2}) while the lowest values were found on Fernandina (with 0.05 ind m^{-2}). *T. depressus* showed significant density patterns associated with depth, suggesting that it prefers shallow habitats less than 10 m depth (Brandt and Guarderas 2002).

Size Structure, Growth and Gonad Production

Tripneustes depressus is the largest sea urchin species in the Galápagos Islands (Hickman 1998). Size structure data for *T. depressus* is available for more sampling periods than for any other sea urchin species in the archipelago. Population surveys done in 1996 showed that the size frequency distribution had two peaks: (1) 7.0 cm, and (2) 12.0 cm. In 1997, the size-frequency distribution had a single mode (12.5 cm) and a mean size of 11.5 cm. Consequently the former surveys may have presented a bimodal shape due to a smaller sample size (Luna 2000). In 1998, population surveys conducted in the western islands of the archipelago, showed a mass recruitment event (Brandt and Guarderas 2002). In 1999, intensive sampling was conducted around three islands of the archipelago (Española, Floreana and San Cristóbal). The results showed a bimodal size frequency distribution for each island (Brandt and Guarderas 2002). On the other hand, the information for the 2000–2001 periods indicates that the mean test diameter around the archipelago was 10.0 ± 0.04 cm, with a range of 1–14 cm. The size frequency distribution had two peaks (5.0 and 9.5 cm). Test diameter was significantly different ($H = 499.9$, $p < 0.0001$, $df = 6$) between regions/islands (Brandt and Guarderas 2002)

Guarderas (2000) determined growth parameters for *T. depressus* using modal progression of size-frequency distributions of monthly data (September 1998 to September 1999) from four sites in Santa Cruz Island (El Barranco, Barco Hundido, Muelle de la Estación Charles Darwin and Punta Núñez). Data from all sites were pooled to estimate population parameters. Analyses and processing of length data were conducted using the Length Frequency Distribution Analysis (LFDA Version 4.01), from the FISAT (FAO/ICLARM) package. Growth rate K ($1 - L(t)/L\infty$) declines with time as the diameter of the animal reaches the asymptotic length. Estimation of growth parameters (K, $L\infty$) of the Von Bertalanffy growth curve was done followed by definition of grid boundaries using the methods of Powell-Wetheral and ELEFAN (Electronic Length Frequency Analysis). The asymptotic diameter (L_∞) ranged from 144.7 mm in the Punta Núñez site to 178.8 mm Barco Hundido. In the same way, the body growth rate coefficient (K) varied from 0.50 in the Barco Hundido site to 0.90 in the Punta Núñez site. The parameters estimated from all sites were $L_{\infty} = 182.3$ mm and $K = 1.38$. The growth function of Von-Bertalanffy used in this study has often been used to calculate growth parameters of fish and many marine invertebrates, including sea urchins. However, other models such as Richards, Jolicour, Tanaka and Gompertz described better growth curves of sea urchins. All these models require information obtained from capture–recapture methods of marked individuals. Therefore it is necessary to carry out direct studies of growth for *T. depressus* and to compare with the growth parameters obtained in this study (Guarderas 2000). Growth rates for *T. depressus* were also estimated in laboratory by monthly measurements of test diameters of different cohorts. Growth rates ranged from 4 to 10 mm per month with an average of 5 mm per month (Guarderas 2000).

In relation to gonad production, Luna (2000) estimated gonad index from bimonthly collections of 20 urchins from Caamaño islet (Santa Cruz Island) between January 1997 and January 1999. This study showed that there was no clear evidence for a distinct reproductive annual cycle for *T. depressus*. It seems that there is spawning activity throughout the year. However, two peaks were observed during the sampling period; one in November 1997 and the other in June 1998 (Luna 2000). On the other hand, Francisco and Smith (unpublished data) determined that the mean gonad index for *T. depressus* was (±SE) 6.09 ± 0.36. This information was obtained of twenty individuals collected haphazardly at approximately 6 m depth at 12 sites on seven islands. Field sampling was conducted between September 2000 and April 2001.

Diet and Feeding Habits

Guarderas (2000) studied the diet composition of *T. depressus* in Caamaño islet, Santa Cruz Island from January 1998 to March 1999. Stomach contents were sampled monthly. The diet consisted mainly of algae (98 %), although remains of sponges, and unidentified invertebrates (2 %) were also found. For the algae, red filamentous algae were the most abundant (49 %) and most frequent functional group (99 %) in its diet (Guarderas 2000). There was temporal variation of different algal functional groups in the diet, which suggests that food availability could determine the food of *T. depressus*. In addition, laboratory observations showed that *T. depressus* in high density and starvation conditions (with low food availability) are cannibalistic. These findings suggest that *T. depressus* tends to be more a generalist than a specialist (Guarderas 2000).

Encope micropora

Size Structures and Allometric Scaling

Differences between paired structures sometimes occur normally, but random variation can give insight into developmental stability and environmental stress.

In clypeasteroid sand dollars, the modified ambulacra, or petals, are among the most conspicuous bilaterally paired structures. Lawrence et al. (1998) studied the bilateral symmetry of the petals in the Ecuadorian sand dollar *Encope micropora* L. Agassiz, 1841 versus others two species of sand dollars (considering large and small specimens): (1) *Mellita tenuis* collected at Mullet Key, Florida, USA (Scutellina, same sub-order of *E. micropora*); and (2) *Arachnoides placenta* collected at South Mission Beach, Queensland, Australia (Clypeasterina, different sub-order of *E. micropora*). Analysis of the anterior and posterior petals of *E. micropora* suggested they were asymmetrical. This fluctuating degree of asymmetry for some petal pairs might be a means to evaluate environmental stress. Knowing the type of asymmetry is necessary to understand its origin, because certain environmental conditions and forces can affect test form in sea urchins (Ellers and Telford 1992; Ellers 1993).

Other Sea Urchins

Distribution and Density

Pinela and Calvache (1997) studied the distribution (Morisita index), and density of *Echinometra vanbrunti* in four different localities along the southwest coast of Ecuador. They found that this sea urchin had an aggregated distribution at all sites, and that its densities ranged from 0.87 to 8.80 ind m^{-2}. The abundance of this species along mainland Ecuador is remarkably higher than in the Galápagos. Brandt and Guarderas (2002) reported that *E. vanbrunti* presented densities of 0.3 ind m^{-2} in the Galápagos Archipelago. In addition, Wellington (1975) reported that *E. vanbrunti* was abundant in the southern, central (including the eastern side of Isabela Island) and northern regions (with 4.38 ind m^{-2}), but rare in Fernandina and on the western side of Isabela Island where *Caenocentrotus gibbosus* (Fig. 6.4h) was the dominant species. Three other sea urchins (*D. mexicanum*, see Fig. 6.3e; *Toxopneustes roseus*, see Fig. 6.3f; and *Centrostephanus coronatus*) were also reported by Brandt and Guarderas (2002). However their abundances were significantly much lower (<0.01 ind m^{-2}) than the other sea urchin species found there. General findings of the Galápagos Marine Biodiversity Baseline, including density patterns of sea urchins, were also published by Edgar et al. (2004). Lawrence and Sonnenholzner (2004) documented the occurrence of eight species of echinoids at seven sites with distinct habitats in central Galápagos. The most diverse assemblage was at Itabaca Channel with *E. galapagensis*, *L. semituberculatus*, *T. depressus*, *T. roseus*, *C. coronatus*, *E. vanbrunti*, *C. gibbosus*, and *D. mexicanum*. All had densities <2.3 ind m^{-2}. The assemblages of species are similar to those reported for the Florida Keys and the Hawaiian Islands (Kier and Grant 1965; Ebert 1971).

Sea Stars

Luidia (Platasterias) latiradiata

Distribution

The living genus *Platasterias* is a surviving somasteroid and *Luidia* is a surviving basal asteroid (Fell 1963). *Luidia (Platasterias) latiradiata,* based upon its skeletal morphology, is considered a "living fossil" (Fell 1963; Caso 1970; Blake et al. 2000). The biogeographical distribution of this species has been extended to Ecuador. It had been found close to the coast of México and Central America (Caso 1970, 1976; Solís-Marín 1998). Villamar and Cruz (1983) found large specimens on a sandy bottom in the intertidal zone at Playas, General Villamil, in the mainland of Ecuador (Fig. 6.1). In México, this species have been reported by incidental mortality due to shrimp fisheries, where this assumes an impact that can cause a decrease in their populations. For more information, review the México chapter.

Diet and Feeding Habits

Villamar and Cruz (1983) reported *L. latiradiata* fed on crustaceans (i.e. cumaceans, miscidaceans), gastropods (*Olivella semistriata*), and bivalves (*Donax mancorensis*).

Phataria unifascialis and *Pharia pyramidata*

Distribution and Density

Distribution of sea star species in shallow waters along the mainland of Ecuador is essentially unknown. Sonnenholzner and Lawrence (2002) conducted a brief survey focused on two sea stars, *P. pyramidata* (Fig. 6.3h), and *P. unifascialis* (Fig. 6.4 g–h) at Los Frailes, Machalilla National Park, Manabí. They made sixty-one transects (2 m wide × 25 m long "double" quadrants, a total area of 3,100 m^2). The density of *P. unifascialis* and *P. pyramidata* was 0.017 ind m^{-2} and 0.0006 ind m^{-2}, respectively. These species were found mainly on open rocky flat bottoms among the rocks, on or under boulders and dead coral.

Arm Size

Sonnenholzner and Lawrence (2002) measured *P. unifascialis* and *P. pyramidata* in the southern and northern embayment at Los Frailes, Manabí. The arm radius of *P. unifascialis* ranged from 40 to 80 mm, with most being 60–70 mm. The arm radius of a small distinct cohort was 90–95 mm. The arm length of *P. pyramidata* was between 93 and 121 mm.

6.3.2 *Immune Response*

Sea Urchins

It is well known that high temperatures have caused a widespread mortality of key species, coral bleaching in many regions of the world, a decrease in number, density and biomass of species with high immigration of predators and a decrease in number of grazers (Epstein et al. 1998). Considering the ecological role of sea urchins as key species in shallow coastal and marine rocky environments, and the paucity of knowledge of their immune response during strong natural disturbances, Sonnenholzner (2001) documented in detail the immune response of the sea urchin *E. vanbrunti* during El Niño episode of 1997–1998 and the cold phase La Niña of 1999–2000. He made observations under laboratory (in vitro) and natural conditions (in vivo, where specimens were collected along a gradient of intertidal rocky pools at Anconcito, Santa Elena, see Fig. 6.1). Four coelomocyte types were identified in the perivisceral coelom of *E. vanbrunti*: (1) the phagocytic amoebocytes, which were the only phagocytic cells studied in this experiments (Fig. 6.5a, b); (2) spherule cells, divided into colorless (with intense amoeboid movements, and colorless cytoplasmic granules, see Fig. 6.5c), and red cells (similar to colorless spherules, but with red granules, see Fig. 6.5c); and vibratile cells (flagellated and small rounded cells, see Fig. 6.5d). The effect of three different temperatures: 19 ± 1 °C (low); 24 ± 1 °C (medium); and 29 ± 1 °C (high) on the immune response of *E. vanbrunti* was evaluated using the perivisceral coelom fluid and four immune assays: (1) antibacterial activity from the plasma (clearance efficiency); (2) total hemocytes count (THC); (3) differential hemocytes count (DHC); and (4) microbicidal activity (oxidative activity by nitroblue tetrazolium reduction related to phagocytosis generated by stimulated phagocytes in phorbol 12-myristate 13-acetate) (Sonnenholzner 2001). Both hemograms (THC and DHC) were useful assays related to the microbicidal activity. At the high temperature in the laboratory conditions and during the mid phase of the El Niño 1997–1998 warming event, four remarkable aspects of the phagocytes were observed in natural conditions: (1) phagocytes were very motile cells and spread rapidly, accompanied by a complete change in morphology (Fig. 6.5a, b); (2) the aggregation-clotting process was accelerated (1.0 ± 0.5 min) at high temperature conditions (>24 °C); (3) the oxidative activity increased ("respiratory burst"); and (4) production of spheres-cellular masses with formation of microvilli (Fig. 6.5b). A significant difference in clearance efficiency was observed with different temperatures. This study showed that hemograms (DHC and THC), clearance efficiency, and phagocytosis are reduced in *E. vanbrunti* at high temperature conditions. These data are important because they provide baseline values for tropical echinoid species in their natural environments that are affected by strong environmental stress episodes. Thus, it is an excellent opportunity for study and verifying through these immune indicators the health condition of this potential bioindicator (as *E. vanbrunti*). Nonetheless, these findings indicate that

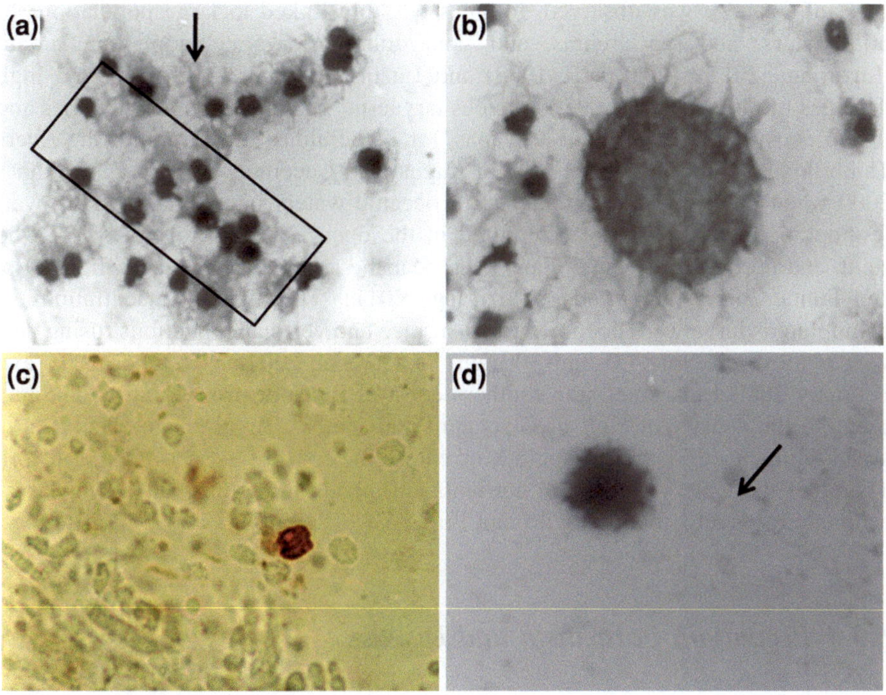

Fig. 6.5 Photomicrography of four types of coelomocytes identified in the perivisceral coelom of the sea urchin *Echinometra vanbrunti*: **a** phagocytic amoebocytes in two different morphological phases: a bladder or petaloid phase (into the square), and a filiform phase (*black arrow*); **b** aggregate of filiform phacogytic amoebocytes; **c** *red* and *colorless* granulocytes; and **d** flagellate cell, *arrow* indicate long flagellum. Photos taken by J. Sonnenholzner

more experimental studies are necessary to reach a definitive conclusion about the possible effect of temperature in the immune response in the tropical sea urchin *E. vanbrunti*.

6.3.3 *Larval Ecology*

In general, the larvae of echinoderms are an integral part of marine zooplankton collections in the Ecuadorian waters, but their abundance and distribution have been reported as rare (Luzuriaga 2011). Luzuriaga (2011) determined areas of distribution and abundance of larvae of echinoderms (mainly *ophiopluteus*) based on the analysis of time series data of zooplankton from fixed coastal stations (from 1996 to 2009) and oceanographic cruises (from 2004 to 2009). In some regions and years an increase of the number of larvae of echinoderms along the continental coast of Ecuador was observed, in areas possibly linked to large populations of brittle stars, and sea urchins, and sand dollars (larva echinopluteus larvae) (Luzuriaga 2011). Luzuriaga (1993) and Luzuriaga et al. (2007) reported high densities of pluteus larvae (>200 pluteus larvae m^{-3}) along the inner estuaries and inland ecosystems of "Mataje-Cayapas" at Esmeraldas, at the mouth the rivers Tambillo and Najurungo (Fig. 6.1), in August and December 1979 and November 2000 when there is a large influence of oceanic waters into estuaries. However, few ophiopluteus larvae were reported in the open sea in October 2003, at the south and north of the Ecuadorian coast, near the archipelago of Jambelí, El Oro, and Punta Galera, Esmeraldas (Luzuriaga 2011). The higher concentrations of ophiopluteus larva were documented in surface and vertical collections (using nets with a mesh size of 200 and 300 μm) in October 2004, November 2005 and February 2009 cruises near the southern coast of Ecuador, from Puntilla de Santa Elena to the estuary of Gulf of Guayaquil and Jambelí Archipelago. This area is greatly influenced by the Humboldt Current off the coast of Ecuador (Fig. 6.6 a–j). Therefore the waters are rich in nutrients and plankton biomass that favor larval echinoderm development (Coello and Prado 1999; Luzuriaga and Elías 1999).

6.3.4 Predation, Parasitism and Disease

Sea Cucumbers

Sonnenholzner and Molina (2005) provided information on the host–parasite relationship that occurs along the southeastern and western bioregions of the Galápagos Islands by examining two families of aspidochirote holothurians: (1) Holothuridae: *Holothuria portovallartensis* (n = 1557) and *H. theeli* (n = 983), and (2) Stichopodidae: *Isostichopus fuscus* (n = 850) and *Stichopus horrens* (n = 150). *Holothuria portovallartensis* and *H. theeli* were infected by the same gastropod parasite *Balcis panamensis*. The prevalence of infection was high, with 95 and 79 %, respectively. The maximum number of parasites (adult and juvenile) in the same host was eight and six, respectively. Both *I. fuscus* (n = 850) and *S. horrens* (n = 150) had a low frequency of parasitism by *Melanella falcata*.

Sand Dollars

Echinoderms are subject to predation by crabs, gastropods, lobsters, fish, and other species, and their populations are affected by biotic diseases caused by a variety of parasites and microorganisms (Jangoux 1990). In particular clypeasteroids are preyed upon by gastropods that drill holes into the coelomic cavity, although the attack is not always successful (Moore 1956; Gerace and Lyndsay 1992; McClintock and Marion 1993). Disease symptoms of populations undergoing mass mortalities include lesions and necrotic tissue of the body wall caused by vibrios

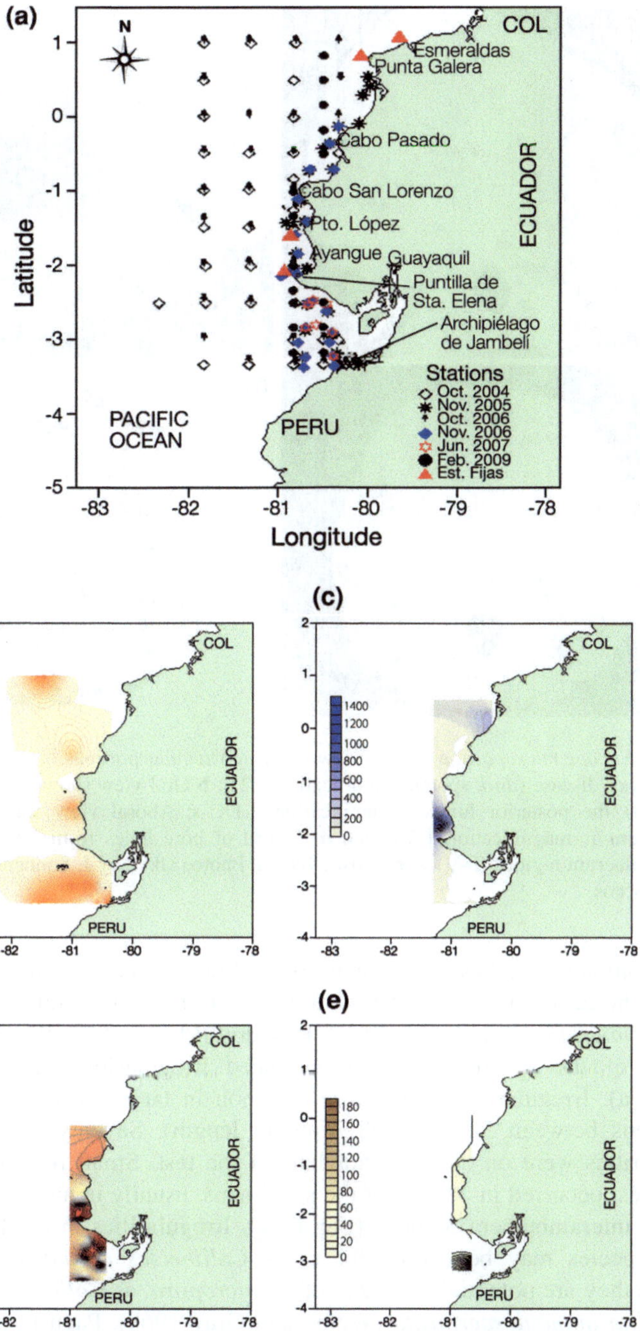

Fig. 6.6 Location of stations 1996–2011 **a** Abundance of larva ophiopluteus larvae 100 m^{-3} during research cruises, in surface waters: **b** October 2004, **c** November 2005, **d** November 2006, and **e** February 2009; taken from Luzuriaga (2011)

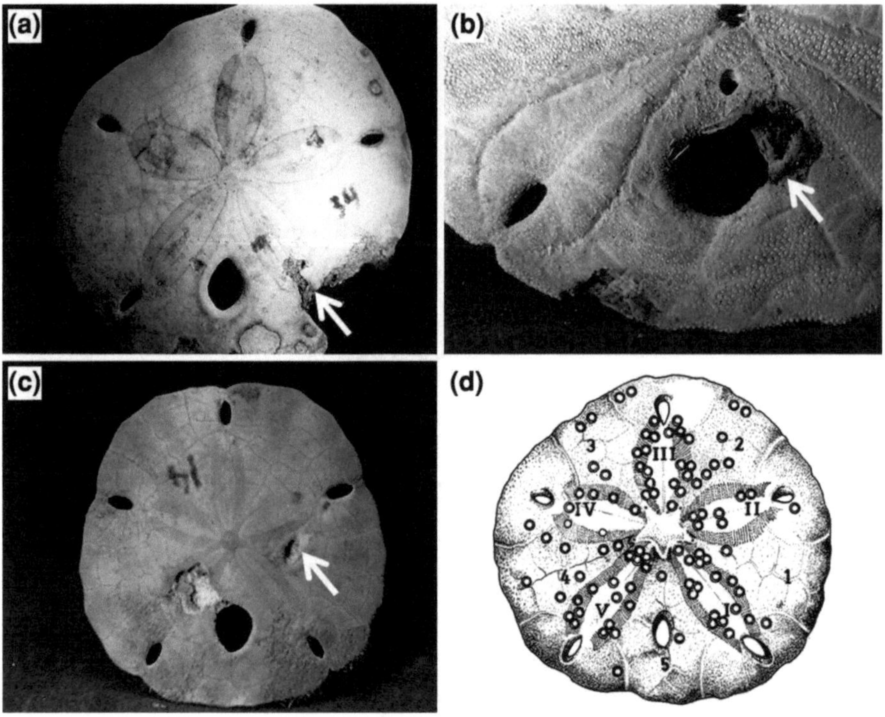

Fig. 6.7 *Encope micropora*. **a** Aboral view showing irregular portions of the test. Note also the indication of disease (*dark* spots), magnification 1.2X; **b** Oral view showing diseased portion of the test at the posterior lunule, magnification 2.1X; **c** Aboral view showing bore hole in ambulacrum 1, magnification 1.2X; and **d** Record of bore holes from 500 specimens, where interambulacrum regions (I–V) were most affected. Photos taken by J. Sonnenholzner, and figure by M. Burgos

and aeromonads (Gilles and Pearse 1986). Sonnenholzner and Lawrence (1998) studied the occurrence of predation and disease in a population of the sand dollar *Encope micropora* on the coast of Ecuador (Playas de Villamil). Thirty-eight percent of the specimens had pronounced irregularities on the test margin (Fig. 6.7a). Irregularities were more common in large individuals (81 % of the specimens between 52.0 and 98.5 mm in length). Seventy-four percent of the irregularities were on the posterior half of the test. Small rounded holes on the aboral test occurred in 48.4 % of the specimens, usually in ambulacra I, IV and V and the interambulacra 3 and 4 (Fig. 6.7c). Irregularities in the test of this sand dollar species may be caused for crabs *Callinectes arcuatus* and *C. toxotes*, because they are potential predators of *E. micropora*, as *Callinectes sapidus* preys on *Mellita quinquiesperforata* (Weihe and Gray 1968). Pagurid hermit crabs are other potential predators. The bore holes found in the tests indicate gastropod attacks (Fig. 6.7c, d). The gastropod *Northia pristis* and *Polinices* sp. are common in the area.

Eighty-seven percent of specimens were diseased (Fig. 6.7b). Symptoms were most frequent on the oral side (65 %) and in the posterior lunules (89 %). Lesions extended to the adjacent posterior interambulacrum (Fig. 6.7b). These lesions were found in the larger specimens (46.4 to 106.0 mm in length), with a mode (38 %) at 62.7 mm. Some specimens had small black spots with surrounding lesions on the oral surface (Fig. 6.7a, b). The authors indicated that oils from fishing boats and commercial crafts are common in this area, and these conditions probably lead to lesions of the epithelium of the test of *E. micropora* that subsequently became infected.

Sea Urchins

Sonnenholzner et al. (2009) studied the effects of predation by fish (the Mexican hogfish *Bodianus diplotaenia*), and lobsters (red spiny lobster *Panulirus penicillatus*, and rocky lobster *Scyllarides astori*) on the herbivore *E. galapagensis*. The authors sampled 10 highly fished and 10 (putatively) lightly fished shallow rocky reefs in the southeastern area of the Galápagos Islands. *Eucidaris galapagensis* showed a negative association with non-coralline algae. In addition, pencil urchins were less abundant where there were many predators indicating an indirect positive association between predators and non-coralline algae occurred. Fishing appeared to affect this trophic cascade. The spiny lobster *P. penicillatus*, the slipper lobster *S. astori*, and the hogfish *B. diplotaenia* were significantly less abundant at highly fished sites. Sea urchin density was higher at highly fished sites. Non-coralline algae were nearly absent from highly fished sites, where a continuous carpet of the anemone *Aiptasia* sp. occurred. The algal assemblage was mainly composed of encrusting coralline and articulated calcareous algae.

Sonnenholzner et al. (2011) conducted an experiment for measuring lobster predation on *Eucidaris galapagensis* in Galápagos. Six slipper lobsters *Scyllarides astori* (19.0–26.2 cm total length, 0.50–0.91 kg) and six red spiny lobsters *Panulirus penicillatus* (25.0–28.0 cm total length, 0.60–1.13 kg) were used as predators. The sea urchin *E. galapagensis* (36.7–51.6 mm test diameter, average 44.9 mm) was offered as prey to each lobster. Two treatments were used for each species of lobster and a control treatment of sea urchins with no lobsters. Sea urchin survivorship was measured for 3 days. Spiny lobsters ate three sea urchins and attacked an additional four urchins. Slipper lobsters did not interact with sea urchins. This experiment identified spiny lobsters as potential predators of urchins. There are many other important predator–prey interactions in the system (Witman et al. 2010). Slipper lobsters might consume sea urchins in the wild, even though they did not prey on them in the laboratory. Predation in aquaria does not necessarily indicate predation rates or preferences in the field only that these predators are capable of eating the prey offered to them. Still, observations in aquaria are wholly consistent with the limited field observations of trophic interactions (Sonnenholzner et al. 2011).

Although the taxonomy and distribution of echinoderms in the Galápagos Islands are well known, their biological interactions with parasitism are not. Echinoderms are intensively parasitized by eulimid gastropods (Pearse and Cameron 1991), but there has been little research about the eastern Pacific eulimid species and their ecology (Skoglung 2004). Sonnenholzner and Molina (2005) provided information on the sea urchin–eulimids host–parasite relationship along the southern and western regions of the Galápagos Islands. A total of 1,855 specimens of *E. galapagensis* were examined. Of them, 925 individuals were mainly affected by *Sabinella shaskyi*. The maximum number of parasites in a host was 63 (both adult and juvenile parasites), seven on the same spine. Between 25 % and 42 % of the host spines were affected. *Eucidaris galapagensis* (n = 930) had another parasite, *Pelseneeria* sp., which crawled freely on the host test and was commonly found attached but not on or inside the spines as *S. shaskyi* was. *Pelseneeria* sp. affected the gonads of the host in weight (a low gonad index) and color (female gonads were pale). The infestation reached high percentages (95 %) and the maximum number of parasites on the same host was five (adult parasites only). At night, snails occasionally occurred on the substrate near sea urchins (J. Sonnenholzner, pers. obs.), suggesting that snails may migrate between sea urchins. It is unknown how *S. shaskyi* affects urchins at the population level, but loss and reduction of spine number (Metz 1994; Sonnenholzner and Molina 2005) would appear to increase susceptibility to predation and energetic costs for spine repair (Heatfield 1971; Denny and Gaylord 1996).

Parasites dominate food web links (Lafferty et al. 2006). For this reason, Sonnenholzner et al. (2011) took advantage of the variation in fishing pressure to investigate complex indirect effects on two eulimid snails that parasitize the most abundant sea urchin *E. galapagensis* in the Galápagos. Past work in the Galápagos suggests that fishing reduces lobster and fish densities and this relaxation of predation pressure indirectly increases sea urchin densities, creating the potential for complex indirect interactions between fishing and parasitic snails. To measure indirect effects of fishing on these parasitic snails, the spatial relationships among sea urchins, parasitic snails, commensal crabs, and large sea urchin predators (hogfish and lobsters) was investigated. Parasitic snails had higher densities at sites where sea urchins were abundant, probably due to increased resource availability. Commensal crabs that shelter under sea urchin spines, particularly the endemic *Mithraculus nodosus*, preyed on the parasitic snails in aquaria. Snails were less abundant at field sites where these crabs were common. In aquaria, hogfish and lobsters readily ate crabs, but crabs under sea urchin spines were protected from predation. This would lead to a facultative mutualism between commensal crabs and sea urchins. In the field, fishing appeared to indirectly increase the abundance of sea urchins and their commensal crabs by reducing predation pressure from fish and lobsters. Fished sites had fewer snails per sea urchin, probably due to increased predation from commensal crabs. However, because fished sites also tended to have more sea urchins, there was no significant net effect of fishing on the number of snails per square meter. These results suggest fishing can have complex indirect effects on parasites by altering food webs.

Sea Stars

Arm regeneration in starfish is related to sub-lethal predation. Sonnenholzner and Lawrence (2002) collected few starfish with signs of sub-lethal predation at Los Frailes, Manabí. They found that one of the 54 specimens collected in their study had four arms and two others had regenerating arms. They observed that the parrotfish *Scarus perrico* is the predator of the tan starfish *P. unifascialis*. Nonetheless, the strength of the body wall, and the availability of food may affect the susceptibility to predation (Lawrence 1987). The low incidence of wounded or regenerating arms in the population of *P. unifascialis* suggests parrotfish are less inclined to feed on it, probably due to its strong body wall.

6.3.5 Interacting Effects of Echinoderms on Benthic Communities

Sea Urchins

Of all benthic consumers, sea urchins are known to be one of the most influential taxa affecting community structure (Lawrence 1975). Sea urchins have the potential to influence the distribution, relative abundance and species composition of algae and other sessile invertebrates (Kitching and Ebling 1961, Paine and Vadas 1969), thereby determining the biomass, diversity and productivity of marine communities. In Ecuador, studies on this topic have been done only in the Galápagos Islands, with the pioneer work of Peter W. Glynn in the late 1970s. In an assessment of coral growth versus coral attrition from grazing, Glynn et al. (1979) showed that *E. galapagensis* reduced coral growth in the Galápagos. He reported individuals of *E. galapagensis* grazing heavily on live hermatypic pocilloporid corals, including *Pocillopora damicornis*, *P. elegans*, *P. capitata* and also *Pavona clavus*. Given that urchins were more abundant along the reef edge, grazing was more intense there. The study suggested *E. galapagensis* might limit reef growth, especially by interfering with the development of reef frame and its lateral expansion. By comparing Panamá to the Galápagos, Glynn et al. (1979) attributed the lower buildup of coral in the Galápagos to a much heavier sea urchin grazing and suggested this difference was due to a lower predation of sea urchins in the Galápagos.

Almost a decade later, Glynn (1988) showed that the distribution and population densities of *E. galapagensis* changed dramatically following the 1982–83 El Niño disturbance. Once abundant in off-reef and reef edges areas, sea urchins moved to the center of the dead reef frameworks of *Porites* and *Pavona*, causing bioerosion rates that ranged from 49 to 99 g $CaCO_3$ m^{-2} day^{-1}. Before 1983, *E. galapagensis* at high densities grazed extensively on live pocilloporid corals (Glynn et al. 1983). However after the El Niño event these corals were locally extinct in Floreana and the remnant non-pocilloporid corals, with less than 0.1 % of cover, were avoided by the urchins. Glynn (1988) suggested that the destructive effect of sea urchins were the result of erosion of reef structures rather than direct mortality of surviving corals. The impact of *E. galapagensis* on habitats other than coral reefs was described by Brandt (2003). This study was done at a depth of 5 m off Caamaño Islet (immediately to the south of Santa Cruz Island, outside Academy Bay). It consisted on removing all *E. galapagensis* from a 6 × 6 m treatment plot. After 12 weeks macroalgal biomass was 20 times higher in the treatment than in the control plot, which had densities of ~40 ind m^{-2}. In addition, benthic diversity, including sessile invertebrates and algal species, was higher in the plot devoid of *E. galapagensis*, while plots with high densities of the sea urchin were characterized by crustose coralline algae and bare rock. Similar to Glynn (1988), who reported that coral erosion rates by urchins were slower when the damselfishes *Stegastes beebei* and *Stegastes arcifrons* were present, Brandt (2003) also noted that the abundance of the green foliose algae *Ulva* sp. was maintained in plots where both urchins and damselfishes were abundant. The role of damselfishes in mediating the impact of urchins thus seems important. Irving and Witman (2009) explored the interactions between the yellow-tail damselfish *Stegastes arcifrons* with *E. galapagensis* and with the "green" sea urchin *L. semituberculatus.* They found that the ability of sea urchins to switch habitats by removing algal turf to create coralline barrens was dependent on the absence of damselfish as well as the species of urchin. In the presence of damselfish, neither species of urchin caused any significant change to algal assemblages. However, in their absence, only *L. semituberculatus* was able to switch the habitat towards greater cover of encrusting algae by grazing filamentous turfs. In addition, one of their results contrasts with previous findings about the impact of *E. galapagensis* on the abundance of macroalgae described above. Irving and Witman (2009) found that removal of *E. galapagensis* from algal communities dominated by encrusting coralline algae had no effect on algal community structure and abundance, while the removal of *L. semituberculatus* resulted in a significant decrease of encrusting coralline algae and an increase in the abundance of green and other erect algae. Since the densities of *E. galapagensis* and *L. semituberculatus* manipulated by Irving and Witman (2009) were different, they concluded that relatively high densities of *L. semituberculatus* impact algal habitats, but relatively low densities of *E. galapagensis* do not. Nevertheless they noted that the densities manipulated were the same as natural densities at the study site. The study of the impacts of sea urchins on benthic communities has greatly improved our understanding about the effects of their grazing and bioerosion in the Galápagos. However more research is needed with other species of sea urchins, such as *T. depressus* and *D. mexicanum* in the Galápagos, and also with sea urchin species on the coast of Ecuador.

6.4 Fisheries and Aquaculture

6.4.1 Sea Cucumbers

Isostichopus fuscus

Sea cucumbers received little or no attention from the scientific community in Ecuador prior to the beginning of commercial harvesting of *I. fuscus* in the early 1990s (Aguilar et al. 1993; Richmond and Martínez 1993; Sonnenholzner 1997). Since then, most studies focused on the biology of this species and other potentially commercial species (such as *S. horrens* and *H. theeli* (Sonnenholzner 2003; Mercier et al. 2004; Hearn 2005, Hearn and Pinillos 2006; Toral-Granda and Martínez 2007), on the social and political implications of the fishery, especially in the Galápagos Islands (Bremner and Pérez 2002), and on the potential for aquaculture (Mercier et al. 2004). In addition, the Charles Darwin Foundation in Galápagos undertook a long term subtidal rocky reef monitoring program to obtain information on the size and abundance of the sea cucumbers found in the Galápagos Islands was recorded (Murillo et al. 2002a). No studies have been done in Ecuador mainland.

About 40 species of commercial sea cucumbers, including *I. fuscus* are harvested throughout the world's oceans (Conand 2004). The fishery of *I. fuscus* has been relatively well documented since the start of its legal fishery in 1999. This fishery began in the mainland Ecuador in 1988. Four years later it was centered mainly on the western side of the Galápagos Archipelago because the sea cucumber population rapidly collapsed on the mainland. This caused the transfer of fishing activity to migrate to the Galápagos Islands (Aguilar et al. 1993; De Paco et al. 1993; Camhi 1995; Powell and Gibbs 1995). At least 100 fishers, mainly from continental Ecuador, began exploiting sea cucumbers in Galápagos in 1992 (Shepherd et al. 2004) although the first signs of sea cucumber harvesting were detected in 1991 (Carranza-Barona and Andrade-Echeverría 1996). Nevertheless the incipient fishery lacked any form of regulations or technical guidelines for its sustainable management (Martínez 2001). After several years of unregulated exploitation, an experimental fishery was opened in 1994, with a catch quota of 500,000 individuals over a 2-month period. However, the fishing season was closed after 6 weeks, with an estimated catch of 8–12 million sea cucumbers (De Miras et al. 1996). Ecuador's General Fisheries Direction reported that ten companies were exporting dried sea cucumbers. Data showed that sea cucumber fishery was less than 0.04 t (fresh total weight) from 1983 to 1988 and 50.3 t from 1989 to 1995. The highest catch (30 t) was reported in 1995. These products were all exported, 67 % to the U.S. and 33 % to Taiwan. The maximum value paid per net kg in Ecuador was USD$30 (Sonnenholzner 1997). Although it is not the most valued sea cucumber species (Conand 2004), this species has been recognized as an important food item in the Chinese market. Sonnenholzner et al. (1997) determined the proximate composition of the body wall of *I. fuscus* collected in April 1993 at the eastern coast of the Fernandina Island along the Bolívar Channel from Punta Mangle to Punta Espinoza. The concentrations (per cent wet weight) of ash (1.0–1.5 %), protein (2.0–4.0 %), fat (3.1–3.6 %), and carbohydrates (<1.0 %) were all low as the water content was high (93.5 %). Carbonate and chloride concentrations were very low (<1.0 %). This indicates the body wall of *I. fuscus* is not a high-quantity protein food source.

The *I. fuscus* fishery was banned from 1995 to 1998. In 1998, the Galápagos Marine Reserve (GMR) was created, along with a novel management system based on the participation of local stakeholders (Galápagos National Park Service, Charles Darwin Research Station, tourism, fishing and local guide sectors). One of the first issues addressed by the new participatory system was to agree upon management measures to re-open the sea cucumber fishery in 1999. Since then, this fishery has been characterized by annual negotiations, conflicts, inconsistencies and a deterioration of the resource which has resulted in its collapse (Shepherd et al. 2004). At first, local fishers remained focused on whitefish and lobster, but as the sea cucumber fishery became established, this figure rose to a maximum of 1,229 active fishers in 2000. After this year, the number declined to just 436 active fishers in 2007 (Murillo and Reyes 2007). In the first official fishing season (1999) 4.4 million sea cucumbers were harvested, 4.9 million in 2000, 2.7 million in 2001, 8.3 million in 2002 and 5.0 million in 2003 (Toral-Granda et al. 2006). From 2002 onwards quotas were set in place which generally outstripped the capacity of the resource, and catches declined steadily until the fishery was closed in 2006. After pressure from fishers and local politicians, it was re-opened in 2007. Overall, more than 25.3 million sea cucumbers (over 6800 t live weight) have been extracted legally from the GMR since 1999 (Murillo et al. 2002a, b). Since its re-opening in 1999, the fishery for *I. fuscus* in the Galápagos Islands has been regulated by a minimum size limit for both fresh and processed sea cucumbers (among other regulatory measures). The size limits are not regarded as a useful tool by the fishers because they believe that sea cucumbers change length too easily to be accurately measured. Nonetheless, local fishers mostly agree, in principle, to have minimum legal size limits rather than none at all. Both management authorities and scientific bodies concur that this regulation is not as effective as others.

Sea cucumber density displayed an irregular pattern with high values of >0.4 ind m^{-2} between 1995 and 1997, but with very low values in 1999 (only 0.13 ind m^{-2}). No surveys were made in 1998. However, after 1999, there was a rapid recovery to prior levels, peaking in 2002, after which densities dropped rapidly and showed no further signs of recovery. At the same time, catch per unit effort (CPUE) measured by onboard observers, increased from 90 to 136 ind diver h^{-1} (between 1999 and 2002, respectively), and then dropped to 55 ind diver h^{-1} by 2005. The decline of *I. fuscus* populations has led to the illegal exploitation of three other, less valuable, sea cucumber species in the GMR: *Stichopus horrens*, *Holothuria kefersteinii* and *H. atra* (Hearn and Pinillos 2006; Toral-Granda et al. 2006).

Several methods, such as molecular phylogeny, morphology, visual, and skeletal features, have been used to try to identify different species of sea cucumbers while live or fresh (Uthicke and Benzie 2003; Uthicke et al. 2004). Concerned with the illegal fishery of *I. fuscus* in the GMR, Toral-Granda (2005) studied the shapes and combination of microscopic spicules from the skeletal components in the body wall and tube feet of *I. fuscus*. Spicules are shaped like tables and C-shaped rods in the body wall and perforate supporting tables in the tube feet plates and perforate supporting tables. This provides a means to diminish the illegal trade in this species, and support conservation efforts. Processed *I. fuscus* looks similar to some other species and can be almost impossible to distinguish from them. This includes *S. horrens*, which is harvested illegally together with *I. fuscus* (Arellano 2004). This methodology can be used in a quick, precise and easy identification. It is cost-effective and can be done in a timely manner by custom officers to distinguish this species from others.

Isostichopus fuscus has been intensively fished along the mainland and around the Galápagos Islands. Therefore its populations have thus been severely depleted over the past decade. Mercier et al. (2004) first attempted to breed *I. fuscus* in land-based installations on the coast of Ecuador. They studied spawning, fertilization, larval rearing, disease control and juvenile growth. Data pooled from monthly trials conducted over 3 years indicate that, under optimal conditions, juveniles can be grown to a length of ca. 8 cm in length in 3.5 months. The survival rate is typically between 30 and 50 %. Furthermore, preliminary experiments showed that growth of young *I. fuscus* in abandoned shrimp ponds is a promising option. Overall, this study demonstrates that *I. fuscus* can be reared in captivity. Thus, this provides an alternative to fisheries, or a way to maintain sustainable harvests and eventually contribute to the restoration of the natural populations.

6.5 Threats

6.5.1 Anthropogenic and Environmental Effects

The fuel tanker Jessica grounded on Shiavioni reef at the entrance of Wreck Bay, San Cristóbal Island (Galápagos Islands) on 16 January 2001. The vessel carried approximately 240,000 gallons of fuel oil consisting of 160,000 gallons of Diesel Oil #2 (DO#2) and 80,000 gallons of Intermediate Fuel Oil 120 (IFO 120, or bunker fuel). By 29th January, most of the remaining 180,000 gallons of fuel oil (75,000 gallons of IFO 120 and 105,000 gallons of DO#2) had escaped from the hull and dispersed to waters within the archipelago. The results of the Galápagos Coastal Oil Survey indicate that environmental contamination from the Jessica spill was broadly and thinly dispersed throughout the southwestern archipelago. Survey work confirmed that bunker fuel went ashore at sites as remote as the coastlines of Fernandina and Isabela and in sensitive habitats such as mangroves (Edgar et al. 2002a; Marshall and Edgar 2003).

Intertidal and shallow subtidal sites on shores affected by oil in San Cristóbal and Santa Fé Islands had been sampled prior to the oil spill. These data were compared with data collected at the same sites 1 month later and at reference locations that had not received oil. No clear effects of oil were detected, indicating that impacts on shore communities of seaweeds, marine invertebrates and fishes were probably slight. Increases in the density of several algal, sea urchin, hydroid and fish taxa were detected adjacent to the wreck, and were probably caused by oil pollution related to the Jessica spill (Edgar et al. 2002a).

The greatest oil pollution of shoreline contamination occurred on Floreana and Isabela Islands. Lower levels of oil pollution were found at Santa Fé, Santa Cruz, and San Cristóbal Islands and secondary contamination (oiled debris and foam) was found on Española and Fernandina Islands. Only one survey site contained sufficient coverage of oil to be termed "moderate". All other sites had "light" or "very light" surface oil (Lougheed et al. 2002).

Along the coast of mainland Ecuador, the absence of historical baseline data for the invertebrate community is a real problem for determining impacts of this kind of event on invertebrates.

Fishing has indirect effects on ecosystems, particularly through trophic cascades (Sala et al. 1998; Pinnegar et al. 2000; Jackson et al. 2001; Bascompte et al. 2005), in which sea urchins are important grazers. For instance, because sea urchins can structure reef communities (Schiel and Foster 1986; Hughes et al. 1987; Pearse and Hines 1987), the removal of sea urchin predators (large lobsters and fish) by artisanal fishing can increase the abundance of sea urchins. They can then overgraze algae (Tegner and Dayton 1981; Tegner and Levin 1983; Siversten 2006). Overgrazing probably magnified impacts of the El Niño 1982–1983 through a cascade of indirect effects involving population expansion of populations of grazing sea urchins.

Edgar et al. (2010) indicated that shallow reef habitats across the central Galápagos Archipelago underwent major transformation at the time of the severe El Niño 1982–1983 warming event with consequences that persist today (Bensted-Smith 2002). Heavily grazed reefs with crustose coralline algae (urchin barrens) replaced former macroalgal and coral habitats, resulting in large local and regional declines in biodiversity. Also, the range of corals decreased. Their ranges and their populations are now largely confined to localized areas in the west (Glynn 1994) and the north of the archipelago (Banks et al. 2009). One echinoderm, the 24-rayed sunstar *Heliaster solaris*, has not been sighted since 1983 (Cairns 1991; Hickman 1998; Edgar et al. 2010). It is considered probably extinct in coastal environments in Galápagos as probably extinct, because it has not been seen for >28 years (Red List IUCN 2008). Marine protected areas with adequate enforcement are predicted to ameliorate but not eliminate ecosystem impacts caused by increasing thermal anomalies associated with El Niño and global climate change (Robinson 1985; Edgar et al. 2010).

6.6 Concluding Remarks and Recommendations

6.6.1 Biodiversity

1. A diverse and complex set of marine ecosystems exist along the coast of mainland Ecuador and the Galápagos Islands. These include rocky shores, sandy beaches, belts of sandy banks, estuaries, mangrove forests, deep sea mounts, and cliffs associated with pocket beaches. All these systems provide a high variety of subtidal and intertidal habitats for a wide range of organisms. Nevertheless information about echinoderms is limited and dispersed. A continuous, long-term monitoring program, which integrates all habitat types and latitudinal variation, is needed in order to understand how environmental conditions influence echinoderm distribution and diversity. In addition, almost nothing is known about echinoderms of bathyal and abyssal zones, especially in the vicinity of the Galápagos Archipelago.
2. Most studies in the inner and outer Gulf of Guayaquil have focused on estuarine production and mangrove ecology. No information on the echinoderm fauna inhabiting this complex and large ecosystem is available, especially on species associated with rocky, sandy and muddy bottoms, such as sand dollars, sea biscuits, sea stars, brittle stars, and sea cucumbers. Viable population sizes should be determined for key groups of taxa and species. They should be monitored throughout their distribution range to quantify changes in population size.
3. No studies have focused on the relationship between climate change and biodiversity, neither for Ecuador mainland nor for the Galápagos Islands. The taxonomic information available is largely at the most basic descriptive level. There is little information about the phylogeny of species and their geographical distributions.
4. Scientists should identify priority areas and taxonomic groups of interest for completing biological inventories. They should encourage the development of a geographic information system for detecting areas of high diversity and design species/area curves for predicting extinction patterns to use to formulate policies for conservation and economic development (Cruz et al. 2003).

6.6.2 Ecology

1. Population dynamics of key sea urchin species should be continuously monitored, because many of the ecological changes in Galápagos might have been exacerbated indirectly through trophic cascades, via loss of predation and disease on *E. galapagensis* (Sonnenholzner et al. 2011).

2. In general, very little information about plankton ecology is available over in Ecuador. There is a lack of information on echinoderm larval ecology, dispersion, and its development along the mainland Ecuador and the Galápagos Islands. This field of research must be developed, because it is important for the future. Key life history traits of stenotopic species need to be known in order to determine their capacity to respond to extreme changes in temperature and salinity in their environments. Additionally, samples of plankton and larvae of echinoderms in the Reference Collections are not being well maintained. There is a need to create a GIS-Database not only for biological but also for oceanographic data.
3. Studies on meiobenthos on the mainland of Ecuador and Galápagos should be continued. However it is necessary to carefully choose study sites for long-term research to monitor changes in biodiversity.
4. As suggested by Okey et al. (2004) for the Galápagos Islands, mass-balanced trophic models (Ecopath/Ecosim/Ecospace) for primary and secondary production for rocky reefs, sandy and muddy beaches, and estuaries environments along the mainland of Ecuador should be constructed. This is necessary to detect those habitats that may act as sinks for oceanic plankton, because reef systems and estuarine environments may be strongly coupled with physical oceanographic conditions. This information should be available for the early life stages of different species of echinoderms in Ecuador.

6.6.3 Fisheries

1. Although several factors contribute to variability in communities of algae, fishes, lobsters, sea urchins, sea urchins parasites, and crabs, in Galápagos, evidence suggests that artisanal fishing has both direct and cascading effects throughout the community (Sonnenholzner et al. 2009, 2011; Edgar et al. 2010). Nonetheless, the magnitude, rate and severity of these changes are still unknown for echinoderms (Sonnenholzner et al. 2011), and community structure (Ruttenberg 2001) because most studies have been observational. More experimental work is needed in order to understand top-down control in the Galápagos.
2. Capture-recapture studies for the sea urchin *T. depressus* (tagged specimens) should be conducted to estimate basic population parameters, such as, age, growth rate, natural mortality, and movement patterns (Sonnenholzner 2011). These studies, combined with other population surveys, should provide baseline data to develop an adequate strategy of coastal management for a potential planned fishery of *T. depressus* in the mainland and the Galápagos Islands.

3. Techniques need to be developed for production of commercial juvenile sea cucumbers (i.e. *I. fuscus*), and sea urchins (i.e. *T. depressus*) through laboratory cultures for aquaculture and/or repopulation.
4. A better understanding of the social and political aspect of the sea cucumber fishery is required. A key external stakeholder in the overexploitation of sea cucumbers in Galápagos is the Asian market. Imaginative solutions are required in order to escape the current situation, whereby social pressure forces the fishery to open after the slightest hint of recovery. This jeopardizes both the long term recovery of the resource as well as the economic stability of the fishers.

6.6.4 Threats

1. Impacts can be grouped into six main categories: (a) fishery exploitation; (b) global climate change; (c) invasions of exotic species; (d) pollution; and (e) loss of habitat due to increased human population of the coastal zones both in Ecuador mainland and in the Galápagos Islands.
2. The collateral effects of the sea cucumber fishery in the Galápagos Islands are not fully understood. What is the damage to biodiversity of illegal fishing camps on pristine islands? To what extent does the presence of a licensed sea cucumber merchant in the islands open a door for the exploitation of other protected species such as sea horses, sharks and sea lions? Socially and economically, who has really benefited from this fishery?
3. Future research should address the following main questions considering the immune response of two sea urchins species: *E. vanbrunti* (at the mainland) and *E. galapagensis* (at the Galápagos Islands) to establish the exact chain of causation that could kill different sea animals after low-level oil contamination, or from global warming change (as was suggested by Romero and Wikelsky 2002). Is the immune response playing a direct role like bio-predictors under a stress response?

Acknowledgments We would like to express our thanks to Manuel Cruz (formerly at the Instituto Oceanográfico de la Armada del Ecuador), Soledad Luna, Kate Clark and Priscilla Martínez (Instituto Ecuatoriano NAZCA de Investigaciones Marinas), Luis Molina (Charles Darwin Foundation, CDF), and Manuel Burgos (Universidad de Especialidades Espíritu Santo, UEES) for their assistance. We also thank Angel Chiriboga, Xavier Romero, and CDF for photographical material included in this manuscript.

References

Abbott DP (1966) Factors influencing the zoogeographic affinities of the Galápagos inshore marine fauna. In: Bowman RI (ed) Proceedings of symposium of the Galapagos International Scientific Project, University of California Press, Berkeley, pp 108–112

Agassiz A (1892) Reports on an exploration off the west Coast of Mexico, Central and South America, and off Galápagos Island, in Charge of Alexander Agassiz. by the U.S. Fish Commission Steamer "*Albatross*" during 1891, Lieut. Commander Z. L. Tanner, U. S. N. Commanding. I. *Calamocrinus diomedae* Agassiz; A new stalked crinoid. With notes on the apical system and the homologies of echinoderms. Mem Mus Comp Zool Harvard Coll 17:1–95

Agassiz A (1898) Preliminary report on the echini. Dredging operations of the Albatross. Bull Mus Comp Zool 32:71–86

Agassiz A (1904) The Panamic deep sea Echini. Report XXXII on an exploration off the west coasts of Mexico, Central and South America, and off the Galápagos Islands. Mem Mus Comp Zool Harvard Coll 31:1–243

Aguilar F, Chalen X, Castro F, Sonnenholzner J, Herrera M (1993) Evaluación del recurso pepino de mar al este de la isla Fernandina. Documento Reservado para la Subsecretaria de Recursos Pesqueros del Ecuador, Quito, Ecuador

Allauca S (1990) Presencia de la corriente costanera ecuatoriana. Acta Oceanogr Pac 6:10–17

Arellano S (2004) Informe de decomiso de *Stichopus horrens*. Informe Interno a la Unidad de Recursos Marinos, Servicio Parque Nacional Galápagos, Santa Cruz, Galápagos, Ecuador

Avilés B (1984) Identificación y distribución de los equinodermos en la provincia del Guayas. Doctorate Thesis, Univ de Guayaquil, Guayaquil, Ecuador

Banks SA (1999) The use of AVHRR data in determining sea surface temperature zonation and variability across the Galápagos Marine Reserve. MSc thesis, Southampton Oceanographic Centre, Southampton

Banks SA (2002) Ambiente físico. In: Danulat E, Edgar GJ (eds) Reserva Marina de Galápagos. Línea Base de la Biodiversidad. Fundación Charles Darwin/Servicio Parque Nacional Galápagos, Santa Cruz, Galápagos, Ecuador, pp 22–35

Banks SA, Vera M, Chiriboga A (2009) Establishing reference points to assess long-term change in zooxanthellate coral communities of the northern Galápagos coral reefs. Galápagos Res 66:41–72

Barham EG, Gowdy RW, Wolfson FH (1973) *Acanthaster* (Echinodermata: Asteroidea) in the Gulf of California. Fish Bull 71:927–942

Bascompte J, Melian CJ, Sala E (2005) Interaction strength combinations and the overfishing of a marine food web. Proc Nat Acad Sci U S A 102:5443–5447

Bensted-Smith R (2002) A biodiversity vision for the Galápagos Islands. Charles Darwin Foundation and World Wildlife Fund, Puerto Ayora

Blake DB, Janies DA, Mooi R (2000) Evolution of starfishes: morphology, molecules, development, and paleobiology. Introduction to the symposium. Amer Zool 40:311–315

Bonham K, Held EE (1963) Ecological observations on the sea cucumber *Holothuria atra* and *H. leucospilota* at Rongelap Atoll, Marshall Islands. Pac Sci 17:305–314

Boone L (1933) Scientific results of cruises of the yachts "Eagle" and "Ara", 1921–1928. Coelenterata, Echinodermata, and Mollusca. Bull Vanderbilt Mar Mus 4:1–217

Brandt M (2003) Influencia del erizo lapicero *Eucidaris thouarsii* en la estructura de la comunidad submarina de la cara norte del islote Caamaño, Islas Galápagos. Licenciatura Thesis, Univ San Francisco de Quito, Ecuador

Brandt M, Guarderas P (2002) Erizos de Mar. In: Danulat E, Edgar GJ (eds) Reserva Marina de Galápagos. Línea Base de la Biodiversidad. Fundación Charles Darwin/Servicio Parque Nacional Galápagos, Santa Cruz, Galápagos, Ecuador, pp 396–418

Bremner MPH, Perez J (2002) A case study of human migration and the sea cucumber crisis in the Galápagos Islands. AMBIO J Hum Environ 31:306–310

Brusca RC (1980) Common intertidal invertebrates of the Gulf of California. Univ Arizona Press, Tucson

Cairns SD (1991) A revision of the ahermatypic Scleractinia of the Galápagos and Cocos Islands. Smith Contr Zool 504:1–32

Calder W (1984) Size, function, and life history. Harvard Univ Press, Oxford

Camhi M (1995) Industrial fisheries threaten ecological integrity of the Galápagos Islands. Conserv Biol 9:715–724

Cañadas L (1983) El mapa bioclimático y ecológico del Ecuador. Banco Central del Ecuador, Quito

Carranza-Barona CC, Andrade-Echeverría M (1996) Retrospectiva de la pesca de pepino de mar a nivel continental. Reporte Interno. Comisión Permanente para las Islas Galápagos, Fundación Charles Darwin and ORSTOM, Quito, Ecuador

Caso ME (1948) Contribución al conocimiento de los equinodermos de México. II. Algunas especies de equinoideos litorales. An Inst Biol UNAM 19:183–231

Caso ME (1958) Contribución al conocimiento de los holoturoideos de México. III. Algunas especies de Holoturoideos litorales de la costa pacífica de México. An Inst Biol UNAM 28:309–338

Caso ME (1961) Los Equinodermos de México. PhD thesis, Fac Cien, Univ Nac Aut México, México DF, México

Caso ME (1962) Estudios sobre equinodermos de México. Contribución al conocimiento de los equinodermos de la islas Revillagigedo. An Inst Biol UNAM 33:293–330

Caso ME (1965) Estudio sobre Equinodermos de México. Contribución al conocimiento de los Holoturoideos de Zihuatanejo y de la Isla de Ixtapa (primera parte). An Inst Biol UNAM 36:253–291

Caso ME (1970) Contribución al conocimiento de los Asterozoa de México. Situación taxonómica actual, morfología externa y datos ecológicos de *Platasterias latiradiata* Gray. An Inst Biol UNAM Ser Cienc Mar Limnol 41:1–62

Caso ME (1976) El estado actual del estudio de los equinodermos de México. An Cent Cien Mar Limnol UNAM 3:1–56

Caso ME (1978) Los Equinoideos del Pacífico de México. Parte 1. Ordenes Cidaroidea y Aulodonta; Parte 2. Ordenes Stiridonta y Camarodonta. An Cent Cien Mar Limnol UNAM Publ Esp 1:1–244

Caso ME (1979) Los Equinodermos de la Bahía de Mazatlán, Sinaloa. An Cent Cien Mar Limnol UNAM 6:197–368

Chávez FP, Brusca RC (1991) The Galápagos Islands and their relation to oceanographic processes in the tropical Pacific. In: James MJ (ed) Galápagos marine invertebrates. Plenum Press, New York, pp 9–33

Chiriboga A, Ruiz D, Banks S (2011) CDF checklist of Galápagos Echinoderms. In: Bungartz F, Herrera H, Jaramillo P, Tirado N, Jímenez-Uzcategui G, Ruiz D, Guézou A, Ziemmeck F (eds) Charles Darwin foundation Galápagos species checklist. Fundación Charles Darwin, Puerto Ayora, pp 1–27

Clark HL (1902) Papers from the Hopkins Standford Galápagos expedition, 1898–1899. XII Echinodermata. Proc Wash Acad Sci 4:521–531

Clark HL (1917) Ophiuroidea. Report XVIII and XXX on the scientific results of the expedition of the "*Albatross*" to the tropical Pacific 1899–1900 and 1904–1905. Bull Mus Comp Zool Harvard Univ 61:429–453

Clark HL (1920a) Asteroidea. XXXII Report on the scientific results of the expedition to the eastern tropical pacific, in charge of Alexander Agassiz, by the U. S. Fish. Comission Steamer "*Albatross*", from October 1904 to March 1905, Lieut. Commander L. M. Garrett, U.S.N., Commanding. Bull Mus Comp Zool Harvard Coll 39:73–113

Clark HL (1920b) Holothuroidea. XXXIII. Reports on the scientific results of the expedition to the eastern tropical pacific, in charge of Alexander Agassiz, by the U. S. Fish. Comission Steamer "Albatross", from October 1904 to March 1905, Lieut. Commander L. M. Garrett, U.S.N., Commanding. Mem Mus Comp Zool Harvard Coll 39:119–154

Clark AH (1939) Echinoderms (other than holothurians) collected on the Presidential Cruise of 1938. Smithson Miscell Collec 98:1–22
Clark HL (1940) Eastern Pacific expeditions of the New York Zoological Society. XXI. Notes on Echinoderms from the west coast of Central America. Zoologica 25:331–352
Clark HL (1948) A report on the Echini of the warmer eastern Pacific, based on the collections of the Velero III. Allan Hancock Pac Exp 8:225–352
Coello D, Prado M (1999) Variabilidad del fitoplancton y microzooplancton frente a la costa ecuatoriana durante 1995–1999. Bol Esp INP, Ecuador, pp 72–108
Conand C (2004) Present status of world sea cucumber resources and utilization: an international overview. In: Lovatelli A, Conand C, Purcell S, Uthicke S, Hamel JF, Mercier A (eds) Advances in sea cucumber aquaculture and management. FAO Technical Paper 463, Rome, pp 13–24
Cox A (1983) Ages of the Galápagos Islands: patterns of evolution in Galápagos organisms. In: Bowman RI, Berson M, Leviton AE (eds) Patterns of evolution in Galápagos organisms. Am Ass Advan Sci, San Francisco, pp 11–24
Cruz M, Gaibor N, Mora E, Jimenez R, Mair J (2003) The known and unknown about marine biodiversity in Ecuador (continental and insular). Gayana 7:232–260
De Miras C, Andrade M, Carranza C (1996) Evaluación socioeconómica de la pesca experimental de pepino de mar en Galápagos. Fundación Charles Darwin para las Islas Galápagos/ ORSTOM, Quito
De Paco C, Hurtado M, McFarland C, Martínez P, Reck G, Richmond R (1993) Evaluación de la pesquería de pepinos de mar en las Islas Galápagos, Ecuador. Informe para la Unión Mundial para la Naturaleza (UICN) como resultado de la Misión realizada a solicitud de la Fundación Charles Darwin para las Islas Galápagos, Santa Cruz, Galápagos, Ecuador
Deichmann E (1938) Holothurians from the western coast of Lower California and Central America, and from the Galápagos Islands; Eastern Pacific Expeditions of the New York Zoological Society XVI. Zoologica 23:361–387
Deichmann E (1941) Deichmann E (1941) The Holothuroidea collected by the Velero III during the years 1932 to 1938. Part I. Dendrochirota. Allan Hancock Pac Exp 8:61–195
Deichmann E (1958) The Holothuroidea collected by theVelero III and IV during the years 1932 to 1954. Part II. Aspidochirota. Allan Hancock Pac Exp 11:253–349
Denny M, Gaylord B (1996) Why the urchin lost its spines: hydrodynamic forces and survivorship in three echinoids. J Exp Biol 199:717–729
Ebert TA (1971) A preliminary quantitative survey of the echinoid fauna of Kealakekua and Honaunau Bays. Hawaii. Pac Sci 25:112–131
Ebert TA (1980) Relative growth of sea urchin jaws: an example of plastic resources allocation. Bull Mar Sci 30:467–474
Edgar GJ, Kerrison L, Shepherd S, Toral-Granda V (2002a) Impacto del derrame del Jessica en la plantas y animales intermareales y submareales someros. In: Edgar GJ (ed) Impactos biológicos del derrame del Jessica en los ambientes de Galápagos. SPNG/CDRS, Santa Cruz, Galápagos, pp 13–14
Edgar GJ, Kerrison L, Shepherd S, Toral-Granda V (2002a) Impacto del derrame del Jessica en la plantas y animales intermareales y submareales someros. In: Edgar GJ (ed) Impactos biológicos del derrame del Jessica en los ambientes de Galápagos. SPNG/CDRS, Santa Cruz, Galápagos, pp 13–14
Edgar GJ, Banks SA, Fariña JM, Calvopiña M, Martínez C (2004) Regional biogeography of shallow reef fish and macro-invertebrate communities in the Galápagos Archipelago. J Biogeogr 31:1107–1124
Edgar G, Banks SA, Brandt M, Bustamante RH, Chiriboga A, Earle SA, Garske LE, Glynn PW, Grove JS, Henderson S, Hickman CP, Miller KA, Rivera F, Wellington GM (2010) El Niño, grazers and fisheries interact to greatly elevate extinction risk for Galápagos marine species. Global Change Biol 16:2876–2890
Ellers O (1993) A mechanical model of growth in regular sea urchins: predictions of shape and a developmental morphospace. Proc R Soc Lond B 254:123–129

Ellers O, Telford M (1992) Causes and consequences of fluctuating coelomic pressure in sea urchins. Biol Bull 182:424–434

Enfield D (1975) Oceanografía de la región norte del frente ecuatorial: aspectos físicos. INOCAR, Dep Cien Mar, Div Oceanogr Fís, Guayaquil, Ecuador

Epstein P, Sherman B, Spanger-Siegfried E, Langston A, Prased S, Muckay B (1998) Marine ecosystems: emerging diseases as indicators of change. Harvard Medical School, Boston

Feldman GC (1985) Satellites, seabirds and seals. In: Robinson G, Del Pino EM (eds) El Niño in the Galápagos Islands: the 1982–1983 event. Charles Darwin Foundation, Quito, Ecuador, pp 125–130

Fell HB (1963) The phylogeny of sea-stars. Phil Trans R Soc (B) 246:381–435

Fisher WK (1911) Asteroidea of the North Pacific and adjascent waters. Part 1: Phanerozonia and Spinulosa. Bull US Nat Mus 76:1–419

Geist D (1996) On the emergence and submergence of the Galápagos Islands. Noticias de Galápagos 56:5–9

Gerace DT, Lindsay W (1992) Cassis in captivity: an ongoing research project. In: Gerace DT (ed) Proceedings of 4th Symposium natural history of the Bahamas, Bahamian Field Station, San Salvador, pp 59–66

Gilles KW, Pearse JS (1986) Disease in sea urchins *Strongylocentrotus purpuratus*: experimental infection and bacterial virulence. Dis Aquat Org 1:105–114

Glynn PW (1988) El Niño warming, coral mortality and reef framework destruction by echinoid bioerosion in the Eastern Pacific. Galaxea 7:129–160

Glynn PW (1990) Coral mortality and disturbances to coral reef in the Tropical Eastern Pacific. In: Glynn PW (ed) Global ecological consequences of the 1982–83 El Niño-Southern Oscillation. Elsevier, Amsterdam, pp 55–126

Glynn PW (1994) State of coral reefs in the Galápagos Islands: natural vs anthropogenic impacts. Mar Poll Bull 29:131–140

Glynn PW, Wellington GM, Birkeland C (1979) Coral reef growth in the Galápagos: limitation by sea urchins. Science 203:47–48

Glynn PW, Wellington GM, Wells J (1983) Coral and coral reefs of the Galápagos Islands. University of California Press, Berkeley

Guarderas AP (2000) Dinámica poblacional y ecología trófica del erizo blanco *Tripneustes depressus* (Echinodermata: Echinoidea) en Santa Cruz, Galápagos. Licenciatura Thesis, Pontificia Univ Católica del Ecuador, Quito, Ecuador

Harris MP (1969) Breeding season of sea-birds in the Galápagos Islands. J Zool 159:145–165

Hearn A (2005) Population dynamics of the exploited sea cucumber *Isostichopus fuscus* in the western Galápagos Islands, Ecuador. Fish Oceanogr 14:375–385

Hearn A, Pinillos F (2006) Baseline information on the warty sea cucumber *Stichopus horrens* in Santa Cruz, Galápagos, prior to the commencements of an illegal fishery. SPC Beche-De-Mer Bull 24:3–10

Heatfield BM (1971) Growth of the calcareous skeleton during the regeneration of spines of the sea urchin *Strongylocentrotus purpuratus* (Stimpson): a light and scanning electron microscope study. J Morph 134:57–90

Herrero-Pérezrul MD, Reyes-Bonilla H, García-Domínguez F, Cintra-Buenrostro CE (1999) Reproduction and growth of *Isostichopus fuscus* (Echinodermata: Holothuroidea) in the southern Gulf of California, Mexico. Mar Biol 135:521–532

Hickman CP Jr (1998) A field guide to sea stars and other echinoderms of Galápagos (Galápagos Marine Life Series). Sugar Spring Press, Lexinton

Hughes TP, Reed DC, Boyle MJ (1987) Herbivory on coral reefs: community structure following mass mortalities of sea urchins. J Exp Mar Biol Ecol 113:39–59

Idrovo F, Sonnenholzner J (1994) Distribución y abundancia de *Tripneustes depressus* Agassiz 1863, y *Diadema mexicanum* Agassiz 1863 (Echinodermata: Echinoidea) en Punta Los Frailes, Parque Nacional Machalilla, Manabí, Ecuador. Licenciatura Thesis, Universidad de Guayaquil, Guayaquil

Irving AD, Witman JD (2009) Positive effects of damselfish override negative effects of urchins to prevent a habitat switch. J Ecol 97:337–347
Jackson JBC, Kirby MX, Berger WH, Bjorndal KA, Botsford LW, Bourque BJ, Bradbury RH, Cooke R, Earlandson J, Estes JA, Hughes TP, Kidwell S, Lange CB, Lenihan HS, Pandolfi JM, Peterson CH, Steneck TS, Tegner MJ, Warner RR (2001) Historical overfishing and the recent collapse of coastal ecosystems. Science 293:629–638
James MJ (ed) (1991) Galápagos marine invertebrates–taxonomy, biogeography and evolution in Darwin's Islands. Plenum Press, New York
Jangoux M (1990) Diseases of echinodermata. In: Kinne O (ed) Disease of marine animals 3. Biologische Anstal Helgoland, Hamburg, pp 439–567
Jiménez R (1987) Antecedentes históricos del fenómeno El Niño. Boletín de la OMM 36:192–194
Kier PM, Grant RE (1965) Echinoid distribution and habitats. Key Largo Coral Reef Preserve, Florida. Smithson Misc Collect 6:149
Kitching JA, Ebling FJ (1961) The ecology of Lough Ine. XI The control of algae by *Paracentrotus lividus* (Echinoidea). J Anim Ecol 383:373–383
Kohtsuka H, Arai S, Uchimura M (2005) Observation of asexual reproduction by natural fission of *Stichopus horrens* Selenka in Okinawa Island, Japan. SPC Beche-De-Mer Bull 22:23
Lafferty KD, Dobson AP, Kuris AM (2006) Parasites dominate food web links. Proc Nat Acad Sci U S A 103:11211–11216
Lawrence JM (1975) On the relationship between marine plants and sea urchins. Oceanogr Mar Biol An Rev 13:213–286
Lawrence JM (1987) Functional biology of Echinoderm. Croom Helm, London and Sidney
Lawrence JM, Sonnenholzner J (2004) Distribution and abundance of asteroids, echinoids, and holothuroids in Galápagos. In: Heinzeiller T, Nebelsick JH (eds) Echinoderms, München. Proc 11th Int Echinoderm Conf. Taylor and Francis Group, London, pp 239–244
Lawrence JM, Pomory CM, Sonnenholzner J, Chao CM (1998) Bilateral symmetry of the petals in *Mellita tenuis*, *Encope micropora*, and *Arachnoides placenta* (Echinodermata: Clypeasteroidea). Invert Biol 117:94–100
Lawrence JM, Pomory CM, Sonnenholzner J, Hiratsuka Y, Molina L (2010) Scaling of Aristotle's lantern in sea urchins *Lytechinus* and *Tripneustes*. In: Harris LG, Böttger SA, Walker CW, Lesser MP (eds) Echinoderms: Durham. Proceedings of the 12th international echinoderm conference. CRC Press, Taylor and Francis group, Balkema, Leiden, pp 219–224
Lessios HA, Kessing BD, Robertson DR, Paulay G (1999) Phylogeography of the pantropical sea urchin *Eucidaris* in relation to land barriers and ocean currents. Evolution 53:806–817
Lougheed LW, Edgar GJ, Snell HL (eds) (2002) Biological impacts of the Jessica oil spill on the Galápagos environment: Final Report v.1.10. Charles Darwin Foundation, Puerto Ayora, Galápagos, Ecuador
Ludwig H (1894) Reports on an exploration off the west coast of México, Central America and South America, and off the Galápagos Islands, in charge of Alexander Agassiz, by the U.S. Fish Commission steamer "Albatross", during 1891. XII. The Holothurioidea. Mem Mus Comp Zool XVII:1–183
Ludwig H (1905) Asteroidea in: explorations of "Albatross" in Tropical Pacific, 1891 and 1899–1900. Mem Mus Comp Zool XXII:1–290
Luna S (2000) Distribución poblacional y ciclo reproductivo del erizo de mar blanco *Tripneustes depressus* (Echinodermata: Echinoidea) en las Islas Galápagos. Licenciatura Thesis, Univ San Francisco de Quito, Quito, Ecuador
Lütken CF, Mortensen T (1899) Reports on an exploration off the west coast of Mexico, Central America and South America, and off the Galápagos Islands, in charge of A. Agassiz, during 1891 on the "Albatross". XXV. The Ophiuridae. Mem Mus Comp Zool XXIII:93–208
Luzuriaga M (1993) Distribución del zooplancton en los manglares de la región norte de la provincia de Esmeraldas, Ecuador. Rev Cien Mar Limnol 3:59–87
Luzuriaga M (2011) Distribución y abundancia de larvas Ophiopluteus en el mar ecuatoriano. Informe interno, Instituto Nacional de Pesca, Guayaquil, Ecuador

Luzuriaga M, Elías E (1999) Variaciones poblacionales del zooplancton e ictioplancton en el mar ecuatoriano durante el período 1994 a 1999. Bol Esp INP, Ecuador, pp 109–132

Luzuriaga M, Solano F, Cevallos D (2007) Bioecología de huevos y larvas de peces *Cetengraulis mysticetus* (chuhueco), *Opisthonema* sp (pinchagua) y anchoas en la Reserva Ecológica Manglares Cayapas-Mataje. Acta Oceanogr Pac 14:101–114

Mair J, Mora E, Cruz M (2002) Manual de campo de los invertebrados bentónicos marinos: Moluscos, Crustáceos y Equinodermos de la zona litoral ecuatoriana. Univ Guayaquil, Guayaquil, Ecuador

Malmquist DL (1991) The past as the key to the present: taphonomy and paleoecology of the Urvina Bay uplift. In: James MJ (ed) Galápagos marine invertebrates: taxonomy biogeography and evolution in Darwin's islands. Plenum Press, New York, pp 393–411

Maluf LY (1988a) Composition and distribution of the Central eastern Pacific echinoderms. Nat Hist Mus LA County Tech Rep 2:1–242

Maluf LY (1988b) Biogeography of the central eastern shelf echinoderms. In: Burke RD, Mlademov PV, Lambert P, Parsley RL (eds) Echinoderm biology. AA Balkema, Rotterdam, pp 389–398

Maluf LY (1991) Echinoderm fauna of the Galápagos Islands. In: James MJ (ed) Galápagos marine invertebrates: taxonomy, biogeography and evolution in Darwin's islands. Plenum Press, New York, pp 345–367

Marshall PA, Edgar GJ (2003) The effect of the Jessica grounding on subtidal invertebrate and plant communities at the Galápagos wreck site. Mar Poll Bull 47:284–295

Martínez PC (1999) Estado poblacional y biología reproductiva del pepino de mar *Stichopus fuscus* en las islas Galápagos. Informe Técnico, Fundación Charles Darwin, Santa Cruz, Galápagos, Ecuador

Martínez PC (2001) The Galápagos sea cucumber: risk or an opportunity for conservation? SPC Beche-De-Mer Bull 14:22–23

Martínez PC, Toral MV, Bustamante RH (1998) Population and reproductive biology of the sea cucumber *Isostichopus fuscus* in the Galápagos Islands. In: Mooi R, Telford M (eds) Echinoderms: San Francisco. AA Balkema, Rotterdam, p 479

Massin C, Doumen C (1986) Distribution and feeding of epibenthic holothuroids on the reef flat of Laing Island (Papua New Guinea). Mar Ecol Prog Ser 31:185–195

McClintock JB, Marion KR (1993) Predation by the king helmet (*Cassis tuberosa*) on six-holed sand dollars (*Leodia sexiesperforata*) at San Salvador, Bahamas. Bull Mar Sci 52:1013–1017

Mercier A, Battaglene C, Hamel JF (1999) Daily burrowing cycle and feeding activity of juvenile sea cucumbers *Holothuria scabra* in response to environmental factors. J Exp Mar Biol Ecol 236:69–87

Mercier A, Ycaza-Hidalgo R, Hamel JF (2004) Aquaculture of the Galápagos sea cucumber, *Isostichopus fuscus*. In: Lovatelli A, Conand C, Purcell S, Uthicke S, Hamel JF, Mercier A (eds) Advances in sea cucumber aquaculture and management. FAO, Rome, pp 347–358

Metz G (1994) Some observations on *Sabinella shaskyi* Waren, 1992. Festivus 26:125–126

Moore DR (1956) Observations of predation on echinoderms by three species of Cassidae. Nautilus 69:73–76

Morante-Carballo FE (2004) Las zeolitas de la costa de Ecuador (Guayaquil): geología, caracterización y aplicaciones. PhD Thesis, Univ Politécnica de Madrid, Madrid

Moreano H (1983) Interacción océano-atmosfera sobre la zona costera del Ecuador. Acta Oceanogr Pac 2:1–11

Murillo JC, Reyes H (2007) Evaluación de la pesquería 2007 de pepino de mar *Isostichopus fuscus* en la Reserva Marina de Galápagos. Informe Fundación Charles Darwin/Parque Nacional Galápagos, Santa Cruz, Galápagos, Ecuador

Murillo JC, Martínez P, Toral MV, Hearn A (2002a) Pepino de mar. In: Danulat E, Edgar GJ (eds) Reserva Marina de Galápagos. Línea Base de la Biodiversidad. Fundación Charles Darwin/Servicio Parque Nacional Galápagos, Santa Cruz, Galápagos, Ecuador, pp 176–196

Murillo JC, Vizcaíno J, Nicolaides F, Moreno J, Espinoza E, Chasiluisa C, Andrade R, Born B, Villalta M, Yépez M, Molina L (2002b) Informe técnico final de la pesquería del pepino de mar (*Isostichopus fuscus*) en las Islas Galápagos 2001: Análisis comparativo con las pesquerías 1999 y 2000. Informe Interno, Estación Científica Charles Darwin y Servicio Parque Nacional Galápagos, Santa Cruz, Galápagos, Ecuador, pp 1–39

Muromtsev AM (1963) The principal hydrological features of the Pacific Ocean. Jerusalem Post Press, Jerusalem

Okey TA, Banks S, Born AF, Bustamante RH, Calvopinia M, Edgar GJ, Espinoza E, Farina JM, Garske LE, Reck GK, Salazar S, Shepherd S, Toral-Granda V, Wallem P (2004) A trophic model of a Galápagos subtidal rocky reef for evaluating fisheries and conservation strategies. Ecol Model 172:383–401

Paine RT, Vadas RL (1969) Calorific values of benthic marine algae and their postulated relationship to invertebrate food preference. Mar Biol 4:79–86

Paredes R (2010) Erizos de mar: Dinámica poblacional y reclutamiento de larvas dentro de la isla San Cristóbal en el Archipiélago de Galápagos-Ecuador. Licenciatura Thesis, Univ San Francisco de Quito, Quito, Ecuador

Pawson DL, Ahearn CA (2001) Bathyal echinoderms of the Galápagos Islands. In: Barker M (ed) Echinoderms 2000. AA Balkema, Rotterdam, pp 41–45

Pearse JS, Cameron RA (1991) Echinodermata: Echinoidea. In: Giese AC, Pearse JS, Pearse VB (eds) Reproduction of marine invertebrates (vol VI): Echinoderms and Lophophorates. The Boxwood Press, Pacific Grove, pp 514–662

Pearse JS, Hines AH (1987) Long-term population dynamics of sea urchins in a central California kelp forest: rare recruitment and rapid decline. Mar Ecol Prog Ser 39:275–283

Peters RH (1983) The ecological implications of body size. Cambridge University Press, Cambridge

Pinela M, Calvache M (1997) Distribución, densidad y biometría de *Echinometra vanbrunti* (Echinodermata: Echinoidea) en la zona intermareal de la costa sudoeste de la provincia de El Guayas. Licenciatura Thesis, Univ de Guayaquil, Guayaquil, Ecuador

Pinnegar JK, Polunin NVC, Francour P, Badalamenti F, Chemello R, Harmelin-Vivien ML, Hereu B, Milazzo M, Zabala M, D'anna G, Pipitone C (2000) Trophic cascades in benthic marine ecosystems: lessons for fisheries and protected-area management. Environ Conserv 27:179–200

Piu M (1998) Resumen de las acciones de patrullaje del Servicio Parque Nacional Galápagos para las pesquerías ilegales (1996–1997). Informe Galápagos 1997–1998. Editores asociados, Fundación Natura and WWF, Puerto Ayora, Santa Cruz, Galápagos, Ecuador, pp 17–19

Piu M (2000) La Reserva Marina de Galápagos: un resumen de las acciones de vigilancia y control de la pesca ilegal, 1998 y 1999. Informe Galápagos 1999–2000. Editores Asociados, Fundación Natura and WWF, Quito, pp 11–14

Pomory CM, Foret TW, Hill SK, Lawrence JM (1998) Characteristics of a population of Holothuria floridana from Florida. In: Mooi R, Telford M (eds) Echinoderms: San Francisco. AA Balkema, Rotterdam, pp 507–511

Powell JR, Gibbs JP (1995) A report from Galápagos. TREE 10:351–345

Reichenbach N, Holloway S (1995) Potential for asexual propagation of several commercially important species of tropical sea cucumber (Echinodermata). J World Aqua Soc 26:272–278

Reyes-Bonilla H, Herrero-Pérezrul MD (2003) Population parameters of an exploited population of *Isostichopus fuscus* (Holothuroidea) in the southern Gulf of California, México. Fish Res 59:423–430

Richmond R, Martínez PC (1993) Sea cucumber fisheries in the Galápagos Islands: biological aspects, impacts and concerns. The World Conservation Union (IUCN) Technical Report, Santa Cruz, Galápagos, Ecuador

Roberts D (1979) Deposit-feeding mechanisms and resource partitioning in tropical holothurians. J Exp Mar Biol Ecol 37:43–56

Robinson G (1985) Influence of the 1982–83 El Niño on Galápagos marine life. In: Robinson G, Del Pino EM (eds) El Niño en las Islas Galápagos: El evento de 1982–1983. Fundación Charles Darwin, Galápagos, Quito, Ecuador, pp 153–190

Roff JC, Taylor ME, Laughren J (2003) Geophysical approaches to the classification, delineation and monitoring of marine habitats and their communities. Aqua Conser Mar Fresh Ecosys 13:77–90

Romero LM, Wikelski M (2002) Severe effects of low-level oil contamination on wildlife predicted by the corticosterone-stress response: preliminary data and a research agenda. Spill Sci Technol Bull 7:309–313

Ruttenberg B (2001) Effects of artisanal fishing on marine communities in the Galápagos Islands. Conserv Biol 15:1691–1699

Sala E, Boudouresque CF, Harmelin-Vivien M (1998) Fishing, trophic cascades, and the structure of algal assemblages: evaluation of an old but untested paradigm. Oikos 82:425–439

Schiel DR, Foster MS (1986) The structure of subtidal algal stands in temperate waters. Oceanogr Mar Biol An Rev 24:265–307

Schoppe S (2000) Sea cucumber fishery in the Philippines. SPS Beche-De-Mer Bull 13:10–12

Sharp DT, Gray IE (1962) Studies on factors affecting the local distribution of two sea urchins, *Arbacia punctulata* and *Lytechinus variegatus*. Ecology 43:309–313

Shepherd SA, Martínez P, Toral-Granda MV, Edgar GJ (2004) The Galápagos sea cucumber fishery: management improves as stocks decline. Environ Conserv 31:1–9

Sierra R (1999) Propuesta preliminar de un sistema de clasificación de vegetación para el Ecuador continental. Proyecto INEFAN-GEF-BIRF y EcoCiencia. Quito, Ecuador

Siversten K (2006) Overgrazing of kelp beds along the coast of Norway. J Appl Phycol 18:599–610

Skoglung C (2004) New distributional records for Panamic province Eulimidae (Gastropoda). Festivus 36:95–98

Sloan NA, Von Bodungen B (1980) Distribution and feeding of the sea cucumber, *Isostichopus badionotus* in relation to shelter and sediment criteria of the Bermuda platform. Mar Ecol Prog Ser 2:257–264

Snell HM, Stone PA, Snell HL (1995) Geographical characteristics of the Galápagos Islands. Noticias de Galápagos 55:18–24

Solís-Marín FA (1998) Catálogo de los equinodermos recientes de México (Fase I). Universidad Nacional Autónoma de México. Instituto de Ciencias del Mar y Limnología. Informe final SNIB-CONABIO proyecto No. G010. México, D.F

Sonnenholzner J (1997) A brief survey of the commercial sea cucumber *Isostichopus fuscus* (Ludwig, 1875) of the Galápagos Islands, Ecuador. SPS Beche-De-Mer Bull 9:12–15

Sonnenholzner J (2001) Desarrollo de pruebas inmunitarias en *Echinometra vanbrunti* (Echinodermata: Echinoidea) para la identificación de biomarcadores de estrés ambiental". Doctorate Thesis, Univ Guayaquil, Guayaquil, Ecuador

Sonnenholzner J (2003) Seasonal variation in the Food composition of *Holothuria theeli* (Holothuroidea: Aspidochirotida) with observations on density and distribution patterns at the central coast of Ecuador. Bull Mar Sci 73:527–543

Sonnenholzner J (2011) Crecimiento y validación de la edad del erizo morado *Strongylocentrotus purpuratus* (Stimpson 1857) en condiciones naturales y de laboratorio". Ph.D. Thesis. Univ Autónoma de Baja California, Ensenada, México

Sonnenholzner J, Cajas L (1993) Dieta del pepino de mar *Isostichopus fuscus*. In: Sonnenholzner J (ed) Contribución al estudio del pepino de mar *Isostichopus fuscus* de las Islas Galápagos. Reporte Interno, Instituto Nacional de Pesca, Guayaquil, Ecuador, pp 18–21

Sonnenholzner J, Lawrence JM (1998) Disease and predation in *Encope micropora* (Echinoidea: Clypeasteroida) at Playas, Ecuador. In: Mooi R, Telford M (eds.). Echinoderms: San Francisco. AA Balkema, Rotterdam, pp 829–833

Sonnenholzner J, Lawrence JM (2002) A brief survey of the echinoderms communities of the central and southern marine-coastal wetlands of the continental coast off Ecuador. Bol Ecotrópica 36:23–35

Sonnenholzner J, Molina L (2005) Parasitic eulimid gastropods in echinoderms of the Islas Galápagos, Ecuador. Festivus 37:85–88

Sonnenholzner J, Camba N, Lawrence JM (1997) Commercial fishing and organic composition of *Isostichopus fuscus* Ludwig, 1875 from the Galápagos Islands. SPS Beche-De-Mer Bull 9:37

Sonnenholzner J, Ladah LB, Lafferty KD (2009) Cascading effects of fishing on Galápagos rocky reef communities: reanalysis using corrected data. Mar Ecol Progr Ser 375:209–218

Sonnenholzner J, Lafferty KD, Ladah LB (2011) Food webs and fishing affect eulimid parasitism of the sea urchin *Eucidaris galapagensis* in the Galápagos. Ecol: doi:10.1890/11-0559.1

Stevenson M (1970) Circulation in the Panama Bight. J Geoph Res 75:659–672

Strickland DL (1968) The shallow water sea urchins of the Galápagos Islands. In: Stanford oceanographic expedition 17 eastern tropical pacific equatorial current system and Galápagos archipelago, pp 241–252

Tanaka Y (1958) Feeding and digestive processes of *Stichopus japonicus*. Bull Fac Fish 9:14–28

Tegner MJ, Dayton PK (1981) Population structure, recruitment and mortality of two sea urchins (*Strongylocentrotus franciscanus* and *S. purpuratus*) in a kelp forest. Mar Ecol Prog Ser 5:255–268

Tegner MJ, Levin LA (1983) Spiny lobsters and sea urchins: analysis of a predator–prey interaction. J Exp Mar Biol Ecol 73:125–150

Thurman HV (1996) Essentials of Oceanography. Prentice Hall, New Jersey

Toral-Granda MV (1996) Biología reproductiva del pepino de mar *Isostichopus fuscus* en la Isla Caamaño, Santa Cruz, Galápagos. Licenciatura Thesis, Univ del Azuay, Cuenca, Ecuador

Toral-Granda MV (2005) The use of calcareous spicules for the identification of the Galápagos sea cucumber *Isostichopus fuscus* on the international market. SPC Beche-de-mer Bull 22:3–5

Toral-Granda MV, Martínez P (2007) Reproductive biology and population structure of the sea cucumber *Isostichopus fuscus* (Ludwig, 1875) (Holothuroidea) in Caamaño, Galápagos Islands, Ecuador. Mar Biol 151:2091–2098

Toral-Granda MV, Murillo JC, Piu M, Nicolaides F, Moreno J, Reyes H, Castejón M, Hearn A (2006) La pesquería de pepino de mar (*Isostichopus fuscus*) en la Reserva Marina de Galápagos en el año 2005. In: Hearn A (ed) Evaluación de las pesquerías en la Reserva Marina de Galápagos Informe Compendio 2005. Fundación Charles Darwin/Parque Nacional Galápagos, Puerto Ayora, Galápagos, Ecuador, pp 1–45

Toropova C, Meliane I, Laffoley D, Matthews E, Spalding M (eds) (2010). Global ocean protection: present status and future possibilities. Brest, France: Agence des aires marines protégées, Gland, Switzerland, Washington, DC and New York, USA: IUCN WCPA, Cambridge, UK: UNEP-WCMC, Arlington, USA: TNC, Tokyo, Japan: UNU, New York, USA: WCS, pp 1–96

Uthicke S (1997) The seasonality of asexual reproduction in *Holothuria atra*, *Holothuria edulis* and *Stichopus chloronotus* (Holothuroidea: Aspidochirotida) on the Great Barrier Reef. Mar Biol 129:435–441

Uthicke S (2001) Influence of asexual reproduction on the structure and dynamics of *Holothuria (Halodeima) atra* and *Stichopus chloronotus* populations of the Great Barrier Reef. Mar Fresh Res 52:205–215

Uthicke S, Benzie JAH (2003) Gene flow and population history in high dispersal marine invertebrates: mitochondrial DNA analysis of *Holothuria nobilis* (Echinodermata: Holothuroidea) populations from the Indo-Pacific. Mol Ecol 12:2635–2648

Uthicke S, O'Hara TD, Byrne M (2004) Species composition and molecular phylogeny of the Indo-Pacific teatfish (Echinodermata: Holothuroidea) Beche-De-Mer fishery. Mar Fresh Res 55:837–848

Villamar F, Cruz M (1983) Presencia de *Platasterias latiradiata* Gray en aguas ecuatorianas. Acta Oceanogr Pac 2:169–177

Weihe SC, Gray IE (1968) Observations on the biology of the sand dollar *Mellita quinquiesperforata* (Leske). J Elisha Mitchel Sci Soc 84:315–327

Wellington GM (1975) The Galápagos Coastal marine environment. A resource report to the Department of National Parks and Wildlife, Quito, Ecuador

Wellington GM (1984) Marine environment and protection. In: Perry R (ed) Key Environments: Galápagos. Pergamon Press, Oxford, pp 247–263

Witman JD, Dayton PK (2001) Rocky subtidal communities. In: Bertness MD, Gaines SD, Hay ME (eds) Marine community ecology. Sinauer Associates, Sunderland, pp 339–366

Witman JD, Brandt M, Smith F (2010) Coupling between subtidal prey and consumers along a mesoscale upwelling gradient in the Galápagos Islands. Ecol Monogr 80:153–177
Wyrtki K (1965) Surface currents of the Eastern Tropical Pacific Ocean. Inter-American Tropical Tuna Com Bull 9:33–68
Wyrtki K (1966) Oceanography of the Eastern Pacific Ocean. Oceanogr Mar Biol An Rev 4:33–68
Wyrtki K (1974) Sea level and the seasonal fluctuations of the equatorial currents in the western Pacific Ocean. J Phys Oceanogr 4:91–103
Wyrtki K (1985) Water displacements in the Pacific and the genesis of El Niño cycles. J Phys Oceanogr 12:984–988
Yamanouchi T (1942) Food habits of *Stichopus japonicus*. Jap J Zool 54:344–346
Yamanouchi T (1956) The daily activity rhythms of the holothurians in the coral reef of Palao Islands. Publ Seto Mar Biol Lab 5:347–362
Ziesenhenne FC (1935) A new brittle star from the Galápagos Islands. Allan Hancock Pac Exp 1:1–4

Chapter 7
Echinoderms from Venezuela: Scientific Recount, Diversity and Distribution

César Lodeiros, Alberto Martín, Vanessa Francisco, Nicida Noriega, Yusbelly Díaz, Jeny Reyes, Orangel Aguilera and José Alió

C. Lodeiros (✉)
Instituto Oceanográfico de Venezuela, Universidad de Oriente y Fundación para la Investigación y Desarrollo de la Acuicultura del Edo. Sucre, Cumaná, 6101, Venezuela
e-mail: cesarlodeirosseijo@yahoo.es

A. Martín · Y. Díaz
Laboratorio de Crustáceos Peracáridos. Centro de Biodiversidad Marina. Departamento de Estudios Ambientales - INTECMAR, Universidad Simón Bolívar, Caracas, 1080, Venezuela
e-mail: amartinz@usb.ve

Y. Díaz
e-mail: yusdiaz@usb.ve

V. Francisco
Laboratorio de Ecología de Ecosistemas de Arrecifes Coralinos, CINVESTAV-IPN Unidad Mérida, 97310, Yucatán, Mexico
e-mail: vanewagen@gmail.com

N. Noriega
Laboratorio de Biología Marina y Laboratorio de Bioquímica, Universidad Simón Bolívar, Caracas, 1080, Venezuela
e-mail: nicidanoriega@gmail.com

J. Reyes
Laboratorio de Zoología de Invertebrados, Universidad del Zulia, Maracaibo 4001 y Fundación para la Investigación y Desarrollo de la Acuicultura del Edo. Sucre, Cumaná, 6101, Venezuela
e-mail: jenyreyesl@yahoo.es

O. Aguilera
Museu Paraense Emilio Goeldi, Coordenação de Ciências da Terra e Ecologia, Departamento de Geociências Av. Perimetral, 1901 Terra Firme, Belém, PA, CEP: 66077-830, Brazil
e-mail: orangel.aguilera@gmail.com

J. Alió
Centro de Investigaciones Agrícolas de Sucre y Nueva Esparta, Instituto Nacional de Investigaciones Agrícolas, Cumaná, Estado Sucre, 6101, Venezuela
e-mail: jalio@inia.gob.ve

J. J. Alvarado and F. A. Solís-Marín (eds.), *Echinoderm Research and Diversity in Latin America*, DOI: 10.1007/978-3-642-20051-9_7,

7.1 Introduction

Venezuela is located in northern South America between 00°45′ and 15°40′N, and between 59°47′ and 73°22′W (Fig. 7.1). Its coastal waters include the Caribbean Sea and the west central Atlantic Ocean, which gives it a special biogeographical position. Ecosystems in the Caribbean Sea include sandy beaches, rocky shores, seagrass meadows, coral reefs, coastal lagoons, soft bottom communities and mangroves, which results in a high environmental variability as a consequence of periodic coastal upwellings due to the trade winds. The Atlantic zone is mainly influenced by the Orinoco River which flows north towards the Caribbean Sea and promotes a brackish water environment in the coastal areas of its delta and the Gulf of Paria, a situation that increases the environmental variability. On the other hand, the 3,964 km coastline (68 % in the Caribbean Sea, 21 % in the Atlantic Ocean and 11 % in the islands) is very diverse both in geology and topography, and a great diversity of geomorphologies coexist including brackish lakes, lagoons, bays, peninsulas and other geologic formations. Such environments, together with 14 archipelagos, encompass multiple ecosystems resulting in one of the most diverse areas in the world (Aguilera et al. 2003; Miloslavich et al. 2005; Miloslavich and Klein 2008).

Sea surface temperature (SST) along the Venezuelan coastline is mainly affected by upwelling, and presents yearly ranges between 20 and 29°C with daily maximum variations of 2.5°C (Penchaszadeh et al. 2000). There are 12 upwelling zones that could be identified along the Venezuelan coast using images from the "Advanced Very High Resolution Radiometer (AVHRR)" to determine the SST. In western Venezuela, the upwelling zones are located to the east of the Guajira Peninsula; in the Gulf of Venezuela and in Punto Fijo coast, western Paraguaná Peninsula; in Cabo San Román and in Puerto Cumarebo, eastern Falcón state. Upwelling is not frequent along the central coast of Venezuela. There are several upwelling systems in the eastern coast between Puerto La Cruz and Cumaná, to the north of the Araya Peninsula and Margarita Island, and the central coast of Sucre state, near Carúpano. Coastal upwelling is frequent in all these locations, associated with the relatively high speed and predominance of the trade winds during the first 6 months of the year. The intensity and propagation of the upwelling is considerable and extends beyond 2,00,000 km^2, contributing to establish the most productive zone of the Caribbean Sea. Offshore from the Paria Peninsula and to the north of the island of Trinidad, upwelling occurs during limited periods of variable intensity. The cold water masses occasionally form eddies that migrate northeast, probably due to the movement of less dense warm water masses from the Gulf of Paria (Müller-Karger and Varela 1989; Penchaszadeh et al. 2000).

All these characteristics that result from environmental variability in the Venezuelan coasts have allowed the establishment of 13 ecoregions according to their ecological, physical and oceanographic characteristics (Fig. 7.1; Miloslavich and Klein 2008).

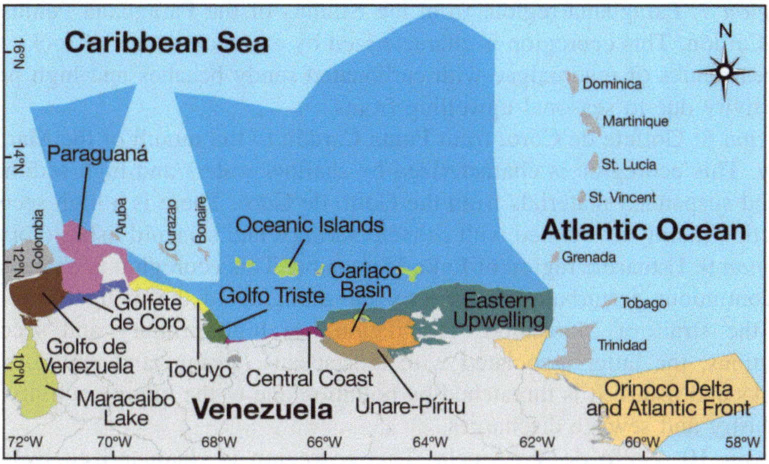

Fig. 7.1 Marine ecoregions of Venezuela (Miloslavich and Klein 2008)

Ecoregion 1: Delta of the Orinoco River and Atlantic Front. This ecoregion is characterized by estuarine and marine ecosystems with a large sedimentary load, high turbidity, sandy and muddy bottoms, with a well developed mangrove community.

Ecoregion 2: Eastern upwelling region, which encompasses the area from the eastern Paria Peninsula to Mochima Bay. This ecoregion is characterized by a shallow continental shelf with coarse sandy sediments, a large rocky shore and few coralline reefs. The region is highly productive due to the upwelling fronts. It includes the islands of Margarita, Coche, Cubagua, Los Frailes, Los Hermanos and Los Testigos Archipelago.

Ecoregion 3: Axis Unare-Píritu-Tacarigua, from Barcelona to Cabo Codera. This ecoregion is characterized by the presence of large coastal lagoons with mangrove community and long sandy beaches, with a large sediment load from the Unare and Tuy rivers. There is a large secondary production in the region, mainly of crustaceans and bivalves.

Ecoregion 4: Central coast, from Cabo Codera to Patanemo Bay. This ecoregion is characterized by a narrow coastal fringe with rocky shores and abundant sandy beaches of coarse sediments. The coral communities are associated with rocky substrates, as well as with some mangrove communities and seagrass.

Ecoregion 5: Golfo Triste, from Patanemo Bay to the mouth of the Tocuyo River. This ecoregion is characterized by a relatively large and shallow continental platform, rocky shores and well developed coralline barriers, particularly in the National Parks Morrocoy and San Esteban. The region has a moderate secondary production.

Ecoregion 6: Tocuyo region, from the Tocuyo River to the isthmus of the Paraguaná Peninsula. This ecoregion is characterized by sandy beaches and rocky cliffs with a high sediment load from the Rivers Tocuyo, Hueque and Ricoa, and low water transparency.

Ecoregion 7: Paraguaná region, from the isthmus of the Paraguaná Peninsula to Punta Cardón. This ecoregion is characterized by emergent zones of rocky shores and great banks of macroalgae with carbonated sandy beaches and high primary productivity due to seasonal upwelling fronts.
Ecoregion 8: Golfete de Coro, from Punta Cardón to the mouth of the Maracaibo estuary. This ecoregion is characterized by shallow waters and high sedimentary load and suspended materials from the Golfo de Coro. There is a high secondary production mainly associated with penaeid shrimps and sciaenid fishes (corvinas).
Ecoregion 9: Estuarine region of Lake Maracaibo. This ecoregion is characterized by a continuous perturbation caused by the dredging of the navigation channel along the strait of Maracaibo. The region has high primary and secondary productions, the latter associated with crustaceans (shrimp and crabs), bivalve mollusks and fishes. It is threatened by pollution due to the intense activity of the oil industry and sewage discharges.
Ecoregion 10: Golfo de Venezuela. This ecoregion is characterized by a wide coastal platform with shallow bottoms (<50 m). It has a high primary and secondary production of shrimp and fishes due to the upwelling fronts. Bottom communities are dominated by detritivorous organisms.
Ecoregion 11: Oceanic Islands which includes all emerged land with the exception of the Eastern Upwelling Region. This ecoregion is characterized by islands and archipelagos, with reefs, seagrass and well developed mangrove areas. It has a high biodiversity. It is a zone shared by marine turtles and birds.
Ecoregion 12: Oceanic area including the Exclusive Economic Zone of Venezuela, an area with depths beyond 200 m. The biodiversity of this ecoregion is little known. It has zones with moderately high primary productivity due to their proximity to upwelling coastal zones.
Ecoregion 13: Cariaco Trench, the depression of the continental shelf below 150 m deep in front of Barcelona and Mochima Bay in eastern Venezuela. This ecoregion represents a special ecosystem with maximum depths close to 1,400 m and anoxic water masses below 250 m. The productive processes are based mainly on chemosynthesis.

In Venezuela, echinoderms have been poorly studied in contrast to other groups of marine organisms like fish, crustaceans (decapods and peracarids) and mollusks, which have been grouped, classified and analyzed taxonomically (Rodríguez 1980; Valdez and Aguilera 1987; Cervigón 1994; Lodeiros et al. 1999; Díaz and Martín 2001; Martín et al. 2002; Martín and Díaz 2003, 2005, 2007; Díaz et al. 2005). Taxonomic and ecological studies on local echinoderms are scarce and limited, in spite of the fact that they represent an important component of the biota in the marine coastal zones (Penchaszadeh 2003). A compilation of studies on echinoderms of Venezuela is presented here, which represents an effort of researchers from several national and international institutions, whose analyses are a current description of the status of these marine invertebrates.

7.2 Research

Along the Venezuelan coast there are few studies pertaining to species of the Phylum Echinodermata, and most have focused mainly on the northern Caribbean Sea region. Studies made in the Atlantic region have been financed by the Venezuelan oil industry (PDVSA) and include the Environmental Base Line of the Platform in the Orinoco River delta (Martín et al. 2007) and the Venezuelan Atlantic Front (Gómez et al. 2005).

7.2.1 Taxonomy

The first studies dealing with the Phylum Echinodermata (Families: Astropectinidae, Luidiidae and Echinasteridae) in Venezuela, came from the Belgian research cruise "*Mercator*" between 1935 and 1936, which followed a trajectory between Florida and Venezuela (Engel et al. 1960). Furthermore, Ummels (1963) and Jangoux (1978) published studies on asteroids where they incorporated locations from the country.

At a regional level, the work on echinoderm taxonomy was initiated in the 1960s with the publication of Zoppi de Roa (1967), who made an exhaustive identification of the species belonging to the different classes along the Venezuelan coast, from where she reported 65 species. On the other hand, Work (1969) reported 27 species for Los Roques Archipelago National Park distributed as follows: eight echinoids, 11 ophiuroids, three asteroids and five holothuroids. Later, Boada (1970) conducted a general study on the taxonomy of echinoderms, but focused in the zone near Cumaná. Belmonte (1985) studied the asteroids, echinoids and ophiuroids present in Margarita Island, while Amaro (2009) made a revision of the echinoderms of the Gulf of Cariaco, and reported 50 species, nine of which belonged to the Class Asteroidea, 17 Ophiuroidea, 12 Echinoidea and 12 Holothuroidea.

Perhaps the holothuroid has been the most highly studied echinoderm group of Venezuela (Mago 1971; Martínez 1971, 1973, 1975, 1982, 1986, 1987a, b, 1988, 1989, 1991a, b; Martínez and Mago 1975), particularly due to their commercial and ecologic interest. In what concerns to the remaining classes, Martínez (1969) evaluated organisms belonging to the Family Brissidae (Class Echinoidea); while Flores and Martínez (1970) described the status of the Family Goniasteridae (Class Asteroidea) and Francisco and Pauls (2008) made a bibliographic compilation of the Order Clypeasteroida (Class Echinoidea) reporting ten species, six of which belong to the Family Clypeasteridae and four to the Family Mellitidae.

7.2.2 Ecology

Studies on populations and ecology of echinoderms have been made primarily in the central-western and eastern coasts of the country. Most refer to abundance and distribution of the species in the Class Echinoidea, Holothuroidea and Ophiuroidea. With respect to the Class Echinoidea, six of the seven species known for the Caribbean Sea have been reported locally (Hendler et al. 1995): *Lytechinus variegatus* (Lamarck, 1816), *Tripneustes ventricosus* Lamarck, 1816, *Diadema antillarum* Philippi, 1845, *Arbacia punctulata* (Lamarck, 1816), *Echinometra lucunter* (Linnaeus, 1758) and *Echinometra viridis* Agassiz, 1863. These species are distributed along the coast from Falcón to Sucre and Nueva Esparta states (Cruz-Motta 2007). The sand dollar, *Mellita quinquiesperforata* (Leske, 1778) has been reported from the coasts of Golfo Triste (Penchaszadeh and Molinet 1994). Most records have been made between 0 and 50 m depth, especially on rocky substrates and seagrass (Cruz-Motta 2007). For the Atlantic region, most species are irregular urchins, among which are *Agassizia excentrica* Agassiz, 1869, *Brissopsis elongata* Mortensen, 1907, *Clypeaster euclastus* Clark, 1941 and *Schizaster orbignyanus* A. Agassiz 1880 (Martínez 1969; Francisco and Pauls 2008).

Diadema antillarum is distributed along the entire continental coast and oceanic islands of Venezuela, associated primarily with coral communities. The abundance of this species in some coastal locations, before the massive mortality registered in 1984, was similar to those found in the rest of the Caribbean Sea. However, only in the eastern zone of the country is there information at the population level before and after the demise that this species underwent in the entire Caribbean Sea, which also affected the populations present in Venezuela (Pauls 1982; Noriega et al. 2006). In Mochima Bay, eastern Sucre state, the average density before the massive mortality varied between 0.28 and 4 ind m^{-2} (Pauls 1982). After the mass mortality, the average density decreased to 0.15 ind m^{-2} in 2000 and 0.47 ind m^2 in 2002 (Noriega et al. 2006). On the other hand, there have been records of densities from 0.28 ind m^{-2} in 2002 to 1.05 ind m^{-2} in 2003 in Oricao Bay, and from 0.84 ind m^{-2} (2002) to 0.74 ind m^{-2} (2003) in Chichiriviche de la Costa, both locations in the central coast of Venezuela (Noriega et al. 2006). The populations of *D. antillarum* in Morrocoy National Park and Cuare Wildlife Refuge showed the highest average densities at Playuela (0.43 ind m^{-2}) and Cayo Sur (0.95 ind m^{-2}), while the remaining areas showed densities <0.1 ind m^{-2}. The density found in Playuela in 2003 is lower than the one reported before the mass mortality event (0.58–3.64 ind m^{-2}) (Noriega et al. 2006). In Morrocoy National Park, Weil et al. (1984) reported average densities between 2.6 and 3.7 ind m^{-2} on three reefs (Pescadores, Playuela and Caiman). The highest densities (4.0–7.0 ind m^{-2}) were found in the shallow and frontal zones of the reefs, characterized by a high structural complexity of the substrate. Low densities (1.5–4.0 ind m^{-2}) were found in deep zones of the reef platform with low to intermediate structural complexity. Densities of *D. antillarum* in Los Roques Archipelago National Park, specifically in Herradura Reef, varied between 0.22 and 0.23 ind m^{-2} (Noriega et al. 2006). Currently, there are no data about the population densities of *D. antillarum* in these or other areas of the country.

Other publications about the distribution and density of Venezuelan echinoids have been mainly related with their association to seagrass, *Thalassia testudinum*. Noriega et al. (2002) showed the relation between the density of *L. variegatus* and several parameters of the seagrass (percent cover, density and length of the leaves, density of the roots and total biomass) in Mochima National Park. Total biomass was the most influential parameter on density (0.6–3.6 ind m^{-2}) and distribution of the species. At Margarita Island, Gómez (2002, 2003) also found a significant correlation between the abundance of *L. variegatus* and the density of *T. testudinum*, reporting maximum densities from 38 to 52 ind m^{-2}.

The association of ophiuroids to specific habitats like sponges, corals, octocorals and algae has been described by some authors (Boffi 1972; Hendler and Miller 1984; Hendler and Littman 1986). Ophiuroids have been reported from the central-western region of Venezuela to the Orinoco River delta, with 52 identified species. The Family Amphiuridae is the most diversified one, being found both on the benthos of shallow and deep water (Borges 2006), while Ophiodermatidae and Ophiocomidae inhabit primarily shallow water. Villamizar and Laughlin (1991), studied the animal communities associated to two species of dominant sponges (*Aplysina archeri* and *A. lacunosa*), characterized by different degrees of structural complexity, in a coral reef of Los Roques Archipelago National Park. Mollusks had the largest number of species in both sponges (76.51 %); crustaceans had the largest percent in number of individuals (67.27 %), while fish and ophiuroids had the largest percent in dry weight (50.45 and 42.68 %, respectively).

In Cubagua Island, Nueva Esparta state, Hernández (2010) reported 24 species of ophiuroids associated primarily with hard bottom, seagrass, aggregations of the bivalve *Arca zebra* and the macroalga *Halimeda opuntia*. In the Gulf of Cariaco, revision of the echinoderms resulted in 50 species, 34 genera, 22 families and four classes: nine species of asteroids, 17 species of ophiuroids, 12 species of echinoids and 12 species of holothuroids (Amaro 2009).

Studies have been made on the trophic ecology of species of Asteroidea, Echinoidea and Holothuroidea. Studies on asteroids have focused mainly on the composition of the diet and trophic interactions. In Golfo Triste, Falcón state, Bitter (1984) found that *Astropecten marginatus* is a non-selective consumer, although it shows a clear preference for mollusks, especially bivalves (33 %) and gastropods (50 %). Similar results were found by Penchaszadeh (1973) with the asteroid *Astropecten brasiliensis,* in which mollusks represented 80 % of the stomach contents of the studied individuals. In relation to the trophic ecology of *Astropecten brasiliensis riensis* and *A. articulatus*, Bitter and Penchaszadeh (1983) found the trophic niches of both species overlap, although competition would be reduced by specialization on the size of the preferred prey. In both species the diet was mainly mollusks, with gastropods as the most important prey item. The gastropod *Finella dubia* accounted for 50 % of the diet of *A. brasiliensis riensis* and the gastropod *Olivella myrmecoon* was 26.5 % of the diet of *A. articulatus.*

Penchaszadeh and Lera (1983) found in their study of the diet of three species of asteroids of the genus *Luidia*, that *L. barimae* is mainly a predator of ophiuroids (85.8 % of its diet); while *L. senegalensis* feeds mainly on bivalves (49.1 %) and *L. clathrata* on ophiuroids (35.3 %) and gastropods (24.4 %). Penchaszadeh and Molinet (1983) evaluated the diet of the asteroid *Tethyaster vestitus*, and found that it is a specialist predator of *Astropecten* spp. (which represented 55.8 % of the total number of prey items). Other prey items of *T. vestitus* were *L. clathrata* and ophiuroids (0.8 and 3.8 % of the diet, respectively). Bitter et al. (1980) studied the trophic interactions between two asteroids (*A. riensis* and *T. vestitus*), and found that echinoderms are the most important group in the diet of *T. vestitus*, with asteroids representing 55.74 % of its diet and the remaining groups 44.20 %. Among the latter, echinoids represented 12.30 %, crinoids 27.87 %, holothuroids 1.64 % and ophiuroids 2.4 % of its diet. The diet of *A. brasiliensis riensis*, was composed of gastropods (67.8 %), pelecypods (20 %), crustaceans (4.71 %) and scaphopods (2.23 %). *A. riensis* is an important element within the trophic structure of the community studied, representing the second and third level of consumers while *T. vestitus* represents the fourth level.

In Mochima Bay, Sucre state, the asteroid *Oreaster reticulatus* behaves mainly as a predator of the sea urchin *L. variegatus*. However, some observations suggest that it could be an opportunistic microphage, with densities between 0.0075 and 0.2 ind m^{-2} (Martín et al. 2001). The diet of *A. marginatus* at the mouth of the Orinoco River and the Gulf of Paria was mainly bivalve mollusks and gastropods (Ortega et al. 2011).

In the Caribbean Sea, many echinoids feed exclusively upon *T. testudinum* and its epibionts (Lawrence et al. 2007). In La Tortuga Island, even though data on density of *T. ventricosus* are not known, its stomach contents consisted primarily of leaves of *T. testudinum* with epibionts although the macroalgae *Acanthophora spicifera, Hypnea cornuta* and *Cladophora* spp. were found, showing their possible importance on the abundance and distribution of this species (Barrios and Reyes 2009). These authors also indicate that the presence of calcareous algae like *Hydrolithon* sp. and *Jania* sp. in the diet of *T. ventricosus* showed that they could be important items in the calcium metabolism associated with formation of the test. On the other hand, Montealegre and Gómez (1999) determined the stomach contents of *L. variegatus* at Margarita Island, observing mainly remnants of *T. testudinum* and smaller amounts of the macroalgae *Ulva lactuca* and diatoms.

With respect to holothuroids, Sambrano et al. (1990) characterized the ingested items of *Isostichopus badionotus* and *Holothuria mexicana* from two types of substrate (with calcareous and terrigenous origins) in *T. testudinum* meadows of the western region of Venezuela. They found that the mean size of ingested species differs significantly between these two holothurian species for both types of substrate. The analyses of the stomach contents of *Duasmodactyla seguroensis, Euthyonacta solida, Ocnus pygmaeus, O. surinamensis, O. suspectus, Thyonella sabanillaenis and Stolus cognatus* from the northeastern region and some oceanic islands, showed that the diet of holothuroids of the Order Dendrochirotida include: planktonic and benthic diatoms, dinoflagellates and copepods as well as filamentous cyanobacteria whereas species of the Order Aspidochirotida (*I. badionotus, Holothuria grisea, H. occidentalis, H. pseudofossor* and *H. surinamensis*) consume detritus consisting of calcareous sand or quartz, coral fragments, bivalves shells, test and spines fragments of echinoderms, sponge spicules, fish scales and amorphous material. Holothuroids of the Order Apoda scrape the surrounding ground with their feathered tentacles and consume mainly benthic diatoms and remnants of green and brown algae (Martínez 1989, 1991a, b).

7.2.3 Reproduction

The biology and reproductive behavior of echinoderms in Venezuela is poorly known. However, some studies suggest that temperature represents a key exogenous factor in the onset of gametogenesis. In the northeastern region of Venezuela, hydrographic and nutrient availability in coastal water masses are variable due to the upwelling phenomena, which promote temperature fluctuations mostly during the first half of the year (with variations between 20 and 28 °C) (Ferraz-Reyes 1987). Montealegre and Gómez (2005) indicate that *L. variegatus* exhibits the highest spawning activity in June and July in coincidence with an increase of water temperature. Espinoza et al. (2008), in spite of not finding a clear influence of environmental factors on the reproduction processes of *E. lucunter* and *L. variegatus*, indicated that low temperature (23–24 °C) seems to be an initiating factor of gametogenesis and high temperature (26–28 °C) of maturation of gametes of *L. variegatus*.

Muñoz (1980), Muñoz and Ellietes (1982, 1983) and Brey (1992), evaluated the effect of different toxic substances (ferrous sulphate and sodium hypochlorite), as well as temperature on embryonic development, fertilization and reproductive processes in *E. lucunter*. Furthermore, Muñoz (1980) evaluated the fertilization of eggs under variable thermal conditions (25, 30, 35 and 40 °C), finding a greater motility of spermatozoa at 30–35 °C, a reduction in the percent of fertilization at 40 °C and a moderate reduction in the number of fertilized eggs at 25 °C. Brey (1992) studied the effect of the ratio sperm/eggs and the time of post-dilution of sperm in sea water on the fertilization success. Toxicity assays with Cr^{6+} ($K_2Cr_2O_7$), Zn^{2+} ($ZnCl_2$) and Cu^{2+} ($CuSO_4$) found that zinc and chromium are less toxic than copper (LD50 for copper between 36.1 and 119.3 ppb).

One aspect that is gaining importance on the eastern coast of Venezuela is knowledge about larval and juvenile development of echinoid species with potential for aquaculture. Some species in this category are *L. variegatus* (Butriago and Lodeiros 2005; Butriago et al. 2005; Gómez and Gómez 2005; Domínguez et al. 2007), *T. ventricosus* (Fernández 2005) and *A. punctulata* (García et al. 2005), in which the effect of salinity and food type were evaluated and are discussed in Sect. 7.5 of this chapter.

In holothurians, Palazón (2001, 2010) studied the reproductive biology of *I. badionotus* from the islands of Margarita, Coche and Cubagua, in northeastern

Venezuela. He found that, in general, male:female sex ratio does not deviate significantly from 1:1 and there were 2 % hermaphrodites. Sexual maturity was reached at a horizontal diameter above 147 mm (55 g). Individuals with mature gametes were found during the entire year, which suggests that reproduction is continuous, although with a greater synchrony in spawning during the period from September to November when water masses stratify and temperature is higher. Fecundity is high, with an average number of eggs per individual of 1,417,330 (SD 151,186), but sometimes exceeded 5 million eggs.

7.2.4 Bioerosion and Herbivory

With respect to bioerosion by echinoderms, Weil (1980) evaluated the impact of *D. antillarum* on coral reefs at Mochima National Park and Díaz et al. (1982) and González (2005) evaluated it in Los Roques Archipelago. Losada et al. (1988) manipulated densities of *D. antillarum* in the intake channel of Planta Centro thermal electric plant, located at Punta Morón-Golfo Triste, and reported a significant increment of bare substrate by grazing at densities that varied from 0.170 ind m^{-2} and bare space of 102.6 cm^2 to more than 1,000 cm^2 bare space when the density reached 0.310 ind m^{-2}.

Another aspect, though less known, is the effect of grazing by echinoids as biocontrol agents. Lodeiros and García (2004) described for the first time in the Gulf of Cariaco, Sucre state, *L. variegatus* and *E. lucunter* as control agents of fouling in culture baskets and bivalve shells in suspended cultivation systems. *L. variegatus* produced a 75 % decrease in the fouling of the culture baskets and 71 % that of the shells of *Pinctada imbricata*, suggesting the usefulness of this method as a natural control of fouling in these systems.

7.2.5 Physiology and Biochemistry

There are few physiological studies of echinoderms in Venezuela. Studies with holothurians have focused mainly on the ion-osmotic response to a great variety of chemical compounds. The first studies were initiated in the 1970s by Zanders (1970), Madrid (1971), Zanders and Herrera (1974) and Madrid et al. (1976), who evaluated the effect of the salinity change of sea water upon the coelomic and intracellular ionic composition of *I. badionotus*. Zanders (1970) studied the histology, ionic composition and permeability to sodium of the intestine of *I. badionotus*. The results showed that the organisms is isosmotic to sea water, but levels of different ions in the tissues could differ from the latter, like potassium which reaches intracellular concentrations 10–30 times higher than in the external environment. Tissue water represents 80–90 % of the wet weight of the intestine and 60–80 % of the tissue water corresponds to extracellular liquid. Zanders and

Herrera (1974) evaluated the distribution and ion flow in cells of the intestine and stomach muscle of *I. badionotus*. They found that potassium concentration in intestinal cells was 240 μ-equiv g^{-1} cell and 252 μ-equiv g^{-1} cell in those of the stomach; sodium was 246 and 296 μ-equiv g^{-1} cell, while chloride was 223 and 302 μ-equiv g^{-1} cell, respectively. In muscle cells, potassium, sodium and chloride concentrations were 348, 119 and 127 μ-equiv g^{-1} cell, respectively. Madrid et al. (1976) found that intracellular ionic concentrations increase in direct relation to the concentration of the external environment and the ionic concentrations of the coelomic liquid of *I. badionotus* equilibrates with 80 and 120 % sea water between 4 and 6 h.

Studies made by Herrera and Plaza (1981a, b), Jurisic (1982), Herrera and López (1983), Jurisic et al. (1983), Sambrano et al. (1990), Conde et al. (1991), Egea et al. (1991), Herrera (1991), Foglietta (1994), Foglietta and Herrera (1996), Herrera and Foglietta (1998) and Herrera et al. (2000) focused their research on the ion-osmotic response of the respiratory structures of *I. badionotus* and *Holothuria glaberrima*. Herrera and Plaza (1981b), found that the respiratory tree and muscles of *H. glaberrima* respire at a rate of 635.5 ± 10 μ l O_2/h × g dw and 137.4 ± 6.3 μ l O_2/h × g dw, respectively, in response to the concentration of sodium in artificial sea water. They indicate that the substitution of sodium by chlorine, lithium or Tris (pH 8.0) does not modify the oxygen consumption of the respiratory trees structures. The low respiratory rate of muscles could be due to limitations of oxygen diffusion in the tissue. Metabolic conditions could be involved in the respiratory response of tissues to pH. Herrera and López (1983) evaluated the effect of sodium concentration in the external medium and the intracellular osmotic condition of the respiratory tree of *H. glaberrima*. They subjected the cells of the respiratory trees to sodium concentrations reduced isosmotically (60, 30, 15, 6 and 0 % of normal sea water). The analyses showed an increasing loss of sodium, potassium and chloride while maintaining the cytoplasm isosmotic with the external medium. This indicates that the maintenance of isosmoticity with the external medium assures the constancy of intracellular volume. The intracellular concentration of potassium was regulated within relatively narrow limits in spite of the changes in cell volume. In sea water with 6 and 30 % sodium, intracellular sodium concentration changed only by 1.5 times which suggests a certain degree of regulation. Herrera and Foglietta (1998) evaluated the effect of ouabain (which blocks the Na–K pump) on the ion-osmotic properties of cells of the respiratory trees of *I. badionotus*. Individuals were exposed to five concentrations of sea water: 17.5, 21.0, 26.25, 35.0 and 42 ‰ in the presence or absence of ouabain. They found that control and ouabain-treated cells responded as perfect osmometers to variations in sea water concentration implying that intracellular organic osmolyte content also remained unchanged despite changes in sea water concentration. In order to test if cells of marine stenohaline invertebrates can increase in volume, Herrera et al. (2000) evaluated the effect of five different solutions of artificial sea water with isosmotic sodium and KCl (100, 83, 71, 60 and 50 %), with and without ouabain, in fragments of the respiratory tree of *I. badionotus*. They found immobilization of extracellular sodium, depolarization of the membrane, an osmometric increase of the cells and a decrease of potassium, sodium and chloride when the cells were treated with the same dilutions of sea water with distilled water. Jurisic et al. (1983) studied the role of chloride in the intracellular isosmotic regulation in cells of the respiratory tree of *H. glaberrima* exposed to different concentrations of chloride (100, 60 and 24 % of normal), finding a small decrease in the intracellular potassium content. However, as a consequence of the 50 % reduction in cell volume, the intracellular concentration of potassium increased approximately by twofold, while that of sodium decreased considerably. Respiratory rate of tissues increased with the reduction of external chloride, which reveals an increase in the metabolic work needed for the sodium and potassium pump against the ever increasing electrolytic gradient of these ions. Foglietta (1994) studied the ion-osmotic response to changes in salinity of the external medium in *I. badionotus* pre-incubated in 57, 100 and 137 % sea water and the effect of ouabain in pre-incubated individuals at 100 % sea water. She found an inverse linear relationship between the relative intracellular space to incubation concentrations. These results indicate that cells show perfect osmometric behavior. On the other hand, the intracellular concentration of K^+ remained unchanged, that of Na^+ decreased with external salinity, and that of Cl^- decreased slightly when incubated above and below 83 % sea water and ouabaine.

Foglietta and Herrera (1996) performed studies similar to those of Foglietta (1994), without the use of ouabain. They evaluated the ion-osmotic response of the respiratory tree of *I. badionotus* pre-incubated in hyper-, iso- and hypo-osmotic sea water. The cells and tissues of individuals pre-incubated in 20 and 48 % sea water responded in a similar way to those pre-incubated in 35 % sea water. The intracellular content of potassium was well regulated, remaining constant in spite of changes in sea water concentration; that of sodium decreased with extracellular concentration, while that of chloride increased in concentrations above and below of 29 %. *I. badionotus* seemed to be more tolerant to hyposaline than to hypersaline sea water, which suggests that normal sea water represents the upper limit of its salinity range. The osmotic condition of the cell and the response of the tissue do not seem to be modified by the variation in the concentration of the means in which it was pre-incubated.

In contrast to the studies just described, Conde et al. (1991) evaluated the patterns of disintegration of fecal material of *H. mexicana* and *I. badionotus*. In *I. badionotus,* the feces completely disintegrate around 8 h, suggesting a high probability of dispersion in comparison with feces of *H. mexicana* which remain on the substrate even after 25 h. The authors suggest that bacterial activity and fungal growth are responsible for the disintegration of the fecal material of *I. badionotus*, while the disintegration of the feces of *H. mexicana* is associated with the activity of motile marine organisms.

There are few studies at the biochemical level. Urbaneja (1968) determined steroidal and lipid contents in the sea star *Echinaster sentus*. Cedeño (1971) identified the fatty acids of the holothurian, *Ludwigothuria mexicana*. Salazar et al. (1994) made histochemical studies of cells of the echinoids, *L. variegatus* and *E. lucunter*. Malavé (1995) characterized and quantified the lipid and fatty acid content in the gonads of *L. variegatus* and *E. lucunter*. He found that the main lipids present in both species were phospholipids and triglycerides, and in smaller concentration free cholesterol, free fatty acids and ester of cholesterol. The higher triglyceride and saturated fatty acid values present in the gonadic tissue of both species coincided with their sexual maturity. The values of saturated fatty acids were, in decreasing order of proportion, 16:0, 14:0, 17:0 and 18:0, while the unsaturated fatty acids were represented by the mono-unsaturated fatty acids (16:1, 18:1 and 22:1) and the polyunsaturated fatty acids (20:5ω3, 18:3 y 20:3). Díaz et al. (1996) determined the total protein content and electrolytes in the hemal lymph of *L. variegatus* and *E. lucunter*, finding that Na^+ concentration varied from 466.42 to 470.30 meq l^{-1}, Cl^- from 372.78 to 739.2 meq l^{-1}, and K^+ from 5.73 to 3.25 meq l^{-1}. Total protein concentration varied from 0.53 to 0.58 %.

7.3 Diversity and Distribution

The information in this section has been compiled from various bibliographic sources which include studies made in Venezuela (Ummels 1963; Zoppi de Roa 1967; Martínez 1969, 1971, 1973, 1975, 1982, 1986, 1987a, b, 1988, 1989, 1991a, b; Boada 1970; Flores and Martínez 1970; Mago 1971; Penchaszadeh 1973, 2003; Martínez and Mago 1975; Bitter et al. 1980; Bitter and Penchaszadeh 1983; Penchaszadeh and Lera 1983; Penchaszadeh and Molinet 1983, 1994; Bitter 1984; Belmonte 1985; Noriega 1998; Francisco and Pauls 2008; Amaro 2009; Hernández 2010; Ortega et al. 2011) and in the Caribbean Sea (Jeannet 1928; Engel 1939; Engel et al. 1960; Ummels 1963; Boffi 1972; Jangoux 1978; Hendler and Miller 1984; Hunte et al. 1986; Scheibling and Mladenov 1998; Donovan 2001; Toral-Granda 2008) that provide information on echinoderm species found in Venezuelan waters as well as the data base available in the web page of GBIF (Biodiversity occurrence data provided by www.gbif.net, 2009-12-10). All echinoderm species from such studies have been placed in systematic order in the appendix. The distribution of systematic groups as classes, families and species was established considering the characterization of the Venezuelan ecoregions (Miloslavich et al. 2005; Miloslavich and Klein 2008).

Two hundred and thirty four species of echinoderms have been reported for the Venezuelan waters, with 97.87 % of them along the Caribbean Sea coast and only 2.13 % of them along the Atlantic coast (Table 7.1). For all coasts, the Asteroidea is the most diverse with respect to the number of species (26.92 %), followed by the Holothuroidea (22.65 %); while the Crinoidea is the least diverse (6.84 %). It is important to emphasize, that in spite of a smaller diversity of species along the Atlantic coast, *Tethyaster vestitus* (Say, 1825) (Asteroidea), *Leptopentacta deichmannae* Domantay, 1953 (Holothuroidea), *Paraster doederleini* Chesher, 1972 (Echinoidea) and *Amphiura otteri* Ljungman, 1872 (Ophiuroidea) have only been reported for this ecoregion.

Most records of echinoderms in Venezuelan waters have been made from depths between 0 and 50 m (47.42 %), followed by the range 51–199 m (26.12 %), 200–499 m (8.25 %) and depths >500 m (18.21 %). Now, considering the distribution of echinoderms in the ecoregions described (Table 7.1, Fig. 7.2), 29.17 % of the known species are reported for the Eastern upwelling zone ecoregion; followed by the Oceanic islands ecoregión (23.33 %) and the Central coast ecoregion (10.42 %). The smallest number was found in the Gulf of Venezuela ecoregion (1.67 %). There are no reports of echinoderms within the Lake Maracaibo ecoregion, most probably due to the low salinity.

The Class Crinoidea includes 16 species, 15 genera, nine families and four orders (Table 7.1), with members mainly in the Central Coast (23.91 %), Oceanic Islands (21.74 %) and Eastern Upwelling ecoregions (19.57 %). They have not been found in the Unare-Píritu, Golfete de Coro and Oceanic area ecoregions. Comasteridae and Antedonidae showed a wider distribution, being present in seven and five ecoregions, respectively. Families like Thalassometridae, Pentacrinidae and Bathycrinidae, have only been reported for a single ecoregion, Oceanic Islands, Central Coast and Tocuyo, respectively. Forty percent of the crinoids reported are found in the depth range 0–50 m, followed with 32 % in the range 51–199 m. The Charitometridae has the largest depth range, with representatives from 51 m to more than 500 m deep. Tropiometridae, Colobometridae, Comasteridae and Antedonidae are only reported for shallow waters, in the depth range 0–50 m for the first, and the remaining in the range 0–199 m. Bathycrinidae and Atelecrinidae show the deepest records (>500 m).

The Class Asteroidea is the most diverse, with 63 species, 40 genera, 15 families and six orders (Table 7.1), being mostly represented in the Eastern upwelling (22.92 %), Oceanic Islands (22.92 %) and Paraguaná (11.81 %) ecoregions; with the smallest representation in Unare-Píritu ecoregion (0.69 %, a single record). The Astropectinidae and Luidiidae have the widest distribution along the coast, with representatives in ten and nine ecoregions, respectively. Families like Brisingidae, Asteriidae, Goniopectinidae and Benthopectinidae, are only present in a single ecoregion, the first in Oceanic Islands ecoregion, the second in Tocuyo ecoregion and the last two in the Eastern upwelling region. It has been calculated that 30.53 % of the asteroid records has been made within 51–199 m, followed by depths greater than 500 m with 27.37 % of the records. Members of Astropectinidae and Goniasteridae were present in all depth intervals. Furthermore, the Oreasteridae, Asterinidae, Ophidiasteridae and Luidiidae, are only reported for shallower waters, within 0–50 m for the first, and the remaining from 0 to 199 m. Brisingidae, Asteriidae, Zorasteridae, Goniopectinidae and Pterasteridae are only recorded at depths greater than 200 m.

The Class Ophiuroidea consisted of 52 species, 31 genera, 14 families and two orders, being mostly represented in the Eastern Upwelling (34.07 %) and Oceanic Islands ecoregions (21.98 %), and has not been found in the Cariaco Trench, Tocuyo, Golfo de Venezuela and Oceanic area ecoregions. The Amphiuridae and Ophiuridae have the widest distribution with representatives in seven ecoregions, while the Asteroschematidae, Ophiochitonidae, Ophiolepididae, Ophiomixidae

Table 7.1 Number of echinoderm taxonomic groups in the marine ecoregions of Venezuela

Ecoregion		Crinoidea	Asteroidea	Ophiuroidea	Echinoidea	Holothuroidea	Total
Delta of the Orinoco River and Atlantic Front	Orders	1	3	2	3	2	9
	Families	1	4	4	4	2	15
	Genera	1	7	6	6	2	22
	Species	2	12	6	7	2	29
Eastern upwelling region	Orders	3	5	2	10	3	23
	Families	6	10	10	17	8	51
	Genera	8	18	17	26	22	91
	Species	9	33	31	33	35	141
Axis Unare-Píritu-Tacarigua	Orders	0	0	0	0	0	0
	Families	0	0	0	0	0	0
	Genera	0	0	0	0	0	0
	Species	0	0	0	0	0	0
Central coast	Orders	0	1	1	2	2	6
	Families	0	1	4	2	3	10
	Genera	0	1	6	2	4	13
	Species	0	1	8	2	5	16
Golfo Triste	Orders	3	2	3	8	0	16
	Families	6	5	8	12	0	31
	Genera	10	5	10	15	0	40
	Species	11	8	14	17	0	50
Tocuyo region,	Orders	2	3	1	6	1	13
	Families	4	6	2	8	2	22
	Genera	6	8	2	12	2	30
	Species	6	16	2	15	2	41
Paraguaná region	Orders	0	0	0	0	0	0
	Families	0	0	0	0	0	0
	Genera	0	0	0	0	0	0
	Species	0	0	0	0	0	0

(continued)

Table 7.1 (continued)

Ecoregion		Crinoidea	Asteroidea	Ophiuroidea	Echinoidea	Holothuroidea	Total
Golfete de Coro	Orders	1	3	2	5	1	12
	Families	1	5	5	6	1	18
	Genera	2	10	8	10	1	31
	Species	2	17	8	13	1	41
Estuarine region of lake Maracaibo	Orders	0	1	1	1	0	3
	Families	0	2	2	2	0	6
	Genera	0	2	2	2	0	6
	Species	0	5	2	2	0	9
Golfo de Venezuela	Orders	2	1	0	2	0	5
	Families	2	2	0	2	0	6
	Genera	2	2	0	2	0	6
	Species	2	4	0	2	0	8
Oceanic Islands	Orders	0	0	0	0	0	0
	Families	0	0	0	0	0	0
	Genera	0	0	0	0	0	0
	Species	0	0	0	0	0	0
Oceanic area	Orders	1	6	3	10	4	24
	Families	6	11	9	18	8	52
	Genera	9	30	13	25	15	92
	Species	10	33	20	27	22	112
Cariaco Trench	Orders	0	0	0	0	0	0
	Families	0	0	0	0	0	0
	Genera	0	0	0	0	0	0
	Species	0	0	0	0	0	0
Orders		4	6	2	10	4	26
Families		9	15	14	22	12	72
Genera		15	40	31	38	34	158
Total species		16	63	52	50	53	234

C Crinoidea, *A* Asteroidea, *O* Ophiuroidea, *E* Echinoidea, *H* Holothuroidea

Fig. 7.2 Sampling of sea urchins in the eastern upwelling ecoregion, in the Lobos island, northern Araya peninsula Sucre state, Venezuela

and Ophionereididae are only reported for a single ecoregion. Records of ophiuroids show that 70.91 % of the species are found in shallow waters between 0 and 50 m deep, with the least number of reports at depths between 200 and 499 m (3.64 %). Most of the reported families of ophiuroids in Venezuela are only found at depths between 0 and 50 m (five families), while the Ophiactidae, Ophiolepididae and Ophiuridae are reported at depths greater than 500 m.

The Class Echinoidea is represented by 50 species, 38 genera, 22 families and 10 orders, with most members in the Eastern Upwelling (26.19 %), Oceanic Islands (21.43 %) and Central Coast (13.49 %) ecoregions, and no records in the Oceanic area region. The Clypeasteridae, Brissidae and Schizasteridae showed a wider distribution along the coast, with representatives in eleven, eight and seven ecoregions, respectively. Six families (Aeropsidae, Cassidulidae, Echinoneidae, Echinidae, Hemiasteridae and Loveniidae) are reported for a single ecoregion, the Oceanic Islands ecoregion for the first three and the remaining in the Eastern Upwelling ecoregion. Most echinoids (45.76 %) in Venezuela occupy shallow waters, between 0 and 50 m deep, followed by the range 51–199 m (42.37 %). The echinoid families reported in this study are only present at most in two depth ranges, most from 0 to 199 m (eight families). Only the Echinothuriidae was found in two depth ranges, 200–500 m and beyond 500 m. The other families (13) were only present in one depth range. The Echinometridae, Echinoneidae, Mellitidae and Scutellidae, are reported between 0 and 50 m deep, while the Aeropsidae, Aspidodiadematidae, Hemiasteridae, Loveniidae and Saleniidae were only found in depths greater than 500 m.

The Class Holothuroidea consisted of 53 species, 34 genera, 12 families and four orders, with most members in the Eastern upwelling (46.58 %) and Oceanic Islands (30.14 %) ecoregions. Members of this class have not been found in the Cariaco Trench, Central Coast, Tocuyo, Golfete de Coro and Golfo de Venezuela ecoregions. The Family Holothuridae shows the widest distribution along the coast (being found in five ecoregions), while the Chiridotidae, Deimatidae, Molpadiidae, Psychropotidae and Psolidae were only present in a single one. Most records of holothurians (69.64 %) are from the depth range 0–50 m. There are no records from the range 200–499 m. Eight families of holothurians are only found in shallow waters down to a maximum depth of 199 m. Representatives of four families (Deimatidae, Molpadiidae, Psychropotidae and Synallactidae) have only been found at depths deeper than 500 m.

7.4 Paleontology

Paleontological studies of echinoderms in Venezuela were initially developed in the 1920s, as a supplementary component during the initiation of petroleum exploration. The lack of knowledge at that time about the Venezuelan sedimentary basins prevented the stratigraphic location of areas with fossil accumulations as they are known today (Fig. 7.3). In addition, the limited attention to echinoderms, being considered among the so called miscellaneous groups due to their apparent lack of importance as paleo-stratigraphic markers, decreased the interest for research on this group of fossils.

Early geological expeditions to Venezuela by P. Christ, C. Wiedenmayer, Vonderschmitt and H.G. Kugler between 1921 and 1926 provided some specimens for this study. Also, J. Nance in 1941 and D. Rusk in 1954, among several other anonymous contributors, added specimens to the invertebrate survey from Venezuela.

Later taxonomic studies, based upon the first collections, led Jeannet (1928), Cooke (1941, 1961), Anisgard (1954), Von der Osten (1957), Weisbord (1969), Hoover (1981) and Mooi and Peterson (2000) to the formal classification of the first fossil species of Venezuelan echinoderms. Mihaljević et al. (2010) placed the study of fossil echinoderms in the paleoecological and paleobiogeographical context, which emphasizes the importance of the study of fossil echinoderms for the paleoenvironmental interpretation.

It should be emphasized, that the data provided in geological order in the following section of paleoenvironmental descriptions are based on the Stratigraphic Lexicon of Venezuela of the Venezuelan Ministry of Energy and Mines (Ministerio de Energía y Minas 1997). This is also referred to in the web site (http://www.pdv.com/lexico) as Venezuelan Geologic Code, and has been developed from Venezuelan fossil echinoderms references and records based on our own field in situ collections (Fig. 7.4) and specimens from museum collections.

7.4.1 Permian

The earliest record of echinoderms in Venezuela is a species of undetermined Cidaroida collected in the Palmarito Formation of Zulia state (Hoover 1981). This corresponds to a paleoenvironment of transitional sedimentation between shallow waters of variable salinity, and deeper waters with constant marine salinity.

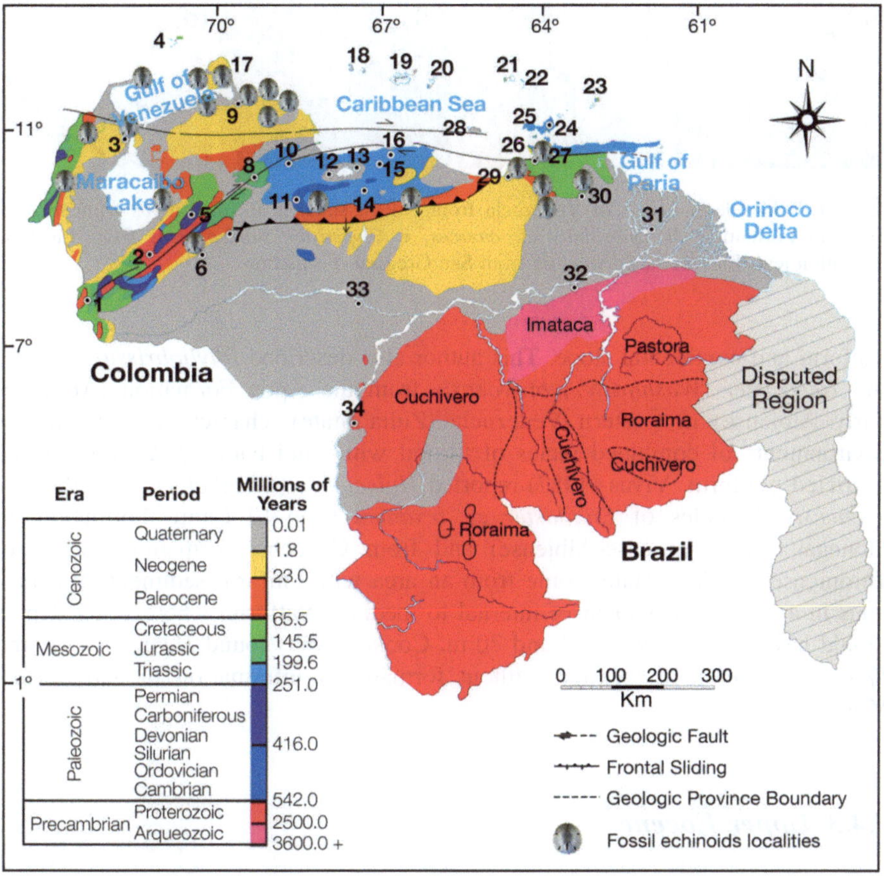

Fig. 7.3 Geological map of Venezuela showing the fossil echinoid localities. *1* San Cristóbal, *2* Mérida, *3* Maracaibo, *4* Los Monjes, *5* Trujillo, *6* Barinas, *7* Guanare, *8* Barquisimeto, *9* Coro, *10* San Felipe, *11* San Carlos, *12* Valencia, *13* Maracay, *14* San Juan de Los Morros, *15* Los Teques, *16* Caracas, *17* Península de Paraguaná, *18* Las Aves, *19* Los Roques, *20* La Orchila, *21* La Blanquilla, *22* Los Hermanos, *23* Los Testigos, *24* Porlamar, *25* Margarita, *26* Península de Araya, *27* Cumaná, *28* La Tortuga, *29* Barcelona, *30* Maturín, *31* Tucupita, *32* Ciudad Bolívar, *33* San Fernando de Apure, *34* Puerto Ayacucho

7.4.2 Lower Cretaceous

The reports of Cooke (1961) show the presence of *Tylocidaris* sp. and *Stereocidaris* sp. in the Barranquin Formation (Hauteriviense-Barremiense) of northeastern Venezuela (Sucre state), an area with a shallow marine paleoenvironment, and *Tetragramma* sp. in rocks without formational information. For the Chimana Formation (Albiense), also in eastern Venezuela (Anzoátegui state). Cooke (1961) reported *Pseudananchys* sp. from a paleoenvironment characterized by external

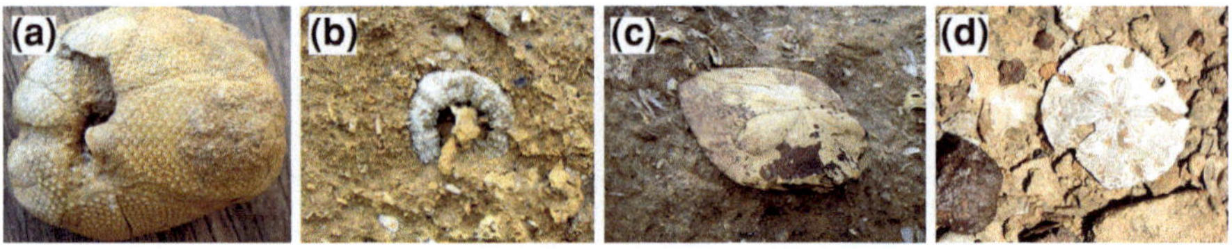

Fig. 7.4 Echinoderm fossils of Venezuela from Falcón state: **a** *Moira atropos* from the San Gregorio Formation, **b** *Lytechinus* cf. *euerces*, **c** *Clypeaster subdepressus* from La Vela Formation and **d** *Melitella falconensis* from San Gregorio Formation

platform and continental slope. This author also described *Phyllobrissus zulianus* and *Heteraster* (*Washitaster*) *bravoensis* from the Apon Formation (Aptiense-Early Albiense) in western Venezuela (Zulia State), characterized by marine environments of combined facies of littoral with sand bars, open platform and restricted platform. Vivas (1987) reported *Holectypus* (*Coenholectypus*) *planatus aponensis*. Samples of *Enallaster* cf. *texanum* from El Cantil Formation, in Monagas state (Aptiense-Albiense) and from Capacho Formation (Albiense-Turoniense) in Zulia state, come from an area with uniform sedimentary conditions in a paleoenvironment of internal to medium platform, where water depths should have been between 10 and 70 m. Cooke (1961) found *Hemiaster* sp. and *Epiaster whitei* in a location without formational information in Río Caripe, Monagas state.

7.4.3 Upper Eocene

In the Misoa Formation (a paleoenvironment with deltaic to coastal marine characteristics) at Trujillo state, *Eupatagus* sp. was found, whereas *Eupatagus clevei* was recorded by Cooke (1961) from rocks without formational information from Lara state. *Oligopygus rotundus*, was collected from rocks without formational information from Guárico State (Cooke 1961), and *Oligopygus christi* and *Pauropygus* sp. were found in rocks without formational information from Barinas State (Mihaljević et al. 2010). The latter authors describe *Oligopygus nancei*, from Caratas Formation (Tinajitas Member) from a reef environment with shallow water in Monagas state.

7.4.4 Oligocene

Weisbord (1969) found *Eucidaris tribuloides* in rocks without formational information from Nueva Esparta state. The stratigraphic record of this species extends until the Recent.

7.4.5 Lower Miocene

Jeannet (1928) described *Brissoma vonderschmitti*, *Opissaster kugleri* and *Opissaster* sp., which were collected in the upper series of Agua Salada Formation, in Falcón state. It is an hemipelagic deposit with paleoenvironmental variations between a deltaic complex and deep waters.

7.4.6 Middle Miocene

All species found for this period come from formations and places in Falcón state (Jeannet 1928; Cooke 1961). Jeannet (1928) reported *Clypeaster* sp. and *Pericosmus* sp. in rocks of the Capadare region; *Clypeaster kugleri*, *Echinolampas* sp. and *Heteroclypeus wiedenmayeri* from rocks without formational information in the vicinity of Caño León, probably La Vela Formation. This should have been a sedimentary environment in shallow waters, with salinity below normal sea water and restricted communication with the open ocean. Mihaljević et al. (2010) reported that *Encope michelini*, *Encope wiedenmayeri*, *Encope vonderschmitti* and *Encope* spp. were found in rocks without formational information in Boca de Hueque, probably of Punta Gavilán Formation from Pliocene, characterized by the sedimentation in a marine environment, with depths higher than 100 m within an external sublittoral zone. These authors also reported *Eurhodia falconensis* from rocks without formational information of La Guarava-Santa Clara (probably of Punta Gavilán Formation); *Echinolampas lycopersicus* in rocks without formational information from El Bejuco, in Valle de Araurima; and undetermined species of *Encope* spp., from rocks without formational information from La Caridad and Mirimire. Finally, *Plagiobrissus lamberti* was described in rocks without formational information from Guaidima to Riecito.

7.4.7 Upper Miocene

Mihaljević et al. (2010) found *Agassizia clevei*, *Antillaster lamberti*, *Brissopsis antillarum*, *Clypeaster subdepressus*, *Lytechinus* cf. *euerces*, and *Prionocidaris* sp. in La Vela Formation of Falcón state. These authors also found *Clypeaster rosaceus* in rocks of Guajira peninsula, probably associated to Castillete Formation of Zulia state, from deposits formed in environments of brackish to shallow marine waters, and in Margarita Island (probably from the Cubagua Formation, a member of Las Tejitas, at Nueva Esparta state). Cooke (1961) and Jeannet (1928) reported *A. lamberti* from Ojo de Agua Formation of Falcón state, a paleoenvironment of deposits in brackish to shallow marine waters, not far from the coast.

7.4.8 Pliocene

Mihaljević et al. (2010) reported *Lovenia* sp. *Rhynobrissus rostratus*, *Schizaster eurynotus*, *Tripneustes tobleri*, *Brissopsis antillarum*, and *Cassidulus* (*Cassidulus*) *falconensis* from Punta Gavilán Formation of Falcón state. Cooke (1961) found *Encope emarginata*, *E. secoensis* and *Moira atropos*, in San Gregorio Formation of Falcón state, a place characterized for having been a shallow littoral marine environment, with ostreid reefs and small waves. This author also reported *Agassizia excentrica*, *A. scrobiculata*, *arbacia punctulata*, *Encope kugleri*, *Litechinus variegatus* and *Melitella falconensis* from San Gregorio Formation in Falcón state. Finally, this author indicated that *Pericosmus stehlini*, and *Plagiobrissus grandis*, occur in Paraguaná Formation in Falcón state, a place which similarly shows a littoral marine paleoenvironment with shallow waters.

7.4.9 Pleistocene

Weisbord (1969) recorded *Brissopsis* cf. *atlantica*, *Echinolampas* sp., *Echinometra lucunter*, *Encope emarginata* Leske, 1778, *E. secoensis* Cooke, 1961, *Leodia divinata* and *M. atropos*, from the Cabo Blanco Group of Vargas state, but the author did not define the formational units. The Cabo Blanco Group with marine fossils is made up by Playa Grande and Mare Formations. The paleoenvironment of both corresponds to a littoral shallow water marine one, with coastal lagoons and moderate surf zone. *Clypeaster ravenelii*, *Meoma ventricosa* and *Encope kugleri*, were recorded for the Cumaná Formation by Macsotay et al. (2009), and its paleoenvironment corresponds to a tropical marine one with quiet and shallow waters with reefs.

7.4.10 Recent

Agassizia excentrica, *A. punctulata*, *Clypeaster rosaceus*, *C. subdepressus*, *Encope emarginata*, *E. secoensis*, *Lytechinus* cf. *euerces*, *Melitella falconensis*, *Moira atropos* and *Prionocidaris* sp., have been registered as fossils in Venezuela (Mihaljević et al. 2010).

The geochronological descriptions of the echinoderm species, as expressed by Mihaljević et al. (2010), show that the fossil echinoderms of Venezuela were most diverse during Middle Miocene, in contrast to the rest of the Caribbean Sea where they were most diverse during Eocene (Kier 1977; McKinney et al. 1992; Donovan 2001). However, it should be noted that the oceanographic and environmental changes that took place in the Caribbean Sea, as a result of the elevation of the Panama isthmus and the separation of the Pacific and Atlantic oceans during

Pliocene, resulted in the extinction and renovation of different invertebrate faunas (Jackson et al. 1993; Cheetham and Jackson 1996; O'Dea et al. 2007; Johnson et al. 2009; Aguilera et al. 2010).

Paleontological records are lacking for the upper Cretaceous, Paleocene and lower and middle Eocene. This bias could probably be associated to sampling problems or underdevelopment of taxonomic work in the units of these eras which have fragmentary remnants of echinoderms without a posterior classification (e.g. Navay Formation of upper Cretaceous: Sánchez and Lorente 1977; Morán Formation of Paleocene: Bushman 1965; Misoa Formation of lower and medium Eocene: Miller and Collinson 1951).

There is a clear preferential bathymetric distribution of recent echinoderms in the western tropical Atlantic and Caribbean Sea (Lessios 2005; Francisco and Pauls 2008). Families like Echinometridae, Clypeasteridae, Mellitidae and Toxopneustidae are found with greater frequency in littoral zones of shallow waters down to a depth of 402 m. This is in extreme contrast to families like Echinothuriidae, Aspidodiadematidae, Pedinidae, Hemiasteridae, Pourtalesiidae, Urechinidae, Aeropsidae and Lovenidae whose preferential depth is between 740 and 5,200 m. Amongst these extremes, the families Schizasteridae, Cidaridae and Diadematidae are distributed with some variation on the platform and continental slope. This fact shows the usefulness of echinoderms as indicators of paleoenvironments, as suggested by Kroh and Nebelsick (2003) and Mihaljević et al. (2010). This agrees with the environment of sedimentation of the formations which hold the co-generic taxa shown in the results described before. In the case of extinct species whose habitat is unknown, it is necessary to use the faunistic associations and lithology of the sediment types to support any paleobathymetric hypotheses.

7.5 Extraction, Fishing and Aquaculture

Fisheries of echinoderms are centered on sea urchin and sea cucumbers. They represent a product of high cost, since sea urchin gonads and the body wall of sea cucumbers (bêche-de-mer) are considered delicacies and aphrodisiacs. Consumption of sea urchin gonads is primarily centered in Europe and Asia. Consumption of sea cucumber body walls is primarily found in Asia and the southwest Pacific.

It is currently considered that most sea urchin and sea cucumber fisheries have reached their maximum exploitation rates. In contrast, their aquaculture is in expansion, particularly that of the sea cucumber *Apostichopus japonicus*. FAO statistics (http://www.fao.org/corp/statistics/en/) show that echinoderm extraction has decreased from 140,000 t at the end of 1990s to 90,000 t in 2007, mainly due to a reduction of catches because of overexploitation of natural stocks. However, their cultivation, practically nonexistent in the twentieth Century, started in 2003 with a production of 37,000 t and reached 85,000 t in 2007, for a value of US$ 300

Fig. 7.5 Sea urchin cone: test with several gonads of the sea urchin *Lytechinus variegatus* prepared for consumption in Margarita island, Venezuela

million. Thus, the demand of the Asiatic countries is supplied both by aquaculture and by importation of sea cucumbers fished in other regions.

7.5.1 Extraction and Fishing

The consumption of sea urchins in Venezuela is known since prehispanic times (Pauls 1992). However, sea urchins in general are not highly appreciated in the Venezuelan diet. There is only an artisanal exploitation of *L. variegatus* in Margarita Island, Nueva Esparta state, without an official control for the exploitation of this resource. Local people gather the gonads of several sea urchins and place them into one of the tests, giving it a conic form (Fig. 7.5). The cones are roasted and sold later informally for approximately one US$ each (Gómez 1999). In other zones of Venezuela, like the surroundings of Mochima National Park, this species is abundant and could be commercially exploited under the administration of the National Institute of Fisheries and Aquaculture (INSOPESCA) and the Office of National Parks. However, until now, the product is neither commercialized nor exploited. *Litechinus varieatus* generally inhabits beds of the seagrass, *T. testudinum.* An uncontrolled exploitation could have negative effects on the marine ecosystem.

There is no exploitation of the sea urchin *E. lucunter*, although it is occasionally consumed by tourists and some local people. Since *E. lucunter* is associated with rocky and coralline substrate of shallow waters where it can reach a relatively high abundance, it could also be threatened by an uncontrolled exploitation. However, there are not studies that suggest how its extraction and administration could be regulated.

The sea urchin *T. ventricosus* has been fished in the Caribbean Sea, particularly in Barbados, where its natural banks were overexploited decades ago (Scheibling and Mladenov 1998). This species is also found in Venezuelan waters, more associated with zones beyond the continental coastline. Its abundance is so low that not even a subsistence fishery is possible.

The exploitation of sea cucumbers in Venezuela is centered on two species, *H. mexicana* and, particularly, *I. badionotus*, and has been generally illegal. Nevertheless, under pressure from dealers and Asiatic fishers for establishing a sea cucumber fishery in Venezuela, by middle 1990s an experimental permit was issued by Servicio Autónomo de los Recursos Pesqueros y Acuícolas (SARPA, currently INSOPESCA) of Venezuela. The permit was intended to be for one-year. It established a maximum weekly quota of 200 kg per vessel, as well as a requirement for continuous supervision and evaluation of the resource (Conde 1997; Rodríguez and Pauls 1998). However, the conditions of this permit were not properly fulfilled, since the information on the extracted amounts was considered neither adequate nor trustable. This led to the suspension of the permit after 6 months, but the illegal exploitation of sea cucumbers has continued. Acccording to Rodríguez and Pauls (1998), had the exploitation of holothurians been continued in that form, the damage caused to these resources along the eastern coasts (where fishing was permitted) could have produced major effects on the populations. These authors, using a report from Marcano and Marcano (1993) about the average dry weight ind^{-1}, showed that landings could have surpassed 42,000 kg y^{-1}, equivalent to more than 3 million individuals.

Some results from this exploitation are provided by León and Cabrera (1994). They found abundance indices of ca. 4 ind m^{-2} in the vicinity of Coche Island and 2 ind m^{-2} near Cubagua Island, which are considered high compared to other zones in the Caribbean Sea (González et al. 2005). According to León and Cabrera (1994), both species of sea cucumbers were distributed in depths from 2 to 40 m, but their exploitation was concentrated between 10 and 15 m due to restrictions of the breathing equipment used for the fishing process. They also found that *I. badionotus* reaches sexual maturity at a length above 180 mm, although Palazón (2001) reported sexual maturity at smaller size (body length of 147 mm).

The Ministry of Agriculture and Livestock established a general closed season for sea cucumbers in Venezuelan waters until the needed technical studies for the evaluation of the resources are made (Gaceta Oficial de la República de Venezuela 14-07-1997, MAC-DM 195, 10-07-1997). Nevertheless, illegal fishing is active and there have been reports of Venezuelan sea cucumbers exported to Japan in 2005 (Toral-Granda 2008). In spite of the importance and uniqueness of the reports by León and Cabrera (1994) and González et al. (2005) pertaining to the evaluation of the sea cucumber resources of Venezuela, it is difficult to reach adequate conclusions from them that would allow administration strategies for these resources. Hence, there is need to generate new knowledge to establish an adequate legal framework for the protection and exploitation of sea cucumbers,

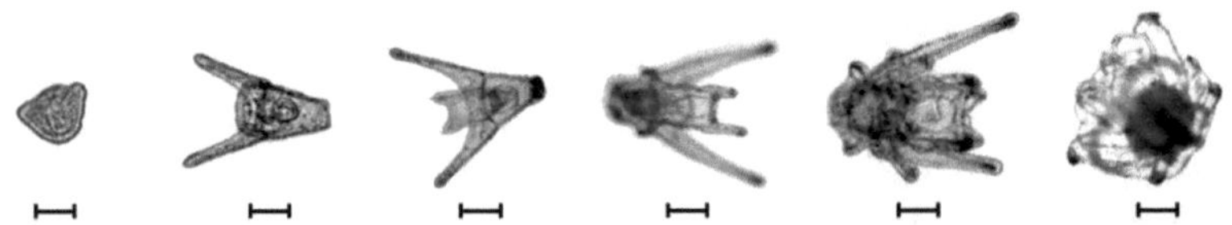

Fig. 7.6 Major stages of larvae of *Lytechinus variegatus*: prisma larva, with 2 arms, with 4 arms, with 8 arms and competent larva. The *bar* in each photo represents 10 μm at 100x magnification (Butriago and Lodeiros 2005)

supported by technical evaluations in different regions of the country. The presence of a continuous illegal fishing of sea cucumbers in the Venezuelan coasts, as revealed from sources in the coastal communities, impoundments made by authorities and our own "in situ" observations, make it urgent to establish an evaluation of these resources along the Venezuelan coast. In this regard, considering that previous studies were limited to a depth of 15 m, future studies should cover explorations to deeper areas.

7.5.2 Aquaculture

The sea urchins (*L. variegatus*, *E. lucunter*, *T. ventricosus* and *A. punctulata*) have been studied with the purpose of increasing production by aquaculture (Lawrence and Balzhin 1998; García et al. 2005; Astudillo et al. 2005; Barrios and Reyes 2009). Furthermore, there is also interest in Venezuela for using the eggs and larvae in sensitivity studies of toxic and contaminant compounds (Esclapés 1999). This supports the importance of establishing techniques to induce spawning in these species and production of larvae.

Most attention for aquaculture has been given to *L. variegatus*. Studies have investigated mass production of larvae (Fig. 7.6), postlarvae and juveniles (Butriago and Lodeiros 2005; Butriago et al. 2005) with the goal to establish initial larval density and the microalgae diet appropriate for cultivation, as well as to determine the conditions leading to metamorphosis and to estimate the postlarval growth rate. These experiments showed that it was possible to mass produce competent larvae with a length of 650 μm after 12–13 days with densities of 0.25–1 larva ml^{-1}. The microalga *Rhodomonas* sp. was a convenient diet for optimal larval growth and survival. Metamorphosis can be induced with biofilms containing benthic diatoms (*Navicula* sp. and *Amphora* sp.) incrusted on plastic layers obtained from sliced disposable beverage bottles (Fig. 7.7), reaching 100 % 96 h after the appearance of competent larvae. Nevertheless, a diet on benthic microalgae is not sufficient for a good larval development, hence a supplementary food source is necessary. For this purpose, the macroalga, *U. lactuca*, supports a considerable growth. However, supplementary addition of feed pellets for marine shrimp to the diet, increases survival (85–86 %). These results suggest studies

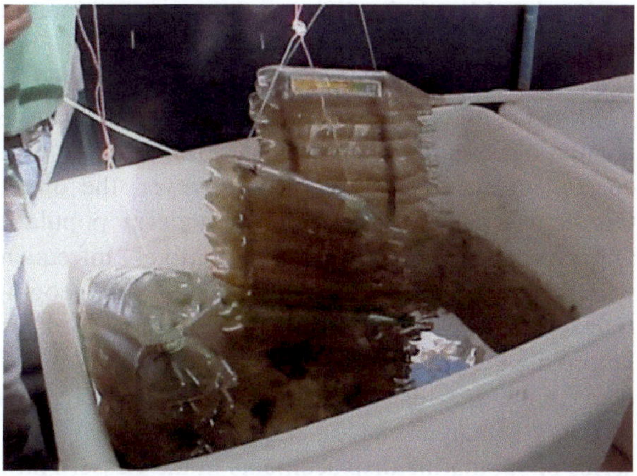

Fig. 7.7 Sliced plastic bottles with biofilms of benthic microalgae to induce the larval metamorphosis of *Lytechinus variegatus* (Butriago and Lodeiros 2005)

should continue to optimize the initial stages of culture, as well as for the following growth stages. Nutritional research should be emphasized to establish the appropriate diet for growth and the production of gonads.

Studies in Venezuela concerning culture of *E. lucunter*, *A. punctulata* and *T. ventricosus* have focused on conditions affecting larval development. García et al. (2005) studied the effect of salinity (30 and 40 ‰) and type of microalga on the larval development of *A. punctulata,* and suggested the use of the microalgae *Isochrysis galbana, Chaetoceros gracilis* and *Chaetoceros calcitrans* for raising larvae. They also found that using 40 ‰ salinity reached the greatest survival, while in 30 ‰ salinity the growth rate was maximized. Astudillo et al. (2005) concluded that *E. lucunter* is a species with excellent characteristics for cultivation because it has short embryonic (16 h) and larval (12 days) periods and good survival rates when fed a monoculture of *C. gracilis*. This microalga also supports larval development of *T. ventricosus*, which completes larval development to metamorphosis in 24 days after fertilization at temperatures from 23 to 29°C (Fernández 2005). Survival is greater at lower temperatures.

Studies to increase production of sea cucumbers by culture have not been done in Venezuela despite the high product market value and the detritivorous condition that make their culture attractive.

7.6 Threats to Echinoderms in Venezuela

The echinoderms in Venezuela are not free from threats experienced by marine organisms nowadays, like global warming (Byrne et al. 2009), increased UV radiation associated with a reduction in the ozone layer (Lesser and Barry 2003) and acidification of the oceans produced by an increase in the CO_2 concentration (Kurihara and Shirayama 2004).

At a local level, the different threats on echinoderm biodiversity can be divided according to their origin, as natural threats and those generated by human intervention. Considering the natural threats, during El Niño years most parts of Venezuela showed a delay in the onset of the rainy season and a reduction in the amount of precipitation (Lozada 2002). This can produce an increase in salinity of coastal waters in some zones. However, the direct effect of El Niño-Southern Oscillation phenomena upon echinoderm populations has not been described. Other threats, like hurricanes, are relatively infrequent in Venezuela in comparison with the rest of the Caribbean Sea due to the geographical position of the country (Díaz-Martín 1995). There are no current records of their direct effects on populations of echinoderms. From a geological point of view, Venezuela is located above a major fault zone, which produces relatively frequent earthquakes. In spite of this, "tsunamis" have been rare (Grasses 1994). However, identification of a series of submarine displacements in the Lesser Antilles reveal the vulnerability of the Venezuelan coasts to this type of phenomenon (Theilen-Willige 2006).

It has been estimated that the most important potential natural threat to echinoderms in Venezuela is the large load of sediments carried by rivers during the rainy season (May–November). This phenomenon can generate mass mortalities of marine organisms (Goodbody 1961) and causes losses in the diversity of echinoderms, particularly in coral reefs (Vázquez-Domínguez 2003). In this sense, it is important to emphasize the influence of the Orinoco River on the patterns of salinity, currents, nutrient concentration and suspended material in the Atlantic coast of the country and the eastern Caribbean Sea whose fluctuations are directly related to the dry-rainy season regime (Penchaszadeh et al. 2000).

Regarding threats by diseases, there are reports that epidemics and the number of emergent diseases are rising worldwide in corals and other marine organisms (Richardson et al. 1998; Harvell et al. 1999; Green and Bruckner 2000). In Venezuela, the echinoid *D. antillarum* did not escape from the mass mortality registered in the general Caribbean Sea between 1983 and 1984 (Lessios et al. 1984; Hunte et al. 1986; Lessios 1988). The abundances of *D. antillarum* recorded between 2002 and 2003 in different localities of the country show that they have not reached yet the population density levels reported before the mortality period of 1984 (Noriega et al. 2006), although our own observations in Mochima Bay and the Gulf of Cariaco show a high density of individuals with signals of recovery of the populations.

In addition to disease, an 80 % reduction in the coralline cover has been reported in the Caribbean Sea (Gardner et al. 2003). In Venezuela coral cover has decreased at Sombrero Key in Morrocoy National Park (Cróquer and Bone 2003). This could also compromise the abundance of echinoderms like the echinoid *E. lucunter*, generally associated with the terrace and the reef crest (Abbott et al. 1974) and *D. antillarum* which also occur in these types of environments.

Regarding anthropogenic threats, it is important to show that intense petroleum exploitation and processing in Venezuela, with a heavy traffic of vessels and a network of oil pipes, the chances for spills of oil and other derivatives represent a latent and constant threat. This is particularly serious during the phase of transportation towards storage places due to the eventual metal fatigue and rupture of pipes, pumps and tanks (López et al. 1999). Populations of echnoids like *E. lucunter* could be reduced by 80 % after an oil spill in the coasts of Panamá (Jackson et al. 1989). Furthermore, oil spills (crude or refined) significantly affect the early larval stages of echinoderms (Jangoux 1984). These effects vary according to the composition of the hydrocarbon (Allen 1971).

More than 50 % of the human population in Venezuela is concentrated in or near the coasts, increasing sedimentation considerably because of infrastructure developments as well as the pollution by untreated sewage (Díaz-Martín 1995). Other contaminants, sub-products of the industry and agriculture, can also become a threat to the fauna in coastal marine environments of Venezuela (Díaz-Martín 1995; Penchaszadeh et al. 2000). Unfortunately, Venezuela does not have effective and widespread sewage treatments and most contaminants reach the sea without previous processing. This can produce high oxygen demand, eutrophication, accumulation of heavy metals and other processes (Penchaszadeh et al. 2000). In this sense, concentrations of some heavy metals, in particular mercury, have been measured in asteroids like *Luidia clathrata, L. bareimaes, L. senegalensis* and *Tethyaster vestitus* in the zone of Golfo Triste. Concentrations ranged between 3.8 and 8.87 μg g dry weight^{-1} in the gonads (Iglesias and Penchaszadeh 1983). Concentrations of the metals Al, Cu, Mn, Ni, Pb, and Zn were measured in the eviscerated digestive tissue of two species of holothurians (*I. badionotus* and *H. mexicana*) in Morrocoy National Park. Concentrations of Cu, Ni, Pb and Zn were higher than in the surrounding sediments (Laboy-Nieves and Conde 2001).

Industrial trawl fishing, recently prohibited by the Venezuelan government (actually in March 2009), was considered a threat to echinoderms. However, echinoderms represented a minor portion (<1%) of the incidental capture in the trawls in eastern Venezuela (Marcano et al. 2001), although their representation in the catches from trawl fisheries has been little studied in other parts of the world (Alverson et al. 1994; Kelleher 2005). Unpublished observations from one of the authors (JA) indicate echinoderm discards by industrial trawlers in eastern Venezuela included echinoids, asteroids, ophiuroids and, in a lesser proportion, holothuroids and crinoids. Most of them, with the exception of ophiuroids due to their fragile structure, were returned alive to the ocean. Hill and Wasemberg (1990) examined the fate of discards from shrimp trawlers in Torres Strait, northern Australia, and recorded an 88 % survival rate in the asteroid, *Astropecten* sp. However, the trawl gear can modify the habitat and generate a reduction in the density of populations of echinoderms and other groups of benthic animals (Bergman and Hup 1992). In this sense, the interdiction of industrial trawling in Venezuelan waters could have reduced the threat to a considerable number of organisms whose capture was intended or not for consumption. It could also have prevented a possible suspension of contaminants present in the sediment, like DDT, PCBs and PAHs, reducing their incorporation into the trophic chains through the plankton (Schiff 2000). The interdiction only affected industrial trawling, because artisanal trawling is still actively practiced in shallow areas of eastern Venezuela. For example, trawl fishing of the turkey wing, *A. zebra*

Fig. 7.8 Samples obtained from a bivalve mollusk dredge with a large number of echinoderms, Araya Peninsula, northeastern Venezuela

(Mollusca, Bivalvia, Acidae), which recently generated more than 50,000 t y^{-1}, has an important component of echinoderms in the incidental catch, which is extracted and discarded maimed or dead, or their habitat destroyed by the low selective fishing gear (Fig. 7.8).

Unfortunately, estimating the echinoderm discards in fisheries has been difficult. In this sense, Kelleher (2005) mentions that both fishers and observers tend to focus on commercial species and recognized animals. There is a tendency to group tunicates, sponges, echinoderms, hermit crabs, worms and corals with jellyfish and perceive such biomass as debris, rather than as organic material. These non-commercial animals are frequently ignored and not recorded as discards during studies. This biomass tends to be omitted from estimates of discards. Even in countries that prohibit discards, allowance is made for discarding of non-commercial species (such as in Iceland). Many of these animals also pose practical problems of measurement of the biomass but may constitute significant large proportion of the total biomass harvested by trawls (Prena et al. 1999).

Considering the direct fishing of echinoderms, there exist evidence of consumption of sea urchins in the Venezuelan territory since prehispanic times (Pauls 1992), although this activity is currently restricted to artisanal fishers of *L. variegatus* in Margarita Island (Gómez 1999, 2002) and has reduced local populations (L. León, Universidad de Oriente, pers. comm.). This situation could be worse in the future because the fishing of *L. variegatus* is still unregulated. There is an interdiction of fisheries of sea cucumbers, which has apparently reduced this activity. However, illegal fishing continues, mainly of *I. badionotus*, and there are reports of large impoundments within Mochima National Park (Rodríguez and Pauls 1998). Unfortunately, there are no current official figures about the amount or species of sea cucumbers illegally fished.

Even though there is no strong tradition for consumption of sea urchins or sea cucumbers in Venezuela, the threat of unregulated exploitation of these resources is present due to the increased demand of Asian markets, the lack of evaluation of their populations and absence of norms that regulate their local fisheries and commerce.

A possible menace that requires more attention is the marketing of live ornamental species for sea water aquaria, about which there are no official data. Also, some species are dried and sold as "souvenirs". These include the starfishes *O. reticulatus* and *Astropecten articulatus* and, in a lesser degree, the sand dollar (called "locha de mar" in Venezuela), *Mellita quinquiesperforata*. Sellers can be seen in the more visited beaches of Margarita Island and also during the high touristic seasons in Chichiriviche (Falcón state) in western Venezuela as well as in Mochima (Sucre state) and Coche Island (Nueva Esparta state) in eastern Venezuela. This activity is not regulated but is expected to increase with the touristic development and will become an environmental problem worldwide (Lunn et al. 2008).

7.7 Recommendations

In spite of the heterogeneity of the ecoregions in Venezuela, the fact that some of them have proposals for management and conservation, and the importance of echinoderms in them, studies on echinoderms are still few. More intense research about their biodiversity and ecology is necessary.

Although the present analysis has attempted to provide information on the different classes, families and species of the extant echinoderms in some ecoregions of Venezuela, the distribution of echinoderms requires incorporation of more detailed spatial analyses that include the entire Venezuelan coastline. It is expected that they will generate information about their density and other population parameters, including growth, reproduction, feeding and their relation to the biotic and abiotic characteristics of the environment. It is important to evaluate the environmental factors that could influence their presence in a certain location, including studies about their vulnerability to climatic change. This information will be of great importance, not only for the scientific knowledge and a better description the autoecologic (individual) and synecologic (community) relations, but also for development of strategies leading towards the sustainable production of echinoderms as a resource and their administration, as well as the search of alternatives to increase their production through aquaculture.

Some studies have considered the echinoderms as model species for ecotoxicologic analyses, as in the case of echinoids (Esclapés 1999). However, there has been no continuity in these types of evaluations, which are very important because the Venezuelan coastline is influenced by many anthropogenic influences that have produced major changes to many habitats.

The studies on sea urchin aquaculture in Venezuela have been encouraging, having completed the techniques for juvenile production. It is recommended that such techniques be optimized and concentrate further studies on post-juvenile growth. Because the gonad is the marketed product from sea urchins, it is necessary to perform studies on the optimization of it production. In contrast with sea urchins, there has been no research on sea cucumbers aquaculture in Venezuela, despite their high market value and their detritivorous condition which makes them attractive for cultivation. Considering that techniques for sea cucumber aquaculture are being optimized in other countries, particularly in Asiatic regions, we should take advantage of this technological expansion to set up culture packages for Venezuelan species. This could also bring environmental benefits, because it would decrease the current pressure on echinoderm resources from illegal fisheries. In this sense, population studies on sea cucumber resources are necessary for their administration. For this, the gathering of information on illegal fishing is important, such as extracted species, the quantification of the landings and the identification of places of extraction and commercialization.

The incidental fishing of echinoderms as bycatch also merits study, in particular the effect of trawls for the ark shell (*A. zebra*) and other fishing gears on echinoderm populations in different zones of Venezuela.

It is possible that the list of echinoderms in Venezuela be larger than the one presented here. In particular, the benthos of the deep ocean of Venezuela has been little studied. Most expeditions covering this area have been made by institutions of other countries where a great amount of information is deposited. Since access to these sources is usually difficult, it is necessary to prepare catalogues of echinoderms with information about their biology, ecology, distribution and fisheries to improve access to knowledge about local echinoderms and with it the promotion of their studies in Venezuela. These catalogues should include paleontological information of the species, produced from studies under a geologic and a paleobiogeographical context. This would allow the ordination in time and space of the ancestral forms with evolutionary continuity, as well as the origin, distribution and ecology of this fauna in the Caribbean Province.

Acknowledgments The collaboration of L. León is appreciated for critical reading of the section on fisheries, as well as that S. Paul for providing bibliographic material. J. Reyes, N. Noriega and Y. J. Díaz, collaborated in the present study during their doctoral studies. The Fundación para la Investigación y Desarrollo de la Acuicultura del Estado Sucre (FIDAES) partially contributed in the research described in the section on aquaculture. This is contribution N° 18 of the Fundación para la Investigación y Desarrollo de la Acuicultura del estado Sucre (FIDAES).

References

Abbott DP, Ogden JC, Abbott IA (1974) Studies on the activity pattern, behavior and food of the echinoid *Echinometra lucunter* (L.) on beachrock and algal reefs at St. Croix, US Virgin Is. West Indies Lab Spec Publ 4:1–11

Aguilera M, Azocar A, González E (2003) Biodiversidad en Venezuela, vol I and II. Fundación Polar and Ministerio de Ciencia y Tecnología, Caracas

Aguilera O, Rodrigues de Aguilera D, Vega FJ, Sánchez-Villagra MR (2010) Mesozoic and Cenozoic decapod crustaceans from Venezuela and related trace-fossil assemblages. In: Sánchez-Villagra MR, Aguilera OA, Carlini AA (eds) Urumaco and Venezuelan palaeontology—The fossil record of the Northern neotropics. Indiana Univ Press, Bloomington, Indiana, pp 103–128

Allen H (1971) Effects of petroleum fractions on the early development of a sea urchin. Mar Poll Bull 2:138–140

Alverson DL, Freeberg MK, Murawski SA, Pope JG (1994) A global assessment of fisheries bycatch and discards. FAO fisheries technical paper N° 339, Rome

Amaro M (2009) Echinodermata del golfo de Cariaco, Venezuela. Bol Inst Oceanogr Univ Oriente 38:137–142

Anisgard HW (1954) An echinoid from the Eocene of western Venezuela. J Paleontol 28:830–835

Astudillo D, Rosas J, Velásquez A, Cabrera T, Maneiro C (2005) Crecimiento y supervivencia de larvas de *Echinometra lucunter* (Echinoidea: Echinometridae) alimentadas con las microalgas *Chaetoceros gracilis* e *Isochrysis galbana*. Rev Biol Trop 53(Suppl 3):337–344

Barrios J, Reyes J (2009) Hábitos alimenticios de *Tripneustes ventricosus* (Lamack 1816) (Echinodermata, Echinoidea) en isla la Tortuga, Venezuela. Foro Iberoam Rec Mar Acuicult II:583–589

Belmonte JL (1985) Contribución al estudio de los equinodermos (asteroideos, equinoideos y ofiuroideos) de la plataforma Margarita - Coche - tierra firme. Estado Nueva Esparta, Isla de Coche, Isla de Margarita, Islas de Venezuela, región Oriental. Associate degree Thesis, Instituto Universitario de Tecnología del Mar, Punta de Piedras, Isla de Margarita

Bergman MJN, Hup M (1992) Direct effects of beam trawling on macrofauna in a sandy segment in the southern north sea. ICES J Mar Sci 49:5–11

Bitter R (1984) Composición de la dieta de *Astropecten marginatus* (Echinodermata: Asteroidea). Golfo Triste, región Centro-Occidental Bol Inst Oceanogr Univ Oriente 23:169–176

Bitter R, Penchaszadeh PE (1983) Ecología trófica de dos estrellas de mar del genero *Astropecten* coexistentes en Golfo Triste, Venezuela. Stud Neotrop Fauna E 18:163–180

Bitter R, Molinet R, Penchaszadeh PE (1980) Interacción trófica entre dos estrellas de mar (*Astropecten riensis* y *Tethyaster vestitus*) en Golfo Triste, Venezuela. Bol Inst Oceanogr Sao Paulo 29:61–63

Boada LM (1970) Estudio taxonómico sobre algunos equinodermos de la costa oeste de Cumaná. Cumaná, Estado Sucre, región Oriental. Bachelor's degree Thesis, Univ Oriente, Cumaná

Boffi E (1972) Ecological aspects of ophiuroids from the phytal of S.W. Atlantic ocean warm waters. Mar Biol 15:316–328

Borges M (2006) Taxonomia, distribuição e biología reprodutiva de Ophiuroidea (Echinodermata) das Regiões Sudeste e Sul do Brasil. PhD Thesis, Univ Estadual Paulista, Sao Paulo

Brey L (1992) Ensayo de toxicidad con gametos del erizo de mar *Echinometra lucunter* (Echinometridae).Bachelor's degree Thesis, Univ Simón Bolívar, Caracas

Bushman JR (1965) Geology of the Barquisimeto area. Final report. Bol Geol Minist Minas e Hidrocarb 6:311–336

Butriago E, Lodeiros C (2005) Producción de larvas, postlarvas y juveniles del erizo verde-blanco del Caribe *Lytechinus variegatus* (Echinodermata: Echinoidea) bajo condiciones de cultivo. Rev Biol Trop 53(Suppl 3):319–328

Butriago E, Lodeiros C, Lunar K, Indorf F, Frontado K, Moreno P, Vasquez Z (2005) Larval production for the sea urchin *Lytechinus variegatus* (Echinodermata: Echinoidea). Effect of initial larval density. Aquacult Int 13:359–367

Byrne M, Ho M, Selvakumaraswamy P, Nguyen HD, Dworjanyn SA, Davis RA (2009) Temperature, but not pH, compromises sea urchin fertilization and early development under near-future climate change scenarios. Proc R Soc B 276:1883–1888

Cedeño G (1971) Identificación de ácidos grasos en holotúroideos por cromatografía de gas. Bol Inst Oceanogr Univ Oriente 10:9–14

Cervigón F (1994) Peces Marinos de Venezuela, vol I–IV. Ed Ex Libris, Caracas

Cheetham AH, Jackson JBC (1996) Speciation, extinction, and decline of arborescent grown in neogene and quaternary cheilostome Bryozoa of tropical America. In: Jackson JBC, Budd AF, Coates AG (eds) Evolution and environment in tropical America. Univ Chicago Press, Chicago, pp 205–233

Conde JE (1997) Una historia de Sushi, Sedimentos y Cocina. Holotureos o pepinos de mar. Cien Hoy 39:1–4

Conde JE, Díaz H, Sambrano A (1991) Disintegration of holothurian fecal pellets in beds of the seagrass *Thalassia testudinum*. J Coastal Res 7:853–862

Cooke CW (1941) *Oligopygus nancei*, a new echinoid from Venezuela. J Paleontol 15:305–306

Cooke CW (1961) Cenozoic and Cretaceous echinoids from Trinidad and Venezuela. Smith Misc Coll 142:1–35

Cróquer A, Bone D (2003) Las enfermedades en corales escleractínidos: ¿Un nuevo problema en el arrecife de Cayo Sombrero, Parque Nacional Morrocoy, Venezuela? Rev Biol Trop 51:167–172

Cruz-Motta JJ (2007) Spatial analysis of intertidal tropical assemblages associated with rocky shores in Venezuela. Cien Mar 33:133–148

Díaz YJ, Martín A (2001) New records of amphipods (Crustacea: Amphipoda) from shallow water on the Caribbean coast of Venezuela. Rev Biol Trop 49:1271–1276

Díaz C, Villamizar E, Alvarez B, Cugat J, Riera RD, Weil E, Laughlin R (1982) Efecto bioerosivo de diferentes densidades del erizo negro *Diadema antillarum* Philippi en el Archipiélago de Los Roques. Dependencias Federales, Archipiélago de Los Roques, Islas de Venezuela. XXXII Convención Anual de la AsoVAC, Univ Simón Bolívar, Caracas. Acta Cient Venez 33:25

Díaz TA, Villafranca S, Arocha J (1996) Determinación de proteínas y electrolitos en la hemolinfa de los erizos *Lytechinus variegatus* y *Echinometra lucunter*, en la ensenada de Turpialito, Golfo de Cariaco, III. Congr Cien Univ Oriente, Maturín

Díaz YJ, Guerra-García JM, Martín A (2005) Caprellids (Crustacea: Amphipoda: Caprellidae) from shallow waters of the Caribbean coast of Venezuela, with a key for species identification. Org Biodiv Evol 3:1–25

Díaz-Martín D (1995) Prioridades de Conservación de las Áreas Marino Costeras de Venezuela. Informe Público de FUDENA, Caracas, Fondo Mundial para la Naturaleza (WWF) y FUDENA

Domínguez A, Rosas J, Velásquez A, Cabrera T, Mata E (2007) Desarrollo, supervivencia y crecimiento del erizo *Lytechinus variegatus* (Lamarck, 1816) (Echinodermata: Echinoidea) alimentado con microalgas a dos salinidades y temperaturas diferentes. Rev Biol Mar Oceanogr 42:49–57

Donovan SK (2001) Evolution of Caribbean echinoderms during the Cenozoic. PALAEO 166:177–192

Egea R, López I, Herrera FC (1991) Respuesta osmótica, en ausencia del cloruro, del árbol respiratorio y músculo del equinodermo *Isostichopus badionotus*. XLI Convención Anual de la AsoVAC, Univ Zulia, Maracaibo. Acta Cient Venez 42:59

Engel H (1939) Echinoderms from Aruba, Curacao, Bonaire and Northern Venezuela. Capita Zool 8:1–11

Engel H, Croes AF, Schroevers WGF (1960) Les Asterides récoltées par l'expédition du navire-école belge *Mercator* 1935–1936. Bull K Belg Inst Nat Wet 36:1–15

Esclapés M (1999) Protocolos estándares para bioensayos de toxicidad con especies acuáticas y terrestres. Versión 2.0. Gerencia General de Tecnología, INTEVEP, Caracas

Espinoza R, Reyes JL, Himelman JH, Lodeiros C (2008) Actividad reproductiva de los erizos *Lytechinus variegatus* y *Echinometra lucunter* Echinodermata: Echinoidea) en relación con factores ambientales en el Golfo de Cariaco, Venezuela. Rev Biol Trop 56(Suppl 3):341–350

Fernández S (2005) Supervivencia y crecimiento del erizo *Tripneutes ventricosus* (Lamarck) alimentadas con las microalgas *Chaetoceros gracilis* y la mezcla *Isochrysis galbana-Chaetoceros gracilis* en condiciones controladas y no controladas. Bachelor's degree Thesis, Univ Oriente, Boca de Río

Ferraz-Reyes E (1987) Productividad primaria del Golfo de Cariaco Venezuela. Bol Inst Oceanogr Univ Oriente 26:97–110

Flores C, Martínez A (1970) Status de la familia Goniasteridae (Asteroidea: Phanerozonia) en la región Oriental de Venezuela. Bol Inst Oceanogr Univ Oriente 9:3–8

Foglietta LM (1994) Estudio de la respuesta ion-osmótica de *Isostichopus badionotus* Echinodermata-Holothuroidea. Bachelor's degree Thesis, Univ Simón Bolívar, Caracas

Foglietta LM, Herrera FC (1996) ion-osmotic response of respiratory trees of the holothurian *Isostichopus badionotus* Selenka preincubated in hyper-, iso- and hypo-osmotic sea water. J Exp Biol Ecol 202:151–164

Francisco V, Pauls SM (2008) Especies del Orden Clypeasteroida (Echinodermata: Echinoidea) de las costas de Venezuela. Rev Biol Trop 56(Suppl 3):215–228

García M, Rosas J, Hernández I, Velásquez A, Cabrera T (2005) Supervivencia y crecimiento larval de *Arbacia punctulata* (Echinodermata: Echinoidea) alimentada con cinco microalgas a dos salinidades. Rev Biol Trop 53(Suppl 3):329–336

Gardner TA, Cote IM, Gill JA, Grant A, Watkinson AR (2003) Long-term region-wide declines in Caribbean corals. Science 301:958–960

Gómez A (1999) Los recursos marinos renovables del estado Nueva Esparta, Venezuela: Biología y pesca de las especies comerciales. Tomo I. Invertebrados y algas. Gráficas Capriles, Caracas

Gómez A (2002) Abundancia de erizo *Lytechinus variegatus* (Lamarck) en la costa norte, este y oeste de la Isla de Margarita (Venezuela). Acta Cien Venez 53:15–20

Gómez A (2003) Relación diámetro-peso y proporción cromática del erizo *Lytechinus variegatus* (Echinoidea: Toxopneustidae) en las islas de Margarita y Cubagua, Venezuela. Rev Biol Trop 51:83–86

Gómez O, Gómez A (2005) Desarrollo embrionario y larval de *Lytechinus variegatus* (Echinoidea: Toxopneustidae) en condiciones de laboratorio en la Isla de Margarita-Venezuela. Rev Biol Trop 53(Suppl 3):313–318

Gómez MG, Capaldo M, Yanes C, Martín A (eds) (2005) Frente Atlántico venezolano. Investigaciones geoambientales. Ciencias Ambientales, Tomo I. PDVSA-Fondo Ed Fundambiente, Caracas

González M (2005) Estudios ecológicos del impacto por erizos sobre la dinámica de un arrecife coralino en el Parque Nacional Archipiélago Los Roques: Bioerosión. Bachelor's degree Thesis, Univ Central de Venezuela, Caracas

González L, León L, Cabrera T, Rengel J (2005) Descripción preliminar de la pesquería de holoturios, *Isostichopus badionotus* (Selenka, 1867) en la zona oriental de Venezuela. VI Congr Venez Ecol, Maracaibo

Goodbody I (1961) Mass mortality of a marine fauna following tropical rains. Ecol 42:150–155

Grasses J (1994) Venezuela. Amenazas Naturales. Terremotos, Maremotos y Huracanes. Academia de Ciencias Físicas, Matemáticas y Naturales de Venezuela. Cámara de Aseguradores de Venezuela, Caracas

Green EP, Bruckner AW (2000) The significance of coral disease epizootiology for coral reef conservation. Biol Conserv 96:347–361

Harvell CD, Kim K, Burkholder JM, Colwell RR, Epstein PR, Grimes DJ, Hofmann EE, Lipp EK, Osterhaus AD, Overstreet RM, Porter JW, Smith GW, Vasta GR (1999) Emerging marine diseases-climate links and anthropogenic factors. Science 285:1505–1510

Hendler G, Littman B (1986) The ploys of sex: relationships among the mode of reproduction, body size and habitats of coral-reef brittlestars. Coral Reefs 5:31–42

Hendler G, Miller J (1984) *Ophioderma devaneyi* and *Ophioderma ensiferum*, new brittlestars species from the Western Atlantic (Echinodermata: Ophiuroidea). Proc Biol Soc Wash 97:442–461

Hendler G, Miller JE, Pawson DL, Kier PM (1995) Sea stars sea urchins and allies: Echinoderms of Florida and the Caribbean. Smith Inst Press, Washington, DC

Hernández I (2010) Ofiuroideos (Echinodermata: Ophiuroidea) de la Isla de Cubagua, Bachelor's degree Thesis Univ Oriente, Cumaná, Venezuela

Herrera FC (1991) La respuesta ion-osmótica en equinodermos y celenterados. IV Congr Latinoam Cien Mar (COLACMAR), Univ Católica del Norte, Coquimbo

Herrera FC, Foglietta LM (1998) Ouabain and the ionosmotic properties of echinoderm cells: from steady-state partial double Donnan to passive Donnan. J Exp Mar Biol Ecol 231:239–254

Herrera FC, López I (1983) Relationship between external sodium concentration and intracellular osmotic effectors in respiratory tree of *Holothuria glaberrima*. Comp Bioch Physiol 74A: 267–274

Herrera FC, Plaza YM (1981a) The role of extracellular sodium in isosmotic intracellular regulation in the respiratory tree of *Holothuria glaberrima*. Comp Bioch Physiol 68A:373–382

Herrera FC, Plaza YM (1981b) Extracellular cations and respiration of tissues of *Holothuria glaberrima*. Comp Bioch Physiol 70:27–32

Herrera FC, Herrera MI, López I (2000) Further studies on the partial bouble donnan is isosmotic KCl solution isotonic with cells of respiratory trees of the holothurian *Isostichopus badionotus* Selenka. J Exp Mar Biol Ecol 247:139–152

Hill BJ, Wassenberg TJ (1990) Fate of discards from prawn trawlers in Torres Strait. Aust J Mar Freshw Res 41:53–64

Hoover PR (1981) Paleontology, taphonomy and paleoecology of the Palmarito Formation (Permian of Venezuela). Bull Amer Paleontol 80:1–138

Hunte W, Cote I, Tomascik T (1986) On the dynamics of the mass mortality of *Diadema antillarum* in Barbados. Coral Reefs 4:135–139

Iglesias N, Penchaszadeh P (1983) Mercury in sea stars from Golfo Triste, Venezuela. Mar Poll Bull 14:396–398

Jackson JBC, Cubit JD, Keller BD, Batista V, Burns K, Caffey HM, Cladwell RL, Garrity CD, Getter CD, González C, Guzmán HM, Kaufmann KW, Knap AH, Levings SC, Marshall MJ, Steger R, Thompson RC, Weil E (1989) Ecological effects of a major oil spill on panamanian coastal marine communities. Science 243:37–44

Jackson JBC, Jung P, Coates A, Collins S (1993) Diversity and extinction of tropical American mollusks and emergence of the Isthmus of Panama. Science 260:1624–1626

Jangoux M (1978) Les étoiles de mer (Echinodermata: Asteroidea) récoltées par le M/V Calamar au large des Guyanes et du Venezuela. Bull Zool Mus Univ Amsterdam 6:93–101

Jangoux M (1984) Diseases of echinoderms. Helgolander Meeresunters 37:207–216

Jeannet A (1928) Contribution á l'étude des échinides tertiaires de la Trinité et du Venezuela. Mém Soc Paléontol Suisse 48:1–49

Johnson KG, Sánchez-Villagra M, Aguilera O (2009) The Oligocene-Miocene transition on coral reefs in the Falcón Basin (NW Venezuela). Palaios 24:59–69

Jurisic M (1982) Papel del cloruro en regulación isosmótica intracelular en *Holothuria glaberrima*. Bachelor's degree Thesis, Univ Católica Andrés Bello, Caracas

Jurisic M, Roque S, López I, Herrera FC (1983) Role of chloride in isosmotic intracellular regulation in *Holothuria glaberrima*. Comp Bioch Physiol 76A:831–837

Kelleher K (2005) Discards in the world's marine fisheries. An update. FAO fisheries technical paper. No. 470, Rome

Kier PM (1977) The poor fossil record of the regular Echinoid. Paleobiol 3:168–174

Kroh A, Nebelsick JH (2003) Echinoid assemblages as a tool for palaeoenvironmental reconstruction: an example from the early Miocene of Egypt. PALAEO 201:157–177

Kurihara H, Shirayama Y (2004) Effects of increased atmospheric CO_2 on sea urchin early development. Mar Ecol Prog Ser 274:161–169

Laboy-Nieves EN, Conde JE (2001) Metal levels in eviscerated tissue of shallow-water deposit-feeding holothurians. Hydrobiol 459:19–26

Lawrence J, Balzhin A (1998) Life-history strategies and the potential of sea urchins for aquaculture. J Shellf Res 17:1515–1522

Lawrence J, Lawrence AL, Watts SA (2007) Feeding, digestion, and digestibility. In: Lawrence JM (ed) Edible sea urchins: biology and ecology. Development in aquaculture and fisheries. Elsevier, Amsterdam, pp 135–158

León L, Cabrera T (1994) Informe de avance sobre la pesquería del pepino de mar en las costas del nororiente de Venezuela. Informe técnico, Escuela de Ciencias Aplicadas del Mar, Universidad de Oriente, Caracas

Lesser MP, Barry TM (2003) Survivorship, development, and DNA damage in echinoderm embryos and larvae exposed to ultraviolet radiation (290–400 nm). J Exp Mar Biol Ecol 292:75–91

Lessios HA (1988) Mass mortality of *Diadema antillarum* in the Caribbean: what have we learned? Ann Rev Ecol Syst 19:371–393

Lessios HA (2005) Echinoids of the Pacific waters of Panama: status of knowledge and new records. Rev Biol Trop 53(Suppl 3):147–170

Lessios HA, Cubit JD, Robertson DR, Shulman MJ, Parker MR, Garrity SD, Levings SC (1984) Mass mortality of *Diadema antillarum* on the Caribbean coast of Panama. Coral Reefs 3: 173–182

Lodeiros C, García N (2004) The use of sea urchins to control "fouling" during suspended culture of bivalves. Aquacult 231:293–298

Lodeiros C, Marín B, Prieto A (1999) Catálogo de moluscos marinos de las costas nororientales de Venezuela. Clase Bivalvia, Edición APUDONS, Cumaná

López R, Núñez O, Morales F, Calderón C, Liewald W (1999) Aplicación de sistemas oxidantes avanzados en el tratamiento de aguas residuales de la industria petrolera (sistema FENTON). Acta Cien Venez 50:75–80

Losada F, Martín A, Feragotto W, Alamo C (1988) Interacciones biológicas en el canal de toma de la Planta Termoeléctrica del Centro en Punta Morón, Venezuela. Ecotrop 1:55–70

Lozada BI (2002) Estudio de la influencia de los fenómenos El Niño y La Niña en la precipitación de la región Centro-Occidental de Venezuela. Rev Brasileira Agrometeorol 10:317–322

Lunn KE, Villanueva-Noriega MJ, Vincent ACJ (2008) Souvenirs from the sea: an investigation into the curio trade in Echinoderms from Mexico. Traffic Bull 22:19–32

Macsotay O, Vivas V, Gil M (2009) Historia Geológica del Golfo y la fosa de Cariaco, Venezuela Nororiental, evolución de la anoxia. Bol Inst Oceanogr Univ Oriente 48:89–108

Madrid E (1971) Acomodación a variaciones en la salinidad del agua de mar de los electrolitos intracelulares y del líquido celómico de la holoturia *Isostichopus badionotus* Selenka. Bachelor's degree Thesis, Univ Central de Venezuela, Caracas

Madrid E, Zanders IP, Herrera FC (1976) Changes in coelomic fluid and intracellular ionic composition in holothurians exposed to diverse sea water concentrations. Comp Bioch Physiol 54A:167–174

Mago A (1971) Contribución al conocimiento de los holoturoideos (Holothuroidea: Echinodermata) de la región Oriental de Venezuela. Bachelor's degree Thesis, Univ Oriente, Cumaná
Malavé M (1995) Caracterización y cuantificación del contenido lipídico y de ácidos grasos en dos especies de erizos de mar: *Lytechinus variegatus* y *Echinometra lucunter*. Región Oriental. Bachelor's degree Thesis, Univ Oriente, Cumaná
Marcano J, Marcano A (1993) Informe de inspección y reconocimiento: 08/1071993. Reporte técnico, Unidad Estatal de Desarrollo Agropecuario de Nueva Esparta, Ministerio de Agricultura y Cría, La Asunción
Marcano LA, Alió JJ, Novoa D, Altuve DE, Andrade G, Álvarez RA (2001) Revisión de la pesca de arrastre en Venezuela. FAO Fisheries Circular 974:330–378
Martín A, Díaz YJ (2003) La fauna de anfípodos (Crustacea: Amphipoda) de las aguas costeras de la región oriental de Venezuela. Bol Inst Esp Oceanogr 19:327–344
Martín A, Díaz YJ (2005) Crustáceos peracáridos de la región atlántica de Venezuela: biodiversidad y taxonomía. In: Gabriela MA, Capaldo M, Yanes C, Martín A (eds) Frente Atlántico venezolano. Investigaciones geoambientales. Ciencias Ambientales. Tomo I. PDVSA-Fondo Ed Fundambiente, Caracas
Martín A, Díaz YJ (2007) Biodiversidad de crustáceos peracáridos en el delta del río Orinoco, Venezuela. Rev Biol Trop 55:87–102
Martín A, Penchaszadeh P, Atienza D (2001) Densidad y hábitos alimentarios de *Oreaster reticulatus* (Linnaeus 1758) (Echinodermata, Asteroidea) en praderas de fanerógamas marinas en Venezuela. Bol Inst Esp Oceanogr 17:203–208
Martín A, Ortiz M, Díaz Y (2002) Nuevos registros de crustáceos anfípodos (Gammaridea) colectados en las costas del Caribe venezolano. Bol Invest Mar Cost 31:15–24
Martín A, Malavé L, Sánchez D, Aparicio R, Arocha F, Bone D, Bolaños-Jiménez J, Castañeda J, Cárdenas J, Carbonini A, Díaz Y, Guada H, Klein E, Lazo R, Lemus A, Lentito M, Lira C, Lodeiros C, López R, Marín B, Martínez G, Marquez B, Marquez R, Molinet R, Morales F, Posada J, Prieto A, Riera A, Rodríguez C, Ramirez A, Señor W, Solana P, Severeyn H, Spinelo P, Valera E, Yanes C, Zoppi E (2007) Línea base ambiental Plataforma Deltana. In: Martín A, Bone D (eds) Petróleos de Venezuela SA. Universidad Simón Bolívar, Caracas
Martínez A (1969) Contribución al conocimiento de la familia Brissidae (Echinoidea: Spatangoidea) en Venezuela. Bol Inst Oceanogr Univ Oriente 8:57–63
Martínez A (1971) Nuevos registros para la fauna de equinodermos de Venezuela y el Surinam. Inst Oceanogr Venezuela, Univ Oriente, Cumaná, Trabajo de Ascenso
Martínez A (1973) Contribución al estudio de los holoturoideos de Venezuela. Bol Inst Oceanogr Univ Oriente 12:41–50
Martínez A (1975) Estudio comparativo sobre la distribución de Holoturoideos en la región Nor-Oriental de Venezuela. Inst Oceanogr Venezuela, Univ Oriente, Cumaná, Trabajo de Ascenso
Martínez A (1982) Nuevos registros de holoturoideos (Holothuroidea: Dendrochirota y Apoda) de la región insular del oriente venezolano. Bol Inst Oceanogr Univ Oriente 2:113–121
Martínez A (1986) Equinodermos de Isla de Aves, Venezuela. Dependencias Federales, Isla de Aves, Islas de Venezuela. Bol Inst Oceanogr Univ Oriente 25:195–213
Martínez A (1987a) Equinoideos y Asteroideos de Venezuela. Bol Inst Oceanogr Univ Oriente 26:153–164
Martínez A (1987b) Presencia del género *Lissothuria verril* (Holothuroidea: Echinodermata) en la costa norte del Estado Sucre, Venezuela. Bol Inst Oceanogr Univ Oriente 26:165–172
Martínez A (1988) Holoturoideos (Echinodermata: Holoturoidea) de la región Nor-Oriental de Venezuela y algunas dependencias federales. Inst Oceanogr Venezuela, Univ Oriente, Cumaná, Trabajo de Ascenso
Martínez A (1989) Holoturoideos (Echinodermata, Holothuroidea) de la región nor-oriental de Venezuela y algunas Dependencias Federales. Bol Inst Oceanogr Univ Oriente 28:105–112
Martínez A (1991a) Holoturoideos Dendrochirotida (Holothuroidea: Echinodermata) I: Familia Cucumaridae. Bol Inst Oceanogr Univ Oriente 30:31–40
Martínez A (1991b) Holoturoideos Dendrochirotida (Holothuroidea: Echinodermata) I: Familia Sclerodactylidae. Bol Inst Oceanogr Univ Oriente 30:41–46

Martínez A, Mago A (1975) Contribución al conocimiento de los holoturoideos (Holothuroidea: Echinodermata) de la región oriental de Venezuela. Bol Inst Oceanogr Univ Oriente 14:187–197

McKinney ML, McNamara JK, Carter BB, Donovan SK (1992) Evolution of paleogene echinoids: a global and regional view. In: Prothero DR, Berggren WA (eds) Eocene–Oligocene climatic and biotic evolution. Princeton Univ Press, Princeton, pp 349–367

Mihaljević M, Klug C, Aguilera O, Wyss P, Lüthi T, Sánchez-Villagra MR (2010) Diversity of Caribbean echinoids including new material from the Venezuelan neogene. Palaeontol Electr 13:1–36

Miller JB, Collinson C (1951) A new species of *Hercoglossa* from Maracaibo. J Paleontol 25:377–379

Miloslavich P, Klein E (2008) Ecoregiones marinas del Caribe venezolano. In: Klein E (ed) Prioridades de PDVSA en la conservación de la biodiversidad en el Caribe venezolano. Petróleos de Venezuela, SA–Univ Simón Bolívar–TNC, Caracas, pp 16–19

Miloslavich P, Klein E, Martín A, Bastidas C, Marin B, Spiniello P (2005) Status of knowledge of marine biodiversity in Venezuela. In: Miloslavich P, Klein E (eds) Marine biodiversity in the Caribbean. DEStech Publications, pp 109–136

Ministerio de Energía y Minas (1997) Léxico Estratigráfico de Venezuela (Tercera Edición). Bol Geol 12:1–828

Montealegre S, Gómez A (1999) Aspectos biológicos de *Lytechinus variegatus* (Echinoidea: Toxopneustidae) en tres localidades al sur de la Isla de Margarita, Venezuela. Estado Nueva Esparta, Isla de Margarita, Islas de Venezuela, región Oriental. 29na Reunión Asoc Lab Mar Carib (ALMC), Cumaná

Montealegre QS, Gómez A (2005) Ciclo reproductivo de *Lytechinus variegatus* (Echinoidea: Toxopneustidae) en el sur de Isla Margarita, Venezuela. Rev Biol Trop 53(Suppl 3):305–312

Mooi R, Peterson D (2000) A new species of *Leoida* (Clypeasteroida: Echinoidea) from the neogene of Venezuela and its importance in the phylogeny of mellitid sand dollars. J Paleontol 74:1083–1092

Müller-Karger F, Varela R (1989) Influjo del Río Orinoco en el Mar Caribe: Observaciones con el CZCS desde el espacio. Memorias Sociedad Ciencias Naturales La Salle, Caracas 186:361–390

Muñoz G (1980) Efectos de las variaciones térmicas sobre el proceso reproductivo del erizo de mar (*Echinometra lucunter*). In: Penchaszadeh PE (ed) Biología, ecología y cultivo de organismos acuáticos. Editorial Equinoccio, Universidad Simón Bolívar, Caracas, pp 209–224

Muñoz G, Ellietes J (1982) The effect of ferrous sulfate and sodium hypochlorite on fertilization and development of *Echinometra lucunter*. In: Lawrence JM (ed) Echinoderms: Proc Int Echinoderms Conf, Tampa Bay, pp 525–527

Muñoz G, Ellietes J (1983) Efectos de la temperatura, del sulfato ferroso e hipoclorito de sodio sobre la fecundación y el desarrollo en el erizo de mar *Echinometra lucunter*. In: Penchaszadeh PE (ed) Ecología del ambiente marino costero de Punta Morón (Termoeléctrica Planta Centro, Estado Carabobo, Venezuela). Intecmar - Univ Simón Bolívar, Caracas, pp 391–422

Noriega N (1998) Caracterización de las comunidades de erizos presentes en praderas de *Thalassia testudinum* y comunidades coralinas de la Bahía de Mochima, Parque Nacional Mochima. Bachelor's degree Thesis, Univ Central de Venezuela, Caracas

Noriega N, Cróquer A, Pauls S (2002) Población de *Lytechinus variegatus* (Echinoidea: Toxopneustidae) y características estructurales de las praderas de *Thalassia testudinum* en la Bahía de Mochima, Venezuela. Rev Biol Trop 50:49–56

Noriega N, Pauls M, Del Mónaco C (2006) Abundancia de *Diadema antillarum* (Echinodermata: Echinoidea) en las costas de Venezuela. Rev Biol Trop 54:793–802

O'Dea A, Jackson JBC, Fortunato H, Smith JT, D'Croz L, Johnson KG, Todd JA (2007) Environmental change preceded Caribbean extinction by 2 million years. Proc Nat Acad Sci 104:5501–5506

Ortega I, Martín A, Díaz D (2011) Distribución, parámetros poblacionales y dieta de *Astropecten marginatus* (Asteroidea: Astropectinidae) en el Atlántico venezolano. Rev Biol Trop 59:53–69

Palazon J (2001) Reproducción del pepino de mar, *Isostichopus badionotus* Selenka (Echinodermata: Holothuroidea). Book of abrstracts, IX Congr Latinoam Cien Mar, San Andrés, Colombia

Palazon J (2010) Casual hermaphoditism in a population of gonochoric sea cucumber, *Isostichopus badionotus* Selenka (Echinodermata: Holothuroidea) of north eastern Venezuela. In: Book of abrstracts, 7th european conference on Echinoderm, Göttingen

Pauls SM (1982) Estructura de las comunidades coralinas de la Bahía de Mochima, Venezuela. MSc Thesis, Univ Oriente, Cumaná

Pauls SM (1992) Erizos de mar en yacimientos indígenas en el Cayo Dos Mosquises Sur, Archipiélago de Los Roques. Reporte Técnico, Fundación Científica Los Roques, Caracas

Penchaszadeh PE (1973) Comportamiento trófico de la estrella de mar *Astropecten brasiliensis*. Ecología 1:45–54

Penchaszadeh PE (2003) Equinodermos. In: Aguilera M, Azocar A, González-Jiménez E (eds) Biodiversidad en Venezuela, vol I-II. Fundación Polar and Ministerio de Ciencia y Tecnología, Caracas, pp 514–521

Penchaszadeh PE, Lera ME (1983) Alimentación de tres especies tropicales de *Luidia* (Echinodermata, Asteroidea) en Golfo Triste, Venezuela. Carib J Sci 19:1–6

Penchaszadeh PE, Molinet R (1983) Diet of sea star *Tethyaster vestitus* Say (Astropectinidae) in the Golfo Triste (Venezuela). Bull Mar Sci 33:187–191

Penchaszadeh PE, Molinet R (1994) Population ecology of the sand dollar *Mellita quinquiesperforata latiambulacra* Clark, 1940 on the west-central coast of Venezuela. In: Bruno D, Guille A, Féral JP, Roux M (eds) Echinoderms through Time. Balkema, Rotterdam, pp 827–835

Penchaszadeh P, León C, Alvarez H, Bone D, Castellano P, Castillo M, Díaz Y, García-Guadilla MP, Lemus M, Losada F, Martín A, Miloslavich P, Parede C, Pérez D, Sebastiani M, Stecconi D, Roa V, Villamizar A (2000) Venezuela. In: Sheppard CR (ed) Seas at the milenium: an environmental evaluation, vol I. Pergamon, Oxford, pp 643–659

Prena J, Schwinghamer P, Rowell TW, Gordon DCJr, Gilkinson KD, Vass WP, McKeown DL (1999) Experimental otter trawling on a sandy bottom ecosystem of the Grand Banks of Newfoundland: analysis of trawl bycatch and effects on epifauna. Mar Ecol Prog Ser 181:107–124

Richardson LL, Goldberg WM, Carlton RG, Halas JC (1998) Coral disease outbreak in the Florida keys: plague type II. Rev Biol Trop 46:187–198

Rodríguez G (1980) Crustáceos decápodos de Venezuela. Instituto Venezolano de Investigaciones Científicas, Caracas

Rodríguez E, Pauls SM (1998) Sea cucumbers fisheries in Venezuela. In: Mooi M, Telford AS (eds) Proceedings of the 9th international confernce on Echinoderms. Balkema, Rotterdam, pp 513–516

Salazar V, Mata M, Montes A, Arredondo L, Rodulfo F (1994) Estudio histoquímico de células inmunológicamente activas de dos especies de Echinoidea: *Lytechinus variegatus* y *Echinometra lucunter* región Oriental. XLIV Convención An AsoVAC, Univ Nac Francisco de Miranda, Coro. Acta Cien Venez 45:45

Sambrano A, Díaz H, Conde JE (1990) Caracterización de la ingesta en *Isostichopus badionotus* (Salenka) y *Holothuria mexicana* Ludwig (Echinodermata: Holothuroidea). Carib J Sci 26:45–51

Sánchez TM, Lorente MA (1977) Paleoambiente del Miembro Quevedo (Formación Navay) en las proximidades de Santa Bárbara. Soc Venez Geol 1:107–133

Scheibling R, Mladenov P (1998) Distribution, abundance and size structure of *Tripneustes ventricosus* on traditional fishing grounds following the collapse of the sea urchin fishery in Barbados. In: Burke R, Mladenov P, Lambert P, Parsley R (eds) Echinoderm biology. Balkema, Rotterdam, pp 449–455

Schiff KC (2000) Sediment chemistry on the mainland shelf of the Southern California Bight. Mar Poll Bull 40:268–276

Theilen-Willige B (2006) Tsunami Hazards in Northern Venezuela. Science of Tsunami Hazards 25:144–159

Toral-GrandaV (2008) Population status, fisheries and trade of sea cucumbers in Latin America and the Caribbean. In: Toral-Granda V, Lovatelli A, Vasconcellos M (eds) Sea cucumbers. A global review of fisheries and trade. FAO fisheries and aquaculture technical Paper. No. 516. Rome, pp 213–229

Ummels F (1963) Asteroids from the Netherlands Antilles and other Caribbean localities. Stud Fauna Curaçao 15:72–101

Urbaneja MJ (1968) Determinación de compuestos esteroidales y lípidos en la estrella de mar *Echinaster sentus*, región Oriental. Bachelor's degree Thesis, Univ Oriente, Cumaná

Valdez J, Aguilera O (1987) Los peces del Golfo de Venezuela. Fondo Editorial CONICIT, Caracas

Vázquez-Domínguez E (2003) Diversity and distribution of crustaceans and echinoderms and their relation with sedimentation levels in coral reefs. Rev Biol Trop 51:183–193

Villamizar E, Laughlin R (1991) Fauna associated with the sponges *Aplysina archeri* and *Aplysina lacunosa* in a coral reef of the Archipiélago de Los Roques, National Park, Venezuela. In: Reitner J, Kaupp H (eds) Fossil and recent sponges. Springer, Berlin, pp 522–542

Vivas V (1987) Bioestratigrafía del Cretáceo de la región Bergantín-Santa Inés, estado Anzoátegui, Venezuela nororiental. Bol Geol Ministerio de Minas y Hidrocarburos 16:3–128

Von der Osten E (1957) A fauna from the Lower Cretacous Barranquin Formation of Venezuela. J Paleontol 31:571–590

Weil E (1980) Papel del erizo *Diadema antillarum* Philippi en la regulación de la estructura de las comunidades coralinas. Bachelor's degree Thesis, Univ Central de Venezuela, Caracas

Weil E, Losada F, Bone D (1984) Spatial variations in density and size of the echinoid *Diadema antillarum* Philippi on some Venezuelan coral reefs. Bijdragen tot de Dierkunde 54:73–82

Weisbord NE (1969) Some late Cenozoic Echinoidea from Cabo Blanco, Venezuela. Bull Am Paleontol 56:277–371

Work R (1969) Systematic ecology and distribution of the mollusk of Los Roques, Venezuela. Bull Mar Sci 19:614–711

Zanders IP (1970) Distribución de electrolitos y movimiento de sodio en el intestino de la holoturia *Isostichopus badionotus* Selenka. Región Oriental. Bachelor's degree Thesis, Univ Oriente, Cumaná

Zanders IP, Herrera FC (1974) Ionic distribution and fluxes in holothurian tissues. Comp Bioch Physiol 47A:1153–1170

Zoppi de Roa E (1967) Contribución al estudio de los equinodermos de Venezuela. Acta Biol Venez 5:267–333

Chapter 8
Echinoderms of Peru

Yuri Hooker, Elba Prieto-Rios and Francisco A. Solís-Marín

8.1 Introduction

8.1.1 Geographic and Geologic Characteristic of the Peruvian Coast

The Peruvian littoral extends 3,080 km. It goes from the border with Ecuador (approx. 3°23′S) to the south (18°S) at the border with Chile. The Peruvian coast faces the South East Pacific Ocean. Its coast is characterized by arid conditions,

Y. Hooker (✉)
Laboratorio de Biología Marina, Departamento de Ciencias Biológicas y Fisiológicas, Universidad Peruana Cayetano Heredia, Av. Honorio Delgado 430, Urb. Ingeniería, S.M.P, Lima, Peru
e-mail: yuri.hooker@upch.pe

Y. Hooker
Unidad Marino Costera, Servicio Nacional de Áreas Naturales Protegidas (SERNANP), Ministerio del Ambiente, Calle Diecisiete No 355, Urb. El Palomar - San Isidro, Lima, Peru

E. Prieto-Rios
Facultad de Ciencias Biológicas, Biología, Universidad Nacional Mayor de San Marcos, Mesa de partes. Ciudad Universitaria de San Marcos, Av. Venezuela s/n, Lima 1, Peru

E. Prieto-Rios
Posgrado en Ciencias del Mar y Limnología, Instituto de Ciencias del Mar y Limnología (ICML), UNAM, Apdo. Post. 70-305, 04510, Mexico, D.F., Mexico

F. A. Solís-Marín
Colección Nacional de Equinodermos "Ma. E. Caso Muñoz", Laboratorio de Sistemática y Ecología de Equinodermos, Instituto de Ciencias del Mar y Limnología (ICML), Universidad Nacional Autónoma de México (UNAM), Apdo. Post. 70-305, 04510, Mexico, D.F., Mexico
e-mail: fasolis@cmarl.unam.mx

J. J. Alvarado and F. A. Solís-Marín (eds.), *Echinoderm Research and Diversity in Latin America*, DOI: 10.1007/978-3-642-20051-9_8,

especially south of 13°45′S where the Atacama Desert starts. The coastal deserts are crossed by fertile valleys formed by rivers of the western slopes of the Andes, which carry water mostly in summer months (December–April). These valleys are vegetated oasis surrounded by extremely arid deserts that may lack any rainfall for years. The desert plains of the Peruvian coast are limited to the east by the Andes, which has heights over 6,700 m. North of 6°S, the coast receives minor seasonal rains during the summer months, which allow the development of a peculiar dry coastal forest. This extends to the border with Ecuador, where it becomes greener and humid because it is located in a rainy area. At the northern tip of the Peruvian coast is a mangrove forest which has a length of 20 km. This ecosystem extents from Ecuador to the Tumbes River delta in Peru and is the southern reach of continuous mangroves in the eastern Pacific. A small, isolated mangrove forest occurs south of Tumbes, at the mouth of the Rio Piura (5°30′S), the most southern reach of the ecosystem in Peru.

The general morphology of the coastline along the Peruvian coast lacks great geographical variants (Fig. 8.1). The most noticeable changes of the coast are the Sechura Bay-Illescas Peninsula (5°38′–6°54′S) and the Paracas Peninsula (13°51′S). There are few large bays, such as San Juan de Marcona, San Nicolas, Independencia, Paracas, Ancon, Tortugas, Samanco, Ferrol, Sechura, and Paita. Along the Peruvian coast are 77 islands (94.36 km^2). The biggest ones are Lobos de Tierra off Lambayeque (16 km^2); the island of San Lorenzo, off Callao (the largest island with 16.48 km^2) and the island of San Gallan (9.32 km^2) and La Vieja (11 km^2) off Paracas. Geologically the littoral is formed primarily by rocky coastlines of sedimentary origin, igneous rocks, and extensive sandy beaches (Sandweiss et al. 1996). Petersen et al. (1972) described the geomorphology of the Peruvian coast. North of Piura and south of Tumbes, plains are formed by an elevated sedimentary platform, which dominates the coastline (western shelf) with sediments of Upper Cretaceous and Tertiary origin. This area has creeks and water run-off, formed during El Niño events which bring intensive rainfall. Between Paita and Punta Illescas, the rocks are metamorphic, igneous, and slate. In Lambayeque and part of La Libertad, desert plains, wide beaches and coastal dunes dominate the coastline. Only offshore do some Paleozoic rocks reach the surface, forming the Lobos de Tierra and Lobos de Afuera islands. The coast of Ancash is dominated by igneous rocks from the Andes mountain range, which at this area approach the coastline, forming one of the most variable geomorphologic sectors of the coast, with numerous bays, islands, emerging rocky areas, and cliffs.

The coastline at Lima province is composed of a variety of geological features that include sedimentary coastal terraces, multiple intrusions from igneous rocks from the Andes that penetrate the sea, sandy plains, and wide valleys associated with numerous rivers of seasonal flow. At Ica province, from Paracas to the south, a geological complex of sedimentary rocks of diverse origin and age dominates the coastline, penetrating towards the sea on the Paracas Peninsula. The eastern Andes mountain range has wide desert plains up to 80 km in width, with spurs of low height that penetrate into the sea. Erosion has formed some high islands such as San Gayan and La Vieja (Independencia). In this coastline of exceptional

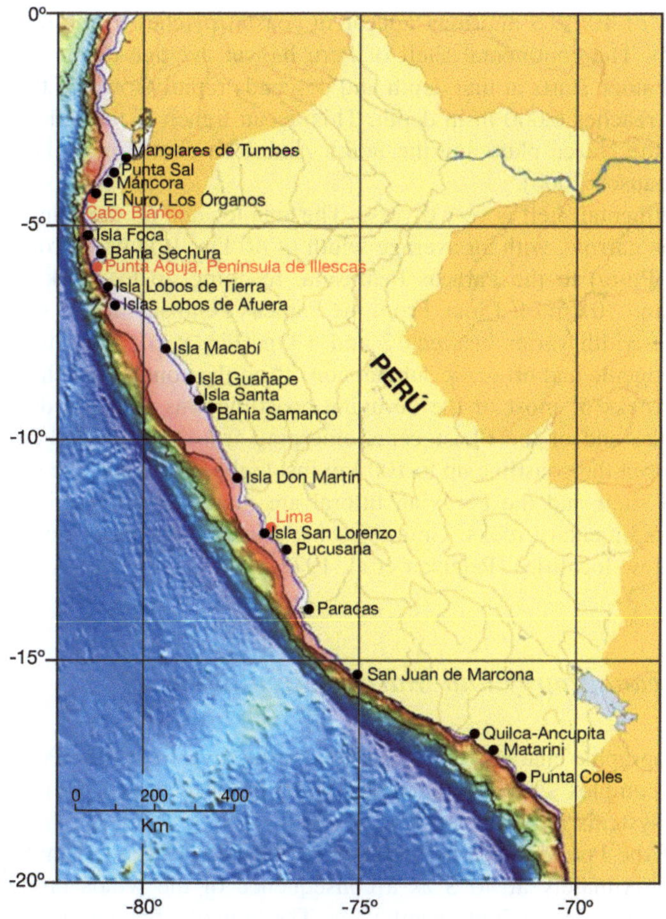

Fig. 8.1 Bathymetry of the Peruvian territorial sea and localities where shallow water echinoderms have been collected by Universidad Peruana Cayetano Heredia (UPCH). Isobath 1 (*blue line*) 50 m; isobath 2 (*red area*) 200 m; isobath 3 (*yellow area*) 1,000 m. Maximum depth 6,280 m

characteristics, emerging sedimentary deposits offer a rich register of fossils that date to the Carboniferous. Further to the south, at the Arequipa region to near the Chilean border, the Andes Mountains extend closer to the sea, forming edges dominated by recent igneous rocks that form cliffs, deep coves, and islands, interspersed by valleys with seasonal rivers. There are long sandy beaches and coastal dunes, forming a coastline with high geomorphologic diversity. This in turn generates great habitat diversity that favors the development of complex biological communities. The marine sediments between 7° and 10°30′S are mainly sand, which dominate the wide continental shelf. Sand also appears in patches towards the south. Mud with high organic content covers the continental shelf and slope south of 10°30′S and may appear in coastal patches to the north (Delgado et al. 1987). The continental shelf of Peru has an average depth of 200 m. The continental slope starts at that depth and descends rapidly towards the Peru ocean trench that reaches 6,000 m in depth. This ocean trench is formed at the edge of contact of the Nazca plate and the South American continental plate (Meschede and Barckhausen 2000).

The continental shelf is very diverse. The area located off the Tumbes and Piura provinces is narrow, with an average width of 40 km. It is wide from the Illescas Peninsula (Piura) to the Paracas Peninsula, reaching up to 140 km in width at Chimbote and 70 km off Lima. From the Paracas Peninsula to the Chilean border the average width varies between 2 and 4 km (Fig. 8.1). Geotectonic blocks as well as pyrogenic and orogenic submersions form the continental shelf. The group of marine rocks of most of the Peruvian coastline was elevated due to tectonic action by the end of the Pleistocene, emerging from the sea and penetrating in many places at the coastline up to 100 km inside the continent. The rocks from the continental shelf and the Peruvian littoral are made primarily of Paleozoic and Mesozoic sedimentary rocks. Granite rocks, granodiorite rocks, and gneiss rocks occur at a few localities (Petersen et al. 1972).

8.1.2 Oceanography and Marine Circulation

The oceanographic characteristics and the marine diversity of the Peruvian sea are ruled by a complex system of currents, which produce one of the most important upwelling systems in the world.

Most of the Peruvian sea is influence by the Peru Current (Humboldt), which begins approximately at 40°S as a consequence of the winds produced by the anticyclonic gyre of the South Pacific Sea. The flow of a branch of this current, the coastal Peruvian Current, is pushed by the trade winds. It follows the topography of the South American coast until 6°S (Illescas Peninsula) where it deviates to the east. This current is characterized by temperate waters through the length of its route, with temperatures between 13 °C and 18 °C according to year, season and geographic position (Fig. 8.2). These temperatures are much colder than expected because it is found in the tropics, close to the equator. This is caused by the intense upwelling that carries cold waters rich in nutrients to the sea surface. These waters have a stable salinity around 35 ‰. As suggested by its name, this current flows along the coast with a width that varies between 96 and 160 km with a depth normally below 200 m. The other branch of the Peru Current, The Peruvian Oceanic Current, moves from south to north, to the east of the meridian 82°W and reaches depths of 700 m. This oceanic current also has temperatures that are lower than those of tropical oceanic waters (17–20 °C) due to its subantarctic origins (Pizarro 2001; Tarazona et al. 2003).

In the Pacific Ocean, over the equatorial line, the Equatorial Counter Current moves towards the east until it collides with the South American continent.

A branch of this stream deviates and moves to the south along the coast of Ecuador and arrives to the north of Peru, where it is named El Niño Current. This stream of warm waters flows up to Cabo Blanco (4°15′S–81°14′W) where it turns towards the west to form part of the Equatorial Current. The area between Cabo Blanco and Punta Illecas is recognized as a region of tropical water masses where waters coming from the north mix with a branch of the Peruvian Current that maintains a weak flow towards the north (Hooker 2010).

Beneath the mentioned surface currents, there are other subsurface current. The Equatorial countercurrent (or Cromwell Current) flows from west to east towards Ecuador. A branch deviates and flows to the southeast to near the Peruvian north coast between 5° and 8°S. It flows under the Peruvian Current and in opposite direction (from north to south). It is close to the coast between Paita and Punta Falsa (5–6°S), flowing between 50 and 300 m in depth. It moves away from the coast to be integrated into the Subsurface Peruvian Current and continues to the south (Cromwell 1953). One of the most important characteristics of the current is a high concentration of oxygen, higher than 1.0 ml^{-1} at 100–200 m (Morón 2000) permitting processes that will be discussed later on Pequeño (2000) discusses the Intermediate Antarctic Water masses in deeper waters. In southern Chile they are rather shallow, going deeper to the north, reaching a depth of 1,500 m at approximately 35°S and even deeper to the north. He also mentions that off Chile, at depths of 3,500–4,200 m, the relatively salty, but very cold, Deep Antarctic Currents flows towards the north over the Pacific sea floor, crossing probably the Chilean-Peruvian trench. This facilitates the displacement of ichthyofauna of Antarctic origin towards less than 20°S, off the Peruvian coast. Pequeño (2000) also mentions that some authors (like Petersen et al. 1972) described the presence of a mass of intermediate water of the southeast Pacific sea or Intermediate Antarctic waters (between 500 and 1,500 m deep) located off Chile, Peru and Ecuador

8.1.3 General Ecological Characteristic

Peru is situated in a peculiar geographical area where temperate waters coming from the south and tropical waters from the north meet, generating a diverse and dynamic ecological system. Furthermore, the recurrent phenomenon called 'El Niño' makes the ecological processes even more complex, altering the abundance and distribution of marine organisms. The Peruvian sea is considered one of the more productive of the world because an intense upwelling system in most of the littoral develops a productivity that has made Peruvian fishing industry one of the most important in the world. The upwelling system is closely related to the Peru Current that flows continuously from south to north, pushed by the trade winds. Upwelling transports waters from depths of 50–150 m towards the surface in areas close to the coast, carrying nutrients accumulated over the bottom of the continental platform. Upwelling waters are characterized by low concentrations of oxygen and high concentrations of nutrients (Zuta and Guillén 1970; Strub et al. 1998).

The arrival of these nutrients to the surface generates intense primary productivity of around 1 kg C m^{-2} y^{-1} (Tarazona et al. 2003). The high abundance of plankton is the base of a trophic chain where the Peruvian anchoveta (*Engraulis ringens*) is the dominant species and the primary food of enormous populations of guano birds whose droppings (guano) are accumulated over the islands and capes. Wind action spreads some of the guano, fertilizing the marine waters even more (Palomares et al. 1987).

The influence of the upwelling reaches up to a depth of 100 km out into the ocean (Tarazona et al. 2003). Its influence is also reflected in coastal waters. The nutrients allow great beds of brown macroalgae as well as other algae to develop. They are used by herbivorous organisms, such as several species of sea urchins and gastropod mollusks.

Nevertheless, the primary productivity does not always favor the abundance of organisms. In the area occupied by the Peruvian Current, oxygen decreases rapidly, reaching hypoxic conditions (<1 ml l^{-1}) at depths below 20 m (Rosenberg et al. 1983). In waters between 100 m and 500 m in depth, oxygen concentration usually is almost zero (<0.1 ml l^{-1}) so diversity there is at a minimum. The cause is the great abundance of plankton in surface waters that, after death, generate a constant organic rain that decomposes as it sinks. The bacterial activity, both in the water column as well as over the organic sediments accumulated on the sea bottom, consumes the dissolved oxygen and generates a minimum oxygen layer that extends the length of the Peruvian Current (Lama et al. 2009).

Outside the continental shelf oxygen reappears in the water column at depths below 50 m carried by the Subsurface Peruvian Current. Over the bottom, at 600 m, biological diversity increases, reaching higher values at 1,000 m in depth, where oxygen-rich Antarctic Intermediate Waters occur (Lama et al. 2009).

In northern Peru, the tropical oceanographic conditions result in totally different ecological characteristics than those occurring south of 6°S. Tropical currents coming from the north maintain surface temperatures at coastal areas usually between 20 and 25 °C (Fig. 8.2). This allows the establishment of a purely tropical ecosystem with a high diversity. Hooker (2009) indicates they include more than 70 % of the total richness of the Peruvian littoral at shallow depths. This region, especially in the northern tip close to the border with Ecuador, is the only place where the influence of low salinity waters is significant due to the great volume of freshwater coming from Ecuadorian rainy areas, generating mangrove forests between the delta of the Tumbes River and the Zarumilla River that mark the border with Ecuador. Sediments carried out by these rivers also influence coastal diversity because waters become murky close to the mouths of the rivers. The effect of suspended sediment is dissipated to the South, where clearer waters are found between 3°52′ and 4°15′S. Species richness increases, especially on rocky reefs. It should be pointed out that Peruvian marine waters do not have any coral reefs and there is only one record of a hermatypic coral species.

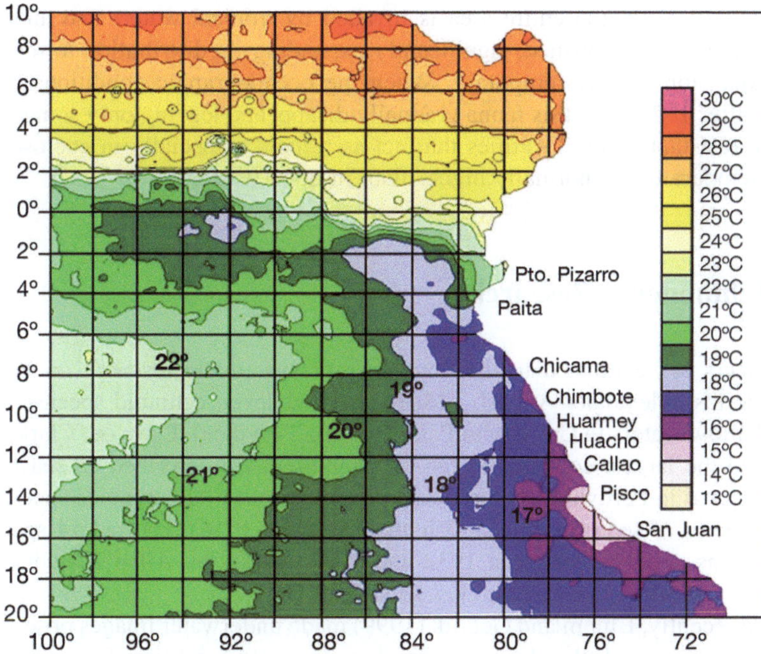

Fig. 8.2 Sea surface temperature distribution in the Peruvian territorial sea during winter during a year of "normal" oceanic conditions (11 August 2004) (*source* NAVO/NOAA)

In the entire sector between the border with Ecuador and 4°15′S the layer of oxygen that exists south of 6°S cannot be detected. There, life is present in all bathymetric gradients, maintaining high richness even below 1,000 m depth. The area called "Banco de Mancora", located about 40 km off Zorritos, a submarine mountain plateau between 90 and 500 m in depth, has high richness exists and some fish and invertebrate species seem to maintain their population centers (Edgar et al. 2004).

Another important factor affecting richness in the Peruvian littoral is the El Niño phenomenon. During these oceanographic events, trade winds push and weaken the Peruvian Current, causing the current to lose strength. As a result, tropical waters invade the Peruvian coast, reaching even to the Chilean north. The entire temperate waters fauna is seriously affected by temperature increases up to 10 °C at some points of the coast. Moreover, when upwellings decreases or disappear, nutrients and primary productivity is notably reduced. This affects macroalgae and filter feeders and the system collapses. In the tropical area of Peru, the water temperature in the littoral reaches 30 °C (Fig. 8.2) and heavy rainfall over the desert brings enormous quantities of sediments to the sea. This makes the sea completely turbid and the submarine rocky formations become clogged due to the sediments, killing the benthic fauna and transforming the coastal sea bottom into mud. The effect of the sediment is even perceptible 2 years after the event, where compact mud is found inside rock crevices and in the shells of dead barnacles.

During El Niño, the entire area is invaded by tropical waters and an aquatic tropical fauna is established, significantly expanding its distribution to the south. This expansion is only temporary. When the oceanographic conditions become normal after 1–2 years, this fauna gradually disappears, leaving only some species that are resistant inside localities that act as refugia, especially in enclosed bays where temperature is normally higher than in open waters.

8.2 Echinoderm Research

Few papers are dedicated to the taxonomy or systematic research on Peruvian echinoderms. Alexander Aggasiz (1881) reported seven echinoid species, mostly from shallow waters in Callao and Paita (at Paita Province). Later on, Clark (1910) published the largest echinoderm list ever reported. He described 25 seastar species collected in Peruvian territorial seas, 10 ophiuroids, 12 echinoids species, and seven species of sea cucumbers. Deichmann (1941, 1958) reported 14 species of holothurians collected between 1932 and 1938 during the Allan Hancock expeditions to the American Pacific (*Velero III* and *IV*).

More recently, Bluhm and Gebruk (1999) made underwater images observations at depths between 4,140 and 4,160 m for three Peruvian localities. They reported 11 species of deep sea holothurians and some non identified sea cucumbers. Hooker et al. (2005) reported 39 echinoderm species at Islas Lobos de Afuera (Lambayeque): seven Asteroidea, eight Ophiuroidea, 11 Echinoidea and 13 Holothuroidea. Prieto-Rios (2010) reported 22 species of sea cucumbers for Peruvian waters, including both shallow and deep areas. Finally, Prieto-Rios et al. (2011) reported *Florometra magellanica* (Bell, 1882) in Piura between 360–814 m, the first crinoid species ever recorded for Peruvian waters.

8.3 Diversity of Echinoderms and Distribution

The Peruvian littoral is divided into two main areas included in two biogeographical provinces (Spalding et al. 2007): Warm Temperate South-Eastern Pacific (Peruvian Province) and Tropical Eastern Pacific (Panamic Province). The former is associated with the Peruvian Current with a high level of endemism. The Temperate Eastern Pacific, on the other hand, is associated with warm waters coming from the north, mostly by a branch of the South Equatorial Current. The fauna of this province is widely distributed from the Gulf of California (Mexico) to north of Piura region, Peru (4°15′S). The area between Cabo Blanco (4°15′S) and Punta Aguja (6°S) is a transition zone between these two provinces (Hooker 2009).

Peruvian echinoderms are represented by 215 species: Crinoidea (1 species), Asteroidea (64 species, distributed in seven orders and 22 families), Ophiuroidea (42 species, distributed in two orders and 11 families), Echinoidea (35 species, distributed in seven orders and 14 families), and Holothuroidea (73 species, distributed in six orders and 16 families). Only 17 species are found in the Temperate South Eastern Province, 64 in the Tropical Eastern Pacific, four are found in both provinces. Seventy eight are deep water species. That means that 20 % of the 84 shallow water species are temperate, 75 % are tropical and 5 % are both temperate and tropical.

When dealing with biogeography and the distribution of Peruvian echinoderms, it is very important to use information gathered in normal oceanographic years. This is because many tropical species widen their geographical distribution southward in El Niño years (every 7–12 years). The species that extend their distribution in El Niño years do so only temporarily, generally for less than 2 years, disappearing later from the colonized area. Our information has been gathered during several years of observations and collections in about 20 localities on the Peruvian coast.

8.3.1 Temperate South Eastern Pacific

South of Punta Aguja, around the Illescas Peninsula (6°S), shallow water echinoderm richness is low, with most species belonging to the Warm Temperate South Eastern Province and endemic to the Peru Current. In this area no differences in species composition are found in the intertidal zone and abundance variations are not significant. However, in the sublittoral zone differences do appear in the distribution of some species, especially those that are mainly Chilean.

Species representative of the rocky intertidal and upper sublittoral are *Stichaster striatus* (Fig. 8.3b), *Heliaster helianthus* (Fig. 8.3d), *Tetrapygus niger* and *Pattalus mollis*. It is common to find juveniles of *T. niger* in rocky walls exposed to wave action feeding on calcareous crustose algae. *P. mollis* occurs inside mussel colonies. They use their tentacles to capture juvenile mussels, small crustaceans, and snails detached from their colonies by wave action (Fig. 8.3h). The seastars *S. striatus* and *H. helianthus* feed in the intertidal below the mussel belt where they are not exposed to air during extended periods of low tide. The brittle star *Ophiactis kroeyeri* lives inside mussel colonies by the thousands. A study carried out in Santa Island, Ancash (Hooker et al. 2011) found an average of 27,000 ind m^{-2} inside the mussel colonies.

The representative species in the sandy intertidal and upper sublittoral is *Athyonidium chilensis.* It lives buried in sand. It has a U-shaped body. The tentacles extend out on the sand and the cloaca shows just at the sediment surface. This species feeds on suspended material from the water column.

In shallow subtidal waters the widely dominant species are the sea urchins *T. niger* and *Caenocentrus gibbosus* (Fig. 8.4b) and *Lytechinus semituberculatus* (Fig. 8.5c). Along the Peruvian central coast both species are equally abundant. *Tetrapigyus niger* lives in shallower waters, usually over 5 m in depth, while *C. gibbosus* is abundant from less than 1 m to about 10 m in depth. This species is

Fig. 8.3 Echinoderms from Temperate South Eastern Pacific (Peruvian Province): **a** *Patiria chilensis* (Lütken, 1859); **b** *Stichaster striatus* Müller and Troschel, 1840; **c** *Luidia magellanica* Leipoldt, 1895; **d** *Heliaster helianthus* (Lamarck, 1816). **e** *Loxechinus albus* (Molina, 1782); **f** *Arbacia spatuligera* (Valenciennes, 1846); **g** *Pseudocnus dubiosus* (Semper, 1868); **h** *Pattalus mollis* Selenka, 1868. (Pictures by Y. Hooker)

Fig. 8.4 Echinoderms from the Tropical Eastern Pacific (Panamic Province): **a** *Astropecten armatus* Gray, 1840; **b** *Phataria unifascialis* (Gray, 1840); **c** *Paulia horrida* Gray, 1840; **d** *Ophiocoma aethiops* Lütken, 1859; **e** *Ophiothela mirabilis* Verrill, 1867; **f** *Ophiothrix magnifica* Lyman, 1860; **g** *Eucidaris thouarsii* (Agassiz & Désor, 1847); **h** *Astropyga pulvinata* (Lamarck, 1816). (Pictures by Y. Hooker)

Fig. 8.5 Echinoderms from the Tropical Eastern Pacific (Panamic Province): **a** *Centrostephanus coronatus* (Verrill, 1867). **b** *Caenocentrotus gibbosus* (L. Agassiz & Desor, 1846); **c** *Lytechinus semituberculatus* (L. Agassiz & Desor, 1846); **d** *Mellitella stokesii* (L. Agassiz, 1841); **e** Dense aggregation of *Neothyone gibber* (Selenka, 1867); **f** *Cucumaria flamma* Solís-Marín & Laguarda-Figueras, 1999; **g** *Holothuria* cf. *inhabilis*; **h** *Isostichopus fuscus* (Ludwig, 1875). (Pictures by Y. Hooker)

numerous along the coasts of central and northern Peru, mainly at Foca island in the middle of the biogeographic transition area. It is very abundant at this locality, covering most of the rock surfaces. To the north, where echinoderm diversity increases, this sea urchin is rare. It is also very numerous in the bays of Sechura, Lobos de Tierra island, Ancash (mainly Samanco) and Pucusana. This species, which extends north to Colombia, seems to have its higher concentrations not only in the area of its biogeographic transition but in the northern bays of the Warm Temperate South Eastern Province. These are all closed and therefore warmer than the surrounding oceanic waters, environments similar to that found in the transition area.

In deeper areas, usually below 10 m, the sea urchin *Arbacia spatuligera* becomes abundant. It prefers organic remains of decomposing algae and deposited organic matter as food, especially in the interface between rocks and sand bottoms (Fig. 8.3f). It is also numerous in soft bottoms with organic matter. *Loxechinus albus* becomes abundant from Paracas southwards (Fig. 8.3e). It is found mainly in areas where kelps are present since it feeds on pieces of algae that fall to bottom and that are picked with the tube feet. Old fishermen of the area affirm that *L. albus* used to cover extended areas and gave a reddish color to the bottom where they lived. When commercial exploitation of the species began, it became scarce in most of its distribution area, remaining abundant only in remote localities or in protected places, such as Punta San Juan in Marcona.

The seastar *S. striatus* prefers the intertidal and wave exposed well oxygenated shallow waters, while *H. helianthus* and *Luidia magellanica* are found in deeper waters (Fig. 8.3c). The last two species are active, voracious predators. *Heliaster helianthus* is an effective biological control of turbinid snails and mussels, as well as other small animals. *Luidia magellanica* may also feed on larger prey, including *A. spatuligera* and *T. niger*. Old fishermen say *T. niger* was not as abundant as it is today and that now it is out of control, covering large areas areas in some locations. In the 90s *T. niger* was a major prey of the wrasse *Semicossiphus darwini* (Y. Hooker personal observations). This fish is now almost extinct. This may be one of the causes of the overpopulation of the sea urchin.

Patiria chilensis, a small seastar, occurs along the entire Warm Temperate South Eastern Province of Peruvian waters (Fig. 8.3a). The ophiuroid *Ophiactis kroeyeri*, is a species that inhabits a wide diversity of habitats and depths in high concentrations. High amounts of organic matter are required to support its vast aggregations. It is usually found inside colonies of intertidal mussels, below rocks with accumulations of detritus, inside the exhalant canals of large sponges, and mainly at depths greater than 60 m because decrease oxygen reduces its abundance. North of Lima to Paita Bay (Ancash), *Ophiothrix magnifica* dominates muddy bottoms. This species is a northern element of the Warm Temperate South Eastern Province as well as the transition zone to the Tropical Province. The sea cucumber *Pseudocnus dubiosus* (Fig. 8.3g) dominates muddy, hypoxic, unstable bottoms, forming large aggregations. It is the most representative member of this habitat.

There are many doubts about the taxonomy of some Peruvian brittle stars like *O. kroeyeri*. It probably includes at least two species that have not yet been distinguished. More detailed studies are required.

8.3.2 Tropical Peruvian Eastern Pacific

Seventy five percent of the intertidal Peruvian echinoderm species are concentrated at the northern tip of Peru along less than 150 km of coastline. North of Cabo Blanco (4°15′S; Fig. 8.1), all species are typical of the Tropical Eastern Pacific, with the exception of *C. gibbosus*, *Pentamera chierchia* and *P. chiloensis*, which are widely distributed. *Pentamera* spp. are even found in cold waters and cannot be considered members of any given province.

On the other hand, *Ophiothrix magnifica*, a brittle star common at north Peru (Fig. 8.4f), enters the Peru Current to Paracas and is successfully established in enclosed shallow bays where warmer waters allow the temporary presence populations. This species is not considered widely distributed since it appears only in refugia where environmental conditions differ from the surrounding areas. The region between El Ñuro and Punta Sal is where the best example of the Tropical Eastern Pacific echinoderm community is found in Peru. Rocky reefs are scarce in northern Peru, represented mainly by flat strata of sedimentary rocks. In this area some taller reefs appear, mainly at Punta Sal. Further north, the diversity is reduced because of the sediment load and turbidity caused by the discharge of adjacent rivers.

In the intertidal, the most common species are *Ophioderma panamensis, Ophiocoma aethiops* (Fig. 8.4d), *O. magnifica*, *Arbacia incisa* and *Echinometra vanbrunti*, mainly in tide pools and below rocks. In the subtidal zone the most abundant species are *Eucidaris thouarsii* (Fig. 8.5g), and *Phataria unifascialis. Pentaceraster cumingi* used to be abundant, but it has become rare since it is collected to be sold as souvenirs. The most common sea cucumber is *Cucumaria flamma* (Fig. 8.5f) which populates the higher portions of reefs. Under the rocks, *Holothuria arenicola* and *H. impatiens* are frequent species, together with *O. panamensis* and *O. aethiops* (Fig. 8.4d). Another common species, *O. magnifica*, may be found in all available cryptic habitats, including rocks, algae, sponge, ascidians, and gorgonians. *Ophiothela mirabilis* is a brittle star that is always associated with gorgonians, where it is attached to their branches in large numbers (Fig. 8.4e).

Astropyga pulvinata (Fig. 8.4h), a large sized sea urchin, is a very rare species in Peru. It has been reported in two localities: north of Punta Sal, where there is a rather large population composed mostly by large adults, and at Lobos de Afuera, where a large group of juveniles on a rocky reef were found (Hooker et al. 2005). Some other rare species are *Toxopneustes roseus*, known in Peru only from scarce reports at El Ñuro, Isla Foca, and Islas Lobos de Afuera. *Paulia horrida* is a rare seastar found through its whole distribution, reported in Peru only from Punta Sal (Fig. 8.4c). This differs from the status of the seastar *Nidorellia armata*, which was frequent but has become close to extinction due to extraction for the souvenir market.

Holothuria cf. *inhabilis* (Fig. 8.5g), a remarkable large sized sea cucumber (about 30 cm in length in life), is frequent on the soft bottoms surrounding rocky reefs, where it feeds on organic sediments. This species, reported as a possible new species when it was collected at Islas Lobos de Afuera (Hooker et al. 2005), should be reevaluated in order to clarify doubts concerning its taxonomy identity.

On shallow sandy bottoms the dominant echinoderms are the seastars *Astropecten* spp., with *A. armatus* as the most frequent (Fig. 8.4a). The sand dollar *Encope micropora* has also been reported from this habitat, but seems to be more frequent in the biogeographic transition region. *Luidia bellonae* is rarely reported in deeper waters (10–20 m) along the north Peruvian coast. However, it is fairly common in Islas Lobos de Afuera. There are no records of this species south of Islas Lobos de Afuera. Nevertheless, Clark (1910), Anonymous (2010) and Morales-Montecinos (2011) reported it from central and southern Peru. In waters below 50 m depth, *L. superba* is the dominant seastar, together with *Clypeaster europacificus*. Both species are frequently caught in artisanal fishing nets.

8.3.3 Tumbes Mangroves

Northern Peruvian mangroves possess a typical Tropical Eastern Pacific fauna but the reported faunal diversity is low. The most representative species is *Mellitella stokesii*, a very common sand dollar in sandy tide pools and intertidal beaches in the interface between the mangroves and the sea (Fig. 8.4d). Another abundant species is the sea urchin *Agassizia scrobiculata*. Dense populations occupy the bottom of tidal creeks at depths between 0.5 and 2 m at low tide. The brittle star *O. magnifica* is also frequent, mainly between mangrove roots and inside sponges and zoanthids. Another ophiuran, an unidentified species of the genus *Ophiogrammus*, has been recorded in the muddy sand bottoms. A bright-orange morph of *Luidia brevispina* is common on sandy beaches around mangroves, different from the grey specimens known from Lobos de Afuera.

8.3.4 Biogeographic Transition Area Temperate South Eastern Pacific-Tropical Eastern Pacific

As mentioned earlier, the transition area between these biogeographic provinces is between Cabo Blanco (4°15′S) and Punta Aguja (6°S). This also includes the islands Lobos de Tierra and Lobos de Afuera. Echinoderms of both provinces are found in this refuge area as well as clear differences in abundance. For example, *O. magnifica* is extremely abundant in almost every habitat, with the exception of mud and fine sand bottoms. Another abundant species is *Neothyone gibber*, especially at Lobos de Tierra where it covers 100 % of the vertical walls of large rocks (Fig. 8.5e). Intertidal cliffs are dominated by *H. helianthus*. The sea cucumber *Psolidium dorsipes* is found in the intertidal zone. In the subtidal, *P. unifascialis* (Fig. 8.4b) has been rarely found on rocks. *Athyonidium chilensis* appears occasionally below them.

Isostichopus fuscus (Fig. 8.4h) has been found throughout the area with low densities. This sea cucumber is not as affected here by over-exploitation as it has been in northern Peru. On the other hand, Lobos de Afuera is the only location where *Tripneustes depressus* has been reported. This species was very abundant there until 2005. It is said that fishermen working in the island extracted them for their own use and for commercialization. This has decreased the populations below commercial levels.

The islands of Lobos de Afuera possess species from both provinces. They may be considered relatively tropical. The relative abundance of the species differs from coastal waters (Hooker et al. 2005). Knowledge about the Lobos de Afuera echinoderm fauna (Hooker et al. 2005), fishes (Hooker 2009) and sponges are more related to the Galapagos Islands than to the coastal tropical waters of northern Peru (Fig. 8.6).

8.3.5 Deep Sea Waters

The diversity of echinoderms in Peruvian deep waters is very high, reaching 46 % of the total reported species for Peru. Richness has been reported mostly by old marine cruises. The information was collected and analyzed by Maluf (1988). The most recent reports come from deep water cruises organized by the Instituto del Mar de Peru (IMARPE). These cruises collected a large number of species that should increase our knowledge of the diversity of Peruvian deep waters.

Prieto-Rios (2010) published the first report for Peru of the sea cucumbers *Molpadia* and *Caudina californica* at depths of 1,412 and 996 m, respectively. Morales-Montecinos (in preparation) has the first collection for Peru of the sea stars *Thrissacanthias penicillatus*, *Ctenodiscus crispatus*, *Pectinaster agassizi*, *Ceramaster leptoceramus*, *Lophaster furcilliger*, and *Hymenaster quadrispinosus*. Recently, Prieto-Rios et al. (2011) published the first report of a crinoids for Peru, adding *F. magellanica* (Bell, 1882) (360–814 m) to the biodiversity list of deep sea echinoderms. Some other species of sea cucumbers and seastar are now under revision and brittle stars and more crinoid species are still awaiting a proper taxonomic revision.

8.4 Aquaculture and Fisheries

Historically echinoderms were of interest to Prehispanic Peruvian cultures. The sea cucumber *A. chilensis*, locally called "ancoco", has been exploited and locally consumed by the Mochica culture in the Lambayeque area ($\sim$6°50′S) for at least

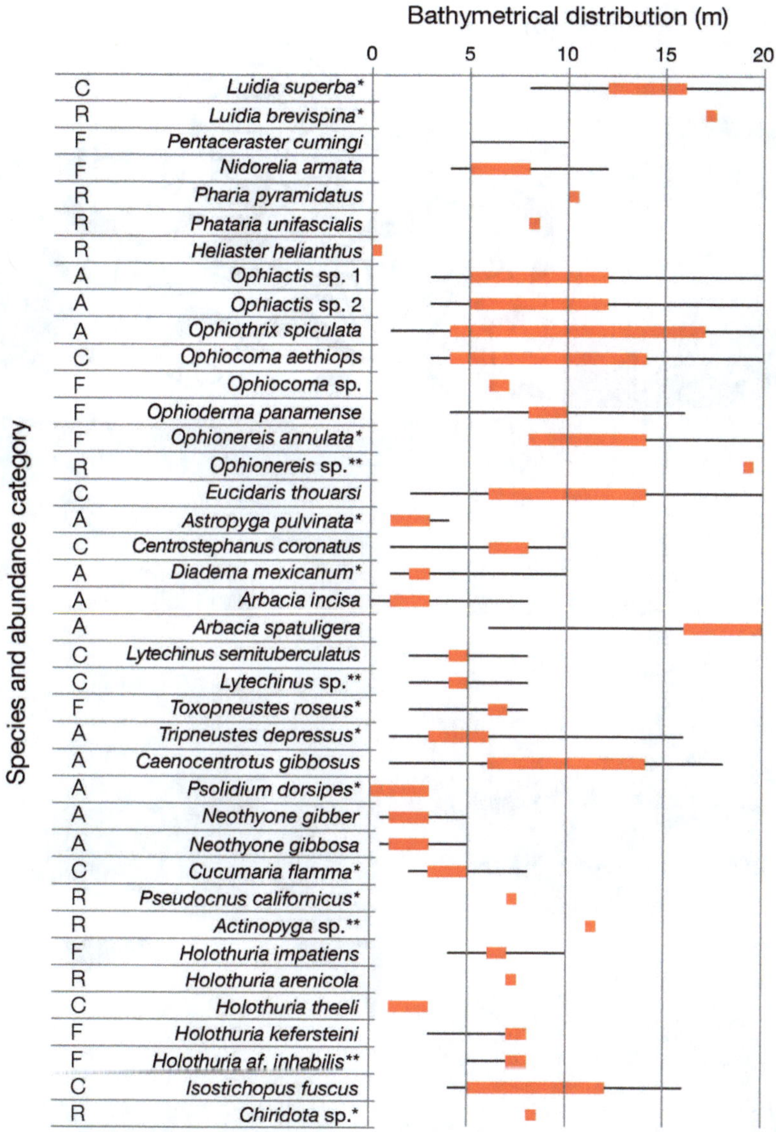

Fig. 8.6 Bathymetric ranges and abundance (to 20 m deep) of the Lobos de Afuera echinoderms in June 1999. *R* Rare (1–2 individuals); *F* Frequent (3–10 individuals); *C* Common (11–50 individuals); *A* Abundant (more than 50 individuals) (from Hooker et al. 2005)

2,000 years. Starfish were also important in the ancient Peruvian cultures, as ritual offerings to the gods. This practice is still kept in the rites carried by the local chamanes. Evidence for this is seen in their representation in different artistic objects like pottery and paintings from the old Moche culture (Fig. 8.7a, b).

Fig. 8.7 **a** Painted Moche ceramic (2000 years old) depicting a seven armed seastar (from Vergara and Sánchez 1996); **b** Moche iconography depicting a fishing scene including some seastar; **c** dry seastar included in different handcrafts; **d** and **e** *Pattallus mollis* landing at Chimbote Pier; **f** Ancoco (*Pattallus mollis*) processed, and **g** Sea urchin (*Loxechinus albus*) processed and ready to sell at the markets in Lima and Ilo (Pictures by Y. Hooker)

Table 8.1 Landing information, total catch (t) of echinoderms per area in the region of Ancash-Tacna, Peru, from 2001 to 2007 (from Sueiro 2009)

Common name	Scientific name/year	2001	2002	2003	2004	2005	2006	2007	Total
Sea urchin	*Loxechinus albus*	2111	1948	2059	1387	3010	80	1365	11960
Sea cucumbers ("mixture of species")	*Pattalus mollis-Holothuria* sp.	5	190	8	1	21	86	567	978
Seastar	*Stichaster* sp.	5	0	0	0	0	0	0	5

Dry specimens of *Stichaster striatus* and *Heliaster helianthus* are used in these rites. Both species are still sold as souvenirs (Fig. 8.7c) or in herb and mystic objects (chamaneria) markets.

Exploitation of echinoderms in Peru is not older than 20 years. It began with exportation of the red sea urchin *Lytechinus albus* gonads and later with some species of sea cucumbers to Japan. The first sea cucumber formally and legally exploited was *I. fuscus* in 1995, possibly when the fishing of the species began to be controlled or even sold in some areas of the Galapagos Islands. Sueiro (2009) mentioned the problem of exploitation of *I. fuscus* in the Galapagos and the regulation measures taken for its protection. Although *I. fuscus* is an economically important marine product in Peru (class "A" product), no protection measures are taken in its fishery. Catches are sent to Ecuador immediately with no information on the catch provided to authorities. Local fishing divers have provided information that *I. fuscus* was eliminated in less than 3 years after fishing began from areas in Peru, Lobos de Afuera and the region between Cabo Blanco and Punta Mero, north of Punta Sal. Legal exportation of sea cucumbers began officially at the end of the 1990s. Catches were principally composed of *P. mollis* and *Holothuria theeli*. The last species became common after El Niño 1997–1998, with large landings in Huacho (Elliott et al. 2005) and Chimbote. The last year this sea cucumber was reported was 2002. Since that it has been replaced in fishing by *P. mollis* captures (Fig. 8.7d–f).

Toral-Granda et al. (2003) and Sueiro (2009) published the catch and exportation statistics of Peruvian sea cucumbers with a "mixture of species", probably including *A. chilensis* (Tables 8.1 and 8.2).

Sueiro (2009) published data that included landings of the sea urchin *L. albus*, a species that was always important for local consumption (Fig. 8.7g). It has been taken to the edge of collapse, being very scarce nowadays. The Minister of Production has established bans several times in order to protect and restore this resource.

Tripnesutes depressus is a sea urchin that reaches large sizes at Lobos de Afuera. Local fishermen reported that it was abundant. However, the presence of purchasers resulted in its illegal and accelerated fishing, with the consequent depletion of its populations. IMARPE's landing records from Lambayeque report it as "*L. albus*". However that species does not occur there, being restricted to south of Paracas.

Table 8.2 Total export data (t) of Peruvian sea cucumbers to Hong Kong (dry, salted or pickled product) 1999-September 2005 (from Toral-Granda 2008)

	1999	2000	2001	2002	2003	2004	2005	Total (t)
Peru	4.1	7.3	3.8	1.8	8.3	19.9	31.0	76.2

Another kind of commerce for the Peruvian echinoderms that is usually not taken into account which affects seriously populations of some species is the souvenir market (Fig. 8.7c). Commerce of dry specimens of *S. striatus* and *H. helianthus* in some tourist places of central and southern Peru is common. These species are also used to make handcrafts. Some species are also sold dry on the tropical north beaches, mainly Mancora. The list includes *N. armata, P. cumingi, P. pyramidatus, P. unifascialis, C. europacificus* and *E. micropora.* The first two species have become close to extinction because they were never abundant and the reefs where they live are not common. *C. europacificus* is accidentally caught by artisanal trawling boats. The effect of this on its populations is unknown.

Sueiro (2009) mentions a trade exportation of 5 tons of *S. striatus* in the year 2001, but the objective of this commerce is unknown. The populations of this seastar have been reduced due to its multiple usages given (ornaments, souvenirs, mystic rituals, etc.). More recently most local universities and schools use them for their teaching laboratories because they are easily collected in the intertidal.

No special effort has been made to develop echinoderm aquaculture techniques in Peru. Nevertheless, the IMARPE's Ilo Aquaculture Laboratory has been working successfully on reproduction of *L. albus*. It is possible that in a short time the aquaculture of this sea urchin will be developed for both market and scientific interests.

8.5 Echinoderm Threats

There is no national strategy for the conservation and management of echinoderms except for the sea urchin *L. albus*. Because new Marine Protected Areas (MPA) in Peru were created, in 2010 and 2011, the country is now faces new challenges. The Servicio Nacional de Areas Naturales Protegidas del Peru currently preparing Red List files of threatened species in Peruvian MPAs. It is expected that several echinoderm species will be included. The following list shows some of the proposed species and their status following the criteria of the International Union for Conservation of Nature (IUCN).

1. *Nidorellia armata*: Critically endangered.
2. *Pharia pyramidatus*: Endangered.
3. *Phataria unifascialis*: Almost threatened.
4. *Stichaster striatus*: Vulnerable.

5. *Clyepaster europacificus*: Insufficient data.
6. *Loxechinus albus*: Vulnerable.
7. *Tripneustes depressus*: Endangered.
8. *Athyonidium chilensis*: Almost threatened.
9. *Isostichopus fuscus*: Endangered.
10. *Pattalus mollis*: Almost threatened.

8.6 Recommendations and Concluding Remarks

Peru has 215 species of echinoderms inhabiting its territorial sea. Crinoidea (1 species), Asteroidea (50 species), Ophiuroidea (32 species), Echinoidea (23 species) and Holothuroidea (53 species). This compilation is important to make recommendations: It is necessary to do more ecological studies. The situation previously described regarding publications about ecology of echinoderms in Peru provides evidence of an unequal and limited development of ecological studies related echinoderms. There are no papers that analyze echinoderm communities in Peru. It is important to have more information about their density and other population parameters, including growth, reproduction, feeding and their relation to biotic and abiotic conditions in the environment. This would make it possible to evaluation of environmental factors that could affect their presence in a certain location and indicate their vulnerability to climate change. This information will be of great importance for the basis for developing strategies leading towards sustainable production of echinoderms as a resource and their administration, as well as the search of alternatives to increase their production through aquaculture.

One of the main problems in Peru, as well as in neighboring countries, is the lack of taxonomists who can work full time reviewing and identifying specimens in biological collections. Deep water specimens as well as brittle stars should be reviewed. This should increase greatly the number of species reported from Peru. Complete knowledge of the echinoderm fauna will allow ecological research directed towards understanding the function of each species in the ecosystem allow researchers to monitor climate change associated with El Niño, and the distribution and biogeography of the species.

Controlled production of commercial species should be developed in order to reduce the impact of fishing on natural resources. Creation of new MPAs should guarantee conservation of exploited species.

Acknowledgments To Dr. Tina Kameya Kameya, Unidad de Investigaciones en Biodiversidad, Instituto del Mar del Peru (IMARPE), for providing access to IMARPE's echinoderm collection.

References

Agassiz A (1881) Report of the echinoidea dredged by the H. M. *Challenger* during the year 1873–76. Report of the scientific results of the voyage of the H. M. S. *Challenger* during the years 1873–1876. Zoology 3:1–321

Anonymous (2010) Cuarto Informe Nacional sobre la aplicación del Convenio de Diversidad Biológica años 2006–2009. Capítulos I, II, III, IV, Apéndices, y Anexos áreas Protegidas. Bicentenario de Desarrollo Estratégico de recursos Naturales. Ministerio del Ambiente, Dirección General de Diversidad Biológica, Lima, Perú

Bluhm H, Gebruk A (1999) Holothuroidea (Echinodermata) of the Peru Basin—Ecological and taxonomic remarks based on underwater images. Mar Ecol 20:167–195

Clark HL (1910) The echinoderms of Peru, vol 52, Bulletin of the Museum of Comparative Zoology, Harvard University pp 321–358

Cromwell T (1953) Circulation in a meridional plane in the central equatorial Pacific. J Mar Res 12:196–213

Deichmann E (1941) The Holothurioidea collected by the Velero III during the years 1932 to 1938. Part I. Dendrochirota. Allan Hancock Pac Exp 8:61–195

Deichmann E (1958) The Holothurioidea collected by theVelero III and IV during the years 1932 to 1954. Part II. Aspidochirota. Allan Hancock Pac Exp 11:253–349

Delgado C, Gomero R, Salzwedel H, Flores L, Carvajal G (1987) Sedimentos superficiales del margen continental peruano: un mapa textural. IMARPE Bol 11:182–190

Edgar GJ, Banks S, Fariña JM, Calvopiña M, Martínez C (2004) Regional biogeography of shallow reef fish and macro-invertebrate communities in the Galapagos archipelago. J Biogeogr 31:1107–1124

Elliott RW, González R, Ramírez A (2005) Bancos naturales de invertebrados en Hornillos y El Colorado (Huacho Región, Lima). IMARPE, Laboratorio Costero de Huacho. Huacho-Región Lima

Hooker Y (2009) Nuevos registros de peces costeros tropicales para el Perú. Rev Peru Biol 16:33–41

Hooker Y (2010) Expediente de creación de 3 nuevas Áreas Marinas Protegidas en el norte tropical del Perú. Informe SERNANP, Lima

Hooker Y, Solís-Marín FA, Lleellish M (2005) Equinodermos de las Islas Lobos de Afuera (Lambayeque, Perú). Rev Peru Biol 12:77–82

Hooker Y, Ubillus O, Heaton J, Garcia O, García M (2011) Evaluación de Objetos de Conservación y Zonificación de Isla Santa. Ancash Rev Áreas Mar Prot Perú 3:1–77

Lama P, Lavika G, Jensena MM, Vossenbergb J, Schmidb M, Woebkena D, Gutiérrez D, Amanna R, Jettenb MSM, Kuypersa MMM (2009) Revising the nitrogen cycle in the Peruvian oxygen minimum zone. Proc Nat Acad Sci 106:4752–4757

Maluf LY (1988) Composition and distribution of the Central eastern Pacific echinoderms. Natural History Museum of Los Angeles County Technical Report 2:1–242

Meschede M, Barckhausen U (2000) Plate Tectonic Evolution of the Cocos-Nazca Spreding Center. In: Silver EA, Kimura G, Shipley TH (eds) Proc Ocean Drill Program Sci Res 170:1–10

Morales-Montecinos M (2011) Revisión taxonómica de los asteroideos del mar del Perú. In Professional Thesis, Univ Nacional Mayor de San Marcos, Lima

Morón OA (2000) Características del ambiente marino frente a la costa Peruana. Bol Inst Mar Perú 19:179–204

Palomares LM, Muck P, Mendo J, Chuman E, Gomez O, Pauly D (1987) Growth of the Peruvian anchoveta (*Engraulis ringens*), 1953 to 1982. In: Pauly D, Sukayama I (eds) The Peruvian anchoveta and its upwelling ecosystem: three decades of changes, vol 15. ICLARM Studies and Reviews pp 117–141

Pequeño G (2000) Delimitaciones y relaciones biogeográficas de los peces del Pacifico suroriental. Estud Oceanol 19:53–76

Petersen G, Mújica MR, Sánchez R (1972) Historia Marítima del Perú. El Mar: Gran Personaje. Lima: Marina de Guerra del Perú, vol 1. Instituto de estudios histórico-marítimos del Perú pp 1–474

Pizarro L (2001) Comportamiento oceanográfico del mar peruano. IMARPE, Lima

Prieto-Rios E (2010) Taxonomía de Holothuroidea (Echinodermata) del mar del Perú. Professional Thesis, Univ Nacional Mayor de San Marcos, Lima

Prieto-Rios E, Valdés-de Anda M, Solís-Marín FA, Laguarda-Figueras A (2011) Primer registro de *Florometra magellanica* (Bell 1882) (Echinodermata: Crinoidea) para el Perú. Rev Peru Biol 18:245–248

Rosenberg R, Arntz WE, Chuman E, Flores LA, Carbajal G, Finger I, Tarazona J (1983) Benthos biomass and oxygen deficiency in the upwelling system off Perú. J Mar Res 41:263–279

Tarazona J, Gutiérrez D, Paredes C, Indacochea A (2003) Overview and challenges of marine biodiversity research in Peru. Gayana 67:206–231

Sandweiss DH, Richardson JBIII, Reitz EJ, Rollins HB, Maasch KA (1996) Geoarchaeological evidence from Peru for a 5000 years B.P. onset of El Niño. Science 273:1531–1533

Spalding MD, Fox HE, Allen GR, Davidson N, Ferdaña ZA, Finlayson M, Halpern BS, Jorge MA, Lombana A, Lourie SA, Martin KD, McManus E, Molnar J, Recchia CA, Robertson J (2007) Marine Ecoregions of the World: a bioregionalization of coastal and shelf areas. Bioscience 57:573–583

Strub PT, Mesías JM, Montesino V, Rutllant J (1998) Coastal ocean circulation off western South America. In: Robinson AR, Brink KH (eds) The Global Coastal Ocean. The Sea, vol. 11. Interscience, New York, pp 273–313

Sueiro JC (2009) Estudio de la actividad extractiva y de comercialización de invertebrados bentónicos en el litoral sur del Perú. ICON-INSTITUT GmbH Private Sector, Proyecto Asistencia Técnica para los Planes Operativos Sectoriales (POS), Lima

Toral-Granda V (2008) Population status, fisheries and trade of sea cucumbers in Latin America and the Caribbean. In: Toral-Granda V, Lovatelli A, Vasconcellos M (eds) Sea cucumbers. A global review of fisheries and trade. FAO Fisheries and Aquaculture Technical Paper. No. 516. Rome, pp 213–229

Toral-Granda V, Martínez P, Hearn A, Vega S (2003) Estado poblacional del pepino de mar (*Isostichopus fuscus*) en la Reserva Marina de Galápagos: Análisis comparativo de los años 1999–2002. Parque Nacional Galápagos, Charles Darwin Foundation

Vergara-Montero E, Sánchez-Vera M (1996) Mitografía mochica. Universidad Nacional de Trujillo, Trujillo

Zuta S, Guillén O (1970) Oceanografía de las Aguas Costas del Perú. Bol Inst Mar Perú Callao 2:157–324

Chapter 9
Echinoderm from Brazil: Historical Research and the Current State of Biodiversity Knowledge

Carlos R. R. Ventura, Michela Borges, Lúcia S. Campos, Leticia V. Costa-Lotufo, Carolina A. Freire, Valéria F. Hadel, Cynthia L. C. Manso, José R. M. C. Silva, Yara Tavares and Cláudio G. Tiago

9.1 Introduction

9.1.1 General Geographic and Oceanographic Features of the Brazilian Coast

Brazil has a long coastal zone which extends from 4°30′N to 33°44′S with a total area of approximately 1,550,000 km^2. It is divided in 16 marine sedimentary basins (Fig. 9.1). One half of this area (about 770,000 km^2) extends up to 400 m

C. R. R. Ventura (✉)
Museu Nacional, Universidade Federal do Rio de Janeiro,
Rio de Janeiro- RJ, 20940-040, Brazil
e-mail: ventura@acd.ufrj.br

M. Borges
Museu de Zoologia, Universidade Estadual de Campinas,
Campinas-SP, 13083-863, Brazil
e-mail: borgesm@unicamp.br

L. S. Campos
Departamento de Zoologia, Instituto de Biologia,
Universidade Federal do Rio de Janeiro, Rio de Janeiro-RJ, 21941-971, Brazil
e-mail: campos-lucia@biologia.ufrj.br

L. V. Costa-Lotufo
Departamento de Fisiologia e Farmacologia,
Universidade Federal do Ceará, Fortaleza-CE, 60430-270, Brazil
e-mail: lvcosta@secrel.com.br

C. A. Freire
Departamento de Fisiologia, Universidade Federal do Paraná,
Curitiba-PR, 81531-990, Brazil
e-mail: cafreire@ufpr.br

J. J. Alvarado and F. A. Solís-Marín (eds.), *Echinoderm Research and Diversity in Latin America*, DOI: 10.1007/978-3-642-20051-9_9,

depth, while the other half (about 780,000 km^2) ranges from depths of 400 to 3,000 m (IBAMA 2008). Along its 8,698 km of shoreline, including all geographic features like indentations (e.g. bays) and capes, it comprises tropical and subtropical environments (IBAMA 2008).

Brazilian's continental shelf is variable in extension. It is larger in the north, ranging from 148 km at the coast of Amapá State to 293.3 km at the mouth of the Amazon River. It narrows in the northeastern region, ranging from 37 to 55 km. From the southeast to the south it expands gradually to 185 km (IBAMA 2008). The Brazilian Exclusive Economic Zone has about 4.2×10^6 km^2 which includes different estuarine and marine ecosystems, such as mangroves, rocky shores, coral reefs, sandy and muddy bottoms in shallow water and deep sea environments. Rocky and calcareous bottoms are common along the northeastern coast while sandy and muddy bottoms prevail along the southeastern and southern coasts (IBAMA 2008).

Two main marine currents flow on the continental margins of Brazil: the North Brazil Current (NBC) and the Brazil Current (BC) (Stramma 1991; Silveira et al. 2000). Both currents are formed at 10°S and are associated with the western border of the South Atlantic Subtropical Gyre (SASG) as two branches of the South Atlantic Equatorial Current (SAEC) (Silveira et al. 2000). The NBC flows northwards while the BC flows southwards along the continental margin up to the Argentine Basin (33° to 38°S), where it meets the Malvinas Current (Brazil-Malvinas Confluence) (Gordon 1988; Silveira et al. 2000). Therefore, the BC prevails on the Brazilian coast.

The first 3 km of the water column of the BC is formed by five water masses: the Tropical Water (TW), the South Atlantic Central Water (SACW), the Antarctic

V. F. Hadel
Centro de Biologia Marinha, Universidade de São Paulo,
São Sebastião-SP, 11600-000, Brazil
e-mail: vafhadel@usp.br

Cynthia L. C. Manso
Laboratório de Invertebrados Marinhos, Universidade Federal de Sergipe,
Itabaiana-SE, 49500-00, Brazil
e-mail: cynthia@phoenix.org.br

J. R. M. C. Silva
Departamento de Biologia Celular e do Desenvolvimento,
Instituto de Ciências Biomédicas, Universidade de São Paulo, São Paulo-SP,
05508-99, Brazil
e-mail: jrmcs@usp.br

Y. Tavares
Faculdade de Ciências, Filosofia e Letras de Paranaguá,
Universidade Estadual do Paraná, Pontal do Paraná-PR, 83203-280, Brazil
e-mail: ytavares@ufpr.br

C. G. Tiago
Centro de Biologia Marinha, Universidade de São Paulo,
São Sebastião-SP, 11600-000, Brazil
e-mail: clgtiago@usp.br

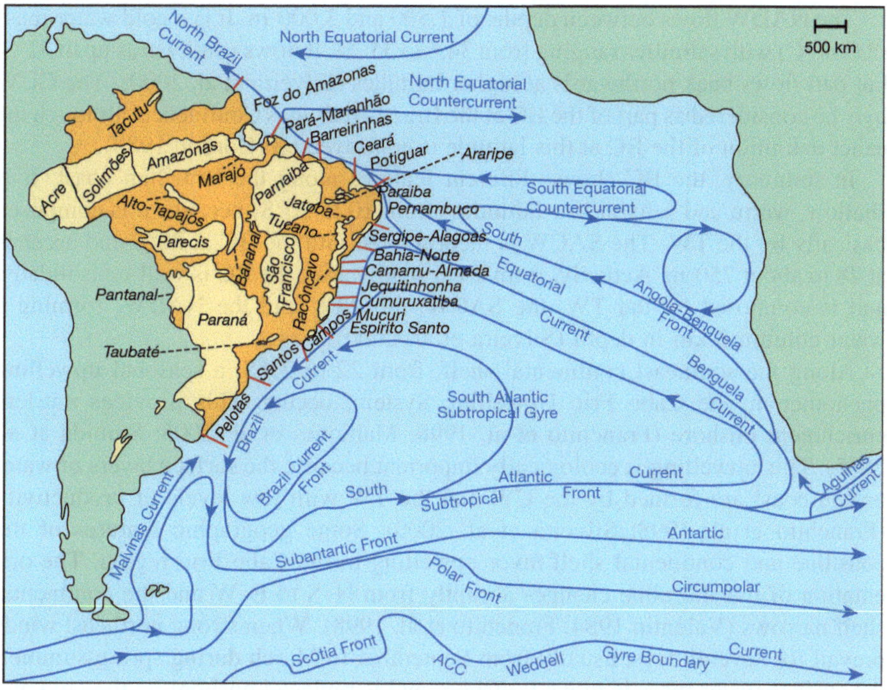

Fig. 9.1 Principal marine currents and Brazilian sedimentary basins in the South Atlantic Ocean

Intermediate Water (AIW), the North Atlantic Deep Water (NADW), and the Circumpolar Upper Water (CUW) (Silveira et al. 2000).

The TW is superficial, warm and salty due to the strong solar radiation and evaporation which characterizes the Tropical Atlantic (Silveira et al. 2000). While the TW is carried to the south by the BC, it is mixed with the Coastal Water (CW) current which is less warm and salty. Consequently, the TW is characterized by temperatures higher than 20 °C and salinities higher than 36 ‰ along the southeastern coast (Silveira et al. 2000).

The SACW is characterized by temperatures between 6 and 20 °C and salinities between 34.6 and 36 ‰. It is formed by the sinking of a water mass at the Atlantic Subtropical Convergence. When it meets the southern border of the SASG, it spreads to the northwest forming the South Atlantic and Benguela Currents. Afterwards it returns to the South America continental margin, carried by the SAEC. Part of the SACW flows to the north towards the equator while the other branch flows southwards below the TW (Silveira et al. 2000).

The AIW flows between depths of 700 and 750 m and is characterized by temperatures ranging from 3 to 6 °C and salinities ranging from 34.2 to 34.5 ‰ (Silveira et al. 2000). The AIW divides into two branches that flow northwards at 28°S and southwards at 25°S (Silveira et al. 2000).

The NADW flows between depths of 1,500 and 3,000 m. It is a cold water mass (3 to 4 °C) with salinities ranging from 34.6 to 35 ‰. It flows southwards up to 32°S, but part flows back northwards at higher latitudes (Silveira et al. 2000). The CUW may be considered as part of the BC at the Brazil-Malvinas Confluence, although the exact definition of the BC at this latitude is controversial (Gordon 1988).

In summary, the BC shows different features along the Brazilian coast. It is shallow, warm and salty at low latitudes (from 10 to 20°S) because it is composed basically by the TW. The SACW is incorporated into the BC at 20°S and ascends at 28 to about 750 m. Actually, from 25°S southwards, the BC is well-consolidated and is composed by the TW, the SACW, the AIW and the NADW, forming a water column 3 km in depth (Silveira et al. 2000).

Along the southeast continental shelf, from 22 to 23°S, a seasonal upwelling phenomenon, the Cabo Frio Upwelling System, occurs. This provides nutrient enrichment onshore (Franchito et al. 1998; Mahiques et al. 2005; Sumida et al. 2005). This upwelling is ecologically important because the surface layers of water on the coast are formed by the CW and the TW with low levels of productivity (Franchito et al. 1998; Silveira et al. 2000). Some geographic features of the coastline and continental shelf favor upwelling in the Cabo Frio region. The orientation of the shoreline changes abruptly from N–S to E–W and the continental shelf narrows (Valentin 1984; Franchito et al. 1998). When strong northeast winds prevail for several days, usually from September to March during spring-summer, the surface water mass is moved offshore and is replaced onshore by the cold and enriched SACW. When cold fronts reach this region from June to August in the winter, the S-SW surface winds blow for a few days and the upwelling cycle is interrupted. Therefore, the seasonality of the upwelling at Cabo Frio area is associated with the movements of the SACW from offshore to onshore and vice versa (Valentin 1984; Franchito et al. 1998).

9.2 Echinoderm Research

9.2.1 Historical Remarks of Echinoderm Research in Brazil

Studies on echinoderms in Brazil have always been characterized by a general approach in which they are considered only as part of the benthic fauna in a given area. The history of such studies can be traced back to 1648 when Marcgrave (1942) reported the first echinoderm species of the Brazilian coast. In his "Brazilian Natural History" the echinoderms were included in the fourth book, which registered the fishes, crustaceans and zoophytes. As zoophytes he listed two sea stars, probably *Luidia senegalensis* (Lamarck, 1816) and *Oreaster reticulatus* (Linnaeus, 1758), a brittle star, probably of the genus *Gorgonocephalus* Leach, 1815, and two sand dollars, both probably *Mellita quinquiesperforata* (Leske, 1778).

The following records were made in the nineteenth century between 1867 and 1898. Among them, three publications must be highlighted: the catalog written by Rathbun in 1879, the report written by Ludwig in 1882, and, 16 years later, the check-list of echinoderms from the southern coast of Brazil (Santa Catarina State) written by Müller in 1898 (Tiago 1998; Hadel et al. 1999; Tiago and Ditadi 2001; Ventura et al. 2006).

The first studies on Brazilian echinoderm fossils were also published in the nineteenth century. In 1878, Derby described a Devonian ophiuroid from the Ponta Grossa Formation (Paraná Basin—southern Brazil), and in 1887, White described 13 species of Cretaceous echinoids from the northeast region, at Sergipe State.

In the twentieth century, paleontological research was extended to other basins: in the middle-west region, at Goiás and Mato Grosso States (Clarke 1913; Löfgreen 1937; Erichsen and Löfgreen 1940; Barbosa 1949; Almeida 1954; Beurlen 1957; Del'Arco et al. 1982; Bosetti and Quadros 1996), and in the north region, in the Amazon Basin (Ferreira and Fernandes 1989).

Taxonomic studies on Brazilian coastal echinoderms were improved after the second half of the twentieth century, especially due to the efforts of Luiz Roberto Tommasi from the Oceanographic Institute of the University of São Paulo (Instituto Oceanográfico da Universidade de São Paulo). Tommasi has studied the five classes of echinoderms morphologically, and provided the first identification keys (Tommasi 1957, 1958a, b, 1960, 1962, 1963, 1964a, b, c, 1965a, b, c, d, 1966a, b, 1967a, b, 1968, 1969a, b, c, 1970a, b, c, d, 1971a, b, c, d, 1972, 1974a, b, 1985, Tommasi and Lima-Verde 1970; Tommasi and Abreu 1974; Tommasi and Oliveira 1976; Tommasi and Aron 1987, 1988; Tommasi and Hecht 1988; Tommasi et al. 1988a, b). Due to his pioneer contributions, his studies were the basis for subsequent taxonomic research in Brazil. In this period, some other studies are also relevant, such as those published by Brito (1959, 1960a, b, c, d, e, f, 1962, 1968, 1971, 1979, 1982), and, afterwards, those by Ávila-Pires (1983); Albuquerque (1986); Monteiro (1987); Monteiro et al. (1992) and Manso (1988a, b, c, d, e, 1989, 1993).

From 1990 onwards there was an increase in echinoderm knowledge, such as studies on fauna (Hadel et al. 1999; Netto et al. 2005; Ventura et al. 2007), community structure (Ventura 1991; Castro et al. 1995; Ventura and Fernandes 1995; Sumida and Pires-Vanin 1997), reproduction and feeding (Ventura et al. 1994, 1997, 1998, 2001; Hadel 1997; Junqueira 1998; Ventura 1998; Guerrazzi 1999, Alves et al. 2002, Carvalho and Ventura 2002, Tavares et al. 2004, Ventura and Barcellos 2004; MacCord and Ventura 2004); development (Hadel 1997; Vellutini and Migotto 2010; Contins and Ventura 2011), growth (Freire et al. 1992), local distribution (Mac Cord and Duarte 2002), environmental education (Hadel and Tiago 2004) and taxonomy (Tiago 1998; Tiago and Ditadi 2001; Borges et al. 2002a; Hopkins et al. 2003).

In 2009, Brazilian researchers gathered for the 1st Brazilian Symposium on Echinoderms (Ventura et al. 2009) and again in 2012 for the second symposium. On these two occasions, information on several current and ongoing research was presented, as well as needs and suggestions to improve the knowledge of Brazilian echinoderms.

In short, research on echinoderms in Brazil was developed on several subjects, mostly on species of the classes Asteroidea, Echinoidea, Ophiuroidea and Holothuroidea. Few studies have been made on crinoids, probably due to their low diversity along the Brazilian coast.

9.2.2 Echinoderm Paleontological Research

Brazilian echinoderms are found from the Palaeozoic to the Tertiary basins. Ten species of Crinoidea, two of Stylophora, three of Blastoidea, 63 of Echinoidea, three of Asteroidea, one of Ophiuroidea and two of Holothuroidea have been recorded (Tommasi 1999; Scheffler 2007). There are difficulties in increasing knowledge of extinct echinoderms, particularly due to the large distances among the basins and the few paleontologists devoted to the study of echinoderm fossils. Besides, very little is known of the origins and geological history of macro and megafaunal biodiversity in the deep-sea, including echinoderms (Campos et al. 2010). Even though it is recognized today that species richness of deep-sea is comparable to, or even exceeds, that of continental shelves, the difficulties in finding identifiable macro-fossils in deep-sea cores limits any assessment of the geological history of specific benthic groups, including echinoderms (Gray et al. 1997, Smith and Stockley 2005). Smith and Stockley (2005) combined evidence from echinoid phylogenetic trees and molecular studies and suggested deep-sea nutrient enhancement over a long geological time has had a key role in the development of its biodiversity.

According to Scheffler (2007), research on Devonian echinoderms (Ponta Grossa Formation, Paraná Basin) started at the end of the nineteenth century with the studies published by Derby (1878) on the Ophiuroidea, and the monograph of Clarke (1913) on the Asteroidea. The first references on Crinoidea were made by Löfgreen (1937) with findings from Goiás State, and by Erichsen and Löfgreen (1940) with material from Mato Grosso State. Lange (1943) identified the first crinoid from Paraná Basin while Petri (1948a, b) described 18 levels of the crinoids and cystoids from Ponta Grossa Formation in the same Basin. Caster (1954a, b) described two new species of Stylophora. Crinoid fragments were recorded by Barbosa (1949), Almeida (1954), Del'Arco et al. (1982) and Bosetti and Quadros (1996) in Mato Grosso State, and Beurlen (1957) in Goiás State. Ferreira and Fernandes (1985) reported the crinoids from the Devonian of Brazil and Ferreira and Fernandes (1989) recorded the crinoid species *Erisocrinus* sp. from the Carboniferous and the genera *Ctenocrinus*, *Exaesyodiscus*, *Laudonomphalus* and *Monstrocrinus* from the Devonian of the Amazon Basin.

In the twenty-first century studies on the Devonian echinoderms were intensified with new descriptions and taxonomic reviews such as those published by Alves and Fernandes (2001) about *Encrinaster pontis*, and Scheffler et al. (2002), who described parts of some crinoid. In addition, several papers describing the crinoids and blastoids from the Ponta Grossa Formation were published by Scheffler (2003, 2004, 2006) and Scheffler and Fernandes (2003, 2007a, b). Descriptions of crinoid species from the Maecuru Formation (Devonian of the Amazon Basin) were presented by Scheffler et al. (2006).

The Cretaceous echinoderms from Brazil were identified for the first time by White (1887) with the description of 13 echinoid species collected by the Geologic Commission of the Empire of Brazil from the Cretaceous of the Sergipe-Alagoas Basin. Derby (1907) reported a little echinoid near the Maraú Peninsula, in the Camamu Basin, Bahia State. In the 1920s and 1930s, Maury (1925, 1930, 1934a, b, 1937) studied the basins from the northeast of Brazil. She described echinoids from the Sergipe-Alagoas, Potiguar and Pernambuco-Paraíba Basins, and the asteroid *Uraster janettae* from the Sergipe-Alagoas Basin. She also revised the paper published by White in 1887 adding two new species. Oliveira (1957) described a new species of *Hemiaster* from Pernambuco-Paraíba Basin.

Few decades later, some authors studied the echinoid fauna from the sedimentary basins of the northeast and north of Brazil from the 1950s to the 1980s, e.g. Santos (1958a, b) who recorded one species of *Salenia* sp. in the Camamu Basin; Santos and Cunha (1959) Santos (1960a, b) who found new species in the Potiguar Basin; Beurlen (1961, 1964a, b, 1966), who made systematic reviews, identified two new echinoids species from the Araripe Basin and a new echinoid species from the Pernambuco-Paraíba Basin. Tinoco (1963) reported fragments of ophiuroids, echinoids, asteroids and holothuroids from Maria Farinha Formation, Tertiary from Pernambuco-Paraíba Basin. Brito (1964, 1980, 1981a, b, c) gave descriptions of the echinoid fauna from the Cretaceous and the Tertiary.

Smith and Bengtson (1991), who published a review with 39 species of echinoids from the Sergipe-Alagoas Basin, made some observations about echinoids from the other Cretaceous Basins of northeastern Brazil. They described two new genera (*Colliclypeus* and *Acriaster*) and a new species (*Acriaster sergipensis*) from the Sergipe-Alagoas Basin. They also concluded that the Brazilian echinoid fauna has great affinities with the West African fauna and a connection with the Texan and Mexican fauna.

Ferré et al. (1996) described species of the roveocrinids crinoids from the Turonian (Cretaceous) of the Sergipe-Alagoas Basin, and Ferré and Bengtson (1997) recorded *Roveocrinus spinosus* from the Turonian for the same basin. According to Morais (1998), Kraatz-Korschlau was the first paleontologist to publish, in 1900, the presence of the echinoid fauna from the Oligo-Miocene of the Pirabas Formation, in the north of Pará State. Carvalho (1926) pointed out similarities between echinoid fauna of Fortaleza Island and Bragança County, Pará State. Santos (1958a, b) recognized an endemic fauna in the Oligo-Miocene of the Pirabas Formation. This fauna consisted of clyperasteroids and cassidulids, and great similarities were found with the Oligo-Miocene Caribbean fauna, suggesting echinoid dispersion to northern Brazil before the Miocene. Santos (1957) was the first to describe cidaroids and spatangoids from the Pirabas Formation and commented on the similarity of the cidarids with the genera from Indo-Pacific Province.

Brito (1979) studied the Brazilian Cenozoic clyperasteroids and described a new species from the Pirabas Formation, *Clypeaster paraensis*. Brito (1981d) reported a new species of clyperasteroid, *Abertella complanata*, from the Pirabas Formation and suggested the presence of *Cidaris antarctica* (Ortmann, 1900). Brito (1986) redescribed *Abertella complanata* in greater detail. Fernandes and Morais (1994) reported *Rhyncholampas candidoi* from the Pirabas Formation in Salinópolis, Pará State. Martinez and Mooi (1997) transfered *Karlaster pirabensis* to the genus *Abertella*.

Morais and Fernandes (1997) recorded for the first time a regular echinoid in Brazilian Cenozoic rocks. Morais (1998) studied the taxonomy and paleoecology of regular echinoids from Pirabas Formation.

Collections in the twenty first century resulted in several publications on taxonomy, paleoecology and paleobiogeography of the echinoderms fauna from the Cretaceous and Tertiary Brazilian Basins (Cooke 1955; Seeling et al. 2000; Manso 2001, 2003a, b, 2006; Souza-Lima and Manso 2000; Manso and Souza-Lima 2003a, b, c, 2005, 2007, 2010; Rodovalho et al. 2003; Manso and Hessel 2007; Manso and Andrade 2008; Manso and Lemos 2008). All this fossil material was deposited at the Phoenix Paleontological Foundation (Fundação Paleontológica Phoenix) in Sergipe State.

9.2.3 Echinoderm Physiological Research

Echinoderm physiology has been the focus of intensive research in the last century. However, interest in this area of research has decreased significantly in the last decades.

The first studies on the effect of acetylcholine on holothuroid muscles were conducted at the Oswaldo Cruz Institute (Rio de Janeiro State) by Moussatché in 1949. Classical studies conducted at the University of São Paulo focused on muscle biochemistry in holothuroids (Sawaya 1951; Ancona-Lopez 1964) and echinoids (Sawaya and Ancona-Lopez 1964; Bianconcini et al. 1982, 1985), and on gas exchange in ophiuroids (Petersen 1976). In the last decade, the physiology of excretion and ionic regulation has been investigated at the Federal University of Paraná (Vidolin et al. 2002, 2007; Santos-Gouvea and Freire 2007). The main focus of this research has been on the interspecific differences in the homeostasis of ions of the perivisceral coelomic fluid of three echinoids (*Echinometra lucunter*, *Lytechinus variegatus* and *Arbacia lixula*), one holothuroid (*Holothuria grisea*) and one asteroid (*Asterina stellifera*). Another target of research has been the investigation of how echinoderms deal with the different ions. The current evidence that echinoderm epithelia treat different ions in distinct ways led to the conclusion that they are not very homogeneous and identical in their osmo-ionic-conforming strategies. These studies pointed out that *E. lucunter* shows reasonable euryhanility because it can tolerate at least a 30 % reduction in seawater salinity for 5 days; however it is not able to maintain large ion gradient, especially of Na^+ and Cl^-, with respect to external seawater (Freire et al. 2011). On the other hand, *L. variegatus* is able to take up Na^+ by active transport, showing selective permeability or ion regulation, somewhat different from *A. lixula*. Therefore, *L. variegatus* was termed as "hyper-conformer" and *A. lixula* as "iso-conformer" (Vidolin et al. 2007). In situ experiments with the holothuroid *H. grisea* and the asteroid *A. stellifera* indicated that these intertidal species may have a mechanism to detect exposure to air, possibly reducing body wall permeability to avoid intense water loss (Vidolin et al. 2002).

Studies on the immunological response of sea urchins have been conducted at the University of São Paulo (e.g., Silva et al. 2001; Shimada-Borges et al. 2005), described later in this chapter. In addition, research on the ecology and biology of echinoderms yielded many contributions on the function of reproduction and development (e.g., Alves et al. 2002; Carvalho and Ventura 2002; MacCord and Ventura 2004; Tavares and Borzone 2006; Cinelli et al. 2007; Lima et al. 2009; Vellutini and Migotto 2010; Contins and Ventura 2011). For instance, there is a significant imbalance in sex ratio in a population of the fissiparous sea star *Coscinasterias tenuispina* where no females have been found (Alves et al. 2002). Although fission occurred throughout the year, it was most frequent in the winter (July, August and September) when the extremely low tides occur during the daytime.

Echinoderms have also been studied in Brazil in the sister-area of physiology, biochemistry, and are also common subjects of toxicological studies in environmental agencies. The biochemistry of sulfated oligo- and polysaccharides from echinoderms with anticoagulant and antithrombotic activities (Mourão et al. 1998) and the reproductive features of these organic substances (e.g., Cinelli et al. 2007, 2010) have been the subject of intensive research for more than two decades at the Federal University of Rio de Janeiro. The antithrombotic activities of a polysaccharide (a fucosylate, chondroitin-sulfate-like polysaccharide) extracted from the body wall of the holothuroid *Holothuria grisea* and its chemically modified derivatives were tested using a static thrombosis model in rabbits. The native organic extract has an effective antithrombotic action. It prevents thrombosis completely at a dose of 1.5 mg kg^{-1} after 10 min stasis, and also, in a dose-dependent manner, it reduces thrombosis. The anticoagulant or antithrombotic effects are suppressed by partial defucosylation and desulphation of the polysaccharide (Mourão et al. 1998). The dynamic of polysaccharide sulfated presented in the jelly egg coats of the sea urchin *E. lucunter* shows that: (1) a great amount of sulfated polysaccharide is present in mature oocytes; (2) a disulfated galactans (2, 6- disulfated galactan) is produced in the ovary firstly and (3) the 6-sulfate ester is removed when the polysaccharide concentration increases in the egg jelly coat. In conclusion, the ovary of *E. lucunter* synthesizes a specific sulfated galactan isotype composed of two subunits (2, 6-disulfated units), but its egg jelly coat has only 2-sulfated residues (Cinelli et al. 2010). These compounds (i.e. sulfated polysaccharides) are important due to their species specificity in inducing the acrosome reaction in the fertilization process.

In conclusion, few investigators are devoted to the study of echinoderm physiology in Brazil. This is a field that should receive more attention in the future. Physiological knowledge of echinoderm species, both adults and larvae, is essential for any initiative for the conservation of these invertebrates in a global changing world.

9.2.4 Echinoderm Immunological Research

Brazilian research with echinoids includes physiological and innate immune response of the tropical sea urchins *L. variegatus* and *E. lucunter*, as well as the Antarctic *Sterechinus neumayeri*. General reviews of the innate immune response in Brazilian echinoderms can be found in Silva et al. (2007) and Tajima et al. (2007).

The stage at which phagocytosis can first be characterized in the embryos of the sea urchin *L. variegatus* was investigated by microinjection of the yeast *Saccharomyces cerevisiae* into its blastocoele. The starting point of phagocytic activity reflects a biological capacity for distinguishing between self and non-self. Thus the phagocytosis of yeast by mesenchymal cells beginning in the mid-gastrula stage in *L. variegatus* may indicate the moment of acquisition of "identity" (self) in this organism. Comparative aspects of embryo and adult phagocytes in *L. variegatus* have been discussed (Silva 2000). A review on the role of the phagocytes on embryos was carried out focusing on the following topics: (1) the search for the origin and first identification of well known adult phagocytes in embryos, including their role after induced injuries; (2) the search for the occurrence of phagocytosis in embryos and its role during their physiological development; (3) the search for phagocytosis in embryos, as a tool to study identity and self recognition. It is possible to verify that different cell types are able to undertake phagocytosis under a variety of different stimuli and that the nature of phagocytosis also varies widely (Silva 2002).

Induced inflammatory processes in the sea urchins *L. variegatus* and *E. lucunter* were demonstrated in vitro and in vivo studies using either yeast or India ink confirming endocytic activity of free ferritin labeled perivisceral phagocytic amoebocytes (Mangiaterra and Silva 2001; Faria and Silva 2008).

In recent years, studies on effects of oil contamination on sea urchins have been conducted on Antarctic and tropical species. They showed that crystalloids can be biomarkers for the analyses of low levels of water-soluble fractions of oil (WSF). The Antarctic sea urchin *S. neumayeri* from regions free of pollution was exposed for 2, 5, 10 and 15 days to different levels of WSF. This study characterized histological alterations in crystalloids of *S. neumayeri* that could be used as biomarkers for oil contaminants with a simple and inexpensive protocol (Borges et al. 2010).

Global warming is a reality and its effects have been evaluated on the effect of temperature increase on the innate immune system of the Antarctic sea urchin

S. neumayeri. This study demonstrated the use of red spherule cells as biomarkers for heat stress because they increase in numbers under acute stress. In addition, a significant increase in phagocytic indexes was observed at 2 °C coinciding with a significant increase of iron intranuclear crystalloids at the same temperature. Moreover, alterations in cell adhesion and spreading were observed. Hsp70 expression was not found in this study and ultrastructural analysis of coelomocytes did not show any significant difference. This was the first study of the innate immune response alterations with an increase in temperature (Branco et al. 2011).

Some investigations compared Antarctic echinoderms with tropical species. One of these studies determined the oxygen consumption in two species of sea urchins, the tropical *L. variegatus* and the polar *S. neumayeri*. In sealed respirometry *L. variegatus* had a higher consumption of oxygen than *S. neumayeri*. This is probably due to diet and temperature. The relation of O:N between *S. neumayeri* and *L. variegatus* was the same, showing that the substrate metabolized of the two animals is similar (Borges et al. 2012).

Phagocytosis was studied in vitro at 0 °C using the coelomic fluid of the Antarctic sea star *Odontaster validus* and the types of amoebocytes involved were identified. This was the first study of this kind on an Antarctic invertebrate. Rates of phagocytosis were similar to, or higher than, previously reported for temperate sea stars. However, this conclusion must be considered with caution because of the scarcity of data and the different methods used. On the other hand, the data do suggest that one of the animals' main defenses against infection may be well adapted to low temperatures in *O. validus*. The phagocytosis and germicide capacity of *S. cerevisiae* by phagocytic amoebocytes (PA) of the Antarctic sea star *O. validus* was also studied in vivo at 0 °C (Silva and Peck 2000; Silva et al. 2001).

9.2.5 Echinoderm Ecotoxicological Research

Bioassays are a valuable approach to evaluate and predict harmful effects of pollutants on the biota at different levels of organization (Environmental Canada 1992, 2011; Nascimento et al. 2002; Losso and Volpi-Ghirardini 2010). In this context, echinoderms have a great importance on ecotoxicological and environmental risk assessment studies.

Toxicity tests for environmental quality analyses in Brazil began in the 1970s with the standardization by the São Paulo State Environmental Agency (Companhia Ambiental do Estado de São Paulo—CETESB) of tests using freshwater organisms (Resgalla and Laitano 2002; Zagatto 2006). Afterwards, standard protocols for marine organisms were published (CETESB 1991, 1992, 1999; Resgalla and Laitano 2002; Zagatto 2006). Since then, toxicity tests with sea urchin embryos have been used extensively in environmental pollution monitoring (Prósperi 1993, 2002; Mastroti 1997; Abessa et al. 2003; Nilin et al. 2007).

Among the most common sea urchin species of the Brazilian coast are *E. lucunter* and *L. variegatus*. Both are commonly used in ecotoxicological assays (Prósperi 1993, 2002; Mastroti 1997; Nascimento et al. 2002; Prósperi and Araújo 2002; Abessa et al. 2003; Nilin et al. 2007; Perina et al. 2011). For instance, Perina et al. (2011) showed that embryotoxicity experiments using *L. variegatus* have shown different effects of two antifouling agents (tributyltin or TBT, and triphenyltin or TPT) and two booster biocides (Irgarol and Diuron). Embryonic-larval effects of these two antifouling agents were quite different, i.e. development was interrupted at the blastula and gastrula stages at two TBT concentrations (1.25 and 2.5 $\mu g\ l^{-1}$, respectively) while the pluteus stage was reached at the same concentrations of TPT. Different ecotoxicological effects were found depending on concentrations of the two biocides: higher concentrations of Irgarol interrupted embryonic development at the blastula-gastrula stages while the pluteus stage was reached under exposure of Diuron at same concentrations. Other studies showed that embryonic-larval development of *E. lucunter* and *L. variegatus* are affected by chronic toxicity of a chemical mixture of light-stick flags (Cesar-Ribeiro and Palanch-Hans 2010) and that the chronic toxicity of the sediment in the Guanabara Bay (Rio de Janeiro) affects the development of *L. variegatus* (Maranho et al. 2010). Short term chronic toxicity tests with embryos of those species were standardized by CETESB (1999) and more recently by the Brazilian Association for Technical Standards (Associação Brasileira de Normas Técnicas—ABNT) (ABNT 2006, 2012).

Acute and chronic toxicity assays employing aquatic organisms at different life stages have been extensively used for environmental monitoring programs. Gametogenesis and embryonic development are some of the aspects that can be measured. They deserve special attention due to the effects they can cause at the population and community levels (Cherr et al. 1992, 1993; Kroeker et al. 2010; Stumpp et al. 2011).

The ecological relevance of echinoderm communities and the vast knowledge on their embryology make these animals the best choice for laboratory and field studies on environmental hazards (Bowmer et al. 1986; Briggs and Wessel 2006; Stumpp et al. 2011). Several species of Echinoidea, Ophiuroidea and Asteroidea have been used, especially sea urchins, which account for the higher number of species and assays described and validated by environmental regulatory agencies (Environmental Canada 1992, 2011; EPA 1995; ABNT 2006, 2012; Nilin et al. 2008). Among the most studied sea urchin species are *Paracentrotus lividus*, *A. lixula*, *L. variegatus*, *E. lucunter* and *Strongylocentrotus purpuratus*.

The use of sea urchin embryos in short-chronic toxicity assays for environmental monitoring and toxicity assessment has been very significant all over the world. More recently this model proved to be suitable also for the evaluation of the biological effects of global warming and ocean acidification (Clark et al. 2009; Dupont et al. 2010; Kroeker et al. 2010; Stumpp et al. 2011). Investigations of the effects of global warming on development of echinoderms with different developmental strategies (i.e. planktothrophic and lecitothrophic development) are currently starting in Brazil (C.R.R. Ventura, pers. comm.). Sea-urchin embryos have been used also for the evaluation of cytotoxicity, genotoxicity, teratogenicity and antimitotic activity of natural and synthetic compounds and on the development of new medicines and pharmacological tools (Kobayashi 1973; Fusetani 1987; Zúñiga et al. 1995; Saotome et al. 1999; Costa-Lotufo et al. 2002; Hansen et al. 2003; Cummings and Kavlock 2004). For instance, the potential cytotoxicity of a diterpene (kaurenoic acid), a natural compound isolated from the wood of a tropical forest tree (i.e. oleo-resin of *Copaifera langsdorffii*), shows nonspecific cytotoxicity. This kaurenoic acid disrupts the embryonic-larval development of *L. variegatus*, inhibits tumor cell growth in mammals and causes hemolysis of mouse and human erythrocytes (Costa-Lotufo et al. 2002).

Many factors contribute for the increasing use of sea urchin embryos as test-organisms, including their worldwide distribution, their key role on intertidal and subtidal communities and the ease with which adults of many species can be collected in their natural habitat. It is also easy to obtain large amounts of ova and sperm for in vitro fertilization, and the embryonic development is rapid and highly successful (Nilin el al. 2008). However, for environmental quality assessment there are some additional factors that have to be considered, such as salinity, ammonia and sulfite concentrations, pH and temperature of the water (Nipper 2000; Carr et al. 2006).

9.2.6 Brazilian Echinoderm Deep-Sea Research

Echinoderms live in a wide variety of marine habitats, exert a considerable influence on benthic communities, especially in the deep-sea where they may represent more than 80 % of the total megafaunal biomass in certain areas, playing an important role in complex food webs (Haedrich et al. 1980; Sibuet and Lawrence 1981; Smith and Hamilton 1983; Gage et al. 1983, 1985; Gooday and Turley 1990; Amon and Herndl 1991; Gage and Tyler 1991; Campos et al. 1994; Hendler et al. 1995; Borges et al. 2002a; Campos et al. 2010).

During the past 23 years, substantial deep-sea research has been carried out in Brazil (Tavares 1999; Lavrado 2006; Campos et al. 2010). The first records of the Brazilian bathyal and abyssal fauna were compiled at the end of the nineteenth century as a result of the *Hassler*, *Challenger* and *Albatross* expeditions (Tavares 1999). In 1987, the French-Brazilian Marion Dufresne (MD55) expedition sampled in the Vitoria-Trindade seamounts region down to a depth of 5,500 m in order to investigate the faunistic transition from Cabo Frio (where the sub-Antarctic SACW upwelling occurs) to the Abrolhos slope (Guille and Ramos 1987; Tavares 1999). Several echinoderm species, especially ophiuroids, sea stars and sea cucumbers, have been sampled during the MD55 campaign (Guille and Ramos 1987).

Other efforts provided valuable deep-sea material from the Brazilian continental margin towards the end of the twentieth century. The INTEGRADO Project sampled from 60 to 600 m depth at São Paulo State coast, and later the Evaluation

Program of the Sustainable Yield of Living Resources from the Brazilian Exclusive Economic Zone (Programa de Avaliação do Potencial Sustentável de Recursos Vivos na Zona Econômica Exclusiva—REVIZEE) sampled down to 800 m in the South region and deeper than 1,000 m at the Southeast one (Pires-Vanin 1993; Lana et al. 1996; Sumida and Pires-Vanin 1997; Amaral et al. 2004; Ventura et al. 2006). Diversity of echinoderm fauna from the Brazilian Southeastern coast is higher between depths of 20 and 100 m and decreases gradually up to 400 m. However, higher frequencies of species are found between depths of 400 and 600 m and diversity increases to similar values at depths of 100–200 m (Ventura et al. 2006). Borges et al. (2002b, 2004a, b) explored ophiuroid diversity and Ventura et al. (2006) expanded the list to other echinoderm classes found at depths down to 2,000 m. Most biodiversity data originated from the South and Southeast upper slope, some stations from the lower slope and a few from the abyss, although a few records were also established in the Northeast (Sumida and Pires-Vanin 1997; Amaral et al. 2004; Amaral and Rossi-Wongtschowski 2004; Lavrado 2006; Lavrado and Brasil 2010; Lavrado et al. 2010a, b; Amaral and Jablonsky 2005; Capítoli and Bemvenuti 2006; Capítoli and Bonilha 2010).

Studies on the effect of anthropogenic activities at the Brazilian continental margin have enhanced our knowledge on benthic communities, especially at Campos Basin down to nearly 3,000 m (Santos et al. 2009; Campos et al. 2010; Lavrado et al. 2010a, b). Echinoderms have been one of the most abundant invertebrate groups sampled during the megabenthic campaigns at Campos Basin (Rio de Janeiro State), representing approximately 45 % of the total megafauna (Lavrado et al. 2010a, b). During February (summer) and August (winter) 2003, 4,392 specimens were collected by trawling. The Ophiuroidea were the most abundant amongst the echinoderms (n = 3,281) followed by the Asteroidea (n = 936) and Holothuroidea (n = 158). The Echinoidea were very rare (n = 17), and no Crinoidea were collected. The abundance of ophiuroids and asteroids was higher at a depth of 1,100 m, and had the tendency to decrease with increasing depth. But this essentially happened because of the presence of a few dominant species from both classes [Ophiuroidea (*Ophiura ljungmani*, *Ophiophycis* sp.1) and Asteroidea (*Plutonaster bifrons* and *Nymphaster arenatus*)]. The higher expected species values were found at a depth of 1,300 m.

According to Campos et al. (2010), 39 species (Appendix) were identified at Campos Basin (CB) during the Campos Basin Deep-Sea Environmental Project/ PETROBRAS. (Caracterização Ambiental de Águas Profundas a Bacia de Campos/CENPES/PETROBRAS). Amongst the ophiuroids, the species *Asteroschema* (*A. arenosum*), *Histampica* (*H. duplicata*), *Ophiuraster* (*Ophiuraster* sp.), *Ophiosphalma* (*O. armigerum*), *Ophiacantha* (*O. aristata*, *O. metallacta* and *O. anomala*) were new records for Brazilian waters. Eight asteroid species from five families (Astropectinidae, Benthopectinidae, Goniasteridae, Zoroasteridae and Brisingidae) were identified: *Astropecten* sp. 1; *Plutonaster bifrons*; *Cheiraster (Cheiraster) sepitus*; *Ceramaster* sp. 1; *Nymphaster arenatus*; *Pseudarchaster gracilis gracilis*; *Zoroaster fulgens*; and *Brisinga* sp. 1. *Pseudarchaster gracilis gracilis* and *Cheiraster (Cheiraster) sepitus* were new records for Brazil (Campos et al. 2010). Sea cucumbers represented 3.6 % of the total echinoderms collected. Most were identified to species level, except for a morpho-species classified as Synallactidae morphotype 1. The most abundant species were *Mesothuria verrilli*, *Molpadia blakei*, *Deima validum validum* and *Molpadiodemas porphyrus*.

Barboza et al. (2011) used metadata from several echinoderm studies and data from the Brazilian continental margin in order to evaluate the potential relationship between the faunas from South America and the Antarctic region, taking into account the proximity of both continents and the fact that they were the last to separated from Gondwana around 34 million years ago. These authors recorded 602 echinoderm species from both regions, 101 of them (~17 %) were shared between the continents. Around 47 % of these are typically from the deep-sea, most deep-sea records being from Brazil.

More recently, other wide spread programs have been carried out by the energy companies in different regions off the Brazilian continental margin. Many new echinoderm records are expected in the near future. Taking into account the length of the Brazilean coastline (~8,000 km), and the fact that around 70 % of its exclusive economic zone, defined between 19.3 km (12 NM) and 321.8 km (200 NM) off the coast, is within the slope and abyss zones (Miloslavich et al. 2011), much work is still needed in order to enhance our knowledge on Brazilian deep sea echinoderms.

9.2.7 Ecology and reproduction

Ecological studies of echinoderms from the Brazilian shallow waters concern spatial distribution patterns (Borzone 1991; Ventura and Fernandes 1995; Sanchez-Jérez et al. 2001; MacCord and Duarte 2002; Capítoli and Benvenutti 2006), population size structure (Ventura and Fernandes 1995; Junqueira et al. 1997), feeding biology (Lima-Verde and Matthews 1969; Ventura et al. 1994; 2001; Calil et al. 2009) and ecological association (Ribeiro et al. 2003; Clavico et al. 2006).

MacCord and Duarte (2002) described the dispersion pattern and the size structure of the crinoid *Tropiometra carinata* along the southeastern coast of Brazil (São Sebastião Channel, São Paulo). This species shows a clumped distribution (especially for smaller individuals), with a preference for attachment to the lateral surface of rocks. Adults are predominant in rocky environment and recruits among algae. The size structure of the population was stable throughout one year.

Regarding sea stars, Ventura and Fernandes (1995) studied the bathymetrical distribution and population size structure of paxillosid species in the upwelling environment of the continental shelf of Cabo Frio. Five species were found but three were mostly frequent for three years: *Astropecten cingulatus*, *A. brasiliensis* and *Luidia ludwigi scotti*. Their bathymetric distribution in this soft-bottom environment is very distinct, according to the grain size of the sediments: *A. brasiliensis* occurs at a depth of 30 m (from very well to well-sorted medium sand) and a depth of 45 m (from well to moderated-well sorted medium sand);

A. cingulatus and *L. ludwigi scotti* are most frequent at a depth of 60 m (from well to moderated-well sorted fine sand) and also at a depth of 45 m. Analysis of population size structure of these three species showed that individuals of the same size range did not occur at the same depth, especially for *Astropecten* species.

Studies on feeding biology are more frequent for paxillosid sea stars, particularly the genera *Astropecten* and *Luidia* because these species show intra-oral digestion and the stomach content analysis provides reliable information about their diet. In general, studies have shown that species of *Astropecten* (*A. cingulatus*, *A. marginatus* and *A. brasiliensis*) and *Luidia* (*L. senegalensisi* and *L. ludwigi scotti*) play a wide range of trophic roles in soft-bottom environments along the Brazilian coast. They can be non-selective deposit feeders (feeding on organic matter from sediments), non-selective predators (feeding on the most frequent prey in the environment, generally mollusks and crustaceans) and also selective predators (showing preferences for some species of prey, especially when there are competitors like other sea stars). Ventura et al. (2001) reported that *A. brasiliensis* and *A. cingulatus* can shift their food niche breadth when they occur together. Also specimens of both species of same size range have broader food niche at depths where they do not occur together.

Calil et al. (2009) carried out field experiments to test the effectiveness of predation of the sea star *Asterina stellifera* in a rocky-shore environment in the south of Brazil. They increased the density of *A. stellifera* in order to evaluate its predation pressure on the community. Their results showed that this starfish has a small effect on the barnacle population (*Chthamalus bisinuatus*) and has a minimal effect on this rocky-shore community of a subtropical region, in contrast to other asteroid species.

Studies on reproduction of echinoderms, especially of common sea urchins and sea stars species from Brazilian shallow waters, have been carried out. Regarding echinoids, in general, continuous gametogenesis occurs throughout the year, but in some cases a long spawning period of three or four months has been detected. That is the case for *E. lucunter,* whose spawning time occurs from December to June in populations from northeastern, southeastern and southern coasts (Tavares et al. 2004; Ventura et al. 2004; Lima et al. 2009).

9.3 Echinoderm biodiversity

9.3.1 Distribution Patterns and Biogeographic Remarks

Echinoderm fauna in Brazilian waters is composed essentially of species with three principal distribution patterns: 1) species ranging from North–South Carolina (USA) or from Caribbean to Argentine; 2) species ranging from North–South Carolina or Florida (USA) to Cabo Frio or São Paulo (Southeastern Brazilian coast); and 3) species ranging from Cabo Frio to the Rio de la Plata mouth (Uruguay/Argentine). However, very few studies on biogeography of echinoderms have been done in Brazil, and even in South America or worldwide (Ghiold and Hoffman 1986; Price et al. 1999; O'Hara and Poore 2000; Martinez 2008). This is probably due to the relatively low number of researchers dedicated to this kind of study, particularly in Brazil, and lack of reliable information on taxonomy and distribution ranges. Although information on the occurrence and distribution of echinoderm species in Brazil still seems to be insufficient regarding the great length of the country's coastline. New records and new species have been described recently as emphasized later in this chapter.

These three patterns of distribution are much correlated with the hydrographical features (i.e. marine currents and water masses) of the Brazilian coast, as described above (see Introduction). Changes in species composition occur markedly at the Cabo Frio region, where upwelling often brings sub-Antarctic water (the South Atlantic Central Water, SAWC) to shallow depths. Therefore, there are features of a temperate environment at a such usually tropical latitude (22°S), which could explain the distribution pattern of some species from Cabo Frio to Uruguay (Rio de la Plata) and the occurrence of tropical and subtropical fauna at Cabo Frio, such as the sea stars *Astropecten cingulatus* (tropical), *Astropecten brasiliensis* (subtropical), *Echinaster* (*Othilia*) *guyanensis* (tropical), *Echinaster* (*Othilia*) *brasiliensis* (subtropical), *Asterina stellifera* (subtropical) and the sea urchin *Paracentrotus gaimardi* (subtropical).

Historically, the South Atlantic Ocean was formed during the Cretaceous (ca. 135 My) when South America was separating from Africa and Antarctica/Australia (the so-called Gondwana paleo-continent). Therefore, species composition on the Brazilian continental shelf should be similar to those at the margins of Africa and Antarctica/Australia and different from the marine fauna of north of the South America ("Tethyan"), because there is no marine connection between the south and the north regions of the Atlantic (Floeter et al. 2009). South America and Africa became effectively separated by a deep-sea barrier during the Paleocene (ca. 60 My). About the same time, a connection between the marine biotas of Brazil and the Caribbean was formed. During the Eocene (ca. 50 My), Australia and Antarctic became separated and also the latter and South America. Each step of the history of South American west coast might mark the evolution of the Brazilian marine biota, at least in shallow water (Floeter et al. 2009). In general, this seems to be true for echinoderms and other marine organisms because the three patterns of species distribution are similar for different taxa. Furthermore, the echinoderm fauna of Brazilian waters is currently very similar to those of the Caribbean, as for other groups of marine invertebrates. This similarity may also be due to the effects of relatively recent dispersal capacity and colonization, as well as ancient extinctions of species at the Brazilian continental shelf.

Comparing the number of species described worldwide and for Brazilian waters, one can notice that there are not as many species of Brazilian echinoderms as expected from the great length of the country's coastline. This is probably due to the low number of researchers dedicated to the study of echinoderms in Brazil. Therefore, information on the occurrence and distribution of echinoderm species in Brazil is insufficient.

There are about 700 extant species of Crinoidea, of which only 16 have been reported for Brazilian waters. Of the 1,800 extant species of Asteroidea, only 77 reported for Brazil. Of the 2,000 extant species of Ophiuroidea, 153 have been reported for the Brazilian coast. Of the 900 extant species of Echinoidea, 52 have been reported for the Brazilian coast. Of the approximately 1,250 extant species of Holothuroidea, 49 reported for Brazil.

9.3.2 Crinoidea, Asteroidea and Echinoidea

The remarkable low diversity of echinoderms of Brazilian waters may be due to the low number of researchers who have been dedicated to the study of the phylum. Considering the great length of the country's coastline, information on the occurrence and distribution of extant echinoderm species in Brazil seems to be insufficient, especially for crinoids, asteroids and echinoids. Little information about earlier studies on crinoids, asteroids and echinoids are available. However, some information about ophiuroids and holothuroids are emphasized on the following topics.

9.3.3 Ophiuroidea

The first report about ophiuroids in Brazil appeared in the 17th century, when Marcgrave (1942) recorded a brittle star, probably from the genus *Gorgonocephalus* Leach, 1815. Ljungman (1867) established the genus *Amphipholis* based on the species *Amphipholis januarii* from the coast of Rio de Janeiro State. Several other studies were carried out in Brazil, mainly on taxonomy and geographical distribution (Brito 1960e; Tommasi 1965a, b, 1967, 1970b, c, d, 1971a, b, c, d; Lima-Verde 1969; Manso 1988a, b, 1989; Absalão 1990). Tommasi (1970b, c, d) published a survey including 65 species of ophiuroids known to that time. Up to the end of the 1980s, eight new species from the Brazilian coast were described and new occurrences were recorded (Tommasi 1971a, b, c, d, Tommasi and Abreu 1974; Tommasi and Oliveira 1976; Albuquerque 1986; Monteiro 1987; Manso 1988a, b, c). In 1999 Tommasi published an identification key for 98 species and 38 genera of Brazilian ophiuroids.

More recently, several other studies were carried out in different regions of Brazil. Alves and Cerqueira (2000) investigated the echinoderms found on beaches of Salvador (Bahia State), recording 15 species of ophiuroids. Albuquerque et al. (2001) described two new species of the family Amphiuridae: *Amphiura callida* and *Amphiodia habilis*. Borges et al. (2002a), in a study about the ophiuroids of the continental shelf off southern and southeastern Brazil, recorded 25 species, including two new records, at depths down to 800 m. Manso (2004) recorded 10 species of ophiuroids in Camamu Bay (Bahia State), at depths down to 50 m.

Ventura et al. (2006) recorded the occurrence of the family Asteronychidae and reported a total of 15 species from the coast of Rio de Janeiro to Bahia States from depths down to 1,800 m. Borges et al. (2006) described the species *Ophiomisidium tommasii* (Ophiuridae) collected at depths down to 800 m, and Borges and Amaral (2007) recorded four new occurrences for Brazil, including the first record of the family Ophiochitonidae. Manso et al. (2008) collected 23 species at Todos os Santos Bay and Aratú Bay (Bahia State), at depths down to 62 m. Gondim et al. (2008) recorded 13 species at Cabo Branco Beach (Paraíba State). Barboza (2010) and Gondim et al. (2010) recorded for the first time the occurrence of the genus *Gorgonocephalus* (*G. chilensis*) and the species *Ophionereis dolabriformis*, respectively.

At the present time, 153 species have been recorded (Appendix), despite the fact that knowledge on Brazilian Ophiuroidea, especially from the continental margin and the deep sea, is yet quite scarce (Tommasi 1999).

Ophiuroids were one of the most abundant macrofauna groups collected at different bathymetric ranges, from the intertidal down to a depth of 1,600 m. Especially in regions deeper than 200 m, new species and many new records of occurrence have been reported (Borges et al. 2006, Borges and Amaral 2007, Campos et al. 2010).

One of the most recent programs of faunal survey was REVIZEE. Between Cabo de São Tomé (Rio de Janeiro State) and Arroio Chuí (Rio Grande do Sul State), more than 16,000 Ophiuroidea specimens were collected between depths of 60 and 810 m with 32 species recorded (Amaral et al. 2004). On the Biota Marine Benthos project (Biodiversidade Marinha do Estado de São Paulo), which sampled from the intertidal down to a depth of 50 m, 29 species of Ophiuroidea were recorded and approximately 2,500 specimens were collected (Borges et al. 2011). Species of the family Amphiuridae predominate in waters shallower than 50 m, although banks of ophiuroids are scarcely found at this bathymetric range. These studies recorded herds of ophiuroids deeper than 500 m for at least three species of the family Ophiuridae: *Ophimisidium tommasii*, *Ophiura clemens* and *Ophiura ljungmani* (Monteiro 1987, 1990; Tommasi et al. 1988a, b; Absalão 1990; Sumida 1994; Heitor 1996; Petti 1997; Capítoli and Monteiro 2000; Borges 2001).

Although the high diversity of Amphiuridae in shallow and deeper waters, population densities are low when compared to those of Ophiuridae in deeper regions. Some genera, such as *Ophiura*, *Ophiomusium*, *Ophiomastus,* and *Ophiomisidium,* are dominant in these regions.

On the inner continental shelf, *Hemipholis elongata* (Ophiactidae), an infaunal and gregarious species, was frequently found together with other species such as *Microphiopholis atra* and *Amphiodia planispina*, from the beach down to a depth of about 50 m, similarly to the findings reported by Hendler et al. (1995) in the Caribbean region. These authors stated that the species occur frequently on poorly oxygenated unconsolidated bottoms, but also on oyster banks and coral reefs. Boffi (1972) recorded only young individuals of *H. elongata* in the phytal of the northern coast of São Paulo State, and reported that adults probably live preferentially on sublittoral soft bottoms.

Several studies in the southeastern and southern regions stated that *Amphipholis squamata*, *Ophiactis lymani*, *Ophiactis savignyi*, and *Ophiothrix angulata* occurred associated with algae, on rocky shores, and in the unconsolidated sublittoral. According to Chao and Tsai (1995), these species preferentially inhabit biological substrates such as sponges, algae, and corals. Morgado and Tanaka (2001) recorded them associated with colonies of the bryozoan *Schizoporella errata*, on the Brazilian southeastern coast. Boffi (1972) analyzed the ophiuroids associated with 23 species of macroalgae on the northern coast of São Paulo State, and found a predominance of *Ophiactis lymani* and *Amphipholis squamata*, which occurred on almost all the algae and were the most frequent and highest-density species. According to Boffi (1972), *Ophiactis savignyi* and *Ophiothrix angulata* preferentially inhabit sponges, although not exclusively. *Ophiothrix angulata* is a frequent species in shallow depths on the Brazilian coast, and may extend to a depth of about 540 m. *Amphipholis squamata* is a cosmopolitan and well-studied species that inhabits several types of substrates, from sand to crevices, macroalgae, sponges, corals, and aggregations of polychaetes. Boffi (1972) reported that this species occurred from the intertidal zone to a depth of 5 m.

Epifaunal species such as *Ophioplocus januarii* and *Ophioderma januarii*, collected in the unconsolidated sublittoral, such as muddy, sandy, and gravel bottoms, are common along the northern coast of São Paulo State from the intertidal to depths of 180 and 100 m, respectively (Monteiro et al. 1992). *Ophioplocus januarii* occurs in the transition zone between the inner and outer continental shelf along the São Paulo State coast (Monteiro et al. 1992; Pires-Vanin 1993).

Among the Amphiuridae, some species were most commonly collected during the REVIZEE/South Score and Biota/FAPESP programs, such as *A. squamata* (coast and algae), *Microphiopholis atra* and *Amphiura joubini* (unconsolidated sublittoral). These last two are typically infaunal species, with morphological adaptations, such as a soft disc and long slender arms, to live in unconsolidated bottoms. *Microphiopholis atra* is common along the Brazilian southeastern coast, in depths to 100 m. *Amphiura joubini*, although frequent on the southeastern-southern coast, is a subantarctic species with a wide bathymetric distribution (from shallow water to a depth of 3,000 m). Its occurrence in this region is possibly a result of the penetration of colder water masses, such as the South Atlantic Central Water (SACW) (Tommasi 1985). According to Pires-Vanin (1993), *A. joubini* is characteristic of an environmental gradient in the transition zone between the inner and outer continental shelf. Also, *Amphiura flexuosa* is present in the zone of environmental transition between the inner and outer shelf, as recorded by findings of REVIZEE and Biota/FAPESP programs.

Capítoli and Monteiro (2000) reported that *Hemipholis elongata*, *Amphiodia planispina*, *Ophiactis lymani*, and *Amphioplus lucyae* form an association of shallow-water species (<30 m), along the extreme southern coast of Brazil. On the other hand, *Amphiura joubini* and *Amphiura complanata* form an association between depths of 30 and 100 m.

Species of the family Ophiuridae are most commonly collected above a depth of 100 m. Sumida and Pires-Vanin (1997) reported that *Ophiura ljungmani* has a wide distribution, although it occurs preferentially at depths above 100 m. According to these authors, there is a clear relationship between depth and species composition, which changes significantly at depths between 300 and 500 m off southeastern Brazil. At the shallower depths, changes in species composition occur more gradually. This differentiation in species composition along the bathymetric gradient was observed by several investigators on the southeastern coast of Brazil, for different taxonomic groups such as Ophiuroidea (Borges 2001), Mollusca (Miyaji 2001; Quast 2003; Arruda 2005), and benthic macrofauna in general (Sumida and Pires-Vanin 1997; Amaral et al. 2004). Borges (2001), analyzing the ophiuroids of the REVIZEE/South Score program, found a group of species at depths to 100 m, another between 100 and 300 m, and a third below 300 m.

Several studies in the southeastern and southern continental shelf of Brazil (Monteiro 1987; Pires-Vanin 1993; Sumida 1994; Petti 1997; Borges 2001) highlighted some groups of species of Ophiuroidea frequently found: (1) species such as *O. lymani*, *O. savignyi*, *A. squamata* and *O. angulata* that are common along the coast, especially ocuring associated with other invertebrate (sponges and octocorals); (2) species such as *O. januarii*, *H. elongata*, *M. atra,* and *Ophiophragmus luetkeni* that occur at depths between 5 and 50 m; (3) species such as *Amphiura flexuosa*, *A. joubini* and *Ophiothrix rathbuni* that occur below a depth of 30 m, although more frequently near depths of 100–200 m; and (4) species of greater depths (>200 m), such as *Ophiomastus satelitae*, *Ophiura clemens*, *O. ljungmani*, *Ophiomusium acuferum*, *Ophiomusium eburneum*, *Ophiostriatus striatus* and *Ophiomisidium tommasii.*

9.3.4 Holothuroidea

Data on Brazilian Holothuroidea appeared at the beginning of the nineteenth century, published by European and North American scientists who identified echinoderms from the Brazilian coastal waters collected on scientific expeditions (Tiago and Ditadi 2001). However, this group has been the most neglected among the echinoderms and even intertidal species known for over 40 years remain poorly studied.

According to Clark (1907), Selenka's monograph (1867) was the pioneering study of the anatomy and modern systematics of the Holothuroidea. Selenka listed 212 species known to that time, 49 of which were new to science. One year later Selenka (1868) added seven other species to his list, although none of the 219 species were reported for Brazilian waters. Indeed, the first record of Brazilian Holothuroidea is in Verrill (1868). He published an article on the Radiata Collection sampled by the geologist Charles Frederick Hartt, who accompanied

Louis Agassiz on his expedition to Brazil. Hartt explored the coral reefs at Santa Cruz and Porto Seguro region (Bahia State) and, among the marine specimens he sampled, were two species of sea cucumbers. One of them was *Thyone braziliensis*, described by Verril as *Thyone (Sclerodactyla) braziliensis*. The other one was *Chiridota rotifera*, recorded for the first time for Brazil by Verril as *Chirodota rotiferum*.

Later Rathbun (1879) provided the first list of Brazilian echinoderms, with four species of sea cucumbers, 11 species of sea urchins, 13 species of sea stars, 32 species of brittle stars, and four species of sea lilies (Tiago and Ditadi 2001). He gathered all the species collected by the Geological Commission of Brazil from 1875 to 1877. He also added the species previously recorded by other scientists, like Ljungman and Fritz Müller. When he recorded the geographic distribution of the sea star *Echinaster brasiliensis*, he stressed that the term "Rio de Janeiro" was commonly used to designate Brazil in general at that time, rendering some confusion about the distribution of many species. The Holothuroidea he recorded were: *C. rotifera* (as *C. rotiferum*), *T. braziliensis* (as *T. (Sclerodactyla) braziliensis*), *Thyonella* sp. and *Holothuria* sp.

In 1881, Ludwig examined the animals collected at the Brazilian coast by E. van Beneden, recording the occurrence of *C. rotifera* and describing two new species of Holothuroidea: *Phyllophorus parvus* (as *Thyonidium parvum*) and *Protankyra benedeni* (as *Synapta benedeni*). In 1882, he also published a list of the echinoderms collected by E. van Beneden at the Brazilian coast (Tommasi 1957).

The specimens collected by the *Challenger* Expedition along the coast of Bahia State were studied by Théel (1882, 1886a), who described *Thyone pervicax*, *Pentacta pygmaeus* (as *Colochirus pygmaeus*), and *Lissothuria braziliensis* (as *Psolus braziliensis*).

Ludwig (1881) and Théel (1886a, b) contributed by describing new species and findings of sea cucumbers in Brazil. However, it was only in 1897 that Hermann von Ihering listed the first marine invertebrates, among them echinoderms, for the coast of the State of São Paulo, specifically at São Sebastião Island. However, Ihering did not identify a single species. That task was performed by Luederwaldt (1929a, b), who cited the occurrence of *Holothuria grisea* at that island (Tiago and Ditadi 2001).

Deichmann (1930) described Holothuroidea sampled over a huge area, from southern Brazil to Cape Cod (Massachussetts, USA). She gave detailed information about the 103 species found, with 12 reported for the Brazilian coast: *Holothuria grisea, Phyllophorus seguroensis, P. occidentalis, P. parvus, Thyone cognita, T. suspecta, T. pervicax, T. belli, Pentacta pygmaeus, Thyonepsolus braziliensis, Synaptula hydriformis* and *Protankyra benedeni*.

Although mentioned by earlier authors such as Verrill (1868), Ludwig (1881), and Théel (1886a, b), the Apoda *Chiridota rotifera* was not reported for Brazil in Deichmann's monograph (Tiago and Ditadi 2001). Later on Luederwaldt (1929a, b) recorded the occurrence of *H. grisea* at São Sebastião Island (São Paulo State).

Systematic research on Holothuroidea by Brazilian biologists is somewhat recent, beginning with Ancona-Lopez (1957), on specimens sampled at Pernambuco, Rio de Janeiro and São Paulo States. In her publication she listed *H. grisea*, *Stichopus badionotus*, *Thyone braziliensis*, *Chiridota rotifera* and *Synaptula secreta*. Tommasi (1957) also published a list of Holothuroidea sampled down to a depth of 15 m along the coast of São Paulo State, including *Synaptula hydriformis*, *H. grisea* and *S. badionotus*.

In 1969, Tommasi also published a list with 31 species of Holothuroidea, as well as a survey of one hundred years of the literature about the group (Tommasi, 1969a). Nine new species were added to the Brazilian fauna in the following thirty years (Tiago and Ditadi 2001).

Other authors also published lists on the Brazilian Holothuroidea, among them Sluiter (1910), Krau (1950), Heding and Panning (1954), Ancona-Lopez and Sawaya (1955), Ancona-Lopez (1962, 1964, 1965a, b), Tommasi (1959, 1969a, b, 1971b, d, 1972, 1974a), Brito (1960f, 1962), Deichmann (1963), Lima-Verde (1969), Freitas et al. (1973), Tommasi and Oliveira (1976), Tommasi and Aron (1988), Tommasi et al. (1988a,b), Freire and Grohmann (1989), Hendler et al. (1995), Hadel (1997), Pires-Vanin et al. (1997), Hadel et al. (1999) and Brites et al. (2011).

Our knowledge of the southeastern Atlantic Holothuroidea is poor, particularly for deep-sea species. Hyman (1955) observed that "interest in this group appears relegated to taxonomic specialists", while Hendler et al. (1995) pointed out that holothurians are far from being attractive animals. Besides, their densities in the Atlantic waters are low compared to the tropical Indo-Pacific. Moreover, the Atlantic species are generally small, dull in colour and are not as easily seen as are coral reef's species.

From Verrill (1868) to the present, 49 valid names of Holothuroidea species have been ascribed to the 7,367 km of the Brazilian coast (see Appendix). At first this may seem to indicate a low occurrence of sea cucumbers on the Brazilian coast, but it properly reflects the lack of systematic studies of these echinoderms by Brazilian biologists (Tiago and Ditadi 2001).

Species with a wide geographical distribution predominate among the Brazilian Holothuroidea, with most species reaching at least to the Caribbean. Endemic species are also well-represented: *Psolus vitoriae*, *Thyone montoucheti*, *Euthyonidiella dentata*, *Cucumaria manoelina*, *Ocnus braziliensis*, *Trachytione crassipeda*, *Leptosynapta brasiliensis* and *Synaptula secreta* (Ancona-Lopez 1957; Tiago and Ditadi 2001).

Although São Sebastião (São Paulo State) is recognized as the southern limit of the geographical distribution of many Brazilian sea cucumbers, some species, like *I. badionotus* and *H. grisea,* have been found at Santa Catarina State, some 750 km south of São Sebastião (Tiago and Ditadi 2001). However, the number of species described for the Brazilian littoral, as well as their geographical distribution, should be reviewed taking into account possible synonymies.

9.4 Aquaculture and Fisheries

Although the collection of echinoderms for human consumption is illegal (Hadel et al. 1999), aquaculture enterprises to culture Brazilian echinoderms are absent in the country (Campos and Lima 2000; Ventura et al. 2006). This may be attributed to a cultural heritage from the first colonizers, mainly of Portuguese and Spanish origins. In the first centuries of European occupation of the Brazilian territory, the main purpose was to collect as much natural resources as possible and take them back to their countries of origin. This attitude towards nature still persists and the only marine and brackish animals cultured commercially are shrimp. Therefore, despite their high economic value on foreign markets, echinoderms are still ignored in Brazil as a potential food resource (Campos and Lima 2000).

9.5 Echinoderm Threats

Nineteen of the 627 threatened species listed in the Brazilian Red List (Machado et al. 2008) are echinoderms: fourteen species of Asteroidea (*Asterina stellifera, Astropecten brasiliensis, A. cingulatus, A. marginatus, Coscinasterias tenuispina, Echinaster (Othilia) brasiliensis, E. (Othilia) echinophorus, E. (Othilia) guyanensis, Linckia guildingii, Luidia clathrata, L. ludwig scotti, L. senegalensis, Narcissia trigonaria, Oreaster reticulatus*), three species of Echinoidea (*Cassidulus mitis, Eucidaris tribuloides, Paracentrotus gaimardi*) and two species of Holothuroidea (*Isostichopus badionotus, Synaptula secreta*). Most of these species live in shallow waters and are easily collected.

Machado et al. (2005, 2008) pointed out several threats that menace Brazilian echinoderms, among them habitat destruction or alteration, eutrophication, and erosion. Anthropogenic impacts are caused by the increased occupation of coastal regions. Landfills, for instance, cause sediment transport to shallow sea waters, changing the physical and chemical properties of the substrate. The sea lily *Tropiometra carinata*, for instance, was once abundant on the exposed rocky shores of the coast of São Paulo State. Due to anthropogenic impacts, however, population densities of these animals are decreasing steadily (Hadel et al. 1999). Oil spills and discharge of domestic or industrial sewage are also side effects of coastal municipaltie's development (Machado et al. 2008).

The impact caused by touristic enterprises, including diving, is recognized as threats to several species of benthic animals, including sea stars, sea urchins and crinoids (Machado et al. 2008).

Accidental by-catch by trawling boats fishing for shrimp is also listed as one of the major menaces for echinoderms, especially for asteroids and echinoids that live on sandy and muddy bottoms (e.g. *A. brasiliensis, A. cingulatus. A. marginatus, L. clathrata, L. ludwigi scotti, L. senegalensis* and *Encope emarginata*). The number of benthic species accidentally captured by trawl-nets with doors may amount to 96 %, with *A. marginatus* always present in the nets (Branco 1999; Branco and Verani 2006a, b). Fishing gear also damages the animals indirectly as they pass over them and disturb the substrate physically (Micael et al. 2009). In the southeast region, fishing of sea-bob-shrimp (*Xiphopenaeus kroyeri*) with bottom trawl nets results in heavy by-catch due to its low selectivity (Coelho et al. 1986; Branco 1999; Branco and Fracasso 2004; Carniel 2008; Bail et al. 2009). According to Bail et al. (2009), 0.04 kg of echinoderm fauna are captured by trawl nets for each kg of sea-bob-shrimp. Branco (1999) estimates that 7.13 ton d^{-1} of by-catch fauna, including fishes, crustaceans, mollusks and echinoderms are captured on the southern Brazilian coast, amounting to 1,711.7 ton y^{-1}.

Either caught or not by the nets, echinoderms may be severely damaged by them. Among the damage listed by Micael et al. (2009) are: injuries caused by demersal trawls passing over the bottom, damage or autotomy of body parts due to physical contact with fishing gear, interactions with other species in the catch or handling by the crew, crushing when dropped on to the deck, punctures in the epidermis, and bacterial infection due to trauma. Echinoderms are menaced also by the impacts caused by trawls and dredges by scraping or resuspension of the substratum (Jennings and Kaiser 1998).

Sea urchins and sea cucumbers have been fished commercially all over the world (Micael et al. 2009) but in Brazil it is illegal to collect native fauna except fishes, mollusks and crustaceans captured for human consumption. Even for research projects, special permission from environmental agencies is needed. Predatory collection for aquarium trade, souvenirs, and decorative pieces, food markets, and other commercial purposes are threats listed for several species of Brazilian echinoderms in addition to those listed by Machado et al. (2008), including the sea star *Asterina stellifera*, the sea urchins *L. variegatus* and *E. lucunter* and the sea cucumber *H. grisea*. Sea stars, *O. reticulatus* for instance, is illegally and extensively collected all over the Brazilian coast and sold as decorative pieces, souvenirs or religious objects used in some local rituals (Hadel et al. 1999; Campos and Lima 2000). Alves et al. (2006, 2009/2010) created the term zoohandicraft to designate workmanship products made of marine macrofauna. When interviewed by the authors, the craftsmen claimed no knowledge on the threats the animals were under. Along the northeast coast of Brazil, Alves et al. (2009/2010) described zoohandicraft production as the only or the main economic source of many people with strong familiar hereditary tradition.

Because it is illegal to collect echinoderms, except for research purposes with a special permission, the capture of those animals occurs in an artisanal way. This is particularly true for sea urchins whose mature female gonads, known as roe, are prepared as a delicacy. Although not menaced at present, it is known that considerable numbers of *E. lucunter*, *L. variegatus* and *P. gaimardi* have been collected by hand during low tides or by divers at localities like Ilhabela Island (São Paulo State). Because there are no external differences between males and females, and between mature and immature females, a large number of individuals are killed before the consumers find the desired specimens, resulting in high levels of mortality. Moreover, killing of mature females, the ones that would produce new generations, endangers even more those species' populations. These threats apply also to the sea cucumbers *I. badionotus* and *H. grisea*. Those echinoderms are reported as a food source for some local populations on southern, southeastern and northeastern Brazilian coasts (Hadel et al. 1999; Campos and Lima 2000; Mendes et al. 2006).

About 17 % of the animals in the aquarium global trade are echinoderms (Micael et al. 2009). Among the species commonly found in Brazilian home aquaria, and at public or private ones, are *A. stellifera*, *E.* (*Othilia*) *brasiliensis*, *O. reticulatus*, *E. tribuloides* and *I. badionotus*. Most of those animals die before they reach the stores. Even after that, most of them survive only for short periods and have to be replaced. Therefore, more and more specimens have been collected, endangering even more the species already menaced by other threats.

Some echinoderm are threatened by low density populations, limited or discontinuous distribution, or because they occur on vulnerable environments. The endemic sea cucumber *Synaptula secreta* was described for one locality only, at Segredo Beach (São Sebastião, São Paulo State). There are no data on the reproductive strategy of this species, but its distribution pattern suggests that they brood their young in the coelom, like other Apoda species such as *Synaptula hydriformis* and *Chiridota rotifera*. The brooding sea urchin *C. mitis*, that keeps their short-term larvae on the female's body, is also threatened by their low dispersion capacity and, in consequence, restricted distribution (Amaral et al. 2008). The sea stars *L. guildingii* and *C. tenuispina* have discontinuous distribution patterns and low population densities, many of which probably persist only by asexual reproduction. This reproductive strategy decreases genetic variability and gene flow among populations, rendering them even more vulnerable to environmental impacts.

Besides the threats already stated here there are others worth mentioning. As benthic animals echinoderms are susceptible to pollutants stored in marine sediments and to biomagnification processes (Camargo 1982; Hadel et al. 1999; Carnevali et al. 2001; Sugni et al. 2007; Micael et al. 2009). Some species, like *L. variegatus* and *E. lucunter,* have been used for ecotoxicological tests according to the standard protocol ABNT-NBR 15350 (Brazilian Association for Technical Standards; ABNT 2006, 2012) because their embryos are a reliable source of material for biochemical analyses (Micael et al. 2009). *Arbacia lixula*, another sea urchin species, has been suggested also as potential test organism for ecotoxicological analysis by Máximo et al. (2008). Although the fact that they are not regarded as severely threatened species, the use of their embryos and larvae as experimental tools is leading to an increasing depletion of wild populations in some regions along the Brazilian coast.

Studies on the reproductive biology, population dynamics and genetics of most species are needed as well as on the taxonomy of those with high morphological variations, such as *Echinaster* species and *P. gaimardi*. In order to achieve such knowledge, which depends on long term studies, it is essential to protect these species through governmental laws that restrict and control collection.

9.6 Concluding Remarks

Although research on echinoderms from Brazilian waters is still scarce, current advances in several fields of knowledge are remarkable. The principal bottlenecks seem to be the current low number of researchers dedicated to the phylum and the lack of integrative actions to promote information exchange and collaborative studies. However, there has been a relevant increase in the number of researchers working in different subject areas during the last two decades.

In fact, an optimistic feeling is corroborated by some evidences. For instance, although there are few taxonomists in Brazil, the zoological collections were improved during the last ten years due to some governmental programs of biodiversity inventory and protective laws that obligate petroleum companies to provide environmental analyses before any exploitation. These facts have induced the needs for more qualified people, which stimulated the academic institutions to have more students working on projects of biodiversity, ecotoxicology and physiology in their Post-graduate Programs. This interest has also generated increasing studies on immunology, biochemistry, ecology, reproduction and development of echinoderms.

In addition, the Brazilian Symposium of Echinoderms has convened twice in the past five years (the first in 2008 and the second in 2012). The second Latin American meeting on echinoderms (II Congreso Latinoamericano de Equinodermos, CLE) will take place in Brazil in 2013. These are important initiatives to enhance exchange of knowledge and to stimulate collaborative studies.

In summary, it is well-known that much time and effort are necessary to train new researchers and to produce knowledge, but some robust steps have taken place as described earlier in this chapter. Echinoderm studies in Brazil have to be stimulated more and integrative research has to be encouraged. The state of echinoderm knowledge seems to have an exciting and successful future in Brazil, like in the whole Latin America.

Acknowledgments We are grateful to the editors of this volume (Juan José Alvarado and Francisco Solís Marín) for the invitation to take part in this volume. We are also indebted to John M Lawrence and the editors for their critical comments and suggestions that improved this chapter.

References

Abessa DMS, Pereira CDS, Zaroni LP, Gasparro MR, Sousa ECPM (2003) Sensibilidade de *Lytechinus variegatus*: comparação entre três populações do Estado de São Paulo. Resumos II Congresso Brasileiro de Pesquisas Ambientais. Santos, CD-Rom 49-51

ABNT (Associação Brasileira de Normas Técnicas) (2006) Ecotoxicologia aquática – Toxicidade crônica de curta duração – Método de ensaio com *Lytechinus variegatus* (Echinodermata: Echinoidea). Norma NBR15350:2006. Rio de Janeiro

ABNT (Associação Brasileira de Normas Técnicas) (2012) Ecotoxicologia aquática – Toxicidade crônica de curta duração – Método de ensaio com ouriço-do-mar (Echinodermata: Echinoidea). Norma NBR15350:2012. Rio de Janeiro

Absalão RS (1990) Ophiuroid assemblages off the Lagoa dos Patos outlet, southern Brazil. Ophelia 31:133–143

Albuquerque MN (1986) *Ophiuroidea* Gray, 1840 (Echinodermata) da Plataforma Continental do Norte e Nordeste do Brasil. PhD Thesis, Universidade de São Paulo, São Paulo

Albuquerque MN, Campos-Creasey LS, Guille A (2001) Two new species of Amphiuridae (Echinodermata, Ophiuroidea) from the southeastern coast of Brazil. Zoosystema 23:591–602

Almeida FFM (1954) Geologia do centro-leste mato-grossense. Boletim Dep Nac Produção Mineral Div Geol Mineral 150:1–97

Alves OFS, Cerqueira WRP (2000) Echinodermata das praias de Salvador (Bahia, Brasil). Rev Bras Zool 17:543–553

Alves SLS, Fernandes ACS (2001) Considerações sobre a presença de *Encrinaster pontis* (Clarke, 1913) (Echinodermata-Ophiuroidea-Oegophiurida) na Formação Ponta Grossa (Devoniano da Bacia do Paraná. R. Resumos XVII Congr Brasileiro Paleontol, Rio Branco

Alves SLS, Pereira AD, Ventura CRR (2002) Sexual and asexual reproduction of *Coscinasterias tenuispina* (Echinodermata: Asteroidea) from Rio de Janeiro, Brazil. Mar Biol 140:95–101

Alves MS, Silva MAS, Pinto SL (2009/2010) Perfil sócio-econômico dos atores envolvidos na produção e comercialização de zooartesanato em Recife, Pernambuco - Brasil. Rev Nordestina Zool 4:97-104

Alves MS, Silva MAS, Melo JrM, Paranaguá MN, Pinto SL (2006) Zooartesanato comercializado em Recife, Pernambuco, Brasil. Rev Bras Zoocien 8:99-109

Amaral ACZ, Jablonski S (2005) Conservation of marine and coastal biodiversity in Brazil. Conser Biol 19:625–631

Amaral ACZ, Rossi-Wongtschowski CLDB (eds) (2004) Biodiversidade Bentônica das Regiões Sudeste e Sul do Brasil - Plataforma Externa e Talude Superior. Série Documentos Revizee: Score-Sul. Instituto Oceanográfico - USP, São Paulo

Amaral ACZ, Lana PC, Fernandes FC, Coimbra JC (2004) Caracterização do ambiente e da macrofauna bentônica. Parte I. In: Amaral ACZ, Rossi-Wongtschowski CLDB (eds) Biodiversidade bentônica das regiões sudeste e sul do Brasil - Plataforma Externa e Talude Superior. Série Documentos Revizee: Score-Sul. Instituto Oceanográfico - USP, São Paulo, pp 11-46

Amaral ACZ, Ribeiro CV, Mansur MCD, Santos SB, Avelar WEP, Matthews-Cascon H, Leite FPP, Melo GAS, Coelho PA, Buckup GB, Buckup L, Ventura CRR, Tiago CG (2008) A situação de ameaça dos invertebrados aquáticos no Brasil. In: Machado ABM, Drummond GM, Pagllia AP (eds) Livro vermelho da fauna brasileira ameaçada de extinção, vol I., – InvertebradosMMA/Fundação Biodiversitas, Brasília/Belo Horizonte, pp 157–167

Amon RMW, Herndl GJ (1991) Deposit feeding and sediment: 1. Interrelationship between *Holothuria tubulosa* (Holothurioidea, Echinodermata) and the sediment microbial community. PSZN I Mar Ecol 12:163–174

Ancona-Lopez AA (1957) Sobre holotúrias do litoral sul brasileiro. Bol Fac Filos Ciên Let Univ São Paulo, Zool 21:5–54

Ancona-Lopez AA (1962) Dendrochirotae de São Sebastião. Arq Mus Nac 52:117–124

Ancona-Lopez AA (1964) Holothurioidea. In: Vanzolini PE (ed) História natural de organismos aquáticos do Brasil: bibliografia comentada. FAPESP, São Paulo, pp 275–286

Ancona-Lopez AA (1965a) Sobre *Pentacta peterseni* spec. nov. (Echinodermata - Holothurioidea). Bol Fac Filos Ciên Let Univ São Paulo, Zool 25:315–320

Ancona-Lopez AA (1965b) Contribution to the ecology of the Holothuroidea of the coast of São Paulo State. An Acad Bras Ciên 37:171–174

Ancona-Lopez AA, Sawaya P (1955) Holotúrias de Recife. Ciên Cult 7:166

Arruda EP (2005) Mollusca Bivalvia (Veneroida) da costa sudeste do Brasil. PhD Thesis, Inst Biociên, Univ São Paulo, São Paulo

Ávila-Pires TCS (1983) Contribuição ao estudo do gênero *Echinaster* Müller & Troschel, 1840 (Echinodermata, Asteroidea) no litoral brasileiro. An Acad Bras Ciên 55:431–448

Bail GC, Branco JO, Freitas Jr F, Lunardon-Branco MJ, Braun JRR (2009) Capítulo 18 - Fauna acompanhante do camarão sete barbas, na Foz do Rio Itajaí-Açú e sua contribuição na diversidade de crustáceos e peixes do ecossistema Saco da Fazenda. In: Branco JO, Lunardon-Branco MJ, Bellotto VR (eds) Estuário do Rio Itajaí-Açú, Santa Catarina: caracterização ambiental e alterações antrópica. Ed UNIVALI, Itajaí, pp 284-312

Barbosa O (1949) Contribuição à geologia Brasil-Bolívia. Mineração e metalurgia 13:271–278

Barboza CAM (2010) Variabilidade espacial das associações de ofiuróides ao longo de um gradiente de contaminação no Canal da Cotinga, Baía de Paranaguá. Master Thesis, Univ Fed Paraná, Brazil, Paraná

Barboza CAM, Moura RB, Lanna AM, Oackes T, Campos LS (2011) Echinoderms as Clues to Antarctic - South American Connectivity. Oecol Australis 15:86–110

Beurlen K (1957) Contribuição ao conhecimento do Devoniano inferior da zona subandinada Bolívia. Bol Dep Nac Prod Min, Div Geol Mineral 101:1–5

Beurlen K (1961) Observações geo-paleontológical no Cretáceo do Rio Grande do Norte e Ceará. Com descrição de amonóides. Coleção Mossoroense sér B 58:1–12

Beurlen K (1964a) A fauna do calcáreo Jandaíra da região de Mossoró (Rio Grande do Norte). Pongetti Ed (Coleção Mossoroense), Rio de Janeiro

Beurlen K (1964b) Notas preliminares sobre a fauna da Formação Riachuelo: IV. Echinóides. Univ Fed Pernambuco, Recife, Bol Geol 4:37–38

Beurlen K (1966) Novos equinóides do Cretáceo do Nordeste do Brasil. An Acad Bras Ciên 38:455–464

Bianconcini MSC, Mendes EG, Valente D (1985) The respiratory metabolismo of the lantern muscles of the sea urchin *Echinometra lucunter* L. – I. The respiratory intensity. Comp Biochem Physiol 80A:1–4

Bianconcini MSC, Medeiros LF, Medeiros LO, Valente D, Mendes EG (1982) The effect of extraction media on the enzyme-activities in the lantern muscles of a sea-urchin. An Acad Bras Ciên 54:729–731

Boffi E (1972) Ecological aspects of ophiuroids from the phytal of S.W. Atlantic Ocean warmwaters. Mar Biol 15:316–328

Borges M (2001) Taxonomia, distribuição e biologia reprodutiva de Ophiuroidea (Echinodermata) das Regiões Sudeste e Sul do Brasil. PhD Thesis, Universidade Estadual Paulista, São Paulo

Borges M, Amaral ACZ (2007) Ophiuroidea (Echinodermata): quatro novas ocorrências para o Brasil. Rev Bras Zool 24:855–864

Borges M, Monteiro AMG, Amaral ACZ (2002a) Taxonomy of Ophiuroidea (Echinodermata) from the continental shelf and slope of the southern and southeastern Brazilian coast. Biota Neotrop 2:1–69

Borges M, Monteiro AMG, Amaral ACZ (2004a) Filo Echinodermta. In: Amaral ACZ, Rossi-Wongtschowski CLDB (eds) Biodiversidade Bentônica da Região Sudeste-Sul do Brasil-Plataforma Externa e Talude Superior. Mus Nac, Rio de Janeiro, pp 156–160

Borges M, Monteiro AMG, Amaral ACZ (2004b) Classe Ophiuroidea. In: Amaral ACZ, Rossi-Wongtschowski CLDB (eds) Biodiversidade Bentônica da Região Sudeste-Sul do Brasil-Plataforma Externa e Talude Superior. Série Documentos REVIZEE - Score Sul. Instituto Oceanográfico USP, São Paulo, pp 156-160

Borges M, Monteiro AMG, Amaral ACZ (2006) A new species of *Ophiomisidium* (Echinodermata: Ophiuroidea) from the continental shelf and slope off southern Brazil. J Mar Biol Ass UK 86:1449–1454

Borges JCS, Branco PC, Pressinotti LN, Silva JRMC (2010) Intranuclear crystalloids of sea urchins as a biomarker for oil contamination. Polar Biol 33:843–849

Borges M, Yokoyama LQ, Amaral ACZ (2011) Ophiuroidea. In: Amaral ACZ, Nallin SAH (eds) Biodiversidade e ecossistemas bentônicos marinhos do Litoral Norte de São Paulo. Sudeste do Brasil, UNICAMP/IB, Campinas, pp 280–288

Borges JCS, Porto-Neto LR, Mangiaterra MBCD, Jensch-Junior BE, Silva JRMC (2002b) Phagocytosis in vivo and in vitro in the Antarctic Sea Urchin *Sterechinus neumayeri* (Meissner) at 0 °C. Polar Biol 25:891–897

Borges JCS, Silva JRMC, Rocha AJS, Jensch-Junior BE, Pressinotti LN, Passos MJACR, Gomes V, Branco PC, Van Ngan P (2012) Metabolic differences between tropical (*Lytechinus variegatus*) and polar (*Sterechinus neumayeri*) echinoderms. Pesqui Antártica Bras 5:71–79

Borzone CA (1991) Spatial distribution and growth of *Mellita quinquiesperforata* (Leske, 1778) on a sandy beach of southern Brazil. Nerítica 7:87–100

Bosetti EP, Quadros R (1996) Contribuição ao estudo dos Lingulida (Brachiopoda: Inarticulata) da Formação Ponta Grossa, Devoniano, Bacia do Paraná, Município de Chapada dos Guimarães, Mato Grosso, Brasil. An Simp Sul Americano do Siluriano-Devoniano: Estratigrafia e Paleontologia 1. Ponta Grossa, Ponta Grossa, pp 167-191

Bowmer T, Boelens RGV, Keegan BF, O'neill J (1986) The use of marine benthic 'key' species in ecotoxicological testing: *Amphiura filiformis* (O.F. Muller) (Echinodermata: Ophiuroidea). Aquat Toxicol 8:93–109

Branco JO (1999) Biologia do *Xiphopenaeus kroyeri* (Heller, 1862) (Decapoda: Penaeidae), analise da fauna acompanhante e das aves marinhas relacionadas a sua pesca, na região de Penha, SC – Brasil. PhD Thesis, Cent Ciên Biol Saúde, Univ Fed São Carlos, São Carlos

Branco JO, Fracasso HAA (2004) Ocorrência e abundância da carcinofauna acompanhante na pesca do camarão sete-barbas, *Xiphopenaeus kroyeri* Heller (Crustacea, Decapoda), na Armação do Itapocoroy, Penha, Santa Catarina, Brasil. Rev Bras Zool 21:295–301

Branco JO, Verani JR (2006a) Análise quali-quantitativa da ictiofauna acompanhante na pesca do camarão sete-barbas, na Armação do Itapocoroy, Penha, Santa Catarina. Rev Bras Zool 23:381–391

Branco JO, Verani JR (2006b) Capítulo 11 - Pesca do camarão sete-barbas e sua fauna acompanhante, na Armação do Itapocoroy, Penha, SC. In: Branco JO, Marenzi AWC (eds) Bases ecológicas para um desenvolvimento sustentável: estudos de caso em Penha, SC edn. UNIVALI, Itajaí, pp 153–170

Branco PC, Pressinotti LN, Borges JCS, Iunes RS, Kfoury Jr JR, Silva MO, Gonzalez M, Santos MF, Peck LS, Cooper EL, Silva JRMC (2011) Cellular biomarkers to elucidate global warming effects on antarctic sea urchin *Sterechinus neumayerii.* Polar Biol 35:221-229

Briggs E, Wessel GM (2006) In the beginning. Animal fertilization and sea urchin development. Develop Biol 300:15–26

Brites AD, Hadel VFH, Tiago CG (2011) Crinoidea, Asteroidea, Echinoidea, Holothuroidea. In: Amaral ACZ, Nallin SAH (eds) Biodiversidade e ecossistemas bentônicos marinhos do Litoral Norte de São Paulo. Sudeste do Brasil, UNICAMP/IB, Campinas, pp 273–279

Brito IM (1959) Sobre uma nova *Clypeaster* do Brasil. Avulso Cent Est Zool Fac Nac Filos 1:1–8

Brito IM (1960a) Asteróides e equinóides brasileiros das coleções do Centro de Estudos Zoológicos. Actas y Trabajos 1er Cong Sudamericano Zool. La Plata 2:61–68

Brito IM (1960b) Clypeasteróides do Rio de Janeiro. Avulso Cent Est Zool Fac Nac Filos 3:1–10

Brito IM (1960c) Os equinóides regulares do Rio de Janeiro. Avulso Cent Est Zool Fac Nac Filos 4:1–8

Brito IM (1960d) Asteróides dos Estados do Rio de Janeiro e São Paulo. Parte I - Forcipulata e Phanerozonia. Avulso Cent Est Zool Fac Nac Filos 5:1–13

Brito IM (1960e) Os ofiuróides do Rio de Janeiro. Parte I - Ophiothrichidae, Ophiochitonidae e Ophiactidae. Avulso Cent Est Zool Fac Nac Filos 6:1–4

Brito IM (1960f) Holothuroides do Rio de Janeiro. Parte I - Aspidochirota e Apoda. Avulso Cent Est Zool Fac Nac Filos 7:1–8

Brito IM (1962) Ensaio de catálogo dos Echinodermata do Brasil. Avulso Cent Est Zool Fac Nac Filos 13:1–11

Brito IM (1964) Equinóides cretácicos do Est. da Bahia. Esc Geol Univ Fed Bahia, Publ Avulsas 1:1–11

Brito IM (1968) Asteróides e equinóideos do Estado da Guanabara e adjacências. Bol Mus Nac 260:1–57

Brito IM (1971) Contribuição ao conhecimento dos equinodermos da Ilha da Trindade, Brasil. Arq Mus Nac 54:261–265

Brito IM (1979) Clipeasteróides cenozóicos do Brasil. An Acad Bras Ciên 51:729–740

Brito IM (1980) Os Equinóides fósseis do Brasil: I. Os endocíclicos. An Acad Bras Ciên 52:569–590

Brito IM (1981a) Os Equinóides fósseis do Brasil: II. Holectypoides e cassiduloides. An Acad Bras Ciên 53:513–527

Brito IM (1981b) Os equinóides fósseis do Brasil. III. Espatangóides, exceto H*emiasteridae*. An Acad Bras Ciên 53:569–578

Brito IM (1981c) O Gênero *Hemiaster* no Cretáceo do Brasil (Echinoidea, Spatangoidea). An Cong Latino-Americano Paleontol 2. Porto Alegre 1:399–414

Brito IM (1981d) Contribuição a Paleontologia do Estado do Pará. A ocorrência de *Abertella complanata* (Echinoidea, Clyperasteroidea) na Formação Pirabas. Bol Mus Paraense Emilio Goeldi nov sé Geol 23:1–8

Brito IM (1982) A ocorrência de *Clypeaster durandi* Cherbonnier no Brasil (Echinoidea, Clypeasteroidea). An Acad Bras Ciên 54:365–368

Brito IM (1986) *Abertella complanata* (Echinoidea, Clyperasteroidea) da Formação Pirabas (Pará, Brasil). Bol Mus Paraense Emilio Goeldi nov sé Geol 28:1–4

Calil P, Rocha RM, Freire RA, Roper JJ (2009) The role of *Asterina stellifera* (Echinodermata: Asteroidea) as a predator in a rocky intertidal communbity in Southern Brazil. Zool 26:279–287

Camargo TM (1982) Papers from the Echinoderm Conference. 10. Changes in the echinoderm fauna in a polluted area on the coast of Brazil. Aust Mus Mem 16:165–173

Campos LS, Lima RPN (2000) Equinodermocultura: criação de equinodermos, um grupo de importância comercial subestimada no Brasil. Panor Aqüicul 10:27–34

Campos LS, Tyler PA, Gage JD, John AWG (1994) Vertical flux coupled to the diet and seasonal life history of the deep-sea echinoid *Echinus affinis*. Deep-Sea Res 41:369–388

Campos LS, Barboza CAM, Alcântara PF, Moura RB, Frensel R, Wanderley P (2010) Echinodermata. In: Lavrado HP, Brasil ACS (eds) Biodiversidade da Região Oceânica Profunda da Bacia de Campos: Megafauna e Ictiofauna Demersal. SAG Serv, Rio de Janeiro, pp 235–292

Capítoli RR, Bonilha LE (1991) Comunidades bentônicas. In: Vooren CM (ed) Projeto Talude. FIPEC Relat final, FURG, Rio Grande, pp 79–92

Capítoli RR, Monteiro AMG (2000) Distribuição e abundância de ofiuróides na plataforma interna do extremo sul do Brasil. Atlântica 22:41–56

Capítoli RR, Bemvenuti C (2006) Associações de macroinvertebrados bentônicos de fundos inconsolidados da plataforma continental e talude superior no extremo sul do Brasil. Atlântica 28:47–59

Car RS, Biedenbach JM, Nipper M (2006) Influence of potentially confounding factors on sea urchin porewater toxicity tests. Arch Environ Contam Toxicol 51:573–579

Carnevali MDC, Galassi S, Bonasoro F, Patruno M, Thorndyke MC (2001) Regenerative response and endocrine disrupters in crinoid echinoderms: arm regeneration in *Antedon mediterranea* after experimental exposure to polychlorinated biphenyls. J Exp Biol 204:835–842

Carniel VL (2008) Interação de aves costeiras com descartes oriundos da pesca artesanal no litoral centro-sul paranaense. Setor de Ciên Biol, Univ Fed Paraná, Curitiba, Master Thesis

Carvalho PE (1926) Reconhecimentos Geológicos entre a Costa Atlântica de Maracanã e Pirabas e a Estação Experimental na E. F. de Bragança, Rio de Janeiro. Bol Dep Nac Prod Min, Diret Geol Mineral 15:115–124

Carvalho ALPS, Ventura CRR (2002) The reproductive cycle of *Asterina stellifera* (Möbius) (Echinodermata: Asteroidea) in the Cabo Frio region, southeastern Brazil. Mar Biol 141:947–954

Caster KE (1954a) A new carpoid echinoderm from the Paraná Devonian. An Acad Bras Ciên 26:123–147

Caster KE (1954b) A Devonian Placoccystoid Echinoderm from Paraná, Brazil. In: Lange FW (ed) Paleontologia do Paraná. Comissão de Comemorações do Centenário do Paraná, Curitiba, pp 137–148

Castro CB, Echeverria CA, Pires DO, Mascarenhas BJA, Freitas SG (1995) Distribuição de Cnidaria e Echinodermata no infralitoral de costões rochosos de Arraial do Cabo. R. J. Rev Bras Biol 55:471–480

Cesar-Ribeiro C, Palanch-Hans MF (2010) Chronic toxicity test with sea urchin *Echinometra lucunter* and *Lytechinus variegatus* (ECHINODERMATA: ECHINOIDEA), exposed to light-stick - flag pasternoster used for fishing surface longline. Braz J Oceanogr 58:71–75

CETESB (Companhia de Tecnologia de Saneamento Ambiental) (1991) Água do mar - Teste de toxicidade aguda com Artemia: método de ensaio. Norma Técnica L5.021. São Paulo

CETESB (Companhia de Tecnologia de Saneamento Ambiental) (1992) Água do mar: Teste de toxicidade aguda com *Mysidopsis juniae* Silva, 1979 (Crustacea: Mysidacea). Norma Técnica L5.251. São Paulo

CETESB (Companhia de Tecnologia de Saneamento Ambiental) (1999) Água do mar: Teste de toxicidade crônica de curta duração com Lytechinus variegatus Lamarck, 1816 (Echinodermata, Echinoidea) – método de ensaio. Norma Técnica L5. 250. São Paulo

Chao SM, Tsai CC (1995) Reproduction and population dynamics of the fissiparous brittle star *Ophiactis savignyi* (Echinodermata, Ophiuroidea). Mar Biol 124:77–83

Cherr GN, Summers RG, Baldwin JD, Merrill JB (1992) Preservation and visualization of the sea urchin blastoceolic extracellular matrix. Microbiol Res Tech 2:11–22

Cherr GN, Higashil RM, Shenker JM (1993) Assessment of Chronic Toxicity of Petroleum and Produced Water Components to Marine Organisms. Univ California, Davis-Bodega Mar Lab U.S. DOI. Available from NTIS PB94-171642. OCS Study, MMS 94-0035, Davis, California

Cinelli LP, Andrade L, Valente AP, Mourão PAS (2010) Sulfated -L-galactans from the sea urchin ovary: Selective 6-desulfation as eggs are spawned. Glycobiol 20:702–709

Cinelli LP, Castro MO, Santos LL, Garcia CR, Vilela-Silva ACES, Mourão PAS (2007) Expression of two different sulfated fucans by females of *Lytechinus variegatus* may regulate the seasonal variation in the fertilization of the sea urchin. Glycobiol 17:877–885

Clark D, Lamare M, Barker M (2009) Response of sea urchin pluteus larvae (Echinodermata: Echinoidea) to reduced seawater pH: a comparison among a tropical, temperate, and a polar species. Mar Biol 156:1125–1137

Clark HL (1907) The apodous holothurians. A monograph of the Synaptidae and Molpadiidae. Smith Contr Knowledge 35:1–231

Clarke JM (1913) Fósseis Devonianos do Paraná. Monografia, Serv Geol Mineral Bras 1:1–353

Clavico EEG, Muricy G, Da Gama BAP, Batista D, Ventura CRR, Pereira RC (2006) Ecological roles of natural products from the marine sponge *Geodia corticostylifera*. Mar Biol 148:479–488

Coelho JAP, Puzzi A, Graça-Lopes R, Rodrigues ES, Preto Jr O (1986) Análise da rejeição de peixes na pesca artesanal dirigida ao camarão sete-barbas (*Xiphopenaeus kroyeri*) no litoral do Estado de São Paulo. Bol Inst Pesca 13:51-61

Contins M, Ventura CRR (2011) Embryonic, larval, and post-metamorphic development of the sea urchin *Cassidulus mitis* (Echinoidea; Cassiduloida): an endemic brooding species from Rio de Janeiro, Brazil. Mar Biol 158:2279–2288

Cooke CW (1955) Some Cretaceous echinoids from the Americas. U.S. Geol Surv prof pap 264-E:87-112

Costa-Lotufo LV, Cunha GMA, Farias PAM, Viana GSB, Cunha KMA, Pessoa C, Moraes MO, Silveira ER, Gramosa NV, Rao VSN (2002) The cytotoxic and embryotoxic effects of Kaurenoic acid, a diterpene isolated from Copaifera langsdorffi oleo-resin. Toxicon 40:1231–1234

Cummings AM, Kavlock RJ (2004) Function of sexual glands and mechanism of sex differentiation. J Toxicol Sci 29:167–178

Deichmann E (1930) The holothurians of the Western part of the Atlantic Ocean. Bull Mus Comp Zoöl Harvard Coll 71:43–226

Deichmann E (1963) Shallow water holothurians known from the Caribbean waters. Stud Fauna of Curacao and other Carib Islands 14:100–118

Del'Arco J, Silva RH, Tarapanoff I, Freire FA, Pereira LGM, Souza SL, Luz DS, Palmeira RC, Barros, Tassinari CCG (1982) Geologia. In: Projeto Radambrasil: Levantamento de recursos naturais. Folha SE.21, Corumbá, 27:25-160

Derby OA (1878) A Geologia da Região Diamantífera da Província do Paraná. Arq Mus Nac 3:89–96
Derby OA (1907) The sedimentary belt of the coast of Brazil. J Geol 15:218–237
Dupont S, Dorey N, Thorndyke M (2010) What meta-analysis can tell us about vulnerability of marine biodiversity to ocean acidification? Est Coast Shelf Sci 89:182–185
Environmental Canada (1992) Biological Test Method: Fertilization assay using echinoids (sea urchins and sand dollars). Environmental Protection Series-EPS 1/RM/27, Environmental Canada, Ottawa
Environmental Canada (2011) Biological test method: fertilization assay using echinoids (sea urchins and sand dollars). Environmental Protection Series-EPS 1/RM/27. Canada. Ottawa
EPA (United States Environmental Protection Agency) (1995) Short-term methods for estimating the chronic toxicity of effluents and receiving waters to west coast marine and estuarine organisms. Report EPA/600/R-95-136. US Environmental Protection Agency, Cincinati
Erichsen AI, Löfgreen A (1940) Geologia de Goiás a Cuiabá. Bol Dep Nac Prod Min, Div Mineral Geol 102:40
Faria MT, Silva JRMC (2008) Innate immune response in the sea urchin *Echinometra lucunter* (Echinodermata). J Invert Pathol 98:58–62
Fernandes JMG, Morais MHC (1994) Uma nova espécie de cassidulóide (Echinodermata:Echinoidea) da Formação Pirabas (MiocenoInferior), Pará, Brasil: *Ryncholampas candidoi* n.sp. Bol Mus Paraense Emílio Goeldi Sér Ciên T. Terra 6:53–58
Ferré B, Bengtson P (1997) An articulated roveacrinid from the Turonian of the Sergipe Basin, Brazil. In: Bengtson P, Greiling R, Schweizer V (eds) IAS Reg Europ Meet Sedimen 18. Gaea heidelbergensis, Heidelberg 3:128-129
Ferré B, Berthou PY, Bengtson P (1996) Apport des crinoides rovéacrinidés à la stratigraphie du Crétacé Moyen du bassin de Sergipe (nordeste, Brésil). Strata 8:101–103
Ferreira CS, Fernandes ACS (1985) Nota sobre alguns crinóides do Devoniano da Amazônia. An Acad Bras Ciên 57:1–39
Ferreira CS, Fernandes ACS (1989) Crinóides do devoniano do Brasil. Res Congr Bras Paleontol, Curitiba11:23
Foeter SR, Soares-Gomes A, Hajdu E (2009) Biogeografia marinha. In: Pereira RC, Soares-Gomes A (eds) Biologia, Marinha edn. Interciência, Rio de Janeiro, pp 421–441
Franchito SH, Oda TO, Rao VB, Kayano MT (1998) The effect of coastal upwelling on the sea-breeze circulation at Cabo Frio, Brazil: a numerical experiment. An Geophys 16:866–881
Freire CAO, Grohmann PA (1989) *Leptosynapta brasiliensis*: a new species of synaptid holothurian (Echinodermata) from a sandy beach in southeastern Brazil. Rev Bras Zool 6:719–723
Freire CA, Santos PJP, Fontoura NF, Magalhães RAO, Grohmann PA (1992) Growth and spatial distribution of *Cassidulus mitis* (Echinodermata: Echinoidea) on a sand beach in south-eastern Brazil. Mar Biol 112:625–630
Freire CA, Santos IA, Vidolin D (2011) Osmolality and ions of the perivisceral coelomic fluid of the intertidal sea urchin *Echinometra lucunter* (Echinodermata: Echinoidea) upon salinity and ionic challenges. Zoologia 28:479–487
Freitas JC, Ancona-Lopez AAA, Galvão JG (1973) Sobre *Semperothuria surinamensis* (Ludwig, 1875) (Holothuroidea - Aspidochirota) da Ilha das Cabras (litoral de São Paulo). C. Ciên Cult 25:359
Fusetani N (1987) Marine metabolites which inhibit development of echinoderm embryos. In: Scheur PJ (ed) Bioorganic Marine Chemistry. Springer-Verlag, Heidelberg, pp 61–92
Gage JD, Tyler PA (1991) Deep-Sea Biology - A Natural History of Organisms at the Deep-Sea Floor. Cambridge Univ Press, Cambridge
Gage JD, Billett DSM, Jensen M, Tyler PA (1985) Echinoderms of the Rockall Trough and adjacent areas. 2. Echinoidea and Holothuroidea. Bull Brit Mus Nat Hist (Zool) 48:173–212
Gage JD, Pearson M, Clarck AM, Paterson GLJ, Tyler PA (1983) Echinoderms of the Rockall Trough and adjacent areas. 1. Crinoidea, Asteroidea and Ophiuroidea. Bull Brit Mus Nat Hist (Zool) 45:263–308

Ghiold J, Hoffman A (1986) Biogoegraphy and biogeographic history of Clypeasteroid echinoids. J Biogeogr 13:183–206

Gondim AI, Manso CLC, Alonso C (2010) First record of *Ophionereis dolabriformis* (Echinodermata: Ophiuroidea: Ophionereididae) for the Brazilian coast. Mar Biodiv Rec 3:1–4

Gondim AI, Lacouth P, Alonso C, Manso CLC (2008) Echinodermata da Praia do Cabo Branco, João Pessoa, Paraíba, Brasil. Biota Neotrop 8:151–159

Gooday AJ, Turley CM (1990) Responses by benthic organisms to inputs of organic material to the ocean floor: a review. Phil Trans Roy Soc London 331:119–138

Gordon AL (1988) The south Atlantic: an overview of results from 1983-88 research. Oceanogr (Nov):12-17

Gray JS, Poore GCB, Ugland KI, Wilson RS, Olsgard F, Johannessen Ø (1997) Coastal and deep-sea benthic diversities compared. Mar Ecol Prog Ser 159:97–103

Guerrazzi MC (1999) Padrões populacionais e ecologia de forrageamento de *Echinaster brasiliensis* Müller & Troschel (Echinodermata: Asteroidea) no Canal de São Sebastião, SP. PhD Thesis, Inst Biociên, Univ Est Paulista, Rio Claro

Guille A, Ramos JM (1987) Terres Australes et Antarctiques Françaises. Mission de Recherche. Les repports des campagnes à la mer. MD 55/Brésil à bord du "Marion Dufresne". Reproduzido por INSTAPRINT SA n° 87-03

Hadel VF (1997) Reprodução e comportamento de *Chiridota rotifera* (Echinodermata, Holothuroidea) em laboratório. PhD Thesis, Inst Biociên, Univ São Paulo, São Paulo

Hadel VF, Tiago CG (2004) Echinoderms for all. In: Heinzeller T, Nebelsick JH (eds) Echinoderms: München. AA Balkema, Leiden, pp 25–26

Hadel VF, Monteiro AMG, Ditadi ASF, Tiago CG, Tommasi LR (1999) Echinodermata. In: Migotto AE, Tiago CG (eds) Biodiversidade do Estado de São Paulo, Brasil: síntese do conhecimento ao final do século XX –, vol 3., Invertebrados MarinhosFAPESP, São Paulo, pp 260–271

Haedrich RL, Rowe GT, Polloni PT (1980) The megabenthic fauna in the deep sea south of New England, USA. Mar Biol 57:165–179

Hansen E, Eilertsen HC, Ernstsen A, Genevière AM (2003) Anti-mitotic activity towards sea urchin embryos in extracts from the marine haptophycean *Phaeocystis pouchetii* (Hariot) Lagerheim collected along the coast of northern Norway. Toxicon 41:803–812

Heding SG, Panning A (1954) Phyllophoridae. Eine Bearbeitung der Polytentaculaten Dendrochiroten Holothurien des Zoologischen Museums in Kopenhagen. Spolia Zool Mus Hauniensis 13:1–209

Heitor SR (1996) Composição e distribuição de Echinodermata na plataforma continental da região da Bacia de Campos (RJ), Master Thesis, Inst Oceanogr, Univ São Paulo, São Paulo, Brasil

Hendler G, Miller JE, Pawson DL, Kier PM (1995) Sea Stars, Sea Urchins, and Allies: Echinoderms of Florida and the Caribbean. Smith Inst Press, Washington DC

Hopkins T, Fontanella FM, Ventura CRR (2003) Morphological diagnosis of three Brazilian sea stars of the genus *Echinaster* (sub-genus Othilia). In: Féral JP, David B (eds) Echinodem Research 2001. Sweets & Zeitlinger, Lisse, pp 97–103

Hyman LH (1955) The invertebrates (Vol 4): Echinodermata. The coelomate Bilateria, McGraw-Hill, New York

IBAMA (Instituto Brasileiro do Meio Ambiente e dos Recursos Naturais Renováveis) (2008) Macrodiagnóstico da Zona Costeira e Marinha do Brasil. MMA, Brasília

von Ihering H (1897) A Ilha de São Sebastião. Rev Mus Paulista 2:129–172

Jennings S, Kaiser MJ (1998) The effects of fishing on marine ecosystems. Adv Mar Biol 34: 203–351

Junqueira AOR (1998) Biologia populacional de Lytechinus variegatus (Lamarck, 1816) em habitats contrastantes do litoral do Rio de Janeiro, Brasil. PhD Thesis, Inst Biociên, Univ São Paulo, São Paulo

Junqueira AOR, Ventura CRR, Carvalho ALPS, Schmidt AJ (1997) Population recovery of the sea urchin *Lytechinus variegatus* in a seagrass flat (Araruama Lagoon, Brazil): the role of recruitment in a disturbed environment. Invert Rep Develop 31:143–150

Kraatz-Koschlau KA, Huber J (1900) Zwischen ocean und Guamá. Beitrag zur Kenntniss des Staates Pará. Mem Mus Hist Nat Ethnographia, Belém 2:1–34

Krau R (1950) Observações sobre os equinodermas da Baía de Guanabara. Mem Inst Oswaldo Cruz 48:357–362

Kobayashi N (1973) Studies on the effects of some agents on fertilized sea urchin eggs, as a part of bases for marine pollution bioassay. Publ Seto Mar Biol Lab 21:109–114

Kroeker KJ, Kordas RL, Crim RN, Singh GG (2010) Meta-analysis reveals negative yet variable effects of ocean acidification on marine organisms. Ecol Let 13:1419–1434

Lana P, Camargo MG, Brogim RSE, Isaac VJ (1996) O bentos da costa brasileira. Avaliação crítica e levantamento bibliográfico (1858-1996). FEMAR, Rio de Janeiro

Lange FW (1943) Novos fósseis Devonianos do Paraná. Arq Mus Paranaense 3:215–231

Lavrado HP (2006) Caracterização do ambiente e da comunidade bentônica. In: Lavrado HP, Ignácio BL (eds) Biodiversidade Bentônica da Região Central da Zona Econômica Exclusiva Brasileira. Museu Nacional, Rio de Janeiro, pp 19–64

Lavrado HP, Brasil ACS (eds) (2010) Biodiversidade da Região Oceânica Profunda da Bacia de Campos: Megafauna e Ictiofauna Demersal. SAG Serv, Rio de Janeiro

Lavrado HP, Brasil ACS, Curbelo-Fernandez MP, Campos LS (2010a) Aspectos gerais da macrofauna bentônica da Bacia de Campos. In: Lavrado HP, Brasil ACS (eds) Biodiversidade da Região Oceânica Profunda da Bacia de Campos: Macrofauna. SAG Serv, Rio de Janeiro, pp 19–27

Lavrado HP, Brasil ACS, Curbelo-Fernandez MP, Campos LS (2010b) Aspectos gerais da fauna demersal da Bacia de Campos. In: Lavrado HP, Brasil ACS (eds) Biodiversidade da Região Oceânica Profunda da Bacia de Campos: Macrofauna. SAG Serv, Rio de Janeiro, pp 23–29

Lima EJB, Gomes PB, Souza JRB (2009) Reproductive biology of *Echinometra lucunter* (Echinodermata: Echinoidea) in a Northeast Brazilian sandstone reef. An Acad Bras Ciên 81:51–59

Lima-Verde JS (1969) Primeira contribuição ao inventário dos equinodermas do nordeste brasileiro. Arq Ciên Mar 9:9-13

Lima-Verde JS, Matthews HR (1969) On the feeding habits of the sea star *Luidia senegalensis* (Lamarck) in the State of Ceará. Arq Ciên Mar 9:173-175

Ljungman A (1867) On Nagranyaarter of Ophiuroider. Ofveraigt af Kongl Vetenskaps-Akademiens Forhandiingar 23:163–166

Löfgreen A (1937) Notas sobre o Devoniano do estado de Goyaz. In: Oliveira EP (ed) Relat anual do Diretor PARA 1936. Serviço Geológico e Mineralógico,

Losso C, Volpi-Ghirardini A (2010) Overview of ecotoxicological studies performed in the Venice Lagoon (Italy). Environ Int 36:92–121

Ludwig H (1881) Über eine lebendiggebärende Synaptide und zwei andere neue Holothurienarten der brazilianischen Küste. Arch Biol 2:41–58

Ludwig H (1882) Verzeichruiss der Von Prof Dr Ed van Beneden an der Küste von Brasilien Gesammelten Echinodermen. Mém couronnés Mém sauvants étrangers 44:1-26

Luederwaldt H (1929a) Resultados de uma excursão scientifica á Ilha de São Sebastião no litoral do Estado de São Paulo e em 1925. Rev Mus Paul 16:1–79

Luederwaldt H (1929b) Errata, addições e modificações aos "Resultados de uma excursão scientifica á Ilha de São Sebastião em 1925". Rev Mus Paulista 16:1011–1019

MacCord FS, Duarte LFL (2002) Dispersion in population of *Tropiometra carinata* in the São Sebastião Channel, São Paulo State, Brazil. Est Coast Shelf Sci 54:219–225

MacCord FS, Ventura CRR (2004) Reproductive cycle of the endemic cassiduloid *Cassidulus mitis* (Echinoidea: Cassiduloida) on the Brazilian coast. Mar Biol 145:603–612

Machado ABM, Martins CS, Drummond GM (eds) (2005) Lista da fauna brasileira ameaçada de extinção. Fundação Biodiversitas, Belo Horizonte

Machado ABM, Drummond GM, Pagllia AP (eds) (2008) Livro vermelho da fauna brasileira ameaçada de extinção. Vol. I - Invertebrados, MMA/Fundação Biodiversitas, Brasília/Belo Horizonte

Mahiques MM, Bícego MC, Silveira ICA, Souza SHM, Lourenço RA, Fukumoto MM (2005) Modern sedimentation in the Cabo Frio upwelling system, Southeastern Brazilian shelf. An Acad Bras Ciên 77:535–548

Mangiaterra MBBCD, Silva JRMC (2001) Induced inflammatory process in the sea urchin (*Lytechinus variegatus*) J Invert. Biol 120:178–184

Mastroti RR (1997) Toxicidade e biodegradabilidade de tensoativos aniônicos em água do mar. Universidade de São Paulo, São Paulo, Master Thesis

Manso CLC (1988a) Uma nova espécie de *Ophiactis* (Echinodermata: Stelleroidea) da costa sudeste do Brasil. Rev Bras Biol 48:375–379

Manso CLC (1988b) Sobre a ocorrência de *Ophionema intricata* Lütken, 1869 e *Ophiophragmus brachyatis* H.L. Clark, 1915 no Brasil (Echinodermata, Ophiuroidea, Amphiuridae). Rev Bras Biol 48:965–968

Manso CLC (1988c) Sobre a ocorrência de *Amphilepis sanmatiensis* Bernasconi and D'Agostino, 1975 na costa sul do Brasil (Echinodermata: Ophiuroidea). Rev Bras Biol 48:371–373

Manso CLC (1988d) Ophiuroidea: situação pré-operacional nos sacos de Pirarquara, região sob influência da descarga da Central Nuclear Almirante Álvaro Alberto (CNAA). Rev Bras Biol 48:75–82

Manso CLC (1988e) Ofiuróides da plataforma continental brasileira. Parte I: Rio de Janeiro (Echinodermata, Ophiuroidea). Rev Bras Biol 48:845–850

Manso CLC (1989) Os Echinodermata da plataforma continental interna entre Cabo Frio e Saquarema, Rio de Janeiro, Brasil. Rev Bras Biol 49:355–359

Manso CLC (1993) Ofiuróides da plataforma continental brasileira. Parte II: Norte do Estado do Rio de Janeiro, Estado do Espírito Santo, Sul do Estado da Bahia e Bancos Royal Charlotte, Hostpur e Davis (Echinodermata: Ophiuroidea). Rev Bras Biol 53:189–195

Manso CLC (2001) Paleogeografia dos equinóides (Echinodermata) do Aptiano-Albiano do nordeste do Brasil. Res XVII Congr Bras Paleontol, Rio Branco:104

Manso CLC (2003a) Paleoecologia, Paleobiogeografia e Sistemática dos Equinóides do Aptiano-Albiano (Cretáceo) do Brasil. PhD Thesis, Univ Fed Bahia, Salvador

Manso CLC (2003b) Os equinóides (Echinodermata) da Bacia Potiguar (RN): Estado da Arte. Res Reunião An Reg Soc Bras Paleontol:17

Manso CLC (2004) Echinodermata da Bacia de Camamu, Bahia, Brasil. Biol Geral Exp 5:19–25

Manso CLC (2006) Primeiro registro de Goniasteridae (Echinodermata:Asteroidea) no Cretáceo do Brasil: Turoniano da Bacia Potiguar. Geociên 25:255–260

Manso CLC, Andrade EJ (2008) Equinóides do Turoniano (Cretáceo Superior) de Sergipe, Brasil. Geociên 27:319–327

Manso CLC, Hessel MH (2007) Revisão sistemática de *Pygidiolampas araripensis* (K. Beurlen, 1966) (Echinodermata: Cassiduloida) da bacia do Araripe, nordeste do Brasil. Geociên 26:271–277

Manso CLC, Lemos ACC (2008) Os Echinoidea (Echinodermata) da localidade Catete Velho I (Cretáceo) em Sergipe, Nordeste do Brasil. Rev Bras Paleontol 11:129–138

Manso CLC, Souza-Lima W (2003a) Os equinóides (Echinodermat) fósseis da bacia de Pernambuco-Paraíba: Estagio atual do Conhecimento. Res Reunião An Reg Soc Bras Paleontol, Natal:18

Manso CLC, Souza-Lima W (2003b) O equinóide *Douvillaster* Lambert, 1917, na Formação Riachuelo, bacia de Sergipe. Rev Bras Paleontol 5:29–37

Manso CLC, Souza-Lima W (2003c) O registro do equinóide *Hemiaster zululandensis* Besaire & Lambert, 1930, no Cretáceo (Albiano Superior) de Sergipe. Rev Bras Paleontol 6:61–67

Manso CLC, Souza-Lima W (2005) Os equinóides *Pygorhynchus colombianus* (Cooke, 1955) e *Pseudholaster altiusculus* (White, 1887) no Neoalbiano da bacia de Sergipe, Brasil. Rev Bras Paleontol 8:229–238

Manso CLC, Souza-Lima W (2007) *Holosalenia bahiensis*, um novo equinóide (Echinodermata) do Albiano da Bacia de Camamu, Brasil. Rev Bras Paleontol 10:229–238

Manso CLC, Souza-Lima W (2010) *Cuscuzispina riachuelensis* new genera and new species, a new echinoid from the Cretaceous (Albian) of Sergipe-Alagoas Basin, Brazil. Arq Mus Nac 68:41–46

Manso CLC, Alves OFS, Martins LR (2008) Echinodermata da Baía de Todos os Santos e da Baía de Aratú (Bahia, Brasil). Biota Neotrop 8:179–196

Maranho LA, Abreu IM, Santelli RE, Cordeiro RC, Soares-Gomes A, Moreira LB, Morais RD, Abessa, DMS (2010) Acute and chronic toxicity of sediments samples from Guanabara Bay (RJ) during the rainy period. Braz J Oceanogr 58(Suppl IV SOB):77-85

Marcgrave J (1942) Historia Natural do Brasil. Imprensa Oficial do Estado, São Paulo

Martinez S (2008) Shallow water Asteroidea and Ophiuroidea of Uruguay: composition and biogeography. Rev Biol Trop 53(suppl 3):205–214

Martinez S, Mooi R (1997) *"Karlaster" pirabensis* from the brazilian Miocene is a species of the *Abertella* (Scutellina, Echinidea), not an mophorasterid. Bol Res Congr Bras Paleontol 15, São Pedro:61

Maury CJ (1925) Fósseis terciários do Brasil, com descripção de novas formas cretáceas. Monografia 4, Serviço de Geologia e Mineralogia do Brasil

Maury CJ (1930) O Cretaceo da Parahyba do Norte. Monographia 8, Serviço de Geologia e Mineralogia do Brasil

Maury CJ (1934a) Fossil Inverterbrata from northeastern Brazil. Bull Am Mus Nat Hist 67: 123–179

Maury CJ (1934b) *Lovenilampas*, a new echinoidean genus from the Cretaceous of Brazil. Bull Am Mus Nat Hist 744:1–5

Maury CJ (1937) O Cretáceo de Sergipe. Monographia 11, Serviço de Geologia e Mineralogia do Brasil

Máximo MV, Mottola LSM, Resgalla JrC (2008) Sensibilidade do ouriço *Arbacia lixula* (Echinodermata: Echinoidea) em testes de toxicidade. J Braz Soc Ecotoxicol 3:47-52

Mendes FM, Marenzi AWC, Di Domenico M (2006) Population patterns and seasonal observations on density and distribution of *Holothuria grisea* (Holothuroidea:Aspidochirotida) on the Santa Catarina Coast, Brazil. SPC Beche-de-mer Inf Bull 23:5–10

Micael J, Alves MJ, Costa AC, Jones MB (2009) Exploitation and conservation of echinoderms. Ann Rev Oceanogr Mar Biol 47:191–208

Miloslavich P, Klein E, Díaz JM, Hernández CE, Bigatti G, Campos L, Artigas F, Castillo J, Penchaszadeh PE, Neill PE, Carranza A, Retana, MV, Díaz de Astarloa JM, Lewis M, Yorio P, Piriz ML, Rodríguez D, Yoneshigue-Valentin Y, Gamboa L, Martín A (2011) Marine biodiversity in the Atlantic and Pacific coasts of South America: knowledge and gaps. PLoS ONE 6:e14631. doi:10.1371/journal.pone.0014631

Miyaji C (2001) Gastrópodes prosobrânquios da plataforma continental externa e talude superior da costa sudeste brasileira. PhD Thesis, Univ São Paulo, São Paulo

Monteiro AMG (1987) Ophiuroidea (Echinodermata) da região de Ubatuba (SP) - Aspectos morfológicos e ecológicos. PhD Thesis, Univ São Paulo, São Paulo

Monteiro AMG (1990) Ophiuroidea (Echinodermata) in the coastal system of the State of São Paulo (Brazil). An II Simp Ecosist Costa Sul e Sudeste Bras: Estrutura, função e manejo. Águas de Lindóia 1:186–201

Monteiro AMG, Reis MO, Pardo EV (1992) Morfologia comparativa e distribuição batimétrica de duas espécies de Ophiuroidea, na região costeira de Ubatuba. Bol Inst Oceanogr São Paulo 40:39–53

Morais MHC (1998) Equinóides regulares da Formação Pirabas (Oligo-mioceno, Pará, Brasil: Sistemática, Master Thesis, Univ Fed Rio de Janeiro, Rio de Janeiro, Taxonomia e Paleobiologia

Morais MHC, Fernandes JMG (1997) Equinóides regulares da Formação Pirabas (Eomioceno, Pará, Brasil. Bol Res Congr Bras Paleontol 15, São Pedro:60

Morgado EH, Tanaka MO (2001) The macrofauna associated with the bryozoan *Schizoporella errata* (Walters) in southeastern Brazil. Scien Mar 65:173-181
Mourão PAS, Guimarães MAM, Mulloy B, Thomas S, Gray E (1998) Antithrombotic activity of a fucosylated chondroitin sulphate from echinoderm: sulphated fucose branches on the polysaccharide account for its antithrombotic action. Brit Jf Hematol 101:647–652
Moussatché H (1949) O músculo das *Holotúrias* (*Holothuria grisea*) como método de ensaio biológico para verificar presença de acetilcolina. Rev Bras Biol 3:525
Müller F (1898) Observações sobre a fauna marinha da costa de Santa Catarina. Rev Mus Paulista 3:31–40
Nascimento IA, Sousa ECPM, Nipper M (2002) Métodos em Ecotoxicologia Marinha. Aplicações no Brasil. Ed Artes Gráficas e Indústria Ltda, São Paulo
Netto LF, Hadel VF, Tiago CG (2005) Echinodermata from São Sebastião Channel (São Paulo, Brazil). Rev Biol Trop 53(Suppl3):207-218
Nilin J, Castro CB, Pimentel MP, Franklin JrW, Matos RFG, Lotufo TMC, Costa-Lotufo LV (2007) Water toxicity assessment of the Ceará river estuary (Brazil). J Braz Soc Ecotoxicol 2:107-113
Nilin J, Castro CB, Pimentel MF, Franklin JrW, Matos RFG, Costa-Lotufo LV (2008) Evaluation of the viability of a microscale method for the short-term chronic toxicity test using *Lytechinus variegatus* embryos. Pan-Am J Aquat Sci 3:122-129
Nipper M (2000) Current approaches and future directions for contaminant-related impact assessments in coastal environments: Brazilian perspective. Aquat Ecosyst Health Manag 3:433–447
O'Hara TD, Poore GCB (2000) Patterns of distribution for southern Australian marine echinoderms and decapods. J Biogeogr 27:1321–1335
Oliveira PE (1957) Invertebrados cretácicos do fosfato de Pernambuco. Bol Dep Nac Prod Min, Dir Geol Mineral 172:29
Perina FC, Abessa DMS, Pinho GLL, Fillmann G (2011) Comparative toxicity of antifouling compounds on the development of sea urchin. Ecotoxicol 20:1870–1880
Petersen JA (1976) Aspects of gas exchange in ophiuroids from the coast of Brazil. Thalassia Jugoslavica 12:295–296
Petri S (1948a) Contribuição ao Estudo do Devoniano Paranaense. Bol Dep Nac Prod Min 129: 1–120
Petri S (1948b) Transgressões marinhas no Devoniano na América do Sul. Bol Fac Ciên Letras USP:121-128
Petti MAV (1997) Macrofauna Bentônica de Fundos Inconsolidados das Enseadas de Picinguaba e Ubatumirim e Plataforma Interna Adjacente, Ubatuba, São Paulo. PhD Thesis, Universidade de São Paulo, São Paulo
Pires-Vanin AMS (1993) A macrofauna bêntica da plataforma continental ao largo de Ubatuba, São Paulo, Brasil. Pub Esp Inst Oceanogr São Paulo 10:137–158
Pires-Vanin AMS, Corbisier TN, Arasaki E, Möelmann AM (1997) Composição e distribuição espaço-temporal da fauna bêntica no Canal de São Sebastião. Rel Téc Inst Oceanogr Univ São Paulo 41:29–46
Price ARG, Keeeling MJ, O'Callaghan CJ (1999) Ocean-scale patterns of 'biodiversity' of Atlantic asteroids determined from taxonomic distinctness and other measures. Biol J Lin Soc 66:187–203
Prósperi VA (1993) Aplicação de testes de toxicidade com organismos marinhos para análise de efluentes industriais lançados em áreas estuarinas. Univ São Paulo, Escola de Engenharia de São Carlos, São Carlos, Master Thesis
Prósperi VA (2002) Comparação de métodos ecotoxicológicos na avaliação de sedimentos marinhos e estuarinos. PhD Thesis, Univ São Paulo, Escola de Engenharia de São Carlos, São Carlos
Prósperi VA, Araújo MMS (2002) Teste de toxicidade crônica com *Lytechinus variegatus* Lamarck, 1816 e *Echinometra lucunter* Linnaeus, 1758 (Echinodermata: Echinoidea). In:

Nascimento IA, Sousa ECPM, Nipper M (eds) Métodos em ecotoxicologia Marinha: Aplicações no BrasilEd. Artes Gráficas, São Paulo, pp 99–110
Quast M (2003) Moluscos Bivalves (Arcoida e Ostreoida) da costa Sudeste do Brasil. Univ Est Campinas, Campinas, Master Thesis
Rathbun R (1879) A list of the Brazlian echinoderms with notes on their distribution. Trans Conn Acad Arts Sci 5:139–158
JrC Resgalla, Laitano KS (2002) Sensibilidade dos organismos marinhos utilizados em testes de toxicidade no Brasil. Notas Téc Facimar 6:153–163
Ribeiro SM, Omena E, Muricy G (2003) Macrofauna associated to *Mycale microsigmatosa* (Porifera, Demospongiae) in Rio de Janeiro State, Brazil. Est Coast Shelf Sci 57:1–9
Rodovalho N, Gontije RC, Milhomen PS, Lima CCU, Manso CLC (2003) Bacias sedimentares Brasileiras: Bacia de Cumuruxatiba. Phoenix 5:1–4
Sánchez-Jérez CA, Cortez FS, Pereira CDS, Silva SLR (2001) Spatial distribution of the most abundant sea urchin populations on the southeast coast of São Paulo (Brazil). Cien Mar 27:139- 53
Santos MECM (1957) Equinóides miocênicos da Formação Pirabas. Atas Simpósio sobre a biota Amazônica. Geociên 1:407–410
Santos MECM (1958a) Ocorrência de equinóides fósseis no Estado da Bahia. Bol Dep Nac Prod Min, Dir Geol Mineral 14:5
Santos MECM (1958b) Equinóide Miocênicos da Formação Pirabas. *Boletim do Departamento Nacional de Produção Mineral, Divisão de Geologia e Mineralogia*, 179:1-24 + 5 pl
Santos MECM (1960a) *Phymosoma parahybensis* (Maury) um equinóide fossil da formação Itamaracá. Bol Dep Nac Prod Min, Dir Geol Mineral 166:5
Santos MECM (1960b) Equinóides cretácicos do Rio Grande do Norte. Bol Dep Nac Prod Min, Dir Geol Mineral 189:7–26
Santos MECM & Cunha FLS (1959) Sobre *Hemiaster jacksoni* Maury e outros equinóides do Cretáceo Brasileiro. Bol Dep Nac Prod Min, Dir Geol Mineral 186:19
Santos MFL, Lana PC, Silva J, Fachel JG, Pulgati FH (2009) Effects of non-aqueous fluids cuttings discharge from exploratory drilling activities on the deep-sea macrobenthic communities. Deep Sea Res II 56:32–40
Santos-Gouvea IA, Freire CA (2007) Effects of hypo- and hypersaline seawater on the microanatomy and ultrastructure of epithelial tissues of *Echinometra lucunter* (Echinodermata: Echinoidea) of intertidal and subtidal populations. Zool Stud 46:203–215
Saotome K, Sofuni T, Hayashi M (1999) A micronucleus assay in sea urchin embryos. Mutat Res 446:121–127
Sawaya P (1951) Sensibilidade do músculo longitudinal radial de holotúria à acetilcolina (efeito da acetilcolina). Ciên Cult 3:41–42
Sawaya P, Ancona-Lopez AA (1964) Fisiologia dos animais marinhos. In: Vanzolini PE (ed) História natural de organismos aquáticos do Brasil: bibliografia comentada. FAPESP, São Paulo, pp 19–29
Scheffler SM (2003) Nova Ocorrência de cálice de Blastoidea na Formação Ponta Grossa (Devoniano, Bacia do Paraná), Estado do Paraná. Brasil. Paleontol Destaque 44:40
Scheffler SM (2004) Os crinoides e blastóides da Formação Ponta Grossa (Devoniano, Bacia do paraná), Estado do Paraná, Master Thesis, Univ Fed Rio de Janeiro, Rio de Janeiro, Brasil
Scheffler SM (2006) Novos equinodermos do Devoniano paranaense (Formação Ponta Grossa, Bacia do Paraná). Bol Res PALEO, Curitiba:26
Scheffler SM (2007) Equinodermas da formação Ponta Grossa (Devoniano, Bacia do Paraná): grupos conhecidos e perspectivas futuras. In: Carvalho IS, Cassab RCT, Schwanke C, Carvalho MA, Sequeira-Fernandes AC, Rodrigues MAC, Salgado de Carvalho MS, Arai M, Queiroz-Oliveira ME (eds) Paleontologia: Cenários da Vida.Vol 1. Ed Interciên Ltda, pp157-166
Scheffler SM, Fernandes AC (2003) Primeira ocorrência de cálice de Blastoidea na Formação Ponta Grossa (Devoniano, Bacia
Scheffler SM, Fernandes AC (2007a) Crinoidea da Formação Ponta Grossa (Devoniano, Bacia do Paraná), Brasil. Arq Mus Nac 65:83–98

Scheffler SM, Fernandes AC (2007b) Blastoidea da Formação Ponta Grossa (Devoniano, Bacia do Paraná), Brasil. Arq Mus Nac 65:99–112

Scheffler SM, Bolzon RT, Azevedo I (2002) Análise morfológica dos crinóides do afloramento Rio Caniú (Formação Ponta Grossa), Estado do Paraná, Brasil. Acta Geol Leopoldinensia 25:65–76

Scheffler SM, Fernandes AC, Medina VM (2006) Crinoidea da Formação Maecuru (Devoniano da Bacia do Amazonas), Estado do Pará, Brasil. Rev Bras Paleontol 9:235–242

Seeling J, Souza-Lima W, Manso CLC (2000) Os fósseis da bacia de Sergipe-Alagoas: Os equinóides. Phoenix 2:1–3

Selenka E (1867) Beiträge zur Anatomie und Systematik der Holothurien. Z Wiss Zool 17:291–374

Selenka E (1868) Nachtrag zur den Beiträgen zur Anatomie und Systematik der Holothurien. Z Wiss Zool 18:109–119

Shimada-Borges JC, Jensch Jr.BE, Gurgel-Garrido PA, Cunha-Silva JRM (2005) Phagocytic amoebocyte sub populations in the perivisceral coelom of the sea urchin *Lytechinus variegatus* (Lamarck, 1816). J Exp Zool A 303:241-248

Sibuet M, Lawrence JM (1981) Organic content and biomass of abyssal holothuroids (Echinodermata) from the Bay of Biscay. Mar Biol 65:143–147

Silva JRMC (2000) The onset of phagocytosis and identity in the embryo of *Lytechinus variegatus*. J Develop Comp Immunol 24:733–739

Silva JRMC (2002) The role of the phagocytes on embryos some morphological aspects. Microscopy Res Tech 57:498–506

Silva JRMC, Peck L (2000) Induced in vitro phagocytosis of the Antarctic starfish *Odontaster validus* (Koehler, 1906) at 0 °C. Polar Biol 23:225–230

Silva JRMC, Hernandez-Blazquez FJ, Porto-Neto LR, Borges JCS (2001) Comparative study of in vivo and in vitro phagocytosis including germicidal capacity in *Odontaster validus* (Koehler, 1906) at 0 °C. J Invert Pathol 77:180–185

Silva JRMC, Vellutini B, Porto-Neto L, Pressinotti LN, Ramos MC, Cooper EL, Hernandez-Blazquez FJ, Jensch-Junior BE, Borges JCS (2007) Resposta Imune Inespecífica de Animais Ectotérmicos Antárticos sob Temperaturas Polares. Oecol Bras 11:110–121

Silveira ICA, Schmidt ACK, Campos EJD, Godoi SS, Ikeda Y (2000) A corrente do Brasil ao largo da costa leste brasileira. Rev Bras Oceanogr 48:171-183

Sluiter CP (1910) Westindische Holothurien. Zoöl Jahr Suppl 11:331–342

Smith AB, Bengtson P (1991) Cretaceous echinoids from north-eastern Brazil. Foss Strata 31:1–88

Smith AB, Stockley B (2005) The geological history of deep-sea colonization by echinoids: roles of surface productivity and deep-water ventilation. Proc Roy Soc B 272:865–869

Smith CR, Hamilton SC (1983) Epibenthic megafauna of a bathyal basin off Southern California: patterns of abundance, biomass, and dispersion. Deep-Sea Res 30:907–928

Souza-Lima W, Manso CLC (2000) Os fósseis da bacia de Sergipe-Alagoas. Os crinóides e asteróides. Phoenix 2:1–2

Stramma L (1991) Geostrophic transport of the South Equatorial Current in the Atlantic. J Mar Res 49:281–294

Stumpp M, Wren J, Melzner F, Thorndyke MC, Dupont ST (2011) CO_2 induced seawater acidification impacts sea urchin larval development I: Elevated metabolic rates decrease scope for growth and induce developmental delay. Comp Biochem Physiolt A Mol Integr Physiol 160:331–340

Sugni M, Mozzi D, Barbaglio A, Bonasoro F, Carnevali MDC (2007) Endocrine disrupting compounds and echinoderms: new ecotoxicological sentinels for the marine ecosystem. Ecotoxicol 16:95–108

Sumida PYG (1994) Associações bênticas da quebra da plataforma e talude superior ao largo de Ubatuba - SP, Brasil. PhD Thesis, Univ São Paulo, São Paulo

Sumida PYG, Pires-Vanin AMS (1997) Benthic Associations of the Shelfbreak and Upper Slope off Ubatuba-SP, South-eastern Brazil. Est Coast Shelf Sci 44:779–784

Sumida PYG, Yoshinaga MY, Ciotti AM, Gaeta AS (2005) Benthic response to upwelling events off the SE Brazilian coast. Mar Ecol Prog Ser 291:35–42

Tajima K, Silva JRMC, Lawrence JM (2007) Disease in sea urchins. In: Lawrence JM (ed) Edible sea urchin: Biology and Ecology. Elsevier, Oxford, pp167-182

Tavares M (1999) The cruise of the Marion Dufresne off Brazilian coast: account of the scientific results and list of stations. Zoosyst 21:597–605

Tavares YAG, Borzone CA (2006) Reproductive cycle of *Mellita quinquiesperforata* (Leske) (Echinodermata, Echinoldea) in two contrasting beach environments. Rev Bras Zool 23: 573–580

Tavares YAG, Kawall HG, Borzone CA (2004) Biochemical changes the gonad in relation to the reproductive cycle of *Echinometra lucunter* and *Arbacia lixula* in southern Brazil. In: Lawrence JM, Guzman O (eds) Sea-Urchin Fisheries and Aquaculture: Proc Int Conf SeaUrchin Fish Aquacul, Puerto Varas, pp 147–155

Théel H (1882) Report on the Holothurioidea. Part I. "Challenger" Sci Res Zool IV:1-176

Théel H (1886a) Report on the Holothuroidea dredged by H.M.S. Challenger during the years 1873-76. Part II. In: Rep Sci Res Voyage HMS Challenger during the years 1873-76 Zool 14:1-290

Théel H (1886b) Report on the Holothuroidea - Report XXX of Reports on the results of dredging, under the supervision of Alexander Agassiz, in the Gulf of Mexico (1877-78), in the Caribbean Sea (1879-80), and along the eastern coast of the United States during the summer of 1880, by the U.S. Coast Survey Steamer "Blake", Lieut.-Commander C.D. Sigsbee, U.S.N., and Commander J.R. Bartlett, U.S.N., commanding. Bull Mus Comp Zoöl Harvard Coll 13:1-21

Tiago CG (1998) Dendrochirotida (Echinodermata: Holothuroidea) no Canal de São Sebastião. PhD Thesis, Univ São Paulo, São Paulo

Tiago CG, Ditadi ASF (2001) Holothurians from the Brazilian coast: a historical survey. In: Barker M (ed) Echinoderms 2000. Swets & Zeitlinger, Lisse, pp 379–384

Tinoco IM (1963) Fragmentos de equinodermas do Paleoceno de Pernambuco. Arq Geol Univ Pernambuco 4:49–63

Tommasi LR (1957) Os equinodermas do litoral de São Paulo. I. Echinoidea, Crinoidea e Holothuroidea do bentos costeiro. Pap Avulsos Dep Zool São Paulo 13:19–44

Tommasi LR (1958a) El genero *Astropyga* Gray, nuevo para América del Sur (Echin. Echinoidea). Neotrop 4:85–87

Tommasi LR (1958b) Os equinodermas do litoral de São Paulo. II. Diadematidae, Schizasteridae, Brissidae, Cidaroidae (Echinoidea) e Asteroidea do bentos costeiro. Contrib Avulsas Inst Oceanogr Univ São Paulo, sér Oceanogr Biol 2:1–39

Tommasi LR (1959) Equinodermas do Estado do Rio de Janeiro. I - Crinoidea, Asteroidea, Equinoidea e Holothuroidea da região compreendida entre o Cabo dos Búzios e Cabo Frio. An Acad Bras Ciên 31:601–604

Tommasi LR (1960) Echinodermata do Estado do Rio de Janeiro. An Acad Bras Ciên 31(4): 601–604

Tommasi LR (1962) Equinodermos del litoral de San Pablo: sobre el hallazgo de *Tripneustes ventricosus*. Neotrop 8:59–60

Tommasi LR (1963) Notas sobre alguns crinoideos del Brasil (Echinodermata). Neotrop 9:95–102

Tommasi LR (1964a) Equinodermes (menos Holothurioidea) recogidos por el barco "Emilia" del IOUSP, en el litoral de Santa Catarina. Neotrop 10:112–113

Tommasi LR (1964b) Equinodermas (exceto Holotúrias). In: Vanzolini PE (ed) História natural dos organismos aquáticos do Brasil. Bibliografia comentada, FAPESP, São Paulo, pp 271–273

Tommasi LR (1964c) Observações sobre equinóides do Brasil. Rev Bras Biol 24:83–93

Tommasi LR (1965a) *Amphiodia repens* (Lyman 1869) sinônimo de *Amphiodia pulchella* Verril 1899. Contr Avulsas Inst Oceanogr Univ São Paulo, sér Oceanogr Biol 7:1–5

Tommasi LR (1965b) Alguns Amphiuridae (Ophiuroidea) do litoral de São Paulo e de Santa Catarina. Contr Avulsas Inst Oceanogr Univ São Paulo, sér Oceanogr Biol 8:1–9

Tommasi LR (1965c) Lista dos crinóides recentes do Brasil. Contr Avulsas Inst Oceanogr Univ São Paulo, sér Oceanogr Biol 9:1–33

Tommasi LR (1965d) Faunistic provinces of the Western South Atlantic littoral region. An Acad Bras Ciên 37:261–262
Tommasi LR (1966a) Distribuição geográfica de alguns equinodermas do Brasil. Rev Bras Biol 26:239–246
Tommasi LR (1966b) Lista dos equinoides recentes do Brasil. Contr Avulsas Inst Oceanogr Univ São Paulo, sér Oceanogr Biol 11:1–50
Tommasi LR (1967a) Sobre dois Amphiuridae da fauna marinha do Sul do Brasil. Contr Avulsas Inst Oceanogr Univ São Paulo, sér Oceanogr Biol 12:1–5
Tommasi LR (1967b) Observações preliminares sôbre a fauna bêntica de sedimentos moles da Baía de Santos e regiões vizinhas. Bol Inst Oceanogr Univ São Paulo 16:43–65
Tommasi LR (1968) Os ofiuróides existentes nas coleções do Museu de Buenos Aires, coletados do La Plata até 42° latitude sul. Pap Avulsos Dep Zool 21:115–124
Tommasi LR (1969a) Lista dos Holothurioidea recentes do Brasil. Contr Avulsas Inst Oceanogr Univ São Paulo, sér Oceanogr Biol 15:1–29
Tommasi LR (1969b) Nova contribuição à lista dos crinóides recentes do Brasil. Contr Avulsas Inst Oceanogr Univ São Paulo, sér Oceanogr Biol 17:1–8
Tommasi LR (1969c) Os equinodermas da região de Ilha Grande, Estado do Rio de Janeiro. PhD Thesis, Fac Filos Ciên Let, Univ São Paulo, São Paulo
Tommasi LR (1970a) Lista dos asteróides recentes do Brasil. Contr Avulsas Inst Oceanogr Univ São Paulo, sér Oceanogr Biol 18:1–61
Tommasi LR (1970b) Os ofiuróides recentes do Brasil e de regiões vizinhas. Contr Avulsas Inst Oceanogr Univ São Paulo, sér Oceanogr Biol 20:1–146
Tommasi LR (1970c) Observações sôbre a fauna bêntica do complexo estuarino-lagunar de Cananéia (SP). Bol Inst Oceanogr Univ São Paulo 19:43–56
Tommasi LR (1970d) Nota sôbre os fundos detríticos do circalitoral inferior a plataforma continental brasileira ao sul do Cabo Frio (RJ). Bol Inst Oceanogr Univ São Paulo 18:55–62
Tommasi LR (1971a) Equinodermes da região entre o Amapá (Brasil) e a Flórida (E.U.A.). I. Crinoidea. Contr Avulsas Inst Oceanogr Univ São Paulo, sér Oceanogr Biol 23:1–6
Tommasi LR (1971b) Equinodermes do Brasil. I. Sôbre algumas espécies novas e outras pouco conhecidas para o Brasil. Bol Inst Oceanogr Univ São Paulo 20:1–21
Tommasi LR (1971c) Equinodermes do Brasil. II. Equinodermes da Baía do Trapandé, situada no complexo estuarino-lagunar de Cananéia, SP. Bol Inst Oceanogr Univ São Paulo 20:23–26
Tommasi LR (1971d) The echinoderms of the Ilha Grande region (RJ, Brasil). Distribution and abundance of six species up to the isobath of 50 m. In: Costlow JrJD (ed) Fertility of the Sea vol. 2. Gordon and Breach Sci Publ, New York, pp 581-592
Tommasi LR (1972) Equinodermes da região entre o Amapá (Brasil) e a Flórida (E.U.A.). II. Echinozoa. Bol Inst Oceanogr Univ São Paulo 21:15–67
Tommasi LR (1974a) Equinodermes do Brasil. III. Observações sobre algumas espécies coletadas durante as viagens do N/Oc. "Almirante Saldanha". Bol Inst Oceanogr Univ São Paulo 23:1–15
Tommasi LR (1974b) Observações sobre a distribuição batimétrica de seis espécies de Echinodermata na região da desembocadura do Rio Doce (ES). Rev Bras Biol 34:187–189
Tommasi LR (1985) Equinodermes da região da Ilha da Vitória (SP). Relat Int Inst Oceanogr, Univ São Paulo 13:1–4
Tommasi LR (1999) Echinodermata Recentes e Fósseis do Brasil. BDT.<http://www.bdt.fat.org.br/zoologia/echinodermata> (Access in: May 28, 2004)
Tommasi LR, Abreu J (1974) Equinodermes do Brasil. IV. Sobre seis espécies novas de Ophiuroidea da Região ao Largo da Ilha Grande (RJ). Bol Inst Oceanogr, Univ São Paulo 23:17–32
Tommasi LR, Aron MA (1987) Equinodermes da Plataforma Continental Do Sudeste do Estado da Bahia. Relat Int Inst Oceanogr, Univ Sao Paulo 19:1–6S
Tommasi LR, Aron MA (1988) Equinodermes da plataforma continental do sudeste do Estado da Bahia. Relat Int Inst Oceanogr, Univ Sao Paulo 19:1–6
Tommasi LR, Hecht SMCC (1988) Sobre a ocorrência do gênero *Tretocidaris* Mortensen, 1903 (Echinoidea) no Atlântico Sul Ocidental. Relat Int Inst Oceanogr, Univ Sao Paulo 20:1–7

Tommasi LR, Lima-Verde JS (1970) Observações sobre *Cassidulus mitis* Krau, 1954 (Cassiduloida, Echinoidea). Bol Inst Oceanogr Univ São Paulo 18:1–9

Tommasi LR, Oliveira E (1976) Equinodermes do Brasil. V. Sobre algumas espécies coletadas durante viagens do N/Oc. "Prof. Besnard". Bol Inst Oceanogr, Univ São Paulo 25:77–100

Tommasi LR, Castro SM, Sousa ECPM (1988a) Echinodermata coletados durante as campanhas oceanográficas do N/Oc. "Almirante Saldanha" no Atlântico Sul Ocidental". Relat Int Inst Oceanogr, Univ São Paulo 21:1–11

Tommasi LR, Cernea MCW, Condeixa MCG (1988b) Equinodermes coletados pelo N/Oc. "Almirante Saldanha" entre 26°59′S e 38°39′S. Relat Int Inst Oceanogr, Univ São Paulo 22:1–11

Valentin JL (1984) Analyses des paramètres hydrologiques dans la remontée de Cabo Frio (Brésil). Mar Biol 82:259–276

Vellutini BC, Migotto AE (2010) Embryonic, larval, and juvenile development of the sea biscuit *Clypeaster subdepressus* (Echinodermata: Clypeasteroida). PLoS ONE 5:e9654. doi:10.1371/journal.pone.0009654

Ventura CRR (1991) Distribuição, abundância e hábito alimentar de Asteroidea (Echinodermata) de fundos inconsolidados da plataforma continental do Cabo Frio, RJ. MSc Thesis, Mus Nac, Univ Fed Rio de Janeiro, Rio de Janeiro

Ventura CRR (1998) Biologia de *Astropecten brasiliensis* Müller & Troschel, 1842 e *Astropecten cingulatus* Sladen, 1889 (Echinodermata: Asteroidea) na região de ressurgência de Cabo Frio, RJ. PhD Thesis, Inst Biociên, Univ São Paulo, São Paulo

Ventura CRR, Barcellos CF (2004) Instantaneous comparison of reproductive and morphological traits of *Paracentrotus gaimardi* along Brazilian coast. In: Lawrence JM, Guzmán O (eds) Sea urchins: fisheries and ecology. DEStech Publ, Lancaster, Proc Int Conf Sea-Urchin Fish Aquacul, pp 156–163

Ventura CRR, Femandes FC (1995) Bathymetric distribution and population size structure of paxillosid seastars (Echinodermata) in the Cabo Frio upwelling ecosystem of Brazil. Bull Mar Sci 56:268–282

Ventura CRR, Grillo MCG, Fernandes FC (2001) Feeding niche breadth and feeding niche overlap of paxillosid starfishes (Echinodermata: Asteroidea) from a midshelf upwelling region, Cabo Frio, Brazil. In: Barker M (ed) Echinoderms 2000. Balkema, Lisse, pp 227–233

Ventura CRR, Junqueira AOR, Fernandes FC (1994) The relation betwen body size and number of prey in starfish (Echinodermata: Asteroidea). In: David B, Guille A, Feral J-P, Roux M (eds) Echinoderms through time. Balkema, Rotterdam, pp 375–380

Ventura CRR, Lima RNP, Carvalho ALPS (2004) Analysis of gametogenesis of *Echinometra lucunter* (Echinoidea: Echinometridae) from a coastal upwelling rocky-shore and an offshore coral reef environments on Brazilian coast. In: Lawrence JM, Guzmán O (eds) Sea urchins: fisheries and ecology. DEStech Publ, Lancaster, Proc Int Conf Sea-Urchin Fish Aquacul, p 120

Ventura CRR, Falcao APC, Santos JS, Fiori CS (1997) Reproductive cycle and feeding periodicity in the starfish Astropecten brasiliensis in the Cabo Frio upwelling ecosystem (Brazil). Invert Reprod Develop 31:135–141

Ventura CRR, Santos JS, Falcão APC, Fiori CS (1998) Reproduction and food intake in *Astropecten cingulatus* (Asteroidea: Paxillosida) in the upwelling environment of Cabo Frio (Brazil). In: Mooi R, Telford M (eds) Echinoderms San Francisco. Balkema, Rotterdam, pp 313–318

Ventura CRR, Lima RPN, Nobre CC, Veríssimo I, Zama PC (2006) Filo Echinodermata. In: Lavrado HP, Ignacio BL (eds) Biodiversidade Bentônica da Região Central da Zona Econômica Exclusiva brasileira. Série Livros 18, Documentos REVIZEE/SCORE-Central, Mus Nac, Rio de Janeiro, pp 339-389

Ventura CRR, Veríssimo I, Lima RPN, Barcellos CF, Oigman-Pszczol SS (2007) Echinodermata. In: Creed JC, Pires DO, Figueiredo MAO (eds) Biodiversidade marinha da Baía da Ilha Grande. Série Biodiversidade, MMA/SBF, Brasília, pp 273–290

Ventura CRR, Cerqueira WRP, Costa-Lotufo LV, Freire CAO, Hadel VF, Manso CLC, Monteiro AMC, Silva JRMC, Tiago CL (2009) Echinodermata. In: Rocha RM, Boeger WAP (eds)

Estado da arte e perspectivas para a Zoologia no Brasil. Ed UFPR, Série Pesquisa 149, Curitiba, pp183-201

Verrill AE (1868) Notice on the corals and echinoderms collected by Prof. C.F. Hartt at the Abrolhos Reefs, Province of Bahia, Brazil, 1867. Trans Conn Acad Arts Sci 1:351–371

Vidolin D, Santos-Gouvea IA, Freire CA (2002) Estabilidade osmótica dos fluidos celômicos de um pepino do mar (*Holothuria grisea*) e de uma estrela-do-mar (*Asterina stellifera*) (Echinodermata) expostos ao ar durante a maré baixa: um estudo de campo. A. Acta Biol Paranaense 31:113–121

Vidolin D, Santos-Gouvea IA, Freire CA (2007) Differences in ion regulation in the sea urchins *Lytechinus variegatus* and *Arbacia lixula* (Echinodermata : Echinoidea). J Mar Biol Ass UK 87:769–775

White CA (1887) Contribuições á Paleontologia do Brasil. Arq Mus Nac 7:1–273

Zagatto PA (2006) Ecotoxicologia. In: Zagatto PA, Bertoletti E (eds) Ecotoxicologia, Aquática edn. Rima, São Carlos, pp 1–14

Zúñiga M, Roa R, Larrain A (1995) Sperm cell bioassay with the sea urchin *Arbacia spatuligera* on samples from two polluted Chilean coastal sites. Mar Poll Bull 30:313–319

Chapter 10
Echinoderm Research in Uruguay

Sergio Martínez

10.1 Introduction

The eastern 220 km of the Uruguayan coastline are considered marine (salinities around 30–35 ‰), and the remaining ca. 350 km correspond to the Río de la Plata estuary. The Río de la Plata is the second largest river system in South America (drainage basin: 3.1 × 106 km^2, sediment load: 91 × 106 t y^{-1}, discharge of an average of 22,000 m^3s^{-1} water to the ocean) (Framiñan and Brown 1996; Guerrero et al. 1997; Syvitski et al. 2005). Along the coast there is a remarkable salinity gradient, with changing fronts, caused by the discharge of fresh water from the Paraná and Uruguay rivers into the Río de la Plata and from it into the Atlantic Ocean (Fig. 10.1a). Winds change the location of the salinity and turbidity fronts (Larrañaga 1894; Nagy et al. 1987). The Río de la Plata is extremely muddy because of mud derived mainly by the Paraná River.

The Atlantic area is influenced by the warm coastal N–S Brazilian Current and by the cold S–N Malvinas (Falkland) Current (Boltovskoy 1966; Podestá et al. 1991; Lentini et al. 2000; Piola et al. 2000). On the platform there are superficial warm waters and deeper cold ones (Fig. 10.1b). The sub-tropical warm waters converge and mix with the cold subantartic waters at about the isobaths of 100 and 200 m. Other factors such as topography, seasonality, and El Niño-Southern Oscillation (ENSO) phenomena also contribute to the littoral environmental conditions (Olson et al. 1988; Podestá et al. 1991; Lentini et al. 2000; Ortega and Martínez 2007; Raicich 2008). In brief, the estuary and the adjunct platform have complex horizontal and vertical structures complicated by a high degree of

S. Martínez (✉)
Departamento Evolución de Cuencas, Facultad de Ciencias,
Iguá 4225, 11400, Montevideo, Uruguay
e-mail: smart@fcien.edu.uy

J. J. Alvarado and F. A. Solís-Marín (eds.), *Echinoderm Research and Diversity in Latin America*, DOI: 10.1007/978-3-642-20051-9_10,

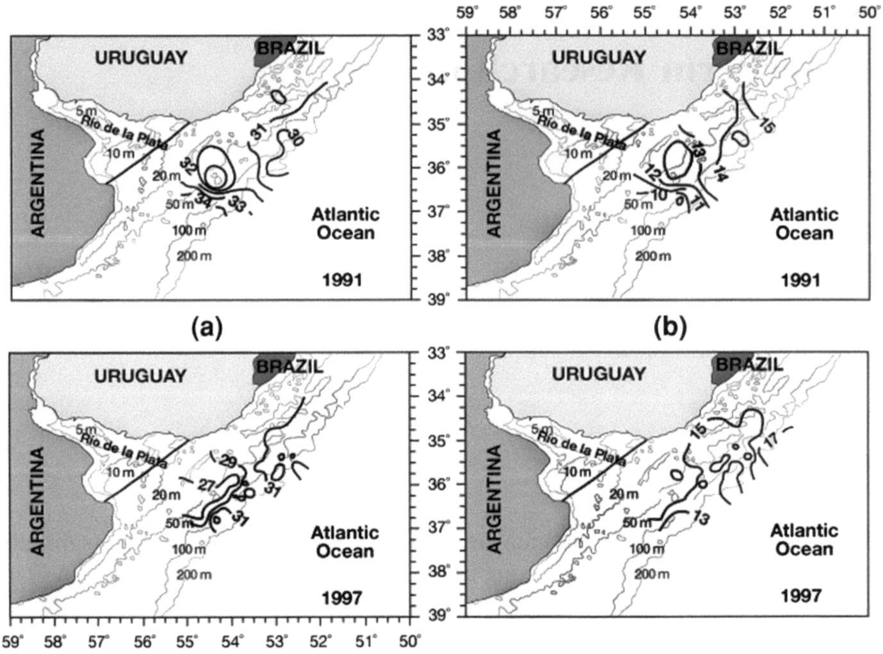

Fig. 10.1 **a** Examples of distribution of salinity (‰) in the Río de la Plata and adjacent platform (from Ortega and Martínez 2007). **b** Examples of distribution of temperature (°C) in the Río de la Plata and adjacent platform (from Ortega and Martínez 2007)

seasonal and inter annual variability. The substrate is dominated by soft sediments, with few areas consolidated or constituted by crystalline rocks (Correia et al. 1996).

The Uruguayan Atlantic coast shows arched (sometimes rather straight), wave dominated sandy beaches of variable size, delimitated by rocky points (Fig. 10.2). They have a wide range of morphodynamic grades, from dissipative to reflective. Coastal morphology is conditioned more by its geological background than by the present dynamics (Brazeiro et al. 2003; Gómez Pivel 2006). Astronomic tides are between 0.4 and 0.6 m, sea level changes induced by the wind being more important (Wells and Daborn 1997; SOHMA 2002).

Jurisdictional issues are regulated by the "Treaty between Uruguay and Argentina concerning the Río de la Plata and the Corresponding Maritime Boundary", signed November 19th 1973 (English version can be consulted in UN 2002, Spanish one in FREPLATA 2011). Apart from the Río de la Plata and Atlantic coastal boundaries, it establishes a "common fishing zone, beyond 12 nautical miles measured from the corresponding coastal baselines, for duly registered vessels flying their flag. Such zone shall be that determined by two arcs of circumferences of a radius of 200 nautical miles, the centers of which are located at Punta del Este (Eastern Republic of Uruguay) and Punta Rasa del Cabo San

Fig. 10.2 Examples of coastal morphology of Uruguay. **a** Lagoon, rocky point, and rather straight beaches at each side. **b** Rocky points and arched beaches in between

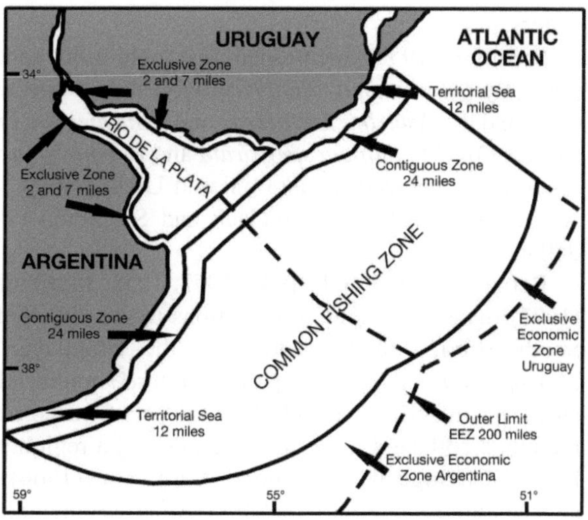

Fig. 10.3 Political limits of Uruguayan waters (modified from FREPLATA)

Antonio (Argentine Republic) respectively" (Fig. 10.3). It also provides a "Joint Technical Commission …which shall be responsible for conducting studies and adopting and co-coordinating plans and measures for the conservation, preservation and rational exploitation of living resources and the protection of the marine environment in the common zone binational organisms…"

Uruguay has a National System of Protected Areas (SNAP). In the Atlantic zone of influence, there are two coastal marine protected areas (from the terrestial coast seaward some miles: Laguna de Rocha and Cabo Polonio). Two additional areas will be formalized soon (Laguna Garzón and Cerro Verde). There are not exclusively marine Protected Areas (SNAP 2010).

In such a variable and unpredictable context as described above, it is not surprising that recent echinoderm faunas are rather poor. But, as mentioned below, the lack of research must be taken into account. Fossil echinoderms, although also not diverse, include key representatives of the extinct echinoid family Monophorasteridae, the sister group of the abundant and widespread Mellitidae. The aim of this chapter is give an overview of the knowledge about extinct and extant echinoderms of Uruguay.

10.2 Research

Echinoderm research is not yet systematically developed in Uruguay. There is little knowledge of the modern fauna. Evidence of this situation is the few papers that have been devoted to Uruguayan echinoderms.

Although Mortensen (1917) mentioned the presence of *Isometra vivipara* in Uruguay, the given coordinates (33°00′S; 51°10′W) correspond to Brazilian waters.

Barattini (1938) mentioned one ophiuroid species (*Ophioceramis januarii*, presently *Ophioplocus januarii*), two asteroid species (*Asterina marginata*, today classified as *Asterina stellifera*, and *Astropecten brasiliensis*) and two echinoid species (*Mellita quinquesperforata* and *Encope emarginata*; Fig. 10.4b). The same species are included in Barattini and Ureta (1961), a book for the general public. A similar book by Klappenbach and Scarabino (1969) mentions only the sand dollar species.

Bernasconi (1941a, b, 1947, 1953, 1955, 1966) and Bernasconi and D'Agostino (1977) made important contributions to the knowledge of the echinoderms of the Southwestern Atlantic, adding to the Uruguayan fauna one asteroid species (*Astropecten cingulatus*; Fig. 10.4a), three ophiuroid species and an echinoid species. This material was included in papers focused on museum collections or about the fauna collected in vessel campaigns at a regional scale (e.g. "Vema" from the Lamont Geological Observatory). Bernasconi (1964) include the species of asteroids and echinoids of Southern South America, including the Uruguayan species, to already known biogeographical entities (e.g. Argentine and Magellan provinces).

In the framework of a faunistic survey, Milstein et al. (1976) added five ophiuroid species to the Uruguayan fauna, and indicated some already known asteroid and echinoid species. Also in a report of a faunistic survey and ecologic study, Juanicó and Rodríguez-Moyano (1976) added three asteroid and two regular echinoid species to the Uruguayan fauna. Parts of these papers include environmental data.

Tommasi (1970a, b, 1974) and Tommasi et al. (1988a, b) published monographs or detailed surveys concerning Brazilian fauna, but included some Uruguayan species found near the borderline between the two countries. In fact, the only Uruguayan holothuroids identified at the species level can be found in these works.

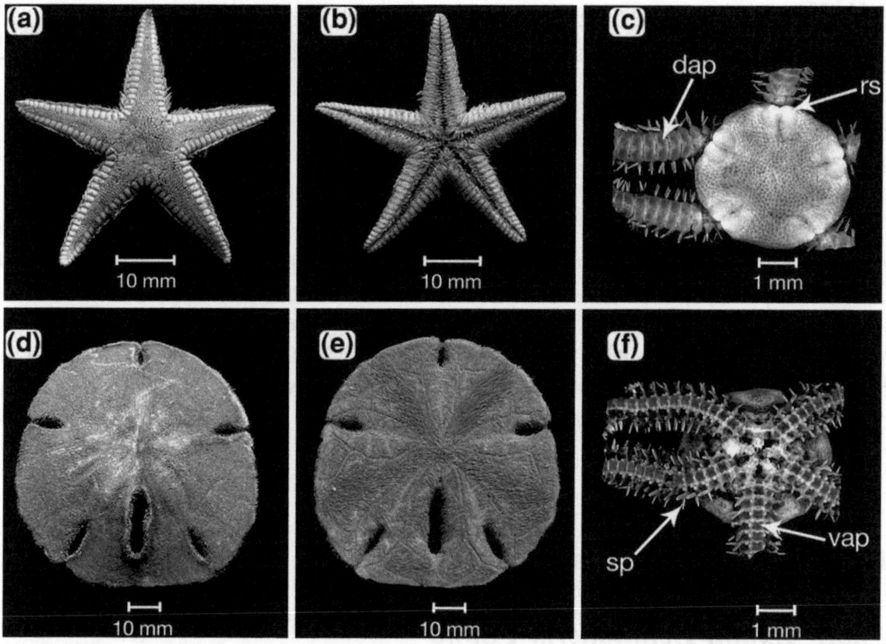

Fig. 10.4 **a** *Astropecten cingulatus*, dorsal view. **b** *A. cingulatus* ventral view. c *Amphiodia* sp., dorsal view (from Carranza et al. 2007). **d** *Encope emarginata*, adapical view. **e** *E. emarginata*, oral view. **f** *Amphiodia* sp., ventral view (from Carranza et al. 2007). Specimens of *A. cingulatus* and *E. emarginata* from S. Martínez's personal collection

Lucchi (1985) identified 17 species of ophiuroids in samples obtained between 10 and 800 m. (continental shelf and slope), being five of them reported for the first time for Uruguay. He and Martínez (2008, with an updated database and including asteroids) made some biogeographical considerations about ophiuroids (see below).

Martínez and Mooi (2005) published a bibliographical review about the fossil and recent sand dollars of Uruguay, recognizing three recent species, two Quaternary, and four Miocene ones. They stressed the importance of the extinct Miocene family Monophorasteridae in the phylogeny of Mellitidae and included some biogeographical observations (see Sect. 3.1).

Scarabino (2006) presents an updated list of the Uruguayan echinoderms reported to a depth of 50 m. He provided a brief account of the characteristics of their habitats. There is only one new record (Carranza et al. 2007) of *Amphiodia* sp. (Fig. 10.4c), an unnamed ophiuroid species different from the other Uruguayan representatives of the genus.

Amaro (1974), Riestra et al. (1992, 1998), Roux and Bremec (1996), Obenat et al. (2001), Giberto et al. (2004), Bremec and Giberto (2004), or Demichelli and Scarabino (2006) have considered Uruguayan echinoderms, adding new localities or habitat characteristics, but without adding new taxa.

Table 10.1 Numbers of echinoderms per taxonomic group in Uruguay

	Crinoidea	Asteroidea	Ophiuroidea	Echinoidea	Holothuroidea	Total
Order	0	3	2	3	3	11
Family	0	4	7	3	3	17
Genus	0	4	15	5	3	27
Species	0	5	24	5	3	37

Puig (1986) found three unidentified holothuroid species and three species of Ophiuroidea in the stomach content of the fish *Micropogonias opercularis*.

Fossil material has been more intensively studied (Martínez 1984, 1994; Martínez and Durham 1988; Mooi et al. 2000; Martínez and Mooi 2005). It is relatively well known but is restricted almost exclusively to Cenozoic echinoids because of their best preservation potential and of the characteristics of the sedimentary units of Uruguay.

10.3 Diversity and Distribution

10.3.1 Extant Echinoderm Fauna

The echinoderm fauna of Uruguay (Table 10.1) is composed of 37 species. Ophiuroidea is by far the richest class with 24 species. Asteroidea has five species; Echinoidea five species; and Holothuroidea three species. There are no species of Crinoidea recorded for Uruguayan waters. Is this species proportion among classes reliable? As commented below, there has been no systematic sampling of the Uruguayan waters, so this question remains open.

Twenty-four of these species have been collected from a depth of less than 50 m, four from between 50 and 100 m, two from between 100 and 200 m, four from 400 m, two from 700–800 m, and one at a wide range of depth (*Amphiura eugeniae*, 46–800 m).

Considering an average depth of 200 m for the shelf break, there are thus only seven species beyond the shelf. Again, it is not clear if this distributional pattern, with species concentrated in the littoral, is reliable, or if it is an artifact caused by the proximity of the coast (easier to collect), and/or by the fact that some collections are a side consequence of commercial fishing (more intense at these depths). Research-oriented faunistic surveys of the Uruguayan waters have been scarce, except for recent years and the collections have not been studied.

Biogeographic considerations have been few. Tommasi (1970a, b) indicated the marine area surrounding the mouth of the Río de la Plata (Rio Grande do Sul State, Brazil, to Buenos Aires Province, Argentina) is an area with reduced diversity. He related this to the latitudinal decrease of temperature, decrease in salinity caused by the discharge of the Río de la Plata, and the relatively cold water mass along the coasts of Rio Grande do Sul.

Lucchi (1985) identified two groups within the ophiuroids she studied, one related to the Malvinas (Falkland) Current and living at a depth greater than 60 m, and the other related to warm waters and living at less a depth than 60 m.

Martínez (2008) identified two distributional patterns of ophiouroid and asteroid species. The first one comprises species with the southern end of their distributional range near Uruguay. This was divided in two sub-groups, depending on whether their distribution extended across the Río de la Plata or not. Species whose distribution is limited by the Río de la Plata, are affected by two main factors in this area: temperature variation of the Subtropical Confluence Zone (ca. 30° to 45°S, see Boltovskoy 1966 for example), where the cold Malvinas (Falkland) and the warm Brazilian currents meet, and the strong variation in salinity and the increase of turbidity of waters caused by fresh water discharge from the Rio de la Plata (Piola et al. 2000, González-Silvera et al. 2004). The "salinity barrier" of the Río de La Plata has been considered an explanation for the truncation of the distribution of other organisms at the Rio de la Plata area, such as mollusks (Scarabino 1977) or sand dollars (Martínez and Mooi 2005). The second group is composed of species distributed within the area of the Argentinean or Patagonian Province (ca. 43°S and ca. 28°S; Scarabino 1977; Boschi 2000), a unit characterized by a complex faunal composition. Tommasi (1970b; ophiuroids), Scarabino (1977; mollusks) and Martínez and del Río (2002; mollusks) suggested this zone may be a true transition zone or "Provinciatone" between the adjacent Magellanic and Brazilian Provinces.

The distribution of sand dollars was studied by Martínez and Mooi (2005) with results and explanations comparable to the sub-groups of asterozoans that do not cross the Rio de la Plata barrier.

10.3.2 Extinct Echinoderm Fauna

There are only eight species of echinoderms in the Uruguayan fossil record, one crinoid species, one ophiuroid species, and six echinoid species. The crinoid and ophiuroid species are from the Devonian.The echinoid species are from the Cenozoic, being abundant in the Late Miocene. During those times (ca. 10 Ma. ago) the Monophorasteridae (sister group of the Mellitidae) were particularly abundant and diverse in what are today Uruguay and Argentina. Two species (*Mellita* sp. and *Encope* sp.) are recorded only for the Quaternary and are still extant in the area.

The following taxonomic arrangement of the fossil species. Suprageneric classification is based on: Crinoidea, Ausich (1998); Holothuroidea, Pawson et al. (2010); Ophiuroidea, Smith et al. (1995) and Shackleton (2005); Echinoidea: Kroh and Smith (2010).

Class Crinoidea J.S.Miller, 1821
Subclass Camerata Wachsmuth and Springer, 1885

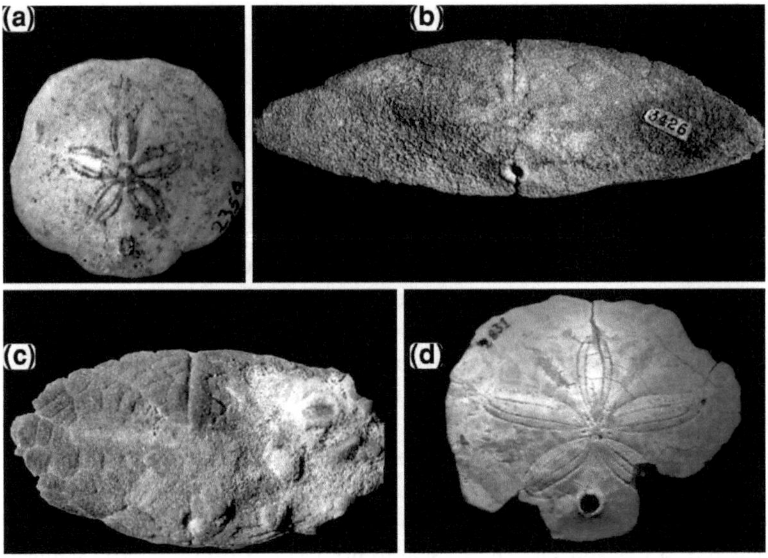

Fig. 10.5 Miocene sand dollars. **a** *Monophoraster duboisi* (x 1.3), FCDP 2359; **b** *Amplaster coloniensis* (x 0.9); holotype MNA-CPO 3426; **c** *Amplaster ellipticus* (x 0.9), paratype MNA-CPO 3425; **d** *Amplaster alatus* (x 1.3), FCDP 2831. From Martínez and Mooi (2005)

Order Monobathrida Moore and Laudon, 1943
Family Melocrinitidae d'Orbigny, 1852
Genus *Ctenocrinus* Bronn, 1840
Ctenocrinus sp.

References: Daners (1990).
Age and Stratigraphic Unit: Early Devonian, Cordobés Fm.
Comments: identified on the basis of columnar plates.

Class Ophiuroidea Gray, 1840
Family Encrinasteridae Schuchert, 1914
Genus *Encrinaster* Haeckel, 1866
Encrinaster pontis (Clarke, 1913)

References: Méndez-Alzola (1938).
Age and Stratigraphic Unit: Early Devonian, Cordobés Fm.
Comments: a single disk. The specimen is lost.

Class Echinoidea Leske, 1778
Order Clypeasteroida L. Agassiz, 1835
Crown group Clypeasteroida
Crown group Scutelliformes
Family Monophorasteridae Lahille, 1896
Genus *Monophoraster* Lambert and Thiéry, 1921
Monophoraster duboisi (Cotteau, 1884)
(Fig. 10.5a)

References: Martínez (1994), Mooi et al. (2000).

Age and Stratigrapic unit: Late Miocene, Camacho Fm.

Comments: Martínez and Mooi (2005) stated. "Prior to these records, all authors (e. g., Goso and Bossi 1966, Figueiras and Broggi 1971) mentioned the presence of *Monophoraster darwini* in Uruguay. In fact, this species is not present in the Miocene of Uruguay and all the known specimens are referable to *M. duboisi*. The species was originally described from the contemporaneous Paraná Formation of Argentina (Cotteau 1884, Mooi et al. 2000)."

Genus *Amplaste*r Martínez, 1984
Amplaster coloniensis Martínez, 1984
(Fig. 10.5b)

References: Martínez (1984, 1994) (partim), Martínez and Durham (1988), Mooi et al. (2000).

Age and Stratigrapic unit: Late Miocene, Camacho Fm.

Comments: This species is endemic to Uruguay.

Amplaster ellipticus Mooi, Martínez and Parma, 2000
(Fig. 10.5c)

References: Mooi et al. (2000).

Age and Stratigrapic unit: Late Miocene, Camacho Fm.

Comments: This species is endemic to Uruguay.

Amplaster alatus (Rossi de García and Levy, 1989)
(Fig. 10.5d)

References: Mooi et al. (2000).

Age and Stratigrapic unit: Late Miocene, Camacho Fm.

Comments: This species was described from the "Patagoniense" strata from Chubut Province (Rossi de García and Levy 1989), but with a controversial age assignment ranging from? Middle Eocene to the? Late Oligocene-Early Miocene (see Mooi et al. 2000).

Family Mellitidae Stefanini, 1912
Genus *Mellita* L. Agassiz, 1841
Mellita sp.

References: Figueiras (1962), Figueiras and Broggi (1967); Martínez and Mooi (2005).

Age and Stratigrapic unit: Holocene, Villa Soriano Formation.

Genus *Encope* L. Agassiz, 1841
Encope sp.

References: Martínez and Mooi (2005)

Age and Stratigrapic unit: Quaternary.

10.4 Aquaculture and Fisheries

There are no echinoderm fisheries or aquaculture currently in Uruguay (see in example Defeo et al. 2009). The effect of fishing of other species (fishes, shrimps) on echinoderm populations has not been evaluated.

10.5 Threats

For most of the Uruguayan coast, baseline studies include echinoderms only incidentally. Hence, it is unknown if species are being affected by anthropogenic activity (e.g., fishing, coastal urbanization), or if the echinoderms are affecting the environment to some extent.

There are not Uruguayan echinoderms considered to be charismatic or bioengineers.

10.6 Recommendations and Concluding Remarks

As shown in this chapter, knowledge of the Uruguayan echinoderm fauna is very poor. Therefore, the priority is to increase substantially all types of research on this group. A reliable inventory of the species must be developed, including basic data such as georeferences, depth, salinity, temperature, sedimentological characteristics, and other species associated with the echinoderms. The lack of national experts on the group undoubtedly has retarded the progress of the knowledge of echinoderms in Uruguay.

Acknowledgments Juan José Alvarado kindly invited me to participate in this book and, along with three anonymous reviewers, made important suggestions on earlier versions of the manuscript. Alejandro Brazeiro provided useful information about the National System of Protected Areas.

References

Amaro J (1974) Noticia sobre una campaña exploratoria a bordo del "Striker" durante el verano de 1971–1972. Bol Com Nac Oceanogr 1:15–17

Ausich WI (1998) Phylogeny of Arenig to Caradoc crinoids (Phylum Echinodermata) and suprageneric classification of the Crinoidea. Paleontol Contrib Univ Kansas 9:1–36

Barattini LP (1938) Equinodermos uruguayos (Contribución al conocimiento de las especies que viven en nuestras aguas). Bol Serv Oceanogr Pesca 1:17–29

Barattini LP, Ureta EH (1961) La fauna de las costas del este (invertebrados). Publicaciones de Divulgación Científica, Museo "Dámaso Antonio Larrañaga". Montevideo, Uruguay

Bernasconi I (1941a) Los equinodermos de la expedición del Buque oceanográfico "Comodoro Rivadavia" A.R.A. Physis 19:37–49
Bernasconi I (1941b) Sobre la distribución geográfica de "Mellita sexiesperforata" (Leske). Physis 19:106–108
Bernasconi I (1947) Una nueva especie de "Mellita" en la República Argentina. Physis 20:117–118
Bernasconi I (1953) Monografía de los equinoideos argentinos. An Mus Hist Nat Montevideo 6:1–58
Bernasconi I (1955) Equinoideos y asteroideos de la colección del Instituto Oceanográfico de la Universidad de San Pablo. Primera contribución. Bol Inst Oceanogr 6:51–77
Bernasconi I (1964) Distribución geográfica de los equinoideos y asteroideos de la extremidad austral de Sudamérica. Bol Inst Biol Mar 7:43–50
Bernasconi I (1966) Los equinoideos y asteroideos colectados por el buque oceanográfico R/V "Vema" frente a las costas argentinas, uruguayas y sur de Chile. Rev Mus Arg Cienc Nat "Bernardino Rivadavia" (Zool) 9:147–17
Bernasconi I, D'Agostino MM (1977) Ofiuroideos del mar epicontinental argentino. Rev Mus Arg Cienc Nat "Bernardino Rivadavia" Inst Nac Inv Cs Nat 5:65–114
Boltovskoy E (1966) La zona de convergencia subtropical/subantártica en el Océano Atlántico (parte occidental) (un estudio en base a la investigación de Foraminíferos indicadores). Serv Hidrogr Nav Arg, Buenos Aires
Boschi EE (2000) Species of decapod crustaceans and their distribution in the American marine zoogeographic provinces. Rev Invest Des Pesq 13:7–136
Brazeiro A, Acha EM, Mianzan HW, Gómez M, Fernández V (2003) Áreas acuáticas prioritarias para la conservación de la integridad ecológica del Río de la Plata y Frente Marítimo. Informe completo. Documento Freplata. http://www.freplata.org/documentos/archivos/documentos_freplata/Prioridad_cons/GeoPriorities.pdf
Bremec CS, Giberto DA (2004) New records of two species of Sabellaria (Polychaeta: Sabellariidae) from the Argentinean biogeographic province. Rev Biol Mar Oceanogr 39:101–105
Carranza A, Borges M, Rodríguez M, Borthagaray AI (2007) Ophiuroidea (Echinodermata) from La Coronilla-Cerro Verde (Uruguay): a new record for the Uruguayan coast. Biota Neotropica 7:103–107
Correia ICS, Villwock JA, Isla FI, López-Laborde J, Jackson JM, Furtado VV, Calliari LJ (1996) ATLAS. Morphology and sedimentology of the Southwest Atlantic coastal zone and continental shelf from Cabo Frio (Brazil) to Península Valdéz (Argentina), Porto Alegre, Ponto UM/UFRGS-IG-CECO
Cotteau MG (1884) Echinides nouveaux ou peu connus. Bull Soc Zool France 7:1–185
Daners G (1990) Presencia de crinoideos (Echinodermata) en la Formación Cordobés (Devónico Inferior, Uruguay). Rev Soc Urug Geol 4:47–48
Defeo O, Horta S, Carranza A, Lercari D, de Álava A, Gómez J, Martínez G, Lozoya JP, Celentano E (2009) Hacia un manejo ecosistémico de pesquerías. Áreas marinas protegidas en Uruguay, Facultad de Ciencias-Dinara
Demichelli M, Scarabino F (2006) Invertebrados bentónicos de La Paloma (Rocha, Uruguay). In: Menafra R, Rodríguez-Gallego L, Scarabino F, Conde D (eds) Bases para la conservación y el manejo de la costa uruguaya. Vida Silvestre Uruguay, Montevideo, pp 523–534
Figueiras A (1962) Sobre nuevos hallazgos de moluscos subfósiles de la Transgresión Querandina. Com Soc Malac Uruguay 1:53–68
Figueiras A, Broggi J (1967) Estado actual de nuestros conocimientos sobre los moluscos fósiles del Uruguay. I Com Soc Malac Uruguay 2:147–186
Figueiras A, Broggi J (1971) Estado actual de nuestros conocimientos sobre los moluscos fósiles del Uruguay. III Com Soc Malac Uruguay 3:131–154
Framiñan MB, Brown OB (1996) Study of the Río de la Plata turbidity front. Part I: spatial and temporal distribution. Cont Shelf Res 16:1259–1282
FREPLATA (Proyecto Protección Ambiental del Río de la Plata y su Frente Marítimo: Prevención y Control de la Contaminación y Restauración de Hábitats) (2011) Tratado del Río de la Plata y

su Frente Marítimo. Montevideo, República Oriental del Uruguay, 19/11/1973.http://www.freplata.org/bases_de_datos/instrumentos_juridicos/internacionales/detalles.asp?id=20
Giberto DA, Bremec CS, Acha EM, Mianzan H (2004) Large-scale spatial patterns of benthic assemblages in the SW Atlantic: the Rio de la Plata estuary and adjacent shelf waters. Estuar Coast Shelf Sci 61:1–13
Gómez Pivel MA (2006) Geomorfología y procesos erosivos en la costa atlántica uruguaya. In: Menafra R, Rodríguez-Gallego L, Scarabino F, Conde D (eds) Bases para la conservación y el manejo de la costa uruguaya. Vida Silvestre Uruguay, Montevideo, pp 35–43
González-Silvera A, Santamaria-del-Angel E, García VMT, García CAE, Millán-Núñez R, Muller-Karger F (2004) Biogeographical regions of the tropical and subtropical Atlantic Ocean off South America: classification based on pigment (CZCS) and chlorophyll-a (SeaWiFS) variability. Cont Shelf Res 24:983–1000
Goso H, Bossi JC (1966) Cenozoico. In: Bossi JC (ed) Geología del Uruguay. Universidad de la República, Montevideo, pp 259–305
Guerrero RA, Acha EM, Framiñan MB, Lasta CA (1997) Physical oceanography of the Río de la Plata estuary, Argentina. Cont Shelf Res 17:727–742
Juanicó M, Rodríguez-Moyano M (1976) Composición faunística de la comunidad de Mytilus edulis platensis d'Orbigny, 1846, ubicada a unas 55 millas al SE de La Paloma. Com Soc Malac Uruguay 4:113–116
Klappenbach MA, Scarabino V (1969) El borde del mar. Nuestra Tierra 2:1–68
Kroh A, Smith AB (2010) The phylogeny and classification of post-Palaeozoic echinoids. J Syst Palaeontol 8:147–212
Larrañaga DA (1894) Memoria geológica sobre la formación del Río de la Plata, deducida de sus conchas fósiles, escrita por los años 1819. An Mus Nac Montevideo 1:3–12
Lentini CAD, Campos EJD, Podestá GG (2000) The annual cycle of satellite derived sea surface temperature on the western South Atlantic shelf. Rev Bras Oceanogr 48:93–105
Lucchi C (1985) Ophiuroidea (Echinodermata) del frente marítimo uruguayo hasta 800 m de profundidad, con claves para su reconocimiento. Contr Depto Oceanogr Fac Hum Cienc 2:115–171
Martínez S (1984) Amplaster coloniensis n.g. n.sp. (Echinoidea: Monophorasteridae) del Mioceno de Uruguay. Mem III Congr Latinoamer Paleont, México 505–508
Martínez S (1994) Bioestratigrafía (Invertebrados) de la Formación Camacho (Mioceno, Uruguay). PhD thesis, Universidad de Buenos Aires, Argentina
Martínez S (2008) Shallow water Asteroidea and Ophiuroidea of Uruguay: composition and biogeography. Rev Biol Trop 56(Suppl 3):205–214
Martínez S, del Río CJ (2002) Late Miocene mollusks from the Southwestern Atlantic Ocean (Argentina and Uruguay): a palaeobiogeographic analysis. Palaeo 188:167–187
Martínez S, Durham JW (1988) La cara oral de Amplaster coloniensis (Echinoidea: Monophorasteridae) (Mioceno Superior, Uruguay). Ameghiniana 25:185–186
Martínez S, Mooi R (2005) Extinct and extant sand dollars (Clypeasteroida: Echinoidea) from Uruguay. Rev Biol Trop 53(Suppl 3):1–7
Méndez-Alzola R (1938) Fósiles devónicos del Uruguay. Bol Inst Geol 24:3–115
Milstein A, Juanicó M, Olazarri J (1976) Algunas asociaciones bentónicas frente a las costas de Rocha, Uruguay. Resultados de la campaña del R/V "Hero", viaje 72-3A. Com Soc Malac Uruguay 4:143–164
Mooi R, Martínez S, Parma SG (2000) Phylogenetic systematics of tertiary monophorasterid sand dollars (Clypeasteroida: Echinoidea) from South America. J Paleontol 74:263–281
Mortensen T (1917) Notocrinus virilis n. g., n. sp. a new viviparous crinoid from the Antarctic Sea. Vidensk Medd Dansk Naturh Foren 68:205–208
Nagy GJ, López-Laborde J, Anastasía L (1987) Caracterización de los ambientes del Río de la Plata exterior (salinidad y turbiedad óptica). Inv Oceanológicas 1:31–56
Obenat S, Ferrero L, Spivak E (2001) Macrofauna associated with Phyllochaetopterus socialis aggregations in the southwestern Atlantic. Vie et Milieu 51:131–139

Olson DB, Podestá GP, Evans RH, Brown OT (1988) Temporal variations in the separation of Brazil and Malvinas Currents. Deep-Sea Res A 35:1971–1990
Ortega L, Martínez A (2007) Multiannual and seasonal variability of water masses and fronts over the Uruguayan shelf. J Coast Res 23:618–629
Pawson DL, Pawson DJ, King RA (2010) A taxonomic guide to the Echinodermata of the South Atlantic Bight USA: 1. Sea cucumbers (Echinodermata: Holothuroidea). Zootaxa 2449:1–48
Piola AR, Campos EJD, Möller OO Jr, Charo M, Martínez C (2000) Subtropical shelf off eastern South America. J Geophys Res 105:6565–6578
Podestá GP, Brown OB, Evans EH (1991) The annual cycle of satellite-derived sea surface temperature in the Southwestern Atlantic Ocean. J Clim 4:457–467
Puig P (1986) Análisis de contenidos estomacales de corvina blanca (Micropogon opercularis) (Sciaenidae, Perciformes). Verano 1984. Publ Com Téc Mixta Frente Mar 1:333–340
Raicich F (2008) A review of sea level observations and low frequency sea-level variability in South Atlantic. Phys Chem Earth 33:239–249
Riestra G, Fabiano G, Foti R, Santana O (1998) Mortandad de organismos bentónicos en la costa atlántica del Uruguay. Com Soc Malac Uruguay 8:73–79
Riestra G, Giménez JL, Scarabino V (1992) Análisis de la comunidad macrobentónica infralitoral de fondo rocoso en Isla Gorriti e Isla de Lobos (Maldonado, Uruguay). Frente Marítimo 11:123–127
Rossi de García E, Levy R (1989) Presencia de Amplaster n.sp. (Echinodermata: Clypeasteroidea) en el Terciario de Patagonia. Actas IV Congr Arg Paleont Bioest, Mendoza 4:89–92
Roux A, Bremec C (1996) Comunidades bentónicas relevadas en las transecciones realizadas frente al Río de la Plata (35°15′S), Mar del Plata (38°10′S) y Península Valdés (42°35′S), Argentina. INIDEP Informe Técnico 11:1–13
Scarabino F (2006) Faunística y taxonomía de invertebrados bentónicos marinos y estuarinos de la costa uruguaya. In: Menafra R, Rodríguez-Gallego L, Scarabino F, Conde D (eds) Bases para la conservación y el manejo de la costa uruguaya. Vida Silvestre Uruguay, Montevideo, pp 113–142
Scarabino V (1977) Moluscos del Golfo San Matías (Provincia de Río Negro, República Argentina). Inventario y claves para su identificación. Com Soc Malac Uruguay 4:177–285
Shackleton JD (2005) Skeletal homologies, phylogeny and classification of the earliest asterozoan echinoderms. J Syst Palaeontol 3:29–114
Smith AB, Paterson GLJ, Lafay B (1995) Ophiuroid phylogeny and higher taxonomy: morphological, molecular and palaeontological perspectives. Zool J Linn Soc 114:213–243
SNAP (Sistema Nacional de Áreas Protegidas) (2010) Plan de mediano plazo 2010/2014. Hemisferio sur, Montevideo
SOHMA (Servicio de Oceanografía, Hidrología y Meteorología de la Armada) (2002) Información oceanográfica que posee el Servicio de Oceanografía, Hidrología y Meteorología de la Armada. Freplata, Proyecto PNUD/GEF-RLA/99/G31, Montevideo. (http://www.freplata.org/documentos/archivos/Documentos_Freplata/Datos_SOHMA.pdf)
Syvitski JMP, Vörösmarty CJ, Kettner AJ, Green P (2005) Impact of humans on the flux of terrestrial sediment to the global coastal ocean. Science 308:376–380
Tommasi LR (1970a) Lista dos asteróides recentes do Brasil. Contr Inst Oceanogr 18:1–16
Tommasi LR (1970b) Os ofiuroides recentes do Brasil e de regiões vizinhas. Contr Avulsas Inst Oceanogr 20:1–146
Tommasi LR (1974) Equinodermes do Brasil. III. Observações sobre algumas espécies coletadas durante as viagens do N/Oc. "Almirante Saldanha". Relatorio Interno Inst Oceanogr 23:1–15
Tommasi LR, de Castro SM, de Sousa ECPM (1988a) Echinodermata coletados durante as campanhas oceanográficas do N/Oc. "Almirante Saldanha" no Atlântico Sul Ocidental. Relatorio Interno Inst Oceanogr 21:1–11
Tommasi LR, Cernea MCW, Condeixa MCG (1988b) Equinodermes coletados pelo N/Oc. "Almirante Saldanha", entre 26°59 e 38°39′S. Relatorio Interno de Inst Oceanogr 22:1–11
UN (United Nations) (2002) Treaty between Uruguay and Argentina concerning the Rio de la Plata and the corresponding maritime boundary, 19 November 1973 (entry into force: 12

February 1974; registration #: 21424; registration date: 17 December 1982. (http://www.un.org/Depts/los/LEGISLATIONANDTREATIES/PDFFILES/TREATIES/URY-ARG1973MB.PDF)

Wells PG, Daborn GR (1997) El Río de la Plata. Una revisión ambiental. Un informe de antecdentes del Proyecto EcoPlata./The Río de la Plata. An environmental overview. An EcoPlata Project background Review. Dalhousie University, Halifax

Chapter 11
Echinoderms from Argentina: Biodiversity, Distribution and Current State of Knowledge

Martín I. Brogger, Damián G. Gil, Tamara Rubilar, Mariano I. Martinez, María Enriqueta Díaz de Vivar, Mariana Escolar, Lucía Epherra, Analía F. Pérez and Alejandro Tablado

---

Martín I. Brogger, Damián G. Gil, Tamara Rubilar These authors contributed equally to this work.

M. I. Brogger (✉) · M. I. Martinez · A. Tablado
Museo Argentino de Ciencias Naturales "Bernardino Rivadavia"-CONICET,
Buenos Aires, Argentina
e-mail: mbrogger@bg.fcen.uba.ar, mmartinez@macn.gov.ar, tablado@macn.gov.ar

D. G. Gil
Instituto de Desarrollo Costero, Universidad Nacional
de la Patagonia San Juan Bosco, Comodoro Rivadavia, Argentina
e-mail: gil_damian@hotmail.com

T. Rubilar · L. Epherra
Centro Nacional Patagónico-CONICET, Puerto Madryn, Argentina
e-mail: rubilar@cenpat.edu.arepherra@cenpat.edu.ar

L. Epherra
e-mail: epherra@cenpat.edu.ar

M. E. Díaz de Vivar
Facultad de Ciencias Naturales, Universidad Nacional
de la Patagonia San Juan Bosco, Puerto Madryn, Argentina
e-mail: mediazdevivar@gmail.com

M. Escolar
Instituto Nacional de Investigación y Desarrollo Pesquero,
Mar del Plata, Argentina
e-mail: marianaescolar@inidep.edu.ar

A. F. Pérez
Facultad de Ciencias Exactas y Naturales, Universidad de Buenos Aires,
Buenos Aires, Argentina
e-mail: analiafperez1@gmail.com

J. J. Alvarado and F. A. Solís-Marín (eds.), *Echinoderm Research and Diversity in Latin America*, DOI: 10.1007/978-3-642-20051-9_11,

11.1 Introduction

Echinoderm research in Argentina began with Professor Irene Bernasconi (1896–1989; Fig. 11.1), at the Argentine Museum of Natural Sciences "Bernardino Rivadavia" (MACN), who dedicated 55 years of her life to study the echinoderms inhabiting the Argentine Sea. Her first studies were focused on Echinoidea of Tierra del Fuego (54°S, 68°W) and Ophiuroidea and Crinoidea of South Georgia islands. In 1924, she published the first work with pictures and drawings of the most common sea urchins (Bernasconi 1924). In 1935, she published her first taxonomic publication about Asteroidea describing new species of the genus *Pteraster* (Bernasconi 1935). From that date, Asteroidea became the focus of her research, describing two new species of the genus *Luidia* (e.g. *Luidia patriae*) (Bernasconi 1941a). She revised the taxonomy of seven families: Pterasteridae (Bernasconi 1937), Luidiidae (Bernasconi 1943), Odontasteridae (Bernasconi 1962a), Goniasteridae (Bernasconi 1963), Ganeriidae (Bernasconi 1964a), Asterinidae (Bernasconi 1973a) and Echinasteridae (Bernasconi 1980). The 1960's were the most productive of her life. She published 15 papers (e.g. Bernasconi 1961, 1962a, b, 1963, 1964a, b, c, 1965a, b, c, 1966a, b), described one new genus (*Vemaster*; Bernasconi 1965a) and four new species (e.g. *Vemaster sudatlanticus*). In 1964, she published the first key for orders, families and genera of Asteroidea from the Argentine Sea (Bernasconi 1964b). Although her major contribution was with sea stars, she also made significant advances in taxonomic research of sea urchins (e.g. Bernasconi 1953) and brittle stars (e.g. Bernasconi and D'Agostino 1977).

Irene Bernasconi was a pioneer in echinoderm research in Argentina. At the age of 72 she travelled on an expedition to Antarctica (1968–1969), being one of the first Argentinean female scientists to explore the white continent (Fig. 11.1). Her work was and still is invaluable for all the research on echinoderms in Argentina. She retired in 1984 and passed away in 1989. We would like to dedicate this chapter to Professor Irene Bernasconi, the first echinoderm specialist in our country.

This chapter reviews all knowledge of Argentine echinoderms up to the present time, available in published papers, meetings, and Licentiate and PhD theses focused on this *Phylum*. Several studies that focus on Antarctic echinoderms are beyond the scope of this chapter, which is restricted to the Argentine Sea (from the Río de la Plata to Tierra del Fuego and Malvinas Islands). Current PhD projects and soon to be published data about chemical, biological, ecological and taxonomic research conducted by the authors and other colleagues are not included in this chapter. Information about depth range, habitat and taxonomic accounts of the species is limited to published information.

Fig. 11.1 Professor Irene Bernasconi, on an expedition to Antarctica (1968–69) and working at the Museum (MACN)

11.1.1 Oceanography and Marine Environments

The Argentine Continental Shelf (ACS; Fig. 11.2) is one of the largest and flattest in the world. This shelf extends from Río de la Plata (RP) (36°00′S, 56°11′W) to Tierra del Fuego (TF) (54°38′S, 64°45′W) (Martos and Piccolo 1988). The ACS has several major coastal embayments: Bahía Blanca bay (BBB) (38°55′S, 62°12′W), San Matías gulf (SMG) (41°29′S, 64°34′W), San José gulf (SJG) (42°19′S, 64°21′W) and Nuevo gulf (NG) (42°78′S, 64°95′W), gulfs which border the headland of Península Valdés (PV) (42°26′S, 63°32′W), and San Jorge gulf (SJOG) (46°19′S, 66°15′W). The ACS (ca. 1,000,000 km^2) is an exceptionally large Neritic Province (Acha et al. 2004). The width of the ACS varies between a few kilometers at 55°S to 850 km along 51°S (Martos and Piccolo 1988). The principal circulation is characterized by the confluence of the cold north-flowing Malvinas Current and the warm south-flowing Brazil Current (Garzoli and Garraffo 1989; Bastida et al. 1992; Brandini et al. 2000; Acha et al. 2004; Lucas et al. 2005).

The Malvinas Current is an energetic north-flowing branch of the Subantarctic Cabo de Hornos Current that affects both coastal and offshore areas. As it moves farther north, it affects only offshore waters (Garzoli and Garraffo 1989; Bastida et al. 1992). The Malvinas Current has seasonal variations, with temperature and

Fig. 11.2 The Argentine Continental Shelf (ACS). Scheme of major upper level currents, Malvinas and Brazil currents and the Brazil-Malvinas confluence (BMC) (adapted from Brandini et al. 2000). It shows the Río de la Plata (RP), San Matías gulf (SMG), San José gulf (SJG), Nuevo gulf (NG), San Jorge gulf (SJOG), the Malvinas Islands (MI) and Tierra del Fuego (TF)

salinity ranging seasonally from 4 to 11 °C and from 33.8 to 34.4 ‰, respectively (Bastida et al. 1992). On the other hand, the Brazil Current is a warm oligotrophic south-flowing branch of the South Equatorial Current (Garzoli and Garraffo 1989; Bastida et al. 1992). Nutrient concentration increases gradually from north to south, with lowest values in the subtropical region coincident with the lowest value of chlorophyll-*a* (Brandini et al. 2000).

The Brazil and the Malvinas Currents converge along the continental margin of South America within latitudes 35°–45°S as the Brazil-Malvinas confluence (BMC) (Garzoli and Garraffo 1989; García et al. 2004). The front shows a substantial spatial and temporal variation forming a large area of intense mixing of subtropical and subantarctic waters (Garzoli and Garraffo 1989; Brandini et al. 2000; Rivas et al. 2006). Maximum annual values of chlorophyll-*a* at the ACS occur in spring and summer, with minimum values in winter (Rivas et al. 2006). This high primary productivity supports a great number of fisheries in the Southwestern Atlantic Ocean (Bremec et al. 2000; Schejter et al. 2002; Acha et al. 2004).

In order to characterize the different coastal regions, we will describe the physical oceanographic features of the coastal fronts (Fig. 11.2).

The Río de la Plata estuary is an extensive and shallow coastal plain estuary at 35°–36°S that receives freshwater from two major rivers (Paraná and Uruguay). The system is characterized by strong vertical stratification. Freshwater flows seaward on the surface while denser shelf water intrudes along the bottom, taking the shape of a salt wedge (Guerrero et al. 1997; Acha et al. 2004; Lucas et al. 2005). Due to lack of osmoregulation in echinoderms, this frontal zone might represent an important geographic barrier for the distribution of coastal echinoderms. Many echinoderm species are distributed off RP estuary, but only in denser and deeper shelf waters of the front.

The front at the coastal zone of El Rincón (39°–41°S, depth < 40 m) encloses an area of 10,000 km^2, showing weak seasonality. The front is characterized by vertical homogeneity due to tidal forcing. A coastal front separates diluted coastal water, from Negro and Colorado rivers, and shelf waters. Salinity gradients increase on the continental shelf due to high saline waters originated in SMG (Guerrero and Piola 1997; Acha et al. 2004; Lucas et al. 2005).

The Península Valdés tidal mixing front is a mesoscale (100–1,000 km) thermal front. The front is observed in spring and summer, which establishes the boundary between stratified waters offshore and a coastal vertically mixed body of water (Acha et al. 2004). The structure of the front is maintained until autumn when stratification of shelf waters decays. The front begins to form in spring (Acha et al. 2004).

Rivas and Ripa (1989) analyzed the temperature and salinity of the Nuevo gulf. They observed no stratification of temperature during winter. However, this condition was not observed during summer when a layer of warmed water overlayed cooler, deeper water. San José gulf is influenced by winds and tidal currents. Therefore, for a great part of the year, this gulf does not show stratification. Only a weak vertical structure occurs during summer (Rivas 1990).

Off southern Patagonia, water masses on the northern extreme of the Drake Passage are diluted. This dilution is caused by an excess of rainfall in the SE Pacific and the continental discharge along the west coast of South America. The flow towards the Atlantic, known as Cabo de Hornos Current, enters onto the continental shelf and contributes to lowering the salinity of the Magellan Strait. Several basins drain along the strait, collecting abundant rainfall during summer and discharging further diluted waters into the ACS. A subsequent diluted plume is then traced 200 km off-shore (100 m depth) and 800 km northward. The plume reaches the southern limit of SJOG and the coastal tidal front (Krepper and Rivas 1979; Acha et al. 2004).

Near the ACS break, the Subantarctic shelf waters of the Malvinas Current produce an important thermohaline front (Martos and Piccolo 1988; Lutz and Carreto 1991). The shelf-break front is a permanent feature that characterizes the border of the shelf. This front has mild gradients in density controlled mainly by temperature in summer and by salinity in winter. The dynamics of the Malvinas Current affects the latitudinal and seasonal variation of the front, moving offshore during summer and onshore during spring and autumn (Olson et al. 1988; Fedulov et al. 1990; Carreto et al. 1995; Acha et al. 2004). The front extends from the Burwood Bank along the shelf-break and around Malvinas Islands (Guerrero et al. 1999) to the Brazil/Malvinas confluence in the North, producing a large region of interrelated fronts (Acha et al. 2004).

The ACS, particularly the Patagonian shelf, is identified as a high primary productivity area (Rivas et al. 2006). The ecology and reproduction of echinoderms inhabiting the coastal, shelf and shelf-break region of the ACS are affected by variations of the primary productivity and oceanographic variables (e.g. Brogger et al. 2004, 2010; Rubilar et al. 2005, 2008a; Marzinelli et al. 2006; Gil and Zaixso 2007; Martinez 2008; Pérez et al. 2008; Gil et al. 2009).

The ACS is characterized by a homogeneous layer of unimodal fine and medium size sand, which is the main component of the sediment. In some areas, this layer is up to 2 m thick (Urien and Ewing 1974; Bastida and Urien 1981). Although some places show a high percentage of gravel, other areas associated with estuaries, gulfs and bays have fine sediments (mud-sand). Hard-bottom areas with rocky shores seem to be scarce (Bastida et al. 1992).

The fauna of the Argentine shelf can be separated into two biogeographic provinces: the Argentine and the Magellanic. These are part of the Subtropical and Subantarctic Regions (Boschi 1979; Balech and Ehrlich 2008). The Argentine province is a warm-temperate subregion extending from the littoral of Buenos Aires southward to the coastal waters of northern Chubut (43–44°S). The depths of the external boundaries range between 40 and 60 m. Water temperature varies latitudinally and seasonally from 8 to 23 °C (Boschi 1979). This subregion also reaches northward to Cabo Frio, (Brazil; 23°S, 42°W) having seasonal fluctuations in the southern area. On the other hand, the Magellanic province is a cold-temperate subregion. At latitudes of 44–45°S it leaves the coastal region following the Malvinas Current, upon reaching 37°S. This province extends about 150-200 km from the continent, with depths greater than 80 m. The northern boundary of this subregion can be tentatively placed at 34–35°S. Water temperature varies seasonally from 4 to 16 °C. This subregion extends to the Pacific Ocean along the southern coastline of Chile (Boschi 1979).

11.2 Diversity and Distribution of Echinoderms

11.2.1 Expeditions to the Argentine Sea

The first campaign in the Argentine Sea was the *Eugenie* (1851–1853), which explored the Río de la Plata and the Magellan Strait. From the collections of this expedition, Ljungman (1867) studied the ophiuroids and Ekman (1925) studied the holothuroids. The *Hassler* Expedition (1871–1872) explored the Brazilian coast southward to the Magellan Strait. Lyman (1875) described the ophiuroids and A. Agassiz (1873) the echinoids of this expedition. One of the most important expeditions was that of the *Challenger* (1872–1876), which established knowledge of the deep sea echinoderms, as well as other deep sea groups in the region. The echinoderms from this famous expedition were described by A. Agassiz (1881; Echinoidea), Lyman (1882; Ophiuroidea), Théel (1882, 1886; Holothuroidea), Carpenter (1884, 1888; Crinoidea), and Sladen (1889; Asteroidea). From 1874 to 1876 the *Gazelle* explored the same region. From the collections of this expedition, Studer examined the ophiuroids (1882), asteroids (1883) and echinoids (1880), while Lampert (1889) examined the holothuroids. The *Albatross* (1888–1904) explored a great part of the world, including the Argentina Sea. Agassiz (1904) studied the echinoids, Ohshima (1915) studied the holothuroids and Clark (1915, 1947) the crinoids.

The vessels *Vettor Pisani* and *Alert*, the expedition Hamburger Magalhaensische Sammelreise (1892–1893), and the Mission scientifique du Cap Horn (1882–1883) explored the southern region around the Magellan Strait, the coast of Patagonia and the Beagle Channel. The ophiuroids and holothuroids were studied by Ludwig (1886, 1898, 1899) and echinoderms in general by Bell (1881) and Perrier (1891).

Between 1897 and 1910, Antarctic expeditions from different countries explored the Argentine waters and Subantarctic Islands. The Belgian Antarctic Expedition 1897–1899, the Scottish National Antarctic Expedition on the SY *Scotia* 1903–1904, the Swedish Antarctic Expedition 1901–1903, the German Antarctic Expedition 1901–1903. Koehler (1902, 1908, 1923) studied the ophiuroids, Hérouard (1906), Ekman (1925) and Vaney (1909) the holothuroids, Andersson (1904), Mortensen (1918), Vaney and John (1939) the crinoids, Ludwig (1903) and Koehler (1923) the asteroids, and Koehler (1902) and Mortensen (1920a) the echinoids. The French Antarctic expeditions during 1903–1905 (on the *Français* vessel) and during 1908–1910 (on the *Pourquoi-Pas? IV* vessel) collected invertebrates. The holothuroids from this campaign were identified by Vaney (1906, 1914). Two British campaigns: RV *Scoresby* (1926–1936) and RV *Discovery* (1927–1932) explored the Southern Ocean. Mortensen (1936) identified the ophiuroids and echinoids, John (1938) identified the crinoids, Fisher (1940) identified the asteroids, and Ahearn (1994) identified the holothuroids.

The R/V *Vema* (1958–1962), from the Lamont Geological Observatory, explored the Argentine Basin. Bernasconi (1966b) studied both echinoids and asteroids. The Campagne de la *Calypso* in 1961–1962, explored 29 stations on the shelf off Uruguay and northern Argentina (33–39°S) and a few samples from Nuevo gulf. Most results were published in special issues of the Annales de l'Institut Oceanographique (Monaco), vols. 44 (1966), 45 (1967), 47 (1969–1970), 49 (1973), and 55 (1979).

The US Antarctic Research Program onboard the USNS *Eltanin* from 1961 to 1966 studied the Magellan region and Malvinas Islands. Agatep (1967), Pawson (1977) and Speel and Dearborn (1983) studied the ophiuroids, holothuroids and crinoids respectively. The FRV *Walter Herwig* in 1966 and 1971 studied the Uruguayean and Argentine Break, and the Malvinas Islands. From this expedition Bernasconi (1973b) studied the ophiuroids, echinoids and asteroids, and Hernández (1982) studied the holothuroids. For the second campaign (1971) of the *Walter Herwig*, only the ophiuroids were studied by Bartsch (1982).

The USNS *Hero* (1969–1971) from the US Antarctic Research Program explored the Argentine shelf. Tommasi et al. (1993) studied the ophiuroids and Larrain Prat (1975) studied the echinoids. There were no reports of other echinoderm classes.

Since 1955, Russian vessels such as RV *Ob* (1955–1958), RV *Akademik Kurchatov* (1971), RV *Dmitriy Mendeleev* (1989) and RV *Akademik Sergei Vavilov* explored the western coast of South America. Abyssal echinoderms associated with the waters of the Argentine Sea were collected (Gebruk 1988, 1990, 1993; Gebruk and Shirshov 1994; Malyutina 2004).

From 1965 to 1972 a Brazilian vessel *Almirante Saldanha* explored the Brazilian, Uruguayean and Argentine waters, from 20°S to 38°S. Tommasi (1965, 1966, 1968, 1969a, b, 1970, 1971, 1974) and Tommasi et al. (1988a, b) described all of the echinoderms collected.

The *Shinkai Maru*, from Japan, made twelve expeditions to the Argentine Sea between 1978 and 1979. Four of these campaigns collected echinoderms from 46 benthic stations. Only the results of holothuroids have been published (Hernández 1982, 1985).

Many Argentinian vessels have explored the Argentine Sea: ARA *Rosales*, ARA *Azopardo*, ARA *Patria*, ARA *San Luis*, CAP *Undine*, ARA *Independencia*, ARA *Pueyrredón*, CAP *Maneco*, ARA *Comodoro Rivadavia*, ARA *Bahía Blanca*, ARA *Capitán Cánepa*, ARA *Zapiola*, *Hero*, *Sta. Maria Magdalena*, *Lamatra*, *Torre Blanca*, ARA *San Juan*, ARA *Madryn*, B/O *Puerto Deseado* and multiple expeditions from the Instituto Nacional de Investigación y Desarrollo Pesquero (INIDEP) (Bernasconi 1937, 1941b, 1953, 1962b, 1964a, b, c, 1972; Bernasconi and D'Agostino 1977; Hernández 1981, 1982, 1987).

11.2.2 Diversity

The echinoderm diversity of Argentina (Table 11.1; Appendix) is composed by 134 species, being Asteroidea the most speciose with 58, followed by Ophiuroidea (32), Holothuroidea (26), Echinoidea (16), and Crinoidea (2). Nevertheless, the taxonomic status of some groups (i.e. Asteroidea, Ophiuroidea and Holothuroidea) needs to be reviewed.

Crinoidea

The stalked crinoids, predominantly associated with bathyal zones (Messing 1997; Améziane and Roux 1997), have not been found in the Argentine Sea. Mortensen

Table 11.1 Numbers of echinoderms from Argentina by taxonomic classification

	Orders	Families	Genera	Species
Crinoidea	1	1	2	2
Asteroidea	6	16	45	58
Ophiuroidea	2	9	20	32
Echinoidea	5	9	12	16
Holothuroidea	3	7	15	26
Total	17	42	94	134

(1917, 1920b) described two viviparous comatulids, *Phrixometra nutrix* and *Isometra vivipara*, from the family Antedonidae. These species are the only two crinoids reported until now from Argentina.

Asteroidea

The systematics and distribution of the class Asteroidea were first studied by Bernasconi (1934, 1935, 1937, 1941a, 1943, 1961, 1962a, b, 1963, 1964a, b, c, 1965a, 1966a, 1972, 1973a, 1980) and later by Tablado (1982) and Tablado and Maytia (1988). These authors reported a total of 58 species of sea stars inhabiting the region. Twenty-eight species are found exclusively along the Magellanic Province, 13 along the Argentine Province, while the others are found in both provinces. Six orders are represented: Paxillosida, Notomyotida, Valvatida, Velatida, Spinulosida and Forcipulatida. The order Valvatida is the most numerous one, with 19 species. Forcipulatida has 14 species and Paxillosida has 11 species. In the order Forcipulatida, Hernández and Tablado (1985) identified three species of the genus *Anasterias*: *A. antarctica, A. minuta* and *A. pedicellaris*, off Ría Deseado ($\sim$48°S). However, according to Clark and Downey (1992) this remains to be settled. Although they recognized that there are morphologically distinct forms among the genus *Anasterias*, the sea stars can belong to different species or they may comprise a very variable species. In the same order, *Allostichaster* has one representative in Argentine waters. According to Clark and Downey (1992) this species is a synonym of *Allostichaster capensis* from South African waters, but this is not certain. The order Velatida is represented by nine species, followed by Spinulosida and Notomyotida with three and two species, respectively.

Ophiuroidea

Knowledge about diversity of brittle stars from Argentina is based on Bernasconi and D'Agostino (1971, 1974, 1977). Thirty-two species of brittle stars have been reported from the region. Only one is an endemic species, *Amphilepis sanmatiensis* from the San Matías gulf (Bernasconi and D'Agostino 1975). The order Euryalida is represented by the family Gorgonocephalidae with two species,

Gorgonocephalus chilensis and *Astrotoma agassizzi*. The order Ophiurida is well represented, with eight families and 18 genera. Family Amphiuridae is the most speciose with 13, of which seven in the genus *Amphiura*. A complete taxonomic revision, based upon additional benthic samples, is needed (Brogger et al. 2008).

Echinoidea

Sixteen known species of sea urchins are present (Mortensen 1910, 1936; Bernasconi 1924, 1925, 1941c, 1947, 1953, 1964c, 1966b, 1970). Two are members of the Subclass Cidaroidea and 12 belong to the Subclass Euechinoidea. Spatangoida is well represented, with eight species, followed by Camarodonta (4), Cidaroida (2), Clypeasteroida (1) and Arbacioida (1). There is a marked predominance of Magellanic species. Some Magellanic species have a wide distribution (e.g., *Pseudechinus magellanicus* and *Arbacia dusfresnii*) reaching low latitudes (~35°S) but only in deeper waters (Bernasconi 1947, 1957).

Holothuroidea

Twenty-six species of holothurians occur in the Argentine Sea (Martinez et al. 2009). Records of holothurians from South of Argentina are scarce, but they increase northward in the Magellanic Province (Pawson 1969). There are a great number of species from cold waters that also occur off Buenos Aires (35°–38°S). At these latitudes, holothurians are found only in deep waters (Bernasconi 1941b; Hernández 1981; Martinez 2008). Holothurians from the Argentine Province are restricted to the North (Deichmann 1947). Overall, cold water species predominate.

Dendrochirotida is well representated by 18 species, followed by Apodida (5) and Molpadiida (3). The fauna has the closest similarity with those of Chile (Pawson 1969) and Antarctica (Deichmann 1947).

The diversity of substrates in ACS provides suitable habitats for several species of holothurians. Some are infaunal, and others are epifaunal on rocky bottoms, while others live on *Macrocystis* sp. holdfasts and fronds or on hard substrata provided by other invertebrates (Bernasconi 1941b; Deichmann 1947; Hernández 1981; Zaixso and Lizarralde 2000; Martinez 2008).

11.3 Research

11.3.1 Biological Studies

Crinoidea

Crinoids are the least species-rich class of the Argentine echinoderms and therefore the least studied among them. Only two species are known from the region. Andersson (1904) and Mortensen (1917, 1920b) studied the brooding habits of *Isometra vivipara* on the Argentine shelf using samples collected by the Sweden South Polar Expedition. Austin H. Clark (1915 *et seq.*) covered the Argentine species in his monograph. Speel and Dearborn (1983) provided important tools for the identification of crinoids from the Southern Ocean.

Tommasi described crinoids from Argentina and Brazil, collected by the expeditions of the Brazilian vessel N/Oc. *Almirante Saldanha* between 26–38°S (Tommasi 1969b; Tommasi et al. 1988a, 1988b). Only one publication (Bernasconi 1934) deals exclusively with Argentine taxa.

Asteroidea

Research has been focused mainly on eight species of sea stars: *Allostichaster capensis, Anasterias minuta, Asterina stellifera, Cosmasterias lurida, Astropecten brasiliensis, Luidia ludwigi scotti, Ctenodiscus australis* and *Diplopteraster clarki.*

Allostichaster capensis (= *A. inaequalis*; Fig. 11.3D) is a fissiparous sea star inhabiting South Atlantic Ocean coasts (Clark and Downey 1992). This sea star has been extensively studied. Pucheta and Urban (1989) focused on trophic behavior. The authors carried out laboratory and in situ observations by SCUBA diving in NG. *Allostichaster capensis* is a relatively specialized predator, feeding on the gastropod *Tegula patagonica* and on the mussels *Lithophaga patagonica* and *Aulacomya atra atra*, selecting this last species over the other prey regardless of their abundance. Rubilar et al. (2005) found that *A. capensis* has a very low percentage of females and displays an annual sexual and asexual reproductive cycle. Fission is most frequent in summer (November-January) and arm regeneration occurs during the rest of the year. In March, gametogenesis begins when gonads appear in the regenerating arms. In August, sexual maturity is achieved. Shedding of gametes does not take place until the arms are almost fully regenerated and individuals are ready to undergo fission. Spawning occurs during September and October followed by a long recovery phase from November to March while arms are in the first stages of regeneration. The pyloric caecum index displays a clear annual cycle and a reciprocal relationship with gonad index, demonstrating the storage function of pyloric caeca and suggesting nutrient transfer. Biochemical analysis revealed that only lipids are translocated from pyloric caeca to gonads at the beginning of gametogenesis (Rubilar et al. 2008a).

Fig. 11.3 Common species of echinoderms from Argentina. **a** The sea star *Cosmasterias lurida*. **b** The brooding sea star *Anasterias* sp. (probably *Anasterias minuta*). **c** The brittle star *Ophioplocus januarii*. **d** The sea urchin *Arbacia dufresnii* and the fissiparous sea star *Allostichaster capensis*. All scale bars = 10 mm

The sea star population appears to be maintained primarily by fission, as only two females of 840 individuals examined were found. Fission in *A. capensis* lasts approximately eight hours. At the beginning of the process the body wall begins to lose its stiffness just before the two halves start to walk away from each other in opposite directions opening a wound across the disc. The oral tissue splits easily, in about four hours. The two halves remain joined only by the aboral tissues and they continue moving apart until the tissue completely breaks. After fission, the two new individuals are usually immobile while the wound heals. The body wall contracts to seal the wound. The tips of the new arms can be observed one week after fission. Individuals had a variable regeneration rate related to body size and number of regenerating arms. Larger individuals showed higher regeneration rates than smaller individuals. Accordingly, individuals regenerating four or five arms had faster regeneration rates than those regenerating three arms (Rubilar et al. 2006).

The effect of fission and regeneration on feeding, energy storage and gonad production was evaluated during the different stages of regeneration in the laboratory, and under differing feeding regimens (Villares 2008; Villares et al. 2009). Split individuals showed the lowest feeding rates, lipids, energy in the pyloric caeca, and gonad production. Regenerating individuals had the highest feeding rate, lipids were allocated to the pyloric caeca and gametogenesis was carried out. Individuals with arms in an advance regeneration state showed an intermediate feeding rate and no allocation of lipids to the pyloric caeca. These individuals spawned, demonstrating that the gonads were mature. Availability of food and healthy nutritional state before fission seem to play key roles in survival. The sea star appears to be extremely well adapted to carry out sexual and asexual reproduction by increasing feeding rates. Regenerating individuals increase their feeding rates to store energy for regeneration, gamete production and nutrient storage. When gonads are mature, individuals have arms in an advanced state of regeneration, and feeding rates decrease since spawning may inhibit feeding. After fission, when capacity to attack prey is reduced, the stored energy is used for survival. Only those individuals that were able to store energy through high feeding rates can survive. During this period, individuals do not spend energy in gamete production, but invest all energy resources in regenerating new arms, which are necessary to increase feeding capacity for the following sexual reproduction period.

The great regenerative capacity of this sea star led Rubilar (2009) to study the cellular mechanism involved in arm regeneration. The sea star regenerates through a morphallatic mechanism (i.e. cellular migration, dedifferentiation and transdifferentiation), followed by a morphogenetic process producing a miniature arm. Then new structures arise between the arm tip and the stump. The morphology of the radial nerve cord and circumoral ring of the sea star was examined by Rubilar et al. (2008b), using antisera raised against the gastropod neuropeptide Tritonia Pedal Ganglia (TPep) (Lloyd et al. 1996). This peptide has been found in motorneurons as well as in neurons located in the pedal ganglia of the gastropod *Tritonia diomedea* and appears to be involved in motor functions. In the sea star, TPep-like immunoreactivity was detected in the hyponeural, ectoneural systems and innervations of the tube feet. In the hyponeural system, the label was observed in numerous neurons and groups of cell bodies in the radial nerve cord as well as in the circumoral ring. In the ectoneural systems, the TPep-like immunoreactivity was observed in isolated neurons as well as in the neuroepithelium.

The oral-brooding sea star *Anasterias minuta* (Fig. 11.3b) is the most abundant intertidal sea star in Patagonia (Salvat 1985; Gil and Zaixso 2007). This species is widely distributed along the Patagonian coast, including the Malvinas Islands, and occurs from the middle intertidal zone to 80 m depth (Bernasconi 1964c). Brooding females bend their arms in order to create a chamber to protect the young. In Ría Deseado (RD; 47°45′S, 65°55′W), brooding occurs during the coldest months of the year, from March to October (Salvat 1985; Gil and Zaixso 2007). Salvat (1985) suggested a biennial oogenic cycle. Gonads are in a recovery phase from April to June after eggs are laid. Gametogenesis begins in July and vitellogenesis continues during the second austral winter; maturity is reached in January. Female gonads might take around 20 months to mature. On the contrary, males display an annual cycle. Spawning occurs once a year during March and April, following a short recovery period in May. Gametogenesis begins again in June, and maturity is reached in February (Salvat 1985). The pyloric caecum index displays a clear annual cycle, with maximum index values occurring before the maximum gonad weight, suggesting nutrient translocation. Except in males, in which a one month lag was observed, no inverse relationship was found between pyloric caeca and gonads (Gil and Zaixso 2007). In Punta Maqueda (46°01′S, 67°01′W), the brooding season of *A. minuta* shows the same pattern (Escudero 1999). The number of juveniles follows a linear regression with female size (Escudero and Zaixso 2003). This sea star is a common predator and prominent member of intertidal and shallow sublittoral communities (Gil 2002; Zaixso 2004; Bertness et al. 2006; Gil and Zaixso 2007). The relationship between predatory and brooding activities in RD was studied by Gil (2002) and by Gil and Zaixso (2007). They found a seasonal feeding cycle with the maximum feeding frequency during the austral summer that was related to the female brooding habits. During brooding season, females with juveniles did not feed. The feeding frequency was also positively correlated with seawater temperature and sea star size, but no differences were found along a shore-level gradient. *Anasterias minuta* consumes a wide range of prey, including mollusks and crustaceans. The diet composition differed among the study sites (GSJO and RD), and it varied with depth and body size. The diversity of prey consumed by this sea star was highest on a semi-exposed coast during the warmest season on the infralittoral fringe. Abundance of prey beneath rocks within tide pools at the infralittoral fringe does not seem to determine diet composition. Generally, the sea star consumes the purple mussel *Perumytilus purpuratus* and the gastropod *Pareuthria plumbea* more than other prey, even when they are not the most abundant prey present. However, at the midlittoral zone of semi-exposed coast, it feeds on prey species according to their abundance. A significant positive correlation was found between the size of the sea star and the size of the most frequent prey. The authors suggested that *A. minuta* could have a variable impact on the community structure since there appears to be an opportunistic feeding strategy, taking unattached, wave-washed mussels when available rather that attached mussels (Gil and Zaixso 2008). An ascothoracid parasite (*Dendrogaster argentinensis*) of *A. minuta* was found at RD (Grygier and Salvat 1984). It is the first ascothoracid described from the South American coast.

The sea star *Asterina stellifera* (= *Patiria stellifer*) is distributed from Cabo Frio (Brazil) to Mar del Plata (MDP) (Argentina; 38°10′S, 57°10′W) in the West Atlantic Ocean (Clark and Downey 1992) and is a very common sea star on MDP shores (Salvat 1975a). The embryonic and larval development was studied by Salvat (1975a). The dipleurula larvae appear 32 h after fertilization and the larva is completely formed after one week. The size of the sea star varied with depth; the largest individuals were found in deeper waters (Salvat 1975b). This sea star has been defined as an omnivore and scavenging feeder (Meretta et al. 2009a). It feeds on macroalgae, bryozoans, tunicates, polyps and biofilm. Between 2006 and 2009, the size structure distribution changed from bimodal to unimodal, thus suggesting that the younger cohort had grown to reach the maximum size and that there was no recruitment during this period. The high density (35 ind m^{-2}) and the trophic range suggest a potential role in the hard bottom community (Meretta et al. 2009b).

Stomach contents of *Astropecten brasiliensis* and *Luidia ludwigi scotti* have been analyzed by Brogger and Penchaszadeh (2008) at Puerto Quequén coast (38°37′S, 58°50′W) between 40 and 60 m depth. *Astropecten brasiliensis* was the most abundant sea star sampled and showed the most diversified diet, with a total of ~28 different items. Bivalves and gastropods were dominant. In contrast, *L. ludwigi scotti* was found in lower numbers and preyed upon eight species. Bivalves, foraminiferans and ophiuroids were the most common prey. There were differences not only in items eaten by the two species, but also in the size of the common prey. This suggests that differences in diet may allow the coexistence of the two predators. Penchaszadeh (1973) also observed mollusks as main prey items for *A. brasiliensis* off Mar del Plata.

The sea star *Cosmasterias lurida* (Fig. 11.3a) shows a continuous distribution from the Argentine shelf (38°S, 57°W) to the Chilean coastline (29°S, 78°W) (Bremec et al. 2000). It is usually found on subtidal coarse bottoms from 0 to 650 m depth (Clark and Downey 1992) in very low densities (Bremec et al. 2000). Reproduction has been studied on pier piles of Almirante Storni in NG (Pastor de Ward et al. 2006). This highly modified habitat had an unusually supply of mussels and large aggregations of sea stars. Gametogenesis takes place from June to January, sexual maturity occurs during the summer, and spawning in April. The gonadal index and pyloric caecum index tended to have an inverse relationship, but this was statistically significant only in males. However, biochemical analyses indicated no transfer of nutrients between pyloric caeca and gonads in either sex. Pastor de Ward et al. (2006) suggested that the high availability of trophic resources in the study area could be the clue to the high allocation of nutrients, the absence of nutrient transfer from the pyloric caeca to the gonads during proliferation and the continuous proliferation of oocytes in *C. lurida* at the Almirante Storni pier.

The brooding sea star *Ctenodiscus australis* is distributed along the east coast of South America, from southern Brazil (23°00′S, 41°40′W) to TF and the Malvinas Islands, between 70 and 700 m depth. It is usually found on sand bottoms, occasionally on coarse sand or sand with stones (Clark and Downey 1992). This is a conspicuous species in the community associated with the shelf-break front (Bremec and Lasta 2002). While brooding, the young (at different stages of development) are hidden under the paxillae, mainly at the base of the arms near the interradius. The youngest stages were entirely hidden by the paxillae. In advanced stages, parts of their bodies protruded between the spines and are easily seen. The youngest stages show a larva typical of an abbreviated development (e.g. a yolky, nonfeeding larva) (Lieberkind 1926). The larvae lack arms and ciliated bands. Nothing is known about internal morphology of the larvae, but they are probably yolky, or has a direct-development mesogen (McEdward and Miner 2001).

The pterasterid *Diplopteraster clarki* has a continuous distribution along the coast of Argentina, from RP to waters off SJOG, between 82 and 600 m depth

(Bernasconi 1973a). This is a brooding sea star with an aboral nidamental chamber, where fifty embryos can be brooded. The size of embryos ranges between 1.5 and 7 mm. Inside the nidamental chamber, embryos are located in specific areas: eggs and smallest embryos are among the paxillae, close to the osculum; the intermediate embryos are, following the edge of the arms, between paxillae and resting on the actinolateral membrane; and the largest embryos (radius = 4–6 mm) are located between paxillae in the interradii. When embryonic development is completed, the juveniles leave the aboral chamber by breaking the suture between the supradorsal and actinolateral membranes. The estimated brooding season is between September and May, from spring until autumn (Bernasconi 1937).

Ophiuroidea

Few publications focus exclusively on this group in Argentina. Bernasconi (1965b, c), Tommasi (1968), and Bernasconi and D'Agostino (1971, 1974, 1975, 1977) published a taxonomic revision for the Argentine region. Bartsch (1982) analyzed benthic samples from the Patagonian shelf and described external and internal characters of 11 species of brittle stars. The first reports of Argentinian brittle stars were provided by Bernasconi (1926, 1934). She documented the presence of *Gorgonocephalus chilensis* and *Amphipholis squamata* in the region.

Zaixso and Lizarralde (2000) examined the distribution of the most frequent echinoderm species of the SJG and southern SMG, referring to four brittle star species. They analyzed species densities by using a point kriging method, and related them to environmental parameters (depth, substrata granulometry, currents and bottom temperature). *Amphiura crassipes* (1–25 ind m^{-2}) was found close to 80 m depth on substrates with high percentages of silt and clay. The increase of abundance was associated with an increase of fine sand substrates and no correlation was found with temperature and currents. *Ophiactis asperula* was the most abundant ophiuroid (25–3,125 ind m^{-2}) in the study area. The density of this species was not related to any particular depth, but the density decreased with increasing summer temperatures and the increase of silt–clay and fine sand substrates. Higher densities were correlated with faster surface currents. *Amphiura magellanica* (5–125 ind m^{-2}) was associated with coarse sand substrates. The abundance was not related to any particular depth or temperature. However, abundance was positively related to faster surface currents. The abundance of *Ophioplocus januarii* (5–125 ind.m^{-2}) was highest at shallow depths and on coarse sand substrates, and increased with summer temperature.

Several aspects of the biology of the brittle star *Ophioplocus januarii* (Fig. 11.3c) in a shallow water population from SJG were studied. Arribas et al. (2008) analyzed presence and abundance of the fauna associated with this ophiuroid species throughout two years. The authors observed that the fauna associated, mainly mollusks, arthropods and annelids, remained stable over the study period (2005–2007). The number of species did not show annual variation. This is consistent with a biocenosis previously described by Zaixso et al. (1998). Growth studies were conducted by Rodríguez et al. (2008, 2009) using calcein as a chemical marker. Growth was observed through a distinct green fluorescent label on the jaws. Coulin et al. (2009) described jaw growth and related it to individual body growth. The authors proposed a power function relationship between the growth of the teeth and dental plate and the body size.

Echinoidea

Sixteen species of sea urchins have been reported from the region. However, there is information available on the general biology and ecology of only six species: *Arbacia dufresnii*, *Pseudechinus magellanicus*, *Loxechinus albus*, *Tripylaster philippii*, *Abatus cavernosus* and *Encope emarginata*.

The sea urchin *Arbacia dufresnii* (= *A. dufresnei*; Fig. 11.3d) is widely distributed around the southern tip of South America, from RP on the Atlantic coast to the north of Puerto Montt (Chile; 41°30′S, 72°56′W) on the Pacific coast, at a depth range of 0–315 m (Bernasconi 1953). This species was associated with shallow waters and areas with coarse sediments (sand) and hard substrates in SJG and NG (Carriquiriborde et al. 1983; Zaixso and Lizarralde 2000; Brogger 2005; Epherra 2009). A new turbellarian species (*Syndesmis patagonica*) was found and described in the gut of *A. dufresnii* from NG. This is the first record of a sea urchin of the genus *Arbacia* as host of an umagillid (Brogger and Ivanov 2010). The diet of *A. dufresnii* was studied by using a qualitative analysis of prey items in gut contents at different sites and across different depths from Buenos Aires to GSJ. *Arbacia dufresnii* is primarily carnivorous (Vásquez et al. 1984), and its diet is determined by the local composition of potential prey (Penchaszadeh and Lawrence 1999). Some aspects of the embryonic development were studied by Bernasconi (1942) and Brogger et al. (2004). Reproduction of *A. dufresnii* was studied at NG (Brogger 2005; Brogger et al. 2010). There are two spawning periods, a first spawning event occurs during spring and a second in summer. Gametogenesis is synchronous between males and females and spawning is synchronized. Epherra et al. (2009a) and Epherra (2009) found that *A. dufresnii* has a biochemical cycle directly related to the reproductive cycle at NG and shows differences between sexes. A similar prolonged spawning period during spring and summer was also observed. The overall biochemical composition, in different organs, was similar to that of other sea urchin species, with a high proportion of protein, followed by lipids and carbohydrates. The population of *A. dufresnii* at NG displays a good nutritional state, as both the gut and the gonads act as storage organs. The reproduction of this species in southern populations (SJOG) has some interpopulation differences in the duration of spawning events and gonadal indices (Epherra et al. 2009b).

Pseudechinus magellanicus (Fig. 11.4b) is the most common sea urchin in southern South America. This species is distributed from 35°S on the Atlantic Ocean coast through the Magellan Strait and throughout TF, to Puerto Montt,

Fig. 11.4 Common species of echinoderms from Argentina. **a** The brooding heart urchin *Abatus cavernosus*. **b** The sea urchin *Pseudechinus magellanicus*. **c** The sea cucumber *Psolidium dorsipes*. **d** The sea cucumber *Pseudocnus dubiosus leoninus*. All scale bars = 10 mm

Chile, on the Pacific Ocean coast. The sea urchin occurs from intertidal to 340 m depth (Bernasconi 1953) and lives in a wide variety of habitats, including tidepools and rocky and mixed gravel-sand bottoms. It can also be found as associated biota on *Macrocystis* kelp and in mussel beds (Carriquiriborde et al. 1983; Bremec and Roux 1997; Adami and Gordillo 1999; Penchaszadeh et al. 2004; Marzinelli et al. 2006). At SJG *P. magellanicus* has a wide distribution, and its abundance appears to be related primarily to sites with strong surface currents, dominated by coarse substrates at intermediate depths (Zaixso and Lizarralde 2000). This distribution seems different from that of *A. dufresnii* and may be related to the consumption of both live food (algae and animals) and drift-material produced by currents (Zaixso and Lizarralde 2000). *Pseudechinus magellanicus* has been noted as a member of various major macrozoobenthos associations in SJG (Carriquiriborde et al. 1983; Zaixso et al. 1998) and SJOG (Roux et al. 1995). A parasitic gastropod (*Pelseneeria sudamericana*) is attached to the test in Isla de los Estados (54°45′S, 63°52′W) (Pastorino and Zelaya 2001). In the Beagle Channel (54°50′S) this species was sexual mature at a diameter of 12 mm. The relative fecundity fluctuated between 70,000 and 100,000 eggs/gonad. The reproductive cycle produced a fully synchronous spawning event between late winter and early spring. A new gametic proliferation may occur during the spring with subsequent partial spawning in the summer (Orler 1992). At NG a main spawning event occurrs in winter and a second minor event during the summer (Marzinelli et al. 2006). The gonad index at NG was negatively correlated with photoperiod and water temperature with a lag period of 3 mo. Egg strain values are high in eggs spawned by small individuals. According to Marzinelli et al. (2006) the jelly coat of eggs of *P. magellanicus* may protect them from the forces experienced during spawning, thus allowing reproductive success.

Diet was studied by Penchaszadeh et al. (2004). The authors suggested an omnivorous behaviour related to different ecological conditions. The diet spectrum of *P. magellanicus* depends mostly on the habitat and biotic community in which they occur. The species has been identified as a grazer in the benthos, associated with *Macrocystis pyrifera* forests in Beagle Channel (Ojeda and Santelices 1984; Vásquez et al. 1984; Adami and Gordillo 1999), and as a secondary consumer in circalittoral mussel banks, feeding on barnacles, polychaetes, bivalve recruits, small gastropods and crustaceans (Penchaszadeh 1979). It can also feed on drifting or detached algae (Castilla and Moreno 1982; Penchaszadeh et al. 2004). According to Penchaszadeh et al. (2004) this species can play an important role in the community structure of the circalittoral beds of *Mytilus edulis platensis*, mainly through intense predation on bivalve spat.

Due to the invasion of the kelp *Undaria pinnatifida* in Argentine Patagonia, the potential role of invertebrate grazers, such as the sea urchins *P. magellanicus* and *A. dufresnii* were preliminarily analyzed by Teso et al. (2009). They found that both species of sea urchins fed on the alga at a low rate. In this context, they are unlikely to play a role in the control of the seaweed because of the high rates of recruitment and growth of *U. pinnatifida.*

The edible sea urchin *Loxechinus albus* has a wide geographic distribution along the Pacific coast, from Ecuador (6°S) to south of TF including the Magellan Strait and Beagle Channel (BC) (Bernasconi 1947, 1953) and Malvinas Islands (Schuhbauer et al. 2010). Growth was analysed by Schuhbauer et al. (2010) in the Malvinas Islands and growth curves were found to be asymptotic with the von Bertalanffy growth model as the best predictor. The authors argue that the relatively short period of optimal water temperatures limited this species' final size compared to populations in regions with higher seasonal temperatures (e.g. central Chilean coasts). Reproductive aspects of *L. albus* populations have been studied along the Chilean Pacific coast (Gutiérrez and Otsú 1975; Zamora and Stotz 1992), in the Magellanic region (Bay-Schmith et al. 1981; Oyarzún et al. 1999), in the Beagle Channel (Orler 1992; Pérez et al. 2008, 2010, 2009) and at the Malvinas Islands (Schuhbauer et al. 2010). The BC population of *L. albus* represents the southernmost limit of distribution. There it is exposed to strong variations in photoperiod, temperature, and marked seasonal fluctuations in primary productivity (Hernando 2006). In BC, a major peak of gonadal index occurred in winter, followed by spawning in spring. A minor gonadal index peak occurred in autumn, explained by an increase in the nutritive phagocytes, a non-gametic component of gonads (Pérez et al. 2010). A similar spawning period was reported for *L. albus* populations of the Magellanic region (Bay-Schmith et al. 1981; Arana et al. 1996; Oyarzún et al. 1999). However, populations of *L. albus* from other sites do not seem to follow this pattern. In the Malvinas Islands, *L. albus* spawns during the warmest time of the year (summer). This allows for greater growth of their pelagic larvae and newly settled juveniles (Schuhbauer et al. 2010). Spawning is delayed along the Chilean coast as latitude increases (Zamora and Stotz 1992; Vásquez 2007).

In the BC population of *Loxechinus albus*, the reproductive cycle may be influenced by photoperiod fluctuation and primary production. In contrast, water temperature was not correlated with gonadal maturation (Pérez 2009; Pérez et al. 2010). Pérez et al. (2008, 2010) made the first attempt to analyze the energetic content of different organs of *L. albus* using calorimetric determinations. High values of gonad energy content suggest the presence of lipids in the gonads. Both gonad energy content and total gonad energy content were higher in females than in males. The reproductive output of *L. albus* was higher in females than males, thus indicating that spawning is energetically more expensive in females (Pérez et al. 2010). The values of gonad energy content and total gonad energy content showed minimum values in winter (ripe stage) and maximum values in spring (spawned stage) (Pérez et al. 2008). Total gonad energy content reached higher monthly average values (50–200 kJ) than total gut energetic content (20–40 kJ). These authors suggested this difference indicates that the gonads constitute the most important storage organs in *L. albus*. Organic stores are built up in the gonads after spawning and then used in gamete production.

The gonads of this sea urchin are a highly valued seafood product and are considered as delicacies in many parts of the world. Pérez et al. (2010) suggested that autumn is the best period to harvest this species since the gonads are in early stages, having the quality (color, texture, and flavor) desired by the market. Malanga et al. (2007) studied the addition of carotenoid in diet in order to improve the organoleptic properties and protection against oxidative damage of gonads. They fed *L. albus* with carotenoid during the pre-reproductive period and during spawning. β-carotenoids, not only provided a desirable color to the product, but also protect their lipid constituents against oxidation. The influence of seasonal factors on oxidative metabolism in the *L. albus* gonads was tested by Malanga et al. (2009). They found that lipid radical content showed non-significant changes with seasons while the content of ascorbyl radicals and -tocopherol were lower in spring and summer as compared to winter. The authors suggested the existence of different profiles of response against oxidative stress of the lipophilic and hydrophilic milieus in *L. albus* gonads.

The sand dollar *Encope emarginata* inhabits the Atlantic Ocean from Colombia to the northern coasts of Argentina. Brogger and Penchaszadeh (2003) described some aspects of population ecology at MDP at 15–25 m depth. Length frequency analysis indicated that *E. emarginata* is characterized by slow growth rates, reaching a smaller size than populations from Brazil. They reported a very low frequency of occurrence of the commensal crab *Dissodactylus crinitichelis* between oral surface spines.

Tripylaster philippii is a spatangoid echinoid known from the region. This species is found in greater depths with a maximum at 135 m. It is associated with substrates having high percentages of silt–clay fraction and very fine sands with a clear predominance of the mud fraction. Despite the existence of suitable substrates in large areas at SJG, this species is absent from those areas, probably because the fine sediments of SJG are not stable enough to allow its establishment, or perhaps summer temperatures are too high (Zaixso and Lizarralde 2000). *Tripylaster philippii* is also found at great depths of NG (Carriquiriborde et al. 1983). At SJOG it is part of an important benthic association with the colonial coelenterate *Renilla* sp. (Roux 2000).

The heart urchin *Abatus cavernosus* (Fig. 11.4a) is another spatangoid that occurs off RP to TF. This species also occurs in the Subantarctic and Antarctic Islands (Bernasconi 1953, 1966b; Schatt and Féral 1991; López Gappa and Sueiro 2007). It inhabits unconsolidated, muddy and sandy sediments and has a wide bathymetric range, from the intertidal to 700 m (Bernasconi 1966b). In RP, this species has a mean density of 63.9 ind m^{-2} and can be found buried in sand, muddy sand, and pebbly sand substrata, from midlittoral to the shallow subtidal, and within soft-bottom tidepools (Dolcemascolo and Zaixso 2000; Gil et al. 2009). According to Gil et al. (2009), *A. cavernosus* is a deposit-feeder that broods its young inside four dorsal ambulacral pouches. Sex ratio was not homogeneous across size classes and females were found to be larger than males. The temporal pattern in the size-adjusted gonad dry weight demonstrated that reproduction in the littoral population of *A. cavernosus* is seasonal and synchronous. Spawning occurs from May to July, and females breed within a 9-month period from May to February, followed by a 1–2 month lag between gonad development and decreasing seawater temperature. Spawning occurred when the seawater temperature decreased and may be influenced by spring and summer periods of sediment enrichment. Juveniles develop directly without formation of larval structures, and are released during the austral summer. The heart urchin exhibited some reproductive characteristics that differ from those of other Subantarctic and Antarctic spatangoids, including relatively large egg size and low fecundity. Tolosano et al. (2006) found that gonopore diameter may be used to sex immature heart urchins, using a breakpoint regression model. The epibiont bivalve *Waldo parasiticus* is usually found among the spines of *A. cavernosus* (Zelaya and Ituarte 2002). According to Zaixso et al. (2003) the number of bivalves per host is not influenced by sea urchin sex, but it is positively related to size. Epibionts were mainly found on the ventral ambulacral areas but can also occupy other areas on the ventral and dorsal side. Maximum abundance of epibionts per host was found during summer and lowest values in winter. The relation between these two species should be tentatively classified as inquiline or commensal.

Holothuroidea

Some holothurian specialists that worked with species from Argentina were Risso (1826), Lesson (1830), Müller (1850), Phillipi (1857), Lampert (1889), Perrier (1904, 1905) and Cherbonnier (1941). More recently Pawson (1964, 1969, 1970) studied some species distributed from Chile to Argentina. Massin (1994), O'Loughlin (2002, 2009), Bohn (2007), O'Loughlin and Ahearn (2008) and O'Loughlin (2009) studied the distribution and taxonomy of many species from Antarctica and near waters with distribution in Argentina. Important advances in taxonomy of holothurians from Argentine waters have been made by Tommasi, through multiple expeditions made with the Brazilian vessel N/Oc. *Almirante Saldanha* (Tommasi 1969a, 1971, 1974; Tommasi et al. 1988a, b).

Bernasconi described some species from the collection of the MACN (1934, 1941b). Hernández (1981) described the holothurians of Puerto Deseado and three members of the genus *Trachythyone* from the southwestern Atlantic, *Trachythyone peruana*, *T. parva* and *T. lechleri* (Hernández 1982); he also described *Trachythyone baja* (Hernández 1987). According to O'Loughlin (2009) this species is a synonym of *T. bouvetensis*. Hernández (1985) studied several aspects of different holothurian species, including a taxonomic description of all species inhabiting the Argentine Sea. He also analyzed the phenetic relationship among the holothurian species found, and the phylogenetic relationships of five genera of the family Cucumariidae, studied the biology and ecology of some intertidal species, publishing data on their distribution. Besides this, he studied the sediment feeding behavior and evisceration as a defense mechanism of *Chiridota pisanii* (Hernández 1985). Martinez et al. (2009) reported the total number of holothurians from Argentina, and developed tools for species recognition in the families Cucumariidae (Fig. 11.4d) and Psolidae (Fig. 11.4c).

Different authors have studied multiple aspects of the reproduction of the brooding dendrochirotid *Psolus patagonicus* (Martinez 2008; Giménez and Penchaszadeh 2010; Martinez et al. 2011). Martinez (2008) studied the ecology and reproduction of a *P. patagonicus* population that inhabits the shelf break at 38°S ($\sim$100 m depth). This species exhibited an annual reproductive cycle with one reproductive event in late summer/autumn (February–March). The maximum oocyte diameter registered in *P. patagonicus* was 900 μm with more or less continuous development and growth of oocytes all through the year. Giménez and Penchaszadeh (2010) estimated the brooding season from February to October and the beginning of the brooding period is consistent with a decrease in the gonadal index of females, larger oocytes and mature spermatozoa in males (Martinez 2008; Martinez et al. 2011). Schejter et al. (2002) reported a seasonal food input in the associated Patagonian scallop community, which is correlated with the reproductive timing of *P. patagonicus*, thus reinforcing the contention that this factor influences the timing of reproduction (Martinez 2008; Martinez et al. 2011).

11.3.2 Argentine Continental Shelf Waters

The study of benthic communities in ACS started four decades ago (Bastida et al. 2007). In some of these studies, echinoderm species showed high abundance and diversity and were used to define different ecological areas along the ACS (Roux et al. 1988, 1993; Bastida et al. 1992).

The ecological information on echinoderms from shelf waters comes from monitoring studies aimed at determining the invertebrate by-catch associated with commercial fisheries. The first faunistic inventory of the Patagonian shrimp (*Pleoticus muelleri*) fishing grounds of SJOG was developed by Roux et al. (1995). The benthic community between 27 and 98 m showed a total of 22 species of echinoderms with a higher abundance in soft than in hard bottoms. *Pseudechinus magellanicus*, together with four species of mollusks and one crustacean were dominant in the benthic assemblage. Other species commonly collected in the gulf were *Henricia obesa*, *Cosmasterias lurida*, *Ophiomyxa vivipara* and *Ophiactis asperula*.

In external shelf areas the assemblage of macroinvertebrates dominated by the Patagonian scallop *Zygochlamys patagonica* has been broadly studied. This community is located along the shelf-break front, following the 100 m isobath, between 37° and 44°S. Echinoderms comprise one of the most important taxa, in terms of biomass and numbers. The northern sector (37–39°S) is characterized by high density of *Ctenodiscus australis*, *Diplasterias brandti*, *Labidiaster radiosus* (Asteroidea), *Gorgonocephalus chilensis*, *O. asperula* (Ophiuroidea) and the echinoid *Austrocidaris canaliculata*. *Sterechinus agassizii* (Echinoidea), *Ophiacantha vivipara* and *O. asperula* (Ophiuroidea) and the sea star *D. brandti* stand out in the southern sector (Bremec and Lasta 2002; Bremec et al. 2006; Schejter et al. 2008).

Echinoderms, mainly sea stars, have a potentially important ecological role in this community. The sea star *Calyptraster* sp. (possibly *Diplopteraster clarki*, according to Escolar and Bremec 2009) appears as the top predator, preying on the sea star *C. australis*, and the gastropods *Odontocymbiola magellanica*, *Adelomelon ancilla* and *Fusitriton magellanicus*. These gastropod species are the main predators of the Patagonian scallop. Other echinoderm species such as *L. radiosus*, *C. australis* and *S. agassizii* are intermediate predators, preying on scallops and brittle stars, scallops being the principal item in the diet of *L. radiosus* (Botto et al. 2006).

Another conspicuous sea star in ACS is *Astropecten brasiliensis*. This is an abundant species in the Buenos Aires Province coastal and shelf areas (Roux 2004). It inhabits soft bottoms (sand-mud) between 7 and 60 m. This species feeds on banks of the mussel *M. edulis platensis*, commercially exploited until 1988 (Penchaszadeh 1980; Ciocco et al. 1998). This sea star is one of the most important predators of mollusks, mostly young gastropods and bivalves (Penchaszadeh 1973). *Astropecten brasiliensis* shows a selective diet that varies with depth and bottom type, limited only by the size of prey. This sea star ingests prey whole; the maximum mussel size recorded in a stomach was 24 mm long. The diet composition of this sea star at *M. edulis platensis* banks varied according to the seasonal recruitment and growth of mussels. When young mussels were absent (late fall and winter), the contribution of cumaceans, other pelecypods and gastropods increased in the gut contents of *A. brasiliensis*.

With respect to echinoderms from deeper areas of the ACS, there are only the faunal inventories of the H.M.S *Challenger*, *Vema* and *Walther Herwig* expeditions (Sladen 1889; Bernasconi 1966b, 1972, 1973b; Bartsch 1982). Recently, samples taken from a submarine canyon at 365 m depth, around 43°35′S and 59°33′W, show that echinoderms are diverse and constitute an important part of the benthic community (18 species, 14 % wet weight). The same species are found in scallop fishing grounds, with the exception of the brittle star *Astrotoma agassizii* (Bremec and Schejter 2010).

11.3.3 Chemical Studies

Secondary metabolites from Echinodermata have been thoroughly investigated for the past four decades. Members of this phylum contain steroidal and triterpernoidal oligoglycosides and saponins, which are widely diffused in terrestrial plants but apparently not so common in marine organisms (Burnell and Apsimon 1983; Minale et al. 1986, 1993, 1995; Stonik and Elyakov 1988; Kalinin et al. 1992, 1996, 2005; D'Auria et al. 1993; Verbist 1993; Stonik et al. 1999; Iorizzi and Riccio 1999; Kalinin 2000; Iorizzi et al. 2001; Chludil et al. 2002b, 2003a, b; Maier and Murray 2006; Maier 2008).

Asteroidea

Asterosaponins are steroidal glycosides which share a 3β-O-sulfated,6α-O-glycosylated $\Delta^{9(11)}$ steroidal skeleton- only one reported structure shows a 12α-hydroxy functionality, with different side chains (Yasumoto and Hashimoto 1965). Twelve asterosaponins have been isolated from four species of sea stars from the Patagonian coast of Argentina: *Luidia ludwigi scotti* from SMG, *Cosmasterias lurida* and *Allostichaster capensis* from NG, and *Anasterias minuta* from SJOG. Analysis of their chemical structures showed six novel asterosaponins: Cosmasterosides A, B, C and D (from *C. lurida*) (Roccatagliata et al. 1994) and Anasterosides A (from *A. minuta* and *A. capensis*) and B (from *A. minuta*) (Díaz de Vivar 2002; Maier et al. 2002) (Fig. 11.5a). The hexaglycoside Anasteroside A showed antifungal activity against the plant pathogenic fungus *Cladosporium cucumericum*. Two additional known saponins were isolated from *C. lurida*: Ophidianoside F, previously isolated from *Ophidiaster ophidianus* (Riccio et al. 1985), as the main component and Forbeside H, previously isolated from *Asterias forbesi* (Findlay et al. 1990). Acanthaglycosides B and C, previously isolated from

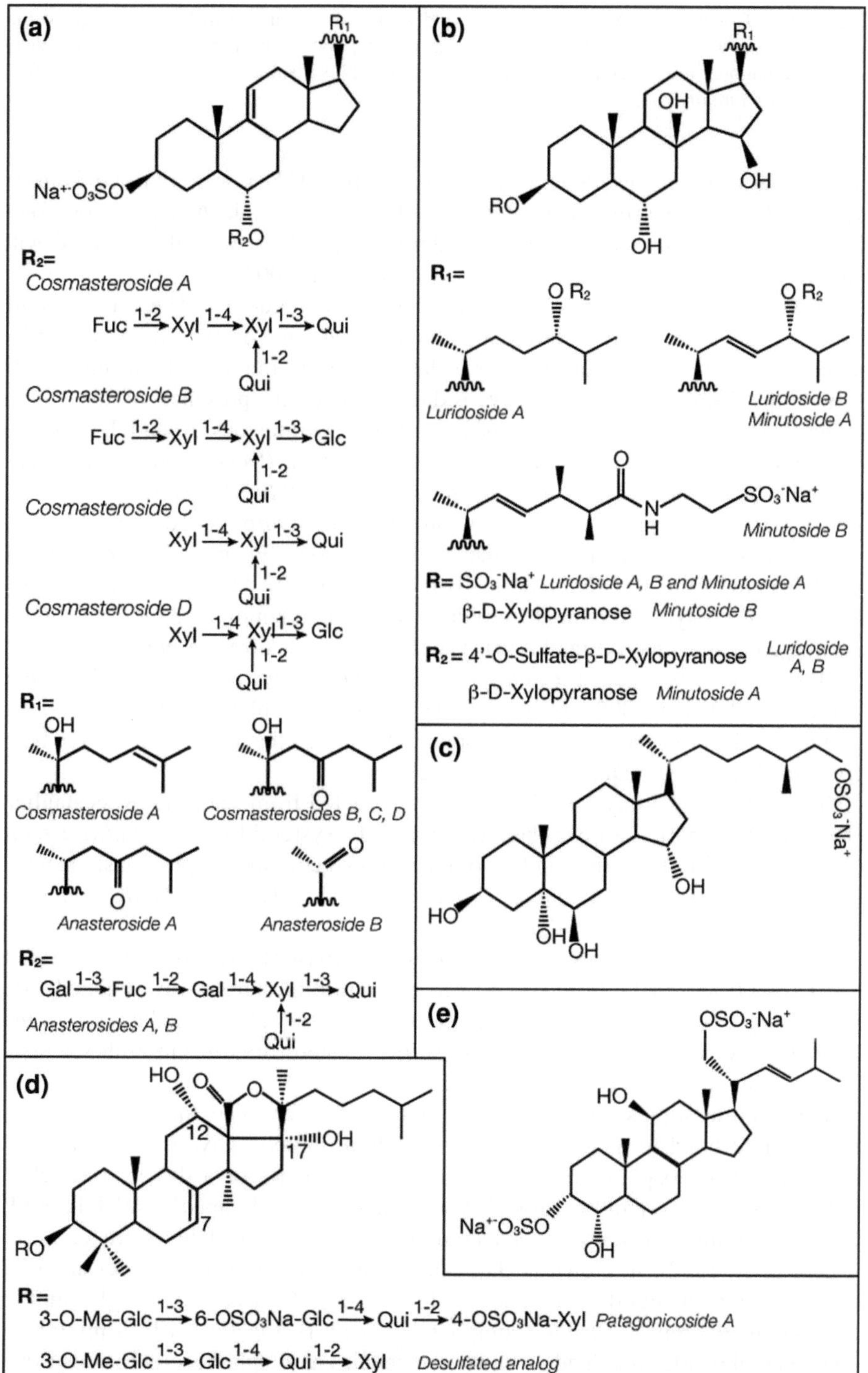
(a)
R1
Na+-O3SO
R2O
R2=
Cosmasteroside A
Fuc 1-2 Xyl 1-4 Xyl 1-3 Qui
1-2
Qui
Cosmasteroside B
Fuc 1-2 Xyl 1-4 Xyl 1-3 Glc
1-2
Qui
Cosmasteroside C
Xyl 1-4 Xyl 1-3 Qui
1-2
Qui
Cosmasteroside D
Xyl 1-4 Xyl 1-3 Glc
1-2
Qui
R1=
OH
Cosmasteroside A
OH
O
Cosmasterosides B, C, D
O
Anasteroside A
O
Anasteroside B
R2=
Gal 1-3 Fuc 1-2 Gal 1-4 Xyl 1-3 Qui
Anasterosides A, B
1-2
Qui
(b)
R1
OH
RO
OH
OH
R1=
O R2
O R2
Luridoside A
Luridoside B
Minutoside A
O
N
H
SO3-Na+
Minutoside B
R= SO3-Na+ Luridoside A, B and Minutoside A
β-D-Xylopyranose Minutoside B
R2= 4'-O-Sulfate-β-D-Xylopyranose Luridoside A, B
β-D-Xylopyranose Minutoside A
(c)
OSO3-Na+
HO
OH
OH
OH
(d)
HO
O
O
12
17
OH
7
RO
(e)
OSO3-Na+
HO
Na+-O3SO
OH
R =
3-O-Me-Glc 1-3 6-OSO3Na-Glc 1-4 Qui 1-2 4-OSO3Na-Xyl Patagonicoside A
3-O-Me-Glc 1-3 Glc 1-4 Qui 1-2 Xyl Desulfated analog

◀ **Fig. 11.5** **a** Cosmasterosides **a**, **b**, **c** and **d**, and Anasterosides **a** and **b**. **b** Sulfated glycosides from *Cosmasterias lurida* and *Anasterias minuta*. **c** New polyhydroxysteroid from *Luidia ludwigi scotti*. **d** Patagonicoside A and its desulfated analogue; **e** Novel sulfated polyhydroxysteroid from *Ophioplocus januarii*

the sea star *Acanthaster planci* (Itakura and Komori 1986), were purified from the sea star *L. ludwigi scotti* (Roccatagliata 1996). Versicoside A, an asterosaponin from *Asterias amurensis versicolor* (Itakura et al. 1983), was also purified with Anasterosides A and B from *A. minuta* (Maier et al. 2002).

Polyhydroxysteroids are also extracted from echinoderms. These compounds show a polyhydroxylated steroidal aglycone and one or two monosaccharide moieties with variability in the hydroxylation pattern, the side chain functionalization, the localization of the glycosidic link and the presence of the sulfate moiety. Two novel steroidal xyloside sulfates (Luridosides A and B) and the known Pycnopodioside C, previously isolated from *Pycnopodia helianthoides* (Bruno et al. 1989), were isolated from the sea star *C. lurida* (Maier et al. 1993). *Anasterias minuta* provided two new sulfated polyhydroxylated steroidal xylosides: Minutosides A and B, together with the known Pycnopodioside B, previously isolated from *P. helianthoides* (Bruno et al. 1989). Minutoside B is the first example of a polyhydroxylated steroidal xyloside containing an amide function in the aglycone side chain (Fig. 11.5b). The three xylosides from *A. minuta* exhibited antifungal activity against *Cladosporium cucumerinum* and *Aspergillus flavus* (Chludil and Maier 2005). Helianthoside, previously isolated from *Heliaster helianthus* (Vázquez et al. 1993) was obtained from the ethanolic extract of the body walls of *A. capensis* (Díaz de Vivar 2002).

Sulfated polyhydroxysteroids were also extracted from sea stars. Roccatagliata et al. (1995) reported a novel sulfated polyhydroxysteroid from *L. ludwigi scotti* (collected at SMG) together with two other known compounds previously isolated from *Myxoderma platyacanthum* (Finamore et al. 1991) and *Luidia maculata* (Minale et al. 1984) (Fig. 11.5c).

The first study on sterols from sea stars of the coast of Argentina was by Seldes and Gros (1985) on *C. lurida*, and later, a mixture of sterols was obtained from *L. ludwigi scotti* by Roccatagliata et al. (1996b). These studies show that 24-methyl-Δ^7-sterols account for ~40 % of the sterol mixture in both sea stars, followed by choles-7-en-3β-ol (more than 20 %).

Glycosphingolipids consist of a sphingoid or a ceramide linked at C-1 by a β-glycosidic bond to a hydrophilic carbohydrate moiety. Glycosylceramides are the simplest glycosphingolipids and consist of a mono- or a disaccharide chain linked to the ceramide (http://www.chem.qmul.ac.uk/iupac/misc/glylp.html). Glucosylceramides (monoglycosylceramides) have been isolated from *C. lurida* (Maier et al. 1998), *A. capensis* (Díaz de Vivar 2002; Díaz de Vivar et al. 2002) and *A. minuta* (Chludil et al. 2003a) (Fig. 11.6b). Two new glucosylceramides were obtained from *C. lurida* (compounds 3 and 4), together with a mixture of related cerebrosides obtained as molecular species and Ophidiacerebrosides C and

D (Maier et al. 1998), and first reported from *Ophidiaster ophidianus* (Jin et al. 1994). *Anasterias minuta* produced a new glucosylceramide: Anasterocerebroside A, together with five other previously reported pure cerebrosides. Isolated as a pure compound, Glucosylceramide 7, previously obtained as molecular species from *C. lurida* and *A. capensis*, was characterized for the first time (Chludil et al. 2003a). The novel Inaequalocerebroside 1 was obtained from body walls and gonads of *A. capensis*. The body walls also contained two other new glucosylceramides as nearly homogeneous mixtures and five known cerebrosides: Phalluside 1, Ophidiacerebrosides C and D (Jin et al. 1994) and compounds 3 and 4, firstly isolated from *C. lurida*. Gonads had two new glucosylceramides (Inaequalocerebrosides 2 and 3), containing the new LCB (2*S*,3*R*,4*E*,15*Z*)-2-amino-4,15-docosadien-1,3-diol and Phalluside 1 as major components. A comparative analysis of the composition of the cerebroside mixtures obtained from body walls and gonads of *A. capensis* showed differences in the fatty acid and LCB moieties: the 9-methyl, $\Delta^{4,8,10}$ triene-LCB and saturated or unsaturated (Δ^{10} or Δ^{15})-24-C fatty acids were the most abundant in the glucosylceramides obtained from body walls, while the octadecatriene and the *cis*-Δ^{15}-22-C LCBs and 2-hydroxyheptadecanoic acid were predominant moieties in cerebrosides from gonads. All fatty acids isolated from body walls had normal chains whereas gonads had fatty acids with *iso* chains as well (Díaz de Vivar et al. 2002).

Ophiuroidea

Secondary metabolites from three species collected at SMG: *Ophioplocus januarii*, *Ophiactis asperula* and *Ophiacantha vivipara* were studied. *Ophioplocus januarii* afforded four sulfated polyhydroxysteroids, one of them is new (Fig. 11.5d). The antiviral activity of the four compounds was tested against herpes simplex virus type 1, Junín virus, respiratory syncytial virus and polio virus. These studies showed the importance of the unsaturation at C-22 for the inhibitory action towards respiratory syncytial virus and at C24-28 for replication of Junín virus (Roccatagliata et al. 1996a). *Ophiactis asperula* and *O. vivipara* did not have sulfated polyhydroxylated steroids as the major secondary metabolites, but showed nucleosides and nucleobases in their ethanolic extracts instead. From *O. asperula*, inosine (19.0 %), 2′-deoxyuridine (10.6 %), thymidine (14.9 %), and the bases uracyil (17.4 %), cytosine (15.9 %) and thymine (22.2 %) were obtained. *Ophiacantha vivipara* afforded inosine (7.1 %), 2′-deoxyinosine (13.4 %), uridine (3.8 %), 2′-deoxyuridine (7.1 %), 2′-deoxycytidine (1.4 %), and thymidine (4.6 %) with the bases uracyl (8.1 %), cytosine (4.8 %), thymine (31.9 %) and hypoxanthine (17.8 %) (Murray et al. 2002). Considering echinoderms, 2′-deoxyribonucleosides had only been previously isolated from the sea stars *A. planci* and *L. maculata* (Komori et al. 1978, 1980). Their importance as systematic markers was suggested by Dematté et al. (1986), since they had only been isolated from tunicates and echinoderms. This is currently under discussion as they have also been purified from two marine sponges (Kondo et al. 1992; Searle and Molinski 1994).

Holothuroidea

Secondary metabolites from holothurians are characterized by the presence of triterpenoidal saponins (holothurins) and can be considered characteristic of the class. Their toxicity (acting as membranotropic towards cell membranes) can be associated with a defensive response of the organism and is the cause of their biological activity (Maier 2008). They have also been used as chemotaxonomic markers since some holothurins are specific for different taxa (see Maier and Murray 2006; Antonov et al. 2008).

Two species of holothurians from the Patagonian coast of Argentina have been studied: *Hemioedema spectabilis* (collected at SJOG) and *Psolus patagonicus* (collected off Bahía Ensenada; 54°50′S, 68°29′W). Patagonicoside A, the major component of the ethanolic extract of *P. patagonicus* was isolated and its structure was determined by Murray et al. (2001) (Fig. 11.5d). Patagonicoside A has revealed a broad spectrum of biological activities: the glycoside and its semi-synthetic desulfated analogue were tested for their antifungal activity against the phytopathogenic fungi *Cladosporium cucumerinum* (Murray et al. 2001), *C. fulvum*, *Fusarium oxysporum* and *Monilia* sp. (Muniain et al. 2008). The intact compound is more effective than the desulfated analogue. It showed an antifungal activity comparable with a commercial fungicide towards *Monilia* sp. Bioactivity assays showed that the presence of the sulphate groups in the oligosaccharide chain plays an important role in the antifungal activity. Patagonicoside A and the desulfated derivative were also tested for antiproliferative, cytotoxic and haemolytic activities and their effect on NF-κB activation (Careaga et al. 2009). Both glycosides showed low cytotoxic activity in A549 tumour cells, which the authors attribute to the uncommon presence of 12α,17α-hydroxyl groups in the aglycone moiety. Both compounds exert an antiproliferative effect although their mechanisms of action do not involve inhibition of NF-κB.

Chludil et al. (2002b) isolated and described two new sulfated triterpene glycosides, Hemoiedemosides A and B from the Patagonian holothurian *H. spectabilis* (Fig. 11.6a). Hemoiedemoside B is a new example of a small number of trisulfated triterpene glycosides from holothurians belonging to the family Cucumariidae. Both glycosides exhibit considerable antifungal activity against the phytopathogenic fungus *C. cucumerinum*, while the semisynthetic desulfated derivative of Hemoiedemoside A is less active.

11.4 Aquaculture and Fisheries

Although fisheries of the sea urchin *Loxechinus albus* has been extensive in Chile, in Argentina it has been captured only sporadically by artisanal fisheries in the Beagle Channel near Ushuaia Bay, especially for the local market, to serve tourists from the Orient, and occasionally exported to Chile.

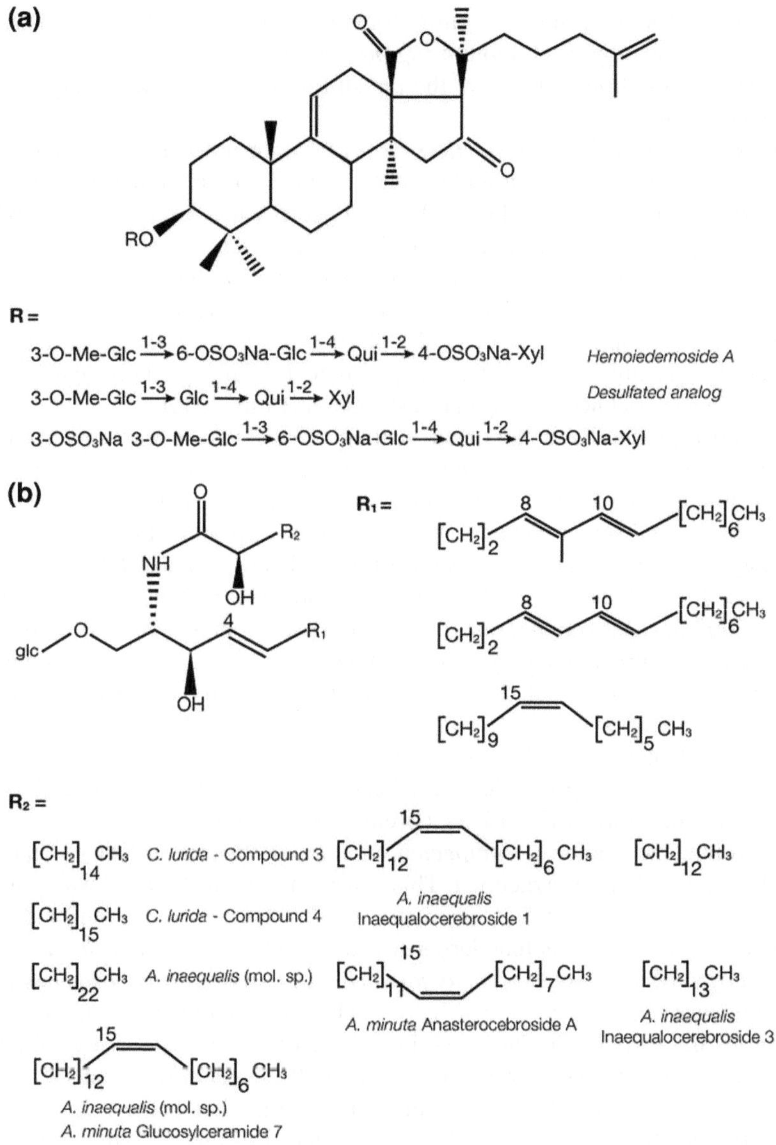

Fig. 11.6 **a** Triterpene glycosides from *Hemioedema spectabilis*. **b** Glucosylceramides from sea stars of the Patagonian coast of Argentina

11.5 Echinoderm Threats

The trawl fishery is considered one of the major factors disturbing the seabed (Kaiser et al. 2006). Otter trawls collect and kill huge amounts of non-target species, physically disturb the seabed, and alter the structure of benthic communities. There are several bottom trawling fisheries in Argentina. Currently, three fisheries have by-catch monitoring programs in order to assess the condition of the disturbed bottoms; these are the red shrimp fishery and the Patagonian and Tehuelche scallop fisheries.

The red shrimp (*Pleoticus muelleri*) fishery has operated in SJOG since 1980. Only 20 % of the species found in the benthic community are retained as incidental catch during the fishery trawls. Echinoderms, together with crustaceans and mollusks, are the main groups of by-catch. *Diplasterias brandti*, *Henricia obesa*, *Pseudechinus magellanicus*, *Arbacia dufresnii*, *Austrocidaris canaliculata*, *Trypilaster philippi*, *Hemioedema spectabilis* and *Gorgonocephalus chilensis* are the echinoderm species caught by the shrimp trawl (Roux et al. 2007); *Diplasterias brandti* being the most frequent. The benthic communities subjected to trawling partially maintain their structure, species richness and spatial distribution (Roux 2008).

The most productive Patagonian scallop (*Zygochlamys patagonica*) fishing grounds in Argentina are associated with the shelf-break frontal system and have been exploited since 1996. Several studies on incidental catches show that brittle stars, sea stars and sea urchins are the dominant taxa in most of the scallop beds; reaching 86 % of the total catch in some of them (Bremec et al. 2006, Escolar et al. 2007; Schejter et al. 2008). The biomass of echinoderms as bycatch increased from 1997 to 2002 in sites with high fishing activity (Escolar et al. 2009). Because many echinoderms are predators, they could be favored by fishery discards. In addition, an increase in biomass of *Ophiactis asperula* (detritivore) was recorded in intensively trawled sites, while *Ophiacantha vivipara* and *Sterechinus agassizii* decreased (Schejter et al. 2008). Although relative abundances of some taxa varied over time, no loss of species has been recorded in the Patagonian scallop community since the start of the fishery (Bremec et al. 2008).

The Tehuelche scallop (*Aequipecten tehuelchus*) fishery in SMG is the main coastal bivalve fishery in Argentina. This fishery started in 1969 and was closed for ten years from 1973 to 1983 (Ciocco et al. 2006). Impact studies on biodiversity reported five species of echinoderms (*Arbacia dufresnii*, *Pseudechinus magellanicus*, *Tripylaster* sp., *Ophioplocus januarii* and *Cosmasterias lurida*) as part of incidental catches. The total density of echinoderms remained stable in sites without fishing activity and increased in sites that were intensively exploited until 2000; the dominant species were *A. dufresnii*, *P. magellanicus* and *O. januarii*. In contrast, echinoderm density had been reduced at sites subjected to intensive and continuous fishing activity. Patches of *A. dufresnii* and *P. magellanicus* were depleted in these fishing grounds (Morsán 2009).

11.6 Concluding Remarks

Until recently studies on echinoderms in Argentina were very limited. During the last century Bernasconi's publications were the principal references for this phylum. In the last ten years there was an increase of studies on the echinoderm fauna of Argentina. This is reflected in recent Doctoral theses (Díaz de Vivar 2002; Pérez 2009; Brogger 2010; Escolar 2010; Rubilar 2010, Epherra in prep., Gil in prep., Martinez in prep., Meretta in prep.) and Licenciatura theses (Gil 2002; Rubilar 2004; Brogger 2005; Marzinelli 2005; Martinez 2008; Villares 2008; Epherra 2009; Rodríguez 2009; Di Giorgio 2011; Pérez-Gallo in prep; Reartes in prep.). These studies are mainly focused on biological and ecological aspects. However, some have focused on chemical, taxonomical, physiological or anatomical aspects. We expect that these will be the beginning of new echinoderm researches in Argentina.

Acknowledgments The present chapter was greatly improved with the valuable comments on the manuscript of five reviewers: J Lawrence, D Pawson, P Lambert, J Pearse, and J Cortés. We would like to thank them for kindly reviewing the manuscript and for their useful suggestions. We are very grateful to all Argentinean researchers who contributed to increase the knowledge about this phylum in the region. A special recognition must be done to the Institutions and the directors for helping us conduct research on echinoderms. Thank you very much to P Penchaszadeh, H Zaixso, C Pastor de Ward, C Bremec, J Calvo, and to the late A Seldes and I Bernasconi.

References

Adami ML, Gordillo S (1999) Structure and dynamics of the biota associated with *Macrocystis pyrifera* (Phaeophyta) from the Beagle Channel, Tierra del Fuego. Sci Mar 63:183–191

Acha E, Mianzan H, Guerrero R, Favero M, Bava J (2004) Marine fronts at the continental shelves of austral South America, physical and ecological processes. J Mar Syst 44:83–105

Agassiz A (1873) The Echini collected in the Hassler expedition. Bull Mus Comp Zool Harvard 3:187–190

Agassiz A (1881) Report on the echinoidea report of the challenger expedition. Zoology 9:1–321

Agassiz A (1904) Reports on an exploration off the west coast of Mexico, Central and south America, and off the Galapagos Islands, in charge of Alexander Agassiz, by the U.S. Fish Commission Steamer "Albatross", during 1891, Lieut. Commander Z.L. Tanner, U.S. Commanding. XXXII. The Panamic deep sea echini. Mem Mus Comp Zool Harvard Coll 31:1–246

Agatep CP (1967) Some Elasipodid holothurians of Antarctic and Subantarctic Seas. Antarctic Res Ser Am Geophys Union, Washington

Ahearn CG (1994) Family Psolidae: new distributional records from the Antarctic. In: David B, Guille A, Feral JP (eds) Echinoderms Through Time, Proc 8th Int Echinoderm Conference, Dijon pp 503

Améziane N, Roux M (1997) Biodiversity and historical biogeography of stalked crinoids (Echinodermata) in the deep sea. Biodiv Conserv 6:1557–1570

Andersson KA (1904) Brutpflege bei *Antedon hirsuta* Carpenter. Wiss Ergebn d Schwed Südpolar-Exped 1901–1903(5):1–7

Antonov AS, Avilov SA, Kalinovsky AI, Anastyuk SD, Dmitrenok PS, Evtushenko EV, Kalinin VI, Smirnov AV, Taboada S, Ballesteros M, Avila C, Stonik VA (2008) Triterpene glycosides from Antarctic sea cucumbers. 1. Structure of liouvillosides A1, A2, A3, B1, and B2 from the sea cucumber *Staurocucumis liouvillei*: new procedure for separation of highly polar glycoside fractions and taxonomic revision. J Nat Prod 71:1677–1685

Arana P, Palma S, Bay-Schmith E, Monardes M, Galvez M (1996) Aspectos biológicos y pesqueros del erizo (*Loxechinus albus*) en la Región de Magallanes. Univ Católica, Valparaiso, Informe Final 23:83–139

Arribas LP, Penchaszadeh PE, Brogger MI (2008) Seasonal variation of assemblage fauna to *Ophioplocus januarii* (Echinodermata: Ophiuroidea) in Golfo San José, Patagonia. Book of abstracts, 5th North Am Echinoderm Conference, Melbourne, Florida

Balech E, Ehrlich M (2008) Esquema biogeográfico del Mar Argentino. Rev Invest Des Pesq 19:45–75

Bartsch I (1982) Ophiuroidea (Echinodermata) from the Patagonian shelf. Mitt Hambur Zool Mus Inst 79:211–250

Bastida R, Urien CM (1981) Investigaciones sobre comunidades bentónicas. Características generales del sustrato (campañas IV, V, X y XI del B/I Shinkai Maru). Contrib INIDEP 338:318–339

Bastida R, Roux A, Martinez DE (1992) Benthic communities of the Argentine continental shelf. Oceanol Acta 15:687–698

Bastida R, Zamponi M, Bremec C, Roux A, Genzano G, Elías R (2007) Las comunidades bentónicas. In: Carreto I, Bremec C (eds) El Mar Argentino y sus recursos pesqueros 5. INIDEP, Mar del Plata, pp 91–125

Bay-Schmith E, Werlinger C, Silva J (1981) Ciclo anual de reproducción del recurso erizo *Loxechinus albus* entre la X y XII Región Informe Final Subsec Pesca. Univ Concepción, Chile

Bell FJ (1881) Account of the Echinodermata collected during the survey of HMS "*Alert*" in the Straits of Magellan and on the coast of Patagonia. Proc Zool Soc Lond 1881:87–101

Bernasconi I (1924) Equinoideos de las costas Argentinas. Physis 7:1–254

Bernasconi I (1925) Resultados de la Primera Expedición a Tierra del Fuego (1921), enviada por la Facultad de Ciencias Exactas, Físicas y Naturales de la Universidad Nacional de Buenos Aires. Equinodermos I. Equinoideos. An Soc Cien Arg 99:1–247

Bernasconi I (1926) Una ofiura vivípara de Necochea. An Mus Nac Hist Nat 34:145–153

Bernasconi I (1934) Los equinodermos de los mares argentinos. Observaciones sobre algunos "erizos de mar", "estrellas de mar" y otras formas comunes en nuestro litoral. Bol Cent Naval 53:1–21

Bernasconi I (1935) Dos nuevas especies de *Pteraster* en la Argentina. Physis 11:482–483

Bernasconi I (1937) Asteroideos argentinos I: Familia Pterasteridae. An Mus Argentino de Cien Nat "Bernardino Rivadavia" 39:167–187

Bernasconi I (1941a) Dos nuevas especies argentinas de *Luidia*. Physis 19:117–118

Bernasconi I (1941b) Los equinodermos de la expedición del Buque Oceanográfico "Comodoro Rivadavia" ARA. Physis 19:37–49

Bernasconi I (1941c) Sobre la distribución geográfica de *Mellita sexiesperforata* (Leske). Physis 19:105–108

Bernasconi I (1942) Primeros estados larvales de *Arbacia dufresnei* (Blv.). Physis 19:305–317

Bernasconi I (1943) Los asteroideos sudamericanos de la familia Luidiidae. An Mus Argentino de Cien Nat "Bernardino Rivadavia" 61:1–20

Bernasconi I (1947) Distribución geográfica de los equinoideos argentinos. An Soc Arg Est Geog 6:97–114

Bernasconi I (1953) Monografía de los equinoideos argentinos. An Mus Hist Nat Montevideo Seg Ser 6:1–58

Bernasconi I (1957) Equinoideos y asteroideos de la colección del Instituto Oceanográfico de la Universidad de San Pablo. Bol Inst Oceanogr 7:119–148

Bernasconi I (1961) Una nueva especie de asteroideo. Neotrop 7:1–2

Bernasconi I (1962a) Asteroideos argentinos. III. Familia Odontasteridae. Rev Mus Argentino Bernardino Rivadavia 9:1–25

Bernasconi I (1962b) *Perknaster denus patagonicus* nueva sub especie de la Argentina. Physis 23:257–258

Bernasconi I (1963) Asteroideos argentinos IV. Familia Goniasteridae. Rev Mus Argentino de Cien Nat Bernardino Rivadavia Cienc Zool 9:1–26

Bernasconi I (1964a) Asteroideos argentinos V. Familia Ganeriidae. Rev Mus Argentino de Cien Nat Bernardino Rivadavia Cienc Zool 9:59–89
Bernasconi I (1964b) Asteroideos argentinos. Claves para los órdenes, familias, subfamilias y géneros. Physis 24:241–277
Bernasconi I (1964c) Distribución geográfica de los Equinoideos y Asteroideos de la extremidad austral de Sudamérica. Bol Inst Biol Mar 7:43–50
Bernasconi I (1965a) Nuevo género y nueva especie abisal de Goniasteridae (Equinodermata, Asteroidea). Physis 25:333–335
Bernasconi I (1965b) Ophiuroidea de Puerto Deseado (Santa Cruz, Argentina). Physis 25:143–152
Bernasconi I (1965c) *Astrotoma agassizii* Lyman, especie vivípara del Atlántico sur. Physis 25:1–5
Bernasconi I (1966a) Descripción de una nueva especie de *Calyptraster*. (Asteroidea, Pterasteridae). Physis 26:95–99
Bernasconi I (1966b) Los equinoideos y asteroideos colectados por el buque oceanográfico R/V "Vema", frente a las costas argentinas, uruguayas y sur de Chile. Rev Mus Argentino de Cien Nat Bernardino Rivadavia. Zool 9:147–175
Bernasconi I (1970) Sobre la validez específica de *Austrocidaris lorioli* (Mortensen, 1903) (Echinodermata, Echinoidea). Physis 30:251–254
Bernasconi I (1972) Nuevas especies de asteroidea: *Bathydiaster herwigi* sp. nov. (Astropectinidae) y *Calyptraster vitreus* sp. nov. (Pterasteridae). Physis 31:9–14
Bernasconi I (1973a) Asteroideos argentinos VI. Familia Asterinidae. Rev Mus Argentino de Cien Nat Bernardino Rivadavia. Hidrobiol 3:333–346
Bernasconi I (1973b) Los equinodermos colectados por el "Walther Herwig" en el Atlántico Sudoeste. Rev Mus Argentino de Cien Nat Bernardino Rivadavia. Hidrobiol 3:287–334
Bernasconi I (1980) Asteroideos argentinos VII. Familia Echinasteridae. Rev Mus Argentino de Cienc Nat Bernardino Rivadavia. Hidrobiol 4:247–258
Bernasconi I, D'Agostino MM (1971) Ofiuroideos argentinos. Claves para los órdenes, subordenes, familias, subfamilias y géneros. Physis 30:447–469
Bernasconi I, D'Agostino MM (1974) Ampliación del área de distribución de *Amphiura crassipes* Ljungman, 1867 (Ophiuroidea, Amphiuridae). Physis 33:135–138
Bernasconi I, D'Agostino MM (1975) Nueva especie de ofiuroideo argentino *Amphilepis sanmatiensis* sp. nov. (Ophiuroidea, Amphilepididae). Physis 34:355–358
Bernasconi I, D'Agostino MM (1977) Ofiuroideos del mar epicontinental argentino. Rev Mus Argentino de Cienc Nat Bernardino Rivadavia. 5:65–114
Bertness MD, Crain CM, Silliman BR, MC Bazterrica, Reyna MV, Hidalgo F, Farina K (2006) The community structure of western Atlantic Patagonian Rocky shores. Ecol Monogr 76:439–460
Bohn JM (2007) *Pseudrotasfer microincubator* gen. et spec. nov., a brooding cucumariid holothurian (Echinodermata: Holothuroidea: Dendrochirotida) from the Burdwood Bank (south-western Atlantic Ocean). Zootaxa 1662:61–68
Boschi EE (1979) Geographic distribution of argentinian marine decapod crustaceans. Bull Biol Soc Wash 3:134–143
Botto F, Bremec C, Marecos A, Schejter L, Lasta M, Iribarne O (2006) Identifying predators of the SW Atlantic Patagonian scallop *Zygochlamys patagonica* using stable isotopes. Fish Res 81:45–50
Brandini FP, Boltovskoy D, Piola A, Kocmur S, Rottgers R, Abreu PC, Lopes RM (2000) Multiannual trends in fronts and distribution of nutrients and chlorophyll in the southwestern Atlantic (30–62°S). Deep-Sea Res I 47:1015–1033
Bremec CS, Roux A (1997) Resultados del análisis de una campaña de investigación pesquera, sobre comunidades bentónicas asociadas a bancos de mejillón (*Mytilus edulis platensis* D'orb.) en costas de Buenos Aires Argentina. Rev Invest Des Pesq 11:153–166
Bremec CS, Lasta ML (2002) Epibenthic assemblage associated with scallop (*Zygochlamys patagonica*) beds in the Argentinian shelf. Bull Mar Sci 70:89–105
Bremec C, Brey T, Lasta M, Valero J, Lucifora L (2000) *Zygochlamys patagonica* beds on the Argentinian shelf. Part I: Energy flow through the scallop bed community. Arch Fish Mar Res 48:295–303

Bremec C, Schejter L (2010) Benthic diversity in a submarine canyon in the Argentine Sea. Rev Chil Hist Nat 83:453–457

Bremec C, Schejter L, Madirolas A, Trípode M (2006) Comunidades de aguas profundas: macrofauna bentónica de un cañón submarino localizado en la plataforma patagónica (43°35′S, 59°33′O). Book of abstracts. IV Jornadas Nac Cien Mar, Buenos Aires

Bremec C, Schejter L, Giberto D, Escolar M, Souto V, Acha M, Mianzán H (2008) Benthos of the Argentinian continental shelf: faunal assemblages, species distribution and richness along a latitudinal gradient Book of abstracts. World Conf Mar Biodiv, Valencia

Brogger MI (2005) Biología reproductiva del erizo verde *Arbacia dufresnii* (Blainville, 1825) en las costas de Golfo Nuevo. Licentiate Thesis. Univ Buenos Aires, Argentina

Brogger MI (2010) Crecimiento, reproducción y alimentación del ofiuroideo *Ophioplocus januarii* (Lütken, 1856) (Echinodermata: Ophiuroidea) en costas de Patagonia, Argentina. PhD Thesis, Univ Buenos Aires, Argentina

Brogger MI, Ivanov VA (2010) *Syndesmis patagonica* n. sp. (Rhabdocoela: Umagillidae) from the sea urchin *Arbacia dufresnii* (Echinodermata: Echinoidea) in Patagonia Argentina. Zootaxa 2442:60–68

Brogger MI, Penchaszadeh PE (2003) Observaciones sobre la ecología poblacional del erizo plano *Encope emarginata* en la región de Mar del Plata. Book of abstracts. V Jornadas Nac Cien Mar, Mar del Plata

Brogger MI, Penchaszadeh PE (2008) Infaunal mollusks as main prey for two sand bottoms sea stars off Puerto Quequén (Argentina). Rev Biol Trop 56(Suppl 3):329–334

Brogger MI, Arribas LP, Martinez MI, Rodríguez MS (2008) Biodiversidad de equinodermos del Mar Argentino: estado actual del conocimiento Book of abstracts. III Congr Nac Conserv Biodiv, Buenos Aires

Brogger MI, Martinez MI, Penchaszadeh PE (2004) Reproductive biology of *Arbacia dufresnei* in Golfo Nuevo, Argentine Sea. Proc Int Conf Sea-Urchin Fish Aquacul 1:165–169

Brogger MI, Martinez MI, Penchaszadeh PE (2010) Reproduction of the sea urchin *Arbacia dufresnii* (Echinoidea: Arbaciidae) from Golfo Nuevo, Argentina. J Mar Biol Ass UK 90:1405–1409

Bruno I, Minale L, Riccio R (1989) Steroidal glycosides from the starfish *Pycnopodia helianthoides*. J Nat Prod 52:1022–1026

Burnell DJ, Apsimon JW (1983) Echinoderm saponins. In: Scheuer PJ (ed) Marine Natural Products: Chemical and Biological Perspectives. Academic Press, New York, pp 287–371

Careaga V, Bueno C, Muniain C, Alché L, Maier MS (2009) Antiproliferative, cytotoxic and hemolytic activities of a triterpene glycoside from *Psolus patagonicus* and its desulfated analog. Chemoth 55:60–68

Carpenter PH (1884) Report on the Crinoidea. First Part.-Stalked Crinoids. Report on the Challenger Expedition vol 11, London

Carpenter PH (1888) Report on the Crinoidea. Second Part.-Comatulæ. Report on the Challenger Expedition, vol 26, London

Carreto JI, Lutz VA, Carignan MO, Cucchi-Colleoni AD, de Marco SG (1995) Hydrography and chlorophyll in a transect from the coast to the shelf-break in the Argentinian Sea. Cont Shelf Res 15:315–336

Carriquiriborde L, Borzone C, Lizarralde Z, Pombo A, Manrique R, Ichazo M (1983) Aspectos biocenológicos del golfo Nuevo (Chubut, Argentina). Centro Nacional Patagónico. Informe Interno, Puerto Madryn

Castilla JC, Moreno CA (1982) Sea urchins and *Macrocystis pyrifera*: Experimental test of their ecological relation in southern Chile. In: Lawrence JM (ed) Echinoderms: Proc 4th Int Echinoderm Conf. Tampa Bay, Balkema Rotterdam, pp 257–263

Cherbonnier G (1941) Note sur une nouvelle holothurie antarctique, *Cucumaria cornuta* sp.nov. Bull Soc Zool France 66:271–276

Chludil HD, Maier MS (2005) Minutosides A and B, antifungal sulfated steroid xylosides from the Patagonian starfish *Anasterias minuta*. J Nat Prod 68:1279–1283

Chludil HD, Seldes AM, Maier MS (2002a) Glycosphingolipids from starfishes and sea cucumbers. Res Advan Lip 2:9–18

Chludil HD, Muniain C, Seldes AM, Maier MS (2002b) Cytotoxic and antifungal triterpene glycosides from the Patagonian sea cucumber *Hemoiedema spectabilis*. J Nat Prod 65:860–865

Chludil HD, Seldes AM, Maier MS (2003a) Anasterocerebroside A, a new glucosylceramide from the Patagonian starfish *Anasterias minuta*. Z Naturforsch 58:433–440

Chludil HD, Murray AP, Seldes AM, Maier MS (2003b) Biologically active triterpene glycosides from Sea cucumbers. Stud Nat Prod Chem 28:587–616

Ciocco NF, Lasta M, Bremec C (1998) Pesquerías de Bivalvos: Mejillón, vieiras (tehuelche y patagónica) y otras especies. In: Boschi EE (ed) El Mar Argentino y sus recursos pesqueros 2. INIDEP, Mar del Plata, pp 143–166

Ciocco NF, Lasta M, Narvarte M, Bremec C, Bogazzi E, Valero J, Orensanz JM (2006) Fisheries and Aquaculture: Argentina. In: Shumway SE, Parsons GJ (eds) Scallops: Biology Ecology and Aquaculture. Elsevier, Amsterdam, pp 1251–1292

Clark AH (1915) A monograph of the existing crinoids: Vol 1. The Comatulids. Part 1. US Nat Mus Bull 82:1–406

Clark AH (1947) A monograph of the existing crinoids. Vol 1. The Comatulids. Part 4b. Superfamily Mariametrida (Concluded. The family Colobometridae) and Superfamily Tropiometrida (Excep the families Thalassometridae and Charitometridae). US Nat Mus Bull 82:1–473

Clark AM, Downey ME (1992) Starfish of the Atlantic. Chapman and Hall, New York

Coulin C, Brogger MI, Penchaszadeh PE (2009) Estudio del desarrollo mandibular y su aplicación en la estimación del crecimiento y edad en *Ophioplocus januarii* (Echinodermata: Ophiuroidea) del Golfo San José, Chubut. Book of abstracts. VII Jornadas Nac Cien Mar. Bahía Blanca

D'Auria MV, Minale L, Riccio R (1993) Polyoxygenated steroids of marine origin. Chem Rev 93:1839–1895

Deichmann E (1947) Shallow water holothurians from the Cabo de Hornos and adjacent waters. An Mus Argentino de Cien Nat Bernardino Rivadavia 42:325–351

Dematté N, Guerriero A, Lafargue F, Pietra F (1986) 2′-deoxynucleoside uronic acids from the ascidian *Aplidium (= Amaroucium) fuscum* (Drasche, 1883). Comp Bioch Physiol 84B:11–13

Díaz de Vivar ME (2002) Aislamiento y elucidación estructural de metabolitos secundarios de las estrellas de mar *Labidiaster annulatus* (Sladen, 1889) y *Allostichaster inaequalis* (Koehler, 1923). PhD Thesis, Univ Nac Patagonia San Juan Bosco, Argentina

Díaz de Vivar ME, Seldes AM, Maier MS (2002) Two novel glucosylceramides from gonads and body walls of the Patagonian starfish *Allostichaster inaequalis*. Lipids 37:597–603

Di Giorgio GV (2011) Regeneración en el ofiuro *Ophioplocus januarii* (Lutken,1956). Licentiate Thesis. Univ Nac Patagonia San Juan Bosco, Argentina

Dolcemascolo A, Zaixso HE (2000) Aspectos ecológicos del erizo espatangoideo *Abatus cavernosus*, en la ria Deseado (Santa Cruz). Book of abstracts. IV Jornadas Nac Cien Mar, Puerto Madryn

Ekman S (1925) Holothurien. Further zoological Results Sweden Antarctic Expedition 1:1–194

Epherra L (2009) Ciclo bioquímico del erizo verde *Arbacia dufresnii* (Blainville, 1825), en costas del Golfo Nuevo. Licentiate Thesis. Univ Nac Patagonia San Juan Bosco, Argentina

Epherra L, Díaz de Vivar ME, Rubilar T, Pastor de Ward CT, Gil DG, Tolosano J, Latorre M, Reartes M (2009a) Ciclo bioquímico del erizo verde *Arbacia dufresnii* (Blainville, 1825) en costas del Golfo Nuevo, Patagonia. Book of abstracts. VII Jornadas Nac Cien Mar, Bahía Blanca

Epherra L, Díaz de Vivar ME, Rubilar T, Pastor de Ward CT, Gil DG, Tolosano J, Latorre M, Reartes M (2009b) *Arbacia dufresnii* (Blainville, 1825): Dieta, estado nutricional y ciclo reproductivo en dos poblaciones de la Patagonia central: Resultados preliminares. Book of abstracts. I Jornadas Patagónicas de Biol, III Jornadas Estudiantiles de Cien Biol, Trelew

Escolar M (2010) Variaciones espacio-temporales en la comunidad de invertebrados bentónicos asociada al frente de talud. Equinodermos como caso de estudio. PhD Thesis, Univ Buenos Aires, Argentina

Escolar M, Bremec C (2009) Familia Pterasteridae (Echinodermata: Asteroidea) en bancos de vieira patagónica en el Mar Argentino. Book of abstracts. VII Jornadas Nac Cien Mar, Bahía Blanca

Escolar M, Diez M, Marecos A, Campodónico S, Vallarino E, Schejter L, Bremec C (2007) *Zygochlamys patagonica* fishery: by-catch data from the "Observers On Board Program" (INIDEP). J Shellfish Res 26:1308

Escolar M, Diez M, Hernández D, Marecos A, Campodónico S, Bremec C (2009) Invertebrate bycatch in Patagonian scallop fishing grounds: a study case with data obtained by the On Board Observers Program. Rev Biol Mar Oceangr 44:369–377

Escudero G (1999) Incubación en *Anasterias minuta* Perrier. Licentiate Thesis, Univ Nac de la Patagonia San Juan Bosco, Argentina

Escudero G, Zaixso HE (2003) Incubación en *Anasterias minuta* Perrier Book of abstracts. V Jornadas Nac Cien Mar, Mar del Plata

Fedulov PP, Remeslo AV, Burykin SN, Pulishchuk JA (1990) Variabilidad de la Corriente de Malvinas. Frente Marítimo 6:121–127

Fisher WK (1940) Asteroidea Discovery Reports. Cambridge Univ Press, Cambridge

Finamore E, Minale L, Riccio R, Rinaldo G, Zollo F (1991) Novel marine polyhydroxylated steroids from the starfish *Myxoderma platyacanthum*. J Org Chem 56:1146–1153

Findlay JA, He ZQ, Blackwell B (1990) Minor saponins from the starfish *Asterias forbesi*. Canad J Chem 68:1215–1217

García CAE, Sarma YVB, Mata MM, García VMT (2004) Chlorophyll variability and eddies in the Brazil–Malvinas Confluence region. Deep-Sea Res II 51:159–172

Garzoli SL, Garraffo Z (1989) Transports, frontal motions and eddies at the Brazil-Malvinas currents confluence. Deep-Sea Res A 36:681–703

Gebruk AV (19889 New taxa of the deep-sea holothurians from the family Elpidiidae (Elasipoda). Zoologicheskij J 67:914–922

Gebruk AV (1990) Deep-sea holothurians of the family Elpidiidae. Nauka, Moscow

Gebruk AV (1993) New data on elasipodid fauna of the South Atlantic and Antarctic. Trudy Inst Okeanologii 127:228–244

Gebruk AV, Shirshov PP (1994) Revision of two deep-sea holothurian genera *Psychrelpidia* Hérouard, 1923 and *Rhipidothuria* Hérouard, 1901 (Elpidiidae). Mitteilungen Hamb Zool Mus Inst 91:147–154

Gil DG (2002) Alimentación en *Anasterias minuta* Perrier. Licentiate Thesis. Univ Nac Patagonia San Juan Bosco, Argentina

Gil DG, Zaixso HE (2007) The relation between feeding and reproduction in *Anasterias minuta* (Asteroidea: Forcipulata). Mar Biol Res 3:256–264

Gil DG, Zaixso HE (2008) Feeding ecology of the subantarctic sea star *Anasterias minuta* within tide pools in Patagonia Argentina. Rev Biol Trop 56(Suppl 3):311–328

Gil DG, Zaixso HE, Tolosano JA (2009) Brooding of the sub-Antarctic heart urchin, *Abatus cavernosus* (Spatangoida: Schizasteridae), in southern Patagonia. Mar Biol 156:1647–1657

Giménez J, Penchaszadeh PE (2010) Brooding in *Psolus patagonicus* (Echinodermata: Holothuroidea) from Argentina, SW Atlantic Ocean. Helgoland Mar Res 64:21–26

Grygier MJ, Salvat MB (1984) *Dendrogaster argentinensis*, new species, a South American sea-star parasite (Crustacea: Ascothoracida). Proc Biol Soc Wash 97:43–48

Gutiérrez J, Otsu I (1975) Periodicidad en las variaciones biométricas de *Loxechinus albus* Molina. Rev Biol Mar 15:179–199

Guerrero RA, Piola AR (1997) Masas de agua en la plataforma continental. In: Boschi EE (ed) El Mar Argentino y sus recursos pesqueros. Antecedentes históricos de las exploraciones en el mar y las características ambientales, vol. I. Inst Nac Inves Des Pesq, Mar del Plata pp 107–118

Guerrero RA, Acha EM, Framiñan MB, Lasta C (1997) Physical oceanography of the Río de la Plata Estuary, Argentina. Cont Shelf Res 17:727–742

Guerrero RA, Baldoni A, Benavides HR (1999) Oceanographic conditions at the southern end of the Argentine continental slope. INIDEP Doc Cien 5:7–22

Hernández DA (1981) Holothuroidea de Puerto Deseado (Santa Cruz, Argentina). Rev Mus Argentino de Cien Nat "Bernardino Rivadavia" Hidrobiol 4:151–168

Hernández DA (1982) Holothuroidea des Sudwestatlantiks I. Die Trachythyone-Arten. Mitteilungen Hambur Zool Mus Inst 79:251–261

Hernández DA (1985) Holothuroidea del Mar Argentino. PhD Thesis, Univ Buenos Aires, Argentina

Hernández DA (1987) *Trachythyone baja* sp. n., a new species from Antarctic waters (Echinodermata: Holothurioidea). Mitteilungen Hambur Zool Mus Inst 84:161–165

Hernández DA, Tablado A (1985) Asteroidea de Puerto Deseado (Santa Cruz, Argentina). CENPAT, Argentina, Contribución 104:1–16

Hernando MP (2006) Efectos de la radiación solar sobre el fitoplancton de aguas Antárticas y sub-Antárticas. PhD Thesis, Universidad de Buenos Aires, Buenos Aires

Hérouard E (1906) Holothuries. Resultats du Voyage du SY Belgica 1:1–19

Iorizzi M, Riccio R (1999) Steroidal oligoglycosides from marine sources. In: Ikan R (ed) Naturally ocurring Glycosides. Wiley, West Sussex, pp 311–344

Iorizzi M, De Marino S, Zollo F (2001) Steroidal glycosides from Asteroidea. Curr Org Chem 5:951–973

Itakura Y, Komori T (1986) Steroid oligoglycosides from the starfish *Acanthaster planci* L. Structures of four new oligoglycoside sulfates. Justus Liebigs An Chem 1986:499–508

Itakura Y, Komori T, Kawasaki T (1983) Steroid oligoglycosides from the starfish *Asterias amurensis* [cf.] *versicolor* Sladen, 1 Structural elucidation of a new oligoglycoside sulfate. Justus Liebigs An Chem 1983:2079–2091

Jin W, Rinehart JL, Jares-Erijman E (1994) Ophidiacerebrosides: cytotoxic glycosphingolipids containing a novel sphingosine from a sea star. J Org Chem 59:144–147

John DD (1938) Crinoidea. Discovery Rep 18:121–222

Kaiser MJ, Clarke KR, Hinz H, Austen MVC, Somerfield PJ, Karakassis I (2006) Global analysis of response and recovery of benthic biota to fishing. Mar Ecol Progr Ser 311:1–14

Kalinin VI (2000) System-theoretical (holistic) approach to the modelling of structural-functional relationships of biomolecules and their evolution: an example of triterpene glycosides from sea cucumbers (Echinodermata, Holothuroidea). J Theor Biol 206:151–168

Kalinin VI, Volkova OV, Likhatskaya GN, Prokofieva NG, Agafonova IG, Anisimov MM, Kalinovsky AI, Avilov SA, Stonik VA (1992) Hemolytic activity of triterpene glycosides from Cucumariidae family holothurians and evolution of this group of toxins. J Nat Toxins 1:17–30

Kalinin VI, Anisimov MM, Prokofieva NG, Avilov SA, Afiyatullov SS, Stonik VA (1996) Biological activities and biological role of triterpene glycosides from holothuroidea (Echinodermata). In: Jangoux M, Lawrence JM (eds) Echinoderm Studies. AA Balkema, Rotterdam pp 139–181

Kalinin VI, Silchenko AS, Avilov SA, Stonik VA, Smirnov AV (2005) Sea cucumbers triterpene glycosides, the recent progress in structural elucidation and chemotaxonomy. Phytochem Rev 4:221–236

Koehler R (1902) Expédition Antarctique Belge. Résultats du Voyage du S.Y. Belgica en 1897–1898–1899 sous le commandement de A. De Gerlache de Gomery. Rapports scientifiques. Zoologie. Échinides et Ophiures

Koehler R (1908) Asteries, ophiures et echinides de l'expedition antarctique national ecossaisse. Trans R Soc Edinb 46:529–649

Koehler R (1923) Asteries et Ophiures. Further zool. Res Swed Antarct Exped 1901–1903(1):1–145

Komori T, Sakamoto K, Matsuo J, Sanechika Y, Nohara T, Ito Y, Kawasaki T, Schulten HR (1978) IUPAC Int Symp Chem Nat Prod 2:120

Komori T, Sanechika Y, Ito Y, Matsuo J, Nohara T, Kawasaki T (1980) Biologisch aktive glykoside aus Asteroidea, I. Strukturen eines neuen cerebrosidgemischs und von nucleosiden aus dem seestern *Acanthaster planci*. Justus Liebigs An Chem 5:653–668

Kondo K, Shigemori H, Ishibashi M, Kobayashi J (1992) Aplysidine, a new nucleoside from the Okinawan marine sponge *Aplysina* sp. Tetrahedron 48:7145–7148

Krepper CM, Rivas AL (1979) Análisis de las características oceanográficas de la zona austral de la Plataforma Continental Argentina y aguas adyacentes. Acta Oceanogr Argentina 2:55–82
Lampert K (1889) Die während der Expedition SMS "Gazelle" 1874–1876 von Prof. Dr. Th. Studer gesammelten Holothurien. Zool Jahrb (Abt Syst Geogr Biol Thiere) 4:806–858
Larrain-Prat AP (1975) Los equinoideos regulares fosiles y recientes de Chile. Gayana Zool 35:1–189
Lesson RP (1830) Centurie zoologique ou choix d'animaux rares. Paris
Lieberkind I (1926) *Ctenodiscus australis* Lütken. A brood-protecting asteroid. Vidensk Medd Dansk Naiurh Foren 8:183–196
Ljungman A (1867) Ophiuroidea viventia huc usque cognita. Ofvers K Vet Akad Forh 29: 303–336
Lloyd PE, Phares GA, Phillips NE, Willows AO (1996) Purification and sequencing of neuropeptides from identified neurons in the marine mollusk, *Tritonia*. Peptides 17:17–23
López-Gappa J, Sueiro MC (2007) The subtidal macrobenthic assemblages of Bahía San Sebastián (Tierra del Fuego, Argentina). Polar Biol 30:679–687
Lucas AJ, Guerrero RA, Mianzan HW, Acha EM, Lasta CA (2005) Coastal oceanographic regimes of the Northern Argentine Continental Shelf (34–43°S). Estuar Coast Shelf Sci 65:405–420
Ludwig H (1886) Die von G. Chierchia auf der Fahrt der Kgl. Ital. Corvette "Vettor Pisani" gesammelten Holothurien. Zool Jahr 2:1–36
Ludwig H (1898) Die Ophiuren der Sammlung Plate. Zool Jahr 4:750–786
Ludwig H (1899) Ophiuroiden. Hamb Magalhaensische Sammelreise 1:1–28
Ludwig H (1903) Seesterne. Résultats du Voyage du S.Y. Belgica 1897–1899. Rap Sci Zool:1-72
Lutz VA, Carreto JI (1991) A new spectrofluorometric method for the determination of chlorophylls and degradation products and its application in two frontal areas of Argentine Sea. Cont Shelf Res 11:433–451
Lyman T (1875) Zoological results of the Hassler expedition. II. Ophiuroidea and Astrophytidae Illust. Cat Mus Comp Zool Harvard 1:1–193
Lyman T (1882) Report on the Ophiuroidea. report of the scientific research on the challenger expedition vol 5. Zoology 14:1–387
Maier MS (2008) Biological activities of sulfated glycosides from echinoderms. Stud Nat Prod Chem 35:311–354
Maier MS, Murray AP (2006) Secondary metabolites of biological significance from echinoderms, in biomaterials from aquatic and terrestrial organisms. In: Fingerman M, Nagabhushanam R (eds) Biomaterials from aquatic and terrestrial organisms. Science Publishers, Enfield, New Hampshire, pp 559–594
Maier MS, Kuriss A, Seldes AM (1998) Isolation and structure of glucosylceramides from the starfish *Cosmasterias lurida*. Lipids 33:825–827
Maier MS, Chludil HD, Seldes AM (2002) Antifungal steroidal glycosides from the Patagonian starfish *Anasterias minuta*: structure–activity correlations. J Nat Prod 62(2):153–157
Maier MS, Roccatagliata AJ, Seldes AM (1993) Two novel steroidal glycoside sulfates from the starfish *Cosmasterias lurida*. J Nat Prod 56:939–942
Malanga GF, Pérez AF, Calvo J, Puntarulo S (2007) The effects of dietary carotenoid on oxidative damage in the gonads of sea urchin *Loxechinus albus*. V Meet of SFRBM—South American group and V Int Conf Peroxynitrite and Reactive Nitrogen Species, Montevideo
Malanga G, Pérez AF, Calvo J, Puntarulo S (2009) The effect of seasonality on oxidative metabolism in the sea urchin *Loxechinus albus*. Mar Biol 156:763–770
Malyutina M (2004) Russian deep-sea investigations of Antarctic fauna. Deep-Sea Res II 51:1551–1570
Martinez MI (2008) Biología reproductiva de *Psolus patagonicus* Ekman, 1925 (Holothuroidea), asociado a los bancos de vieiras frente a la Provincia de Buenos Aires, en el Mar Argentino. Licentiate Thesis. Univ Buenos Aires, Argentina

Martinez MI, Brogger MI, Penchaszadeh PE (2009) Revisión taxonómica de cuatro especies de dendrochirotidos (Echinodermata: Holothuroidea) del Mar Argentino. Book of abstracts. VII Jornadas Nac Cien Mar, Bahía Blanca

Martinez MI, Giménez J, Penchaszadeh PE (2011) Reproductive cycle of the sea cucumber Psolus patagonicus Ekman 1925, off Mar del Plata, Buenos Aires, Argentina. Invert Reprod Dev 55:124–130

Martos P, Piccolo MC (1988) Hydrography of the Argentine continental shelf between 38° and 42°S. Cont Shelf Res 8:1043–1056

Marzinelli EM (2005) Ecología y biología reproductiva del erizo rojo *Pseudechinus magellanicus* (Philippi, 1857) (Echinoidea: Temnopleuridae) en Golfo Nuevo Patagonia. Licentiate Thesis. Univ Buenos Aires, Argentina

Marzinelli EM, Bigatti G, Giménez J, Penchaszadeh PE (2006) Reproduction of the sea urchin *Pseudechinus magellanicus* (Echinoidea: Temnopleuridae) from Golfo Nuevo, Argentina. Bull Mar Sci 79:127–136

Massin C (1994) Ossicle variation in Antarctic dendrochirote holothurians (Echinodermata). Biology 64:129–146

McEdward LR, Miner BG (2001) Larval and life-cycle patterns in echinoderms. Canad J Zool 79:1125–1170

Meretta PE, Farías NE, Cledón M (2009a) Comportamiento y hábitos alimenticios de *Asterina stellifera* (Möbius, 1859) (Echinodermata: Asteroidea) en el Puerto de Mar del Plata. Book of abstracts. VII Jornadas Nac Cien Mar, Bahía Blanca

Meretta PE, Farías NE, Ventura CRR, Cledón M (2009b) Estructura poblacional y distribución espacial de *Asterina stellifera* (Möbius, 1859) (Echinodermata: Asteroidea) en el Puerto de Mar del Plata. Book of abstracts. VII Jornadas Nac Cien Mar, Bahía Blanca

Messing CG (1997) Living comatulids. In: Waters JA, Marples CG (eds) Geobiology of Echinoderms. Paleontol Soc, London, pp 3–30

Minale L, Pizza C, Riccio R, Squillace-Greco O, Zollo F, Pusset J, Menou JL (1984) New polyhydroxylated sterols from the starfish *Luidia maculata*. J Nat Prod 47:784–789

Minale L, Riccio R, Zollo F (1993) Steroidal oligoglycosides and polyhydroylated steroids from echinoderms. In: Hertz W, Kirby GW, Moore RE, Steglich W, Tamm C (eds) Progress in the chemistry of organic natural products. Springer, New York, pp 65–308

Minale L, Riccio R, Zollo F (1995) Structural studies on chemical constituents of echinoderms. In: Atta-ur-Rahman FZB (ed) Bioactive natural products—studies in natural products chemistry. Elsevier Science Publishers, Amsterdam, pp 43–110

Minale L, Riccio R, Pizza C, Zollo F (1986) Steroidal oligoglycosides of marine origin. In: Imura H, Goto T, Murachi T, Nakajima T (eds) Natural products and biological activities. University of Tokyo Press, Tokyo, pp 59–73

Morsán EM (2009) Impact on biodiversity of scallop dredging in San Matías Gulf, northern Patagonia (Argentina). Hydrobiology 619:167–180

Mortensen T (1910) The Echinoidea of the Swedish South Polar Expedition. Wiss Ergebn Schwed Südpolar Exp 6:1–114

Mortensen T (1917) The Crinoidea of the Swedish Antarctic Expedition. Wiss Ergebn Schwed Südpolar Exp 8:10–15

Mortensen T (1918) The Crinoidea of the Swedish Antarctic Expedition. Wiss Ergebn Schwed Südpolar Exp 8:1–23

Mortensen T (1920a) Echinoidea. Wiss Ergebn Schwed Südpolar Exp 6:1–114

Mortensen T (1920b) The Crinoidea. Wiss Ergebn Schwed Südpolar Exp 6:1–24

Mortensen T (1936) Echinoidea and Ophiuroidea. Discovery Rep 12:109–348

Müller J (1850) Anatomische Studien über die Echinodermen. Arch Anat Physiol 1850:129–155

Muniain C, Centurión R, Careaga VP, Maier MS (2008) Chemical ecology and bioactivity of triterpene glycosides from the sea cucumber *Psolus patagonicus* (Dendrochirotida, Psolidae). J Mar Biol Ass UK 88:817–823

Murray AP, Muniain C, Seldes AM, Maier MS (2001) Patagonicoside A: a novel antifungal disulfated triterpene glycoside from the sea cucumber *Psolus patagonicus*. Tetrahedron 57:9563–9568

Murray AP, Araya E, Maier MS, Seldes AM (2002) Nucleosides and nucleobases from *Ophiactis asperula*, *Ophiacantha vivipara*, and *Gorgonocephalus chilensis*. Biochem Syst Ecol 30: 259–262

O'Loughlin PM (2002) Report on selected species of BANZARE and ANARE Holothuroidea, with reviews of *Meseres* Ludwig and *Heterocucumis* Panning (Echinodermata). Mem Mus Victoria 59:297–325

O'Loughlin PM (2009) BANZARE holothuroids (Echinodermata: Holothuroidea). Zootaxa 2196:1–18

O'Loughlin PM, Ahearn CG (2008) Antarctic and Sub-Antarctic species of *Psolidium* Ludwig (Echinodermata: Holothuroidea: *Psolidae*). Mem Mus Victoria 65:23–42

O'Loughlin PM, Manjón-Cabeza ME, Moya-Ruiz F (2009) Antarctic holothuroids from the Bellingshausen Sea, with descriptions of new species (Echinodermata: Holothuroidea). Zootaxa 2016:1–16

Ohshima H (1915) Report on the holothurians collected by the United States fisheries steamer "Albatross" in the Northwestern Pacific during the summer of 1906. Proc US Nat Mus 48:213–291

Ojeda FP, Santelices B (1984) Invertebrate communities in holdfasts of the kelp *Macrocystis pyrifera* from southern Chile. Mar Ecol Progr Ser 16:65–73

Olson DB, Podesta GP, Evans RH, Brown OB (1988) Temporal variations in the separation of Brazil and Malvinas Currents. Deep-Sea Res A 35:1971–1990

Orler PM (1992) Biología reproductiva comparada de *Pseudechinus magellanicus* y *Loxechinus albus*, equinoideos del Canal Beagle. PhD Thesis, Univ Nac de La Plata, Argentina

Oyarzún ST, Marín SL, Valladares C, Iriarte JL (1999) Reproductive cycle of *Loxechinus albus* (Echinodermata: Echinoidea) in two areas of the Magellan Region (53°S, 70–72°W), Chile. Sci Mar 63:439–449

Pastorino G, Zelaya DG (2001) A new species of the eulimid genus *Pelseneeria* Koehler and Vaney, 1908 (Mollusca: Gastropoda) from Staten Island, Argentina. The Veliger 44:310–314

Pastor de Ward CT, Rubilar T, Díaz de Vivar ME, González-Pisani X, Zárate E, Kroeck M, Morsan E (2006) Reproductive biology of *Cosmasterias lurida* (Echinodermata: Asteroidea) an anthropogenically influenced substratum from Golfo Nuevo, Northern Patagonia (Argentina). Mar Biol 151:205–217

Pawson DL (1964) The Holothuroidea collected by the Royal Society Expedition to Southern Chile, 1958–1959. Pacif Sci 18:453–470

Pawson DL (1969) Holothuroidea from Chile. Rep No. 46 Lund Univ. Chile Exped. 1948–1949. Sarsia 38:121–145

Pawson DL (1970) The marine fauna of New Zealand: sea cucumbers (Echinodermata: Holothuroidea). New Zealand Dep Sci Ind Res Bull 201:1–70

Pawson DL (1977) Molpadiid sea cucumbers (Echinodermata: Holothuroidea) of the southern Atlantic, Pacific and Indian Oceans. Antarctic Res Ser 26:97–123

Penchaszadeh PE (1973) Comportamiento trófico de la estrella de mar *Astropecten brasiliensis*. Ecol 1:45–54

Penchaszadeh PE (1979) Estructura de la comunidad y procesos que la determinan en bancos circalitorales de mejillón *Mytilus platensis* UNESCO. Sem Bentos Atlántico Suroccidental, Montevideo

Penchaszadeh PE (1980) Ecología larvaria y reclutamiento del mejillón del Atlántico Suroccidental, *Mytilus platenses* d'Orbigny. Cah Biol Mar 21:169–179

Penchaszadeh PE, Lawrence JM (1999) *Arbacia dufresnei* (Echinodermata: Echinoidea): A carnivore in Argentinian waters. In: Carnevali C, Bonasoro F, Bonasoro MD (eds) Echinoderm Research 1998. AA Balkema, Rotterdam, pp 525–530

Penchaszadeh PE, Bigatti G, Miloslavich P (2004) Feeding of *Pseudechinus magellanicus* (Philippi, 1857) (Echinoidea: Temnopeluridae) in the SW Atlantic Coast (Argentina). Ophelia 58:91–99

Pérez AF (2009) Reproducción, energética y metabolismo oxidativo del erizo comestible *Loxechinus albus* en el Canal Beagle, extremo sur de su distribución. PhD Thesis, Univ Buenos Aires, Argentina

Pérez AF, Morriconi E, Boy C, Calvo J (2008) Seasonal changes in energy allocation to somatic and reproductive body components of the common cold temperature sea urchin *Loxechinus albus* in a Sub-Antarctic environment. Polar Biol 31:443–449

Pérez AF, Boy C, Morriconi E, Calvo J (2010) Reproductive cycle and reproductive output of the sea urchin *Loxechinus albus* (Echinodermata: Echinoidea) from Beagle Channel, Tierra del Fuego. Polar Biol 33:271–280

Perrier MR (1891) Echinodermes. I Stellérides. Mission Scientifique du Cap Horn, 1882–1883.: Mission Scientifique du Cap Horn, Zoologie

Perrier MR (1904) Holothuries du Cap Horn. Bull Mus Hist Nat 10:13–17

Perrier MR (1905) Holothuries antarctiques du Museum d'Histoire Naturelle de Paris. An Sci Nat (Zool) 9:104–146

Phillipi RA (1857) Vier neue Echinodermen des Chilenischen Meeres. Arch Naturgesh 23:134–130

Pucheta E, Urban G (1989) Estudio sobre la alimentación del asteroideo *Allostichaster inaequalis*. Licentiate Thesis. Univ Nac Córdoba, Argentina

Riccio R, Pizza C, Squillace-Greco O, Minale L (1985) Starfish saponins. Part 17. Steroidal glycoside sulphates from the starfish *Ophidiaster ophidianus* (Lamarck) and *Hacelia attenuate* (Gray). J Chem Soc Perkin Trans 1:655–660

Risso A (1826) Historie naturelle des principales productions de l'Europe meridionales. Paris

Rivas AL (1990) Análisis estacional de la estructura termo-halina en el Golfo San José, Argentina. Geoacta 17:37–48

Rivas AL, Ripa P (1989) Variación estacional de la estructura termo-halina de Golfo Nuevo, Argentina. Geofís Int 28:3–24

Rivas AL, Dogliotti AI, Gagliardini DA (2006) Seasonal variability in satellite-measured surface chlorophyll in the Patagonian Shelf. Cont Shelf Res 26:703–720

Roccatagliata AJ (1996) Aislamiento y elucidación estructural de metabolitos secundarios polares presentes en equinodermos. PhD Thesis, Univ Buenos Aires, Argentina

Roccatagliata AJ, Seldes AM, Maier MS, Iorizzi M, Minale L (1994) Starfish saponins, part 2. Steroidal oligoglycosides from the starfish *Cosmasterias lurida*. J Nat Prod 57:747–754

Roccatagliata AJ, Seldes AM, Maier MS (1995) A novel marine steroidal sulfate from the starfish *Luidia ludwigi*. J Nat Prod 58:1941–1944

Roccatagliata AJ, Maier MS, Pujol CA, Damonte EB, Seldes AM (1996a) Antiviral sulfated steroids from the ophiuroid Ophioplocus januarii. J Nat Prod 59:887–889

Roccatagliata AJ, Maier MS, Seldes AM (1996b) Main sterols from the starfish *Luidia ludwigi*. An Asoc Quim Argentina 84:615–618

Rodríguez MS (2009) Estudio de crecimiento en el ofiuroideo *Ophioplocus januarii* (Echinodermata: Ophiuroidea) del Golfo San José, Patagonia. Licentiate Thesis. Universidad CAECE, Argentina

Rodríguez MS, Penchaszadeh PE, Brogger, MI (2008). Chemical tagging assay on the ophiuroid *Ophioplocus januarii* growth. Fifth North American Echinoderm Conference. Melbourne, Book of abstracts

Rodríguez MS, Brogger MI, Penchaszadeh PE (2009) Análisis del crecimiento del ofiuroideo *Ophioplocus januarii* (Echinodermata: Ophiuroidea) mediante técnicas de marcado químico Book of abstracts. VII Jornadas Nac Cien Mar, Bahía Blanca

Roux A (2004) Estrellas y Erizos. In: Boschi EE, Cousseau B (eds) La vida entre mareas: vegetales y animales de las costas de Mar del Plata, Mar del Plata. Argentina INIDEP, Buenos Aires, pp 285–290

Roux A (2008) Invertebrados bentónicos acompañantes de la captura de langostino patagónico (*Pleoticus muelleri*). Resultados de la campaña de investigación AR-01/08 INIDEP. Informe de Investigación 11:1–8

Roux A (2000) Evaluación del estado de los fondos de pesca de langostino (*Pleoticus muelleri*) a través del análisis de la fauna de la fauna bentónica acompañante, en el Golfo San Jorge y litoral de Chubut, Argentina. Frente Mar 18:143–149

Roux A, Bastida R, Bremec C (1993) Comunidades bentónicas de la plataforma continental argentina. Campañas transección BIP "Oca Balda" 1987/88/89. Bol Inst Oceanogr San Pablo 41:81–94

Roux A, Fernández M, Bremec C (1995) Estudio preliminar de las comunidades bentónicas de los fondos de pesca de langostino patagónico del Golfo San Jorge (Argentina). Cien Mar 21:295–310

Roux A, Bastida R, Lichtschein V, Barreto A (1988) Investigaciones sobre las comunidades bentónicas de plataforma a través de una transecta frente a Mar del Plata. Spheniscus 6:19–52

Roux A, Piñero R, De la Garza J (2007) Guía para la identificación de la fauna asociada a la pesquería del langostino patagónico. INIDEP, Bueno Aires

Rubilar T (2004) Ciclo reproductivo y bioquímico del asteroideo fisíparo *Allostichaster capensis* en Golfo Nuevo. Licentiate Thesis. Univ Nac de la Patagonia San Juan Bosco, Argentina

Rubilar T (2009) Mecanismos de regeneración en equinodermos. Book of abstracts. VII Jornadas Nac Cien Mar. IADO-CONICET, Bahía Blanca

Rubilar T (2010) Morfología del sistema nervioso y regeneración del asteroideo fisíparo *Allostichaster capensis*. PhD Thesis, Univ Nac del Comahue, Argentina

Rubilar T, Pastor de Ward CT, Díaz de Vivar ME (2005) Sexual and asexual reproduction of *Allostichaster capensis* (Echinodermata:Asteroidea) in Golfo Nuevo. Mar Biol 146:1083–1090

Rubilar T, Pastor de Ward CT, Díaz de Vivar ME (2006) Timing of fission in the starfish *Allostichaster capensis* (Echinodermata:Asteroidea) in laboratory. Rev Biol Trop 53(Suppl 3): 299–303

Rubilar T, Díaz de Vivar E, Pastor de Ward C (2008a) Biochemical composition of body compartments during the reproductive cycle of the starfish *Allostichaster capensis* in Patagonia Argentina. Rev Biol Trop 56(Suppl 3):351–360

Rubilar T, Schinder AF, Pastor de Ward CT, Brogger M (2008b) First detection of TPep (Tritonia Pedal Peptide) in the nervous system of the fissiparous starfish *Allostichaster capensis*. Book of abstracts, 5th North Am Echinoderm Conference, Florida Inst Tech, Melbourne

Salvat MB (1975a) Algunos aspectos de la biología de *Patiria stellifer* (Moebius), (Echinodermata, Asteroidea) I Desarrollo Embrionario. Physis 34:49–58

Salvat MB (1975b) Algunos aspectos de la biología de *Patiria stellifer* (Moebius), (Echinodermata, Asteroidea) II Consideraciones Ecológicas. Physis 34:257–268

Salvat MB (1985) Biología de la reproducción de *Anasterias minuta* Perrier (Echinodermata, Asteroidea), especie incubadora de las costas patagonicas. PhD Thesis, Univ Buenos Aires, Argentina

Schatt P, Féral JP (1991) The brooding cycle of *Abatus cordatus* (Echinodermata: Spatangoida) at Kerguelen Islands. Polar Biol 11:283–292

Schejter L, Bremec C, Hernández D (2008) Comparison between disturbed and undisturbed areas of the Patagonian scallop (*Zygochlamys patagonica*) fishing ground "Reclutas" in the Argentine Sea. J Sea Res 60:193–200

Schejter L, Bremec CS, Akselman R, Hernández D, Spivak ED (2002) Annual feeding cycle of the patagonian scallop *Zygochlamys patagonica* (King and Broderip, 1832) in Reclutas Bed (39°S–55°W), Argentine Sea. J Shellfish Res 21:549–555

Schuhbauer A, Brickle P, Arkhipkin A (2010) Growth and reproduction of *Loxechinus albus* (Echinodermata: Echinoidea) at the southerly peripheries of their species range, Falkland Islands (South Atlantic). Mar Biol 157:1837–1847

Searle PA, Molinski TF (1994) Isolation of spongosine and 2′-deoxyspongosine from a Western Australian sponge of the order Hadromerida (Tethyidae). J Nat Prod 57:1452–1454

Seldes AM, Gros EG (1985) Main sterols from the starfish *Comasterias lurida*. Comp Biochem Physiol, B: Comp Biochem 80:337–339

Sladen WP (1889) Report on the Asteroidea. Zoology. Report of the Scientific Research on the Challenger Expedition Vol 30, London

Speel JA, Dearborn JH (1983) Comatulid crinoids from R/V Eltanin cruises in the Southern Ocean. Antarctic Res Ser 38:1–60

Stonik VA, Elyakov GB (1988) Secondary metabolites from echinoderms as chemotaxonomic markers. In: Scheuer P (ed) Bioorganic marine chemistry. Springer, New York, pp 43–86

Stonik VA, Kalinin VI, Avilov SA (1999) Toxins from sea cucumbers (holothuroids): Chemical structures, properties, taxonomic distribution, biosynthesis and evolution. J Nat Toxins 8: 235–248

Studer T (1880) Gazelle-Echinoidea. Monatsber Akad Berlin 1880:861–885

Studer T (1882) Übersicht über die Ophiuriden, welche während der Reise S.M.S. Gazelle um die Erde 1874–1876 gesammelt wurden. Abhandlungen der Preussischen Akademie der Wissenschaften zu Berlin 1:1–37

Studer T (1883) Über der Asteriden, welche während der Reise SMS Gazelle um die Erde gesammelt wurden. Sitzungsberichte der Gesellschaft Naturforschender Freunde zu Berlin 18:128–132

Tablado A (1982) Asteroideos argentinos. Familia Poraniidae. Com Mus Argentino Cien Nat Hidrobiol 2:87–106

Tablado A, Maytia S (1988) Presencia de *Perissasterias polyacantha* H.L. Clark 1923 (Echinodermata, Asteroidea) en el Atlántico sudoccidental. Com Zool Mus Hist Nat Montevideo 12:1–11

Teso SV, Bigatti G, Casas G, Piriz ML, Penchaszadeh PE (2009) Do native grazers from Patagonia, Argentina consume the invasive kelp *Undaria pinnatifida*? Rev Mus Argentino Cien Nat 11:7–14

Théel H (1882) Report on the Holothuroidea 1 Challenger Sci Results. Zoology 4:1–136

Théel H (1886) Report on the Holothurioidea 2 Challenger Sci Results. Zoology 14:1–290

Tolosano JA, Gil DG, Zaixso HE (2006) Diferenciación de sexos en *Abatus cavernosus* (Echinoidea: Echinodermata). Book of abstracts. VI Jornadas Nac Cien Mar, Puerto Madryn

Tommasi LR (1965) Lista dos crinoides recentes do Brasil. Contr Inst Oceanogr Univ São Paulo Ser Oceanogr Biol 9:1–33

Tommasi LR (1966) Lista dos equinoides recentes do Brasil. Contr Inst Oceanogr Univ São Paulo Ser Oceanogr Biol 11:1–50

Tommasi LR (1968) Os Ofiuroides existentes nas coleçoes do Museu de Buenos Aires coletados de La Plata até 42°S. Pap Avulsos de Zool 21:115–124

Tommasi LR (1969a) Lista dos Holothuroidea recentes do Brasil. Contr Avulsas Inst Oceanogr Univ São Paulo Ser Oceanogr Biol 15:1–29

Tommasi LR (1969b) Nova contribuição à lista dos crinóides recentes do Brasil. Contribuições Avulsas do Instituto Oceanográfico, Universidade de São Paulo, série Oceanografia Biológica 17:1–8

Tommasi LR (1970) Lista dos asteroides recentes N/Oc. "Almirante Saldanha" do Brasil. Contr Avulsas Inst Oceanogr Univ São Paulo Ser Oceanogr Biol 18:1–61

Tommasi LR (1971) Equinodermes do Brasil. I. Sobre algumas especies novas e outras pouco conhecidas para o Brasil. Bol Inst Oceanogr São Paulo 20:1–21

Tommasi LR (1974) Equinodermes do Brasil. III Observacoes sobre algunas espécies coletadas durante as viagens do N/Oc. "Almirante Saldanha". Bol Inst Oceanogr São Paulo 23:1–15

Tommasi LR, de Castro SM, de Sousa ECPM (1988a) Echinodermata coletados durante as campanhas oceanograficas do N/Oc. "Almirante Saldanha" no Atlantico sul occidental. Relat interno Inst Oceanogr Univ São Paulo 21:1–11

Tommasi LR, Cernea MCW, Condeixa MCG (1988b) Equinodermes coletados pelo N/Oc. "Almirante Saldanha", entre 26o 59′ S e 38o 39′ S. Relat interno Inst Oceanogr Univ São Paulo 22:1–11

Tommasi LR, Albuquerque M, Monteiro AMG (1993) Ophiuroidea das regioes antártica e subantártica 3. Amphiuridae (Ophiroidea) coletados durante as viagens do USNS "Eltanin" e "Hero" ás regioes antártica e subantártica. Relat interno Inst Oceanogr Univ São Paulo 34:1–12

Urien CM, Ewing M (1974) Recent sediments and environments of South Brazil, Uruguay, Buenos Aires and Rio Negro continental shelf. In: Burk RCA, Drake CL (eds) The geology of continental margins. Springer, Belirn, pp 157–177

Vaney C (1906) Holothuries. Exp Antarctique Française (1903–1905) 5:1–30

Vaney C (1909) Les holothuries de l'expédition antarctique nationale écossaise. Trans R Soc Edinb 46:405–441

Vaney C (1914) Holothuries. Deuxième Expédition Antarctique Française 1908–1910, commandée par le Dr. J. Charcot, Sciences Naturelles: Documents Scientifiques. Masson et Cie, Paris

Vaney C, John DD (1939) Scientific results of the voyage of SV "Scotia", 1902–04: The Crinoidea. Trans R Soc Edinb 59:661–672

Vásquez JA (2007) Ecology of *Loxechinus albus*. In: Lawrence JM (ed) Edible Sea Urchins: Biology and Ecology. Elsevier, Amsterdam, pp 227–241

Vásquez JA, Castilla JC, Santelices B (1984) Distributional patterns and diets of four species of sea urchins in giant kelp forest (*Macrocystis pyrifera*) of Puerto Toro, Navarino Island, Chile. Mar Ecol Progr Ser 19:55–63

Vázquez MJ, Quiñoá E, Riguera R (1993) Helianthoside from *Heliaster heliantus*, an asterosaponin with a C3′-sulphated pyranose. Canad J Chem 71:1147–1151

Verbist JF (1993) Pharmacological effects of compounds from echinoderm. In: Jangoux M, Lawrence JM (eds) Echinoderm Studies. AA Balkema, Rotterdam, pp 111–186

Villares G (2008) Variación de la eficiencia alimenticia con la abundancia de alimento y la regeneración del asteroideo *Allostichaster capensis* (Echinodermata: Asteroidea). Licentiate Thesis. Univ Nac de la Patagonia San Juan Bosco, Argentina

Villares G, Rubilar T, Epherra L, Díaz de Vivar ME, Pastor de Ward CT (2009) Efecto de la fisión y regeneración sobre la alimentación, reproducción y almacenamiento de energía en la estrella de mar fisipara *Allostichaster capensis*. Book of abstracts. VII Jornadas Nac Cien Mar, Bahía Blanca

Yasumoto T, Hashimoto Y (1965) Properties and sugar components of asterosaponin A isolated from starfish. Agr Biol Chem 29:804–808

Zaixso HE (2004) Bancos de cholga *Aulacomya atra atra* (Molina) (Bivalvia: Mytilidae) del golfo San José (Chubut, Argentina): Diversidad y relaciones con facies afines. R. Rev Biol Mar Oceanogr 39:61–78

Zaixso HE, Lizarralde ZI (2000) Distribución de equinodermos en el golfo San José y sur del golfo San Matías (Chubut, Argentina). Rev Biol Mar Oceanogr 35:127–145

Zaixso HE, Gil DG, Tolosano J (2003) Aspectos de la relación entre el bivalvo *Waldo parasiticus* y el erizo *Abatus cavernosus* (Echinoidea) Book of abstracts. V Jornadas Nac Cien Mar, Mar del Plata

Zaixso HE, Lizarralde Z, Pastor C, Gomes-Simes E, Romanello E, Pagnoni G (1998) Distribución espacial del macrozoobentos submareal del golfo San José (Chubut, Argentina). Rev Biol Mar Oceanogr 33:43–72

Zamora S, Stotz W (1992) Ciclo reproductivo de *Loxechinus albus* (Molina 1782) (Echinodermata: Echinoidea) en Punta Lagunillas, IV Región, Coquimbo, Chile. Rev Chil Hist Nat 65:121–133

Zelaya DG, Ituarte C (2002) The identity of *Waldo parasiticus* (Dall, 1976) and description of *Waldo trapezialis* new species (Bivalvia: Galeomatoidea). The Nautilus 116:109–117

Chapter 12
Echinoderms of the Cuban Archipelago

Rosa del Valle-García, Mercedes Abreu-Pérez, Francisco Alonso Solís-Marín and Alfredo Laguarda-Figueras

12.1 Introduction

The Cuban Archipelago is found between 19°50′–23°17′N and 74°08′–84°57′W. It is comprised of the island of Cuba (105,007 km^2), Isla de Pinos (2,200 km^2), and about 4,195 cays and islets (Núñez-Jiménez 1982). The southern coasts border the Caribbean Sea and the northern the Atlantic Ocean and the Gulf of Mexico.

The island of Cuba is the largest of the group comprising "the Greater Antilles". Its coast line is 3,209 km on the north side and 2,537 km on the south side. The archipelago's insular platform is shallow (<30 m in depth) and has a submerged plain relief with an emerged surface that, including the cays and islets (with exception of the island of Cuba), covers an area of 67,831 km^2 (Núñez-Jiménez 1982). According to Ionin et al. (1977), the insular platform may be considered a unique area from geological, tectonic and geomorphological points of view. It has four independent shallow regions surrounded by the islands, cays, little cays and reefs forming archipelagos (Fig. 12.1).

The insular slope is found nearby the exterior edge of the cays. The reefs begin to decline between 20 and 50 m depth (usually at depths between 25 and 30 m). In most of the periphery of the platform it drops nearly vertically down to a depth between 100 and 150 m. This is followed by a zone of smaller slope (60–80°)

R. d. Valle-García (✉) · M. Abreu-Pérez
Instituto de Oceanología, Ministerio de Ciencias
Tecnología y Medio Ambiente (CITMA), La Habana, Cuba
e-mail: rosa.denisse1@hotmail.cominformatica@giron.sld.cu

F. A. Solís-Marín · A. Laguarda-Figueras
Laboratorio de Sistemática y Ecología de Equinodermos,
Instituto de Ciencias del Mar y Limnología, Universidad Nacional Autónoma
de México (UNAM), Apdo. Post. 70-305, 04510, Mexico, D. F, Mexico
e-mail: fasolis@icmyl.unam.mx

J. J. Alvarado and F. A. Solís-Marín (eds.), *Echinoderm Research and Diversity in Latin America*, DOI: 10.1007/978-3-642-20051-9_12,

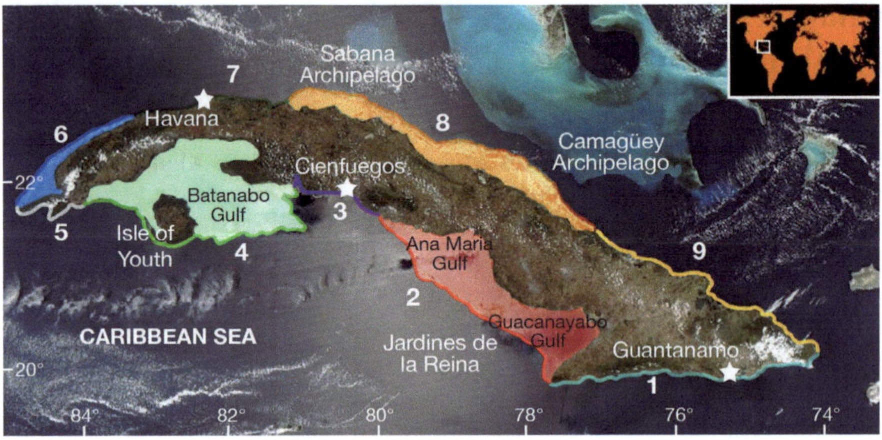

Fig. 12.1 Satellite image of the Cuban Archipelago (Jeff Schmaltz, MODIS, Rapid Response Team, NASA/GSFC), and division of the Cuban marine platform into nine ecoregions (taken from Areces 2002). *1* Southeastern Ecozone (Costa Sur de Oriente), *2* Southcentral Ecozone (Jardines de la Reina Archipelago), *3* Southcentral Ecozone (Coast to the South of macizo de Guamuhaya), *4* Southwestern Ecozone (Archipelago of Los Canarreos), *5* Southwestern Ecozone (Península de Guanhancabibes), *6* Northwestern Ecozone (Archipelago of Los Colorados), *7* Northern Ecozone (Habana-Matanzas), *8* Northcentral Ecozone (Archipelago Sabana-Camagüey), *9* Northeastern Ecozone (Costa Norte de Oriente)

which continues with slight changes far beyond a depth of 600 m. These deep mark the geographic limits of the Cuban Archipelago (Claro 2007).

The salinity and nutrient values of the Cuban platform water show comparatively wide seasonal and spatial variation since there is relative isolation of these oceanic waters, the mean depth is small, and there is a high level of river damming. During the rainy season salinity may drop to an average of 32 ‰ in some regions and under certain conditions even more. On the other hand, in other sites of limited exchange with the ocean, salinity may reach more than 50 ‰ during the dry season (Lluis-Riera 1972, 1977, 1981a, b, 1983, 1984; Alcolado et al. 1998; Claro 2007).

The water circulation over the Cuban platform seems to be determined mainly by the wind, the tide, and the influence of sea currents of the adjacent open sea (Emilsson and Tápanes 1971) whereas the currents system in the oceanic water that surrounds the Cuban Archipelago is extraordinarily complex and variable (Sukhovey et al. 1980; García-Díaz et al. 1991; Victoria et al. 1997; Gallegos et al. 1998).

The horizontal distribution of the primary production in the platform is affected by the hydrologic regime which is highly influenced by the oceanic currents and winds and by the input of inland water. Changes on any of these factors or other factors linked to them (e. g. local variation on the biogenic elements content), may produce local conditions of variable duration resulting in different environmental conditions for phytoplankton (Claro 2007).

12.1.1 Seagrasses

Seagrasses, known locally as "seibadales", are marine angiosperms on unconsolidated sediment bottoms. Six species of angiosperms have been recorded in Cuba: *Halophila decipiens*, *Halophila engelmanni*, *Thalassia testudinum*, *Halodule wrightii*, *Syringodium filiforme*, and *Ruppia maritima*. The greatest amount of primary production on the Cuban platform is due to marine phanerogams, especially *T. testudinum* which generally constitutes more than 70 % of the macro phytobenthos wet biomass on the Cuban platform.

Seagrasses are the main energy entrance pathway for biological and fishery productivity on the Cuban platform. It constitutes a major ecological reserve of matter and energy in biomass. Part of this is exported to the reefs and the ocean and increases productivity in these biotopes. They also stabilize the substrate, preventing the erosion on reefs and nearby beaches. They regulate oxygen concentrations on the sea water and strongly affect local biogeochemical processes. Development of calcareous algae, mainly *Halimeda*, in the seagrass beds produce most of the sand of beaches (Alcolado 2007).

12.1.2 Coral Reefs

Cuban coral reefs appear in the shape of ridges ("restingas"), coral barriers, promontories ("cabezos" or patch reefs), alternating bars with sand canals (bottoms of "camellones" (ridges), or "macizos" and canals), and carpet boulders and rocky terraces, amongst other varied structures (Ionin et al. 1977). More than 98 % of the Cuban marine platform edge is surrounded by frontal reefs, ridges, or barriers (Alcolado 2007).

The seagrass-reef complex is the source of approximately 60 % of the commercial catches of fish of high commercial value. Reefs contain a great diversity of microorganisms, plants and invertebrates that have biologically active substances. These are used or constitute potential resources as drugs or biochemical and experimental reagents of interest. Its barriers or ridges are an effective protection for coastal lines against wave erosion. They also protect coastal towns and buildings. They are extraordinarily attractive for diving tourism, i.e. the reefs of Archipiélago de los Jardines de la Reina (Alcolado 2007) (Fig. 12.1). Recently, Figueredo-Martín et al. (2010) evaluated the environmental quality of coral reefs at Jardines de la Reina and concluded the area was in optimal conservation condition.

Isle of Youth, National Park of the National System of Protected Areas of Cuba (Fig. 12.1), has a marine area of 5.15 km^2 with coral reefs of major species and biotope diversity. Forty two species of algae, three phanerogams, 29 sponges, 21 gorgonians, 35 corals, and 79 fishes have been identified in this area (De la Guardia et al. 2004). The most representative habitats between the coastline and a depth of 25 m are sand bottoms with or without seagrasses, reefs in shallow patches, *Acropora* ridges, rocky esplanades with isolated corals and gorgonians ending in an escarpment or contour. Reefs in deep patches and spurs and groves very close to the wall. In general, the Isle of Youth possesses healthy reefs, although some indicators, such as the damaging effect of a dense growth of algae and cyanophytes over sessile benthic organisms and the scarcity of large herbivorous or carnivorous fish indicate the existence of potential disruptions in the reef (De la Guardia et al. 2004).

12.1.3 Mangrove Swamps

Most of the Cuban Archipelago coasts are lined with mangroves, as well as swampy areas and thousands of coastal lagoons and estuaries. This biotope in Cuba is integrated by four arboreal species: mangle rojo (red mangrove) (*Rhizophora mangle*), mangle prieto (black mangrove) (*Avicennia germinans*), and patabán (*Laguncularia racemosa*), considered true mangroves, and the yana (*Conocarpus erectus*), a false or pseudomangrove. The mangrove swamps occupy approximately 4.8 % of the national territory, 26 % of the forest surface of the country (Alcolado 2007).

Approximately 30 % of Cuban mangroves are being affected by an increase in salinity and nutrient depletion, the result of damming, pollution and illegal deforestation. There has also been an increase in contamination from sugar mills, mining, oil refineries, paper and alcohol industries and cattle farms. Hydrologic changes on the coasts and at sea have been produced by building dams and viaducts over the sea. The effect on coasts that have lost the frontline protection of red mangrove produces a sudden accumulation of sand (Alcolado 2007).

12.2 Echinoderm Research

The earliest researches on echinoderms made in Cuban seas were by the vessels "*Blake*" (1877–1879), "*Fish Hawk*" (1880–1881), "*Albatross*" (1883–1884), "*Vanderbilt*" (1921–1928), "*Eagle*" and "*Ara*" (1921–1930), and the "*Atlantis*" (1938–1939). Leading specialists from Cuba and abroad participated in these voyages and made valuable contributions to echinoderm knowledge, especially for those species inhabiting deep waters off Cuba. Most of the results obtained from those investigations are in the works of Théel (1886), Agassiz (1869, 1878, 1880, 1888), Verrill (1899), Clark (1917, 1931, 1940, 1947, 1950), Clark (1919, 1933, 1941), Deichmann (1930, 1940, 1954, 1963), and Thomas (1963).

Other foreign specialists have made major contributions to the echinoderm knowledge in Cuba, including Boone (1933, 1935), Mortensen, with his masterpiece "*Monograph of Echinoidea*" (1928, 1935, 1940, 1943a, b, 1948a, b, 1950,

1951), Tortonese (1956), Tommasi (1958), Pawson and Fell (1965), Downey (1973), Meyer et al. (1978), Serafy (1979), Pawson and Caycedo (1980), Miller and Pawson (1984), Clark and Downey (1992), and Hendler et al. (1995).

The first publication by a Cuban scientist was by Arango (1877) "*Los Radiados de la Isla de Cuba*". He gave a list of marine invertebrates including the echinoderms, the first record for the phylum in the archipelago. Later, Delgado-Núñez (1917) did his doctoral thesis in Natural Sciences at Universidad de la Habana, with a contribution on the study of the nervous system of the sea urchin *Centrechinus antillarum* (= *Diadema antillarum* Philippi, 1845).

The first identification key for echinoderms of Cuba made by Aguayo (1945) is still of great practical use for scholars. That same year, Sirven-Villuenda (1945) did his doctoral thesis on the subject "Equinoideos litorales de Cuba" ("Littoral Echinoids of Cuba"), and in 1959, Pérez-Farfante (1959) published his doctoral thesis on irregular echinoids of Cuba, presenting dichotomous keys and descriptions for 24 species of shallow and deep waters.

Subsequently there were no publications related to echinoderms until 1965, when the Instituto de Oceanología de la Habana (Oceanology Institute of la Habana) was created and study of echinoderms was resumed. The mission of this institute was to establish the scientific basis to guarantee preservation and sustainable use of coastal and marine resources of Cuban ecosystems and adjacent waters through multidisciplinary research. Since then, major development of marine sciences began in Cuba, under the direction of Dr. Darío Guitar, an internationally renowned and professor at Universidad de la Habana. Consequently, new talents have emerged who acquired different specialties and research cruises have been undertaken especially from the former Soviet Union, Bulgaria and Romania.

During the second half of the twentieth century, joint expeditions were made by Cuban and European institutions and specialists, like the "I y II Expediciones Conjuntas cubano-soviéticas" ("1st and 2nd Cuban-Soviet Joint Expeditions"). Both of these expeditions studied different aspects of echinoderms, including the extraction of bioactive substances to determine their potential use in medicine, especially as anticarcinogens (Archivo Científico, Instituto de Oceanología). To accomplish this, work was done in close relationship with specialists from the Oncology and Radiology Institute at la Habana, Cuba.

As a result of the collections made by these cruises and their identification of the specimens deposited in the collection at Instituto de Oceanología, Levin and Gómez (1975) published a work on the holothurians of shallow waters of Cuba. They described the external morphology and spicules of 17 species of sea cucumbers and also provided data on their population density, habitat and ecology. Nine of these species were recorded for the first time for Cuba.

After this, Suárez (1974) published his work "Lista de los equinodermos cubanos recientes" ("List of the recent Cuban echinoderms"), which provided the scientific name and synonymies of all species recorded for Cuba until that date. It included all five classes of echinoderms and gave new records for the archipelago. She listed 308 species of echinoderms (Crinoidea: 33 species, Asteroidea: 83 species, Ophiuroidea: 116 species, Echinoidea: 67 species, and Holothuroidea:

9 species). It also includes bibliographic references to these species for the Cuban Archipelago and/or the region. This dissertation is considered the first scientific work in which all the taxonomic information existing in Cuba on the Phylum Echinodermata is integrated. It laid the groundwork for the beginning of the different studies that have been made until today.

It was not until the 1980s when the study of echinoderms began in a deeper way. The pioneers in this enterprise where Ángela Corvea and Mercedes Abreu-Pérez, under the guidance of Dr. Osvaldo Gómez, an outstanding pedagogue responsible for the formation of numerous generations of biologists at the Universidad de la Habana.

Dr. Pedro M. Alcolado, a specialist from the Instituto de Oceanología, also made major contributions to the development of this special field in Cuba. In the early 1990s, Dr. José Espinosa began study of echinoderms, although for a short period of time. Four years later M. Sc. Rosa del Valle assumed this position.

Among the main research projects for systematics and/or ecology developed by echinoderm specialists in the benthos department from the Instituto de Oceanología were "Comunidades del Bentos de Punta del Este" ("Benthos communities from Punta del Este"); "El Bentos de la Macrolaguna del Golfo de Batabanó" ("Benthos from the macro-lagoon of Golfo de Batabanó"); "Comunidades macrobentónicas asociadas a tres arrecifes coralinos del Golfo de Batabanó" ("Macrobenthic communities associated with three coral reefs from Golfo de Batabanó"); "Caracterización del bentos de las Bahías de Jururú y Vita" ("Characterization of the benthos from Bahías de Jururú y Vita"); "Protección de la biodiversidad y desarrollo sostenible en el ecosistema Sabana-Camaguey" ("Biodiversity protection and sustainable development in the Sabana-Camagüey ecosystem"); and "Ecología y sistemática de las comunidades de equinodermos de las costas del Caribe Mexicano y Cuba" ("Ecology and systematics of echinoderm communities from the Mexican Caribbean coasts and Cuba") under the framework of the Joint Collaboration Agreement with the Instituto de Ciencias del Mar y Limnología, from the UNAM, Mexico.

Other projects in which echinoderms have been studied are "La Biodiversidad Marina de Cuba" ("Cuban Marine Biodiversity"); "Evaluación del deterioro ocasionado por los vertimientos del emisario submarino de Playa el Chivo, mediante el análisis de las comunidades bentónicas" ("Evaluation of the damage made by the dumping of the submarine emissary of Playa el Chivo, by means of benthic communities analysis"); "Evaluación general del estado ecológico de los arrecifes de Cuba y monitoreo de la estación regional cubana de CARICOMP" ("General evaluation of the ecological status of reefs in Cuba and monitoring of the regional Cuban station of CARICOMP"); "Hacia el uso sostenible del Golfo de Batabanó" ("Towards the sustainable use of the Golfo de Batabanó"); "Análisis de sistema y modelación de escenarios" ("System analysis and scenario modeling"); "Evaluación del Archipiélago de los Canarreos con vista a su categorización dentro del Sistema Nacional de Áreas Protegidas" ("Evaluation of the Archipiélago de los Canarreos with a view to its categorization within the Sistema Nacional de Áreas Protegidas"). As a result of participation in these projects, numerous scientific papers, popular science works, scientific-technical reports and many book chapters have been published.

Among the most relevant results obtained by the echinoderm specialists from the benthos department are the works by Corvea (1986, 1990) that reports three new records of holothuroids for Cuba: *Euthyonidiella destichada* (*as Neophyllophorus destichaudus*), *Synaptula hydriformis*, and *Holothuria thomasi*. Corvea et al. (1990) worked with the echinoderm communities from the Golfo de Batabanó macro-lagoon and presented taxonomic and ecological results for echinoderms from the gulf.

Abreu-Pérez (1983, 1987) published early works on new records of brittlestars for the Cuban waters, including 14 species of shallow waters. Later, Abreu-Pérez (1990) presented an updated list of Cuban ophiuroids that reported 158 species of brittlestars, including seven new records. Espinosa et al. (1997) added new records of three species of echinoderms: *Davidaster rubiginosus*, *Astropyga magnifica*, *Lytechinus variegatus carolinus*. They also give data of bathymetric distribution and exceptional sizes of four other species previously reported for Cuba: *Eucidaris tribuloides*, *Podocidaris sculpta*, *Moira atropos* and *Holothuria thomasi*.

Abreu-Pérez (1997) published "Los asteroideos (Echinodermata) del Archipiélago Cubano" ("The seastars (Echinodermata) of the Cuban Archipelago") that updated the records of the seastar species of the coasts of Cuba. She reported 20 species, two of them being new records: *Asterina harmeyeri* and *Ophidiaster guildingi (= Linckia guildingi)*. In addition, she referred to previous works citing the existence of the seastar species and their bathymetric ranges.

Abreu-Pérez and Del Valle (1998) recorded the sea urchin *Lytechinus williamsi* for Cuba. That same year, Del Valle and Abreu-Pérez (1998) participated in the National Study of Biological Diversity with a chapter related to the Phylum Echinodermata that provided a list of all echinoderms recorded for Cuba and gave the main characteristics of the group and ecological data of interest.

Ibarzábal et al. (1999) published a list of 26 species of echinoderms inhabiting the Jardines de la Reina Archipelago. The class Ophiuroidea was the most diverse with 14 species.

Abreu-Pérez et al. (2000) published an updated list of the sea urchins of the Cuban Archipelago. Del Valle et al. (2000) recorded for the first time the family Chiridotidae for Cuba with the species *Chiridota rotifera*. Fernández (2001) made a new contribution for the class Asteroidea with the species *Copidaster lymani*.

Del Valle et al. (2003) presented the Final Report of the "Inventory of marine flora and fauna from the Sabana-Camagüey Archipelago" that included the list of echinoderm species present in the northern part of the providences of Matanzas, Sancti Spiritus, Villa Clara, Ciego de Ávila, and Camagüey (Archivo Científico del Instituto de Oceanología). That same year, Abreu-Pérez (2003) published five new records of brittlestars for Cuba: *Ophiolepis gemmata*, *Ophiopsila hartmeyeri*, *Ophionereis vittata*, *Ophioderma ensiferum*, and *Ophiactis quinquerradia*, extending the bathymetric range for some of them and increasing the number of known species of brittlestars for Cuba. The next year, Del Valle and Barreras-Rodríguez (2006) reported the sea cucumber *Holothuria princeps* for Cuba.

In the book "Corales Pétreos: Jardines Sumergidos de Cuba" ("Stony corals: underwater gardens of Cuba") Del Valle and Abreu-Pérez (2004) give a general view of the Phylum Echinodermata for Cuban waters, listing the main species of echinoderms associated with stony corals in Cuba. A year later, Abreu-Pérez et al. (2005) and Del Valle et al. (2005) published the catalogue of neritic seastars and brittlestars and a catalogue of neritic sea lilies, sea urchins, and sea cucumbers of the Cuban Archipelago, respectively. In both, they listed 125 species for the neritic zone (down to a depth of 30 m), the bibliographic references that cite them for the country, habitat, depth, distribution in the Western Atlantic, and the collection localities in Cuba. The catalogue of Del Valle et al. (2005) reported two new records of holothuroids for Cuban waters: *Ocnus suspectus* and *Phyllophorus (Urodemella) occidentalis*.

The next year, Del Valle and Barreras-Rodríguez (2006) recorded the holothurian *Holothuria princeps* for Cuba. Soon after, Rodríguez-Barrera et al. (2007) contributed a new record of an echinoid for Cuba, *Paraster doederleini*, increasing the number of species and extending the knowledge on the geographic distribution for this group.

In the book " Marine Biodiversity of Cuba", Del Valle and Abreu-Pérez (2007) present a taxonomic list of all echinoderm species recorded for Cuba, where the distribution of each one is given by ecozones (Areces 2002), as well as the relative abundance of them. For each class, the most remarkable species for each ecosystem present in Cuba are highlighted. In this work they analyzed the status of exploitation of the sea cucumber *Isostichopus badionotus* in the Cuban Archipielago.

In the book "Sabana-Camagüey Ecosystem: current state, progresses and challenges on the protection and sustainable use of Biodiversity", Del Valle et al. (2008) listed 1,370 marine species inventoried in the archipelago. Of these, 1,064 are invertebrates with 71 species of echinoderms. In addition, the areas with the upper and lower species richness values are pointed out. Pertinent recommendations for protection and better knowledge of the marine fauna of the Sabana-Camagüey Archipelago are given. This study found that in the outer zone of the macro-lagoon, where human impact is minor, the seagrasses develop better, the echinoderm communities have a greater number of species and moderated population densities. This was associated with a reduction in percentage of suspension-feeders. In contrast, towards the inner side of the macro-lagoon, where there is little exchange with oceanic waters, seagrasses are rare or absent and the organic matter and silt values are high. Fewer echinoderm species occur and a greater percentage of invertebrate filter-feeders are present.

12.2.1 Other Studies

Echinoderms have been used by other specialists, in different fields of study, for research. Elyakov et al. (1975) made a comparative and chemical composition study of holothurins (glycosides) isolated from nine species of sea cucumbers:

Astichopus multifidus, *Isostichopus badionotus*, *Actinopyga agassizii*, *Holothuria* mexicana, *H. cubana*, *H. arenicola*, *H. grisea*, *H. surinamensis*, and *Euapta lappa*.

Gómez et al. (1980) evaluated the macrobenthos of the Golfo de Batabanó. They provide ecological data of interest related to echinoderms. Berovides-Alvarez and Ortíz-Touzet (1981) studied color polymorphism in the seastar *Oreaster reticulatus*, finding four patterns of coloration: olive green, red, reddish brown and creamy yellow. They concluded that the most frequent pattern is the reddish brown with 49.2 % of presence and the olive green pattern was frequent in juveniles. Herrera-Moreno et al. (1981) used random sampling to study populations of the sea urchin *Diadema antillarum* in front of the rocky zone of the Instituto de Oceanología. Espinosa (1982) analyzed the stomach contents of the predatory seastar *Astropecten articulatus* and *A. duplicatus*. Martínez-Estalella (1984) contributed a list of organisms associated with coral reefs. Beltrán et al. (1988) used the sea urchin *Echinometra lucunter* as an indicator of oil pollution in the rocky littoral adjacent to Bahía de la Habana. Ablanedo et al. (1990) used the same species as an indicator of heavy metals (iron, zinc, manganese, and mercury) pollution. They measured the concentration of these metals in sea urchins gonads and skeletal parts (test, Aristotle's lantern). The results obtained showed that heavy metals accumulate differentially in different body tissues: iron and mercury in gonads, zinc and manganese in skeletal parts. They also observed that accumulation of heavy metal showed seasonal variation.

Ortea et al. (1991) pointed out the 30 most conspicuous species of echinoderms inhabiting the Cuban Archipelago. These included Luidia senegalensis, *A. duplicatus*, *Linckia guildingii*, *O. reticulatus*, Astrophyton muricatum, *Ophiocoma* riisei, *Ophiothrix suensonii, D. antillarum, E. tribuloides, E. lucunter, E. viridis, Tripneustes ventricosus, L. variegatus, Clypeaster rosaceus, Leodia sexiesperforata, Meoma ventricosa ventricosa, Holothuria cubana, H. arenicola, I. badionotus and Stichopus regalis*.

Santana et al. (2001) documented the ingestion of a specimen of *D. antillarum* (30 mm test diameter) by the sea anemone *Condylactis gigantea* in an aquarium. Six stages were involved in this process during a period of 24 h. All of the skeletal material of the sea urchin was discarded by the anemone at the end of digestion. Castellanos-Iglesias et al. (2004) performed the "Reef Check Monitoring" at "Rincón de Guanabo" reef in la Habana. As part of the results, they showed that the population density of *D. antillarum* was extremely low in this reef.

Nodarse-Konnorov (2001) analyzed the ecological structure of *E. lucunter* populations in a coastal reef at Ciudad de la Habana, using a stratified random sampling. The goal was to detect possible natural regularities in abundance and size of the *E. lucunter* over space and time. The authors found that anfractuosity was larger in the middle strata (5–15 m away from the shore), therefore containing the major refuge availability, and with greater densities per month. Depth, an indicator of the hydro-physical stress, increased farther from shore. Smaller individuals were collected in the deeper strata during September and November. For this reason, it was concluded distribution and size of individuals were related.

Ortiz (2001) provided a list of marine, estuarine and semi-terrestrial invertebrates of Playa Cojimar, located in the northern coast of the Province Ciudad de la Habana. It showed that several species, *Oreaster reticulatus* and *Arbacia punctulata* among them, were starting to disappear during a period of 40 years of sampling in that locality.

Later on, De la Guardia et al. (2003) published a work on marine biodiversity in the coastal lagoon "El Guanal", Cayo Largo, Cuba. They made an inventory of the flora and fauna of this site. Four echinoderm species were identified: *O. reticulatus, D. antillarum, Tripneustes ventricosus*, and *Lytechinus variegatus*. The authors concluded that none of the species identified during this study is exclusive to this zone, nor in danger of extinction or overexploitation. Despite the intense nautical activity and dredging work prior to this study, the system's characteristics have allowed the maintenance of a phase, undistinguishable from natural situations.

Due to the importance of the results obtained by the joint expeditions from Harvard and La Habana Universities, on board the B/I "*Atlantis*" during 1938 and 1939; as well as for the difficult access to the original literature on the subject, Lalana et al. (2004) published a taxonomic list (with bathymetric distribution) of 371 species and 5 subspecies of mollusks, crustaceans and echinoderms from that expedition (31 species and one subspecies were echinoderms).

Recently, Martín-Blanco et al. (2011) determined the abundance, distribution, and size structure of *Diadema antillarum* and the influence of their populations on the community structure of algae in Jardines de la Reina Archipelago.

12.3 Diversity and Distribution

Three hundred and fifty seven species of echinoderms have been recorded for Cuba. The Class Crinoidea is represented by 34 species (four orders, 19 families, and 21 genera), the Class Asteroidea by 76 species (six orders, 23 families, and 50 genera), the Class Ophiuroidea by 163 species (two orders, 14 families, and 61 genera), the Class Echinoidea by 63 species (13 orders, 23 families, and 35 genera), and the Class Holothuroidea by 51 species (five orders, 12 families, and 24 genera).

The 34 crinoid species recorded for Cuba are distributed from the littoral zone down to the great depths, as in the species *Coccometra hagenii* which was collected at 1,046 m depth. Only eight species of crinoids have been recorded at depths below 30 m. Conspicuous by their apparent frequency in coral reefs are the species *Davidaster rubiginosus* and *N. discoideus*.

The Class Asteroidea is represented in Cuban waters by 76 species; of these, only 20 (26 %) are distributed down to a depth of 30 m. Conspicuous by their apparent frequency in coral reefs are *Linckia guildingii* and *Ophidiaster guildingi*. Whereas, in sandy substrates with macro-vegetation, the most frequent and abundant seastar is *O. reticulatus*. In bays and coastal lagoons, where sand-mud substrates predominate mostly without vegetation, the nine-arm seastar Luidia senegalensis occurs frequently. In beaches or sandy plains *Astropecten duplicatus* can be observed, with relative frequency. The 56 species left are distributed in the deeper zones. Conspicuous amongst them are *Benthopecten spinosus* and *Pteraster aciculata*, the first recorded at depths between 1,300 and 3,700 m and the second at depths between 260 and 3,712 m.

The Class Ophiuroidea is the best represented in all coastal ecosystems of the Cuban Archipelago, both in species number (163) and appearance frequency and abundance. Of the total of recorded species, 55 have been observed at depths shallower than 30 m. Conspicuous in coral reefs are the genera *Ophiomyxa*, *Astrophyton*, *Ophiocoma*, *Ophioderma*, and *Ophiothrix*. In particulate substrates with macro-vegetation, stand out, by its appearance frequency, the species *Ophiolepis paucispina* and *Ophionereis reticulata*. In particulate substrates without vegetation in bays, coastal lagoons, beaches or sandy plains, *Amphiodia trychna* and *Amphipholis gracillima* are the most frequent.

In semi-enclosed areas with limited water circulation and with major input of organic matter, as is the case for Bahía de Santa Clara (northern-central Cuba), very high densities of the species *Ophiothrix suensonii* (up to 8.84 ind m^{-2}) and *Ophiothrix angulata* (up to 7.57 ind m^{-2}) have been recorded. Both are associated with macroalgae of the genera *Laurencia*. In the mangrove swamps of Cuba, the brittle stars species *Ophiopsila riisei* and *Ophionereis olivacea* are recorded with greater frequency It is interesting that the only species of brittle star that has been recorded in the rocky littoral is *Ophioderma brevicadum* and the two species recorded at greatest depths are *Ophiomusium lymani* and *Homolophiura abyssorum* (584 and 983 m, respectively).

Of the 63 echinoids recorded for Cuba, the best represented orders are Cidaroida, with nine species, Clypeasteroida, also with nine species, and Echinoidea with seven. Twenty two species have been observed at depths less than 30 m. Conspicuous, both by its appearance frequency and abundance (in density and biomass), are *E. lucunter*, *E. viridis*, *E. tribuloides*, *T. ventricosus* and *D. antillarum*.

In particulate sand-mud substrates with macro-vegetation, *Meoma ventriciosa* and *Clypeaster rosaceus* have been recorded in high densities. The latter is also abundant in sandy bottoms. In muddy substrates without macro-vegetation in bays and coastal lagoons, the irregular echinoid *Moira atropos* is the best represented. On beaches and/or protected sandy plains *Leodia sexiesperforata* is the most frequent and abundant. It must be emphasized that prior to the 1980s, *D. antillarum* was very abundant in the rocky littoral of Cuba, where it occurred along with *E. lucunter*. However, after the mass mortality of this sea urchin, *E. lucunter* became the most frequent and abundant species of this ecosystem. The deepest records for the Phylum Echinodermata are for *Pourtalesia miranda* (450–5,850 m) and *Aceste bellidifera* (550–5,220 m).

Among the 51 species of holothurians recorded for Cuba, 26 inhabit depths greater than 30 m. The other 25 species are distributed in the different ecosystems present on the island platform. In coral reefs, 10 species are recorded, of which *Euapta lappa*, *Holothuria impatiens*, and *H. thomasi* inhabit exclusively in this ecosystem. Others, like the sea cucumber *Actinopyga agassizi*, have been found with relative frequency in coral reefs. It can also be collected with some frequency in seagrasses associated with rocks and stones. *Holothuria thomasi* has been observed with less frequency. In seagrasses, the most frequent and abundant species are *H. mexicana* and *H. floridana*. In bays and coastal lagoons *I. badionotus* can be found at times with very high densities. *Holothuria arenicola* is common on beaches and/or sandy plains.

12.4 Ecology

The Workshop of Eco-regionalization and classification of marine habitats in the Cuban platform, celebrated in 2002, brought together Cuban and foreigner specialists in different fields of study. They undertook the task of analyzing the geographic, geomorphologic, and ecological characteristics of the Cuban platform, as well as the distribution of its biotopes. The main objectives of this workshop were to identify the regions that form the marine platform of the Cuban Archipelago and to determine the main biotopes of the Cuban marine ecosystem. As result of the analysis, nine "ecoregions" were established (Areces 2002) (Fig. 12.1), named later by Claro (2007) as "ecozones". It was stated that the main biotopes of the marine ecosystem in Cuba are coral reefs and hard non-reef bottoms (from interior waters); non-consolidated sediments (sand, mud); submerged vegetation (seagrasses and macroalgae, mangrove swamps, coastal lagoons and estuaries, low rocky coasts or with cliffs, and beaches).

Among the ecological studies performed in the island platform of Cuba are the major ones made in Punta del Este (at La Isla de la Juventud, ecozone 4), and the Golfo de Batabanó macro-lagoon (ecozone 4), the echinoderm communities of Sabana-Camagüey Archipelago (ecozone 8), of Jururú and Vita bays (ecozone 9), of Jardines de la Reina Archipelago (ecozone 2) and the northern coast of Habana-Matanzas (ecozone 7) (Fig. 12.1).

An integrative analysis of the ecological results for the Golfo de Batabanó macro-lagoon, Sabana-Camagüey Archipelago (ecozones 4 and 8, respectively) (Fig. 12.1), and the Jururú and de Vita bays (ecozone 9) showed all were dominated by unconsolidated sediments (sand and mud) and submerged vegetation (seagrasses and macroalgae). Both biotopes are present in different proportions (sand-mud, seagrasses-macroalgae), according to the oceanographic and geomorphologic characteristics of each locality.

At Golfo de Batabanó (ecozone 4) (Fig. 12.1), were recorded 50 species of echinoderms (33 brittle stars, four seastars, seven echinoids, and six holothurians). The most important species of brittle stars, due to their wide distribution within that macro-lagoon were *Ophioderma olivaceum*, *Ophiolepis elegans*, and *Ophiostigma isacanthum* (Corvea et al. 1990). The richest and most diverse sites corresponded to those zones where appropriate geomorphologic and oceanographic conditions exist for the development of macro-vegetation and the establishment of diverse and abundant echinoderm communities. Whereas, in some interior zones, where muddy bottoms without vegetation predominate, very few or no species were recorded. The dry biomass values varied between 0 and 4.72 g m^{-2}. The highest ones corresponded to "Cayo Largo del Sur" zone, part of the Northwestern zone of "Isla de la Juventud" (ecozone 4) (Fig. 12.1). The poorest zones in biomass were the Southeast part of Golfo de Batabanó, "La Broa" inlet (ecozone 1) (Fig. 12.1), which suffers from strong erosion in the northern muddy coasts, due to disappearance of the protective red mangrove. This produced a marked and permanent water turbidity that has eliminated the seagrasses throughout the west part of this inlet where some benthic organisms were once abundant (Alcolado 2007).

All over the macro-lagoon, the dominant species in biomass were the sea urchin *L. variegatus*, the seastar *O. reticulatus*, and the irregular echinoid *C. rosaceus*. This study showed *L. variegatus* plays an important role in seagrasses from Golfo de Batabanó, because is an active and abundant micro phytobenthos consumer and represents the main biological energy mobilizing mesh in the ecosystem (Alcolado 1990).

The mega-benthic communities of the Sabana-Camagüey Archipelago (ASC, ecozone 8) were analyzed using a 109 stations network and numerous oceanographic parameters including temperature, current, salinity, particulate organic matter and silt content in sediments (Espinosa and Abreu-Pérez unpublished data; Alcolado et al. 1998). Echinoderms represented by 53 species (29 brittle stars, 12 holothurians, seven echinoids, and five seastars) were recorded in 71 stations just like the Golfo de Batabanó macro-lagoon, the dominant species in number in most stations (18) and with the largest frequency of appearance (32 stations), was *L. variegatus*. Other conspicuous species in terms of numbers were the brittle stars were evident: *O. angulata* (seven stations), *Ophiostigma isacanthum*, *Ophioderma olivaceum*, *Ophiothrix orstedii,* and *O. angulata*. These also were important for being recorded in 20, 19, 18, and 12 stations, respectively.

It is interesting to note the high abundance in number in which the brittle star *O. angulata* occurred. Over two thousand specimens were collected in 1 min of sampling. This seston-feeder species was found on the red algae *Acanthophora* sp. at a station located at the western of Bahía de Santa Clara, where the current provoked by predominant winds (from east to west) accumulates a larger amount of nutrients, producing the so called "pocket" effect which favors this kind of organism (Alcolado 1990).

The greatest species number, as well as the highest Margalef's Richness Index (1951) (R1 = 3.9), was found at station 36. This station is located in an external area where good exchange of water with the ocean exists and the bottom is covered with seagrass. When the 109 stations are combined, taking into account the biotope type, the greatest species number corresponded to seagrass growing on thick aleurite. In addition, it was observed that the brittle stars *I. isacanthum*, *O. angulata*, and *O. oerstedii* were the species appearing on a larger numbers of biotopes.

The effect of salinity was also analyzed. Of the 53 potential species existing in the Sabana-Camagüey Archipelago at salinities of 36 ‰, the number is markedly reduced as salinity rises, down to 10 species at 41 ‰ and only one, the brittle star

Ophiolepis elegans, at a salinity of 42.2 ‰. This is the most salinity-tolerant species reported in Cuban waters.

A major variation in particulate organic matter content was found within the sediments of the interior parts of Sabana-Camagüey Archipelago. The greatest species numbers (33–36) were found at stations with low-medium content of organic matter (2.5–15.5 %). The species number was greatly reduced (11–5) in places with high-medium percentages (15.6–35.0 %) (Espinosa and Abreu-Pérez unpublished data). This observation is very similar to that found by Alcolado et al. (1998) for all mega-zoobenthic organisms he studied.

The sites with greater species richness were those with good water circulation and exchange of interior waters with the open sea, presence of good development of macro-vegetation, salinity values lower than 38 ‰, and moderate particulate organic matter and silt values. The highest species density values were recorded in regions where the geographic configuration favors particle retention (pocket effect) (seston and larvae from benthic colonizers), provided that stress factor values were at the levels indicated above.

In a general sense, the results from studies at Bahías de Jururú and Vita (ecozone 9), Cárdenas and Nuevitas (ecozone 8) (Fig. 12.1) showed the same pattern observed in Sabana-Camagüey Archipelago.

Comparison of the echinoderm list of ASC with the Golfo de Batabanó macro-lagoon and Bahías de Jururú and Vita, using Sorensen's Similarity Index (1948), shows that a high faunistic affinity exists. Integration of the echinoderm species lists from all of the studied sites in Cuba indicates that the echinoderm communities present in soft bottoms total 91 species. Fifty six are brittle stars, 18 holothurians, nine echinoids, and eight seastars.

Another relevant ecological study was performed on three coral reefs in Golfo de Batabanó (Espinosa and Abreu-Pérez unpublished data) (Fig. 12.1). The main characteristics of echinoderm communities present were analyzed in different zones (reef lagoon (L), rear zone (ZT), plain (M), pounding zone (ZE), abrasive rocky esplanade (ER) at 5 m depth, and external slope (PE) at 10, 15, and 20 m).

The "Juan García" reef, located in the western part of Golfo de Batabanó, is the most exposed of the three reefs. It receives a strong influence from enriched water with organic matter from Isle of Youth and the numerous adjacent cays and little cays. The "Cantiles" reef is located south of Golfo de Batabanó, in a more protected zone and is the one receiving the least land influence. The "Diego Pérez" reef, located east of Golfo de Batabanó, is the most protected one and receives a major land influence from Península de Zapata, a region rich in swamps and mangroves. Major differences were found among the three reefs in the ecological structure of echinoderm communities, as well as between the zones that composed them (coastal lagoon, reef plain complex, abrasive rocky esplanade, and external slope at 10, 15, and 20 m)

The number of species in "Juan García" reef was less than in the other two reefs, except on the external slope. These results have an important relationship with different environmental factors acting negatively on the ecological structure of echinoderm communities. Among them are the greater degree of exposure to

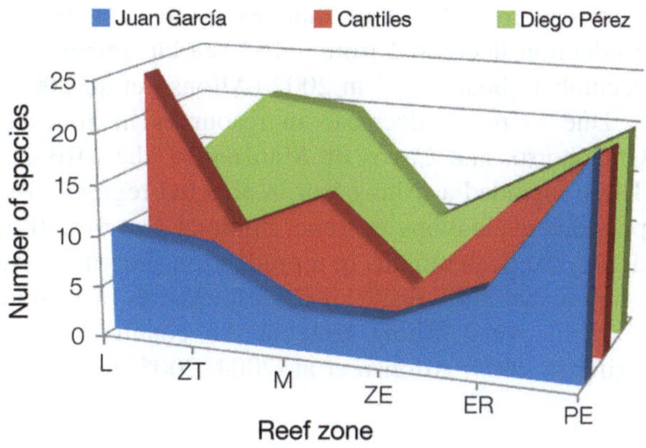

Fig. 12.2 Echinoderm species distribution in the different zones at the coral reefs (Juan García, Cantiles, and Diego Pérez) from Golfo de Batabanó. *L* reef lagoon, *ZT* rear zone, *M* plain, *ZE* pounding zone; *ER* rocky esplanade at 5 m depth; and *PE* external slope at depths of 10, 15, and 20 m

waves and winds present. This became evident with the high values recorded by the Agitation Index (IA) by Alcolado (1990): rear zone (IA = 98), pounding zone (IA = 100), and abrasive rocky esplanade (IA = 99). These harsh conditions, along with the major input of organic matter received from Isle of Youth and the numerous adjacent cays and little cays, create a constant and severe re-suspension of sediment in these zones. This stress factor may turn out to be intolerable for many echinoderm species. In this reef, brittle star species predominated. In some stations, only this class was recorded, especially in the Rear Zone-Plain-Pounding Zone complex (ZT-M-ZE), where the species *Ophiocoma echinata* and *Ophiocoma riisei* dominated. However, on the external slope (at 15, 19, and 20 m depth), where conditions are not so severe, a remarkable increase in species number was observed. This also happened in Cantiles and Diego Pérez reefs (Fig. 12.2).

Valdivia et al. (2004) described representative biotopes of coral reefs from the southern zone of Guanahacabibes Peninsula (western edge of Cuba) and a taxonomic list of marine organisms identified in the zone during 2001. This zone is characterized by the greatest concentration of reef-former organisms and fish in three biotopes: rocky plain with coral and gorgonians (down to 2–4 m), "cabezos" (5–8 m), and "camellones" (ridges) (12–15 m). At a depth of 3 m, and forming bands 20 m wide, sea urchins *E. lucunter* and *E. viridis* form a characteristic biotope of this zone. In addition, 109 algal species, 23 sponges, and 86 fish species typical of coral reefs were identified. They argued that excessive overfishing may be one cause of large fish in the area.

12.5 Aquaculture and Fisheries

Since 1999, the sea cucumber *I. badionotus* has been exploited in Cuba. This activity is in the charge of the Ministerio de la Pesca (Fishing Ministry) (MIP). During the first 2 years of fisheries (1999–2000), more than 3 million individuals were captured. It seems that, as result of these large extractions, catches and production decreased from 1,153 sea cucumbers boat^{-1} d^{-1} in 1999 to 350 sea cucumbers boat^{-1} d^{-1} in 2002 (Alfonso et al. 2003).

Due to major decrease in resources in zones like Bahía de Banes, Banco Chinchorro, and Cayos de Manzanillo, the MIP established several regulations. These included a fishing ban in affected regions, fishing effort limits regarding to previous evaluations by zone, minimal size of 220 mm total length, a fishing ban during the peak months in reproduction (February–April), and a rotation system of catch areas. As a result of these measures, a resource recovery was observed in 2005 in every fishing zone in the country, including the most affected ones (Alfonso 2004; Alfonso et al. 2004, 2005, pers. comm. R. Castelo).

12.6 Threats

The main threats affecting echinoderms of the Cuban Archipelago are closely related to habitat destruction, fundamentally caused by coastal development and/or natural factors (e.g. hurricanes, cyclones, etc.).

The known effect in Cuba is related to the sea urchin *D. antillarum* populations. The mass mortality of *D. antillarum* has been widely documented, not only in Cuba but all along the Caribbean region (Hughes et al. 1999; Alcolado et al. 2004). In many reefs from the Cuban Archipelago, an excessive growth of algae has been observed (Habana littoral; Sabana-Camagüey Archipelago; Largo, Paraíso, and Levisa Cays; and nearby Bahía Honda). This seems to be, in general terms, due to the mass mortality of *D. antillarum* and also because of the obvious low numbers of herbivore fishes.

Alcolado et al. (2004) studied population densities of *D. antillarum* in 199 sites at different reef ridges from the Cuban Archipelago. They showed that, even though *D. antillarum* existed in some sites in ecozones Jardines de la Reina (0.925–21.1 ind m^{-2}) and in Golfo de Batabanó (0.58 a 1.17 ind m^{-2}) with densities ranging from moderate to very high, the recovery of populations of this species is happening slowly. Caballero et al. (2007) mentioned that *D. antillarum* density in "María La Gorda" reef (Bahía de Corrientes) was very low (3.5 ind 10 m^{-2}).

Martín-Blanco et al. (2011) presented the results of a study about abundance, distribution, and size structure of *D. antillarum* in 22 frontal reefs from Jardines de la Reina Archipelago. The authors found that *D. antillarum* densities were large in reef ridges (0.08–2.18 ind m^{-2}) and decreased down to three orders of magnitude, with a maximum density of 0.13 ind m^{-2} on the slope. These authors concluded that the size frequency distribution of *D. antillarum* suggests that recruiting began to be evident approximately 3 years prior to the sampling; indicating that their populations were affected during the past and are now recovering.

Another species that seems to have been affected in some way is the sea urchin *A. punctulata*, for it was highly abundant in the northern coast of Cuba, especially in Bahía de Cárdenas (pers. comm. M. Ortiz). However, it has not recorded for over 20 years, despite numerous exploratory samplings in that bay. Moreover, it was never observed during the several samplings made in the Sabana-Camagüey Archipelago. Nevertheless, some specimens have been observed recently at Provincia de Villa Clara (ecozone 8). High densities have being observed of *A. punctulata* in the north and south coasts of Península de Guanahacabibes (ecozones 5 and 6, respectively) (pers. comm. R. Arias and M. Esquivel).

12.7 Conclusions and Recommendations

The Phylum Echinodermata is among the best studied faunistic groups in the Cuban Archipelago with observations from the seventeenth century until the present. However, knowledge of this group differs among classes. The least studied is Class Crinoidea. There are remarkable differences in knowledge among the platform's nine ecozones. These results will contribute to the knowledge of echinoderm diversity in Cuba, and will be of practical and teaching purposes for scholars interested in the subject.

One recommendation is to increase the scientific formation of young men and women specialized in this group with the aim of ensuring the continuity of studies related to the Phylum Echinodermata. New projects and lines of investigation (genetics and molecular amongst them) should be developed. It is necessary to develop a database and an information system on Cuban marine biodiversity, including echinoderms, and to regulate and promote its use. Existing laws prohibiting echinoderm extraction for ornamental purposes without prior impact studies should be effectively enforced. Environmental education, awareness, and citizens' involvement should be promoted so that the population, especially children and young people learn about echinoderm species inhabiting our seas.

Acknowledgments We want to express our gratitude to †Cynthia Ahearn and David Pawson (United States National Museum of Natural History, Smithsonian Institution) and Gordon Hendler (Los Angeles County Museum) for they support and for always enriching our minds with their wise knowledge. We also thank the Department and Laboratories of the Oceanology Institute of La Havana, Cuba (Oceanology Department and Laboratory), for their unconditional support during all these years. Without them this work would have been impossible.

References

Ablanedo N, González H, Ramírez M, Torres I (1990) Evaluación del erizo de mar *Echinometra lucunter* como indicador de la contaminación por metales pesados. Cuba. Aquat Liv Res 3:113–120

Abreu-Pérez M (1983) Nuevos registros de ofiuroideos (Echinodermata: Ophiuroidea) del Golfo de Batabanó. Cuba. Poeyana 259:1–6

Abreu-Pérez M (1987) Contribución al conocimiento de los ofiuroideos del Golfo de Batabanó. Poeyana 336:1–7

Abreu-Pérez M (1990) Lista actualizada de ofiuroideos cubanos. Poeyana 389:1–13

Abreu-Pérez M (1997) Los asteroideos (Echinodermata) del Archipiélago Cubano. Avicennia 6–7:65–72

Abreu-Pérez M (2003) Cinco nuevos registros de ofiuroideos (Echinodermata: Ophiuroidea) para el Archipiélago Cubano. Cocuyo 12:4–5

Abreu-Pérez M, Del Valle R (1998) Presencia de *Lytechinus williamsi Chesher*, 1968 (Echinodermata: Echinoidea) en aguas del Mar Caribe de Cuba. Avicennia 9:163

Abreu-Pérez M, Solís-Marín FA, Laguarda-Figueras A (2000) Los equinoideos (Echinodermata: Echinoidea) del Archipiélago Cubano. Avicennia 12–13:69–77

Abreu-Pérez M, Solís-Marín FA, Laguarda-Figueras A (2005) Catálogo de los equinodermos (Echinodermata: Asteroidea y Ophiuroidea) nerítico-bentónicos del Archipiélago Cubano. Rev Biol Trop 53(Suppl 3):29–52

Agassiz A (1869) Preliminary report on the echini and seastares dredged in deep watter between Cuba and the Florida Reef. by L. F. de Poutalès. Bull Mus Comp Zool 1:253–308

Agassiz A (1878) Reports on the dredging, under the supervision of Alexander Agassiz, in the Gulf of Mexico by the United States Coast Survey Steamer "*Blake*". Proc U S Coast Surv 1:181–195

Agassiz A (1880) Reports on the results of dredging, under the supervision of Alexander Agassiz, in the Caribbean Sea, 1878–79 and along the Atlantic coast of the United States during the summer of 1880, by the U. S. Coast Survey Steamer "*Blake*" commander J. R. Bartlett, U.S. N., Commanding. IX. Preliminary report on the Echini. Bull Mus Comp Zool Harvard Coll 8:69-84

Agassiz A (1888) Characteristic deep-sea types.-Echinoderms. Three cruises of the United States Coast and Geodetic Survey Steamer "*Blake*" in the Gulf of Mexico in the Caribbean Sea, and along the Atlantic Coast of the United States, from 1877 to 1880. The Riverside Press, Cambrige 2:84–127

Aguayo C (1945) Zoografía de invertebrados. Clave para equinodermos. Cooperativa Estudiantil, La Habana

Alcolado PM (1990) Aspectos ecológicos de la macrolaguna del Golfo de Batabanó, con especial referencia al bentos. El Bentos de la Macrolaguna del Golfo de Batabanó. Ed. Academia, La Habana, In, pp 129–157

Alcolado PM (2007) Diversidad, utilidad y estado de conservación de los biotopos marinos. In: Claro R (ed) La Biodiversidad marina de Cuba. Inst Oceanología, Ministerio de Ciencia, Tecnología y Medio Ambiente, La Habana, Cuba

Alcolado PM, Espinosa J, Martínez-Estalella N, IbarzábaL D, Del Valle R, Martínez-Iglesias JC, Abreu-Pérez M, Hernández-Zanuy A (1998) Prospección del megazoobentos de los fondos blandos del Archipiélago Sabana-Camagüey. Cuba. Avicennia 8:87–104

Alcolado PM, Del Valle R, González S, Caballero H, Rodríguez R (2004) Abundance of the long spine black urchin *Diadema antillarum* in Cuban coral reefs. TNC, NFWF, RSMAS, Univ Miami, Miami USA, The Diadema Workshop

Alfonso I (2004) National Report - Cuba. Convention on International trade in endangered species of wild fauna and flora. Technical workshop Report on the conservation of sea cucumbers in the families Holothuridae and Stichopodidae. Kuala Lumpur, Malaysia, pp 1–3

Alfonso I, Frías MP, Aleaga L, Alonso C (2003) Análisis de la pesquería del pepino de mar *Isostichopus badionotus* en la región sur oriental de Cuba. VI Cong Cien Mar, La Habana:10

Alfonso I, Frías MP, Aleaga L, Alonso C (2004) Current status of the sea cucumber fishery in the south eastern region of Cuba. In: Lovatelli A, Conand C, Purcell S, Uthicke S, Hamel JF, Mercier A (eds) Advances in Sea Cucumber Aquaculture and Management. FAO Fisheries Technical Paper 463, Rome, pp 151–162

Alfonso I, Frías MP, Castelo R, Blás Y, Rodríguez Y (2005) Del mar a la producción: Plan de manejo pesquero del recurso pepino de mar en Cuba. PESCA 2005, La Habana

Arango AMR (1877) Radiados de la Isla de Cuba. La Habana, Cuba, pp 13–19

Areces AJ (2002) Ecoregionalización y clasificación de hábitats marinos en la plataforma cubana: resultados. Inst oceanología, WWF-Canada, Environmental Defense, Centro Nacional de Áreas Protegidas, La Habana

Beltrán J, Ramos I, Ruiz F, Mederos R, Pereiras M (1988) El erizo de mar *Echinometra lucunter* como organismo indicador de la contaminación por petróleo. Rev Inst Invest Trans 11:45–56

Berovides-Alvarez V, Ortíz-Touzet M (1981) Polimorfismo del color en la estrella de mar *Oreaster reticulatus* (Echinodermata: Asteroidea). Rev Inv Mar 1:27–42

Boone L (1933) Scientific results of cruises of the Yachts "Eagle" and "Ara", 1921–1928, Willian K. Vanderbilt Commanding. Coelenterata, Echinodermata and Mollusca. Bull Vanderbilt Mar Mus 4:1–217

Boone L (1935) Scientific results of the World Cruise of the Yacht "Alva", 1931, William K. Vanderbilt Commanding. Bull Vanderbilt Mar Mus 6:241–264

Caballero H, González-Ferrer S, Cobián D, Alvarez S, Alcolado PM (2007) Evaluación AGRRA del bentos en diez sitios de buceo de "María la Gorda", Bahía de Corrientes, Cuba. Rev Inv Mar 28:131–138

Castellanos-Iglesias S, Lopeztegui Castillo A, De la Guardia-Llansó E (2004) Monitoreo Reef Check en el arrecife coralino "Rincón de Guanabo", Cuba. Rev Invest Mar 25:219–230

Clark AH (1917) Four new echinoderms from the West Indies. Proc Biol Soc Wash 30:63–70

Clark AH (1931) A monograph of the existing crinoids. Vol 1. The Comatulids. Part 3. Superfamily Comasterida. US Nat Mus Bull 82:1–816

Clark AH (1940) The family Antedonidae in the west tropical Atlantic. Mem Soc Cub Hist Nat 14:139–160

Clark AH (1947) A monograph of the existing crinoids. Vol 1. The Comatulids. Part 4b. Superfamily Mariametrida (Concluded. The family Colobometridae) and Superfamily Tropiometrida (Excep the families Thalassometridae and Charitometridae). US Nat Mus Bull 82:1–473

Clark AH (1950) A monograph of the existing crinoids. Vol. 1. The Comatulids. Part 4c. Superfamily Tropiometrid (the families Thalassometridae and Charitometridae). US Nat Mus Bull 82:1–383

Clark AM, ME Downey (1992) Seastares of the Atlantic. Nat Hist Mus Pub. Identification Guide 3. Chapman and Hall, London

Clark HL (1919) The distribution of the littoral echinoderms of the West Indies. Publ Dept Mar Biol Carnegie Inst Wash 13:49–74

Clark HL (1933) A handbook of the littoral echinoderms of Porto Rico and the other West Indian Islands. Scientific Survey of Porto Rico and the Virgin Islands 16:1–147

Clark HL (1941) Reports on the scientific results of the "*Atlantis*" Expedition to the West Indies, under the joint auspices of the University of Havana and Harvard University. The echinoderms (other than holothurian). Mem Soc Cubana Hist Nat "Felipe Poey" 15:1–154

Claro R (2007) La Biodiversidad marina de Cuba. Inst Oceanología, Ministerio de Ciencia, Tecnología y Medio Ambiente, La Habana

Corvea A (1986) Nuevos registros de holoturoideos neríticos (Echinodermata) para Cuba. Poeyana 322:1–5

Corvea A (1990) Nuevo registro de holoturoideo (Echinodermata) para Cuba. Poeyana 392:1–6

Corvea A, Abreu-Pérez M, PM Alcolado (1990) Características de las comunidades de equinodermos de la macrolaguna del Golfo de Batabanó. In: Alcolado PM (ed) El Bentos de la Macrolaguna del Golfo de Batabanó. Ed Academia, La Habana, pp 90–99

Deichmann E (1930) The holothurians of the western part of the Atlantic Ocean. Bull Mus Comp Zool Harvard Coll 71:43–226

Deichmann E (1940) Report of the holothurians, collected by the Harvard-Havana Expeditions 1938-1939, with a revision of the Molpadonia of Atlantic Ocean. Mem Soc Cub Hist Nat "Felipe Poey" 11:183–240

Deichmann E (1954) The holothurians of the Gulf of Mexico. US Fish Wildlife Bull 89:381–410

Deichmann E (1963) Shallow water holothurians known from the Caribbean waters. Stud Fauna Curacao Carib Islands 14:1–100

De la Guardia E, González-Sansón G, Aguilar C (2003) Biodiversidad marina en la laguna costera El Guanal, Cayo Largo, Cuba. Rev Inv Mar 24:111–116

De la Guardia E, Angulo J, González-Sansón G, Aguilar C, González-Díaz P (2004) Biodiversidad en la zona de buceo del Parque Nacional de Punta Francés, Isla de la Juventud, Cuba. Rev Inv Mar 25:90–102

Delgado-Núñez E (1917) Contribución al estudio del erizo de mar Centecrinus antillarum. Mem Soc Cubana Hist Nat 2:251–280

Del Valle R, Abreu-Pérez M (1998) Equinodermos. In: Vales MA, Alvarez L, Montes L, Avila A (eds) Estudio nacional sobre la diversidad biológica en la República de Cuba. Programa de Naciones Unidas para el Medio Ambiente, Centro Nacional de Biodiversidad, Instituto de Ecología y Sistemática, CITMA, La Habana

Del Valle R, Abreu-Pérez M (2004) Equinodermos. In: González-Ferrer S (ed) Corales Pétreos. Jardines Sumergidos de Cuba, Edit Acad, Cuba, pp 216–217

Del Valle R, Abreu-Pérez M (2007) Diversidad de organismos: Equinodermos–Filo Echinodermata In: Claro R (ed) La Biodiversidad marina de Cuba. Inst Oceanol, Minist Cien, Tecnol y Medio Amb, La Habana, Cuba, pp 136–140

Del Valle R, Barreras-Rodríguez R (2006) Registro nuevo de holoturoideo (Echinodermata) para el Archipiélago Cubano. Cocuyo 16:18

Del Valle R, Solís-Marín FA, Laguarda-Figueras A (2000) Presencia de *Chiridota rotifera* (Pourtalès, 1851) (Echinodermata: Holothuroidea), en aguas del Archipiélago Cubano. Avicennia 12–13:137

Del Valle R, Solís-Marín FA, Abreu-Pérez M, Laguarda-Figueras A, Durán.-González A (2005) Catálogo de los equinodermos (Echinodermata: Crinoidea, Echinoidea y Holothuroidea) nerítico-bentónicos del Archipiélago Cubano. Rev Biol Trop 53(Suppl 3):9–28

Del Valle R, Abreu-Pérez M, Rodríguez R, F A Solís-Marín, Laguarda-Figueras A, Durán-González A (2008). Equinodermos (Echinodermata) del occidente del Archipiélago Sabana-Camagüey, Cuba. Rev Biol Trop 56(Suppl 3):19–35

Del Valle R, Cano LM, Martínez-Daranas B, Espinosa J, Marcos Z, Díaz-Larrea J, Hernández-Zanuy A, López-Cánovas C, Cantelar K, Alcolado PM, Muiñas RR, González-Ferrer S, Martínez JC, Abreu-Pérez M, Esquivel M, Hernández JL, Pérez-Zayas DM, Hernández M (2003) Informe Final del "Inventario de la flora y la fauna marina del ArchipiélagoSabana-Camagüey (Norte de Matanzas, Sancti Spiritus, Villa Clara, Ciego de Avila y Camagüey)". Archivo Científico del Proyecto PNUD/GEF CUB/98/G32, La Habana

Downey ME (1973) Seastares from the Caribbean and the Gulf of Mexico. Smith Contrib Zool 126:1–158

Elyakov GB, Kuznetsova TA, Stonik VA, Levin VS, Albores R (1975) Glycosides of marine invertebrates IV. A comparative study of the glycosides from cuban sublittoral holothurians. Comp Biochem Physiol 52B:413–417

Emilsson I, Tápanes J (1971) Contribución a la hidrología de la plataforma sur de Cuba. Ser Cub Invest Pesq 6:16–35

Espinosa J (1982) *Astropecten articulatus* y *A. duplicatus* (Echinodermata: Asteroidea) dos importantes depredadores de bivalvos. Poeyana 249:1–12

Espinosa J, Abreu-Pèrez M, Gómez O (1997) Adiciones a la fauna de equinodermos de Cuba. Avicennia 6(7):61–64

Fernández A (2001) Nuevo registro de asteroideo (Echinodermata: Asteroidea) para aguas cubanas. Avicennia 14:133–134

Figueredo-Martín T, Pina-Amargós F, Angulo-Valdés J, Gómez-Fernández F (2010) Buceo contemplativo en Jardines de la Reina: caracterización y percepción sobre el estado de conservación de área. Rev Inv Mar 31:23–32

Gallegos A, Victoria I, Zabala J, Fernández M, Penié I (1998) Hidrología de los estrechos del mar Caribe noroccidental. Rev Inv Mar 9:1–37

García-Díaz C, Chirino AL, Rodríguez JP (1991) Corrientes geostróficas en la ZEE al sur de Cuba. Rev Invest Mar 12:29–38

Gómez O, Ibarzabal D, Silva L (1980) Evaluación cualitativa del bentos en la región suroccidental de Cuba. Cien Cuba 149:1–25

Hendler G, Miller JE, Pawson DL, Kier PM (1995) Sea Stars, Sea Urchins, and Allies: Echinoderms of Florida and the Caribbean. Smith Inst Press, Washington DC

Herrera-Moreno, A, Valdés-Muñoz E, D Ibarzábal (1981) Evaluación poblacional del erizo negro, *Diadema antillarum* Philliphi, mediante un diseño de muestreo aleatorio estratificado y algunos aspectos de su biología. Cienc Biol 6:61–80

Hughes TP, Szmant AM, Steneck RS, Carpenter RC, Miller SL (1999) Algal blooms on coral reefs: what are the causes? Limnol Oceanogr 44:1583–1586

Ibarzábal D, Del Valle R, Abreu-Pérez M, Espinosa J, Cortés R (1999) Listado preliminar de los poliquetos, moluscos y equinodermos marinos del Archipiélago de Los Jardines de la Reina. Cocuyo 9:3–7

Ionin AS, Pavlidis YA, Avello O (1977) Geología de la plataforma insular de Cuba. Ed Nauka, Moscow

Lalana R, Ortiz M, Varela C, Tariche N (2004) Compilación sobre los invertebrados colectados en las expediciones del "*Atlantis*" en el archipiélago cubano. Rev Inv Mar 25:3–14

Levin VS, Gómez O (1975) The shallow water holothurians of Cuba. Acad Sci USSR Acad Sci Cuba 6:55–62

Lluis-Riera M (1972) Estudios hidrológicos del Golfo de Batabanó y de las aguas oceánicas adyacentes. Ser Oceanol Acad Cien Cuba 14:1–49

Lluis-Riera M (1977) Estudios hidrológicos de la plataforma suroriental de Cuba y aguas oceánicas adyacentes. Inf Cien Tec Inst Oceanol Acad Cien Cuba 16:1–29

Lluis-Riera M (1981a) Condiciones hidrológicas de la plataforma nororiental de Cuba, durante febrero de 1976. Inf Cien Tec Inst Oceanol Acad Cien Cuba 161:1–32

Lluis-Riera M (1981b) Informe de datos oceanográficos de la plataforma suroriental de Cuba y aguas oceánicas adyacentes. Ed Academia, La Habana

Lluis-Riera M (1983) Estudios hidrológicos de la plataforma noroccidental de Cuba. Invest Inst Oceanol Acad Cien Cuba 13:1–48

Lluis-Riera M (1984) Estudios hidrológicos de la plataforma nororiental de Cuba (zona D). Ed Academia, La Habana

Martín-Blanco F, Clero-Alonso L, González Sansón G, Pina-Amargós F (2011) Influence of *Diadema antillarum* populations (Echinodermata: Diadematidae) on algal community structure in Jardines de la Reina, Cuba. Rev Biol Trop 59:1149–1163

Martínez-Estalella N (1984) Representantes más comunes de la flora y la fauna en la comunidad coralina cubana. Poeyana 265:1–34

Meyer DL, Messing CG, Macurda DB (1978) Zoogeography of tropical western Atlantic Crinoidea (Echinodermta). Bull Mar Sci 28:412–441

Miller JE, Pawson DL (1984) Holothurians (Echinodermata: Holothuroidea). Mem Hourglass Cruises 6:1–79

Mortensen T (1928) A Monograph of the Echinoidea. I. Cidaroidea, CA Reitzel, Copenhagen

Mortensen T (1935) A Monograph of the Echinoidea. II. Bothriocidaroidea, Melonechinoidea, Lepidocentrotida and Stirodonta. CA Rietzel, Copenhagen

Mortensen T (1940) A Monograph of the Echinoidea. III. 1. Aulodonta. CA Rietzel, Copenhagen

Mortensen T (1943a) A Monograph of the Echinoidea. III. 2. Camarodonta. I. Orthopsidæ,Glyphocyphidæ, Temnopleuridæ and Toxopneustidæ. CA Rietzel, Copenhagen

Mortensen T (1943b) A Monograph of the Echinoidea. III. 3. Camarodonta. II. Echinidæ, Strongylocentrotidæ, Parasaleniidæ, Echinometridæ. CA Rietzel, Copenhagen

Mortensen T (1948a) A Monograph of the Echinoidea. IV. 1. Holectypoida, Cassiduloida. CA Rietzel, Copenhagen

Mortensen T (1948b) A Monograph of the Echinoidea. IV. 1. Holectypoida, Cassiduloida. CA Rietzel, Copenhagen

Mortensen T (1950) A Monograph of the Echinoidea, vol 1. Spatangoida, CA Rietzel, Copenhagen

Mortensen T (1951) A Monograph of the Echinoidea, vol 2. Spatangoida, CA Rietzel, Copenhagen

Nodarse-Konnorov A (2001) Abundancia y distribución del erizo *Echinometra lucunter* (Linnaeus) (Echinodermata, Echinoidea) en un arrecife del litoral norte de Ciudad de la Habana. Rev Invest Mar 22:107–115

Núñez-Jiménez A (1982) Cuba: la naturaleza y el hombre. El archipiélago. Tomo 1. Letras Cubanas, La Habana

Ortea J, De la Cruz A, Lalana R, Ortiz M, Báez M, Suárez AM (1991) Cuba desde el mar. Univ Oviedo, Oviedo, España

Ortiz M (2001) Lista de invertebrados marinos, estuarinos y semiterrestres de la playa Cojímar, en la costa norte de la Provincia ciudad de la Habana. Rev Inv Mar 22:93–102

Pawson DL, Caycedo IE (1980) *Holothuria (Thymiosycia) thomasi* new species, a large caribbean coral reef inhabiting sea cucumber (Echinodermata: Holothuroidea). Bull Mar Sci 30:454–459

Pawson DL, Fell HB (1965) A revised classification of the Dendrochirote Holothurians. Brevioria 214:1–7

Pérez-Farfante I (1959) Los erizos irregulares de Cuba. Rev Islas Univ Las Villas 1:331–372

Rodríguez-Barreras R, Del Valle R, Solís-Marín FA, Abreu-Pérez M (2007) Nuevo registro de equinoideo (Echinoidea: Echinodermata) para Cuba. Rev Inv Mar 28:269–270

Santana I, Varela C, Ortiz M (2001) El insólito caso de la ingestión de un erizo de mar *Diadema antillarum* (Echinodermata, Echinoidea) por la anémona Condylactis gigantea (Cnidaria, Actinaria) en el acuario nacional de Cuba. Rev Inv Mar 22:153–156

Serafy DK (1979) Echinoids (Echinodermata: Echinoidea). Mem Hourglass Cruises 5:1–120

Sirven-Villuenda M (1945) Equinoideos del litoral de Cuba. PhD Thesis, Univ Habana, La Habana

Suárez AM (1974) Lista de equinodermos cubanos recientes. Cien Invest Mar 6:61–72

Sukhovey VF, Korotaev GK, Shapiro NB (1980) Hidrología del Mar Caribe y el Golfo de México. Gidrometeoizdat, Leningrad

Théel H (1886) Report on the Holothurioidea 2. Challenger Sci Results. Zool 14:1–290

Thomas LP (1963) A redescription of the amphiurid brittlestar Ophiocnida cubana A. H. Clark, 1917. Proc Biol Soc Wash 76:217–222

Tommasi LR (1958) Os equinodermos do litoral de Sao Paulo. Contr Avuls Inst Oceanogr Sao Paulo 2:1–27

Tortonese E (1956) Catalogo degli echinodermi della collezione E. Tortonese. Ann Mus Civ St Nat Gen 58:177–233

Valdivia A, de la Guardia E, Armenteros M, González P, Suárez AM, Aguilar C, González-Sansón G (2004) Inventario de los componentes más comunes de la flora y la fauna de algunos arrecifes coralinos de la Península de Guanahacabibes, Pinar del Río, Cuba. Rev Inv Mar 25:113–121

Verrill AE (1899) Report on the Ophiuroidea collected by the Bahama Expedition in 1893. Bull Lab Nat Hist State Univ Iowa 5:1–86

Victoria del Río I, Melo N and R Pérez (1997) Results of Cuban oceanographic research related to global change studies. Proc Conferecne/Workshop in the Caribbean Countries and the Inter-American Institute for Global Change Research

Chapter 13
Recent Echinoderms from Hispaniola

Alejandro Herrera-Moreno and Liliana Betancourt

13.1 Introduction

The Island of Hispaniola is the second largest in the Archipelago of the Greater Antilles. It is located between 17°36′15″N and 19°57′09″N, and 68°19′22″W and 74°41′33″W. Its geographical boundaries are the Atlantic Ocean to the north, the Caribbean Sea to the south, the Mona Channel to the East, which separates it from Puerto Rico, and in the West, the Windward Passage and the Jamaica Channel, which separate it from Cuba and Jamaica, respectively. The 76,480 km^2 of its territory is divided politically into two countries: Dominican Republic to the east and the Republic of Haiti in the west, separated by a land border of 360 km.

The Republic of Haiti has a total area of 27,750 km^2, which includes the mainland territory plus several islands and islets such as Gonave (743 km^2), Tortue (180 km^2), Vache (52 km^2), Cayemites (45 km^2) and Navassa (5.2 km^2). The Dominican Republic has an area of 48,442 km^2 of mainland in addition to several islands and islets such as Saona (117 km^2), Beata (27 km^2) and Catalina (9.6 km^2). Two oceanic banks: Christmas and Silver Banks, with areas of 70 and 150 km^2 respectively, are located to the north of Samana in the Atlantic Ocean.

The coastline of Hispaniola extends for about 3,059 km (1,288 km belong to 16 Dominican coastal provinces and 1,771 km to nine Haitian coastal provinces) containing ecosystems such as sandy beaches, rocky shores, cliffs, estuaries, coastal lagoons and mangrove forests. On the Haitian coast important mangroves areas are reported along the north and northeast coast in the Bays of Fort Liberté, l'Acul, Caracol, the Artibonite estuary and the Islands Les Cayes, Vache, Gonave and Cayemite (Ehrlich et al. 1987). Along the Dominican coast the largest areas of mangroves are found in the Bay of Samaná and Montecristi.

A. Herrera-Moreno (✉) · L. Betancourt
Programa EcoMar, Inc., Santo Domingo, Dominican Republic
e-mail: ongprogramaecomar@yahoo.com

J. J. Alvarado and F. A. Solís-Marín (eds.), *Echinoderm Research and Diversity in Latin America*, DOI: 10.1007/978-3-642-20051-9_13,

The Dominican shelf has a surface area of 8,130 km^2 extending down to 200 m depth, with an average width of 7.5 km. The platform area of Haiti is estimated at 6,683 km^2 and is on average narrower. Important coastal habitats include sandy or muddy bottoms (related to the degree of terrigenous influence), commonly populated with abundant macrophytobenthos, especially seagrasses, followed by coral patches and fringing or barrier reefs. In Haiti coral reefs occur near Vache Island, all around Gonave Island in the central Bay of Port-au-Prince, on the Rochelois Bank, at Cayemites Islands, off the northern coast of the southern peninsula, and from the border with the Dominican Republic in the east to Acul Bay just west of Cape Haitien (Creary et al. 2008). Well preserved reefs are found at Navassa Island (Causey et al. 2000).

In the Dominican Republic important reef areas include the Montecristi National Park barrier reef in the north-west (where the shelf is widest), narrow high-energy reefs in the central region, and the Bávaro-Macao-Punta Cana barrier reef system at the eastern end. Samaná Bay receives many rivers and is the largest estuary of the insular Caribbean. Coastal reefs in the vicinity are poorly developed, but the Christmas and Silver Banks reef systems are about 100 km to the north. To the south, on the Caribbean coast, are the well-studied reefs of National East Park and the adjacent Saona Island. Uplifted carbonate terraces with reefs growing on narrow platforms are present in the west from Catalina Island to beyond Santo Domingo (e.g. Boca Chica and the Submarine National Park La Caleta). Conditions are not good for reefs in the south-west, except on the shallow sheltered shelf east of Cabo Beata at National Park Jaragua (Woodley et al. 2000).

13.2 Research

Echinoderm studies of the waters around Hispaniola Island have a long history. The Haitian collection of David Friedrich Weinland from 1857 (NMNH 2002) and the work on holothuroids by Emil Selenka (1867), with the record of *Actinopyga agassizi* (Selenka, 1867) in Haiti and the description of the type specimen of *Holothuria grisea* Selenka, 1867, are among the earliest contributions.

William More Gabb made collections in the Dominican Republic in 1878 and deposited several specimens of *Echinometra lucunter* (Linnaeus, 1758) in the U.S. National Museum of Natural History (Rathbun, 1886). In December 1878, during the *Blake* Expedition, the echinoid *Salenocidaris varispina* (A. Agassiz, 1880) was collected northwest of Haiti at 2,195 m depth (A. Agassiz, 1880). Dodërlein and Hartmeyer (1910) and Verrill (1915) summarized previous records of asteroids from Hispaniola. Hubert L. Clark (1919) provided the first taxonomic summary of Haitian echinoderms listing more than 20 species of all classes, except crinoids.

In the 1930s, new contributions to the knowledge of echinoderms from Hispaniola include Deichmann (1930) who summarized previous records for the Dominican Republic and Haiti in her review of the holothurians of the western part of the Atlantic Ocean. The handbook of the littoral echinoderms of Porto Rico and the other West Indian Islands (H.L. Clark 1933) also summarized historical information about all classes of echinoderms for Hispaniola, except crinoids, with some new records. Austin H. Clark (1939) listed 19 species of shallow-water echinoderms (between 1 and 2 m depth) collected in two localities in the north of Haiti (Cape Haitien and Tortue Island) during the Smithsonian-Hartford expedition of March 1937, aboard the RV Joseph Conrad.

New collections took place during the 1930s on five expeditions: Parish-Smithsonian in 1930 to Haiti on the yacht *Esperanza*, Norcross-Bartlett in July 1931 to the southwest of Santo Domingo down to 200 m depth on the schooner *Effie Morrissey*, Johnson–Smithsonian Deep-Sea in February 1933 on the Yacht Caroline at depths from 26 to 40 m at twelve stations inside the Bay and around the Peninsula of Samaná (Bartsch 1933), RV *Stranger* in February 1933 in Port-au-Prince and the RV *Atlantis* in April 1939 dredging to 1,170 m deep at one station to the northeast of Haiti (H.L. Clark 1941). These expeditions contributed new records for Hispaniola including two deep-water asteroid species: *Persephonaster patagiatus* (Sladen, 1889) and *Phormosoma placenta sigsbei* (A. Agassiz, 1880). The expeditions of 1931 and 1933 collected the first specimens of crinoids: *Poliometra prolixa* (Sladen, 1881) and *Nemaster rubiginosa* (Pourtalès, 1869).

In the decade of 1960s two expeditions took place. The exploratory fishing vessel *Silver Bay* in October 1963 made collections at six stations between 92 and 348 m deep in the north, northeast and east of the Dominican Republic. The exploratory fishing vessel *Oregon* in June and May, 1965 and December 1969 collected between 11 and 59 m deep at four stations off the north Haitian coast and the northeast Dominican coast. During this period, Parslow and Clark (1963) summarized the zoogeographical distribution of shallow-water (less than 17 m deep) ophiuroids known for the West Indies, including 11 species from Hispaniola, updating the previous list of H.L. Clark (1919). Deichmann (1963) summarized previous records of holothuroids for Haiti. Halpern (1969) mentioned the asteorid *Litonotaster intermedius* (Perrier, 1884) found in the Windward Pass, between Cuba and Hispaniola, at a depth between 1,958 and 3,294 m (Downey 1973).

From January to July 1970 and July 1971 the most prolific expeditions carried out in Hispaniolan waters took place aboard the RV *John Elliot Pillsbury*. Collections were made at seven stations in Haiti, between 31 and 2,545 m deep, and 33 stations in Dominican Republic, between 9 and 3,109 m deep (Fig. 13.1) (Staiger and Voss 1970). Meyer et al. (1978) included the RV *John Elliot Pillsbury* collections in their zoogeographical study of western Atlantic Ocean crinoids, offering the most complete summary of this group, with 19 species listed for Dominican and Haitian waters. In her review of the Order Brisingida for the Atlantic Ocean, Downey (1986) also incorporated Haitian specimens from the Pillsbury expeditions. A.M. Clark (1987) described the type specimen of the asteorid*Henricia downeyae* A.M. Clark, 1987 from a Haitian locality, registered in the catalogue of types of the NMNH (Ahearn 1995). Two more records for Haiti from this period were the holothuroid *Psolus tuberculosus* Théel, 1886 (Miller and Pawson 1984) and the asteroid *Ceramaster grenadensis grenadensis* (Perrier, 1881) (Halpern 1970).

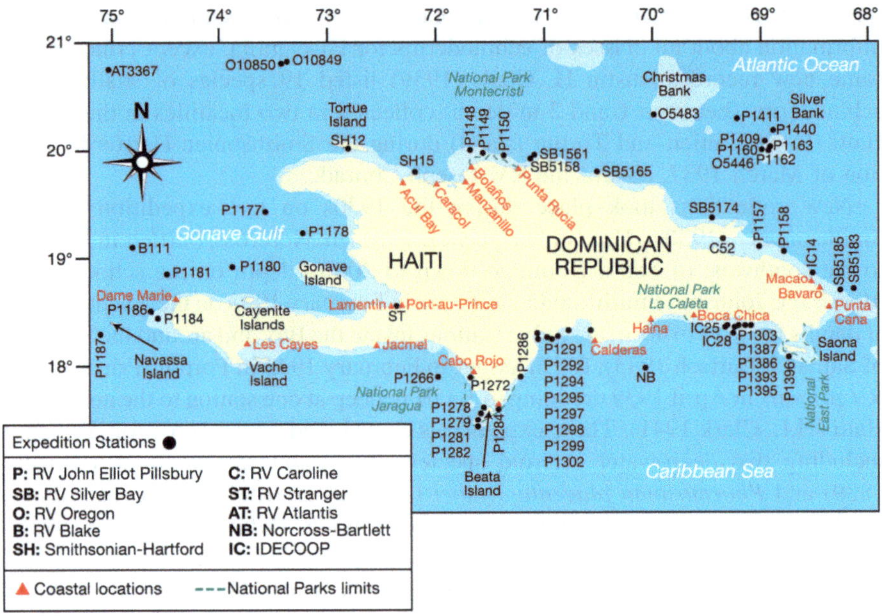

Fig. 13.1 Localities where collections of echinoderms have been made by several expeditions and national and international projects on the shelf and oceanic waters of Hispaniola

All these expeditions increased the collections of the five classes of echinoderms from Hispaniola in international museums with 541 specimens of 132 species located in the U.S. Museum of Natural History (NMNH 2002), Florida Museum of Natural History (FLMNH 2010), Museum of Comparative Zoology at Harvard University (MCZ 2010) and Museum of Natural Sciences of Berlin (ZMB 2010). The RV *John Elliot Pillsbury* expeditions had a decisive role in the enrichment of collections, particularly with deep-sea species, that at the end of the 1970s, 90 % of the crinoid, asteroid, and echinoid species known from Hispaniola were already described and deposited in different museums.

In the Dominican Republic, the creation of the Center of Marine Biology Research (CIBIMA) in 1962 promoted coastal and marine biodiversity research. Between the decades from 1970 to 1990, CIBIMA produced summarized list of echinoids and asteroids (Cicero et al. 1976), ophiuroids (Rathe 1978) and holothuroids (Briones 1983). Multiple reports were finally compiled into a preliminary study on coastal and marine biodiversity of the Dominican Republic (CIBIMA 1992). From these investigations, 242 specimens of 58 species are deposited in the National Museum of Natural History of Santo Domingo (MNHNSD 2008).

In May 1979, the Autonomous University of Santo Domingo, along with the University of Puerto Rico, implemented an expedition on board the RV *Crawford* to make inventories of the southeast Dominican reefs of La Caleta and Catalina and Saona Islands (Williams et al. 1983). Later the IDECOOP expedition

(Fig. 13.1) made collections between 142 and 270 m deep with three new records for the deep-sea Dominican echinoderms: the crinoid *Cenocrinus asterius* (Linnaeus, 1775); the ophiuroid *Astronicda isidis* Lyman, 1872 and the echinoid *Conolampas sigsbei* (A. Agassiz, 1878) (Rivas 1983). Studies carried out by researchers from East Carolina University in Montecristi reefs (Luczkovich 1991) extended the reports of Dominican crinoids with the record of *Tropiometra carinata* (Lamarck, 1816).

The information on Haitian echinoderms comes from old collections, that have already been mentioned, and some more recent studies carried out by foreign institutions. In June 1988, a detailed study at 13 stations on different ecological zones (from the reef lagoon to the frontal reef) of the Arcadines coral reefs was made by the World Wildlife Fund and the Conservation Foundation Wilcox Associates, between 0.3 and 21 m depth (Wilcox et al. 1989). Twelve species of common shallow-water echinoderms were reported. Hendler et al. (1995) summarized the echinoderms from approximately 17 old Haitian records.

With the implementation of the Hispabiota Marina Project (Herrera-Moreno and Betancourt 2012) by Programa EcoMar, Inc. in the Dominican Republic a bibliographical and taxonomic review of echinoderm species was done for the first time, with a historical and insular approach (Herrera-Moreno and Betancourt 2004). A database was created with the results of collections of echinoderms from more than 150 localities of the coastal zone, the shelf and the deep zone of Hispaniola (Fig. 13.1). This project added 60 species to the national inventory of CIBIMA (1992) confirming 123 species for Dominican coastal and marine zone. It also compiled for the first time 79 species for Haiti. A total of 156 species for Hispaniola (Appendix) is discussed in the present report.

13.3 Diversity and Distribution

The numbers of echinoderms in different taxonomic categories for the Dominican Republic, Haiti and Hispaniola are summarized in Table 13.1. There are 22 species of crinoids (18 for the Dominican Republic and eight for Haiti), subdivided into 17 genera, nine families and four orders. About five species of crinoids are found in coral reef environments at depths to 45 m deep including those which, according to Hendler et al. (1995), are more common and accessible to conventional scuba diving. The remaining species are distributed below 100 to 1,033 m deep.

The list of asteroids has 33 known species (18 for Haiti and 21 for the Dominican Republic) subdivided into 24 genera, 11 families and seven orders. Downey (1973) in her review of the asteroids in the Gulf of Mexico and the Caribbean described 95 species and noted that the asteroid fauna from Hispaniola was practically unknown. The bathymetric ranges of the asteroid species known from Hispaniola varies from 0.3 to 3,493 m deep. About eight species are distributed from the surface to 50 m deep primarily from the mangroves to the reef front, with the remaining species found below 50 m deep.

Table 13.1 Summary of the numbers of different taxonomic categories for echinofauna groups of Dominican Republic (DO), Haiti (HA) and Hispaniola (HI)

		Orders	Families	Genus	Species
Crinoidea	HA	3	4	7	8
	DO	4	9	14	18
	HI	4	9	17	22
Asteroidea	HA	7	11	16	18
	DO	4	7	14	22
	HI	7	11	24	33
Ophiuroidea	HA	2	9	12	21
	DO	2	11	15	24
	HI	2	12	20	30
Echinoidea	HA	7	12	15	21
	DO	10	16	31	41
	HI	12	17	36	50
Holothuroidea	HA	3	6	7	11
	DO	3	5	7	18
	HI	3	7	10	21
TOTAL	HA	22	42	57	79
	DO	23	48	81	123
	HI	28	57	107	156

Thirty species of ophiuroids are known for Hispaniola (21 for Haiti and 24 for the Dominican Republic), subdivided into 20 genera, 12 families and two orders. Most species of ophiuroids are distributed in water less than 30 m deep found in seagrass beds and coral reefs. Only four species were collected between 148 and 366 m deep.

Fifty species of echinoids are listed (21 for Haiti and 41 for the Dominican Republic) subdivided into 36 genera (including one species identified only to the genus level), 17 families and 12 orders. Although bathymetric data are not available for all species, the collections from Hispaniola contain the more common shallow-water species from mangrove habitats, seagrass beds and coral reefs. At least about 30 species are distributed from 100 to 2,545 m depth.

The list of holothurians of Hispaniola includes 21 species (11 for Haiti and 18 for the Dominican Republic) with ten genera, seven families and three orders. Most species of holothuroid are common in shallow sedimentary environments. Three species are found between 243 and 1,400 m deep.

13.4 Ecology

In the Dominican Republic, echinoderm species are mentioned in the inventories of invertebrates of seagrass beds and coral reefs from Manzanillo to Punta Rucia in Montecristi (Luczkovich 1991; CIBIMA 1998), Puerto Plata (Herrera-Moreno and

Betancourt 2009), Samaná (Sang and Lysenko 1994; Sang 1996), Punta Cana, Bávaro (CURPOB 2000; Brandt et al. 2003), National East Park (Vega et al. 1997) and Isla Saona in La Altagracia, Catalinita in La Romana, La Caleta in Santo Domingo (Williams et al. 1983), Haina in San Cristóbal (Herrera-Moreno et al. 2009), Las Calderas Bay in Peravia (Almonte 1976), Puerto Viejo in Azua (González et al. 1978) and Jaragua National Park in Pedernales (Weil 2006). In Haiti, inventories with echinoderm species include Lamentin (Beebe 1928), Les Arcadins reefs (FoProBIM 1985; Wilcox et al. 1989) and Navassa (Miller 2003).

Due to the dramatic reduction of black sea urchin *Diadema antillarum* in the Caribbean in the 1980s, the presence of this species is always highlighted in all inventories and some abundance surveys have been done. In the Dominican reefs, Chiappone (2001) estimated an average density of *D. antillarum* of 0.03 ind 50 m^{-2} for the coral reefs in the National East Park, between 9 and 17 m depth. He also estimated 9.15 ind 50 m^{-2} in the Boca Chica reefs between 4 and 20 m depth. Brandt et al. (2003) reported a maximum density of 5.4 ind 10 m^{-2} in the reefs of Punta Cana and Bavaro with the highest average values in the deep forereef (1.2 ind 10 m^{-2}) in relation to the shallow areas of the back-reef (0.5 individuals/10 m^2). For Caracol Bay in Haiti, Hay (1984) reported 20 ind m^{-2}, between 3 and 10 m depth. FoProBIM (FoProBIM 1985) found densities of *D. antillarum* up to 14.5 ind 100 m^{-2} at 5 m depth in the area of Les Arcadines. More recently, abundance values from 0 to 1.5 ind 10 m^{-2} have been reported in the same area by Linton (2003).

However, ecological studies at the level of echinoderm populations or communities seem to be scarce. In the Dominican Republic, Chiappone (2001) conducted a study measuring density and size distribution for five species of echinoids (*D. antillarum*, *Echinometra viridis*, *Echinometra lucunter*, *Eucidaris tribuloides* and *Tripneustes ventricosus*) in the National East Park and Boca Chica. Tewfik et al. (2005) studied the impact of anthropogenic enrichment on the seagrass food web at two locations in southern Dominican Republic (Pedernales and Barahona), involving several species of echinoderms, categorized by them as three consumer functional groups: (a) generalists (regular urchins: *Lytechinus variegatus*, *T. ventricosus*, *D. antillarium*); (b) subsurface deposit feeders (red heart urchin *Meoma ventricosa* and sand dollars) and (c) surface deposit feeders (cushion sea star *Oreaster reticulatus* and sea cucumbers). Hay (1984) compared the activity of grazing fish and sea urchins in the Bay of Caracol in Haiti, on an overexploited coral reef including density data of *Diadema antillarum* and its rate of consumption of *Thalassia testudinum*.

13.5 Aquaculture and Fisheries

Commercial fishing of sea cucumber takes place in the Dominican Republic. According to the Dominican Council of Fisheries and Aquaculture CODOPESCA in 2002 there was an export of 57.6 MT, equivalent to 9 % of seafood exports. The figures for years 2007, 2008 and 2009 indicate much smaller export volumes with 1.27, 2.37 and 1.37 MT, respectively. The trading company for sea cucumbers in the Dominican Republic is NETCO C x A which operates from Juan de Bolaño Beach in Montecristi with CODOPESCA permissions for exportation to the United States. There are no studies that scientifically support this exploitation. In fact, there is no information on the species being harvested, minimum sizes, close seasons or any other descriptive parameters of an organized fishery. The survival of these echinoderm populations is seriously threatened. According to Toral-Granda (2008), between 2002 and 2003, China imported 2,607 kg of dry weight of sea cucumbers from Dominican Republic and 10,680 kg of dry weight from Haiti. In Haiti, the Caribbean SeaFood Company announces online the sale of dried sea cucumbers of the species *Stichopus badionotus*.

13.6 Threats

The threats to echinoderms are not very different from those faced by other marine groups of Hispaniola. The destruction of ecosystems (mainly mangroves, seagrass beds and coral reefs) due to coastal development (human settlements, agriculture, industrial and tourist facilities and/or fishing) and tourism activities (especially diving) is the major threat that affects all groups. In addition to impacts of habitat destruction, some groups of echinoderms are exploited for commercial purposes. These include some species of sea cucumbers used for human consumption and some species of starfish and sea urchins used for sale as aquarium species (SERCM 2004). Other echinoderm species, such as the sand dollar *Mellita quinquiesperforata* and the starfish *Oreaster reticulatus* are collected for handicrafts that are sold in tourist establishments. Some fishermen also use this asteroid species as bait in net traps. No ecological studies or fisheries biology research have been done to support a sustainable exploitation of these resources.

13.7 Recommendations

It is essential to train a group of national specialists from Haiti and the Dominican Republic in the systematic of all echinoderm classes that can work closely together to better understand the echinofauna of Hispaniola. The National Museum of Natural History in Santo Domingo should be involved in the development and review of the marine collections. New investigations in structural ecology and population dynamics of echinoderms must be initiated for those species that may be subject to some kind of extractive use. For exploited species, it is also necessary to conduct fishery biology investigations for the establishment of closed seasons, protected areas from fishing and legal minimum sizes. Fishing regulations and fishery statistics are crucial for a sustainable exploitation of these species. Marine research should continue increasing the knowledge of the echinoderms in new areas of the Dominican and Haitian platforms, including exploration of deeper zones.

References

Agassiz A (1880) Reports on the results of dredging, under the supervision of Alexander Agassiz, in the Caribbean Sea, 1878–79 and along the Atlantic coast of the United States during the summer of 1880, by the U. S. Coast Survey Steamer "Blake"commander J. R. Bartlett, U.S. N., Commanding. IX. Preliminary report on the Echini. Bull Mus Comp Zool Harvard Coll 8:69–84

Ahearn CG (1995) Catalog of the type specimens of seastars (echinodermata: asteroidea) in the national museum of natural history, Smithsonian Institution. Smith Contr Zool 572:1–59

Almonte NC (1976) Bahía de Las Calderas, flora y fauna. Ed Amigo del Hogar, Santo Domingo, República Dominicana

Bartsch P (1933) Station records of the first Johnson–Smithsonian deep-sea expedition. Smith Misc Coll 91:1–31

Beebe W (1928) Beneath tropic seas. A record of diving among the coral reefs of Haiti. GP Putnam's Sons, New York

Brandt ME, Cooper WT, Polsenberg JF (2003) Results of a coral reef survey of Punta Cana, Dominican Republic, with comparisons to past studies and other Caribbean reefs, August 20–25. Report of The National Center for Caribbean Coral Reef Research, Rosenstiel School of Marine and Atmospheric Science, University of Miami, Miami

Briones R (1983) Contribución al estudio de los holoturoideos (Echinodermata: Holothuroidea) de la República Dominicana. Academia de Ciencias, Santo Domingo

Causey B, Delaney J, Diaz E, Dodge D, Garcia JR, Higgins J, Jaap W, Matos CA, Schmahl GP, Rogers C, Miller MW, Turgeon DD (2000) Status of coral reefs in the US Caribbean and Gulf of Mexico: Florida, Texas, Puerto Rico, US Virgin Islands and Navassa. In: Wilkinson C (ed) Status of coral reefs of the world: 2000. Australian Institute for Marine Science, Townsville, pp 239–260

Chiappone M (2001) Coral reef conservation in marine protected areas: a case study of Parque Nacional Del Este Dominican Republic publications for capacity building. The Nature Conservancy, Arlington

CIBIMA (1992) Estudio preliminar sobre la biodiversidad costera y marina de la República Dominicana. Ed Alfa y Omega, Santo Domingo, República Dominicana

CIBIMA (1998) La diversidad biológica de los ecosistemas marinos del Parque Nacional de Montecristi. Reporte técnico del Proyecto GEF-PNUD/ONAPLAN: Conservación y Manejo de Biodiversidad de la Zona Costera de la República Dominicana, Centro de Investigaciones de Biología Marina (CIBIMA), Universidad Autónoma de Santo Domingo (UASD), Santo Domingo, República Dominicana

Cicero J, Rivas V, Bonnelly I (1976) Erizos y estrellas comunes del litoral dominicano. An Acad Cien Rep Dominic 2:73–80

Clark HL (1919) The distribution of the littoral echinoderms of the West Indies. Publ Dept Mar Biol Carnegie Inst Wash 13:49–74

Clark HL (1933) A handbook of the littoral echinoderms of Porto Rico and the other West Indian Islands. Sci Surv Porto Rico Virg Isl NY Acad Sci 16:1–147

Clark AH (1939) Echinoderms of the Smithsonian–Hartford expedition, 1937, with other West Indian records. Proc US Nat Mus 86:441–456

Clark HL (1941) Reports on the scientific results of the "Altantis" expedition to the West Indies, under the joint auspices of the University of Havana and Harvard University. The echinoderms (other than holothurian). Mem Soc Cubana Hist Nat "Felipe Poey" 15:1–154

Clark AM (1987) Notes on Atlantic and other Asteroidea. 5: Echinasteridae. Bull Brit Mus (Nat Hist) Zool 53:65–78

Creary M, Alcolado P, Coelho V, Crabbe J, Green S, Geraldes F, Henry A, Hibbert M, Jones R, Jones-Smith L, Manfrino C, McCoy SMC, Wiener J (2008) Status of coral reefs in the Northern Caribbean and Western Atlantic GCRMN Node in 2008. In: Wilkinson C (ed) Status of coral reefs of the world: 2008. Australian Institute for Marine Science, Townsville, pp 239–252

CURPOB (2000) Cornell Undergraduate Research Program Biodiversisty CURPOB. Cornell Biodiversity Laboratory at Punta Cana. Final Report for June 9 to August 11

Deichmann E (1930) The holothurians of the western part of the Atlantic ocean. Bull Mus Comp Zool Harvard Coll 71:43–226

Deichmann E (1963) Shallow water holothurians known from the Caribbean waters. Stud Fauna Curacao Carib Isl 14:1–100

Dodërlein L, Hartmeyer R (1910) Westindische Seeigel und Seesterne. Zool Jahrbucher 11:145–156

Downey ME (1973) Starfishes from the Caribbean and the Gulf of Mexico. Smith Contrib Zool 126:1–158

Downey ME (1986) Revision of the Atlantic Brisingida (Echinodermata: Asteroidea), with description of a new genus and Family. Smith Contr Zool 435:1–57

Ehrlich MC, Adrien N, Lebeaue F, Lewis L, Lauwereysen H, Lowenthal I, Mayda Y, Paryski P, Smucker G, Talbot J, Wilcox E (1987) Haiti country environmental profile: a field study. USAID Haiti, Port-au-Prince

FLMNH (2010) Florida museum of natural history. Invertebrate zoology master database, Echinodermata. http://www.flmnh.ufl.edu/scripts/dbs/malacol_pub.asp

FoProBIM (1985) Reef check training and coral reef monitoring in Haiti: a preliminary report. Caribbean Coastal Data Centre, Centre for Marine Sciences, University of West Indies, Foundation pour la Protection de la Biodiversite Marine, Port-au-Prince

González Z, Gutiérrez W, Rivas V, Bonnelly I (1978) Informe preliminar sobre la laguna costera de Puerto Viejo, Azua en la República Dominicana. In: Conservación y Ecodesarrollo. Centro de Investigaciones de Biología Marina (CIBIMA). Universidad Autónoma de Santo Domingo, pp 53–93

Halpern JA (1969) Biological investigations of the deep sea. 46. The genus *Litonotaster* (Echinodermata: Asteroidea). Proc Biol Soc Wash 82:129–142

Halpern JA (1970) Goniasteridae (Echinodermata, Asteroidea) of the straits of Florida. Bull Mar Sci 20:193–286

Hay ME (1984) Patterns of fish and urchin grazing on Caribbean coral reefs: are previous results typical? Ecol 65:446–454

Hendler G, Miller JE, Pawson DL, Kier PM (1995) Sea stars, sea urchins, and allies: Echinoderms of Florida and the Caribbean. Smith Inst Press, Washington DC

Herrera-Moreno A, Betancourt L (2004) Especies de equinodermos recientes (Echinodermata: Crinoidea: Asteroidea:Ophiuroidea:Echinoidea y Holothuroidea) conocidas para la Hispaniola. Rev Cien Soc 29:506–533

Herrera-Moreno A, Betancourt L (2009) Impacto de la contaminación sobre los arrecifes coralinos entre Playa Dorada y Cafemba, Puerto Plata. Report from Proyecto EcoMar: Impactos a los arrecifes dominicanos. http://programaecomar.com/Arrecifes_Cafemba.pdf

Herrera-Moreno A, Betancourt L (2012) Hispabiota Marina project, the first inventory of the marine biota from Hispaniola Island. http://programaecomar.com/HISPABIOTAMARINA.htm

Herrera-Moreno A, Betancourt L, Alcolado P (2009) Impacto de la contaminación sobre los arrecifes coralinos al Oeste del Río Haina, San Cristóbal. Report from Proyecto EcoMar: Impactos a los arrecifes dominicanos. http://programaecomar.com/Arrecifes_Haina.pdf

Linton D (2003) Reef check training and coral reef monitoring in Haiti. A preliminary report, Centre for Marine Sciences, UWI

Luczkovich JJ (1991) Marine ecology of the Buen Hombre coast. In: Stoffle RW, Halmo DB (ed) Satellite monitoring of coastal marine ecosystems: a case from the Dominican Republic. East Carolina University, pp 93–141

MCZ (2010) Harvard University, Museum of Comparative Zoology, MCZ Marine Invertebrate Collections. http://collections.mcz.harvard.edu/MarineInvert/MarineInvertSearch.html

Meyer DL, Messing CG, Macurda DB Jr (1978) Zoogeography of tropical western Atlantic Crinoidea (Echinodermata). Bull Mar Sci 28:412–441

Miller MW (2003) Status of reef resources of Navassa Island: cruise report Nov. 2002. NOAA Technical Memorandum NMFS-SEFSC-501

Miller JE, Pawson DL (1984) Holothurians (Echinodermata: Holothuroidea). Memoirs of the Hourglass Cruises 6. Florida Dept. Nat. Res. FL, USA, VII, Part 1:1–79

MNHNSD (2008) Catálogo de equinodermos. Museo Nacional de Historia Natural de Santo Domingo, Santo Domingo

NMNH (2002) Master list: Echinoderms Dominican Republic & Haiti. United States National Museum, Washington DC

Parslow RE, Clark AM (1963) Ophiuroidea of the lesser Antilles. Stud Fauna Curaçao Carribbean Isl 15:24–50

Rathbun R (1886) Catalogue of the collection of recent Echini in the United States National Museum (corrected to july 1, 1886). Proc US Nat Mus 9:225–293

Rathe L (1978) Distribución geográfica de las estrellas frágiles (Subclase Ophiuroidea) de República Dominicana. Trabajo del Curso Métodos de investigación biológica, Centro de Investigaciones de Biología Marina (CIBIMA), Universidad Autónoma de Santo Domingo

Rivas V (1983) Lista de equinodermos recolectados por el Departamento de Pesca del IDECOOP. Contr Centro de Invest Biol Mar 5:107–111

Sang L (1996) Estudio de los arrecifes de coral de la costa Norte de la Península de Samaná. Proyecto inventario de la biodiversidad y caracterización de las comunidades del entorno de la Península y Bahía de Samaná, Centro para la conservación y Ecodesarrollo de la Bahía de Samaná y su Entorno (CEBSE), Santo Domingo

Sang L, Lysenko N (1994) Praderas marinas. In: Caracterización de ecosistemas costeros y marinos en la Bahía de Samaná. Centro para la Conservación y Ecodesarrollo de la Bahía de Samaná y su Entorno (CEBSE), Santo Domingo, pp 47–72

Selenka E (1867) Beitrage zur Anatomie und Systematik der Holothurien. Zeischr Wiss Zool 17:291–375

SERCM (2004) Los Recursos Marinos de la República Dominicana. Subsecretaría de Estado de Recursos Costeros y Marinos/Secretaría de Estado de Medio Ambiente y Recursos Naturales, SERCM/SEMARN, Ed Búho, Santo Domingo

Staiger JC, Voss GL (1970) Narrative of R/V John Elliot pillsbury cruise P-7006 to Hispaniola and Jamaica. Rosentiel school of marine and atmospheric science. University of Miami, Miami

Tewfik A, Rasmussen JB, Mccann KS (2005) Anthropogenic enrichment alters a marine benthic food web. Ecology 86:2726–2736

Toral-GrandaV (2008) Population status, fisheries and trade of sea cucumbers in Latin America and the Caribbean. In: Toral-Granda V, Lovatelli A, Vasconcellos M (eds) Sea cucumbers. A global review of fisheries and trade. FAO Fisheries and Aquaculture Technical Paper. No. 516. Rome, pp 213–229

Vega M, Chiaponne M, Delgado GA, Wright R, Sullivan KM (1997) Evaluación ecológica integral Parque Nacional del Este, República Dominicana. Media Publishing Limited, Bahamas

Verrill AE (1915) Reports on the starfishes of the West Indies, Florida and Brazil. Bull Lab Hist State Univ Iowa 7:1–232

Weil E (2006) Coral, Octocoral and sponge diversity on reefs of the Jaragua National Park, Dominican Republic. Rev Biol Trop 54:423–443

Wilcox E, Deyo T, Gardella A, García R, Glick D, Goneaga C, Medina A, Vicente V (1989) Proposed Les Arcadins National Marine Park resource document. World Wildlife Fund and Conservation Foundation Wilcox Associates

Williams EH, Clavijo I, Kimmel JJ, Colin PL, Díaz C, Bardales AT, Armstrong RA, Bunkley L, Boulon RH, García JR (1983) A checklist of marine plants and animals of the south coast of the Dominican Republic. Carib J Sci 19:39–54

Woodley J, Alcolado P, Austin T, Barnes J, Claro-Madruga R, Ebanks-Petrie G, Estrada R, Geraldes F, Glasspool A, Homer F, Luckhurst B, Phillips E, Shim D, Smith R, Sullivan K, Vega M, Ward J, Wiener J (2000) Status of coral reefs in the Northern Caribbean and Western Atlantic. In: Wilkinson C (ed) Status of coral reefs of the world: 2000. Australian Institute for Marine Science, Townsville, pp 261–285

ZMB (2010) Museum fur Naturkunde, Berlin. Taxonomy and specimen data. http://sesam.gbif-evt3.senckenberg.de/page/index.htm

Chapter 14
Review of Echinoderm Research in Puerto Rico, with the Focus on Biological and Ecological Aspects

Stacey M. Williams, Milena Benavides-Serrato, Jose E. García-Arrarás, Edwin A. Hernández-Delgado and Ruber Rodríguez-Barreras

14.1 Introduction

The island of Puerto Rico is located in the northeastern part of the Caribbean Sea (18°13′N–66°28′W). Puerto Rico is considered an archipelago formed by the main island and other small islands mainly off its eastern and southern coastlines (Fig. 14.1). The eastern shelf is very shallow (typically less than 30 m in depth) and delimited in the north by a chain of small islands and emergent monoliths that extend from Cabezas de San Juan to Culebra island (18°18′N–65°16′W), 27 km off the main island. The island of Vieques [or: Vieques Island] (18°07′N–65°25′W) lies south of the eastern shelf. Numerous mangrove keys and an offshore island (6 km), Caja de Muertos (1.54 km^2), characterize the southern coast of

S. M. Williams (✉)
Centro de Investigación en Ciencias del Mar y Limnología (CIMAR),
Universidad de Costa Rica, San Pedro, San José, CR 11501-2060, USA
e-mail: stcmwilliams@gmail.com

M. Benavides-Serrato
Echinoderm Curator Museo de Historia Natural Marina de Colombia,
Biodiversity Marine Ecosystems (BEM), Instituto de Investigaciones Marinas
y Costeras-INVEMAR, A.A 1016, Santa Marta, Colombia
e-mail: milena_benavides@invemar.org.co

J. E. García-Arrarás · E. A. Hernández-Delgado · R. Rodríguez-Barreras
Department of Biology, University of Puerto Rico,
PO Box 23360 Río Piedras, San Juan, PR 00931-3360, USA
e-mail: jegarcia@hpcf.upr.edu

E. A. Hernández-Delgado
e-mail: coral_giac@yahoo.com

R. Rodríguez-Barreras
e-mail: ruber.rodriguez@hotmail.com

J. J. Alvarado and F. A. Solís-Marín (eds.), *Echinoderm Research and Diversity in Latin America*, DOI: 10.1007/978-3-642-20051-9_14,

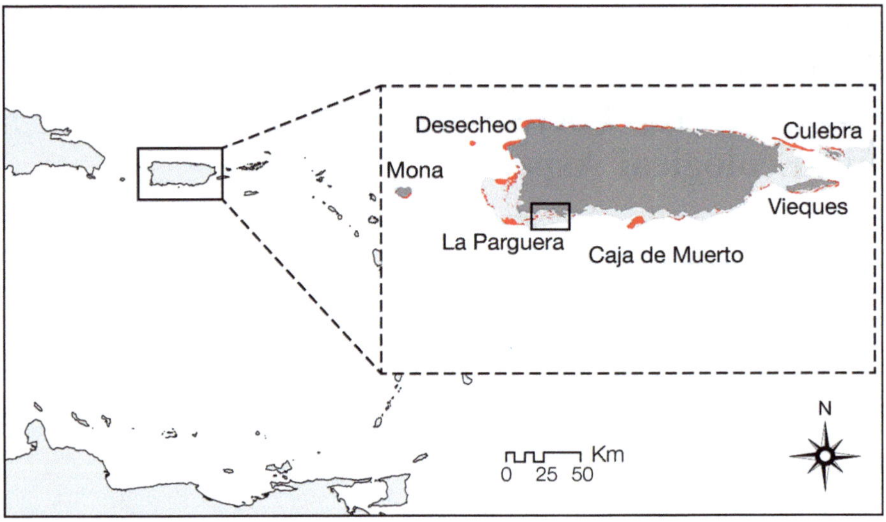

Fig. 14.1 Map of Puerto Rico with outlying islands. Coral reefs are marked in *red* and insular shelf is displayed in light *grey*. Boxed in area identifies La Parguera, Puerto Rico

Puerto Rico. The western shelf is broad and shallow, especially in the southwestern portion. Off the western edge of the shelf in the Mona Passage is Isla de Mona (18°04′N–67°52′W), located 68 km west from the main island with an area of 55 km^2 (Frank et al. 1998). The Mona Passage, which is about 120 km wide, separates Puerto Rico from Hispaniola. The northern coastline is characterized by a very narrow shelf with high energy and sandy beaches. There are other smaller islets surrounding Puerto Rico: Monito and Desecheo on the western side, and Culebrita and Palomino on the eastern side. The coastline of the main island is 580 km. However, if we add the adjacent islands of Vieques and Culebra, the total shoreline is around 700 km.

14.1.1 Coastal Morphology

Sediment discharge from rivers is greater along the northern and western coasts, than the eastern and southern coasts. This is due to the higher annual precipitation levels in the northern and western parts of the island. Abundant sand dunes characterize the northern and northwestern coast. Parts of these are submerged while others have fossilized and formed rocky outcrops and beaches (García-Sais et al. 2005). The eastern coast has sandy deposits with scattered rock formations. Submarine canyons (Morelock et al. 1977) and mangrove islets characterize the southern coast. In Culebra, mangroves grow along the southern while sand beaches dominate the northern side. However, Vieques has an opposite pattern. This is the result of differences in wave action. The northern coast is exposed to more wave action resulting from northeast trade winds and North Atlantic winter storms (Ruiz-Ramos et al. 2011).

14.1.2 Shelf Morphology

Puerto Rico is crossed by two important geological troughs, the Muertos Trough in the south with a depth of 5,000 m and the Puerto Rico Trench, the deepest submarine trough (>8,400 m) in the Atlantic Ocean, located approximately 55 km off the north coast. The ocean depth, shelf, and slope morphology varies greatly in Puerto Rico (Morelock et al. 1977). The shelf off the northern and northwestern coasts is narrow (3 km wide, Fig. 14.1) and is subjected to strong wave action during the winter season (December–March). Along the southern coast, the shelf-edge drops off at a depth of about 200 m with an almost vertical and abrupt steep slope in some areas (García-Sais et al. 2005). Also along the southern coast is a flat platform ($\sim$10 km wide) at a depth of about 30 m located between Salinas and Ponce municipalities. The eastern coast is broad compared to the rest of the island and has a wider shelf (Fig. 14.1). This area is partially protected by a cordillera aligned between Culebra and Fajardo, the Cordillera Reefs System (CRS). The southern coast is usually characterized by lower wave energy. However tropical storms and southern trade winds can affect this area. In the west, the Mona Passage has an average depth of 400 m, but some areas are more than 1,000 m deep. Puerto Rico has 3,370 km^2 of fringing coral reefs surrounding the main island and also Culebra, Vieques, Mona, Monito, and Desecheo (García-Sais et al. 2005).

14.1.3 Oceanographic Conditions

Coastal currents around Puerto Rico are mainly driven by tides and wind. The tide along the north and western coast is semidiurnal and diurnal along the south coast (Morelock et al. 1977). Wind driven currents are dominated by the easterly trade winds, which push surface waters towards the west. Near bottom waters tend to reverse the flow during times of weak easterly winds.

The main objective of this chapter is to describe the species diversity and distribution of echinoderms around Puerto Rico. We also want to present a comprehensive historical account of echinoderm research carried out in Puerto Rico. Lastly, recommendations are given for future ecological echinoderm research and management strategies that seem fit for the conservation of echinoderms in Puerto Rico.

14.2 Echinoderm Research

14.2.1 Taxonomic and Inventories Studies

The study of echinoderms in the Caribbean was initiated at the end of nineteenth and beginning of twentieth centuries, as part of the expeditions HMS *Challenger*, RV *Blake* and US Fish Commission RV *Albatross*. During these expeditions many marine organisms many marine organisms from shallow and deep waters were collected in the Caribbean, including Puerto Rico, Jamaica, Cuba, Bermuda, Tobago, Cayman Island, Virgin Islands, Curaçao, Aruba, Bonaire, Venezuela, and the coast of French Guyana. The main contributions have been the works of Clark (1898, 1901a, b, 1919, 1933, 1941, 1942), (1939), Engel (1939), Cherbonier (1959) and Engel and Aberson (1960). Not only taxonomic characteristics, but also the morphological and ecological aspects of echinoderms were studied. Clark and Downey (1992) was an important contribution to the study of sea stars in the Caribbean Sea, including the Lesser and Major Antilles with the citation of 23 species. The most recent work is by Hendler et al. (1995), who produced an identification guide and life history of 150 echinoderm species from the shallow waters (0–33 m deep) of the Florida Keys, Bahamas and the Caribbean.

The first major report of marine biodiversity of Puerto Rico was done by Clark (1901a), which was based primarily on the extensive collections made by the U.S. Fish Commission steamer *Fish Hawk* in January and February 1899 and by Mr. George R Gray 1900 collection near San Juan. It listed 87 species: three crinoid species, 11 asteroid species, 49 ophiuroid species, 13 echinoid species and 11 holothurian species. This survey produced the first systematic list of marine invertebrates and fish for the island.

The collections of echinoderms made by the New York Academy of Sciences in 1914 were reexamined in a new report (Clark 1921). The revised report contained around 550 specimens. Eleven new species were identified in Puerto Rico. The list was completed and reviewed three decades later by the same author (Clark 1933).

Several other studies on echinoderms have been done in Puerto Rico. For example, Carrera (1974) studied the Class Ophiuroidea (particularly the Amphiuridae family) at La Parguera and found 17 new records for the island. Wolf (1978) studied aspects of the biology of *Astrophyton muricatum* (Lamarck) at Enrique, La Parguera; González-Liboy (1971) emphasized the echinoderms in his study of the macro-invertebrates at the Maní beach, Mayagüez et al. (1976) focused on the role played by the sea urchin *Diadema antillarum* (Philippi) on the composition of corals and sea grass beds at Jobos Bay.

Recently, Benavides-Serrato (2006), based on a review of the literature, museums revision and collections, made the most complete and current taxonomic list of the shallow water (depth from 0 to 100 m) echinoderms from Puerto Rico. One hundred and eight species were observed. Nine species, *Davidaster discoidea* (Carpenter), *Poraniella echinulata* (Perrier), *Ophioderma squamosissimum*

(Lütken), *Ophioderma phoenium* (Clark), *Ophionephthys limicola* (Lütken), *Plagiobrissus grandis* (Gmelin), *Actinopyga agassizi* (Selenka), *Holothuria (Halodeima) floridana* (Pourtalés), and *Synaptula hydriformis* (Lesueur) were considered as first records for Puerto Rico.

14.2.2 Marine Science Collection of Department of Marine Sciences (UPR Mayagüez)

The echinoderm collection of marine invertebrates of the Marine Science Department (MSC, Mayagüez Campus, University of Puerto Rico) is located at Isla Magueyes (La Parguera), and is considered the most complete and important one for Puerto Rico. The invertebrates collection has approximately 4,000 specimens (Weil 2005) belonging to different taxonomic groups, including the Phyla Porifera, Cnidaria, Mollusca, Nemertea, Echinodermata, Arthropoda and Annelida (Class Polychaeta). Echinoderm collection is composed of approximately 74 species, belonging to all five classes.

14.2.3 Reproduction, Larval Biology, Settlement, and Recruitment

Free-spawning is a mechanism that organisms utilize to disperse their genes and increase their population success. Egg fertilization of a free-spawning individual is difficult because of the rapid diffusion and limited life span of gametes in a dynamic ocean (Levitan 1995). The fertilization of the egg is critical for population survival and there are many mechanisms that organisms use to increase fertilization success. Some of these tactics are simultaneous spawning, aggregation, and large population size (Levitan 1995). One way to measure fecundity is to measure the gonad index, which is calculated as gonad weight divided by total weight times 100 (Giese and Pearse 1974). The drop in gonad index can be interpreted as spawning events (Moore et al. 1963). Measurements of fecundity can provide a first estimate of adult contributions to larval populations (Cameron 1986).

Echinoids have a complex life-history, which involves external fertilization of pelagic eggs, planktonic larval development and recruitment into a benthic habitat where it reaches an adult stage and reproduces (Smith 1997). The distribution of larvae is influenced by the dynamics of water motion in the coastal ocean in relation to the location of the spawning sites (Williams and García Sais 2010). Motility depends both on internal, physiological processes and on the physical properties of the fluid environment. The physical properties of water (e.g. density and dynamics viscosity) have a predominant influence on an organism's motion and are considered to be a major force in the evolution of their various modes of locomotion and feeding (Podolsky and Emlet 1993). Questions about larval transport focus largely on: (1) where they go and (2) where they came from (Levin 2006). Assessments of temporal and spatial abundance patterns of pelagic larval stages are relevant for understanding dispersal strategies, spatial connectivity and recruitment dynamics (Williams and García Sais 2010). Larval abundance measurements is a challenging undertaking for echinoderms due to their small size, relatively long pelagic larval life, potential for patchy distributions and the dynamic environment in which they live in (Williams et al. 2009). Larval community characterizations are typically incomplete and may underestimate abundance, especially when using large mesh diameters such as $\geq$202 μm for plankton tows (Williams and García Sais 2010). There are several studies of coastal zooplankton communities of Caribbean systems (Reeve 1970; Youngbluth 1980; McWilliam et al. 1981; Yoshioka et al. 1985) however, these studies do not report on echinoderm larvae from plankton sampling.

The colonization of benthic habitats by marine organisms with planktonic life-stages involves three phases: planktonic development and dispersal, habitat selection and settlement (Keough and Downes 1982). Settlement rate is defined as the rate at which planktonic larvae of benthic organisms establish permanent contact with the substrate. Trying to identify sources and/or sinks of larvae is important to understanding supply to populations (Miller and Emlet 1997). Marine invertebrate larvae may navigate to suitable settlement sites by responding to gradients of environmental stimuli (Kingsford et al. 2002). Recruitment, which is the fourth phase, includes the survival of the organism until observable by an observer (Keough and Downes 1982). Recruitment is a key mechanism shaping and structuring reefal communities (Cameron and Schroeter 1980). Recruitment rather than reproductive output is the key concept for marine species demography and is one of the main determinants of spatial and temporal variations in population sizes (Cowen et al. 2000). Larval availability, substrate selection and juvenile mortality are factors that influence recruitment to a population. There have been only a few larval and recruitment studies of echinoderms in Puerto Rico. Most of these studies focused on Echinoidea.

Studies on the Reproduction, Larval Biology, Settlement, and Recruitment of *Lytechinus variegates* (Lamarck), *Echinometra viridis* (Agassiz), *Echinometra lucunter* (Linnaeus), and *Tripneustes ventricosus* (Lamarck)

Cameron (1983, 1986) measured the reproduction, larval abundance, and size distribution for *Lytechinus variegatus*, *Echinometra viridis*, *E. lucunter* and *Tripneustes ventricosus* in La Parguera, Puerto Rico (Fig. 14.2). The mean gonad index of *L. variegatus* ranged from 0.8 in October 1981 to 7.8 in October 1983. The higher gonad indices occurred during the late summer and early fall months for *L. variegatus* and *Echinometra* spp. This is when the sea surface temperatures were at their highest. Spawning occurred eight times ranging over all seasons of the year for *L. variegatus*. The gonad indices for *Echinometra* spp. reached a peak in August 1983 (10.0). The degree of synchrony was much greater for

Fig. 14.2 Photographs of **a** *Lytechinus variegatus.* **b** *Echinometra viridis.* **c** *Echinometra lucunter.* **d** *Tripneustes ventricosus* (Photos by Carlos Muñoz)

Echinometra spp. than for *L. variegatus*. The reproductive period of the white-spined urchin, *T. ventricosus* in not very synchronous like the other three species and also contrasts by a peak in the coolest time of the year. *Tripneustes ventricosus* displayed gonad indices between 1.2 (July 1982 and June 1983) and 4.4 (May 1982) (Cameron 1986).

All four of these species displayed similar reproductive strategies; they all broadcast gametes which develop into feeding pelagic larvae. Cameron (1986) found that at 23–25 °C *L. variegatus*, *E. lucunter* and *E. viridis* achieve metamorphosis in about two weeks and *T. ventricosus* in about four weeks (Lewis 1958). The seasonal larval occurrence of *L. variegatus* and *Echinometra* spp. coincides with their differences in synchrony. *Lytechinus variegatus* plutei were present in all samples from the spring to fall 1982 and 1983. Plutei of *Echinometra* spp. was only present during the fall reproductive peak. Larval abundances of *L. variegatus*, *E. viridis*, and *E. lucunter* co-varied with the reproductive period. *Tripneustes ventricosus* plutei were only found during August and September 1982. The larval period of *T. ventricosus* is twice as long as the other three species. Summer currents may carry larvae of *L. variegatus* offshore followed by a later return of mature larvae. Larvae of the other three species may be present in the water column at a time when retention is favorable.

Cameron (1986) found *L. variegatus* juveniles (<15 mm in test diameter) at Tres Marias in all four 1982 samples, April 1983 and June 1984. At Caracoles juveniles were present during April and August 1982, April 1983 and June 1984. Settlement of *L. variegatus* occurred throughout the year. The recruitment pattern of this species follows the expected and observed pattern of reproduction and larval occurrence. Juveniles (20 mm in diameter) of *T. ventricosus* found in January or February may have settled the previous September. They most likely settle on the reef face and crest. The larvae of *T. ventricosus* were observed during August 1982, a time that coincides with the expected fall recruitment. Seasonal recruitment was not apparent for *E. lucunter* at Papayo. There were *E. viridis* recruits several times during the year.

Cameron et al. (1985) measured the embryonic response to temperature and salinity by four species of Caribbean sea urchins. They wanted to test if embryos of highly synchronous species of echinoderms are more sensitive to physical factors. They cultured embryos of four species of sea urchins, *L. variegatus, E. lucunter, E. viridis,* and *T. ventricosus* in a 3 × 3 factorial experiment comparing temperatures of 20, 25, and 30 °C and salinities of 31.5, 35, and 38.5 % at La Parguera. The responses to different temperatures and salinities differed among the embryos of the four species. The asynchronous species, *L. variegatus* and *T. ventricosus* exhibit a more gradual response to temperature at cleavage than the synchronous *Echinometra* spp. *Lytechinus variegatus* developed more rapidly at higher salinities and *T. ventricosus* at lower salinities. *Echinometra* spp. are less affected by salinity changes and have a linear response to an increase in temperature. These results support the suggestion that species with a narrower reproductive season have greater responses in development rate of extreme environmental conditions. Salinity proved to be less harmful at all development stages than temperature. Significant mortality occurred during the blastula and early gastrula stages. The development rate of all four species is temperature dependent. At 20 °C the hatching interval was the shortest for *L. variegatus* and longest for *T. ventricosus*; *Echinometra* spp. had an intermediate response. At 25 °C all four species hatched at about the same time, 7.5 h. *Tripneustes ventricosus* had 100 % mortality at extreme temperatures. All four species reached the pluteus stage (18–24 h) at optimum temperature. Development of *L. variegatus* was fastest (18.3 h), while *L. variegatus* was slowest (28.6 h). Most synchronous species (*Echinometra* spp.) were the most responsive to temperature changes.

14.2.4 Reproduction, Larval Biology, and Settlement of *Diadema antillarum*

Lugo (2004) described the population structure and reproductive biology of *D. antillarum* in La Parguera. The average gonad index per month was used to characterize the annual reproductive cycle. The gonad index ranged from 2.9

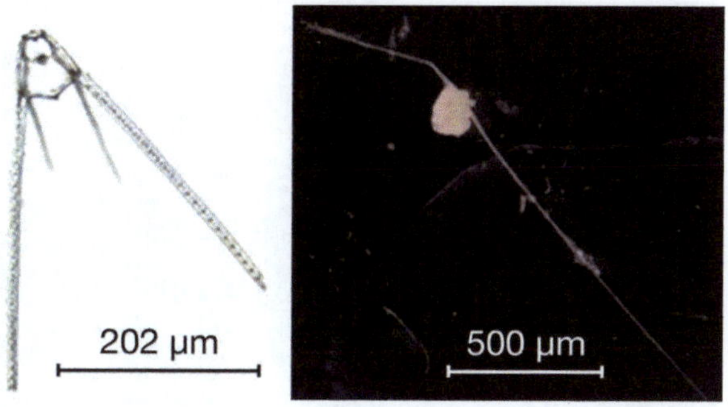

Fig. 14.3 Early (4–8 day-old) and late (~24 day-old) larval stages of *Diadema antillarum*

(June–July 2001) to 11.2 (December 2001). High gonad indices were observed during the late fall (9.5) to winter during times when sea surface temperatures were at its lowest values (26 °C). Lugo (2004) estimated that *D. antillarum* spawn during early spring and early summer. He also found a significant negative relationship between monthly mean gonad indices and sea surface temperature. The smallest sexually mature urchin was 3.8 cm in diameter. The male:female ratio was 0.89:1 during his study.

Williams et al. (2009) measured the temporal variations of early larval stages in neritic waters of La Parguera. They also measured the water current velocities and directions of inshore areas of La Parguera. During April 2005, six drogues were deployed. They also deployed an Acoustic Doppler Current Profiler (ADCP) during April 2005. Only early larval stages (4–8 day-old, Fig. 14.3) of *D. antillarum* were collected in the study. However, Williams and García Sais (2010) observed older larval stages (~24 day old) (Fig. 14.3) of *D. antillarum* at the shelf-edge and oceanic stations.

Early larval stages were collected in 13 of the 16 monthly samples, which indicates that *D. antillarum* spawn throughout the year. Mean larval abundances (ind 100 m^{-3} ± SD) ranged from no individuals in September and October of 2005 and January 2006 to 1,239 ± 309 in April and 1,078 ± 648 in July. There were peaks in larval abundances in April and July 2005. Echinoplutei were patchy within sample replicates. The inconsistent peaks of abundance between years and months may be a result of patchy distribution of these larvae. Williams et al. (2009) did not find a significant correlation between monthly larval abundance and SST. Drogue trajectories and the ADCP time series indicated that the circulation of coastal waters at La Parguera is towards the west-northwest. The flow in La Parguera seems to be influenced by the boundary layers of the coastline and coral reefs. Early dispersal of *D. antillarum* larvae during the first two days is most likely within the insular shelf of La Parguera. It is uncertain if the older larvae continue offshore towards the Mona Passage or follow the shelf contour towards the north along the westcoast of Puerto Rico. The recovering populations of *D. antillarum* in La Parguera may be an important source of larval replenishment for downstream islands.

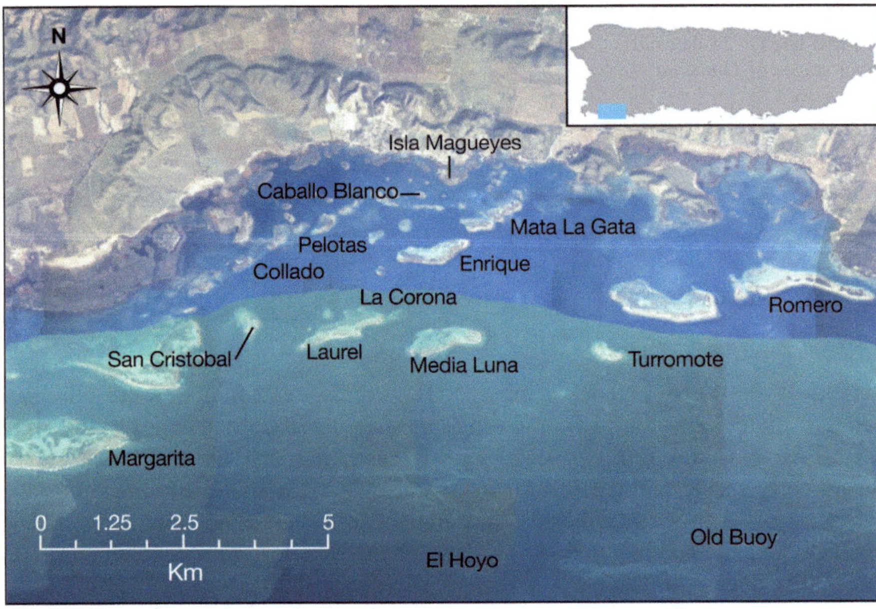

Fig. 14.4 Aerial photograph of La Parguera, Puerto Rico. Major reef systems and Isla Magueyes Marine Station have been identified

Larval settlement of *D. antillarum* was assessed in three separate surveys (Miller et al. 2009; Williams et al. 2010, 2011) in La Parguera, Puerto Rico. Miller et al. (2009) assessed the rates of *D. antillarum* larval settlement at two sites (Enrique and The Buoy) in La Parguera (Fig. 14.4). Settlement was quantified from August 2005 to July 2006 with experimental settlement plates modeled on Bak's (1985) study. During each month plates were examined in the field and all *D. antillarum* settlers were recorded. Williams et al. (2010, 2011) determined the settlement pattern of *D. antillarum* in La Parguera during two separate studies. In 2010, *D. antillarum* settlement was examined at three reefs: an inshore reef (La Pelotas), a mid-shelf reef (Media Luna), and a shelf-edge reef (Old Buoy, "The Buoy") for one year. The study of Williams et al. (2011) included six reefs in La Parguera: Pelotas and Enrique (inner shelf reefs), San Cristobal and Media Luna (mid-shelf reefs), El Hoyo, and Old Buoy (shelf-edge reefs). Settlement was measured for 6 months (Fig. 14.4). Artificial turf was used as the recruitment plates for both studies (Williams et al. 2010, 2011). Plates were inspected for *D. antillarum* recruits in the laboratory.

Mean test size of settlers (mean $\pm$ SD) collected in the studies of Williams et al. (2010, 2011) was 0.90 ± 0.50 mm (Fig. 14.5). The maximum monthly settlement of *D. antillarum* in all studies was ~ 16 ind m^{-2} (Miller et al. 2009), 1,064 ind m^{-2} (Williams et al. 2010), and 265 ind m^{-2} (Williams et al. 2011). The temporal distribution of settlement by *D. antillarum* in the 2011 study was similar to previous observations made by Williams et al. (2010). *Diadema antillarum*

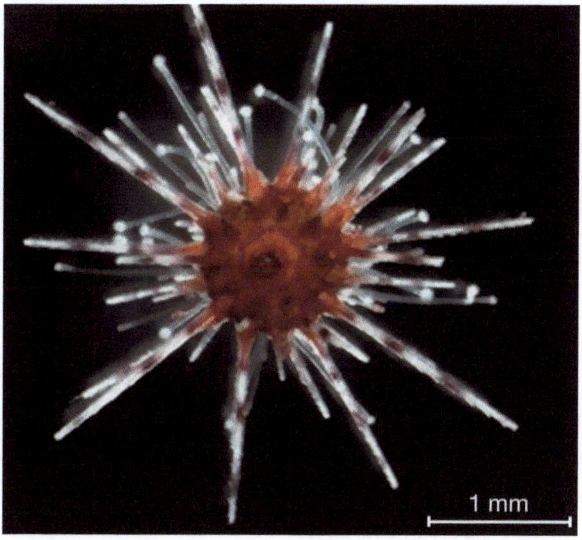

Fig. 14.5 *Diadema antillarum* settler

settlement mainly occurred during the summer (July–September). In 2010, settlers of *D. antillarum* were collected during 12 of the 13 sampled months, confirming year-round reproduction in La Parguera (Lugo 2004; Williams et al. 2009, 2010). Spatial variations of *D. antillarum* settlement were evident. There were significantly more recruits collected at the shelf-edge than at the inshore reefs (Miller et al. 2009; Williams et al. 2010, 2011). Williams et al. (2010) suggested three non-mutually exclusive explanations for the low settlement at the inshore reefs: (1) stochastic variation, (2) a shelf-edge reef larval 'filtering effect' (Gaines et al. 1985), and (3) high abundance of fleshy macroalgae colonizing the artificial turf plates precluded settlement at inshore reefs. However, it was proven that fouling of the substrate was not a major factor precluding the settlement of *D. antillarum* at the inshore sites in La Parguera (Williams et al. 2011). Settler mortality and/or low larval supply were two other possible explanations for the low settlement at inshore sites in La Parguera. Settlement in all studies was concentrated at the shelf-edge where adult *D. antillarum* are rare. Therefore, *D. antillarum* recruitment in La Parguera may be independent of local adult density (Lessios 1988). Post-settlement mortality on *D. antillarum* at the shelf-edge may be high, which explains the absence of adults in this area.

14.2.5 Bioerosion

Sea urchins are considered a major bioeroder on coral reef communities. Griffin et al. (2003) studied the bioerosion of *Echinometra viridis* on coral reef communities in La Parguera, Puerto Rico. The mean size and densities of *E. viridis* were determined at two patch reefs: Palmas and Mario. The daily rate of bioerosion

Fig. 14.6 *Holothuria glaberrima* specimen

(amount of $CaCO_3$ removed) was estimated and measured in fecal pellets produced over 24 h by urchins from depths of 1–3 m. The size ranges of the urchins varied between 2 and 2.5 cm in diameter. They repeated four different treatments 11 times during March and April 1998 with different sea urchins. Substrate (pieces of dead coral) was the same for each treatment. In the different treatments they estimated (1) endolithic and external $CaCO_3$ sediment production by urchins, (2) endolithic bioerosion only, (3) the amount of particulate $CaCO_3$ the sea urchins had in their gut, and (4) control for external sedimentation. Fecal pellets were rinsed with fresh water, dried for 24 h at 70 °C, and then burned in a furnace at 555 °C to eliminate organic matter, and finally at 1,000 °C to determine calcium carbonate ($CaCO_3$). They found that bioerosion rates of *E. viridis* were higher in the shallow areas (1–3 m), areas of high densities. Average densities of *E. viridis* ranged from 0.77 to 62 ind m^{-2} and size ranged from 2.01 to 2.44 cm at both reefs. They found the average daily individual bioerosion rate to be 0.18 g day^{-1} and the mean annual bioerosion rates between 1.14 to 4.14 kg m^{-2} $year^{-1}$. Bioerosion rates measured in this study were lower than that reported for many other species of sea urchins, for example *D. antillarum*, endolithic sponges and parrotfishes. However, they found that bioerosion activity of *E. viridis* had an important effect on the morphology on the reefs and that the high densities of *E. viridis* at Palmas might account for the degradation of coral at this site.

14.2.6 Physiology

Holothuria (Selenkothuria) glaberrima (Selenka) as a Novel Model System for Cellular/Molecular Studies

The laboratory of Dr. José E. García-Arrarás at the University of Puerto Rico has made use of a local echinoderm species, the sea cucumber *Holothuria glaberrima*, as a novel model system to probe the phenomenon of tissue and organ regeneration. *Holothuria glaberrima* (Fig. 14.6) is a species common in costal waters that can be

Fig. 14.7 *Holothuria glaberrima* specimen **a** before and **b** after evisceration

found in rocky shores with high wave action and is usually in close associations with the sea urchin *Arbacia punctulata* (Lamarck). Although the original goal of the laboratory focused on the regeneration of the digestive tract, spin-off research from the laboratory has produced important information on holothurian anatomy, neurobiology, immunology, evolution and genomics among other research areas. Here we will summarize some of the findings that have resulted from this research.

In summary, the studies in *H. glaberrima* have provided information of interest to many research areas other than the field of regenerative biology. Ongoing studies of the transcriptome of nervous and intestinal systems are expected to increase the interest of other colleagues in related fields to expand the number of investigations using not only this species but also other holothurians.

Evisceration

Holothurians are well known to eviscerate their internal organs when submitted to different types of stress. Evisceration in the laboratory is induced by injections of KCl, although it can occur spontaneously in foul water or low oxygen conditions (Fig. 14.7). *Holothuria glaberrima* provides a classic example of this phenomenon by eliminating most of the visceral organs through the cloaca (García-Arrarás et al. 1998). The eviscerated organs include the intestinal system, which breaks at the esophageal-small intestine junction and at the large intestine-cloaca junction. Other organs associated with the intestine, including the hemal system and the right respiratory tree, are also eliminated. Similarly, most of the gonads (if the animal is in the reproductive season) are ejected but the gonadal base remains. The same process has been documented in another local species, *H.* (*Halodeima*) *mexicana* (Ludwig) (unpublished results).

Anatomy

Research into the regeneration process has provided anatomical descriptions of several structures, tissues and organs. The most extensively studied has been the digestive tract, which includes a short esophagus, a long descending and ascending

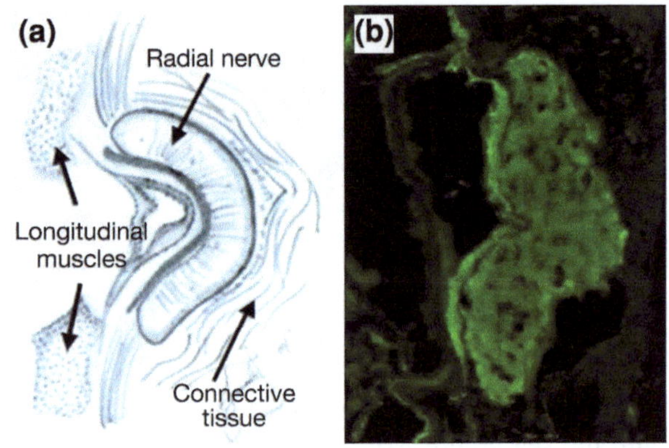

Fig. 14.8 **a** Diagram of a transverse section of one of the five radial nerves found within the sea cucumber body wall.
b Immunohistochemistry of a radial nerve using one of the nervous tissue markers developed in the laboratory (see Díaz-Balzac et al. 2007)

small intestine and a descending large intestine that connects to the cloaca (Díaz-Miranda et al. 1995, García-Arrarás et al. 1998). In addition, many of the investigations provide important information on the neuroanatomy of sea cucumbers, including a description of the nervous system in the digestive tract and tube feet and the discovery of new nerve plexi within the connective tissue, (García-Arrarás et al. 2001; Díaz-Balzac et al. 2007, 2010a, b) thus making *H. glaberrima* a much needed model for understanding neural connections in holothurians.

Neurobiology

Studies on *H. glaberrima* have provided important information on the nervous system. The main focus of these studies has been the development and use of neural markers that serve to identify and characterize components of the nervous system (Fig. 14.8). They have been extensively used in neuroanatomical studies as described above. Many of the markers are antibodies that recognize epitopes in cells of the nervous system. Some of these markers are against vertebrate and invertebrate molecules such as acetylated a-tubulin (García-Arrarás and Viruet 1993), neuropeptides (García-Arrarás et al. 1991a, b; Díaz-Miranda et al. 1995, 1996) or yet unknown proteins (García-Arrarás et al. 2001; Díaz-Balzac et al. 2007; Mashanov et al. 2010b). The work on the nervous system also permitted the characterization of the first holothurian neuropeptide (Díaz-Miranda et al. Díaz-Miranda et al. 1992) and studies on its possible function in the intestinal system (Díaz-Miranda and García-Arrarás 1995). More recently, markers for holothurian glia have been developed and used to characterize this little studied component of the echinoderm nervous system (Mashanov et al. 2009, 2010a).

Intestinal Regeneration

The formation of a new intestine in *H. glaberrima* following evisceration has been the main subject of a series of publications from the García-Arrarás group, including several reviews (García-Arrarás and Greenberg 2010; García-Arrarás and Dolmatov 2010; Mashanov and García-Arrarás 2011). The initial description of the process was published over a decade ago (García-Arrarás et al. 1998). Many of the subsequent articles describe cellular events that take place during the process, including extracellular remodelling (Quiñones et al. 2002), cell migration (Cabrera-Serrano and García-Arrarás 2004; García-Arrarás et al. 2006), myogenesis (Murray and García-Arrarás 2004), cell dedifferentiation (Candelaria et al. 2006), cell division (García-Arrarás et al. 1998), apoptosis (Mashanov et al. 2010a, b) and epithelial to mesenchymal transitions (García-Arrarás et al. 2011). Many of the cellular events are now well understood and have been integrated in a working model (see Mashanov and García-Arrarás 2011, García-Arrarás et al. 2011). In brief, following evisceration cells of the mesentery mesothelium, particularly myocytes, dedifferentiate and migrate as a cell layer toward the free end of the mesentery, where the intestine was previously attached, prior to evisceration. As cells migrate, some undergo proliferation while others ingress into the connective tissue and provide cells for the formation of the intestinal rudiment. This gives rise, initially, to a solid rod of tissue that extends from the oesophagus to the cloaca. Luminal cells from the oesophagus and cloaca migrate into this tissue rod forming the luminal layer and thus, the lumen of the intestine. Once the initial rudiment is formed, cells from the mesothelium differentiate, giving rise to the muscle cells (Murray and García-Arrarás 2004) and possibly to neurons. A comparison to other regenerative processes, such as wound healing of the body wall, has shown that the same cellular events are common to both (San Miguel-Ruiz and García-Arrarás 2007).

In recent years, experiments have aimed at understanding the molecular basis of intestinal regeneration. Many of these experiments are summarized below. The experimental data indicate a large number of genes are associated with intestinal regeneration; over 70 % of the studied sequences showed some type of differential expression between regenerating and normal intestines (Ortiz-Pineda et al. 2009).

Nervous System Regeneration

One of the main research interests of the García-Arrarás laboratory has been the regeneration of the nervous system. The main reason is that this type of regeneration is highly limited in vertebrates and most other chordates. Nonetheless, echinoderms with their amazing regenerative potential should provide excellent models to study nervous system regeneration. Many of the investigations carried out in the laboratory centered on this topic. Thus, neuronal markers were developed and the neuroanatomy of the holothurians was described. Even early experiments on intestinal regeneration (García-Arrarás et al. 1998) were initiated with an interest in the enteric nervous system, that is, the nervous system of the digestive tract. Results have shown that *H. glaberrima* can regenerate its enteric nervous system and that this regeneration occurs concomitantly with the regeneration of the intestine itself (García-Arrarás et al. 1999). Additional experiments were done to demonstrate that the central nervous system of holothurians, the so-called radial nerve cords, could also regenerate (San Miguel et al. 2009). Results showed that following transection, nerve fibers could re-grow and bridge the injury gap forming a new scar-free nerve cord in about a month time. A process of neurogenesis was also documented that produced new neurons within the regenerated cord. Ongoing experiments are aimed at characterizing the genes associated with this regeneration process.

Genomics

The laboratory began to study the molecular basis of intestinal regeneration in the late 1990s. Initially the work was done on a gene-by-gene strategy, where the selected genes were either candidate genes, such as Hox genes (Méndez et al. 2000) or genes identified by differential display techniques comparing regenerating and normal intestines (Roig-López et al. 2001). In this way, several genes of interest were characterized in the holothurian and were shown to be overexpressed during intestinal regeneration. Among these were serum amyloid A (Santiago et al. 2000) and ependymin (Suarez-Castillo et al. 2004).

With the advent of high throughput molecular analyses, the number of genes associated with intestinal regeneration was multiplied. First, three cDNA libraries were prepared from normal and regenerating intestines and over 5,000 EST (Expressed Sequence Tags) were identified (Rojas-Cartagena et al. 2007). These ESTs represent sequences of mRNAs (and therefore gene sequences) that were expressed in the tissues at the time of dissection. This was the first attempt to obtain large genomic information from a holothurian. These ESTs sequences were then probed in a custom-made microarray in order to determine which genes were differentially expressed during the regeneration process (Ortiz-Pineda et al. 2009). As mentioned earlier, a large number of genes ($\sim$73 %) were found to be differentially expressed at $P < 0.05$ probability. At more stringent conditions ($P < .001$) the percentage was still highly significant ($\sim$40 %). These studies have identified holothurian genes possibly associated with wound healing, cell proliferation, differentiation, morphological plasticity, cell survival, stress response, immune challenge, and neoplastic transformation. Among those that have been described are cytoskeletal genes, such as actins, and developmental genes, such as Wnt and Hox genes that show interesting expression profiles during regeneration.

More recently large-scale 454 pyrosequencing techniques have been applied to study regenerative processes and have identified over 2,000,000 ESTs that cluster into $\sim$80,000 contigs (Mashanov et al. in prep.). The tissue used for these analyses was the radial nerve cord of normal and regenerating animals following the transection of their cord. Thus, when the data are fully analyzed, the expectation is to obtain a large percent of the transcriptome of normal and regeneration radial nerve cords.

It is important to emphasize that the sea cucumber *H. glaberrima* has, thus, become the holothurian with the largest number of genes described, and a reference for the genomic analysis of other members of its class.

Immune System

One of the research areas that has benefited indirectly from these regeneration studies is the field of echinoderm immunology. The reason is that regeneration follows the loss of tissues and this loss is usually associated with some traumatic event causing exposure of the animal's internal environment to the outside media. Thus, the immune system is activated to deal with the possible invasion of pathogens. For example, in the case of evisceration, the cloaca ruptures and as the internal viscera are ejected, the coelomic cavity is exposed to the sea water.

The outcome of these events is that activation of the regenerative process could be occurring along with activation of the immune system. Cells and genes that might be associated with regeneration might in reality be involved in the immune response or vice versa. Alternatively the possibility exists that the two events are somehow coupled and that immune associated cells and molecules are also participating in the regenerative process.

Lipopolysaccharide (LPS), a component of the bacterial cell wall, has been used to study if genes associated with regenerative processes are also associated with immune activation. LPS causes a coelomocyte-mediated immune response in *H. glaberrima*, inducing an approximately threefold increase in coelomocyte phagocytic activity (Santiago-Cardona et al. 2003). A large number of immune-related genes were identified from the EST database (Ramírez-Gómez et al. 2008) and many of these were shown to be associated with LPS-induced activation (Ramírez-Gómez et al. 2009).

These studies have prompted a further in depth study into how different pathogens might be able to activate different components of the immune system, in particular different sub-populations of coelomocytes (Ramírez-Gómez et al. 2010). In summary, *H. glaberrima* has also served to study the immune responses in holothurians and has provided important information on fundamental aspects of the cellular and molecular immune response (Ramírez-Gómez and García-Arrarás 2010).

Molecular Evolution

Echinoderms are a key group evolutionary group of animals. They are invertebrate deuterostomes closely related to chordates. As such, they can provide key information on the evolution of genes and their associated roles. Experiments with *H. glaberrima* have already documented the evolution of particular genes. Some of these data can be found associated with genomic studies (Rojas-Cartagena et al. 2007; Ramírez-Gómez et al. 2008; Ortiz-Pineda et al. 2009). Other genes have

Table 14.1 Number of species per classes in Puerto Rico in different kind of ecosystems

Class	# species	Coral reefs	Seagrass beds	Mangroves	Rocky shores	Sandy beaches
Crinoidea	14	11	–	–	–	–
Asteroidea	25	6	2	2	5	12
Ophiuroidea	49	30	20	13	5	10
Echinoidea	16	5	6	1	7	12
Holothuroidea	29	10	18	4	12	17
Total	133	62	46	20	29	51

been studied more in depth and their evolutionary associations have been well-described. This is the case for Hox genes (Méndez et al. 2000, serum amyloid A protein (Santiago et al. 2000) and ependymin (Suarez-Castillo et al. 2007). The latter is an excellent example of how studies in holothurians have served to establish evolutionary patterns. In this case, prior to our study, ependymins were thought to be a vertebrate-specific family. Its characterization in *H. glaberrima* showed it was present in various invertebrate groups.

14.3 Distribution and Diversity of Echinoderms

Based on Hendler et al. (1995), Benavides-Serrato (2006), Pawson et al. (2009) and the database of the National Museum of Natural History (NMNH) in Puerto Rico island, 133 species of echinoderms (Table 14.1) are distributed along different kinds of ecosystems through the island.

According to Weil (2005), Puerto Rico harbors a large diversity of habitats along the coast and on the insular platform. Estimates from recent mapping of the coral reef ecosystems and associated communities and habitats along Puerto Rico by NOAA, indicated a total area of 5,010 km^2. Of these, 757 km^2 were covered by coral reefs and colonized hard bottom habitats (15.1 % of the total area), 625 km^2 by seagrass beds (12.8 %), 97 km^2 by hard bottoms dominated by macroalgae (1.9 %) and 73 km^2 by mangrove forests (1.4 %).

According to (García-Arrarás et al. 1998), the insular shelf of La Parguera, on the southwest coast of Puerto Rico, is characterized by an extensive development of coral reefs, seagrass beds and mangrove forest. The dry, warm and relative stable climate, low wave energy, high water transparency, relatively wide shelf and oligotrophic offshore waters, are some of the factors that contribute to the conditions of the marine ecosystem of La Parguera. Interactions among coral reef, seagrass beds, and mangrove communities provide for a highly productive, structurally complex, and biologically diverse ecosystem. Mean surface salinity is 35.2 %, but varies from 34 to 37 %. Mean temperature is 28.4 °C, varying from 25 °C to 31 °C.

Coral reefs in La Parguera are distributed in three parallel lines: (1) inshore, (2) mid-shelf and (3) outer shelf (Almy and Carrión-Torres 1963, Acevedo and Morelock 1988). The length and the wide range of the reef's depths produce physical, chemical and biological gradients providing a highly productive, structurally complex, and biologically diverse ecosystem (Morelock et al. 1977, Acevedo and Morelock 1988). The most common mangrove species is *Rizophora mangle* (L.) *Laguncularia racemosa* (L. Gaertn. f.) is less common. Extensive sea grass beds are well established in La Parguera. *Thalassia testudinum* (Banks ex König) and *Syringodium filiforme* (Kützing) are the most common and widely distributed species. Also, most extensive sea grass beds are found at a depth of two meter around mangrove islands (García et al. 2003).

With the goal to characterize the echinoderms fauna in two major communities in La Parguera, Benavides-Serrato (2006) collected and described echinoderms associated with coral reefs and seagrass communities during several field trips (February of 2005 to January of 2006) at eleven different localities (Fig. 14.4).

As a result of this study, 32 species were found at La Parguera. Thirty-one species of echinoderms were identified from sampling stations from the seagrass beds. The most abundant classes were Echinoidea (mean abundance varied between 0.22 ind m^{-2} at Caballo Blanco to 5.89 ind m^{-2} at San Cristobal, represented by *L. variegatus, E. lucunter* and *E. viridis*) and Ophiuroidea (mean abundance varied from 0.31 ind m^{-2} at Caballo Blanco to 3.99 ind m^{-2} at San Cristobal, principally represented by *Ophiocoma echinata* (Lamarck), *Ophiactis savignyi* (Müller and Troschel), *Ophiothrix* (*Ophiothrix*) *angulata* (Say) and *Ophiothrix* (*Ophiothrix*) *orstedii* (Lütken).

Twenty-four species of echinoderms were identified from the coral reef community. Ophiuroidea had the highest number of species (14). The composition of the different classes at different depths showed that ophiuroid and echinoid species were widely distributed from depths of 0 to 20 m. Crinoids only appeared at depths greater than the 10 m. On the other hand, asteroids were absent from the shallow water habitats (0–5 m). The crinoids *Davidaster discoidea* and *D. rubiginosa* (Pourtalès) were only found at The Buoy at a depths of 20–30 m.

14.3.1 Asteroidea Distribution

A total of seven species of sea stars within the Order Valvatida have been documented in Puerto Rico. These include *Oreaster reticulatus* (Linnaeus) in the family Oreasteridae. This is probably one of the most significant and ubiquitous sea stars of seagrass communities in Puerto Rico. It has been commonly documented across the entire shelf on seagrass bottoms and sandy patches from nearly sea level to about 25 m depths, but is particularly abundant on shallow water (depth < 10 m) seagrass communities adjacent to Vieques, Culebra Island, La Cordillera, and Fajardo, off northeastern Puerto Rico. Populations have locally disappeared over the last three to four decades from many locations across the main island, probably as a combined impact of long-term habitat decline as a result of chronic water quality degradation. Also, this species was largely exploited in the past by artisans for souvenirs.

Ophidiaster guildingii (Gray), in the family Ophidiasteridae, is very uncommon and has been documented only from shallow water (depth of 2–5 m) fringing and patch reefs off Cabo Rojo and Culebra Island. This is a very cryptic species, usually hidden in reef crevices and overhangs. The common comet sea star, *Linckia guildingi* (Gray), is also a rare species due to its cryptic behavior, but has been documented from coral reefs and rocky reef habitats from Culebra Island, Humacao, Fajardo, Luquillo, Vega Baja, and Cabo Rojo, at depths ranging from about 5 to 20 m. Four species have been also documented in the family Astropectinidae, including *Astropecten articulatus* (Gray), *A. duplicatus* (Koehler), and *A. antillensis* (Lütken). These three species can be typically found buried beneath the surface of sandy bottoms at variable depths, from nearly sea level to more than 20 m. They can be occasionally to commonly observed on sandy and muddy bottoms off Humacao, Naguabo, Ceiba, Fajardo, La Cordillera, Vieques Island and Culebra Island, off eastern Puerto Rico. But *A. antillensis* and *A. articulatus* have also been observed off Patillas in the southeast, and off Luquillo, Río Grande, Carolina, San Juan, and Vega Alta, along the northern coast. In addition, *Astropecten marginatus* (Gray) was also found in high abundance in Luquillo Beach, Puerto Rico and shares the habitat with *Mellita lata* (Clark) (Kenk 1944). The geographic range of *A. marginatus* comprises Venezuela, Trinidad, Guiana, and Brazil. They are located as far south as 27°30′N.

There are also two species within order Spinulosida, family Echinasteridae. The conical spined sea star, *Echinaster sentus* (Say), has been sporadically observed from variable shallow water bottoms (dcpth < 5 m), including seagrass communities, fringing reefs, mixed subaquatic vegetation habitats (algae + seagrass), and rocky outcrops from Culebra Island, La Cordillera, and Fajardo. The thorny sea star, *Echinaster echinoporus* (Lamarck) was observed only once from Culebra Island by Hernández-Delgado on a shallow water (depth < 1 m) transitional rubble-seagrass-mangrove bottom.

14.3.2 Ophiuroidea Distribution

Six species belonging to four families in the order Ophiurida have been documented. These include *Sigsbeia conifera* (Koehler) (family Hemieuryalidae). The rose lace coral brittle star is a highly cryptic, very small and very rare brittle star species apparently only associated with rose lace coral, *Stylaster roseus* (Pallas). Hernandez-Delgado observed only one individual within a reef crevice in Culebra Island during a night dive in 2002 clinging to a *S. roseus* colony at a depth of 2 m. The blunt-spined brittle star, *Ophiocoma echinata* (Lamarck) (family Ophiocomidae) is a common inhabitant of rubble areas in shallow, protected reef lagoons. They typically hide during the day under rocks. The reticulated brittle star,

Ophionereis reticulata (Say) (family Ophionereididae) share nearly similar habitats with *O. echinata*, though they prefer sandy bottoms and hide under reef overhangs and under rock and coral slabs. This is one of the most common species of brittle stars across the Puerto Rican shelf, at depths ranging from nearly sea surface to well beyond 30 m. it prefers sandy bottoms at the edge of coral reefs, under coral overhangs or rubble.

There are also three ubiquitous brittle stars from family Ophiodermatidae, which include *Ophioderma rubicundum* (Lütken), *O. apressum* (Say), and *Ophiothrix suensonii* (Lütken). The first species can be typically found hidden during the day under coral heads and reef crevices but foraging at night across open areas of reef bottoms. During coral mass spawning events they become very active and rapidly climb coral heads, sea fans (*Gorgonia* spp.) and other octocorals approximately two minutes before mass spawning begins. This behavior can be used as an indication of the beginning of spawning and could probably result from some form of chemical reception of signals from corals. In the case of *O. apressum*, its behavior and typical habitat range is fairly similar to that of *O. rubicundum*, while *O. suensonii* is typically hidden during the day within sponges but can be commonly seen active at night, often climbing different octocoral species.

There is also an important species of basket star, *Astrophyton muricatum* (Lamarck) (order Phrynophiurida). It can be commonly found across the region, particularly in coral reefs with abundant sea fan populations and at least moderate water circulation. They often coil into a tight rounded shape attaching to gorgonians or sponges, but open widely to filter plankton at night by orienting its body and spreading its arms perpendicular to the prevailing current.

14.3.3 Echinoidea Distribution

The distribution and abundance of *Echinometra lucunter* and *E. viridis* were examined on seven coral reefs near the southwest of Puerto Rico (McGehee 1992). Urchin distribution was strongly correlated with patterns in water motion, depth, and substrate size among the reefs. However, the two species of *Echinometra* occupy different habitats. *Echinometra lucunter* was most abundant in the shallow forereef zone while *E. viridis* was more common in habitats with less water movement and smaller substrate particle sizes, such as deep forereef and nearshore sites. McGehee (1992) observed that water movement may be the best variable to predict abundance and distribution of the two urchins. As for depth, *E. lucunter* was found in shallow water at depths of 2–3 m, and *E. viridis* was found at depths greater than 3 m. Also, *E. lucunter* can be extremely abundant (e.g., 50–125 ind m^{-2}) on subtidal emergent eolianite platforms along the northern coast (Hernández-Delgado, unpub. data).

As of 1940, *Mellita lata* was ranked the most common echinoderm of Puerto Rico (Kenk 1944). The distribution of this species occurred in the uppermost sediments layers on sandy shallow beaches along the entire circumference of the island. *Mellita lata* was extremely numerous along the north-northeast coast (Luquillo, San Juan, Guayama) and northwest and west coast (Aguadilla, Añasco, Mayagüez). The moves by rhythmical movements of the long spines of the body, marginal spines, interambulacra spines 1, 2, 3, and 4, and posterior lunule spines. Kenk (1944) recorded about 35 strokes of the spines per minute at 26.5 °C. Other members of order Clypeasteroida and family Mellitidae include *Mellita quinquiesperforata* (Leske) and *Leodia sexiesperforata.* Both are also commonly found partially buried under sediments on sandy beaches, but particularly on bottoms with high abundance of organic material (e.g., Naguabo, Humacao, Yabucoa).

Diadema antillarum Distribution

Before the mass mortality *D. antillarum* densities were high at many locations around Puerto Rico, San Juan 13.8 ind m^{-2} (Bauer 1980), La Parguera 12.7 ind m^{-2} (Craft 1975), Guayanilla 3.0 ind m^{-2} (Vicente and Goenaga 1984), and Jobos Bay 2.6 ind m^{-2} (Rivera and Vicente 1976). Lugo (2004) measured the density and size structure of *D. antillarum* at five reefs in La Parguera: Turrumote, Mata la Gata, Cayo Enrique, Laurel, and Media Luna. The median density for all locations was 0.08 ind m^{-2}. The highest mean urchin density (0.13 ind m^{-2}) was at Mata La Gata, followed by Laurel (0.30 ind m^{-2}) and Media Luna (0.15 ind m^{-2}). No adult *D. antillarum* were found at Turromote. The highest mean densities of *D. antillarum* were in the shallow fore-reef zone (depth from 0 to 3 m). He found that the average size ($\pm$ SD) for all sites was 6.71 cm ($\pm$ 0.25). The size varied with depth: 6.95 cm in shallower areas, 6.48 at intermediate depths, and 3.37 cm at deeper reefs.

Shortly after Lugo (2004), Weil et al. (2005) described the decrease in the population of *D. antillarum* from 1984 until 2001 and the population status (abundances, distribution and size structure) after 17 years after the mass mortality in La Parguera. They also found that overall the *D. antillarum* populations were unevenly distributed and low across all sites (0.83–1.55 ind m^{-2}) compared to pre-mortality densities. They also observed more urchins in shallow water areas (depth < 3 m). Urchins were significantly higher in shallow reef platforms (0.89–1.98 ind m^{-2}) (depths from 0 to 3 m), and the upper fore-reef (0.56–2.33 ind m^{-2}) habitats (depths from 3 to 7 m), compared to deeper (depths > 7 m) habitats (0.01–0.88 ind m^{-2}). In contrast, El Atravesado reef had higher urchin densities at the deeper (depths from 8 to 10 m) habitats, along with high abundance of juveniles. A number of factors such as, planktonic settlement, differential morality of juveniles, higher spatial heterogeneity, and presence of refuges could influence *D. antillarum* distribution. Populations in the sea grass mounds were dominated by midsize (6 cm) to large individuals (9 cm). Within reefs, the average size was medium to large. Small sized individuals were more common in shallow reef waters than on the *Thalassia* mounds. They observed a high abundance of juveniles in protected, back reef and lagoonal reef habitats. Higher number of aggregations and a higher number of urchins per aggregation were correlated with low complexity (rugosity) habitats. This supports the idea that this behavior provides protection. They also concluded that recruitment is not limited to highly complex, calcium carbonate habitats. Although average densities were well below pre-mass-mortality densities in Puerto Rico, results of this study indicate that *D. antillarum* seem to be making a slow come back in La Parguera.

More recently, Ruiz-Ramos et al. 2011) measured the population recovery of *D. antillarum* at 26 locations around Puerto Rico during 2003–2004. Seven sites were surveyed in the northern region, five in the western region, four in the southern and eastern regions and six around Culebra Island. Adults were absent in 11 of the 26 localities measured. The highest densities of *D. antillarum* occurred in Culebra and in the northern region. Arrecife El Banderote, Culebra had the highest densities of *D. antillarum* (0.91 ind m^{-2}) followed by Bahía Tamarindo and Bahía Sardinas, with 0.70 and 0.51 ind m^{-2}, respectively. They believe that the continued low densities could be due to a number of factors such as diminished larval supply, recruitment failure, post-recruitment mortality, currents or lack of an appropriate habitat due to chronic environmental degradation. Also, larger individuals were more common in Culebra and the northern region. As for water transparency, the mean value around the island was 6.2 m. Water transparency was greater at the northern sites and sea urchin density was positively correlated with water transparency. The benthic communities of the northern region differed significantly from the other locations. They found the highest mean percent cover of coral was in Culebra (12.6 ± 8.4) and the lowest to the east (0.1 ± 0.3). Cover of macroalgae was greater on reefs in more turbid areas. A positive relationship was observed between *D. antillarum* density and percent coral cover. In conclusion, they found that densities in their study were relatively higher than many other wider Caribbean reef systems. However, they believe that population recovery of *D. antillarum* seems remote.

There is also another important member of order Diadematoida, the magnificent urchin, *Astropyga magnifica* (Clark). This is a very rare species in shallow water areas. It can be often found at depths well below 30 m, but occasionally occurs in shallow water (depth of 1 m) across seagrass communities. There are also two species in the family Echinidae (order Temnopeuroida), *Lytechinus variegatus,* and *Tripneustes ventricosus*. Both species are common on shallow seagrass habitats and protected backreef lagoons. *Lytechinus variegatus* appears to be more abundant on seagrass habitats of the northern coast (e.g., Condado Lagoon, Vega Baja), while *T. ventricosus* is more ubiquitous. Both species tend to cover their test with fragments of algae and seagrass blades. There is a single member of family Cidaridae, the pencil urchin, *Eucidaris tribuloides* (Lamarck). This is a fairly common reef urchin that often hides under sheltered zones It can often be observed on rubble and seagrass areas. There are two members of the family Brissidae (order Spatangoida), *Meoma ventricosa ventricosa (*Lamarck), and *Clypeaster rosaceus* (Linnaeus). *Meoma v. ventricosa* can be commonly found buried under sandy bottoms or sparse seagrass bottoms, while *C. rosaceus* often camouflages its body under seagrass blades, algae and seagrass blade fragments.

14.3.4 Holothuroidea Distribution

There are only a few species of non-pelagic sea cucumbers that are known to execute rapid movements. The species that are able to make rapid movements are members of the order Apodida. *Astichopus multifidus* (Sluiter) is a member of the order Aspidochirotida. Specimens of this group are characterized by dish-shaped tentacles and respiratory trees. The genus *Astichopus* is easily recognized because it is very large and soft and with tube feet uniformly covering both the dorsal and ventral sides of the body. *Astichopus* was most abundant on the northwestern coast, at Crashboat Landing, Puerto Rico. Specimens have been seen at depths of 20–40 m at different times near Isabela, Camuy and Arecibo, Puerto Rico. There were a few individuals collected from La Parguera, Puerto Rico in areas of dense turtle grass (*Thalassia testudinum*). Glynn (1965) stated that *Astichopus* possibly breed in the late autumn or early winter in Puerto Rico. Astichopus is very active. The most frequent means of progression when the animal was left undisturbed is a slow crawl. Glynn (1965) examined the environmental effects on movements. Sudden changes in sea water temperature (3–4 °C) resulted in walking, bounding and rolling movements more effectively than other environmental changes (salinity, oxygen availability, light regimes, etc.). Specimens were attracted to natural and artificial light of low intensity ($\leq$5 foot-candles) and did not demonstrate any clear tendencies to geotactic or thigmotactic behavior. Increased activity of this sea cucumber does not release any toxic substance.

Euapta lappa (Müller) can be also commonly found on shallow (3–10 m) to moderately deep (20–25 m) reefs. They are soft-bodied and are commonly found at the base of reef patches. They usually hide in reef crevices during the day and emerge at dusk, moving around adjacent sandy and rubble bottoms. Several other species in the family Holothuridae (Order Aspirochirotida) can be found across seagrass, sandy and rubble bottoms in northern and eastern Puerto Rico, including *Holothuria mexicana*, *Holothuria* (*Thymiosycia*) *thomasi* (Pawson and Caycedo) and *Actinopygia agassizzi* (Selenka). Also individuals of *Isostichopus badionotus* (Selenka) (family Stichopodidae) have been documented from Culebra Island.

14.3.5 Echinoderm Larval Distribution

Williams and García Sais (2010) described the temporal and spatial distribution patterns of echinoderm larvae in La Parguera, Puerto Rico. Temporal variability of echinoderm larval abundances was examined from a series of monthly tows between Cayo Collado and Cayo San Cristobal in La Parguera. In this study 11,921 echinoderm larvae were collected. Echinoidea larvae were the numerically dominant taxa, representing 52.5 % of the total. Amongst the sea urchin assemblage, 20 % were *D. antillarum* larvae. Ophiuroidea larvae represented 45.6 % of the total. The combined abundance of Holothuroidea and Asteroidea larvae was

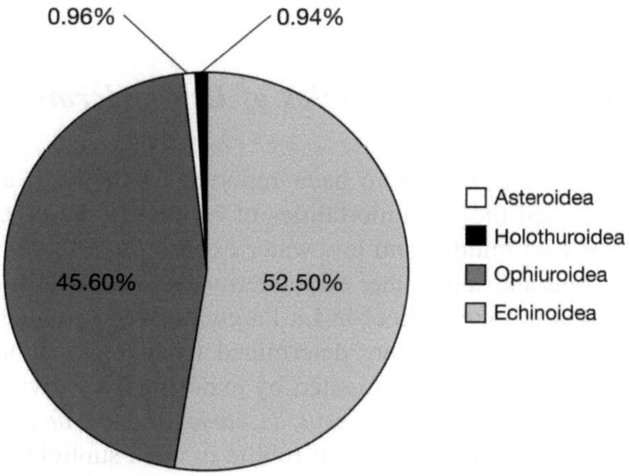

Fig. 14.9 The relative abundance (%) of echinoderm classes to the total larval collection

less than 2 % of the total (Fig. 14.9). Larval composition found in this study was similar to the adult population structure of echinoderms found in la Parguera reported in Benavides-Serrato (2006) study. Echinoidea and Ophiuroidea larvae were collected in all 20 monthly samplings. Holothuroidea and Asteroidea larvae were absent on approximately two-thirds of the samples. A possible explanation for the lack of these larvae may be due to under-sampling associated with low abundance of adults and short larval duration. No seasonal trend of early-stage larvae was observed in this study. However, higher abundances of later-stage larvae of Echinoidea and Ophiuroidea peaked during the summer and fall. Echinoidea, Ophiuroidea and Holothuroidea larval abundances all peaked at the shelf-edge. Larval abundances of Asteroidea showed an increasing pattern of abundance from inshore to offshore, with a peak at the oceanic station. The depth of the substrate at the shelf-edge and the lack of benthic and planktonic predators may account for the higher abundance of echinoderm larvae at this location. Also, internal waves that break at the shelf-edge bring colder, more nutrient rich waters to the surface, which would positively influence zooplankton abundance. They collected more echinoderm larvae using the 64 μm mesh net than with the 202 μm mesh net. Echinoderm larval abundance may be underestimated when 202 μm and/or greater mesh net sizes are used. Larval patchiness also tends to increase in collections when the 202 μm mesh net was used. The larval distribution in the temporal and spatial study was found to be patchy/aggregated. The overall variations and inconsistent peaks may be a result of the patchiness of these larvae.

14.4 Aquaculture and Fisheries

Echinoderms have not been cultured nor fished in Puerto Rico.

14.5 Threats

14.5.1 Mass Mortalities of Echinoderms

Mass mortalities have been reported as far back as 1965, when Glynn (1968) described the mass mortalities of echinoids and other reef flat organisms concurrent with midday and low water exposures in La Parguera, Puerto Rico. All dead urchins and some other echinoderms were counted and removed from the sampling plot at Caracoles Reef in La Parguera over a two day period. Population sizes of living organisms were determined from m^2 quadrat counts. Tolerance of desiccation was also investigated by exposing 25 individuals of each of the five echinoid species (*T. ventricosus, L. variegatus, E. lucunter, Brissus unicolor* (Leske), and *D. antillarum* on coral rubble in open sunlight (unprotected) over a period of six hours. The same desiccation experiment was performed, but this time the sea urchins were provided with some protection. In addition, Glynn (1968) investigated the effects of solar radiation on the sea urchin survival. Fifteen individuals of three species of sea urchins were exposed to sunlight in water baths just deep enough to cover the animal. The most severe mortality of echinoids was during April 14, 1965. *Tripneustes ventricosus* experienced the highest mortality (86.4 %), *L. variegatus* 64 %, *B. unicolor* 25.9 % and *E. lucunter* 1.2 %. *Lytechinus variegatus* and *T. ventricosus* were found dead in shallow water zones and had not been exposed to drying. Before the mass mortality *L. variegatus* had an aggregated mean density of 57.6 ind m^{-2}, *T. ventricosus* 13.6 ind m^{-2}. *Echinometra lucunter* was very abundant, numbering >1,000 individuals. *Diadema antillarum* and *B. unicolor* were scarce but showed a significant increase in densities by spring 1966. There was also a noticeable recovery of *T. ventricosus* (35.6 ind m^{-2}) and *L. variegatus* (539.6 ind m^{-2}) the following fall.

There was two other mass die offs on May 16 and June 28, 1965 reported by Glynn (1968). On those episodes, there were two species of ophiuroids that were affected seriously by extreme temperatures (39 °C*), Ophioderma appressa* (Say) and *Ophiocoma echinata*. During the summer months (May–August) there were a series of minus tides during midday when heat and desiccation approached extreme levels. Glynn (1968) frequently observed temperatures exceeding 32 °C and, during May and June 1965, extreme temperatures as high as 39 °C.

Other echinoderm species involved in these three mass mortalities were: *D. antillarum, Ophiomyxa flaccida (*Say), *Ophiothrix angulata, Ophiocoma riisei* (Lütken), *Ophiactis savignyi, Holothuria* (*Halodeima*) *mexicana,* and *H. (Thymiosycia) arenicola* (Semper). For the desiccation experiment, the loss in weight as a function of exsiccation showed a linear relationship for both protected and unprotected individuals. The loss in weight by *B. unicolor* increased to 70 % from the initial wet weight after 6 h. *Echinometra lucunter* had the highest loss of weight (50 % of initial wet weight after 6 h). All the unprotected urchins died after 3 h of exposure except *E. lucunter*. For the protected urchins, *T. ventricosus* had the highest resistance to desiccation (80 % of the initial wet weight). For the solar radiation experiment, five urchins died after 48 h of recovery, three in the unprotected group and two in the protected group.

The second mass mortality event of echinoderms in Puerto Rico was between 1983 and 1984. Lessios (1995) stated that the 1983–1984 mass mortality of *D. antillarum* was the most extensive and severe die-off ever recorded for a marine invertebrate. Before 1983, the presence of this species was common (22 ind m^{-2}) on coral reefs (Lessios 1995). After the massive die-off, populations were drastically reduced by 95–100 % in many Caribbean locations (Lessios 1995). It was theorized that a water-borne pathogen transported by surface currents was the main cause of this mass mortality (Lessios et al. 1984, Phinney et al. 2001). In addition, one report states that ballast water from ships was also linked with the spread of the urchin-killing pathogen (Bak et al. 1984). The spread of the massive mortality is believed to have originated from Punta Galeta, Panama and proceeded in two different fronts. The first front went east from Panama to Tobago. The second front was faster and went east of Costa Rica, North of Cayman Island, Belize, Mexico, Jamaica, Gulf of Mexico, Florida, and to Bermuda. In February of 1984, an area of 3.5 million square kilometers was affected by this mass mortality (Lessios 1988). Vicente and Goenaga 1984 observed the first diseased *D. antillarum* in January 1984 at Laurel reef, La Parguera, Puerto Rico. Before the mass mortality, densities were reported from 2.6 ind m^{-2} (Jobos Bay) to as high as 13.8 ind m^{-2} (San Juan). In Guayanilla densities before the mass mortality were 3.0 ind m^{-2} and after were as low as 0.13 ind m^{-2} (Vincent et al. 1985).

Bunkley-Williams et al. (1986) reported the third recognized mass mortality of echinoderms in Puerto Rico waters. This study described the isolated mortalities of two sea urchins, *Astropyga magnifica* and *Eucidaris tribuloides*, during the winter of 1984–1985 in the Northwest coast of Puerto Rico. An aggregation of 68 A. magnifica, at depths between 0.6 and 1.2 m deep with dropping spines and rolling in the waves with the surge were observed at Crashboat Beach, Aguadilla on December 8 and 12, 1984. These signs seemed very similar to those witnessed during the *D. antillarum* 1983–84 die-off. *Astropyga pulvinata* (Lamarck) located in deeper waters seemed healthy and there was an absence of sediment on tests. No other species in the vicinity showed signs of disease or stress. Therefore, pollutants were not thought to be a factor in this urchin mortality. In January 1985, there was another mass mortality of *E. tribuloides* at Survival Beach near old Ramey Air Force Base, Aguadilla. Six thousand to 8,000 sea urchins were observed washed up on the beach. These die-offs were during times of rough waves and stormy winter weather. No further mortalities were witnessed. These two mass mortalities of *A. magnifica* and *E. tribuloides* are unexplained events.

Williams (1991) reported mass mortality outbreaks of *T. ventricosus* in Puerto Rico between January and February 1995. The new die offs progressed more slowly (6–8 days) compared to the *D. antillarum* die-off during 1983–84 (3–4 days). One thousand *T. ventricosus* were found dead at La Parguera (Williams et al. 1996). This mortality was not localized but island wide. They also witnessed urchins dying near Humacao and Fajardo. No samples were taken or preserved. There was variation in symptoms. Many urchins were floating near the surface while they still were able to direct spines toward tactile stimulation. The gas production and lost of spines suggests that this species was involved in a disturbance, primary or secondary pathogen. A similar die-off was reported between September and November 1990 in La Parguera (Colon-Jones 1993). This die-off was due to air exposure and/or higher seawater temperatures. The extent, significance and frequency of these die-offs must be better understood in order to identify the cause and their relationship with any other disturbance affecting their habitat.

14.6 Recommendations

The knowledge of echinoderms in Puerto Rico has been scarce and focused principally to ecological studies. Although inventories have been developed, this is incomplete, especially in the north of the island. The focus of research has been mainly located in La Parguera, Puerto Rico. The facilities located on the Isla of Magayues (UPR at Mayagüez) are conveniently located for scientists to carry out research in La Parguera region. Research needs to be extended and increased in other zones of the Archipelago, especially the little islands: Mona, Culebra, Vieques and Desecheo. In order to increase the research of these echinoderms, it is necessary to encourage to students and scientists to study this phylum, especially phylogeny, taxonomy, biogeography and functional aspects.

References

Acevedo R, Morelock J (1988) Effects of terrigenous sediment influx on coral reef zonation in southwestern Puerto Rico. Proc 6th Int Coral Reefs Symp 2:189–194

Almy C, Carrión-Torres D (1963) Shallow-water stony corals of Puerto Rico. Carib J Sci 3:133–162

Bak RPM (1985) Settlement patterns and mass mortalities in the sea urchin Diadema antillarum. Proc 5th Coral Reef Congr, Tahiti 5:267–272

Bak RPM, Carpay MJE, de van Ruyter ED (1984) Densities of the sea urchin *Diadema antillarum* before and after mass mortalities on the coral reefs of Curaçao. Mar Ecol Prog Ser 17:105–110

Bauer JC (1980) Observations on geographical variations in population density of the echinoid *Diadema antillarum* within the western North Atlantic. Bull Mar Sci 30:509–515

Benavides-Serrato M (2006) Taxonomic list of the shallow water echinoderms of Puerto Rico with new information for La Parguera. MSc Thesis, Univ Puerto Rico, Mayagüez

Cabrera-Serrano A, García-Arrarás JE (2004) RGD-containing peptides inhibit intestinal regeneration in the sea cucumber *Holothuria glaberrima*. Dev Dyn 231:171–178

Cameron RA (1983) Dribblers and swampers: Reproductive period and recruitment in Caribbean sea urchins. Am Zool 23:964

Cameron RA (1986) Reproduction, larval occurrence and recruitment in Caribbean sea urchins. Bull Mar Sci 39:332–346

Cameron RA, Schroeter SC (1980) Sea urchin recruitment: effect of substrate selection on juvenile distribution. Mar Ecol Prog Ser 2:243–247

Cameron RA, Boidron-Metairon I, Monterrosa O (1985) Does the embryonic response to temperature and salinity by four species of Caribbean sea urchins parallel the reproductive synchrony? Proc 5th Coral Reef Congr, Tahiti 5:273–278

Candelaria AG, Murray G, File SK, García-Arrarás JE (2006) Contribution of mesenterial muscle dedifferentiation to intestine regeneration in the sea cucumber *Holothuria glaberrima*. Cell Tissue Res 325:55–65

Carrera CJ (1974) The sallow water Amphiurid Brittle Stars (Ophiuroidea: Echinodermata) of Puerto Rico MS Thesis. Univ Puerto Rico, Mayagüez

Cherbonier G (1959) Echinodermes de la Guyane Francaise. (Crinoides, Asterides, Ophiurides, Echinides, Holothurides). Bull Mus Natn Hist Nat Paris 31:168–171

Clark HL (1898) Notes on Bermuda echinoderms. An NY Acad Sci 19:407–413

Clark HL (1901a) The echinoderms of Porto Rico. Bull US Fish Comm 20:231–263

Clark HL (1901b) Bermudan echinoderms. A report on observations and collections made in 1899. Proc Bos Soc Nat Hist 29:339–344

Clark HL (1919) The distribution of the littoral echinoderms of the West Indies. Pabl Carnegie Instn 10:1–240

Clark AH (1921) A monograph of the existing crinoids, vol 1. The Comatulids, Part 2. US Nat Mus Bull 82:795

Clark AH (1939) Echinoderms (other than Holothurians) collected on the Presidential Cruise of 1938. Smithson Miscell Collec 98:1–22

Clark HL (1933) A handbook of the littoral echinoderms of Porto Rico and the other West Indian Islands. Sci Surv Porto Rico Virgin Islands 16:1–147

Clark HL (1941) Reports on the Scientific results of the *Atlantis* expeditions to the West Indies under the joint of the University of Havana and Harvard University. Mem Soc Cubana Hist Nat "*Felipe Poey*" 15:1–154

Clark HL (1942) The echinoderm fauna of Bermuda. Bull Mus Comp Zool Harv 89:367–391

Clark AM, Downey ME (1992) Seastars of the Atlantic. Natural History Museum Publishing Chapman and Hall, London, UK

Colon-Jones DE (1993) Size (age) factors controlling the distribution and population size of the white-spined sea urchin, *Tripneustes ventricosus* (Lamark, 1816) MSc Thesis, Univ Puerto Rico, Mayagüez

Cowen RK, Lwiza KMM, Sponaugle S, Paris CB, Olson DB (2000) Connectivity of marine populations: Open or closed? Science 287:857–859

Craft LL (1975) Aspects of the biology of the crab *Percnon gibbesi* (Milne Edwards) and its commensal association with the sea urchin *Diadema antillarum* Phillipi. MSc Thesis, Univ Puerto Rico, Mayagüez

Díaz-Balzac C, Santacana-Laffitte G, San Miguel-Ruíz J, Tossas K, Valentín-Tirado G, Rives-Sánchez M, Mesleh A, Torres II, García-Arrarás JE (2007) Identification of nerve plexi within the connective tissues of the sea cucumber *Holothuria glaberrima* using a novel nerve-specific antibody. Biol Bull 213:28–42

Díaz-Balzac C, Abreu-Arbelo JE, García-Arrarás JE (2010a) Neuroanatomy of the tube feet and tentacles in *Holothuria glaberrima* (Holothuroidea, Echinodermata. Zoomorphol 129:33–43

Díaz-Balzac C, Mejias W, Jimenez LB, García-Arrarás JE (2010b) The catecholaminergic nerve plexus of Holothuroidea. Zoomorphol 129:99–109

Díaz-Miranda L, García-Arrarás JE (1995) Pharmacological action of the heptapeptide GFSKLYFamide on the muscle of the sea cucumber *Holothuria glaberrima* (Echinodermata). Comp Biochem Physiol 110:171–176

Díaz-Miranda L, Price DA, Greenberg MJ, Lee TD, Doble KE, García-Arrarás JE (1992) Characterization of two novel neuropeptides from the sea cucumber *Holothuria glaberrima*. Biol Bull 182:241–247

Díaz-Miranda L, Blanco R, García-Arrarás JE (1995) Localization of GFSKLYFamide in the sea cucumber *Holothuria glaberrima* (Echinodermata): a light and electron microscopic study. J Comp Neurol 352:626–640
Díaz-Miranda L, Pardo-Reoyo CF, Martínez R, García-Arrarás JE (1996) Galanin-like immunoreactivity in the sea cucumber *Holothuria glaberrima*. Cell Tissue Res 286:385–391
Engel H (1939) Echinoderms from Aruba, Curazao, Bonaire and Northern Venezuela. Capitazool 8:1–12
Engel H, Aberson VT (1960) Ubereinige Echinoderm von el Salvador (C-Amerika) und von den Farallon-Inseln (Kalifornien). Senck Biol 41:363–366
Frank E, Wicks C, Mylroie J, Troester J, Calvin-Alexander E, Carew J (1998) Geology of Isla de Mona, Puerto Rico. J Cave and Karst Stud 60:69–72
Gaines S, Brown S, Roughgarden J (1985) Spatial variation in larval concentrations as a cause of spatial variation in settlement for the barnacle, *Balanus glandula*. Oecol 67:267–272
García-Arrarás JE, Dolmatov IY (2010) Echinoderms; potential model systems for studies on muscle regeneration. Cur Pharma Des 16:942–955
García-Arrarás JE, Greenberg MJ (2010) Visceral Regeneration in Holothurians. Micros Res Tech 55:438–451
García-Arrarás JE, Viruet E (1993) Enteric nerve fibers of holothurians are recognized by an antibody to acetylated a-tubulin. Neurosci Let 157:153–157
García-Arrarás JE, Torres-Avillán I, Ortíz-Miranda S (1991a) Cells in the intestinal system of holothurians (Echinodermata) express cholecystokinin-like immunoreactivity. Gen Comp Endocrinol 83:233–242
García-Arrarás JE, Enamorado-Ayala I, Torres-Avillán I, Rivera V (1991b) FMRFamide-like immunoreactivity in cells and fibers of the holothurian nervous system. Neurosci Let 132:199–202
García-Arrarás JE, Estrada-Rodgers L, Santiago R, Torres II, Díaz-Miranda L, Torres-Avillán I (1998) Cellular mechanisms of intestine regeneration in the sea cucumber, *Holothuria glaberrima* Selenka. J Exp Zool 281:288–304
García-Arrarás JE, Díaz-Miranda L, Torres II, File S, Jiménez LB, Rivera-Bermúdez K, Arroyo E, Cruz W (1999) Regeneration of the enteric nervous system in the sea cucumber, Holothuria glaberrima. J Comp Neurol 406:461–475
García-Arrarás JE, Rojas-Soto M, Jiménez LB, Díaz-Miranda L (2001) The enteric nervous system of echinoderms: Unexpected complexity revealed by neurochemical analysis. J Exp Biol 204:865–873
García-Arrarás JE, Schenk C, Rodriguez-Ramirez R, Torres II, Valentín G, File SK, Candelaria AG (2006) Spherule cells in the echinoderm *Holothuria glaberrima* and their involvement in intestinal regeneration. Dev Dynam 235:3259–3267
García-Arrarás JE, Valentín-Tirado G, Flores JE, Rosa RJ, Rivera-Cruz A, San Miguel-Ruiz JE, Tossas K (2011) Cell dedifferentiation and epithelial to mesenchymal transitions during intestinal regeneration in H. *glaberrima*. Dev Biol 17:11–61
García JR, Morelock J, Castro R, Goenaga C, Hernandez-Delgado E (2003) Puertorican reefs: research synthesis, present threats and management perspectives. In Cortes, J (ed) Latin American Coral Reefs. Elsevier, Berlin, pp 111-130
García-Sais J, Appledoorn R, Bruckner A, Caldow C, Christensen JD, Lilyestrom C, Monaco ME, Sabater J, Williams E, Díaz E (2005) The state of coral reef ecosystems of the commonwealth of Puerto Rico. In: Waddell JE (ed) The state of coral reef ecosystems of the United States and Pacific Freely Associated States. NOAA Technical Memorandum, Silver Spring, Maryland, pp 75–116
Giese AC, Pearse JS (1974) Introduction: general principles. In: Giese AC, Pearse JS (eds) Reproduction of Marine Invertebrates. Academic Press, New York, pp 1–49
Glynn PW (1965) Active movements and other aspects of the biology of Astichopus and Leptosynapta (Holothuroidea). Biol Bull 129:106–127
Glynn PW (1968) Mass mortalities of echinoids and other reef flat organisms coincident with midday, low water exposures in Puerto Rico. Mar Biol 1:226–243

González-Liboy JA (1971) The zonation and distribution of marine beach macro-invertebrates at playa Maní, Mayagüez, Puerto Rico. MSc Thesis, Univ Puerto Rico, Mayagüez

Griffin SP, García RP, Weil E (2003) Bioerosion in coral reef communities in southwest Puerto Rico by the sea urchin *Echinometra viridis*. Mar Biol 143:79–84

Hendler G, Miller JE, Pawson DL, Kier PM (1995) Sea Stars, Sea Urchins, and Allies: Echinoderms of Florida and the Caribbean. Smith Inst Press, Washington

Iliffe TM, Pearse JS (1982) Annual and lunar reproductive rhythms of the sea urchin *Diadema antillarum* (Philippi) in Bermuda. Int J Invert Repro 5:139–148

Kenk R (1944) Distribution and movements of *Mellita lata* and *Astropecten marginatus*. Biol Bull 83:177–187

Keough MJ, Downes BJ (1982) Settlement of marine invertebrates: the role of active larval choices and early mortality. Oecol 54:348–352

Kingsford MJ, Leis J, Shanks A, Lindeman K, Morgan S, Pineda J (2002) Sensory environments, larval abilities and local self-recruitment. Bull Mar Sci 70:309–340

Lessios HA, Robertson DR, Cubit JD (1984) Spread of *Diadema* mass mortality through the Caribbean. Science 226:335–337

Lessios HA (1988) Population dynamics of *Diadema antillarum* (Echinodermata: Echinoidea) following mass mortality in Panama. Mar Biol 99:515–526

Lessios HA (1995) *Diadema antillarum* 10 years after mass mortality: still rare, despite help from a competitor. Proc R Soc Lond B 259:331–337

Levin L (2006) Recent progress in understanding larval dispersal: new directions and digression. Integr Comp Biol 46:282–297

Levitan DR (1995) The ecology of fertilization in free-spawning invertebrates. In: McEdward L (ed) Ecology of Marine Invertebrate Larvae. CRC Press, Boca Raton, pp 123–156

Lewis JB (1958) The biology of the tropical sea urchin *Tripneustes ventricosus* Leske in Barbados, British West Indies. Can J Zool 36:607–621

Lugo MA (2004) Population status of the black sea urchin *Diadema antillarum* (Phillipi) in La Parguera, Puerto Rico, 20 years after the mass mortality event. MSc Thesis, Univ Puerto Rico, Mayagüez

Mashanov VS, García-Arrarás JE (2011) Gut regeneration in holothurians: A snapshot of recent developments. Biol Bull 221:93–109

Mashanov VS, Zueva OR, Heinzeller T, Aschauer B, Naumann WW, Grondona JM, Cifuentes M, García-Arrarás JE (2009) The central nervous system of sea cucumbers (Echinodermata: Holothuroidea) shows positive immunostaining for a mammalian glial secretion. Front Zool 6:11

Mashanov VS, Zueva OR, García-Arrarás JE (2010a) Organization of glial cells in the adult sea cucumber central nervous system. Glia 58:1581–1593

Mashanov VS, Zueva OR, Rojas-Cartagena C, García-Arrarás JE (2010b) Visceral regeneration in a sea cucumber involves extensive expression of *survivin* and *mortalin* homologs in the mesothelium. BMC Dev Biol 10:117

McGehee MA (1992) Distribution and abundance of two species of *Echinometra* (Echinoidea) on coral reefs near Puerto Rico. Carib J Sci 28:173–183

McWilliam PS, Sale PF, Anderson DT (1981) Seasonal changes in resident zooplankton sampled by emergence traps in One Tree Lagoon, Great Barrier Reef. J Exp Mar Biol Ecol 52:185–204

Méndez AT, Roig-López JL, Santiago P, Santiago C, García-Arrarás JE (2000) Identification of Hox gene sequences in the sea cucumber *Holothuria glaberrima* Selenka (Holothuroidea: Echinodermata). Mar Biotech 2:231–240

Miller BA, Emlet RB (1997) Influence of nearshore hydrodynamics on larval abundance and settlement of sea urchins *Strongylocentrotus franciscanus* and *S. purpuratus* in the Oregon upwelling zone. Mar Ecol Prog Ser 148:83–94

Miller MW, Kramer KL, Williams SM, Johnston L, Szmant AM (2009) Assessment of current rates of *Diadema antillarum* larval settlement. Coral Reefs 28:511–515

Moore HB, Jutare T, Bauer JC, Jones JA (1963) The biology of *Lytechinus variegates*. Bull Mar Sci Gulf Carib 13:23–53

Morelock J, Schneidermann N, Bryant WR (1977) Shelf reefs, southwestern Puerto Rico. Reefs and Related Carbonates—Ecology and Sedimentology. Studies in Geology 4. Eds. Stanley H. Frost, Malcolm P. Weiss, and John B. Saunders, Tulsa, Okla.: American Association Petroleum Geologists, pp 17–25

Murray G, García-Arrarás JE (2004) Myogenesis during holothurian intestinal regeneration. Cell Tissue Res 318:515–524

Ortiz-Pineda P, Ramírez-Gómez F, Pérez-Ortiz J, González-Díaz S, Santiago-De Jesús F, Hernández J, Rojas-Cartagena C, Suárez-Castillo EC, Tossas K, Roig-López JL, Méndez-Merced A, Ortiz-Zuazaga H, García-Arrarás JE (2009) Gene expression profiling of intestinal regeneration in the sea cucumber. BMC Genomics 10:262

Pawson D, Vance D, Messing C, Solis-Marin F, Mah C (2009) Echinodermata of the Gulf of Mexico. In: DL Felder, David K (eds) Gulf of Mexico Origin, Waters, and Biota, Volume 1, Biodiversity. Texas A&M Univ Press, pp 1177–1204

Phinney JT, Muller-Karger F, Dustan P, Sobel J (2001) Using remote sensing to reassess the mass mortality of *Diadema antillarum* 1983-1984. Conserv Biol 15:885–889

Podolsky RD, Emlet RB (1993) Separating the effects of temperature and viscosity on swimming and water movement by sand dollar larvae (*Dendraster excentricus*). J Exp Biol 176:207–221

Quiñones JL, Rosa R, Ruiz DL, García-Arrarás JE (2002) Extracellular matrix remodeling and metalloproteinase involvement during intestine regeneration in the sea cucumber *Holthuria glaberrima*. Dev Biol 250:181–197

Ramírez-Gómez F, García-Arrarás JE (2010) Echinoderm immunity. Invert Survival J 7:211–220

Ramírez-Gómez F, Ortíz-Pineda P, Rojas-Cartagena C, Suárez-Castillo E, García-Arrarás JE (2008) Immune-related genes associated with intestinal tissue in the sea cucumber *Holothuria glaberrima*. Immunogenetics 60:57–71

Ramírez-Gómez F, Ortíz-Pineda P, Rivera G, García-Arrarás JE (2009) LPS-induced genes in intestinal tissue of the sea cucumber *Holothuria glaberrima*. PLoS ONE 4:e6178

Ramírez-Gómez F, Aponte-Rivera F, Méndez-Castañer L, García-Arrarás JE (2010) Changes in holothurian coelomocyte populations following immune stimulation with different molecular patterns. Fish Shellfish Immunol 29:175–185

Reeve MR (1970) Seasonal changes in the zooplankton of south Biscayne Bay and some problems of assessing the effects on the zooplankton of natural and artificial thermal and other fluctuations. Bull Mar Sci 20:894–921

Rivera JA, Vicente VP (1976) *Diadema antillarum* (Philippi): Enhancement of coral species diversity in Thalassia beds. Proc Ass Isl Mar Lab Car May 12:18

Roig-López JL, Santiago P, Jiménez B, Santiago C, García-Arrarás JE (2001) Strategies to identify differentially expressed genes during regeneration. In: Barker E (ed) Echinoderms 2000. Swets & Zeitlinger, Lisse pp 49–54

Rojas-Cartagena C, Ortíz-Pineda P, Ramírez-Gómez F, Súarez-Castillo EC, Matos-CruzV, Rodríguez C, Ortíz-Zuazaga H, García-Arrarás JE (2007) Distinct profiles of expressed sequence tags during intestinal regeneration in the sea cucumber *Holothuria glaberrima*. Physiol Genomics 31:203–215

Ruiz-Ramos D, Hernandez-Delgado EA, Schizas NV (2011) Population status of the long-spined urchin *Diadema antillarum* in Puerto Rico, 20 years after a mass mortality event. Bull Mar Sci 87:113–127

San Miguel JE, Maldonado-Soto AR, Garía-Arrarás JE (2009) Regeneration of the radial nerve cord in the sea cucmber *Holothuria glaberrima*. Dev Biol 9:3

San Miguel-Ruiz JE, García-Arrarás JE (2007) Common cellular events occur during wound healing and organ regeneration in the sea cucumber *Holothuria glaberrima*. BMC Dev Biol 7:115

San Miguel-Ruiz JE, Maldonado-Soto AR, García-Arrarás JE (2007) Regeneration of the radial nerve cord in the sea cucumber Holothuria glaberrima. BMC Dev Biol 9:3

Santiago P, Roig-López JL, Santiago C, García-Arrarás JE (2000) Serum amyloid A protein in an echinoderm: Its primary structure and expression during intestinal regeneration in the sea cucumber *Holothuria glaberrima*. J Exp Zool (Mol Dev Evol) 288:335–344

Santiago-Cardona, Pedro, César A. Berríos, Francisco Ramírez and José E. García-Arrarás. (2003) Lipopolysaccharides induce intestinal serum amyloid A expression in the sea cucumber Holothuria glaberrima. Dev Comp Immunol 27:105–110

Smith AB (1997) Echinoderm larvae and phylogeny. Annu Rev Ecol Syst 28:219–241

Suárez-Castillo EC, García-Arrarás JE (2007) Molecular Evolution of the Ependymin Protein Family: A Necessary Update. BMC Evol Biol 15(7):23

Suárez-Castillo E, Medina-Ortíz W, Roig-López JL, JE García-Arrarás (2004) Ependymin, a gene involved in regeneration and neuroplasticity is overexpressed during regeneration in the echinoderm *Holothuria glaberrima*. Gene 334:133–143

Vicente VP, Goenaga C (1984) Mortandades masivas del erizo de mar *Diadema antillarum* (Philippi) en Puerto Rico. Centro para Estudios Energéticos y Ambientales, Div Ecología Marina, Univ Puerto Rico. Reporte CEER-M-195

Weil E (2005) Puerto Rico. In: Miloslavich P and Klein E (eds) Caribbean Marine Biodiversity the known and unknown. Destesch Publications, Lancaster, pp 108

Weil E, Torres JL, Ashton M (2005) Population characteristics of the sea urchin *Diadema antillarum* in La Parguera, Puerto Rico, 17 years after the mass mortality event. Rev Biol Trop 53(Suppl 3):219–231

Williams EH Jr (1991) Threat to black sea urchins. Nature 352:385

Williams SM, García Sais J (2010) Temporal and spatial distribution patterns of echinoderm larvae in La Parguera, Puerto Rico. Rev Biol Trop 58:81–88

Williams LB, Williams EH Jr, Bunkley AG Jr (1986) Isolated mortalities of the sea urchins *Astropyga magnifica* and *Eucidaris tribuloides* in Puerto Rico. Bull Mar Sci 38:391–393

Williams EH Jr, Williams LB, Bruckner RJ, Bruckner AW, Ortiz-Corps EAR, Kerby AB, Jones DEJ (1996) Recurring mass mortalities of the white-spined sea urchin, *Tripneustes ventricosus* (Echinodermata: Echinoidea) in Puerto Rico. Carib J Sci 32:111–112

Williams SM, García Sais J, Capella J (2009) Seasonal variation and spatial distribution patterns of early larval stages of the long-spined sea urchin *Diadema antillarum* in La Parguera, Puerto Rico. Carib J Sci 45:110–117

Williams SM, Yoshioka PM, García Sais J (2010) Settlement pattern of *Diadema antillarum* in La Parguera, Puerto Rico. Coral Reefs. 29:809–812

Williams SM, García Sais J, Yoshioka PM (2011) Spatial variation of *Diadema antillarum* settlement in La Parguera, Puerto Rico. Bull Mar Sci 87:531–540

Wolf TJ (1978) Aspects of the Biology of *Astrophytum muricatum* (Lamarck, 1816) (Ophiuroidea: Gorgonocephalidae). MSc Thesis, Univ Puerto Rico, Mayagüez

Yoshioka PM, Owen GP, Pesante D (1985) Spatial and temporal variations in Caribbean zooplankton near Puerto Rico. J Plank Res 7:733–752

Youngbluth MJ (1980) Daily, seasonal, and annual fluctuations among zooplankton populations in an unpolluted tropical embayment. Est Coast Shelf Sci 10:265–287

Chapter 15
Echinoderms of the Canary Islands, Spain

José Carlos Hernández, Sabrina Clemente, Fernando Tuya, Angel Pérez-Ruzafa, Carlos Sangil, Leopoldo Moro-Abad and Juan José Bacallado-Aránega

J. C. Hernández (✉) · S. Clemente
Departamento de Biología Animal (Ciencias Marinas), Universidad de La Laguna, Biodiversidad, Ecología Marina y Conservación, c/Astrofísico Francisco Sánchez sn, La Laguna, 38206 , Tenerife, Canary Islands, Spain
e-mail: jocarher@ull.es

S. Clemente
e-mail: msclemen@ull.es

F. Tuya
Departamento de Biología, Universidad de Las Palmas de Gran Canaria, Las Palmas de Gran Canaria, Campus Tafira s/n, Las Palmas de G.C., Canary Islands 35017, Spain

A. Pérez-Ruzafa
Departamento de Ecología e Hidrología, Universidad de Murcia, Murcia, Spain

C. Sangil
Departamento de Ecología e Hidrología, Universidad de La Laguna, Biodiversidad, Ecología Marina y Conservación, c/Astrofísico Francisco Sánchez sn, La Laguna, 38206 , Tenerife, Canary Islands, Spain

L. Moro-Abad
Servicio de Biodiversidad del Gobierno de Canarias, Santa Cruz de Tenerife, Canary Islands, Spain

J. J. Bacallado-Aránega
Museo de Ciencias Naturales de Tenerife, c/Fuente Morales s/n, 38004, Santa Cruz de Tenerife (Canarias), Canary Islands, Spain

J. J. Alvarado and F. A. Solís-Marín (eds.), *Echinoderm Research and Diversity in Latin America*, DOI: 10.1007/978-3-642-20051-9_15,

15.1 Introduction

15.1.1 Geographical and Geological Context of the Canary Islands

The Canary Islands are located in the Northeast Atlantic, between 27°–29° N and 14°–18° W. The Archipelago consists of seven major islands, four smaller islets and numerous rocky outcrops (Table 15.1, Fig. 15.1). The islands are arranged longitudinally along 450 km. Fuerteventura Island is only 90 km from the African continent, La Palma is almost 400 km from it (Fernández-Palacios and Martín Esquivel 2001a). The islands have a land surface area of 7,490 km^2 and 1,501 km of coastline. The land surface and length of coastline vary in the different islands; Tenerife is the largest in area and length of coastline (2,034 km^2, 336 km), while El Hierro has the smallest surface (269 km^2) and La Gomera the shortest length of coastline (97 km).

From a geological point of view, the Canary Islands belong to the group of "oceanic islands", they emerge from the ocean floor and have their own structures (insular edifice) completely independent of the continents. In fact, the Canary Islands, except for Lanzarote and Fuerteventura that share the same insular edifice, had independent origins from the others. The size of the islands'edifice is especially relevant since over 80 % of most islands is under water. On average, insular edifices rise from a depth of 3,500 m. The depth is greater than 4,000 m for La Palma and El Hierro (Carracedo 1988; Carracedo et al. 2001).

The origin of the islands dates back to about 20 million years, but there are also important variations in age between islands (Table 15.1). In fact, while in some islands the eruptive periods have ceased several million years ago, in others it has continued up to the present, as indicated by the large number of historical eruptions (<500–600 years old). A good example of this volcanic activity is the recent eruption off El Hierro Island that started in October 2011. The difference in the ages between islands has important implications for the substrate morphology and topography, both inland and underwater, since erosion and remodeling of insular edifices are higher in older islands (Carracedo et al. 2001).

The abyssal plain between the islands is dotted by a large number of submarine mountains and volcanoes. Most rise only a few hundred meters above the ocean floor, but there are two mountains, the "Banco de La Conception" located north of Lanzarote and the "Banco de Amanay" southwest of Fuerteventura, whose summit is less than 100 m deep. In the past, when sea level was lower, these mountains were above sea level and were true islands (Fernández-Palacios 2004). The current submarine volcanism is now considered more important than terrestrial activity, since recent seismic studies and dating of lava flows indicate the presence of active submarine volcanoes in the western region of the archipelago (Carracedo 2001).

Due to the volcanic nature of the islands, their edifices have a significant slope, thus the island platforms (shallow bottoms of gentle slope located between 0 and 100 m depth) are limited in area compared to the land surface (Mitchell 1998; Masson et al. 2002). The extension of the island platforms largely depends on the

Table 15.1 Geographic data of the Canarian Arquipelago, incluiding islets (*). Source: Instituto Geográfico Nacional, Ministerio de Fomento (http://www.ign.es), Carracedo et al. (2001)

Island	Area (km2)	Coastal perimeter (km)	Age (mill. years)
Lanzarote*	885	237	15,5
Fuerteventura*	1.664	310	20,5
Gran Canaria	1.560	252	14,5
Tenerife	2.034	336	7,5
La Gomera	370	97	12
La Palma	708	162	1,5
El Hierro	269	107	0,8

geological age of the islands and on the physicochemical characteristics of erupted materials. The older islands (Fuerteventura, Lanzarote, Gran Canaria, La Gomera) have been exposed longer to the action of erosive agents (waves, wind, runoff) and therefore they have a wider platform than the younger islands (Tenerife, La Palma, El Hierro) (Mitchell et al. 2003). In the case of El Hierro, the youngest island, the edifice has hardly been remodeled and the insular platform is very small or non-existent (Fig. 15.1b). Within each island, the extent of the platforms may differ. They usually show greater development in the northern faces of the islands, which are areas more highly exposed to erosive agents (Mitchell et al. 2003).

Along with erosion and the creation of offshore platforms, other underwater structures and events have been recognized, such as rifts and massive landslides (Urgeles et al. 1999; Masson et al. 2002; Mitchell et al. 2002). Rifts are the growth axes of the islands that constitute their structural basis since they concentrate the greatest amount of volcanic activity. Landslides are sudden and intense events that generate large avalanches, caused by the gravitational effect due to the instability of the insular edifice or by volcanic eruptions. Landslides detected in the archipelago can extend into the ocean for over 100 km and constitute a volume of more than 100 km^3 (Watts and Masson 1995). These are not strange phenomena in oceanic islands, or in the Canary Islands, and they represent typical stages of consolidation and settlement processes of insular edifices (Gee et al. 2001a, 2001b; Mitchell et al. 2002).

The morphology of the coastline and the continental shelf is highly variable throughout the Canarian Archipelago and even in each island. However, several shapes or substrates are more frequent than others, in relation to the geologic age and its interaction with environmental factors (waves, currents, wind, etc.). The coastline on the older islands has been heavily modified. It is homogeneous in shape and hardly scalloped, representing the typical shape of a backward-stepping coast. Beaches of organic origin are abundant, but cliffs and cobble beaches are also frequent. Intertidal platforms, typically of small amplitude, show greater development in the eastern islands. In contrast, in younger western islands where the processes of coastal erosion coexist with volcanism, a more complex littoral, richer in shapes, is found. The coastline is fringed with numerous points and rock

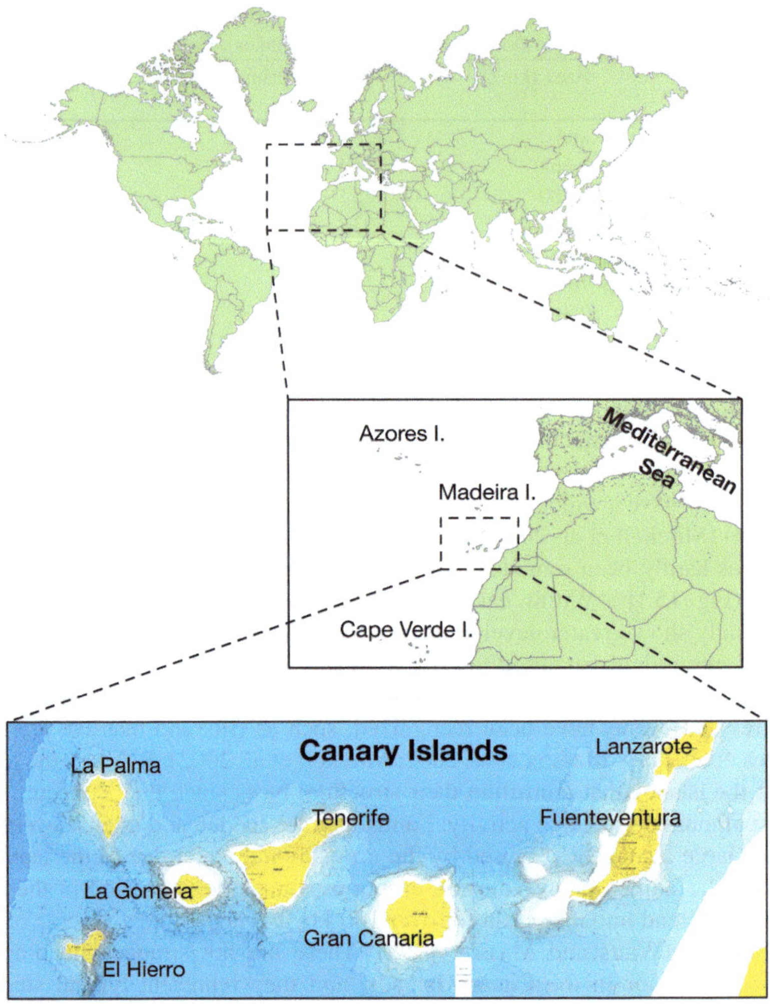

Fig. 15.1 Geographical position and bathimetric map of the Canarian Arquipelago. Island platforms between 0–100 m depth are in *white*. *Source* Estudio Hidrográfico y Oceanográfico de la Zona Económica Exclusiva de Canarias (ZEE-03), Ministerio de Defensa del Gobierno de España

ledges, as well as small shoals, but cliffs are the dominant forms. Beaches are scarce, smaller, and mostly of volcanic origin (Yanes 1990).

In the submerged platforms sandy bottoms dominate, a pattern that increases with depth because the sedimentation is favored by the reduction of hydrodynamics with depth. On the other hand, the percentage of surface occupied by sandy bottoms differs in each island. In older islands sedimentation seems to be favored by the low slope. The heterogeneity (roughness) of the bottoms is particularly high in younger islands, where the steeper slope of their bottoms along with the presence of submarine lava flows creates highly complex environments (Yanes 1990).

15.1.2 Oceanography

From an oceanographic point of view, the Canary Islands are a region of great interest due to the interaction of different phenomena: the upwelling off the African coast, the cold Canary Current, waves and trade winds. The combination of these phenomena creates a complex mosaic of environments throughout the archipelago, with peculiarities in the physical–chemical conditions of its waters. These processes cool the waters of the archipelago that would otherwise have higher temperatures in relation to the latitude (Brito et al. 1984; Aguilera et al. 1994; Fernández-Palacios and Martín-Esquivel 2001a; Fernández-Palacios et al. 2001b). The waters of the archipelago are situated on the eastern limit of the North Atlantic subtropical gyre, occupying an intermediate position between the rich cold upwelling waters of the northwest African coast and the warm and oligotrophic waters of the open ocean, giving rise to the so-called Transitional Coastal Zone (Braun and Molina 1988; Barton et al. 1998; Barton and Arístegui 2004).

The surface seawater temperature ranges from minimum values of 17–19 °C in April-March to maximum values of around 23–25 °C in September–October. This surface temperature regime is slightly cooler than expected at the islands' latitude in the Atlantic Ocean. This is due to the cold Canary Current and the upwelling off the African coast (Braun and Molina 1988; Molina et al. 1996; Hernández-Guerra and Nykjaer 1997; Tejera et al. 2002; Marrero-Díaz et al. 2006).

The upwelling off the nearby African coast has its origin in the action of trade winds from the northeast over the northwestern coasts of Africa. This causes, following Ekman's theory (Ekman1905), the emergence of deep waters perpendicular to the direction of the winds and towards the archipelago. Upwelled seawater is colder, lower in salinity and richer in nutrients compared to surface water. The incidence of this phenomenon over the archipelago has an important spatial and temporal variation. The eastern islands, the closest to it up-welling, are more affected by this process. Its influence on central and western islands is less and mainly due to the incidence of the so-called upwelling filaments. Temporally, upwelling is more regular in those months when the trade winds are more intense and constant (May–August) (Braun and Molina 1988; Nykjaer and Van Camp 1994; Hernández-Guerra and Nykjaer 1997; Tejera et al. 2002; Barton and Arístegui 2004).

The cold Canary Current, which flows NNE–SSW, is a branch of the Gulf Current, a cyclonic system of currents that carries warm waters from the Caribbean Sea to the northeast Atlantic Ocean. As it crosses the Bay of Biscay, this current begins to cool, becoming a cold current when it reaches the Canaries. The average speed of this current is 0.1–0.2 ms^{-1}, but undergoes major seasonal changes reaching minimum values of 0.025 ms^{-1} in summer. Each island acts as a barrier during the flow of the current throughout the archipelago. This causes changes in speed and creates mesoscale phenomena such as cyclonic and anticyclonic eddies that produce sinking and upwelling events of small amounts of waters, and downstream warm wakes in the south and southwest regions of some islands (Braun and Molina 1988; Arístegui et al. 1997; Tejera et al. 2002; Barton and Arístegui 2004).

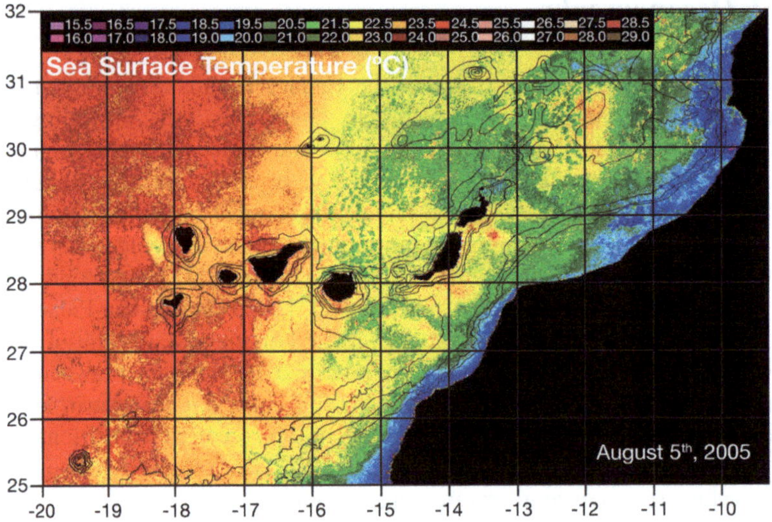

Fig. 15.2 Sea surface temperatura (SST) map of the Canarian Arquipelago and surrounding area. *Source* Dr. Antonio García Ramos, Departamento de Biología de la ULPGC

The different interactions of upwelling and, to a lesser extent, of the cold current of the Canary Islands along the archipelago creates a gradient in temperature (≈ 2 °C), salinity, nutrients and primary production between the eastern and western islands (Fig. 15.2). In this sense, the waters off the eastern islands have more affinity to the waters of the temperate regions, while those off the western islands are more similar to the waters of the tropics (Braun and Molina 1988; Barton et al. 1998; Davenport et al. 2002; Barton and Arístegui 2004).

The open nature of the coasts of the Canary Islands results in a great proportion of the littoral being exposed or semi-exposed to wave action. Bays and coves protected from the waves are much more localized along the islands and are a small part of the coast. The waves that affect the archipelago consist of two substantially different types: wind waves and swells. Wind waves are dominant over the swells. Wind waves, from the NNE and NE, have a mean speed of 18–22 kmh^{-1}. They are produced by the action of trade winds over the seawater surrounding the islands, creating waves that follow the direction and intensity of the winds. The mean annual height of these waves is 1.4 m, with a frequency of 9.5 s^{-1}. Swells from NNW and NW are produced by storms from remote areas of the North Atlantic. They create perturbations that travel along the ocean for distances of up to 3,500 km. This kind of wave movement shows a marked seasonal pattern, being more frequent from October to March. The mean annual height of waves is between 2 and 3 m, with a frequency of 18 s. The prevailing northern component in wave action at the Canary Islands makes the northern coasts of the islands more exposed than the southern coasts (Braun and Molina 1988; Pérez et al. 2003; Yanes et al. 2006).

The climatic regime of the islands is characterized by the predominance of trade winds from the northeast that affects the islands when they are under the influence of the Azores anticyclone. These winds have an average speed of 20–22 km h^{-1}. They are not constant throughout the year, undergoing major seasonal variations. In summer they reach their maximum frequency, occurring around 90–95 % of the days. In fall the Azores anticyclone gets weaker, producing periods of calm. However, it is in winter and spring when there are minimum levels of trade winds, with an incidence of less than 50 % of the days. During this period low pressures from the Atlantic Ocean are frequent, producing winds from the northwest, southwest or southeast, as well as warm and dry invasions of Saharan weather conditions towards the east and southeast (Marzol 1988).

The tidal regime of islands is semidiurnal as in the rest of the Atlantic region, with two high tides and two low tides each day. Usually two consecutive high tides and two low tides have the same amplitude, experiencing a change during the year. The maximum tidal amplitude occurs during the spring and fall equinoxes, reaching about 2.5 m, whereas minimum values are recorded during the summer and winter solstices, not exceeding 0.7 m of amplitude. During the rest of the year the average tidal amplitude usually ranges between 1.2 and 1.3 m, but local variations may occur due to the morphology of the coast, the depth, or to particular climatic conditions of winds and waves in favor or against tidal waves (Braun and Molina 1988; Yanes 1990).

15.1.3 General Characteristics of Communities and Ecosystems

The seabed of the Canary Islands is morphologically highly complex with a rich and diverse biota. The presence of certain biological populations and their structure are determined by the interplay of five factors: insularity, topography of the bottoms, great length of coastline, oceanographic conditions and geographical location. Since the Canary Islands originated from the seabed, isolated from other regions, its biota necessarily came from nearby continents or islands. One could expect to find in the islands the same patterns of communities and ecosystems structures as in nearby regions. However, despite the great dispersal ability of marine species, many species fail to cross ocean barriers. Therefore a biased representation of the biota appears in the archipelago. The volcanic nature of the islands determines the pronounced slope of the bottoms and the small extent of sublittoral platforms that abruptly terminates onto the islands' slopes. This implies that the extension on which benthic communities can develop is spatially very limited. However, the long coastline, along with the rich and diverse coastal morphology, allows the occurrence and development of numerous different habitats and ecosystems. On the other hand, the waters surrounding the islands are oligotrophic that limits production in the environment (Braun and Molina 1988).

From a biogeographic point of view, the geographical location of the archipelago has resulted in the coexistence of both temperate and tropical species (Brito et al. 2001; Sansón et al. 2001; Brito and Ocaña 2004; Tuya and Haroun 2009). It has been noted that the main system of currents that affects the archipelago is the cold Canary Current, a descending branch of the Gulf Current that flows from the north of the Iberian Peninsula towards the southeast. On its way through the European Atlantic coast, it becomes progressively richer in propagule and larvae of different species. Some of these species successfully reach the islands and become established if conditions are favorable. Therefore, most of the species found in the Canary Islands have their origin in the Atlantic-Mediterranean region. To a lesser extent, American tropical and subtropical species have also colonized the Canary Islands through this current. In many cases, the islands constitute the only region in the eastern Atlantic Ocean where these species are found. Although there is no clear system of currents that connects the islands with the tropical southern African coast, some species from this region have colonized the Canaries. Probably southeast African weather conditions occasionally counteract the Canary Current, favoring the arrival of these species. Biota common to the Saharan coast (the region of mainland Africa nearest to the Canary Islands) are scarce, mainly because the oceanographic conditions generated by the upwelling of cold waters near the continent area are very different from the conditions of the Canaries (Aguilera et al. 1994).

This set of variables defines the characteristics displayed by coastal populations, communities and ecosystems. Because the structure, composition and functioning of ecosystems show characters in between those of temperate and tropical systems, they are considered subtropical. The richness, diversity, disparity and originality of the populations are very high and trophic relationships among species are complex in many cases. The size of populations is usually small, due to both the low productivity of the waters and the spatial restrictions of habitats. In this sense, it should be emphasized that all benthic ecosystems, with the exception of abyssal plain ecosystems, lack of continuity throughout the archipelago because they are confined to the insular edifices and seamounts. The surface occupied by benthic ecosystems depends directly on features inherent to the islands' edifices (e.g. coastal ecosystems between 0 and 100 m depth are larger on some islands than in others). In this sense, according to the different availability of habitats and to the oceanographic conditions at the islands, the population size of a species can be different in each of them. On the other hand, given that most species are linked to the bottoms, the non-continuity of ecosystems means that the only way of exchange between populations is by larval dispersal. Since few species (only some species of fish and probably also some cephalopods) have adults that are able to move between islands, genetic connectivity between the islands' populations in turn depends on factors such as the distance between islands, the speed and direction of currents, and characteristics of each species (e.g. dispersal ability of the larvae). The population size of each island is also directly related to different levels of human exploitation. The east to west oceanographic gradient across the archipelago is also responsible for the distribution of biota. In the western islands, with warmer waters, species with tropical affinities are more abundant, while species with temperate affinities are more abundant in the eastern islands of temperate waters (Brito et al. 2001; Sangil et al. 2007). Finally, the proximity of the oceanic waters to the coast, and given that the Canary Islands are among one of the main routes of migratory species (e.g. tunids), the pelagic systems strongly interact with the benthic-littoral systems. Moreover, since great depths are occur near the coast, deep-water species that migrate to shallow areas during their diel rhythms and are temporarily integrated into coastal ecosystems (Aguilera et al. 1994). In this sense, it is also important to emphasize that some deep fauna occur in shallow caves along the coast (Sangil 2007).

15.2 Research

The following section provides a historical perspective on the development of studies on echinoderms in the Canarian Archipelago. We describe the shift in the scientific strategy for the study of this zoological group and as some data to quantity published material.

Although the echinoderm fauna of the Atlantic is well known, much of the research has focused mainly on Western North Atlantic coastal waters. The European Eastern Atlantic coasts have also been well studied (see Koehler 1921; Mortensen 1925, 1927; Høiseater 1990; Hansson 2001), while Macaronesian archipelagos (Azores, Madeira, Salvajes, Canary Islands and Cabo Verde) are less known. Some studies have been done on the coasts of the Azores and Portugal coasts (Barrois 1888; Nobre 1930, 1931, 1938; Chapman 1951; Marques 1983; Pereira 1997). De Jesus and Abreu (1998) studied Madeira although sampling was restricted to soft bottoms between 20 and 100 m depth. Recently, the echinoderm fauna of the shallow rocky bottoms (0–25 m depth) of the Macaronesian archipelagos has been surveyed in the context of the "Macaronesia 2000 Project" promoted by the Museo de la Naturaleza y el Hombre from Tenerife (Pérez-Ruzafa et al. 2002) and Cabo Verde (Pérez-Ruzafa et al. 1999, 2003a, Entrambasaguas 2008; Entrambasaguas et al. 2008).

Echinoderms are one of the groups of marine animals that have received considerable research effort in the Canaries. Because of its economic impact, research interest has primarily concerned, and sometimes even monopolized, by one species: the sea urchin *Diadema* aff. *antillarum*. This species have been recently described and named as *Diadema africana* (Rodríguez et al. in press.). It is worthwhile to note that researchers have used either *D. antillarum* or *D.* aff. *antillarum* to refer to the Eastern-Atlantic species. We recommend the use of *D. africana* when referring to eastern-Atlantic populations and have used this name throughout the text.

15.2.1 Pioneering Studies

The occurrence of echinoderms in the Canary Islands has been noted by a few authors (Greef in Theel 1886; Panning 1939) and by expeditions in these waters.

Expeditions visiting the archipelago at the end of the 19th century and the first half of the 20th include the Challenger, Travaillerur and Talisman, Princesse Alice, Michael Sars, The Java-Sud Africa Expedition. As an example, Barrois (1888) studied the echinoderm faunas of Azores and Madeira and made comparisons with faunas from Canary Islands and Cabo Verde. Mortensen (1943) provided one of the first references on echinoderms of the Canaries where he refers to the presence of long-spined black sea urchins during his visit to Las Palmas (Gran Canaria) at the beginning of 1930s. Reports from other foreign naturalists also emphasized, several decades ago, large densities of this sea urchin in the shallow rocky subtidal of the Canaries (Johnston 1969). This sea urchin (misidentified as *Centrostephanus longispinus*) was even pointed out as "guilty" of the lack of algae in the rocky subtidal through overgrazing (Johnston 1969).

15.2.2 1980s and 1990s-Establishing the Taxonomic Baselines and the First Bionomical, Biogeographical and Ethological Approaches

The first studies carried out by researchers from institutions on the Canary Islands to describe the species composition of the echinoderm fauna are from the early 1980s. A team from the University of La Laguna, led by J.J. Bacallado, implemented the research projects *Bentos I* and *Bentos II* (Bacallado et al. 1982). These works focused on the collection, identification and taxonomic description of the main animal groups of the marine bottoms of the Canary Islands, including echinoderms (Bacallado et al. 1984, 1985). As a result of these projects the original checklist was improved (Bacallado et al. 1988).

The *Bentos* project was the base for studies on holothurians (Pérez-Ruzafa 1984, 1984a, b; Pérez-Ruzafa et al. 1985a, b, 1992a, b; Pérez-Ruzafa and Marcos 1987), including their spatial distribution (Pérez-Ruzafa et al. 1984a), biogeographical relationships (Pérez-Ruzafa et al. 1984b; Pérez-Ruzafa and Marcos 1987) and behavior (Pérez-Ruzafa et al. 1987). A morphological description of spines for eight common species of echinoids was published by Moreno-Batet et al. (1980). Of particular relevance for echinoids, was the publication of a monograph on the main biological characteristics of *Diadema africana* (Bacallado et al. 1987). This work reinforced previous observations of this research group (Brito et al. 1984) on the important role this sea urchin seemed to play in the structure and organization of shallow rocky bottoms of the archipelago. Importantly, this environmental issue took on a social dimension with the publication of the book by Aguilera et al. (1994) about environmental sustainability of the Canaries. These documents constituted the baseline on which subsequent works were carried out, including taxonomic revisions and additions to all classes of echinoderms.

15.2.3 The "recent" Past-An Explosion of Studies on Echinoderm Ecology at the Start of the 21st Century

The end of the 20th is linked to an increase in studies on the ecology of echinoderms from the Canaries. Of special importance is the project *Macaronesia 2000* supported by the Museo de Ciencias Naturales de Tenerife and directed by J.J. Bacallado. This project was done between 1998 and 2006 and was a great advance for echinoderm studies in the Macaronesian islands (Azores, Madeira, Salvage Islands, Canary Islands and Cape Verde). We have to emphasize the importance of the doctoral thesis of L. Entrambasaguas (2009) and other works derived from this project (Pérez-Ruzafa et al. 1999, 2002, 2003a; Entrambasaguas et al. 2008).

These and previous studies on taxonomy had their final zenith with the completion and subsequent publication of the BIOTA book, which compiled all taxonomic citations published until 2003 for each group of marine animals from the Canarian Archipelago (Pérez-Ruzafa et al. 2003b). This document identified 91 species of echinoderms living in the waters of the Canary Islands (including two subspecies): 15 echinoid species, 18 asteroids, nine crinoids, 15 ophiuroids and 34 holothurian species (Sect. 15.3 and Appendix).

Without doubt, the sea urchin *D. africana* was the center of attention for studies on benthic ecology: high density of the sea urchin and the subsequent formation of "urchin barren grounds" (*blanquizales*—white bottoms with no erect macroalgal cover as a result of overgrazing) catapulted the species to the spotlight of benthic research in the Canary Islands. As a result, studies covered a suite of biological and ecological topics: colonization of artificial substrates and interactions of this sea urchin with other elements of the subtidal biota (Herrera 1998), patterns of circadian movements and feeding preferences (Tuya 2002), reproductive biology and genetic diversity (Garrido 2003), gametogenesis, larval development, settlement and recruitment patterns (Hernández 2006), predation as a mechanism that controls populations, including the role of marine protected areas (Clemente 2007), trophic ecology (Cabanillas 2009), and processes determining structure and organization of benthic communities (Ortega 2010). Two projects: "*Estudio de la biología y ecología del erizo Diadema antillarum y de las comunidades algales de sucesión en diferentes zonas de blanquizal del Archipiélago Canario*" and "*Canarias, Costa Viva*", as well as monitoring studies to evaluate the effectiveness of marine protected areas throughout the Canaries, have provided considerable data on the ecology of *D. africana* during 2001–2004. In particular, a large body of research has described the relationships between size and abundance patterns, and the connection between the abundance and size patterns of *D. africana* populations and algal and fish assemblages.

The end of the 20th century saw a drastic shift in the way science was evaluated in Spain: a consolidation of peer-reviewed international indexed journals (included in international databases where journals are ranked according to their scientific relevance) as the main output of scientific publication. Therefore, the publication of articles in peer-reviewed indexed journals on aspects related to the echinoderms

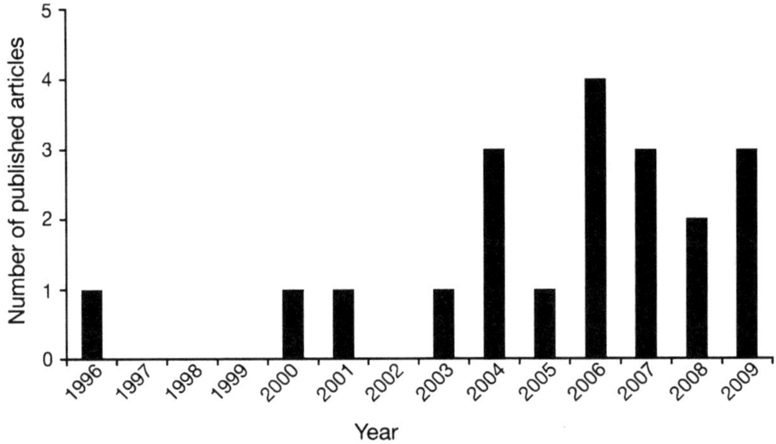

Fig. 15.3 Number of published articles in international, indexed, peer-reviewed journals on aspects related to the sea urchin *Diadema africana* in the Canaries

from the Canaries has increased considerably in the last years (Fig. 15.3). These works include some of the biological and ecological topics derived from the doctoral theses and the projects noted above. Here, we compile these studies on the biology and auto-ecology of *D. africana*, including reproductive biology (Garrido et al. 2000; Hernández et al. 2006a), feeding preferences (Tuya et al. 2001), nocturnal movements and "homing" behavior (Tuya et al. 2004a), seasonality of settlement patterns and its coupling with the oceanographic scenarios of the Canaries (Hernández et al. 2006b, 2010a), as well as technical topics related to the in situ tagging to monitor movement (Tuya et al. 2003; Clemente et al. 2007a) and predation (Clemente et al. 2007b). Another group of studies analyzed the spatial and temporal variability of populations (Tuya et al. 2006a; Hernández et al. 2008a), including the effect of a set of environmental factors (e.g. turbulence (Tuya et al. 2007), the topography of the reef and sedimentation (Clemente and Hernández 2008), respectively human-induced factors such as fishing effort (Hernández et al. 2008a) and the establishment of protection frameworks over the coast (Hernández et al. 2008b). A large body of research analyzed the way *D. africana* affects the composition, abundance and organization of algal assemblages (Tuya et al. 2004b, 2004c, 2005). Two studies analyzed how the populations of this echinoid interact with other parts of the animal biota, including the abundance, diversity and organization of macro-invertebrate assemblages (Tuya et al. 2006b; Ortega et al. 2009). Finally, recent investigations have focused on the way predation, mainly exerted by fishes, shape the abundance and size patterns of *D. africana* in marine protected areas (Clemente et al. 2009, 2010) and non-protected zones (Clemente et al. 2007b). Comparison of protected areas (marine reserves) and zones subject to prominent exploitation that have qualitative and quantitative distinct fish assemblages, in conjunction with previous observations (Tuya et al. 2004b; Hernández et al. 2008a), has been used to demonstrate the key role that predation, chiefly induced by overexploited fish species, exert in the control of populations of *D. africana*. Consequently, overexploitation of sea urchin predators has been directly linked to large densities of *D. africana*, which causes the direct elimination of erect algal beds and results in the prevalence of rocky unvegetated bottoms in the rocky subtidal ("urchin barren grounds"). The disappearance of these algal beds limits benthic primary production, as well as the amount of food and shelter these algal habitats provide for associated invertebrates and fishes. This exacerbates, even more, the deterioration of marine biodiversity in the shallow subtidal waters of the Canarian Archipelago. Last, but not least, there is a positive relation between the increases in densities of *D. africana* and the recent increase in sea water temperature (Hernández et al. 2010a). This work is based on a historical compilation of sea water surface temperature data from 1948 to the present and a continuous collection of data on settlement, recruitment and densities from 2001 to 2008. The major conclusion is that the increase in sea water temperature, coupled with the prevalence of "urchin barren grounds", enhanced the survivorship of *D. africana* recruits. This work predicts an increase in the densities of this sea urchin as well as the extension of "urchin barren grounds" under the predicted scenario of raising temperatures. In conclusion, the rise in sea water temperature has been a relevant factor in conjunction with overfishing of sea urchin predators to enhance densities of the sea urchin.

The high level of social interest about the effect of high densities of *D. africana* has not only increased the publication of scientific manuscripts. Publications have disseminated the characteristics of the sea urchin to society in non-technical language (Brito et al. 2004; Tuya et al. 2004d; Hernández et al. 2009). These publications have tried to raise awareness on the necessity to implement conservation strategies to guarantee the sustainability and conservation of benthic communities in the Canary Islands since high densities of the sea urchins are directly connected with perturbed bottoms. Indeed, increasing interest on preserving the rocky subtidal reefs of the Canaries has been consolidated by the development of the project *Diademar* (Instituto Canario de Ciencias Marinas), which is exploring ways to commercially exploit the high densities of the sea urchin, mainly for human consumption of gonads, as a way of population control. In addition, the *Viceconsejería de Medio Ambiente* and the *Viceconsejería de Pesca* of the Canary Government and several insular administrations (*Cabildos*) have implemented a program to periodically eliminate this sea urchin in large experimental plots across the entire archipelago. The goal is to preserve algal beds (but see Sect. 15.5).

Other echinoderms have not received the same focus of attention within the scientific community. There are only a few studies on other species of sea urchins, such as the genetic diversity (Calderon et al. 2008) and a description of the juvenile phases of *Paracentrotus lividus* (Hernández et al. 2005), and the spatial distribution patterns of *P. lividus* and *Arbacia lixula* across depth gradients in the shallow subtidal (Tuya et al. 2007). Other works focused on some ecological aspects, such the relation between densities of *D. africana* and other species of echinoderms (Ortega et al. 2009), particularly the holothurians (Tuya et al. 2006b).

15.3 Diversity of Echinoderms and Distribution

The echinoderm fauna from the Canary Islands is presently composed of 127 species, nine of Crinoidea, 25 of Asteroidea, 42 of Ophiuroidea, 26 of Echinoidea and 28 of Holothuroidea. Of these, 71 species inhabit the littoral zone, at depths less than 50 m (one of Crinoidea, 16 of Asteroidea, 24 of Ophiuroidea, 13 of Echinoidea and 17 of Holothuroidea) (see Appendix and pictures on Figs. 15.4–8 for the most common species). This biodiversity is similar to that reported for the Mediterranean (147 species in total, 101 in shallow water) or the British Isles (119 and 71 respectively). It is higher than the reported in other archipelagos such as Cabo Verde (66 species in total and 42 in shallow waters) but this is due probably to differences in sampling effort.

The present status of the echinoderm fauna of the Archipelago needs a major revision, because many citations come from old works and indirect references, some species reported could be misidentifications. For example, *Holothuria (Lessonothuria) poli*, was reported for the Canary Islands by Ludwig (Greef, in Theel 1886), Herouard (Tortonese 1965) and for Palma by Tortonese (1965). We have never found the species in our sampling of the different islands. In the same way, the cluster composed of *Holothuria (Holothuria) stellati*, *Holothuria (Holothuria) tubulosa*, *Holothuria (Holothuria) mammata* and *Holothuria (Holothuria) dakarensis*, which have been cited for the archipelago, must be reduced to *Holothuria (Holothuria) mammata* after the taxonomic and genetic review of the group in Atlanto-Mediterranean waters by Borrero-Pérez et al. (2009). According to their results, *H. dakarensis* would be restricted to the Cape Verde Islands and the coast of West Africa, *H. mammata* to the Atlanto-Mediterranean area and *H. tubulosa* to the Mediterranean Sea. *Holothuria stellati* was considered a junior subjective synonym of *H. tubulosa*.

The small holothuroid *Holothuria (Holothuria) helleri* was reported by Pérez-Ruzafa (1984) and Pérez-Ruzafa et al. (1984a) from samples of coral branches at 102 m depth, but the validity of this species must be reviewed as it could be confused with juveniles of other *Holothuria* species.

Some confusion has existed with species with anfiatlantic distributions. Lessios et al. (2001) divided *Diadema antillarum* into Western and Eastern Atlantic clades from a molecular genetic analysis. This could also be the case for the small brooding holothuria *Neocnus incubans*, reported at the Canary Islands as its only locality of its distribution in the Atlantic Ocean (Pérez-Ruzafa et al. 1984a; Pérez-Ruzafa et al. 1984b).

Shallow rocky bottoms are inhabited by few species. The more conspicuous are the sea stars *Ophidiaster ophidianus*, *Coscinasterias tenuispina* and *Marthasterias glacialis*, the holothuroid *Holothuria (Platyperona) sanctori*, and the sea urchins *Diadema africana*, *Arbacia lixula* and *Paracentrotus lividus*. The three sea urchins coexist at all depths down to 30 m, but some vertical zonation can be observed in their distribution (Tuya et al. 2007). *Paracentrotus lividus* is herbivorous and dominates in shallower areas, from 0 to 5 m depth, with algal cover and well

Fig. 15.4 **a** *Antedon bifida* (Pennant, 1777); **b** *Thethyaster subinermis* (Philippi, 1837) (Photo: J. Barquín); **c** *Chaetaster nodosus* Perrier, 1875; **d** *Asterina gibbosa* (Pennant, 1777); **e** *Hacelia attenuata* Gray, 1840; **f** *Narcissia canariensis* (d'Orbigny, 1839); **g** *Ophidiaster ophidianus* (Lamarck, 1816); **h** *Echinaster sepositus* (Retzius, 1783)

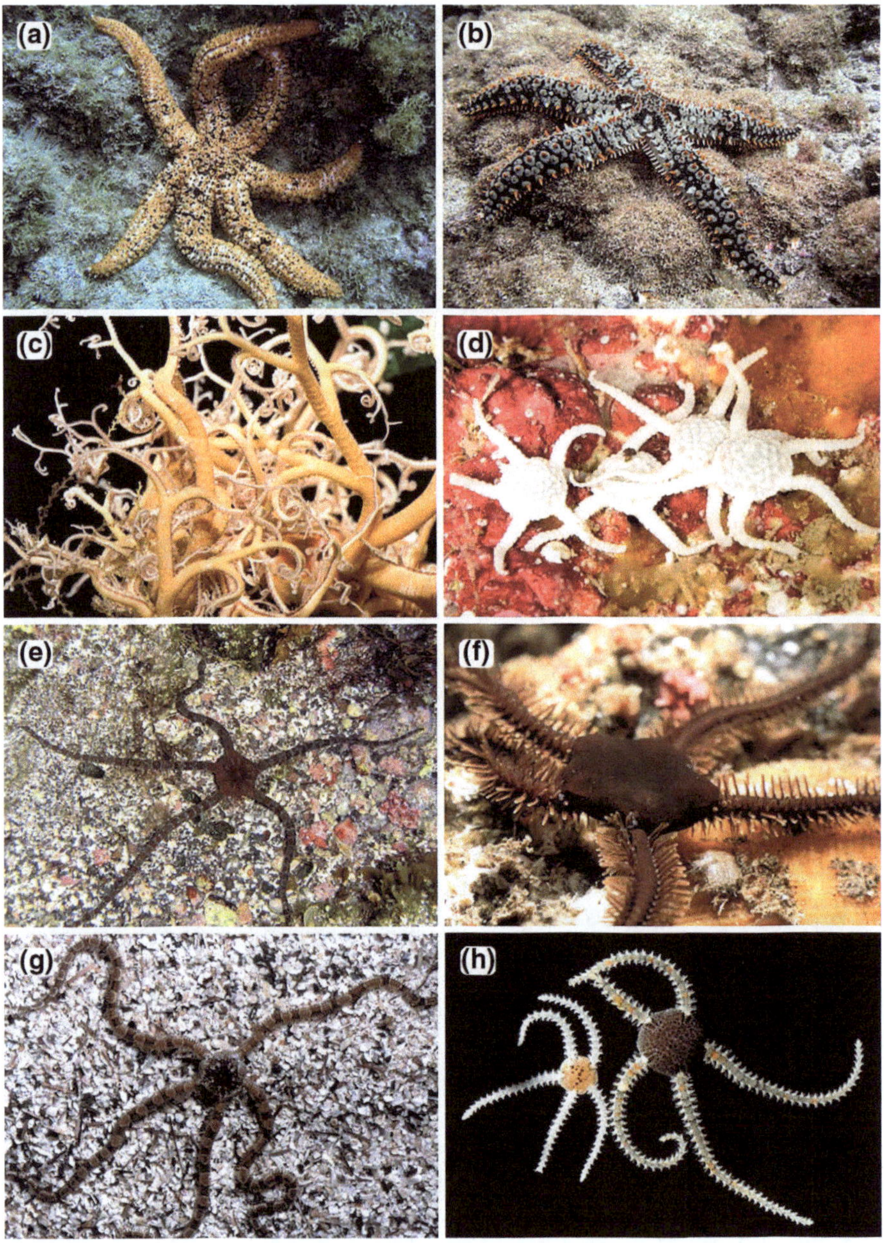

Fig. 15.5 **a** *Coscinasterias tenuispina* (Lamarck, 1816); **b** *Marthasterias glacialis* (Linnaeus, 1758);**c** *Astropartus mediterraneus* (Risso, 1826); **d** *Ophiolepis paucispina* (Say, 1825); **e** *Ophioderma longicaudum* (Retzius, 1789); **f** *Ophiocomina nigra* (Abildgaard, 1789); **g** *Ophiopsila aranea* Forbes, 1845; **h** *Amphipholis squamata* (delle Chiaje, 1829)

Fig. 15.6 **a** *Ophiothrix fragilis* (Abildgaard, 1789); **b–c** *Stylocidaris affinis* (Philippi, 1845); **d** *Centrostephaus longispinus* (Philippi, 1845); **e** *Diadema africana* (Rodríguez, Hernández, Clemente 2010); **f**, **g** *Arbacia lixula* (Linnaeus, 1758); **h** *Coelopleurus floridianus* Agassiz, 1872

Fig. 15.7 **a** *Sphaerechinus granularis* (Lamarck, 1816); **b–c** *Paracentrotus lividus* (Lamarck, 1816); **d** *Genocidaris maculata* Agassiz, 1869; **e** *Echinocyamus pusillus* (Müller, 1776); **f, g** *Brissus unicolor* (Leske, 1778); **h** *Echinocardium cordatum* (Pennant, 1777)

Fig. 15.8 **a** *Neocnus incubans* Cherbonnier, 1972; **b** *Holothuria mammata* Grube, 1840; **c**, **d** *Holothuria arguinensis* Koehler and Vaney, 1906; **e**, **f** *Holothuria sanctori* delle Chiaje, 1823; **g** *Euapta lappa* (Müller, 1850); **h** *Leptosynapta inhaerens* (Müller, 1776)

illuminated environments. *Arbacia lixula* is more abundant at slightly deeper waters (1–15 m) and in sciaphilous assemblages, probably having a slightly higher trophic level than *P. lividus*. Finally, *D. africana* is the most abundant species, reaching densities up to 240 ind. per 100 m^2 covering large extensions of unvegetated rock. It can reach depths up to 100 m. A less abundant species that can be confused with *Diadema* is another long-spined sea urchin *C. longispinus*. The sea star *Hacelia attenuata* can also be found on rocky and detritic bottoms.

On rocky substrata, but close to soft bottoms, the holothurians *H.* (*H.*) *mammata* and *Holothuria* (*Lessonothuria*) *arguinensis* are common. The daily feeding behavior of these species in Canary Islands has been described by Pérez-Ruzafa and Marcos (1987). *Holothuria* (*Vaneyothuria*) *lentiginosa lentiginosa* inhabits rocky and sandy bottoms deeper than 20 m. *Holothuria* (*Panningothuria*) *forskali* has been collected on rocky substrata at 348 m depth at Tenerife Island (Pérez-Ruzafa et al. 1987). This is the southernmost and greatest depth reported for the species.

Pérez-Ruzafa (1984) made the first report of the small incubating holothurian *Neocnus incubans* in the Atlantic from the Canary Islands, it was always found on *Cystoseira* spp. on the upper infralittoral. The small, incubating ophiuroid, *Amphipholis squamata*, is common in photophilous algae.

The ophiuroid *Ophiothrix* (*Ophiothrix*) *fragilis*, *Ophiocomina nigra*, *Ophiomyxa pentagona* and *Ophioderma longicauda* are common under stones and in crevices. The nocturnal anfiatlantic apodid holothurian *Euapta lappa* is common in cryptic habitats and is particularly abundant in the western part of the archipelago where the waters are warmer. Other species of ophiuroids on rocky bottoms of the archipelago are *Ophioconis forbesi*, *Ophioderma appressum*, *Ophiocoma pumila*, *Ophiactis savignyi* and *Ophiactis virens*.

The sea stars *Luidia ciliaris* and *Astropecten aranciacus* and the sea urchin *Echinocardium cordatum* occur on sandy bottoms. The sea urchin *Brissus unicolor*, occurs on sandy bottoms under stones. The sea stars *Astropecten irregularis irregularis* and *Peltaster placenta*, the ophiuroids *Ophiura* (*Dictenophiura*) *carnea* and *Ophiura texturata*, the sea urchins *Genocidaris maculata*, *Schizaster canaliferus*, *Spatangus purpureus*, *Plagiobrissus costai* and the holothuroids H. (*H.*) *mammata* and *Thyone fusus* occur on sandy and muddy bottoms in shallow water and *Eostichopus regalis* and *Stichopus tremulus* in deeper water.

The sea urchin *Echinocyamus pusillus* and the ophiuroid *Ophiura grubei* are common on detrictic bottoms. The ophiuroid *Ophiacantha setosa* can be found on gorgonians.

15.3.1 Biogeographic Relationships of the Echinoderm Fauna of the Canary Islands

The origin of the echinoderm fauna of the Canary Islands is mainly warm-temperate (28 %) and with a wide thermal distribution (26 %) (Fig. 15.9). The strictly

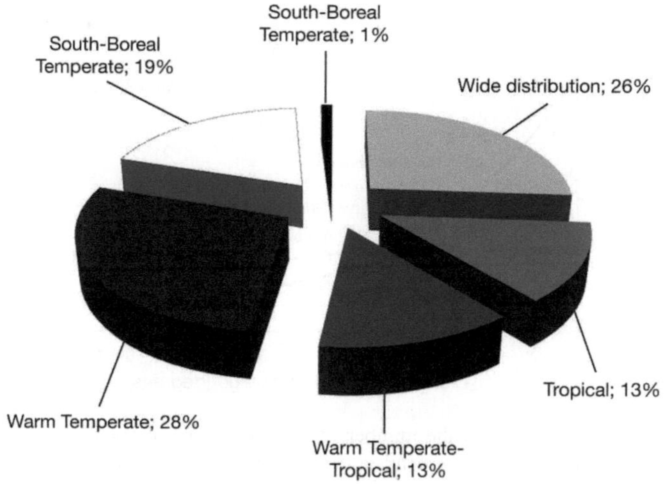

Fig. 15.9 Origin and affinities of the echinoderm fauna of the Canary Islands in relation with the thermal regions in the Atlantic described by Briggs (1996)

south boreal component is represented only by *Mesothuria connectens* reported for the archipelago by Mortensen (1927) from material collected by the Talisman expedition in 1975 at 2,518 m. However, the South-Boreal-Temperate component constitutes the 19 % of the species. It includes the shallow water species *Antedon bifida*, *Luidia ciliaris*, *Asterina gibbosa*, *Echinaster sepositus*, *Henricia sanguinolenta*, *Ophiocomina nigra*, *Echinocyamus pusillus*, *Aslia lefebrei*, *Pawsonia saxicola*, *Stichopus tremulus*, *Holothuria forskali* and *Leptosynapta inhaerens*, and the deep water species *Ophiomitrella clavigera*, *Amphilepis norvegica*, *Ophiactis abysicola*, *Araeosoma fenestratum* and *Araeosoma histrix*. The tropical (13 %) and tropical-warm temperate faunas (13 %) are well represented. Except for *Euapta lappa* which occurs at depths between 10 and 20 m and *Ophiolepis paucispina*, which reaches a maximum depth of 37 m, all strictly tropical species present at the archipelago occur in deep waters or over a wide range of depths. These include *Astropecten hermatophilus* (15–1,500 m), *Narcisia canariensis* (25–155 m), *Oreaster reticulatus* and *Astrophyton muricatum* (0–70), *Ophiomisidium pulchellum* (1,380–1,830 m), *Ophiocoma pumila* (0–368 m), *Ophionereis sexradia* (18–130), *Plesiodiadema antillarum* (651–3,111), *Diadema africana* (0–400), *Bathyplotes pourtalesi* (134–1,477 m) or *Psychropotes semperiana* (1,143–5,600 m). All of these, except *Astropecten hermatophilus*, *Narcisia canariensis* and *Ophionereis sexradia*.

The tropical-warm temperate component is represented mainly by Eastern Atlantic-Mediterranean species as *Astropecten aranciacus*, *Thethyaster subinermis*, *Chaetaster longipes*, *Peltaster placenta*, *Hacelia attenuata*, *Ophidiaster ophidianus*, *Ophiomyxa pentagona*, *Astropartus mediterraneus*, *Ophiacantha bidentata*, *Ophiacantha setosa*, *Ophiacantha valenciannensi*, *Ophiura* (*Dictenophiura*) *carnea*, *Ophiura grubei*, *Ophioderma longicaudum*, *Ophiopsila*

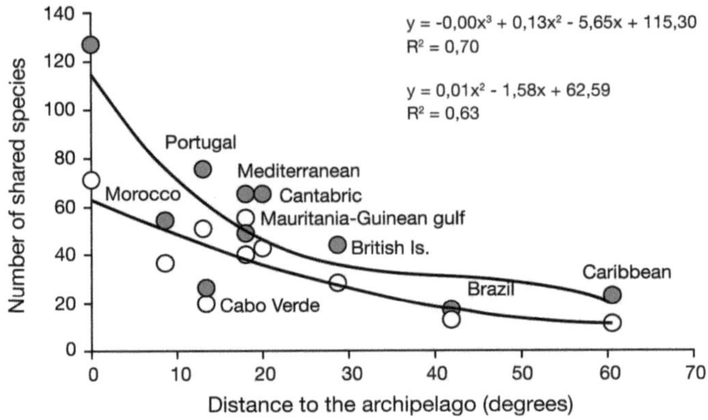

Fig. 15.10 Relationship between the number of shared species of the different biogeographic areas in the Atlantic and the Canary Islands and the distance to the archipelago. This relationship is negative emphasizing the importance of oceanic currents and connectivity to explain the composition of the fauna in oceanic islands. White circles: shallow water species (<50 m); Grey circles: all depths

guineensis, *Ophiactis virens*, *Ophiothrix maculata*, *Centrostephaus longispinus*, *Brissopsis atlantica mediterranea*, *Holothuria* (*Holothuria*) *dakarensis*, *Holothuria* (*Platyperona*) *sanctori*, *Holothuria* (*Vaneyothuria*) *lentiginosa lentiginosa* and *Molpadia musculus*. There are also some anfiatlantic species as *Coscinasterias tenuispina*, *Ophioderma appressum*, *Ophiactis savignyi*, *Arbacia lixula*, *Genocidaris maculata*, *Brissus unicolor*, *Benthodites lingua* and *Psychropotes depressa*.

There are few analyses on the distribution and biogeography of Atlantic echinoderms (Clark and Downey 1992). The relationships of the echinoderm fauna of the Canary Islands with other Atlantic faunas were analyzed for the class Holothuroidea (Pérez-Ruzafa et al. 1984b) and in the context of South-Western Mediterranean echinoderm fauna relationships (Pérez-Ruzafa and Marcos 1987).

The shallow water echinoderm fauna of Canary Islands is a consequence of its oceanography and location in the Atlantic. The number of species shared with other geographical areas is directly related to distance (Fig. 15.10). Therefore, of the 71 species inhabiting the archipelago above 50 m depth, the highest number corresponds to the Mediterranean (55 species), Portugal (50), the Cantabric (43) and Mauritania and Guinea Gulf region (40). The low number shared with Morocco (36) is probably due to the scarcity of works in this region. In fact, the number of species increases to 54 when deep water fauna is considered because most of the deep water expeditions which passed through the archipelago sampled in both areas.

The lower number of shallow water shared species is with the Caribbean (12), Brazil (13), and Cabo Verde Archipelago (20). In the case of the American coasts, this can be explained because of the distance. But in the case of Cabo Verde, the explanation has to do with the oceanographic currents system that precludes the flux of tropical species from south to north in the Eastern Atlantic.

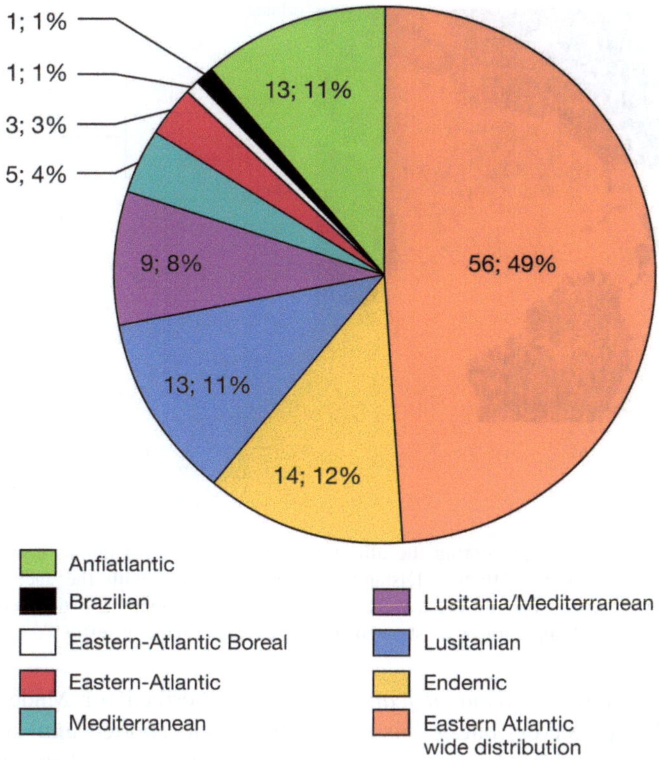

Fig. 15.11 Contribution to the echinoderm fauna of the Canary Islands of the different faunal groups according to the biogeographic regions and provinces described by Briggs (1996). Figures indicate the number of species shared and the percentage of the total

A 56 % of the species present in the archipelago has a wide distribution in the Eastern Atlantic area (Fig. 15.11). Furthermore, 27 species (23 %) are from the Lusitanian province (five from the Mediterranean, 13 from the Atlantic and nine which are present in both areas). Three species (*Astropecten hermatophilus*, *Narcisia canariensis*, *Ophionereis sexradia*) are shared exclusively with the tropical Eastern Atlantic region (from Cabo Verde Islands and the Cape Verde on the mainland to Angola (Briggs 1996). Only one species, the deep water holothurian *Mesothuria connectens*, is exclusively from the Eastern-Atlantic Boreal region. Finally, 13 species (13 %) are characteristic of the Caribbean-West Indian and Brazilian provinces in the Western Atlantic.

Only 13 species have been reported exclusively from the Canarian Archipelago waters: the sea stars *Lonchotaster tartareus*, *Thoracaster cylindratos*, *Colpaster scutigera*, *Freyastera tuberculata*, the ophiuroids *Ophiomusium serranum*, *Ophiomusium vallidum*, *Ophiozonella molesta*, *Ophiura inermis*, *Amphiura grandis*, the sea urchins *Cidaris blakei*, *Salenocidaris rarispina*, *Aceste bellidifera* and the holothurian *Peniagone ferruginea*. All are deep water species occurring at more

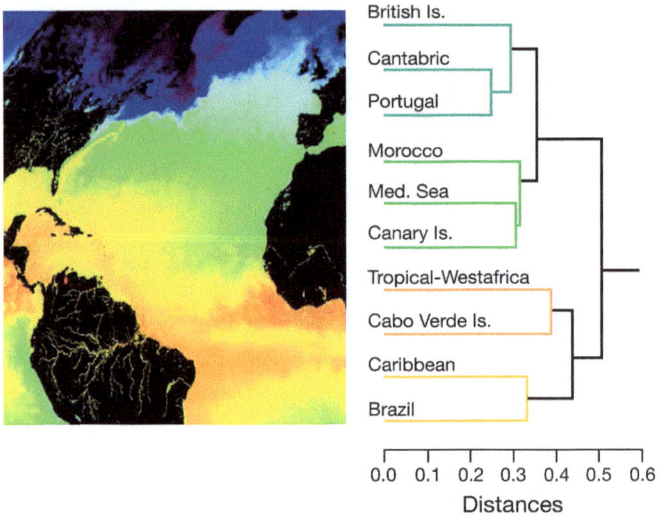

Fig. 15.12 Cluster tree representing the affinities of the echinoderm faunas between different geographic regions in the Atlantic. Distances were constructed with the Jaccard index for presence-absence data. Note the close relationship between distances and differences in mean annual surface water temperatures as showed by the satellite image at left

than 900 m depth. *Diadema africana* has been described for Madeira, Canary Islands and Cape Verde, but as this species had been reported until now as *Diadema antillarum* or *Diadema* aff. *antillarum*, a review of the distribution of the species in the Eastern Atlantic must be done.

The application of the Jaccard index to the echinoderm fauna of these mentioned regions (including shallow and deep sea faunas) produces two main groups (Fig. 15.12). The first includes British Islands, Cantabric Sea and Portugal coasts and the Atlantic coast of Morocco, the Mediterranean and the Canary Islands. The second group links the Caribbean and Brazil on one hand with Cabo Verde and the Mauritanian and Guinean Gulf on the other. These groups represent the four major thermal regions in the Atlantic surface waters: South Boreal, Cold-temperate, Warm-temperate and Tropical (Briggs 1996). The tropical zone is delimited by the 20 °C isotherm for the coldest month.

15.4 Echinoderm Ecology: Implications for Conservation

Echinoids are by far the most abundant group of echinoderms of shallow waters of the Eastern Atlantic Archipelagos. Few detailed ecological studies have focused on other taxa in this region. Comparisons of spatial distribution of sea urchins *A. lixula*, *P. lividus* and *D. africana* show a consistent depth-dependent partitioning pattern on rocky reefs, in which the latter is the competitive dominant species both at the Canary Islands (Tuya et al. 2007) and Madeira Island (Alves et al. 2001). This depth gradient in the vertical distribution of the three sea urchin species has been related to their distinct morphological characteristics, and consequently, their different resistance to natural hydrodynamic forces (Tuya et al. 2007). *Arbacia lixula* and *P. lividus* seem to be morphologically adapted to minimize the chances of being dislodged by high water motion at the intertidal and first few meters of the subtidal and coexist in relatively high densities only in the shallowest depths (<5 m). Densities of *D. africana*, less adapted to support intense water movement, increase with depth. The result is a general pattern in the vertical distribution of these sea urchins in the warm-temperate rocky reefs of the eastern Atlantic Ocean (Alves et al. 2001, Tuya et al. 2007), in which wave exposure and island slope that affects wave energy are major environmental factors determining densities of these key herbivorous sea urchins (Hernández et al. 2008a). Moreover, protected coasts have calmer waters that promote retention of urchin larvae (Hernández 2006). Larvae at exposed sites larvae tend to be carried away. The result is that areas with higher hydrodynamics show greater urchin density in a deep band (15–20 m) just below the algal stand, whereas along sheltered coasts algal beds the urchins occur only at the first meters depth where more food is available (Hernández et al. 2008a).

Because of the biological and ecological features, such as size, motility, feeding preferences or longevity, several urchin species play a key role, even more important than other herbivores, in controlling epibenthic communities and in turn of the overall structure of coastal communities (Lawrence and Sammarco 1982). The main consequence of this overall phenomenon of intense grazing activity is a shift from large areas of rocky reefs covered by complex erect macroalgal beds to overgrazed substrates ('urchin barrens' or 'barren grounds') dominated by encrusting algae and some sessile invertebrates.

The influence of echinoids grazing activity on rocky reefs communities of the eastern Atlantic oceanic islands, particularly in the Canary Islands, have been well studied (Alves et al. 2001, Tuya et al. 2004c; Hernández et al. 2008b). These are relatively small barren grounds produced by *P. lividus* in tidepools, and by *A. lixula* on shallow bottoms and around artificial jetties (JC Hernández pers. obs.). However, barren grounds resulting from the intense grazing activity of *D. africana* are more extensive and commonly spread throughout the entire region (Aguilera et al. 1994; Brito et al. 2004; Tuya et al. 2004c), reaching up to 50 m depth and covering about 75 % of the total littoral rocky bottoms of the Canary Islands (Barquín et al. 2004). Densities of this species can reach more than 12 ind m^{-2}, with severe consequences (Brito et al. 1984, 2004; Alves et al. 2001; Tuya et al. 2004b; Hernández 2006). Field monitoring and experimental manipulations of urchin densities have shown its major effects on algal communities (Hernández et al. 2008a). At moderate densities, sea urchin grazing can stimulate plant production and promote species diversity. Because of this sea urchins should be considered an integral part of ecosystem dynamics (Sammarco 1982; Valentine et al. 1997). However, densities of as few as 4 urchins m^{-2} can remove up to 70–80 % of algal cover (Hernández et al. 2008a) and dramatically reduce macroalgal diversity, resulting in the formation of impoverished sea urchin-dominated areas denuded of all but encrusting algae (Lawrence 1975; Sangil et al. 2006a, 2006b). Compared to typical urchin barrens elsewhere in the world (Chapman 1981; Pearse and Hines 1987; Valentine and Johnson 2005; Guidetti and Dulcic 2007), barrens generated by *D. africana* in the eastern Atlantic appear to be maintained by relatively low numbers of sea urchins (starting at 2 ind m^{-2}). Probably, the oligotrophic condition of waters surrounding the archipelago provides limited nutrient resources that are not sufficient to maintain extremely high urchin densities. However, species movement and food consumption rates, which involve grazing large areas of rocky substrate at night (Tuya et al. 2004a), allow comparatively lower urchin densities to maintain a barren habitat denuded of most erect macroalgae biomass.

Loss of macroalgal beds due to *D. africana* grazing activity produces not only lower species diversity (Herrera 1998; Garrido 2003) but also a lack of habitat suitable for fish feeding and reproduction (Brito et al. 2004; Tuya et al. 2005; Clemente 2007). Macroalgae constitute a major energy source that maintains the food web, and are the main biological engineers on rocky reefs of the Canary Islands, which lie between tropical and temperate waters and has no coral reefs. Consequently, algal cover provides the habitat that supports the whole system. It constitutes a well-known indicator of benthic conservation status (Hernández et al. 2008b). Another good example of a third-order change that occurs as a consequence of the development of urchin barrens is the distribution and abundances of large holothurians, linked to specific type of habitats across shallow-waters of the eastern Atlantic (Tuya et al. 2006b). This fact could be related to variations in the availability of particulate organic matter across habitats. Benthic grazers, such as sea urchins, can transform up to 70 % of the consumed algae to detrital fecal matter (Vadas 1977; Koike et al. 1987; Frantzis and Gremare 1992; Lison de Loma et al. 2000; Mills et al. 2000; Mamelona and Pelletier 2005). High densities of *D. africana* on barren grounds likely produce large amounts of fecal pellets, as a source of fresh particulate organic matter (Lawrence 1975; Wotton 1994; Lawrence and Klinger 2001; Levinton et al. 2002; Mamelona and Pelletier 2005). Consequently, holothurians were significantly more abundant in *D. africana* barrens compared with vegetated habitats, a pattern that was strongest for the numerically dominant species, *Holothuria sanctorii* (Tuya et al. 2006b).

As a major herbivore, *D. africana* is recognized as having a key ecological role that threatens the resilience of marine systems in the eastern Atlantic, with the subsequent establishment of undesired conservation states (Hernández et al. 2008a). The reported increasing frequency of overgrazing events, concomitant with loss of associated ecosystem goods and services, generates a serious environmental problem that constitutes a key threat to coastal marine ecosystems and local communities of the islands. Hence, the necessity of managing the populations of *D. africana* to prevent overgrazing management has become apparent, especially in areas of the Canary Islands where algae-associated services (such as fishing stocks) are important for livelihood sustenance. Algal stands are very restricted in space in the Canary Islands and have shown a temporal regression related to the demographic explosion of the sea urchin during recent years (Clemente 2007; Hernández et al. 2008a). High percentage cover is usually attained only in shallow water stands at exposed sites where hydrodynamics prevent urchin activity (Alves et al. 2001; Tuya and Haroun 2006), at certain marine protected areas (Clemente 2007; Hernández et al. 2008a), or at barely perturbed areas associated with a delayed anthropogenic development such as isolated and less populated islands (Clemente et al. 2009).

Many factors are thought to be involved in the recent expansion of *D. africana* populations and regression of algal beds at the eastern Atlantic Archipelagos. We can indicate an array of environmental factors that promote high urchin densities, such as substrate topography and diversity and certain oceanographic events that can influence sea urchin settlement, recruitment and survival (Hernández et al. 2008a). In addition, several anthropogenic factors, such as the number of islanders and tourists per coastal perimeter, the number of operational fishing boats and the overfishing of the sea urchins' natural predators have been associated with increased urchin abundances (Hernández et al. 2008a; Clemente et al. 2009).

Diadema africana is a termophilus species, which probably constitutes a competitive advantage for its proliferation at the sublittoral latitude of the Canary Islands in contrast to sea urchins of temperate affinities such as *P. lividus*. This has been especially relevant in the last decades, when a trend of progressive sea water warming has occurred in the region as a consequence of global climate change processes (Brito 2008). In fact, a long-term study has detected a clear relationship between sea surface temperature and *D. africana* settlement (Hernández et al. 2010a). Urchin settlement was higher during warmer years. Temperatures above 24 °C produced major settlement pulses and higher post-settlement survival rates in barren areas. However, some variability across habitats indicates resilience of the systems. While sea water temperature is a major factor influencing *D. africana* density in urchin barrens because it enhances post-settlement survival, specific factors in healthy macroalgal beds constrain settlement, increases post-settlement mortality and controls sea urchin populations (Hernández et al. 2010a). The study shows that warmer summer temperatures (temperatures at settlement) have been more frequent since the early 1980′s with a parallel increase in urchin population densities (Hernández et al. 2008a). Assuming climate change predictions (IPCC 2007) for the future are correct, the strong positive relationship between thermal history and post-settlement survival of this ecologically important species suggests that barren grounds will become more extensive in the future as seawater temperatures warm in relation to anthropogenic activities. Nonetheless, mass mortality events, as found for *P. lividus* in 2003 (Girard et al. in press), could became more frequent due to warmer waters. In this sense, we would like to point out that researchers have observed mortality events of *D. africana* in Madeira and Gran Canaria islands in 2010 (F. Alves, M. Kaufman and F. Espino pers. com., respectively). However, it is too soon to know the cause of this recent phenomenon that, initially, seems to be very spatially restricted. In both cases, urchins were infected by the bald sea urchin disease, and possibly originated by the bacteria *Vibrio alginolitycus* (Hernández et al. 2010b).

Regulation of sea urchin populations also depends on certain trophic interactions. Experimental studies have identified key fish species that successfully prey upon *D. africana* and have the potential to control their populations (Clemente 2007; Clemente et al. 2011). This has important implications for coastal management. By controlling herbivore densities, and thus the amount of intact macroalgal cover, predatory fish indirectly aid in preserving ecosystems service or capacity to provide locally important fishing resources and coastal productivity in sublittoral rocky reefs. The predatory fish observed are mainly species very susceptible to fishing (Clemente 2007). Consequently increased fishing activities have been associated with lower numbers of these predatory species and higher urchin densities in the Canary Islands (Clemente et al. 2009; 2011) and elsewhere (Sala et al. 1998; Guidetti 2006; Guidetti and Dulčić 2007). Additionally, fishing decreases the mean size of fish. This adds to the loss of species capable of consuming large sea urchins and resulting in a smaller escape by size from predation that favour their proliferation. The removal of fish predators by severe overfishing in most areas of the Canarian Archipelago have led to an urchin population outbreak driven only by settlement and recruitment in the absence of any kind of predatory or top-down control (Hernández 2006; Clemente 2007). This ecological disequilibrium, mediated by trophic cascade processes, has major consequences for the structure and function of the entire benthic community.

Because top-down effects on *D. africana* at the eastern Atlantic archipelagos depend on specific predatory species and their abundances within the food web, preserving the diversity of fish predators must be a priority of conservation efforts (Clemente et al. 2010). However, at this time this scientific knowledge has not been fully taken into consideration by policy makers and no protection measures have been directly applied to key predators of the sea urchin. There is only one kind of effective management tool in force that can indirectly contribute to the control of sea urchin populations. This is marine protected areas in which fishing restriction measures are implemented to allow fishing stocks to recover (Hernández et al. 2008a). As elsewhere around the world (Sala et al. 1998; Babcock et al. 1999; McClanahan et al. 1999; Guidetti 2006), studies in the Canary Islands have stressed the value of marine protected areas as a tool for ecosystem restoration; not only for fishes and fishing resources that increases predation on urchins and indirectly aids regeneration of benthic communities (Hernández et al. 2008a; Clemente et al. 2009). In fact, the three marine protected areas within the Canary support higher densities of fish predators and higher levels of predation intensity on *D. africana* compared to equivalent non-protected locations (Clemente 2007; Clemente et al. 2011).

Regardless of the complexity of factors involved in regulation of overpopulations of *D. africana* may be achieved by aiding the recovery of the predatory fish community with appropriate fishing bans and protection measures. Current knowledge about the species' population dynamics has critical implications for the conservation of rocky reef ecosystems in the eastern Atlantic Islands. They add to a growing body of research suggesting that the incidence of barrens and their associated impacts are related to human disturbance (Steneck 1998; Jackson et al.

2001). However, active management actions directed to control urchin population have been limited to developing a strategy of manually removing sea urchin by scuba divers, providing evidence that erect macroalgae have the potential to recolonize encrusting macroalgae barrens (Brito et al. 2004), but of doubtful outcome in the long-term. Combined perturbations caused by overfishing and progressive sea water warming are likely to result in an environment increasingly favourable to the sea urchin, which seems difficult to control with such small-scale actions. Conservation policies need to focus efforts on specifically protecting healthy macroalgal bed habitats and the predatory fishes within them, because they are predicted to increase post-settlement mortality of *D. africana*. Such actions could improve ecosystem function by retaining the equilibrium between sea urchins and macroalgae.

15.5 Aquaculture and Fisheries

Pre-hispanic human settlement in the Canary Islands, commondly known as Guanches (Majos-Lanzarote and Fuerteventura; Canarii-Gran Canaria; Guanches-Tenerife; Benahoritas-La Palma; Bimbaches-El Hierro) usually consumed sea urchins. Indications of pre-hispanic human consumption have been discovered in various shells accumulations locations in the islands. However sea urchin species have been identified only at the "Guinea deposit" in Frontera-El Hierro (Martín Oval et al. 1985). The species most consumed was *P. lividus*, probably because of its abundance in intertidal rocky areas. Nowadays, this sea urchin is still consumed by the islanders. However it remains only in particular coastal areas as a tradition and no commercial exploitation exists. Presently, and probably then, this sea urchin is also used as fish bait for fishing traps or to attract fishes to hook lines. In some coastal locations this is the only known use. Today, other sea urchin species, such as *Arbacia lixula*, *Diadema africana* and *Sphaerechinus granularis*, are also used as bait. *Arbacia lixula* and *D. africana* are collected by snorkeling while *S. granularis*, due its deeper distribution, is a bycatch of trammel nets. The number of urchins used for this purpose is not known and is very difficult to monitor due to the diversity of discharging places, many of them located at inaccessible sites.

Paracentrotus lividus is commercially exploited in several European countries such as Ireland, France, Portugal, Italy, Greece and Spain. In the Spanish Peninsula, it is mainly exploited in the northwest (Galicia), north (Cantabria, Asturias y Pais Vasco), northeast (Cataluña), and in the south (Andalucía). Although, it is locally consumed in all these Spanish Comunidades, France is the main consumer of sea urchins in Europe and the second consumer in the world after Japan (Grosjean 2001). At the present, some of the most important fisheries in Europe like the ones in Ireland and Brittany have collapsed (Le Direach 1987; Andrew et al. 2002). In Spain, fishing activities regulations depend on the "Comunidad Autónoma". While Galicia has a well developed sustainable fishing activity (Catoira-Gòmez 1992), others like Asturias, Cantabria and Andalucía have a growing semi-professional activity. However, the general overfishing situation of the traditional resources could favour an increase of sea urchin fisheries in the near future (Thorpe et al. 2000). At this time accurate stock studies are needed to promote appropriate regulations for a sustainable fishing activity and prevent the collapse of this resource.

Human fishing activity effects on *P. lividus* populations, although it has not been studied in the Canary Islands, are evident in many intertidal habitats of the islands compared to remote or marine protected areas locations. In these areas, of human use, *P. lividus* is scarce and is restricted to inaccessible deep cavities in the basaltic volcanic rock. In the Canary Islands, fishing activities in the intertidal are regulated. Only Sport Fishing Licence holders can fish for shellfish during weekends and holidays. No number or close season regulations exist. In our opinion, this is an insufficient measure to regulate shellfish fishing activities that favors overfishing in many intertidal benches. This situation has focused the attention of managers who have recently funded a National Project to optimize and promote *P. lividus* culture in Galicia and in the Canary Islands. This project aims to improve the technical and biological aspect of larval culture and to develop culture systems at different locations along the Spanish coast to provide juveniles for farms or repopulation. This project was supported by the *Ministerio de Medio Ambiente y Medio Rural y Marino through the Junta Asesora de Cultivos Marinos* (JACUMAR). It has contributed to the improvement of spawning and fertilization, larval culture and metamorphosis-settlement.

In addition to human collection, other factors such as temperature and competition also modify the abundance of *P. lividus*. The Canary Islands are the southern-most limit of this temperate species. It is more abundant off the eastern, colder islands. Although, there are no specific studies that assess the temporal abundance of this sea urchin, we have noticed that anomalous warm summers like occurred in 2003 were associated with localized mass mortalities at several intertidal and subtidal locations in Tenerife (Girard et al. in press). The sea water warming trend in the Canarian Arquipelago (Hernández et al. 2010a) will probably increase the frequency of these mortality events in the future, which would decrease the intertidal populations. Bathymetric segregation of *P. lividus* occurs in the littoral area. It is more abundant in shallow waters between 1–5 m (Tuya et al. 2007). This bathymetric segregation is mainly caused by competition with the grazing of the abundant *D. africana* which has greatly reduced food resources and has limited *P. lividus* distribution to shallow macroalgae stands refuges. Otherwise it could reach deeper waters as in the Mediterranean Sea. In summary, the depletion of macroalgae stands by *D. africana*, sea water warming and human collection is modifying the potential distribution of *P. lividus* in the archipelago.

In recent years, due to the outbreak of *D. africana*, local management agencies have become interested in the development of a commercial sea urchin fishery. At the same time, these agencies are proposing reducing activity of *D. africana* at some sites to gain in the conservation status of rocky bottoms of the Canary Islands. However, both strategies have the potential to restore local unproductive barren grounds to algal stands (Brito et al. 2004). These strategies should be combined with other conservation measures such as the exclusion of urchin predators from fishing and protection of algal stand ecosystems. Combining these strategies would ensure that the ecosystem would keep the balance between urchins and algae. Urchin removals alone are inadequate (but see Sect. 15.4). In general, the management of urchin outbreak cases around the world has consisted in urchin removal. However better results have been always obtained when this is done inside a marine protected area (Tegner and Dayton 1977; McClanahan et al. 1996). This kind of management measures has been proposed for the Canary Islands several times (Clemente 2007; Hernández et al. 2008a; Clemente et al. 2009; Clemente et al. 2011).

These combined actions would potentially decrease the frequency and duration of urchin overgrazing but limited knowledge of direct and indirect effects of these ecosystem scale actions makes it difficult to assess the applicability and sustainability of this method. Therefore we call for governmental responsibility and rigor when planning this kind of action. Uncontrolled removal of urchins during the planned fishery could cause urchin populations to "crash" with potential ecological and economic side effects. The recent "boom and bust" trend of sea urchin fisheries around the word (USA, Mexico, Canada) (Andrew et al. 2002; Uthicke et al. 2009), most of them considered nowadays as overexploited, indicates the need for more information on basic population parameters of species when implementing removal actions (Russell 2000). Rigorously evaluating population demographics has, in the case of the species *D. africana*, both ecological and economic significance, especially now that it is beginning to be commercially harvested. We encourage seeking sustainable solutions to the increasing populations of the sea urchin *D. africana*, trying to contribute to the correct management and conservation of the fragile marine resources of the Canary Islands. However, we have to develop standards for the commercial exploitation of this species of what we lack today.

15.6 Recomendations

We have divided our recommendations into three major sections: biodiversity, ecology and fisheries. These constitute the main research lines that currently drive research on echinoderms in the Canarian Institutions (universities and research institutes). Moreover, in each section we have pointed out research gaps that should be considered for future studies in the archipelago.Biodiversity

- Accurate information about spatial distribution of echinoderm fauna is needed to precisely identify species habitats and depths ranges.
- Genetics studies of some taxonomic groups are needed to clarify biogeographical patterns and taxonomic identification problems (e.g. *Arbaciella elegans* is thought to be small *Arbacia lixula*).

- Field explorations should be extened to deeper habitats such as rhodoliths and deep coral communities. These have been poorly studied and probably contain a rich echinoderm fauna.

Ecology

- Population dynamics of key sea urchin species should be monitored, including their influence in triggering transition phases between algal stands and unproductive urchin barren grounds.
- The effectiveness of marine protected areas as tool to prevent and revert urchin barren situations should be studied.
- The effects of sea water warming and acidification on the calcification processes in larvae and adults of echinoderms and its relevance on population dynamics should be analyzed. The influence of environmental parameters is of special importance for species of echinoderms that modify the habitat and whose activity can greatly affect the whole system.
- The magnitude of the recent mortality events of *D. africana* in Maderia and the Canary Islands, and any related potential ecosystem wide effects should be determined.

Fisheries and aquaculture

- Tagging-recapture studies should be done to determine age, growth rate and natural mortality of sea urchins. These studies, combined with other population surveys, will provide baseline data to develop an adequate strategy of coastal management for the planned fishery of *D. africana*.
- Experimental reduction of sea urchin densities inside Marine Protected Areas that are still affected by dense urchin barren grounds should be done to identify effective strategies that could promote ecosystem transitions toward productive macroalgal stands. The potential reconstruction of the trophic structure at marine protected areas, where fishing activities have ceased during several years, may favour the control of newly settled urchins after reduction in density.
- Techniques need to be developed for production of juvenile sea urchin and sea cucumbers through laboratory cultures for aquacture or repopulation.

References

Aguilera F, Brito A, Castilla C, Díaz A, Fernández-Palacios JM, Rodríguez A, Sabaté F, Sánchez J (1994) Canarias, economía, ecología y medio ambiente. Francisco Lemus (ed), Islas Canarias, La Laguna

Alves FMA, Chícharo LM, Serrao E, Abreu AD (2001) Algal cover and sea urchin spatial distribution at Madeira Island (NE Atlantic). Sci Mar 65:383–392

Andrew NL, Agatsuma Y, Bahzin A, Creaser T, Barnes D, Botsford L, Bradbury A, Campbell A, Einnarsson S, Gerring P, Hebert K, Hunter M, Hur S, Johnson C, Kalvass P, Miller R, Moreno C, Palleiro J, Rivas D, Robertson S, Steneck RE, Vadas R, Woodby D, Xiaoqi Z (2002) Status and management of world sea urchin fisheries. Oceanogr Mar Biol Ann Rev 40:343–425

Arístegui J, Tett P, Hernández-Guerra A, Basterretxea G, Montero MF, Wild K, Sandrá P, Hernández-León S, Cantón M, Braun JA, Pacheco M (1997) The influence of island-generated eddies on chlorophyll distribution: a study of mesoscale variation around Gran Canaria. Deep Sea Res 44:71–95

Babcock RC, Kelly S, Shears NT, Walker JW, Willis TJ (1999) Changes in community structure in temperate marine reserves. Mar Ecol Prog Ser 189:125–134

Bacallado JJ, Brito A, Cruz T, Carrillo M, Barquín J (1982) Estudio del Bentos Marino del Archipiélago Canario. Publicaciones restringidas de la Consejería de Agricultura y Pesca, Catálogo preliminar de los invertebrados marinos bentónicos del Archipiélago Canario. Tomo I-II y III. Gobierno de Canarias

Bacallado JJ, Moreno E, Perez-Ruzafa A (1984) Echinodermata (Canary Islands). Provisional check-list. Proceedings of 5th International Echinoderm Conference, Galway 149–151

Bacallado JJ, Pérez-Ruzafa A, Martínez de Saavedra M., Gonzalez JF, Carrillo M, Moreno E, Perez G (1988) New contributions to the echinodermata check-list for the circumcanarian area. In: Burke RD (ed) Echinoderm Biology. Proceedings of 6th Int Echinoderm Conference, Victoria, 1988

Bacallado JJ, Brito A, Cruz T, Carrillo M, Barquín J (1987) Proyecto Bentos II. Anexo: Estudio de la biología del erizo de lima (*Diadema antillarum*). Informes de la Consejería de Agricultura y Pesca del Gobierno de Canarias

Bacallado JJ, Moreno E, Pérez-Ruzafa A (1985) Echinodermata (Canary Islands) provisional Check-list. In: Keegan BF, O'connor B (eds) Echinodermata. AA Balkema, Rotterdam, pp 149–151

Barquín J, Gil-Rodríguez MC, Del Arco Aguilar MJ, González G, Domínguez-Álvarez S, González R, Aldanondo-Ariztizabal N, Rodríguez M, Cruz-Reyes MA, Herrera-Lopez G, Sancho A, Cruz T, O'dwyer J, Martin L, Cansado S (2004) La Cartografia bionomica del litoral de Tenerife: resultados preliminaries. XIII Simp Iberico de Estudio del Bentos Marino. Libro de resúmenes

Barrois T (1888) Liste des échinodermes recueillis aux Açores durante les mois d'Aout et septembre 1887, 109–115

Barton ED, Arístegui J (2004) The Canary Islands coastal transition zone-upwelling eddies and filaments. Prog Oceonogr 62:67–69

Barton ED, Arístegui J, Tett P, Cantón M, Braun J, Hernández-León S, Nykjaer L, Almeida C, Almunia J, Ballesteros S, Basterretxea G, Escánez J, García-Weill L, Hernández-Guerra A, López-Laatzeen F, Molina R, Montero MF, Navarro-Pérez E, Rodríguez JM, Van Lenning K, Vélez H, Wild K (1998) The transition zone of Canary Current up-welling region. Prog Oceonogr 41:455–504

Borrero-Pérez GH, Pérez-Ruzafa A, Marcos C, González-Wanguemert M (2009) The taxonomic status of some Atlanto-Mediterranean species in the subgenus Holothuria (echinodermata: holothuroidea: holothuriidae) based on molecular evidence. Zool J Lin Soc 157:51–69

Braun J, Molina M (1988) El mar, el Atlántico en el área de Canarias. In: Afonso L, Galán FM, Rodríguez-Martín JA (eds) Geografía de Canarias. Interinsular Canaria, Santa Cruz de Tenerife, pp 18–28

Briggs JC (1996) Global Biogeography. Elsevier, Amsterdam

Brito A (2008) Influencia del calentamiento global sobre la biodiversidad marina de las Islas Canarias. In: Afonso-Carrillo J (ed) Naturaleza amenazada por los cambios en el clima Actas de la III Semana Científica Telesforo Bravo. Inst Estudios Hispánicos de Canarias, Puerto de La Cruz, pp 141–161

Brito A, Ocaña O (2004) Corales de las Islas Canarias. Ed Francisco Lemus, La Laguna

Brito A, Falcón JM, Aguilar N, Pascual P (2001) Fauna Vertebrada Marina. In: Fernández-Palacios JM, Martín-Esquivel JL (eds) Naturaleza de las Islas Canarias: Ecología y, Conservación edn. Turquesa, Santa Cruz de Tenerife, pp 219–229

Brito A, Hernández JC, Falcón JM, García N, González-Lorenzo G, Gil-Rodríguez MC, Cruz A, Herrera G, Sancho A, Clemente S, Cubero E, Girad D, Barquín J (2004) El erizo de Lima

(*Diadema antillarum*): una especie clave en los fondos litorales rocosos de Canarias. Makaronesia 6:68–86

Brito A, Cruz T, Moreno E, Pérez JM (1984) Fauna Marina de las Islas Canarias. In: Bacallado-Aránega JJ (ed) Fauna marina y terrestre del Archipiélago, Canario edn. Edirca, Las Palmas de Gran Canaria, pp 42–65

Cabanillas N (2009) Ecología y estatus trófico del erizo de mar *Diadema antillarum* (Philippi, 1985) en los fondos rocosos de las Islas Canarias, Gran Canaria, España. PhD Thesis, Univ de Las Palmas de Gran Canaria, Las Palmas

Calderon I, Giribet EG, Turon X (2008) Two markers and one history: phylogeography of the edible common sea urchin *Paracentrotus lividus* in the Lusitanian region. Mar Biol 154:137–151

Carracedo JC (1988) Marco geográfico. In: Alfonso L, Galán FM, Rodríguez-Martín JA (eds) Geografía de Canarias. Interinsular Canaria, Santa Cruz de Tenerife, pp 10–16

Carracedo JC (2001) Vulcanismo reciente y riesgo volcánico. In: Fernández-Palacios JM, Martín-Esquivel JL (eds) Naturaleza de las Islas Canarias: Ecología y Conservación, edn. Turquesa, Santa Cruz de Tenerife, pp 65–76

Carracedo JC, Badiola ER, Guillou H, De La Nuez J, Torrado FJ (2001) Geology and vulcanology of La Palma and El Hierro, Western Canaries. Est Geól 57:175–273

Catoira-Gòmez JL (1992) La pêche des oursins en Galice, Espagne, pendant la campagne 1990–1991. In: Scalera-Liaci L, Canicattì C (eds) Echinoderm Research. AA Balkema, Rotterdam, pp 199–200

Chapman ARO (1981) Stability of sea urchin dominated barren grounds following destructive grazing of kelp in St. Margaret's Bay, eastern Canada. Mar Biol 62:307–311

Chapman MG (1951) Aspects of the fauna and flora of the Azores IV. Echinodermata. An Mag Nat Hist 12:398–400

Clark AM, Downey ME (1992) Starfish of the Atlantic. Chapman and Hall, New York

Clemente S (2007) Evolución de las poblaciones del erizo *Diadema* aff. *antillarum* en Canarias y valoración de la depredación como factor de control. PhD Thesis, Univ de La Laguna, Tenerife, Islas Canarias

Clemente S, Hernández JC (2008) Influence of wave exposure and habitat complexity in determining spatial variation of the sea urchin *Diadema* aff. *antillarum* (Echinoidea: Diadematidac) populations and macroalgal cover (Canary Islands–Eastern Atlantic Ocean). Rev Biol Trop 56 (Suppl 3):229–254

Clemente S, Hernández JC, Brito A (2007a) An external tagging technique for the long-spined sea urchin *Diadema* aff. *antillarum*. J Mar Biol Ass UK 87:777–779

Clemente S, Hernández JC, Toledo K, Brito A (2007b) Predation upon *Diadema* aff. *antillarum* at barrens grounds in the Canary Islands. Sci Mar 71:745–754

Clemente S, Hernández JC, Brito A (2009) Evidence of the top–down role of predators in structuring sublittoral rocky-reef communities in a marine protected area and nearby areas of the Canary Islands. ICES J Mar Sci 66:64–71

Clemente S, Hernández JC, Rodríguez A, Brito A (2010) Identifying keystone predators and the importance of preserving functional diversity in sublittoral rocky bottoms. Mar Ecol Prog Ser 413:55–67

Clemente S, Hernández JC, Brito A (2011) Context-dependent effects of marine protected areas over predatory interactions. Mar Ecol Prog Ser 437:119–133

Davenport R, Never S, Helmke P, Pérez-Moreno J, Llinas O (2002) Primary productivity in the northern Canary Islands region as inferred from Seawifs imagery. Deep-Sea Res II 49: 3481–3496

De Jesus C, Abreu D (1998) Contribution to the knowledge of the soft bottom echinoderms of Madeira islands. Bol MusMunicipal do Funchal 50:59–69

De Lison Loma T, Harmelin-Vivien M, Naim O, Fontaine MF (2000) Algal food processing by *Stegastes nigricans*, an herbivorous damselfish: differences between an undisturbed and a disturbed coral reef site (La Reunion, Indian Ocean). Oceanol Acta 23:793–804

Ekman W (1905) On the influence of the earth's rotation on ocean currents. Arkiv foer Matematik Astronomi och Fysik 2:1–52

Entrambasaguas L (2008) Estudio faunístico y ecológico de los equinodermos del archipiélago de Cabo Verde. PhD Thesis, Univ Murcia, Murcia

Entrambasaguas L, Pérez-Ruzafa A, García Charton JA, Stobart B, Bacallado JJ (2008) Abundance, spatial distribution and habitat relationships of echinoderms in the Cabo Verde Archipelago (eastern Atlantic). Mar Fresh Res 59:477–488

Fernández-Palacios JM (2004) Introducción a las islas. In: Fernández-Palacios JM, Morici C (eds) Ecología Insular/Island Ecology. AEET/Cabildo Insular de La Palma, pp 21–55

Fernández-Palacios JM, Martín Esquivel JL (2001a) Las islas como experimento de laboratorio. In: Fernández Palacios JM, Martín-Esquivel JL (eds) Naturaleza de las Islas Canarias: Ecología y Conservación. Ed Turquesa, Santa Cruz de Tenerife, pp 39–44

Fernández-Palacios JM, Vera A, Brito A (2001b) Los Ecosistemas. In: Fernández Palacios JM, Martín-Esquivel JL (eds) Naturaleza de las Islas Canarias: Ecología y Conservación. Ed Turquesa, Santa Cruz de Tenerife, pp 157–165

Frantzis A, Gremare A (1992) Ingestion, absorption, and growth rates of *Paracentrotus lividus* (Echinodermata, Echinoidea) fed different macrophytes. Mar Ecol Progr Ser 95:169–183

Garrido M (2003) Contribución al conocimiento de *Diadema antillarum* Philippi 1845, en Canarias. PhD Thesis, Univ de Las Palmas de Gran Canaria, Las Palmas de Gran Canaria

Garrido M, Haroun RJ, Lessios HA (2000) Annual reproductive periodicity of the sea urchin *Diadema antillarum* Phillipi in the Canary Islands. Bull Mar Sci 67:989–996

Gee MJ, Masson DG, Watts AB, Mitchell NC (2001a) Offshore continuation of volcanic rift zones, El Hierro, Canary Islands. J Volcanol Geoth Res 105:107–119

Gee MJ, Watts AB, Masson DG, Mitchell NC (2001b) Landslides and evolution of El Hierro in Canary Islands. Mar Geol 177:271–293

Girard D, Clemente S, Toledo-Guedes K, Brito A, Hernández JC (in press) A mass mortality of subtropical intertidal populations of the sea urchin *Paracentrotus lividus*: analysis of potential links with environmental conditions. Mar Ecol. doi:10.1111/j.1439-0485.2011.00491.x

Grosjean P (2001) Growth model of the reared sea urchin *Paracentrotus lividus* (Lamarck, 1996). PhD Thesis, Univ Libre de Bruxelles, Bruxelles

Guidetti P (2006) Marine reserves reestablish lost predatory interactions and cause community effects in rocky reefs. Ecol Appl 16:963–976

Guidetti P, Dulčić J (2007) Relationships among predatory fish, sea urchins and barrens in Mediterranean rocky reefs across a latitudinal gradient. Mar Environ Res 63:168–184

Hansson HG (2001) Echinodermata. In: Costello MJ, Emblow CS, White RJ (eds) European register of marine species: a check-list of the marine species in Europe and a bibliography of guides to their identification. Collection Patrimoines Naturels, Mus Nat Hist Nat, Paris, pp 336–351

Hernández-Guerra A, Nykjaer L (1997) Sea surface temperature variability of north-west Africa: 1981–1989. Int J Rem Sen 18:2539–2558

Hernández JC (2006) Estrategia reproductiva de la población canaria del erizo *Diadema* aff. *antillarum*, Phillipi, 1845: maduración gonadal, asentamiento larvario y reclutamiento. PhD Thesis, Univ de La Laguna, Tenerife, Islas Canarias

Hernández JC, Clemente S, Brito A (2011) Effects of seasonality on the reproductive cycle of *Diadema* aff. *antillarum* in two contrasting habitats: implications for the establishment of a sea urchin fishery. Mar Biol 158:2603–2615

Hernández JC, Toledo K, Girard D, Clemente S, Cubero E, Brito A (2005) Description of post-larva and juvenile phases of three echinoids from the Canary Islands: *Diadema antillarum* Philippi, 1845, *Paracentrotus lividus* (Lamarck, 1816) and *Arbaciella elegans* Mortensen, 1910. Vieraea 33:385–397

Hernández JC, Brito A, García N, Gil-Rodríguez MC, Herrera G, Cruz-Reyes A, Falcón JM (2006a) Spatial and seasonal variation of the gonad index of *Diadema antillarum* (echinodermata: echinoidea) in the Canary Islands. Sci Mar 70:689–698

Hernández JC, Brito A, Cubero E, Girard D, González-Lorenzo G, Falcón JM (2006b) Temporal patterns of larval settlement of *Diadema antillarum* (echinodermata: echinoidea) in the Canary Islands using an experimental larval collector. Bull Mar Sci 78:271–279

Hernández JC, Clemente S, Sangil C, Brito A (2008a) The key role of the sea urchin *Diadema* aff. *antillarum* throughout the Canary Islands (eastern subtropical Atlantic) in controlling macroalgae assemblages: an spatio-temporal approach. Mar Environ Res 66:259–270

Hernández JC, Clemente S, Sangil C, Brito A (2008b) Actual status of the sea urchin *Diadema* aff. *antillarum* populations and macroalgal cover in marine protected areas compared to a highly fished area (Canary Islands-eastern Atlantic Ocean). Aquat Conser: Mar Freshw Ecosyst 18:1091–1108

Hernández JC, Clemente S, Sangil C, Brito A (2009) La degradación de los fondos marinos de Canarias: crónica de una muerte anunciada. El Indiferente 20:62–71

Hernández JC, Clemente S, Girard D, Pérez-Ruzafa A, Brito A (2010a) Effect of temperature on settlement and post settlement survival in a barrens-forming sea urchin. Mar Ecol Progr Ser 413:69–80

Hernández JC, Clemente S, Brito A, Lorenzo-Morales J, Martín C, Valladares B (2010b) Informe sobre la reciente mortalidad del erizo *Diadema* aff. *antillarum*: estudio preliminar y recomendaciones. Agencia Insular del Mar del Cabildo de Tenerife, Viceconsejería de Medio Ambiente. Grupo de investigación en Biodiversidad, Ecología Marina y Conservación (BIOECOMAC) e Instituto de Enfermedades Tropicales y Salud Pública de Canarias, Departamento de Biología Animal (Ciencias Marinas), Univ de La Laguna, Tenerife

Herrera R (1998) Dinámica de las comunidades bentónicas de los arrecifes artificiales de Arguineguín (Gran Canaria) y Lanzarote. PhD Thesis, Universidad de La Laguna, Tenerife

Høiseater T (1990) An annotated check-list of the Echinoderms of the Norwegian coast and adjacent waters. Sarsia 75:83–106

IPCC (2007) Cambio climático 2007: Informe de síntesis. Contribuciones de los grupos de trabajo I, II y III al Cuarto Informe de evaluación del Grupo Intergubernamental de Expertos sobre el Cambio Climático. IPCC, Ginebra

Jackson JBC, Kirby MX, Berger WH, Bjorndal KA, Botsford LW, Bourque BJ, Bradbury RH, Cooke R, Erlandson J, Estes JA, Hughes TP, Kidwell S, Lange CB, Lenihan HS, Pandolfi JM, Peterson CH, Steneck RS, Tegner MJ, Warner RR (2001) Historical overfishing and the recent collapse of coastal ecosystems. Science 293:629–638

Johnston CS (1969) Studies on the ecology and primary production of Canary Islands marine algae. Proc Int Seaweed Symp 6:213–222

Koehler R (1921) Fauna de France. Echinoderms. Office Central de Faunistique, Fédération Francaise des Sociétés de Sciences Naturelles, Paris

Koike I, Mukai H, Nojima S (1987) The role of the sea urchin, *Tripneustes gratilla* (Linnaeus), in decomposition and nutrient recycling in a tropical seagrass bed. Ecol Res 2:19–29

Lawrence JM (1975) On the relationships between marine plants and sea urchins. Oceanogr Mar Biol Ann Rev 13:213–286

Lawrence JM, Sammarco PW (1982) Effects of feeding on the environment: echinoidea. In: Jangoux M, Lawrence JM (eds) Echinoderm Nutrition. AA Balkema, Rotterdam, pp 499–635

Lawrence JM, Klinger TS (2001) Digestion in sea urchins. In: Lawrence JM (ed). Edible Sea Urchins: Biology and Ecology. Elsevier, Amsterdam, The Netherlands, pp 103–113

Le Direach JP (1987) La peche des oursins en Méditerranée: historique, techniques, legislation, production. In: Boudouresque CF (ed) Colloque Internacional sur *Paracentrotus lividus* et les oursins comestibles. GIS Posidonie, Marseilles, pp 335–362

Lessios HA, Kessing BD, JS Pearse (2001) Population structure and speciation in tropical seas, global phylogeography of the sea urchin *Diadema*. Evolution 55:955–975

Levinton JS, Ward JE, Shumway SE (2002) Feeding response of the bivalves *Crassotrea gigas* and *Mytilus trossulus* to chemical composition of fresh and aged kelp detritus. Mar Biol 141:367–376

Mamelona J, Pelletier E (2005) Green urchin as a significant source of fecal particulate organic matter within nearshore benthic ecosystems. J Exp Mar Biol Ecol 314:163–174

Marques VM (1983) Peuplements benthiques des Azores. I- Echinodermes. Arq Mus Bocage Ser A 2:1–7

Marrero-Díaz A, Rodríguez-Santana A, Machín F, Pelegrí JL (2006) Analytic salinity temperature relations for the upper-termocline waters of the eastern North Atlantic subtropical gire. Sci Mar 70:167–175
Martín-Oval M, Arnay de La Rosa R, Ponte-Lira E, Cerolo-González I, Jiménez-Gómez MC (1985) Estudio preliminar de la fauna del Conchero de Guinea (Frontera, El Hierro). Tabona 6:227–240
Marzol MV (1988) El clima. In: Afonso L, Galán FM, Rodríguez-Martín JA (eds) Geografía de Canarias. Interinsular Canaria, Santa Cruz de Tenerife, pp 157–202
Masson DG, Watts AB, Gee AMJR, Urgelés R, Mitchell NC, Le Bas TP, Canals M (2002) Slope failures on the flanks of the western Canary Islands. Earth-Sci Rev 57:1–35
McClanahan TR, Kamuru AT, Muthiga NA, Gilagabher Yebio M, Obura D (1996) Effects of sea urchin reductions on algae, coral, and fish populations. Conserv Biol 10:136–154
McClanahan TR, Muthiga NA, Kamukuru AT, Machano H, Kiambo RW (1999) The effects of marine parks and fishing on coral reefs of northern Tanzania. Biol Conserv 89:161–182
Mills SC, Peyrot-Clausada A, Fontaine MF (2000) Ingestion and transformation of algal turf by *Echinometra mathaei* on *Tiahura fringing* reef (French Polynesia). J Exp Mar Biol Ecol 254:71–84
Mitchel NC, Masson DG, Watts AB, Gee MJR, Urgeles R (2002) The morphology of submarine flanks of volcanic islands. A comparative study of the Canary and Hawaiian hotspod island. J Volcanol Geoth Res 115:83–107
Mitchell NC (1998) Characterising the irregular coastlines of volcanic ocean islands. Geomorphol 23:1–14
Mitchell NC, Dade WB, Masson DG (2003) Erosion of submarine flanks of the Canary Islands. J Geophys Res 108:1–11
Molina R, Cabanas JM, Laatzen FL (1996) Corrientes e hidrografía en la región canaria. Campaña Canarias 9205. Bol Inst Español Oceanogr 12:43–51
Moreno-Batet E, Bacallado JJ (1978) Contribución al estudio de los equinodermos de las Islas Salvajes. Contribución al estudio de la Historia Natural de las Islas Salvajes. Resultados de la expedición científica "Agamenon 76". Aula de Cultura de Tenerife:195–198
Moreno-Batet E, Fernández-Palacios H, Bacallado JJ (1980) Estudio morfológico de las púas de ocho especies de Equinoideos de las Islas Canarias. Vieraea 10:39–52
Mortensen T (1925) Echinoderms du Maroc et de Mauritanie. Bull Soc Sci Nat du Maroc 5:178–187
Mortensen T (1927) Handbook of the Echinoderms of the British Isles. Oxford University Press, London
Mortensen T (1943) A Monograph of the Echinoidea. III. 3. Camarodonta. II. Echinidæ, Strongylocentrotidæ, Parasaleniidæ, Echinometridæ. CA Rietzel, Copenhagen
Nobre A (1930) Materiais para o estudo da fauna dos Açores. Oporto
Nobre A (1931) Echinodermes de Portugal. VII. Holothuridios. Inst Zool Univ do Porto:133–158
Nobre A (1938) Fauna marinha de Portugal. 2. Aditamento. Mem Mus Zool Univ Coimbra 108:1–49
Nykjaer L, Van Camp L (1994) Seasonal and interanual variability of coastal upwelling along northwest Africa and Portugal from 1981 to 1991. J Geophys Res 99:14197–14207
Ortega L (2010) *Diadema antillarum* (Phillipi, 1845) en Canarias: Procesos determinantes de la estructura y organización de comunidades bentónicas. PhD Thesis, Univ de La Palmas de Gran Canaria, Las Palmas
Ortega L, Tuya F, Haroun RJ (2009) El erizo de mar *Diadema antillarum* Phillipi, 1845 influye sobre la diversidad y composición de la comunidad de mega-invertebrados vágiles en fondos rocosos del Archipiélago Canario. Rev Biol Mar Oceanogr 44:489–495
Panning A (1939) Holothurien von den kanaren und von Dakar. Vidensk Medd Fra Dansk Naturh Foren I Kopenhavn Bd 103:523–546
Pearse JS, Hines AH (1987) Long-term population dynamics of sea urchins in a central California kelp forest: rare recruitment and rapid decline. Mar Ecol Progr Ser 39:275–283
Pereira MOR (1997) Checklist of the littoral echinoderms of the Azores. Açoreana 8:331–337

Pérez-Ruzafa A (1984) Estudio sistemático, ecológico y biogeográfico de la clase Holothurioidea (Echinodermata) en las Islas Canarias. Univ de La Laguna, Tenerife, Licenciatura Thesis

Pérez-Ruzafa A, Marcos C (1987) Observaciones sobre la actividad diaria y la ecología de algunas holoturias (Echinodermata: Holothuroidea) litorales. An Biol 12:79–90

Pérez-Ruzafa A, Marcos C, Bacallado JJ (1984a) Algunas citas nuevas y otras holoturias (holothuroidea, echinodermata) de interés para la fauna de las Islas Canarias (España). Actas do IV Simp Ibérico de Estudos do Benthos Marinho III:277–284

Pérez-Ruzafa A, Marcos C, Bacallado JJ (1984b) Relaciones biogeográficas entre la fauna de holoturias (holothuroidea, echinodermata) de las Islas Canarias y la de otras áreas del Atlántico y la del Mediterráneo. Actas do IV Simp Ibérico de Estudos do Benthos Marinho I:36–45

Perez-Ruzafa A, Bacallado JJ & Marcos Diego C (1985a) Relaciones biogeográficas entre la fauna de holoturias (holothuroidea: echinodermata) de las Islas Canarias y la de otras áreas del Atlántico Norte y la del Mediterráneo. Actas do IV° Simp Ibérico de Estudios do Benthos Marinho, Lisboa 3:36–45

Pérez-Ruzafa A, Bacallado JJ, Marcos Diego C (1985b) Algunas citas nuevas y otras holoturias (Holothuroidea: Echinodermata) de interés para la fauna de las Islas Canarias (España). Actas do IV° Simp Ibérico de Estudios do Benthos Marinho, Lisboa 3:277–284

Pérez-Ruzafa A, Marcos C, Bacallado JJ (1987) Presencia de *Holothuria (Panningothuria) forskali* (echinodermata: holothuroidea) en las islas Canarias. Vieraea 17:361–367

Pérez-Ruzafa A, Marcos C, Bacallado JJ (1992a) Holoturias (echinodermata: holothuroidea) de las islas Canarias, I. Consideraciones generales y Orden Aspidochirotida. Rev Acad Canaria Cien 4:139–162

Pérez-Ruzafa A, Marcos C, Bacallado JJ (1992b) Holoturias (Echinodermata: Holothuroidea) de las islas Canarias, II. Ordenes Dendrochirotida, Elasipodida, Apodida y Molpadida. Rev Acad Canaria Cien 4:163–185

Pérez-Ruzafa A, Entrambasaguas L, Bacallado JJ (1999) Fauna de equinodermos (Echinodermata) de los fondos rocosos infralitorales del archipiélago de Cabo Verde. Rev Acad Canaria Cien 11:43–62

Pérez-Ruzafa A, Entrambasaguas L, Espejo C, Marcos C, Bacallado JJ (2002) Fauna de equinodermos (Echinodermata) de los fondos rocosos infralitorales del archipiélago de Salvajes (océano Atlántico). Rev Acad Canaria Cien 14:277–294

Pérez-Ruzafa A, Entrambasaguas L, Marcos C, Bacallado JJ García-Chartón JA (2003a) Spatial relationships of the echinoderm fauna of Cabo Verde islands: A multiscale approach. In: Féral J-P, David B (eds). Echinoderm Research 2001. Proceedings of the sixth european confernce on echinoderm research.Sweets y Zeitlinger, Lisse. pp 31–39

Pérez-Ruzafa A, Bacallado JJ, Acevedo I, Docoito JR (2003b) División Echinordermata. In: Moro L, Martín JL, Garrido MJ, Izquierdo I (eds) Lista de especies marinas de Canarias (algas, hongos, plantas y animales). Consejería de Política Territorial y Medio Ambiente del Gobierno de Canarias, pp 51–57

Pérez OM, Telfer TC, Ross LG (2003) On the calculation of wave climate for offshore cage culture site selection: a case study in Tenerife (Canary Islands). Aquac Engin 29:1–21

Rodríguez A, Coppard SE, Clemente S, Hernández JC (in press) A new specie of *Diadema* (Echinodermata:Echinoidea:Diadematidae) from the eastern Atlantic Ocean and a *neotype* designation of *Diadema antillarum* (Philippi, 1845). Zootaxa

Russell MP (2000) Spatial and temporal variation in growth of the green sea urchin, *Strongylocentrotus droebachiensis*, in the Gulf of Maine, USA. In: Barker M (ed) Proceedings of 10th International Echinoderm Conference. AA Balkema, Rotterdam, pp 533–538

Sala E, Boudouresque CF, Harmelin-Vivien M (1998) Fishing, trophic cascades, and the structure of algal assemblages: evaluation of an old but untested paradigm. Oikos 82:425–439

Sammarco PW (1982) Echinoid grazing as a structuring force in coral communities: whole reef manipulations. J Exp Mar Biol Ecol 61:31–55

Sangil C (2007) Distribución de la fauna marina en la cueva del Infierno. Vulcania 8:70–78

Sangil C, Sansón M, Afonso-Carrillo J (2006a) Distribución de las comunidades algales de sustratos rocosos en la isla de La Palma (islas Canarias). Proc XIV Simp Ibérico de Estudios en Biol Mar, Barcelona

Sangil C, Afonso-Carrillo J, Sansón M (2006b) Estructura y composición de las comunidades algales sublitorales de los fondos rocosos de la isla de El Hierro (islas Canarias). Proc XIV Simp Ibérico de Estudios en Biol Mar, Barcelona

Sangil C, Sanson M, Afonso-Carrillo J (2007) Biodiversidad y biogeografía de las macroalgas bentónicas de la isla de La Palma. Rev Est Gen de La Palma 3:457–479

Sansón M, Reyes J, Afonso-Carrillo J (2001) Flora marina. In: Fernández-Palacios JM, Martín-Esquivel JL (eds) Naturaleza de las Islas Canarias: Ecología y Conservación. Ed Turquesa, Santa Cruz de Tenerife, pp 193–198

Steneck RS (1998) Human influences on coastal ecosystems: does overfishing create trophic cascades? Trends Ecol Evol 13:429–430

Tegner MJ, Dayton PK (1977) Sea urchin recruitment patterns and implications of commercial fishing. Science 196:324–326

Tejera A, García-Weil L, Heywood KJ, Cantón-Garbín M (2002) Observations of oceanic mesoescale features and variability in the Canary Islands area from ERS-1 altimeter data, satellite infrared imagery and hydrographyc measurements. Int J Rem Sens 23:4897–4916

Theel HJ (1886) Report on the Holothuroidea II. "Challenger" Sci Results. Zoology 14:1–290

Thorpe JP, Solé-Cava AM, Watts PC (2000) Exploited marine invertebrates: genetics and fisheries. Hydrobiology 420:165–184

Tortonese E (1965) Echinodermata. Fauna d'Italia. Vol. 6. Calderini, Bologna

Tuya F (2002) Contribución al conocimiento de aspectos bio-ecológicos de la zona propuesta como reserva marina Gando-Arinaga. PhD Thesis, Univ de Las Palmas de Gran Canaria, Las Palmas

Tuya F, Haroun RJ (2006) Spatial patterns and response to wave exposure of shallow water algal assemblages across the Canarian Archipelago: a multiscaled approach. Mar Ecol Prog Ser 311:15–28

Tuya F, Haroun RJ (2009) Phytogeography of *Lusitanian Macaronesia*: biogeographic affinities in species richness and assemblage composition. Eur J Phycol 44:405–413

Tuya F, Martín JA, Reuss GM, Luque A (2001) Food preferences of the sea urchin *Diadema antillarum* in Gran Canaria (Canary Islands, central-east Atlantic Ocean). J Mar Biol Ass UK 81:845–849

Tuya F, Martín JA, Luque A (2003) A novel technique for tagging the long-spined sea urchin *Diadema antillarum.* Sarsia 88:365–368

Tuya F, Martín JA, Luque A (2004a) Patterns of nocturnal movement of the long-spined sea urchin *Diadema antillarum* (Philippi) in Gran Canaria (the Canary Islands, central East Atlantic Ocean). Helgoland Mar Res 58:1–26

Tuya F, Boyra A, Sánchez-Jerez P, Haroun RJ, Barberá C (2004b) Relationships between rocky-reef fish assemblages, the sea urchin *Diadema antillarum* and macroalgae throughout the Canarian Archipelago. Mar Ecol Progr Ser 278:157–169

Tuya F, Boyra A, Sánchez-Jerez P, Barbera C, Haroun RJ (2004c) Can one species determine the structure of the benthic community on a temperate rocky reef? The case of the long-spined sea-urchin *Diadema antillarum* (echinodermata: echinoidea) in the easter Atlantic. Hydrobiology 519:211–214

Tuya F, Boyra A, Haroun RJ (2004d) Blanquizales en Canarias. Ed Oceanográfica, Telde, Islas Canarias

Tuya F, Boyra A, Sánchez-Jerez P, Haroun RJ (2005) Influence of fishing and functional group of algae on sea urchin control of algal communities in the eastern Atlantic. Mar Ecol Progr Ser 287:255–260

Tuya F, Ortega L, Del Rosario-Pinilla AB, Haroun RJ (2006a) Spatio-temporal variability in a key herbivore, the long-spined black sea urchin (*Diadema antillarum*, echinodermata: echinoidea) in the Canary Islands. J Mar Biol Ass UK 86:791–797

Tuya F, Hernández JC, Clemente S (2006b) Is there a link between the type of habitat and the patterns of abundance of Holothurians in shallow rocky reefs? Hidrobiologia 571:191–199
Tuya F, Cisneros-Aguirre J, Ortega L, Haroun RJ (2007) Bathymetric segregation of sea urchins on reefs of the Canarian Archipelago: role of flow-induced forces. Estuar Coast Shelf Sci 73:481–488
Urgeles R, Masson DG, Canals M, Watts AB, Le Bas T (1999) Recurrent large-scale landsliding on the west flank of La Palma, Canary Island. J Geophys Res 104:331–348
Uthicke S, Schaffelke B, Byrne M (2009) A boom–bust phylum? Ecological and evolutionary consequences of density variations in echinoderms. Ecol Monogr 79:3–24
Vadas RL (1977) Preferential feeding: an optimization strategy in sea urchins. Ecol Monogr 47:337–371
Valentine JP, Johnson SR (2005) Persistence of sea urchin (*Heliocidaris erythrogramma*) barrens on the east coast of Tasmania: inhibition of macroalgal recovery in the absence of high densities of sea urchins. Bot Mar 48:106–115
Valentine JF, Heck KL, Busby J, Webb D (1997) Experimental evidence that herbivory increases shoot density and productivity in a subtropical turtlegrass (*Thalassia testudinum*) meadow. Oecol 112:193–200
Watts AB, Masson DG (1995) A giant landslide on the north flank of Tenerife, Canary Islands. Geophys J Int 119:648–666
Wotton RS (1994) The classification of particulate and dissolved matter. In: Wotton RS (ed). The Biology of Particles in Aquatic Systems. CRC Press, Boca Raton, pp 1–6
Yanes A (1990) Morfología litoral de las islas Canarias Occidentales. Secretariado Publicaciones Universidad de La Laguna, La Laguna
Yanes A, Marzol MV, Romero C (2006) Characterization of sea storms along the coast of Tenerife, the Canary Islands. J Coastal Res 48:124–128

Chapter 16
Latin America Echinoderm Biodiversity and Biogeography: Patterns and Affinities

A. Pérez-Ruzafa, J. J. Alvarado, F. A. Solís-Marín, J. C. Hernández, A. Morata, C. Marcos, M. Abreu-Pérez, O. Aguilera, J. Alió, J. J. Bacallado-Aránega, E. Barraza, M. Benavides-Serrato, F. Benítez-Villalobos, L. Betancourt-Fernández, M. Borges, M. Brandt, M. I. Brogger, G. H. Borrero-Pérez, B. E. Buitrón-Sánchez, L. S. Campos, J. R. Cantera, S. Clemente, M. Cohen-Renfijo, S. E. Coppard, L. V. Costa-Lotufo, R. del Valle-García, M. E. Díaz de Vivar, J. P. Díaz-Martínez, Y. Díaz, A. Durán-González, L. Epherra, M. Escolar, V. Francisco, C. A. Freire, J. E. García-Arrarás, D. G. Gil, P. Guarderas, V. F. Hadel, A. Hearn, E. A. Hernández-Delgado, A. Herrera-Moreno, M. D. Herrero-Pérezrul, Y. Hooker, M. B. I. Honey-Escandón, C. Lodeiros, M. Luzuriaga, C. L. C. Manso, A. Martín, M. I. Martínez, S. Martínez, L. Moro-Abad, E. Mutschke, J. C. Navarro, R. Neira, N. Noriega, J. S. Palleiro-Nayar, A. F. Pérez, E. Prieto-Ríos, J. Reyes, R. Rodríguez-Barreras, T. Rubilar, T. I. Sancho-Mejías, C. Sangil, J. R. M. C. Silva, J. I. Sonnenholzner, C. R. R. Ventura, A. Tablado, Y. Tavares, C. G. Tiago, F. Tuya and S. M. Williams

A. Pérez-Ruzafa (✉) · A. Morata · C. Marcos
Departamento de Biología Animal Biodiversidad, Ecología Marina y Conservación (Ciencias Marinas), Facultad de Biología, Universidad de La Laguna, Tenerife, islas Canarias, Spain
e-mail: angelpr@um.es

A. Morata
e-mail: aixamaria.morata@um.es

C. Marcos
e-mail: cmarcos@um.es

J. J. Alvarado · S. M. Williams
Centro de Investigación en Ciencias del Mar y Limnología (CIMAR), Universidad de Costa Rica, San Pedro, San José, 11501-2060, Costa Rica
e-mail: juanalva76@yahoo.comjuan.alvarado@ucr.ac.cr

S. M. Williams
e-mail: stcmwilliams@gmail.com

J. J. Alvarado and F. A. Solís-Marín (eds.), *Echinoderm Research and Diversity in Latin America*, DOI: 10.1007/978-3-642-20051-9_16,

F. A. Solís-Marín
Colección Nacional de Equinodermos "Ma. E. Caso Muñoz", Laboratorio de Sistemática y Ecología de Equinodermos, Instituto de Ciencias del Mar y Limnología (ICML), Universidad Nacional Autónoma de México (UNAM), Apdo, 70-305 04510, Mexico, D.F., Mexico
e-mail: fasolis@cmarl.unam.mx

J. C. Hernández · S. Clemente
Departamento de Biología Animal (Ciencias Marinas), Universidad de La Laguna, Biodiversidad, Ecología Marina y Conservación, c/Astrofísico Francisco Sánchez sn, La Laguna, Tenerife, Canary Islands 38206 , Spain
e-mail: jocarher@ull.es

M. Abreu-Pérez
Instituto de Oceanología, Ministerio de Ciencias Tecnología y Medio Ambiente (CITMA), , Cuba

O. Aguilera
Museu Paraense Emilio Goeldi, Coordenação de Ciências da Terra e Ecologia, Departamento de Geociências Av. Perimetral, 1901 Terra Firme, CEP, PA, 66077-830, Brazil
e-mail: orangel.aguilera@gmail.com

J. Alió
Centro de Investigaciones Agrícolas de Sucre y Nueva Esparta, Instituto Nacional de Investigaciones Agrícolas, Estado Sucre, 6101, Cumaná, Venezuela
e-mail: jalio@inia.gob.ve

J. J. Bacallado-Aránega
Museo de La Naturaleza y el Hombre de Tenerife, Santa Cruz de Tenerife, islas Canarias, Spain

E. Barraza
Gerencia de Vida Silvestre, Ministerio de Medio Ambiente y Recursos Naturales, San Salvador, El Salvador
e-mail: ebarraza@marn.gob.sv

M. Benavides-Serrato · G. H. Borrero-Pérez
Instituto de Investigaciones Marinas y Costeras—INVEMAR, A.A 1016, Santa Marta, Colombia
e-mail: milena_benavides@invemar.org.co

G. H. Borrero-Pérez
e-mail: giomar_borrero@invemar.org.co

F. Benítez-Villalobos · J. P. Díaz-Martínez
Instituto de Recursos, Universidad del Mar, Carretera a Zipolite Km 1.5, Puerto Ángel, Oaxaca, Mexico
e-mail: fvb@angel.umar.mx

J. P. Díaz-Martínez
e-mail: lobi2@hotmail.com

L. Betancourt-Fernández · A. Herrera-Moreno
Programa EcoMar, Inc., Santo Domingo, Dominican Republic
e-mail: ongprogramaecomar@yahoo.com

M. Borges
Museu de Zoologia, Universidade Estadual de Campinas, Campinas-SP, 13083-863, Brazil
e-mail: borges@unicamp.br

M. Brandt
Department of Ecology and Evolutionary Biology, Brown University, Providence, RI, USA
e-mail: margarita.brandt@gmail.com

M. I. Brogger · M. I. Martínez · A. Tablado
Museo Argentino de Ciencias Naturales "Bernardino Rivadavia"-CONICET, Buenos Aires, Argentina
e-mail: mbrogger@bg.fcen.uba.ar

M. I. Martínez
e-mail: mmartinez@macn.gov.ar

A. Tablado
e-mail: tablado@macn.gov.ar

B. E. Buitrón-Sánchez
Departamento de Paleontología, Instituto de Geología UNAM, Mexico, DF 04510, Mexico
e-mail: blancab@servidor.unam.mx

L. S. Campos
Departamento de Zoologia, Instituto de Biologia, Universidade Federal do Rio de Janeiro, Rio de Janeiro-RJ, 21941-971, Brazil
e-mail: campos-lucia@biologia.ufrj.br

J. R. Cantera K. · R. Neira
Departamento de Biología, Universidad del Valle, A.A. 25360, Cali, Colombia
e-mail: jaime.cantera@correounivalle.edu.co

R. Neira
e-mail: rneira56@yahoo.com

M. Cohen-Renfijo
Centre d'Oceanologie de Marseille, Université de la Méditerranée, Marseille Cedex, France
e-mail: ablocsberg@gmail.com

S. E. Coppard
Smithsonian Tropical Research Institute, PO BOX 0843-03092, Balboa, Ancón, Panamá
e-mail: CoppardS@si.edu

L. V. Costa-Lotufo
Departamento de Fisiologia e Farmacologia, Instituto de Biologia, Universidade Federal do Rio de Janeiro, Rio de Janeiro-RJ, 21941-971, Brazil
e-mail: lvcosta@secrel.com.br

R. del Valle-García
Instituto de Oceanología, Ministerio de Ciencias Tecnología y Medio Ambiente (CITMA), , Cuba
e-mail: rosa.denisse1@hotmail.cominformatica@giron.sld.cu

M. E. Díaz de Vivar
Facultad de Ciencias Naturales, Universidad Nacional de la Patagonia San Juan Bosco, Puerto Madryn, Argentina
e-mail: mediazdevivar@gmail.com

Y. Díaz · A. Martín
Laboratorio de Crustáceos Peracáridos. Centro de Biodiversidad Marina. Departamento de Estudios Ambientales - INTECMAR, Universidad Simón Bolívar, Caracas, 1080, Venezuela
e-mail: yusdiaz@usb.ve

L. Epherra · T. Rubilar
Centro Nacional Patagónico-CONICET, Puerto Madryn, Argentina
e-mail: epherra@cenpat.edu.ar

T. Rubilar
e-mail: rubilar@cenpat.edu.ar

M. Escolar
Instituto Nacional de Investigación y Desarrollo Pesquero, Mar del Plata, Argentina
e-mail: marianaescolar@inidep.edu.ar

V. Francisco
Laboratorio de Ecología de Ecosistemas de Arrecifes Coralinos, Centro de Investigación y de Estudios Avanzados del Instituto Politécnico Nacional, Mérida, Yucatán, Mexico

C. A. Freire
Departamento de Fisiologia, Universidade Federal do Paraná, Curitiba-PR, 81531-990, Brazil
e-mail: cafreire@ufpr.br

J. E. García-Arrarás · E. A. Hernández-Delgado · R. Rodríguez-Barreras
Department of Biology, University of Puerto Rico, Río Piedras, PO Box 23360, San Juan, 00931-3360, PR, USA
e-mail: jegarcia@hpcf.upr.edu

E. A. Hernández-Delgado
e-mail: coral_giac@yahoo.com

R. Rodríguez-Barreras
e-mail: ruber.rodriguez@hotmail.com

D. G. Gil
Instituto de Desarrollo Costero, Universidad Nacional de la Patagonia San Juan Bosco, Comodoro Rivadavia, Argentina
e-mail: gil_damian@hotmail.com

P. Guarderas
GEOcentro, Universidad San Francisco de Quito, Cumbayá, Ecuador
e-mail: pagigv@yahoo.com

V. F. Hadel · C. G. Tiago
Centro de Biologia Marinha, Universidade de Sao Paulo, Sao Sebastiao-SP, 11600-000, Brazil
e-mail: vafhadel@usp.br

C. G. Tiago
e-mail: cltiago@usp.br

A. Hearn
Department of Wildlife, Fish and Conservation Biology, University of California, Davis, CA, USA
e-mail: arhearn@ucdavis.edu

M. D. Herrero-Pérezrul
Centro Interdisciplinario de Ciencias Marinas, Instituto Politécnico Nacional, Ave. IPN S/n Col., Playa Palo de Santa Rita, CP, 23096, La Paz, BCS, Mexico
e-mail: dherrero@ipn.mx

Y. Hooker
Laboratorio de Biología Marina, Departamento de Ciencias Biológicas y Fisiológicas, Universidad Peruana Cayetano Heredia, Av. Honorio Delgado 430, Urb. Ingeniería, S.M.P, Lima, Peru
e-mail: hookery@yahoo.com

Y. Hooker
Unidad Marino Costera, Servicio Nacional de Áreas Naturales Protegidas (SERNANP), Ministerio del Ambiente, Calle Diecisiete No 355, Urb. El Palomar—San Isidro, Lima, Peru

M. B. I. Honey-Escandón · E. Prieto-Ríos
Posgrado en Ciencias del Mar y Limnología, Instituto de Ciencias del Mar y Limnología (ICML), UNAM, Apdo, 70-305 04510, Mexico, D.F., Mexico

C. Lodeiros
Instituto Oceanográfico de Venezuela, Universidad de Oriente y Fundación para la Investigación y Desarrollo de la Acuicultura del Edo. Sucre, 6101Cumaná, Venezuela
e-mail: cesarlodeirosseijo@yahoo.es

M. Luzuriaga
Laboratorio de Plancton e Ictioplancton, Instituto Nacional de Pesca del Ecuador, Guayaquil, Ecuador
e-mail: mluzuriagav@yahoo.com

C. L. C. Manso
Laboratório de Invertebrados Marinhos, Universidade Federal de Sergipe, Itabaiana-SE, 49500-000, Brazil
e-mail: cynthia@phoenix.org.br

S. Martínez
Departamento Evolución de Cuencas, Facultad de Ciencias, Iguá 4225, 11400, Montevideo, Uruguay
e-mail: smart@fcien.edu.uy

L. Moro-Abad
Consorcio Insular de La Reserva Mundial de La Biosfera La Palma, Servicio de Biodiversidad del Gobierno de Canarias, ,

E. Mutschke
Laboratorio de Ecología y Ciencias Ambientales, Universidad de Magallanes, ,
e-mail: erika.mutschke@umag.cl

J. C. Navarro
LECROY S.A., Km 2.5, Avenida Carlos Julio Arosemena, Bodega No.5, Guayaquil, Ecuador
e-mail: jcnavarrogo@gmail.com

N. Noriega
Laboratorio de Biología Marina y Laboratorio de Bioquímica, Universidad Simón Bolívar, Caracas, 1080, Venezuela
e-mail: nicidanoriega@gmail.com

J. S. Palleiro-Nayar
Centro Regional de Investigación Pesquera, Instituto Nacional de Pesca, Apartado postal 1305, Ensenada, BC, México
e-mail: juliopalleiro@yahoo.com.mx

A. F. Pérez
Facultad de Ciencias Exactas y Naturales, Universidad de Buenos Aires, Buenos Aires, Argentina
e-mail: analiafperez1@gmail.com

E. Prieto-Ríos
Facultad de Ciencias Biológicas, Biología, Universidad Nacional Mayor de San Marcos, Mesa de partes. Ciudad Universitaria de San Marcos, Av. Venezuela s/n, Lima 1, Peru

16.1 Introduction

The first attempt to study the diversity and biogeography of echinoderms from Latin America was done by Maluf (1988a, b). She identified 630 species of echinoderms from southern California to southern Peru (Central Eastern Pacific or

J. Reyes
Laboratorio de Zoología de Invertebrados, Universidad del Zulia, Maracaibo 4001 y Fundación para la Investigación y Desarrollo de la Acuicultura del Edo. Sucre, Cumaná, 6101, Venezuela
e-mail: jenyreyesl@yahoo.es

T. I. Sancho-Mejías
Museo de Zoología, Escuela de Biología, Universidad de Costa Rica, San José, Costa Rica
e-mail: sancho.ekin02@gmail.com

C. Sangil
Departamento de Ecología e Hidrología, Universidad de La Laguna, Biodiversidad, Ecología Marina y Conservación

J. R. M. C. Silva
Departamento de Biologia Celular e do Desenvolvimento, Instituto de Ciências Biomédicas, Universidade de São Paulo, São Paulo-SP, 05508-990, Brazil
e-mail: jmcs@usp.br

J. I. Sonnenholzner
Universidad Laica Eloy Alfaro de Manabí, Departamento Central de Investigación, ,
e-mail: jsonnenholzner@hotmail.com

C. R. R. Ventura
Museo Nacional, Universidade Federale do Rio de Janeiro, Rio de Janeiro-RJ, 20940-040, Brazil
e-mail: ventura@acd.ufrj.br

Y. Tavares
Faculdade de Ciências, Filosofia e Letras de Paranaguá, Universidade Estadual do Paraná, Pontal do Paraná-PR, 83203-280, Brazil
e-mail: ytavares@ufpr.br

F. Tuya
Departamento de Biología, Universidad de Las Palmas de Gran Canaria, Las Palmas de Gran Canaria, Campus Tafira s/n, Las Palmas de G.C., Islas Canarias 35017, Spain

A. Durán-González
Laboratorio de Sistemática y Ecología de Equinodermos, Instituto de Ciencias del Mar y Limnología (ICML), Colección Nacional de Equinodermos "Ma. E. Caso Muñoz", Universidad Nacional Autónoma de México (UNAM), Apdo, 70-305 04510, ,

M. B. I. Honey-Escandón
Laboratorio de Sistemática y Ecología de Equinodermos, Instituto de Ciencias del Mar y Limnología (ICML), Colección Nacional de Equinodermos "Ma. E. Caso Muñoz", Universidad Nacional Autónoma de México (UNAM), Apdo, 70-305 04510, Mexico, D.F., Mexico
e-mail: mbihoney@hotmail.com

CEP). Two-thirds of all CEP echinoderms occur on the continental shelf (depth <200 m), with a low level of endemism. She found an increase in species richness from higher to lower latitudes, with peaks of richness in the Gulf of California, Panama and the Galápagos Archipelago, places that also have a higher number of endemic species, and are also areas with more research. The Gulf of California had a greater similarity with the tropics than the Pacific side of the Baja California peninsula and southern California, while the oceanic island of Cocos was more similar to the mainland than the Galápagos and Revillagigedo archipelagos. Moreover, she indicated the presence of five faunal transition zones related to large-scale abiotic parameters. Those transition zones were: (1) the Galápagos Islands, (2) Gulf of Guayaquil, (3) Costa Rica–Panama, (4) the mouth of the Gulf of California and (5) Central Baja California outer coast. Seventy percent of all the 265 CEP shelf restricted species were endemic to the region, 12 % were of a Northeast Pacific affinity (Oregonian and Alaska provinces), 5 % had a Southeast Pacific affinity (Magellanic and Peruvian provinces), and 13 % were widespread species. Most species in this last category were transpacific more than circumtropical, and were species associated with reefs and rocky shores (Maluf 1988b).

On the Atlantic side, Price et al. (1999) analyzed a database of presence/absence of the Asteroidea compiled by Clark and Downey (1992) that was based on records collected over 150 years. They determined geographical patterns of diversity and make comparisons between coastal and deep-sea diversity for 26 regions of the Atlantic. For 349 species, they found a higher degree of endemism in coastal waters and a greater level of similarity between regions with increasing depth. However, the number of species between bathymetrical categories was similar: 199 species at depths between 0 and 200 m, 135 and 169 species between 200 and 500 m, and 198 species at depths >500 m. The greatest number of species (81 species) was in the region between Florida and the Yucatan Peninsula, followed by the Bahamas-Caribbean (77 species) and the Caribbean and Guyana basins together with the Gulf of Mexico region (66 species). This last region possessed the highest number of endemic species of all the Atlantic Asteroidea (14 species). Regions like Guyanas-Cape Frío, Cape Frío-Río de la Plata, Río de la Plata-Tierra del Fuego-Falklands islands and the Canary islands had few endemic species (richness: 48, 22, 43, 26 species respectively; endemism: 1, 1, 8, 0 species respectively). According to Price et al. (1999) the disparity in sampling efforts is an important factor which can influence interpretation of geographic patterns.

Alvarado and Cortés (2004), Alvarado et al. (2008, 2010) and Alvarado (2011) studied the diversity of echinoderms in Central America, covering both Pacific and Caribbean coast. On both coasts the class Ophiuroidea was the richest with 85 species on the Pacific side and 79 on the Caribbean coast. Panama was also the richest country in the region on both coasts, with 253 species on the Pacific coast and 154 species on the Caribbean coast. Alvarado et al. (2010) and Miloslavich et al. (2010) made an extensive review of published records on echinoderms of the

Caribbean and reported a total of 433 species. Ophiuroidea was the richest class with 148 species, followed by Asteroidea (116 species). Mexico and Colombia were the richest countries with 182 and 180 species, respectively. In terms of Caribbean ecoregions, the Southwestern Caribbean was richest (283 species), followed by the Western Caribbean (268 species), the Greater Antilles (248 species), the Southern Caribbean (151 species), and lastly the Eastern Caribbean (79 species).

Miloslavich et al. (2011) analyzed the marine biodiversity of South America (including the Pacific coast of Costa Rica and Panama, and excluding the Caribbean coast from Colombia and Venezuela). The best known groups in the region are fish, mollusks, crustaceans, echinoderms, cnidarians, and macroalgae. They divided the region into five subregions: (1) Eastern Tropical Pacific (223 species of echinoderms, 3.3 % of the total species, 4.4 spp./100 km of the coast, 51 % spp. in OBIS–Ocean Biogeographic Information System), (2) Humbolt Current-Chile and Peru (364 species of echinoderms, 3.6 % of the total species, 5.0 spp./100 km of the coast, 38 % spp. in OBIS), (3) Patagonian shelf-Uruguay and Argentina (207 species of echinoderms, 5.5 % of the total species, 3.7 spp./100 km of the coast, 76 % spp. in OBIS), (4) North, South and East Brazilian shelves (254 species of echinoderms, 2.8 % of the total species, 3.4 spp./100 km of the coast, 60 % spp. in OBIS) and (5) tropical west Atlantic-Venezuelan Atlantic, Guyana, Suriname and French Guyana (107 species of echinoderms, 3.9 % of the total species, 5.7 spp./100 km of the coast, 84 % spp. in OBIS). According to their analysis with the OBIS database, echinoderms have a low level of endemism (3.6 %) and seem to have reached a relatively stable number with few new additions. This could be the result of the lack of taxonomic expertise, limited funding for research, lack of collecting effort, and limited access to sampling sites.

Sthör et al. (2012) presented a global biodiversity analysis of the class Ophiuroidea, reporting 2,064 species. They divided Latin America into three regions: East Pacific (EP), South America (SA) and West Atlantic (WA). West Atlantic was the richest region with 335 species with 60.6% of them endemic to the region, followed by EP with 186 species and 62.9 % endemics, and SA with 124 species and 24.2 % endemics. Moreover, according to their depth strata, the bathyal stratum (depths between 200 and 3,500 m) was the richest in the three regions (229 species in WA, 111 species in EP and 102 species in SA), followed by the shelf stratum (depths between 0 and 200 m) (217 spp. in WA, 92 spp. in EP and 79 spp. in SA). The abyssal (depths between 3,500 and 6,000 m) (16 spp. in WA, 28 spp. in EP and 17 spp. in SA) and the hadal strata (depths > 6,500 m) (0 sp. in WA, 1 sp. in EP and SA) were less rich.

On this chapter we analyze the database presented in the appendix of this book. We investigate the current patterns of diversity by country and by class of echinoderms. In addition, we analyze their biogeographical, depth, and habitat or substratum affinities.

Fig. 16.1 Biogeographical regions in central and south American coasts according to Briggs (1995) and Clark and Downey (1992)

16.2 General Oceanographic Variables of Latin America and the Canary Islands

Traditionally, the area has been divided into five biogeographical regions with nine provinces (Briggs 1995) (Fig. 16.1). Western Atlantic Region, from Bermuda (Southern Florida) to Cape Frío (Brazil), which includes the Caribbean, Brazilian and West Indian provinces. In the Pacific Ocean, the Eastern Pacific Region includes the Mexican, Panamanian and Galápagos Provinces, the Eastern South America Region with the Peru-Chilean Province and Easter Island in the central south Pacific. In the Eastern Atlantic, the Canary Islands belong to the Lusitania Region.

Together they cover a wide biogeographical and climate range (Table 16.1). They are affected by the main currents in the Atlantic and the Pacific oceans (Fig. 16.2). Minimum temperature reaches −2 °C South of the Brazilian region and Eastern South America. The highest temperature is reached in the tropical

Table 16.1 Climatic and trophic variables in the water column in the coastal biogeographical provinces of North, Central and South America and the Canary Islands

	Minimum T (°C)[a]	Maximum T (°C)[a]	Mean T (°C)[a]	ST0 (°C)[b]	ST50 (°C)[b]	MLD (m)[b]	Photic Depth (m)[b]	Maximum Chl ($mg\ m^{-3}$)[a]	Minimum Chl ($mg\ m^{-3}$)[a]	Mean Chl ($mg\ m^{-3}$)[a]	Chl ($g\ m^{-1}$)[b]	Productivity (g $C\ m^{-2}\ d^{-1}$)[b]
Caribean	5.3	31.	25.6	26.9	25.8	23.0	44.2	74.3	0.0	0.6	6.6	0.5
West Indian	2.8	31.2	27.3	26.9	25.8	23.0	44.2	74.3	0.0	0.3	6.6	0.5
Brazilian	−2.0	30.5	24.6	27.6	26.7	14.5	38.7	74.3	0.0	0.4	55.8	1.9
EasternSouth America	−2.00	24.7	10.6	8.5	7.1	101.2	30.0	74.3	0.0	1.7		1.3
Canary Islands	16.79	24.7	20.7	21.2	18.0	28.2	28.7	40.9	0.0	0.2	35.0	2.0
Easter Island				16.7	14.2	23.3	43.3				15.7	0.7
Chilean				16.7	14.2	23.3	43.3				15.7	0.7
Galápagos				27.6	24.2	22.5	53.3				4.3	0.3
Panamanian				27.3	21.0	10.2	39.2				11.3	0.9
Mexican				27.3	21.0	10.2	39.2				11.3	0.9

[a] Mean values obtained from monthly satellite data for the period 1982–1991 for sea surface temperature and 2003–2010 for chlorophyll (http://neo.sci.gsfc.nasa.gov)

[b] Data from Longhurst et al. (1995), Longhurst (1998), compiled by Taeger and Lazarus (2010)

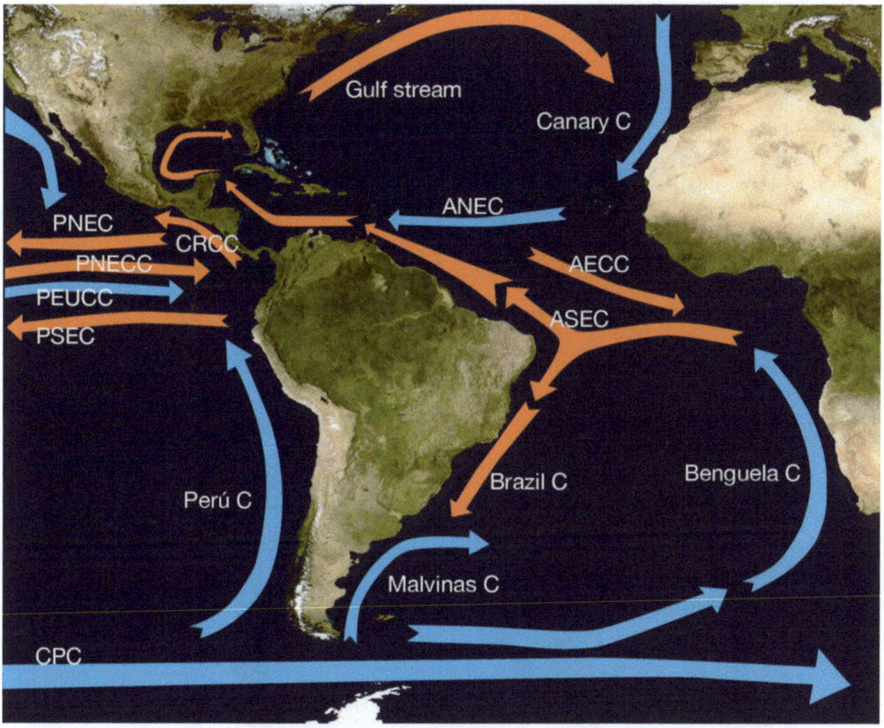

Fig. 16.2 Main current systems affecting the American coasts. *Orange arrows* correspond to warm currents, *blue arrows* correspond to cold currents. *PNEC* Pacific North Equatorial Current, *PNECC* Pacific North Equatorial Countercurrent, *PEUCC* Pacific Equatorial Undercountercurrent, *PSEC* Pacific South Equatorial Current; *CRCC* Costa Rica Costal Current; *CPC* Circumpolar current; *ANEC* Atlantic North Equatorial current; *AECC* Atlantic Equatorial Countercurrent; *ASEC* Atlantic South Equatorial Current

areas of the Pacific and Caribbean, exceeding 31 °C. The thermal range between summer and winter is greater than 25 °C in most areas and reaches 28.4 °C in the West Indies. Canary Islands show less variability, with a range of only 7.8 °C (Fig. 16.3a, b). In general, all the regions are productive. The Brazilian and Canary Islands provinces appear to be among the most productive in the world (1.92 and 2.01 g C $m^{-2}day^{-1}$, respectively) (Taeger and Lazarus 2010). However, in the case of the Canary Islands, this value is reached in the West African upwelling. Productivity in the archipelago is less and maximum chlorophyll concentration does not exceed 40.91 mg m^{-3}. Minimum mean chlorophyll concentration is reached in the Galápagos archipelago, with only 4.29 g m^{-3} (Fig. 16.3c).

The Caribbean, West Indian and Galápagos provinces are characterized by warm waters, with the highest photic depth and with a high mixed layer depth and low chlorophyll concentration and low productivity. The Panamanian and Mexican

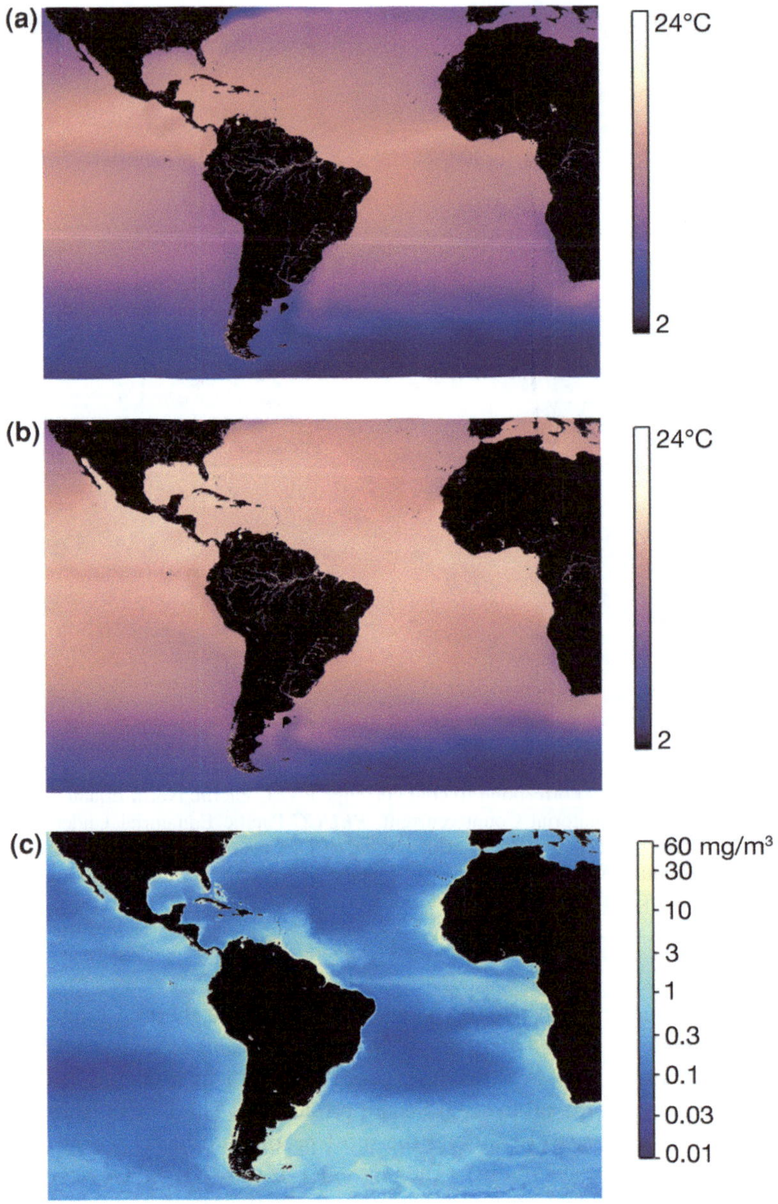

Fig. 16.3 Mean annual minimum (**a**) and maximum (**b**) sea surface temperature (°C) for the period 1982–1992 obtained from monthly satellite images. **c** Mean annual concentration of chlorophyll a (mgm^{-3}) for the period 2003–2010. Data source (http://neo.sci.gsfc.nasa.gov)

provinces share with the former the high temperature and photic depth, but have a narrow mixed layer depth and higher productivity. The Chilean, Easter Island and

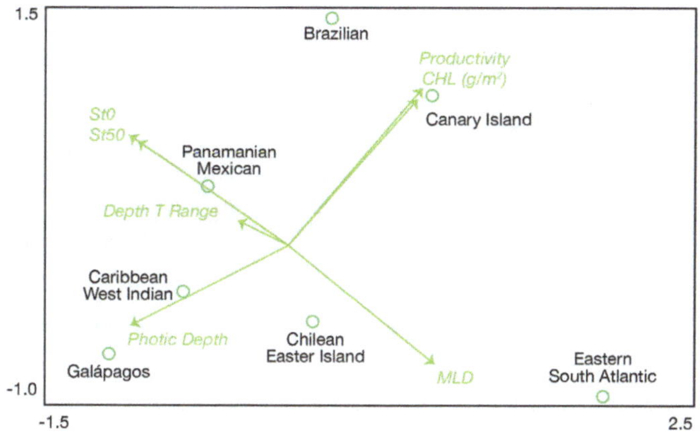

Fig. 16.4 Ordination of North, Central and South American biogeographical regions and environmental variables in a biplot representation of the first axes of the Principal Component Analysis (PCA) performed on the surface water column parameters according to Longhurst (see Table 16.1). Key to abbreviations for environmental variables are in Table 16.1

Eastern South Atlantic provinces are characterized by cold and transparent waters, with the highest mixed layer depth and low productivity (Fig. 16.4).

16.3 Diversity and Biogeographic Affinities

The echinoderm fauna of Latin America and the Canary Islands consist of 1,539 species, 82 species of Crinoidea, 392 species of Asteroidea, 521 species of Ophiuroidea, 242 species of Echinoidea and 302 species of Holothuroidea. Six hundred and twenty-seven species are found only along the Atlantic coasts of Central and South America, 597 species are found only along the Pacific coasts and oceanic islands and 19 species are found only in the Eastern Atlantic, in the Canary Islands.

Species richness is highly variable among the different countries, from 20 species in Easter Island and 23 species along the Caribbean coast of Guatemala to a maximum of 374 species in Cuba (Fig. 16.5a). However, the number of species is highly dependent of the coast length of the countries (Fig. 16.6). The number of species per 10 km of coast is 0.22 in Chile and 66.6 in Malpelo. Among biogeographical provinces (Fig. 16.7) the highest mean value is in the Panamanian province with 11.88 (s.e. ± 7.4) species per 10 km. However, Permanova analyses do not detect significant differences between provinces ($P = 0.255$).

The highest diversity (species richness (S), average taxonomic distinctness (Δ+), total taxonomic distinctness (sΔ+), average phylogenetic diversity (Phi+) and total phylogenetic diversity (sPhi+)) is in Caribbean and West Indian countries (Atlantic coasts of Mexico, Guatemala and Cuba). This suggests they could be the

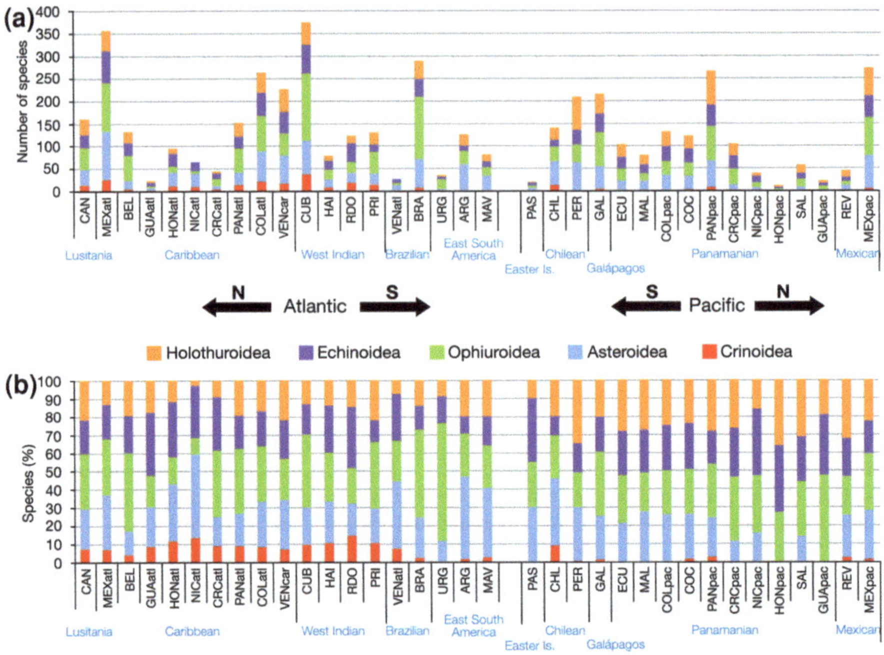

Fig. 16.5 Number of species (*up*) and percentage (*down*) of the five classes of echinoderms along the Pacific and Atlantic coasts of North, Central and South American countries and the Canary Islands. *MEX* Mexico; *REV* Revillagigedo Archipelago; *GUA* Guatemala; *SAL* El Salvador; *HON* Honduras; *NIC* Nicaragua; *CRC* Costa Rica; *COC* Cocos Island; *PAN* Panama; *COL* Colombia; *MAL* Malpelo Island; *GAL* Galápagos Archipelago; *ECU* Ecuador; *PER* Perú; *CHL* Chile; *PAS* Eastern Islands. *BEL* Belize; *VEN* Venezuela; *BRA* Brazil; *URG* Uruguay, *ARG* Argentina; *MAV* Malvinas Islands; *CUB* Cuba; *HAI* Haiti; *RDO* Dominican Republic, *PRI* Puerto Rico; *CAN* Canary Islands; atl: Atlantic; car: Caribbean; pac: Pacific

origin of North Atlantic and Eastern Tropical Pacific echinoderm faunas. Currently, both areas (Caribbean and West Indies) are quite similar in temperature regimes (maximum 31.21 °C in both areas, minimum 5.28 and 2.80 °C, mean 25.61 and 27.25 °C in the Caribbean and West Indian respectively), but differ in chlorophyll concentration (0.64 and 0.28 mg m^{-3} respectively).

On the other hand, there are significant differences (Permanova P = 0.001) in the percentage of the five classes of echinoderms between provinces (Fig. 16.5b). Ophiuroidea is the dominant class in most regions, except in the Chilean province that is dominated by Asteroidea, the Easter Island province that is dominated by Echinoidea and the Mexican province that is dominated by Holothuroidea. Crinoidea is always the least abundant class. Their absence in the Easter Island region and scarcity in the Eastern South America, Galápagos, Panamanian and Mexican provinces should be noted. The absence of Crinoidea has been also emphasized in other Atlantic archipelagos like Cabo Verde (Pérez-Ruzafa et al. 1999). It is also

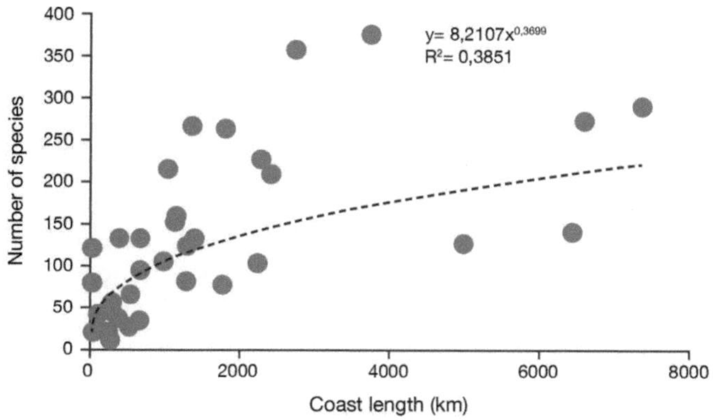

Fig. 16.6 Relationship between the number of echinoderm species and shore length of the Pacific and Atlantic coasts of the North, Central and South American countries and the Canary Islands

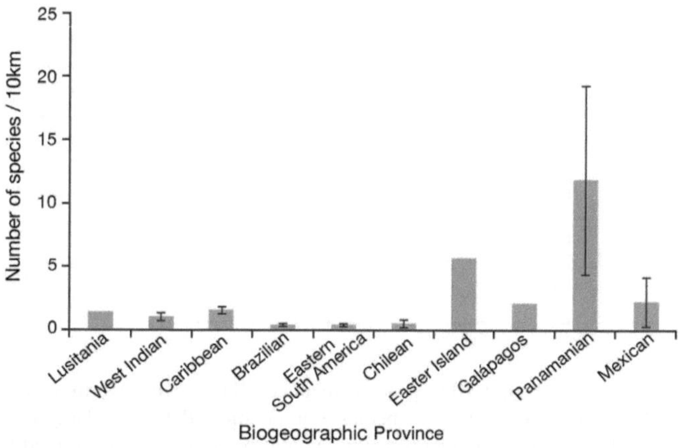

Fig. 16.7 Standardized species richness per 10 km shore length in the biogeographical provinces of North, Central and South American countries and in the Canary Islands. Error *bars* correspond to s.e. of the mean

worthwhile to note the negative relationship that exists between the dominance of holothurians in the faunas and the presence of crinoids (Fig. 16.8).

The echinoderm fauna of the different countries is consistent with their belonging to one of the above mentioned regions and provinces. NMDS analyses, performed with Bray-Curtis similarity calculated on species presence-absence (Bray and Curtis 1957), spatially distributed the countries according to their geographic location and importance to a given province (Fig. 16.9). The plot shows two well-established clusters. One includes the Panamanian, Galápagos and

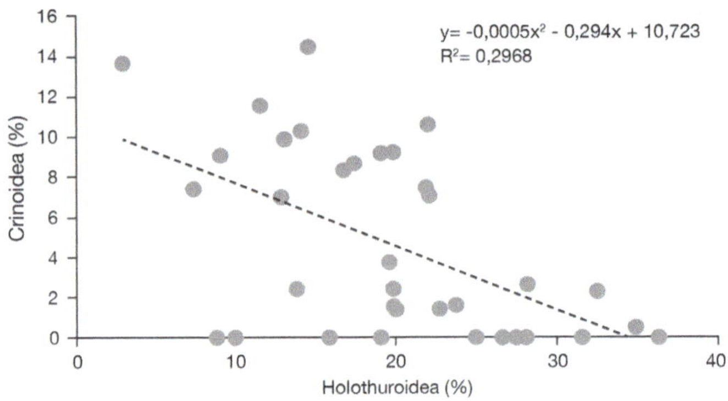

Fig. 16.8 Negative relationship between the percentage of species of crinoids and holothurians of the North, Central and South American countries and in the Canary Islands

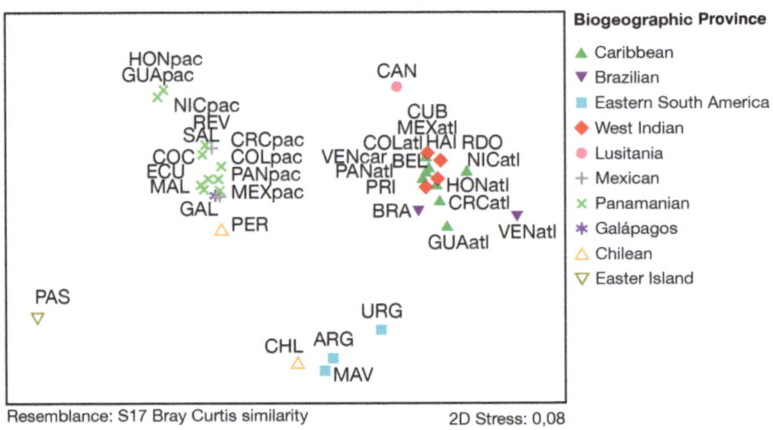

Fig. 16.9 Multivariate analysis nMDS plot of distance-related echinoderm species assemblage structure from Latinamerica based on species composition data and Bray-Curtis similarity measure. The faunal similarity between biogeographical provinces is represented by the relative distance. *MEX* Mexico; *REV* Revillagigedo Archipelago; *GUA* Guatemala; *SAL* El Salvador; *HON* Honduras; *NIC* Nicaragua; *CRC* Costa Rica; *COC* Cocos Island; *PAN* Panama; *COL* Colombia; *MAL* Malpelo Island; *GAL* Galápagos Archipelago; *ECU* Ecuador; *PER* Perú; *CHL* Chile; *PAS* Eastern Islands. *BEL* Belize; *VEN* Venezuela; *BRA* Brazil; *URG* Uruguay, *ARG* Argentina; *MAV* Malvinas Islands; *CUB* Cuba; *HAI* Haiti; *RDO* Dominican Republic, *PRI* Puerto Rico; *CAN* Canary Islands; atl: Atlantic; car: Caribbean; pac: Pacific

Chilean provinces to the tip of the Eastern South American region. Another cluster well separated region includes the Caribbean, West Indian, Lusitanica and Brazilian provinces. Earlier studies (Maluf 1988a, 1991) and results presented here suggest that the echinoderm faunas of the Panamic, Galápagos and the Chilean provinces are biogeographically related. There is a low level of association among echinoderm species from the Chilean fauna and Eastern South American

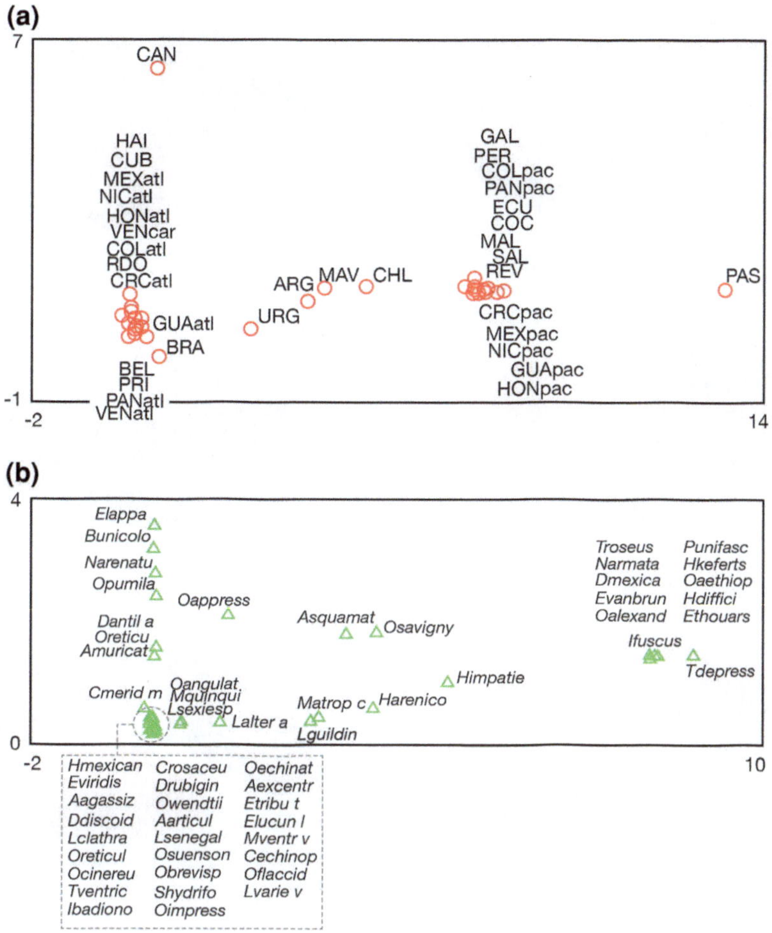

Fig. 16.10 **a** Ordination of countries in the representation of the first axes of the Detrended Correspondence analysis (DCA) performed on the echinoderm species matrix. **b** Ordination of species in the representation of the first axes of the Detrended Correspondence analysis (DCA) performed on the echinoderm species matrix. Only species with a weight >40 % on the axis are represented. *MEX* Mexico; *REV* Revillagigedo Archipelago; *GUA* Guatemala; *SAL* El Salvador; *HON* Honduras; *NIC* Nicaragua; *CRC* Costa Rica; *COC* Cocos Island; *PAN* Panama; *COL* Colombia; *MAL* Malpelo Island; *GAL* Galápagos Archipelago; *ECU* Ecuador; *PER* Perú; *CHL* Chile; *PAS* Eastern Islands. *BEL* Belize; *VEN* Venezuela; *BRA* Brazil; *URG* Uruguay, *ARG* Argentina; *MAV* Malvinas Islands; *CUB* Cuba; *HAI* Haiti; *RDO* Dominican Republic, *PRI* Puerto Rico; *CAN* Canary Islands; atl: Atlantic; car: Caribbean; pac: Pacific

biogeographic region. Chile is closer to Argentina and the Malvines than to Peru. In fact, there are two biogeographical provinces, the Peru-Chilean and the South America or Magellan.

The first two axis of the Detrended Correspondence Analyses (DCA), also performed on country species composition, also show a geographical ordination,

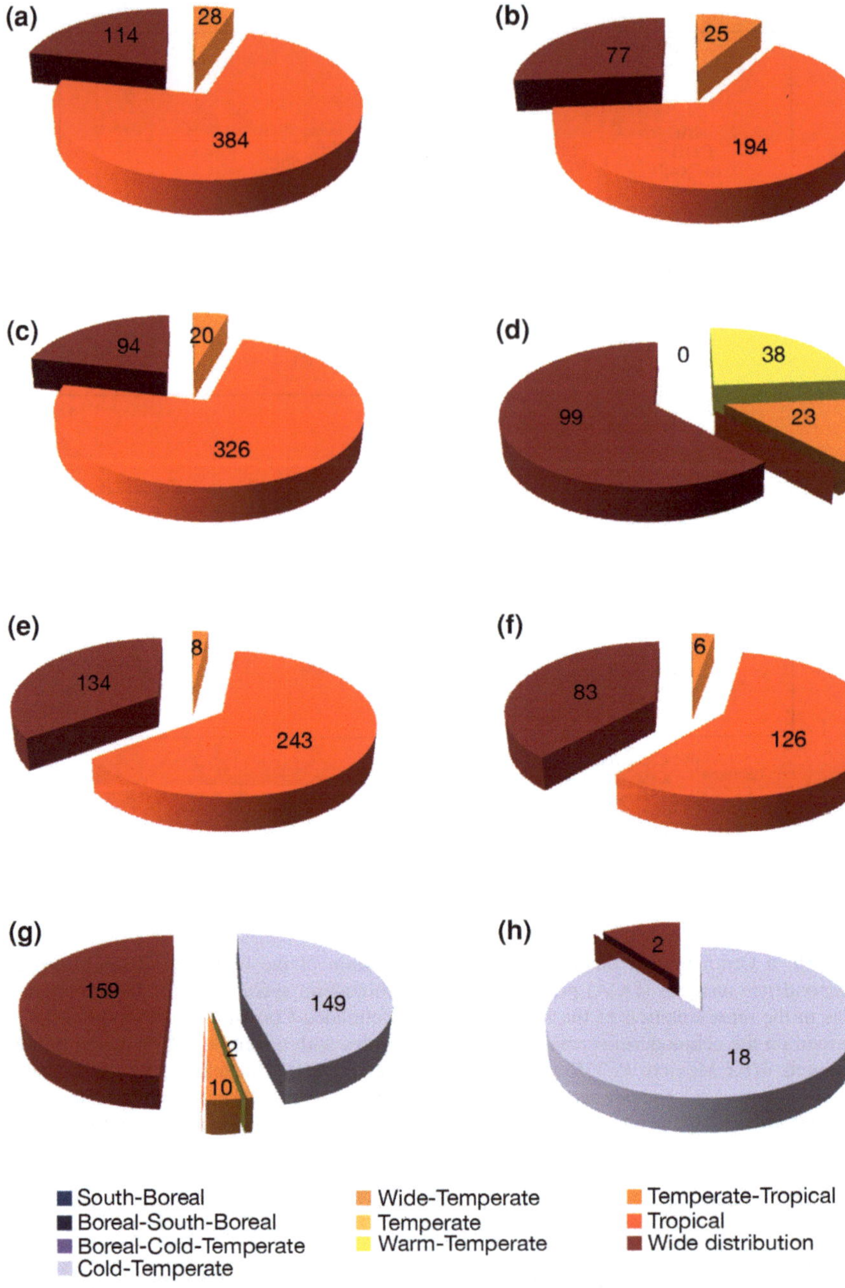

Fig. 16.11 Composition of the echinoderm faunal assemblages of the biogeographical provinces of North, Central and South America and the Canary Islands according to the thermal distribution range of the species inventoried. **a** Caribbean, **b** Brazilian, **c** West Indian, **d** Canary Is. (Lusitania), **e** Panamanian, **f** Galápagos, **g** Chilean, **h** Easter Island

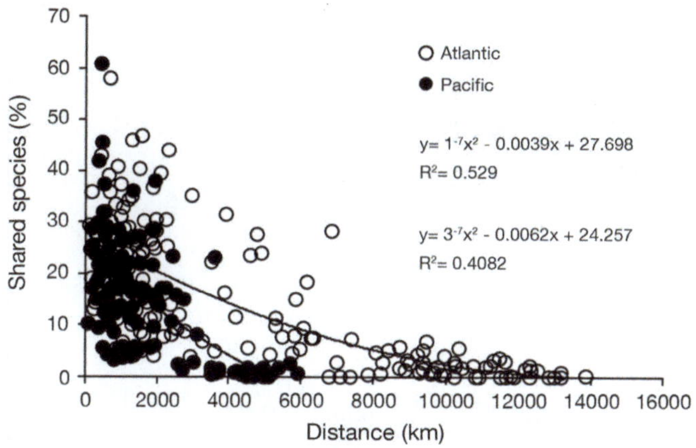

Fig. 16.12 Number of echinoderm species shared by two countries as a function of the distance between them. Distances have been calculated using the geographical coordinates at the *midpoint* of their *shorelines*

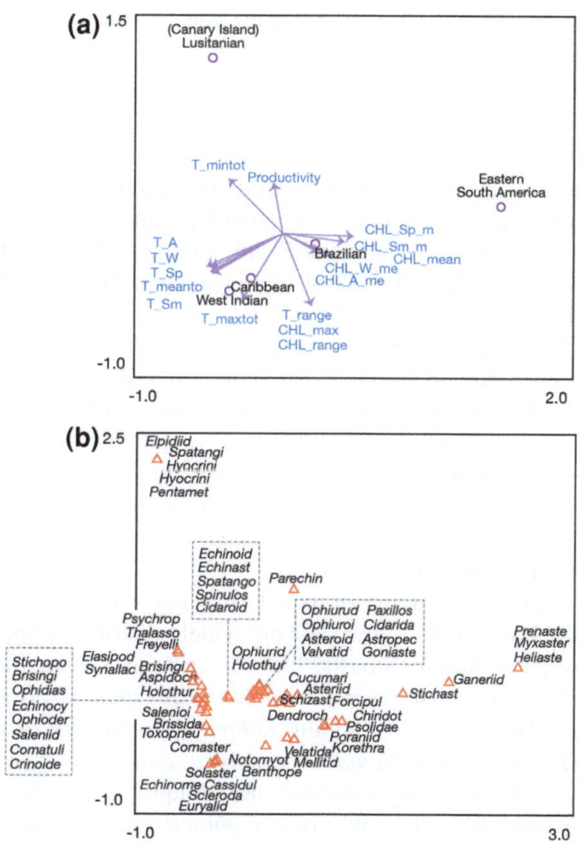

Fig. 16.13 Ordination of samples (biogeographical areas) and environmental variables (**a**) and families (**b**) in the representations of the first axes of the canonical correspondence analysis (CCA) performed on the matrix containing the number of species representing the echinoderm orders and families in the different Atlantic biogeographical provinces

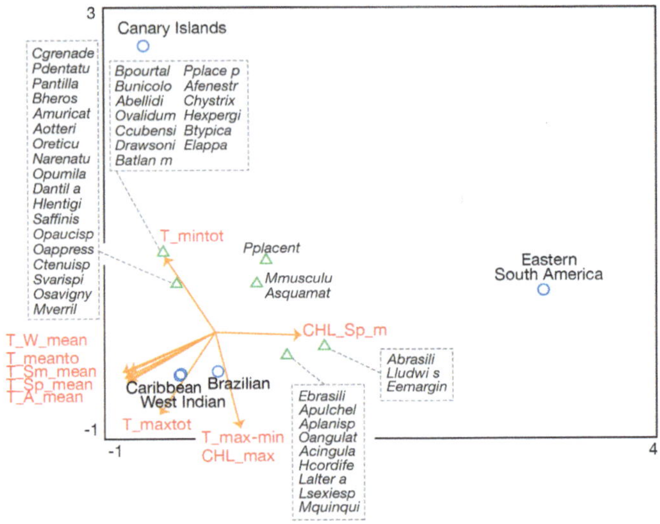

Fig. 16.14 Ordination of samples (biogeographical areas) and environmental variables (**a**) and species (**b**) in the representations of the first axes of the canonical correspondence analysis (CCA) of Atlantic echinoderm species matrix using environmental data as explanatory variables. Key of abbreviations for environmental variables and species are in Table 16.1

with North Atlantic locations on the second axis, Pacific countries on the positive part of the first axis, and Southern Atlantic, and Southern Pacific countries as a transition between the two main groups. The first axis, which determines the separation of Atlantic and Pacific faunas and the gradient throughout the Magellan region explain 29.6 % of total variance. The second axis explains only an additional 3.8 % (Fig. 16.10a).

The species represented in Fig. 16.10b are those that have a weight on the axis higher than 40 %. The positive part of axis 1 is represented by species exclusive to the Pacific and with a wide distribution in this ocean, like the sea urchins *Toxopneustes roseus, Diadema mexicanum Echinometra vanbrunti, Eucidaris thouarsii, Tripneustes depressus*, the sea stars *Nidorellia armata, Ophidiaster alexandri, Phataria unifascialis*, the ophiuroids *Ophiocoma aethiops* and the holothuroids *Holothuria (Halodeima) kefersteini, Holothuria (Platyperona) difficilis, Isostichopus fuscus.*

The left extreme of axis 1 is characterized by exclusively Atlantic species. In the lower part of the axis 2 are widely distributed species and exclusively from the western Atlantic coasts, like the crinoids *Comactinia meridionalis meridionalis, Davidaster discoideus*, the ophiuroids *Ophionereis reticulata*, the sea star *Luidia clathrata*, the sea urchin *Echinometra viridis* or the holothuroids *Holothuria (Halodeima) mexicana, Actinopyga agassizii.*

Between both extremes are ubiquitous species, widely distributed in both oceans, like the ophiuroids *Ophiactis savignyi* and *Amphipholis squamata*, the

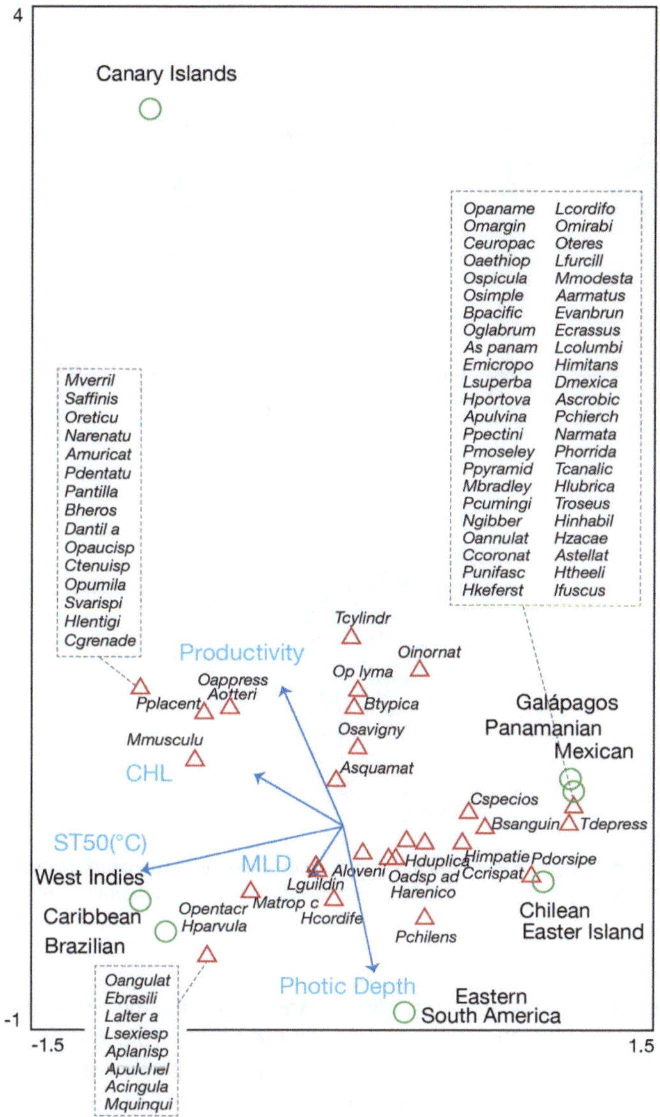

Fig. 16.15 Ordination of samples (biogeographical areas) and environmental variables (**a**) and species (**b**) in the representations of the first axes of the canonical correspondence analysis (CCA) of Atlantic and Pacific echinoderm species matrix using environmental data as explanatory variables. Key of abbreviations for environmental variables and species are in Table 16.1

holothuroids *Holothuria (Thymiosycia) arenicola* and *Holothuria (Thymiosycia) impatiens*, and the sea star *Linckia guildingi*.

Along the axis 2 are anfiatlantic species, shared by the American coasts and the Canary Islands, like the ophiuroids *Astrophyton muricatum, Ophiocoma pumila,*

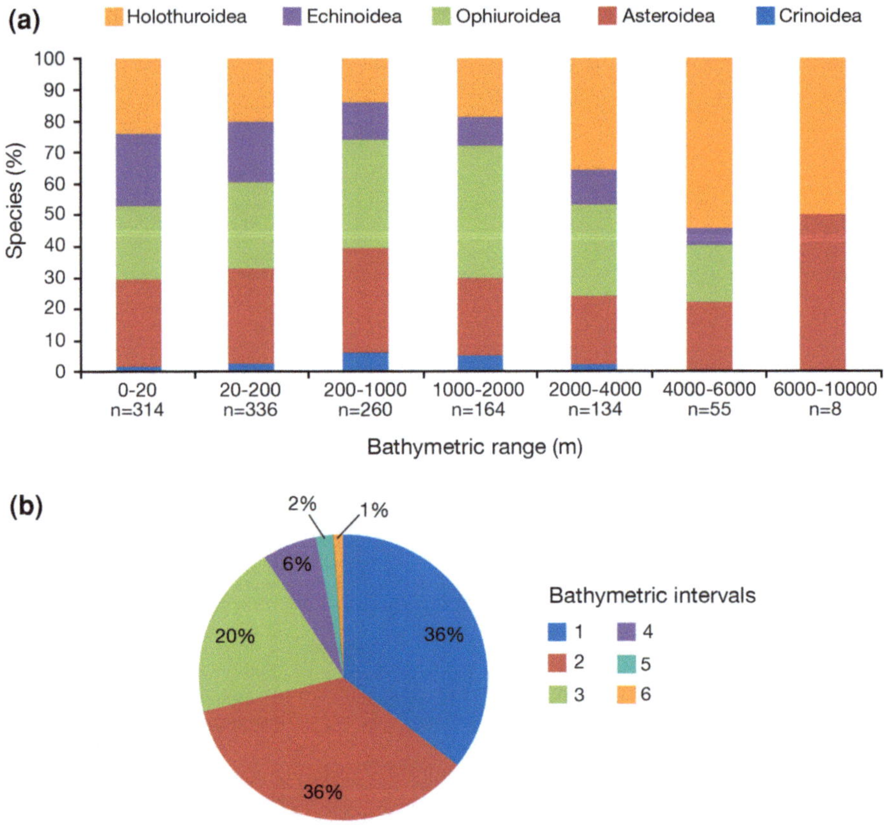

Fig. 16.16 **a** Percentage of echinoderm species per class by bathymetric range in the Pacific taxonomic list and **b** percentage of species present in one to six bathymetric intervals. N = 620 species

the sea stars *Oreaster reticulatus*, *Nymphaster arenatus*, the sea urchins *Diadema antillarum antillarum, Brissus unicolor* and the holothuroid *Euapta lappa*.

Cosmopolitan species are an important component in all the biogeographical regions (Fig. 16.11). Except the Chilean and the Easter Island provinces, which are dominated by cold temperate species. The Central and South American regions are dominated by tropical species and to a lesser extent, temperate species with a wide distribution range. In the Canary Islands, the warm temperate component is also important.

Similarity between country faunas depends on climatic and trophic conditions and on geographical proximity. The number of shared species is highly dependent on the distance separating two countries (Fig. 16.12). The diminishing proportion of shared species between two areas with increasing geographical distance is an obvious feature of natural systems (Hengeveld 1990; Huston 1994; Rosenzweig 1995; Brown and Lomolino 1998; Hubbell 2001).

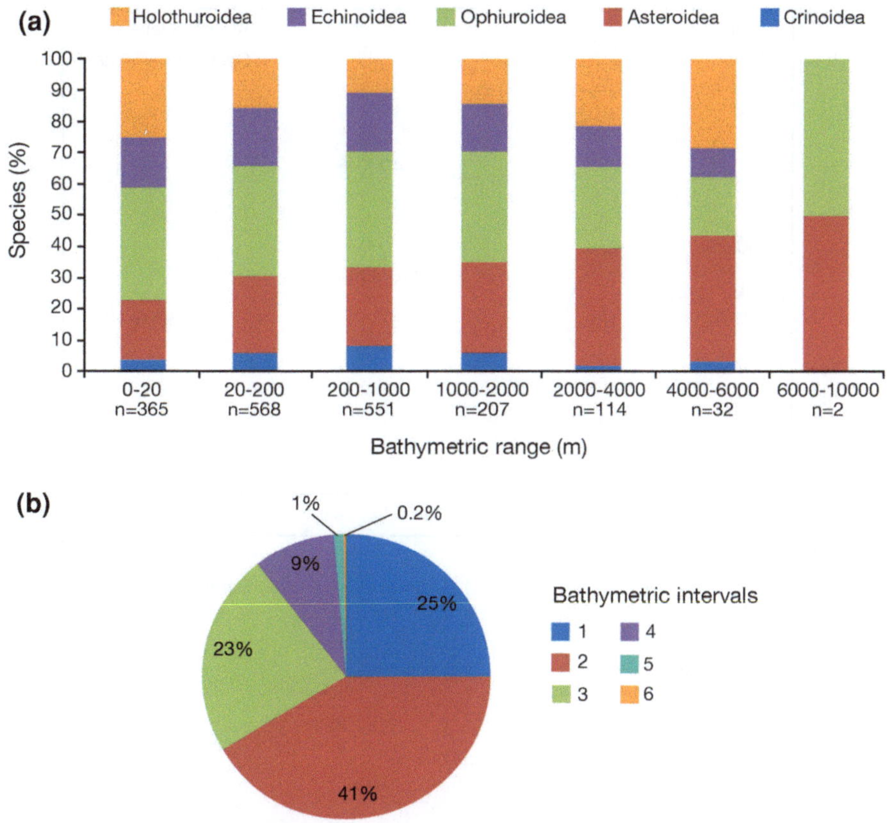

Fig. 16.17 **a** Percentage of echinoderm species per class by bathymetric range in the Caribbean and Atlantic taxonomic list and **b** percentage of species present in one to six bathymetric intervals. N = 836 species

There are more islands in the Atlantic region (i.e. the Lesser Antilles). Geographical distance between islands might be an important source of similarity in terms of species richness and taxonomic composition. Two islands may share the same number of species not because they are similar in area and/or in diversity because they are geographically close (Rosenzweig 1995), which allows individuals to move easily from one island to the other.

Canonical Correspondence analyses performed on the number of species representing the echinoderm orders and families in the different biogeographical provinces (Fig. 16.13) are consistent with the previous non-canonical analyses on species presence-absence data and climatic conditions. Figure 16.13 shows the results obtained for Atlantic provinces. The environmental variables used were mean seasonal and mean, minimum, maximum and range of variation annual values obtained from monthly satellite data for sea surface temperature (1982–1991) and for chlorophyll (2003–2010). Productivity values were obtained

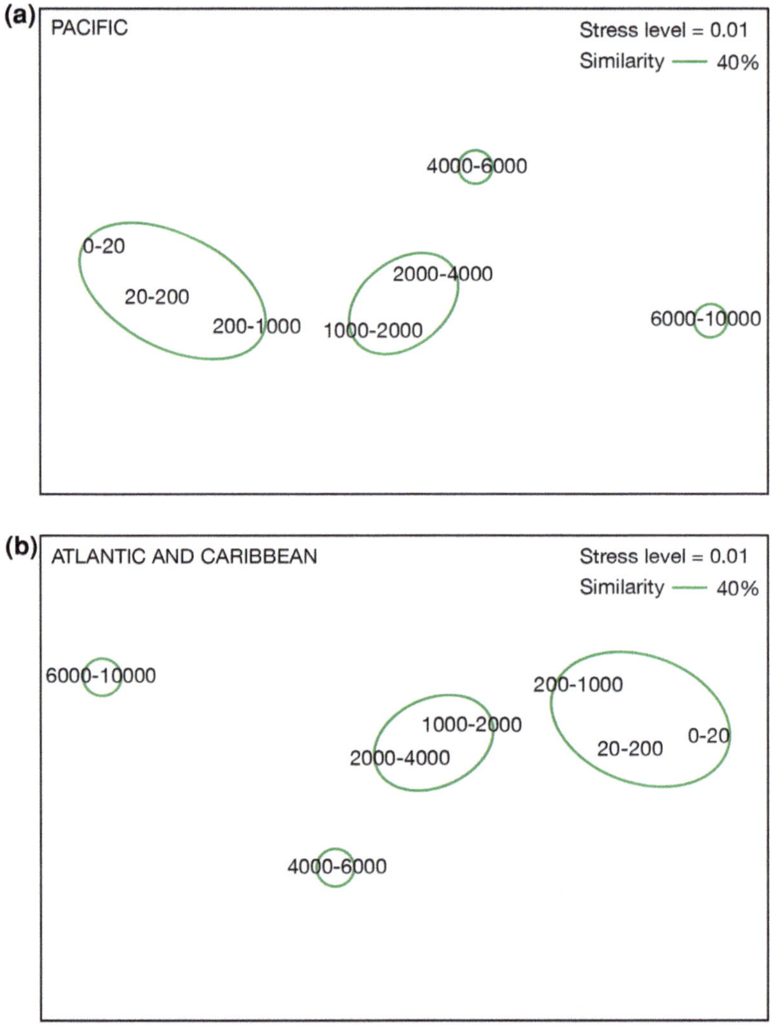

Fig. 16.18 Multivariate analysis nMDS *plot* of distance-related echinoderm species assemblage structure from **a** the Pacific taxonomic list and **b** the Caribbean and Atlantic taxonomic list, based on species presence/absence matrix by bathymetric interval using Bray-Curtis similarity measure

from Longhurst et al. (1995), Longhurst (1998), compiled by Taeger and Lazarus (2010) (Table 16.1).

The first two axes account for 46.6 and 30.3 %, respectively, of the total variance of species-environment relation. The first axis represents a gradient from the Caribbean and West Indies to the south. The positive extreme of the first axis is associated with the Eastern South America region and highest mean chlorophyll concentration in all seasons and the lowest temperatures. It is represented by the

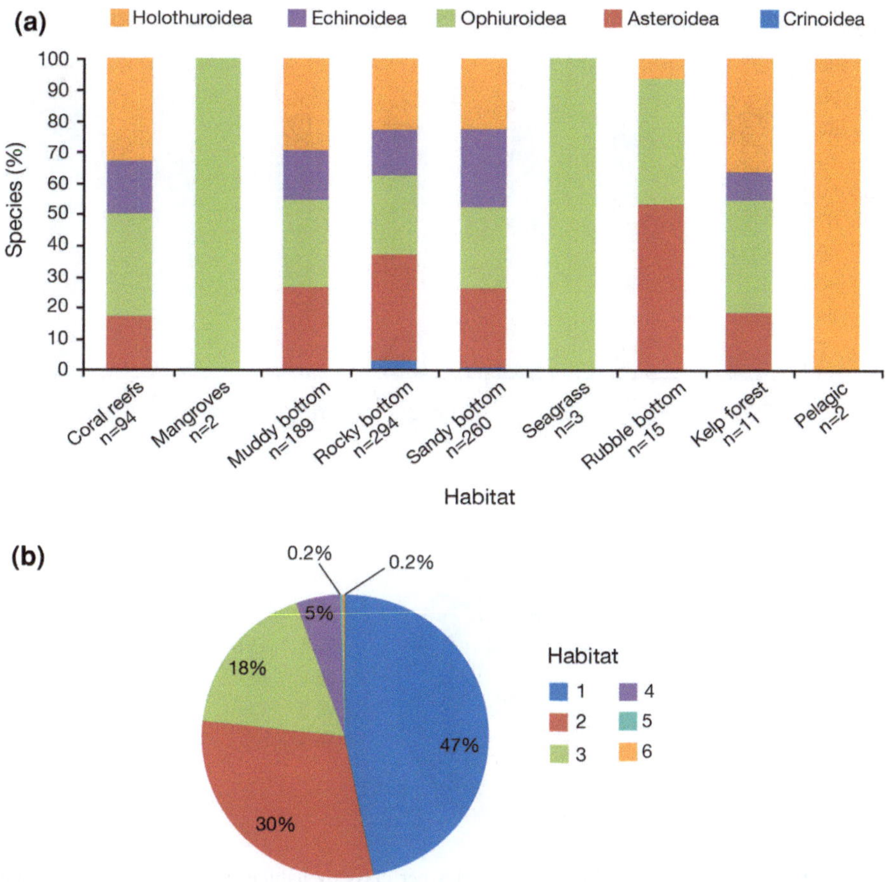

Fig. 16.19 **a** Percentage of echinoderm species per class by habitat type in the Pacific taxonomic list and **b** percentage of species presents in one to seven habitat types. N = 473 species

families Heliasteridae, Stichasteridae, Ganeriidae, Myxasteridae and Prenasteridae. The Myxasteridae is exclusive to the Eastern South America region and the Prenasteridae is shared with the Chilean province.

In the negative part of this axis, associated with highest temperatures and low chlorophyll concentration, are families with a wide distribution in the Atlantic but absent in the Eastern South America region like Echinocyamidae. Some of them like Brisingida, Ophidiasteridae, Saleniidae, Stichopodidae, Ophiodermatidae or the class Crinoidea are even present along Pacific coasts. This indicates their tropical origin and ancient links, prior to closure of the Isthmus of Panama. In general, these families are represented by different species in each ocean. For example, the sea star family Ophidiasteridae, is present in the area with 29 species, 14 in the Pacific, five in the Western Atlantic and nine in the Anfiatlantic. None is shared between the Atlantic and Pacific coasts. Congeneric species are *Ophidiaster*

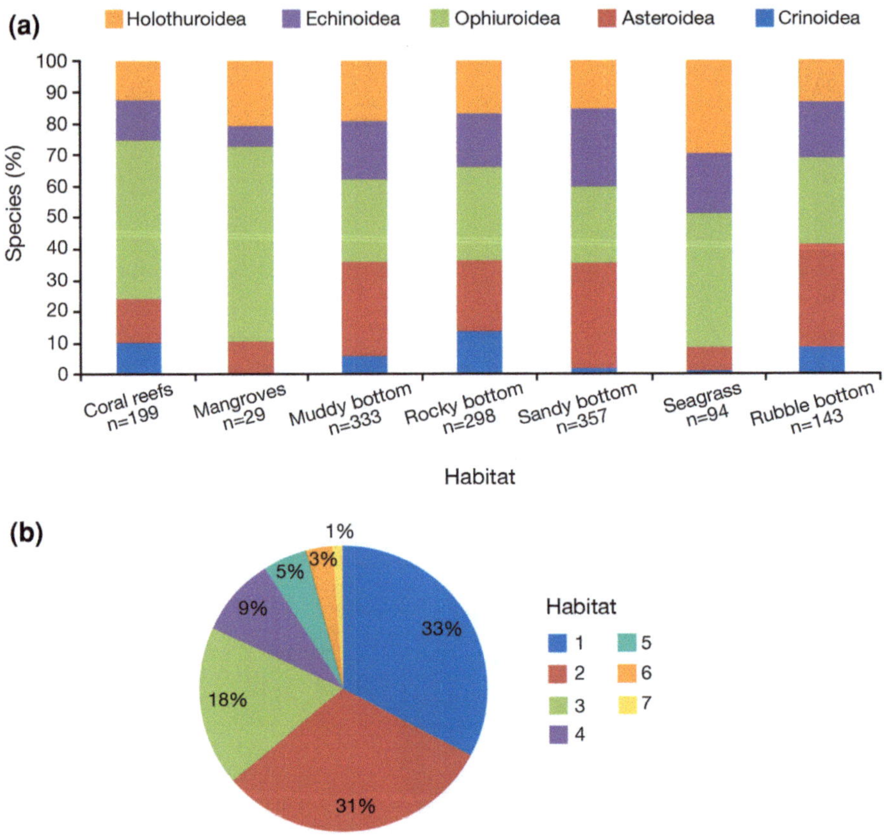

Fig. 16.20 **a** Percentage of echinoderm species per class by habitat type in the Caribbean and Atlantic taxonomic list and **b** percentage of species presents in one to seven habitat types. N = 630 species

alexandri, Ophidiaster bayeri, Tamaria floridae and *Tamaria halperni* in the Western Atlantic, *Ophidiaster agassizi, Ophidiaster ludwigi* along the Pacific coasts, and *Tamaria obstipa* and *Tamaria stria. Ophidiaster guildingii* and *Ophidiaster ophidianus* in the Anfiatlantic. *Narcissia canariensis* and *N. trigonaria* are Anfiatlantic, *Narcissia gracilis* and the subspecies *N. gracilis malpeloensis* occur in the Pacific.

The second axis represents a Western–Eastern gradient. Pentametrocrinidae, Elpidiidae, Hyocrinidae and Spatangidae characterize the positive part of this axis. Pentametrocrinidae is present in the Canary Islands and is not shared with south American regions, Elpidiidae and Hyocrinidae are shared by the Canary Islands and all the Pacific regions except Easter Island, but are absent in the South American coasts. In the same way, Spatangidae is a family shared by the Canary Islands and the Pacific Mexican coast but is absent from the other regions.

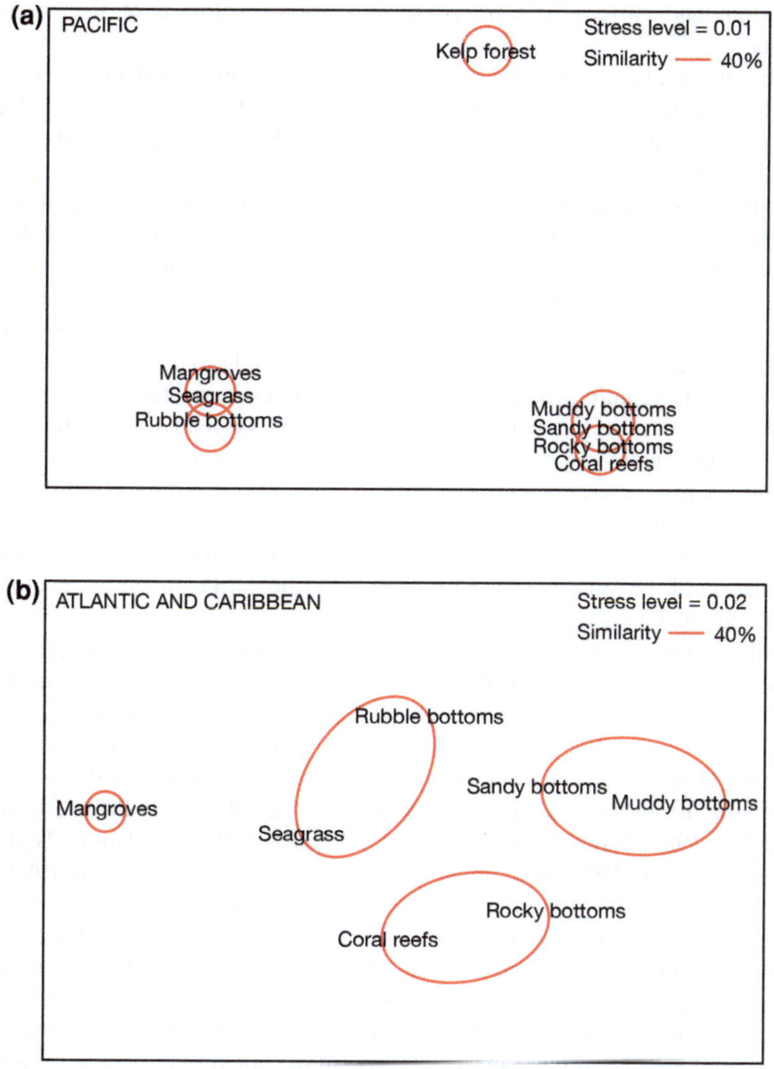

Fig. 16.21 Multivariate analysis nMDS plot of distance-related echinoderm species assemblage structure from the **a** the Pacific taxonomic list and **b** the Caribbean and Atlantic taxonomic list, based on species presence/absence matrix by habitat category using Bray-Curtis similarity measure

The negative part of this axis is represented by the families Solasteridae, Echinometridae, Sclerodactylidae and the order Cassiduloida. All of them are present in the Caribbean, West Indian and Brazilian provinces, and even in the warm Pacific coasts. But they are not present in the Canary Islands or the Eastern South America region. It indicates a common tropical origin, after the separation of western and eastern Atlantic coasts but prior to closure of the Isthmus of Panama.

The same analyses performed on species presence-absence data can be seen in Fig. 16.14 for the Atlantic regions and Fig. 16.15 for Pacific and Atlantic Regions. The figure includes only species fitting more than 50 % of the axis and weighting more than 50 %.

As previously mentioned, the affinities between faunas are a consequence of the combination of climatic and trophic factors, connectivity as a function of distance, current patterns and historical processes. The partition of the variance to see the relative influence of environmental variables and spatial organization (latitude and longitude coordinates) was done following the methodology proposed by Borcard et al. (1992). It shows that the covariation of space and environment reach 62.2 % in the case of Atlantic provinces. The environmental variables alone are responsible for 37.8 % of the variance. The covariation of space and environment of the Atlantic and Pacific regions analyzed together account for only 11.4 % of the variability while environmental variables explain 58.6 % of the variability. There is 14.5 % variation that is purely spatial and an additional 15.5 % is of unexplained variation and stochastic fluctuations or due to historical processes linked to the isolation of the faunas of the Caribbean and tropical Pacific after the closure of the isthmus of Panama.

Furthermore, different environmental factors would be responsible for faunal composition and species distribution at different spatial scales (Barry and Dayton 1991; Levin 1992; Pérez-Ruzafa et al. 2003; Entrambasaguas et al. 2008). Water temperature is considered the most important influence on the global distribution of marine animals and minimum temperature is usually the factor that determines faunal similarities at small geographical scales (Steele 1983; Pérez-Ruzafa and López-Ibor 1988; Pérez-Ruzafa et al. 2003). Factors such as predation (Tegner and Dayton 1981; Sala 1997), settlement and recruitment (Young and Chia 1982; Ebert 1983; Hereu et al. 2004; Hernández et al. 2010), availability of trophic resources (Menge 1992), disease epidemics (Hagen 1999; Dumont et al. 2004), or competitive interactions (Hagen and Mann 1992), physical factors like substrate nature, bottom complexity, depth, wave exposition, etc. (Drouin et al. 1985; Tyler et al. 2000; Entrambasaguas et al. 2008) or harvesting (Pfister and Bradbury 1996; Hasan 2005) are of major importance in explaining small scale species distribution and abundance.

16.4 Bathymetrical and Substrate Distribution

In the Pacific taxonomic list (see appendix) there were 620 species of echinoderms with bathymetric information (Fig. 16.16) that represent 86 % of the species of the list, while in the Caribbean and Atlantic taxonomic list (see appendix) there were 836 species, that represent 92 % of the species on the list (Fig. 16.17). In the Pacific most species were reported at depths between 20 and 200 m (336 species),

30 % in the class Asteroidea (Fig. 16.16a). This class was also the predominant at depths between 0 and 20 m (28 %). The class Ophiuroidea was the predominant at depths between 200–1,000 m (35 %) and 1,000–2,000 m (42 %). The class Holothuroidea was predominant at depths between 2,000–4,000 m (36 %) and 4,000–6,000 m (55 %). Most species on the Pacific list were found only in one or two bathymetric intervals (each 36 %) while just 2 and 1 % were found in five and six bathymetric intervals respectively (Fig. 16.16b). In the Caribbean and Atlantic list most species were also at depths between 20 and 200 m (568 species) (Fig. 16.17a), 35 % in the class Ophiuroidea. This class also predominated at depths between 0–20 m (36 %), 200–1,000 m (37 %) and 1,000–2,000 m (35 %). The class Asteroidea was predominant at depths between 2,000–4,000 m (38 %) and 4,000–6,000 m (41 %). Most of the species on the Caribbean and Atlantic list were found only in two or one bathymetric intervals (41 and 25 %, respectively) while only 1 and 0.2 % were found in five and six bathymetric intervals respectively (Fig. 16.17b).

According to the depth categories in the Pacific and the Caribbean-Atlantic, the similarity analysis showed four groups with 40 % resemblance (Fig. 16.18a, b). The first group is composed by the depths from 0 to 20 m, 20 to 200 m and 200 to 1,000 m. These groups possesses the highest number of species (314, 336, 260 species respectively in the Pacific and 365, 568, 551 species in the Caribbean-Atlantic), Margalef species richness (54.4, 57.5, 46.5 respectively in the Pacific and 61.5, 89.2, 87.1 respectively in the Caribbean-Atlantic) and Shannon diversity (5.7, 5.8, 5.5 respectively in the Pacific and 5.8, 6.3, 6.3 respectively in the Caribbean-Atlantic). The second group is composed by depths from 1,000 to 2,000 m and 2,000 to 4,000 m, that possesses moderate values of number of species (164 and 134 species in the Pacific and 207 and 114 in the Caribbean-Atlantic), Margalef species richness (31.9 and 27.1 in the Pacific and 38.6 and 23.8 in the Caribbean-Atlantic) and Shannon diversity (5.1 and 4.8 in the Pacific and 5.3 and 4.7 in the Caribbean-Atlantic). The other two groups are composed by only one bathymetric interval (4,000–6,000 m and 6,000–10,000 m) and possesses the lowest values of all indices; number of species (55 and 8 species in the Pacific and 32 and 2 in the Caribbean-Atlantic), Margalef species richness (13.4 and 3.3 in the Pacific and 8.9 and 1.4 in the Caribbean-Atlantic) and Shannon diversity (4.0 and 2.0 in the Pacific and 3.4 and 0.6 in the Caribbean-Atlantic). Of the first two groups, the Caribbean-Atlantic always had higher values than the Pacific in all the indices. But in deeper waters (groups 3 and 4), the Pacific was richer than the Caribbean-Atlantic.

There were 473 species of echinoderms (66 % of the total) in the Pacific taxonomic list (see appendix) with habitat or substrate information (Fig. 16.19) In the Caribbean and Atlantic taxonomic list (appendix) there were 630 species (70 % of the total) with habitat or substrate information (Fig. 16.20). Most of the species in the Pacific were on rocky bottoms. Of the 294 species, 30 % are in the class Asteroidea (Fig. 16.19a). This class was also the predominant on rocky and rubble bottom (34 and 53 %, respectively). On coral reefs and in kelp forest, the classes Ophiuroidea (33 and 36 %, respectively) and Holothuroidea (33 and 36 %, respectively) were predominantd. Holothuroidea were predominant on muddy bottoms (30 %). On sandy bottoms, Asteroidea (26 %) and Ophiuroidea (26 %) were predominant. Most of the species in the Pacific list were found in only one habitat (each 47 %) while only 0.2 % were found in five and seven habitats (Fig. 16.19b). In the Caribbean and Atlantic list, most of the species were on sandy bottoms (357 species) (Fig. 16.20a), 33 % in the class Asteroidea. This class also was predominant on rubble (33 %) and muddy (30 %) bottoms. The class Ophiuroidea was predominant on coral reefs (50 %), mangroves (62 %), seagrass (43 %) and rocky bottoms (30 %). Most of the species in the Caribbean and Atlantic lists were found in only one habitat (33 %) while only 3 % and 1 % were found in six and seven habitats respectively (Fig. 16.20b).

There are three groups according to habitat categories in the Pacific (Fig. 16.21a). The first group is composed of coral reefs, muddy, rocky and sandy bottoms. These habitats possess the highest values of total species richness (94–294), Margalef species richness (20.3–51.4) and Shannon Diversity (5.3–5.9). The second group is composed by sea grass, mangroves and rubble bottom. These habitats possess low values of total species richness (2–15), Margalef species richness (1.4–5.2) and Shannon Diversity (0.7–2.7). The third group is composed by the kelp forest that possesses low values of the total species richness (11), Margalef species richness (4.2) and Shannon Diversity (2.3). In the Caribbean-Atlantic, the nMDS indicates the presence of four groups (Fig. 16.21b). The first one is composed of sandy and muddy bottoms that possess the highest values of total species richness (357–333 species respectively), Margalef species richness (60.4–57.1 respectively) and Shannon Diversity (5.9–5.8). The second group is composed of rocky bottoms and coral reefs that possess high to moderate values of total species richness (298–199 species respectively), Margalef species richness (52.1–37.4 respectively) and Shannon Diversity (5.7–5.3 respectively). The third group is composed of seagrass and rubble bottoms with moderate values of total species richness (94–143 species respectively), Margalef species richness (20.5–28.6 respectively) and Shannon Diversity (4.5–4.9 respectively). The last group is composed of mangroves that possess the lowest values of total species richness (29), Margalef species richness (8.3) and Shannon Diversity (3.4).

Acknowledgments We are grateful with all the authors that participated on this contribution. We appreciate the help and extensive work of the students from the Laboratorio de Ecología y Taxonomía de Equinodermos, Colección Nacional de Equinodermos Dra. María Elena Caso, Instituto de Ciencias del Mar y Limnología from the Universidad Autónoma Nacional de México. JJA wants to acknowledge Monica Chavez for preparing all the figures from this chapter and CONACYT and CONICIT for all the economic support during the preparation and analysis of this review.

References

Alvarado JJ (2011) Echinoderm diversity from the Caribbean Province. Mar Biodiv 41:261–285

Alvarado JJ, Cortés J (2004) The state of knowledge on echinoderms of Costa Rica and Central America. In: Heinzeiller T, Nebelsick JH (eds) Echinoderms: München. Proceedings of the 11th international Echinoderm conference. Taylor and Francis Group, London, pp 149–155

Alvarado JJ, Solís-Marín FA, Ahearn C (2008) Equinodermos (Echinodermata) del Caribe Centroamericano. Rev Biol Trop 56(Suppl 3):37–55

Alvarado JJ, Solis-Marin FA, Ahearn C (2010) Echinoderms (Echinodermata) diversity off Central America Pacific. Mar Biodiv 40:45–56

Barry JP, Dayton PK (1991) Physical heterogeneity and the organization of marine communities. In: Kolasa K, Pickett STA (eds) Ecological heterogeneity. Springer, New York, pp 270–320

Borcard D, Legendre P, Drapeau P (1992) Partialling out the spatial component of ecological variation. Ecology 73:1045–1055

Bray JR, Curtis JT (1957) An ordination of the upland forest communities of southern Wisconsin. Ecol Monogr 27:325–349

Briggs JC (1995) Global biogeography. Elsevier, Amsterdam

Brown JH, Lomolino MV (1998) Biogeography. Sinauer Associates, Sunderland

Clark AM, Downey ME (1992) Starfishes of the Atlantic. Chapman & Hall, London

Drouin G, Himmelman J, Béland T (1985) Impact of tidal salinity fluctuations on echinoderm and mollusc populations. Can J Zool 63:1377–1387

Dumont CP, Himmelman JH, Russell MP (2004) Sea urchin mass mortality associated with algal debris from ice scour. In: Heinzeiller T, Nebelsick JH (eds) Echinoderms: München. Proceedings of the 11th international Echinoderm conference. Taylor and Francis Group, London, pp 177–182

Ebert TA (1983) Recruitment in Echinoderms. In: Jangoux M, Lawrence JM (eds) Echinoderm studies, vol I. Balkelma, Rotterdam, pp 169–203

Entrambasaguas L, Pérez-Ruzafa A, García-Charton JA, Stobart B, Bacallado JJ (2008) Abundance, spatial distribution and habitat relationships of echinoderms in the Cabo Verde Archipelago (eastern Atlantic). Mar Fresh Res 59:477–488

Hagen N, Mann K (1992) Functional response of the predators American lobster *Homarus americanus* and Atlantic wolf fish *Anarhichas lupus* to increasing numbers of the green sea uchin *Strongylocentrotus droebachinesis*. J Exp Mar Biol Ecol 159:89–112

Hasan MH (2005) Destruction of a *Holothuria scabra* population by overfishing at Abu Rhamada Island in the Red Sea. Mar Environ Res 60:489–511

Hengeveld R (1990) Dynamic biogeography. Cambridge University Press, Cambridge

Hereu B, Zabala M, Linares C, Sala E (2004) Temporal and spatial variability in settlements of the sea urchin *Paracentrotus lividus* (Lamarck) in the NW Mediterranean. Mar Biol 144:1011–1018

Hernández JC, Clemente S, Girard D, Pérez-Ruzafa A, Brito A (2010) Effect of temperature on settlement and postsettlement survival in a barrens-forming sea Urchin. Mar Ecol Prog Ser 413:69–80

Hubbell SP (2001) The unified neutral theory of Biodiversity and Biogeography. Princeton University Press, Princeton

Huston MA (1994) Biological diversity: the coexistence of species on changing landscapes. Cambridge University Press, Cambridge

Levin SA (1992) The problem of pattern and scale in ecology. Ecology 73:1943–1967

Longhurst AR (1998) Ecological geography of the sea. Academic Press, San Diego

Longhurst A, Sathyendranath S, Platt T, Caverhill C (1995) An estimate of global primary production in the ocean from satellite radiometer data. J Plankton Res 17:1245–1271

Maluf LY (1988a) Composition and distribution of the Central eastern Pacific echinoderms. Nat Hist Mus L A County Tech Rep 2:1–242

Maluf LY (1988b) Biogeography of the central eastern shelf echinoderms. In: Burke RD, Mlademov PV, Lambert P, Parsley RL (eds) Echinoderm biology. AA Balkema, Rotterdam, pp 389–398

Maluf LY (1991) Echinoderm fauna of the Galápagos Islands. In: James MJ (ed) Galápagos marine invertebrates: taxonomy, biogeography and evolution in Darwin's islands. Plenum Press, New York, pp 345–367

Menge BA (1992) Community regulation: under what conditions are bottom-up factors important on rocky shores? Ecology 73:755–765

Miloslavich P, Díaz JM, Klein PE, Alvarado JJ, Díaz C, Gobin J, Escobar-Briones E, Cruz-Motta JJ, Weil E, Cortés J, Bastidas AC, Robertson R, Zapata F, Martín A, Castillo J, Kazandjan A, Ortiz M (2010) Marine biodiversity in the Caribbean: regional estimates and distribution patterns. PLoS ONE 5:e11916. doi:10.1371/journal.pone.0011916

Miloslavich P, Klein E, Díaz JM, Hernández CE, Bigatti G, Campos L, Artigas F, Castillo J, Penchaszadeh PE, Neill PE, Carranza A, Retana MV, Díaz de Astarloa JM, Lewis M, Yorio P, Piriz ML, Rodríguez D, Yoneshigue-Valentin Y, Gamboa L, Martín A (2011) Marine biodiversity in the Atlantic and Pacific Coasts of South America: knowledge and gaps. PLoSONE 6:e14631. doi:10.1371/journal.pone.0014631

Pérez-Ruzafa A, López-Ibor A (1988) Echinoderm fauna from south-western Mediterranean. Biogeographic relationships. In: Burke RD, Mladenov PV, Lambert P, Parsley RL (eds) Echinoderm biology. AA Balkema, Rotterdam, pp 355–362

Pérez-Ruzafa A, Entrambasaguas L, Bacallado JJ (1999) Fauna de equinodermos (Echinodermata) de los fondos rocosos infralitorales del archipiélago de Cabo Verde. Rev Acad Canaria Cien 11:43–62

Pérez-Ruzafa A, Entrambasaguas L, García Charton JA, Bacallado JJ, Marcos C (2003) Spatial relationships of the echinoderm fauna of Cabo Verde islands: A multi-scale approach. In: Féral JP, David B (eds) Echinoderm Research 2001. Sweets & Zeitlinger, Lisse, pp 31–39

Pfister CA, Bradbury A (1996) Harvesting red sea urchins: recent effects and future predictions. Ecol Appl 6:298–551

Price ARG, Keeeling MJ, O'Callaghan CJ (1999) Ocean-scale patterns of 'biodiversity' of Atlantic asteroids determined from taxonomic distinctness and other measures. Biol J Lin Soc 66:187–203

Rosenzweig ML (1995) Species diversity in space and time. Cambridge University Press, Cambridge

Sala E (1997) Fish predators and scavengers of the sea urchin *Paracentrotus lividus* in protected areas of the north-west Mediterranean Sea. Mar Biol 129:531–539

Steele DH (1983) Marine ecology and zoogeography. In: South GR (ed) Biogeography and ecology of the Island of Newfoundland. Junk Publishers, The Hague, pp 421–465

Stöhr S, O'Hara TD, Thuy B (2012) Global diversity of Brittle Stars (Echinodermata: Ophiuroidea). PLoS ONE 3:e31940. doi:10.1371/journal.pone.0031940

Taeger M, Lazarus D (2010) VLIZ maritime boundaries. Mus Natur, Berlin. Geodatabase http://www.vliz.be/vmdcdata/marbound/download.php

Tegner MJ, Dayton PK (1981) Population structure, recruitment and mortality of two sea urchins (*Strongylocentrotus franciscanus* and *S. purpuratus*) in a kelp forest. Mar Ecol Prog Ser 5:255–268

Tyler PA, Young CM, Clarke A (2000) Temperature and pressure tolerances of embryos and larvae of the Antarctic sea urchin *Sterechinus neumayeri*: potential for deep-sea invasion from high latitudes. Mar Ecol Prog Ser 192:173–180

Young CM, Chia FS (1982) Factors controlling spatial distribution of the sea cucumber *Psolus chitonoides*: settling and post-settling behaviour. Mar Biol 69:195–205

Appendix

Solís-Marín FA, JJ Alvarado, M Abreu-Pérez, O Aguilera, J Alió, JJ Bacallado-Aránega, E Barraza, M Benavides-Serrato, F Benítez-Villalobos, L Betancourt-Fernández, M Borges, M Brandt, MI Brogger, GH Borrero-Pérez, BE Buitrón-Sánchez, LS Campos, J Cantera, S Clemente, M Cohen-Renjifo, S Coppard, LV Costa-Lotufo, R del Valle-García, Y Díaz, ME Díaz de Vivar, JP Díaz-Martínez, A Durán-González, L Epherra, M Escolar, V Francisco, CA Freire, JE García-Arrarás, DG Gil, P Guarderas, VF Hadel, A Hearn, JC Hernández, EA Hernández-Delgado, A Herrera-Moreno, MD Herrero-Pérezrul, Y Hooker, MBI Honey-Escandón, C Lodeiros, M Luzuriaga, CLC Manso, A Martín, MI Martinez, S Martínez, L Moro-Abad, E Mutschke, JC Navarro, R Neira, N Noriega, JS Palleiro-Nayar, AF Pérez, A Pérez-Ruzafa, E Prieto-Rios, J Reyes, R Rodríguez, T Rubilar, T Sancho-Mejía, C Sangil, JRMC Silva, JI Sonnenholzner, CR Ventura, A Tablado, Y Tavares, CG Tiago, F Tuya, SM Williams

J. J. Alvarado and F. A. Solís-Marín (eds.), *Echinoderm Research and Diversity in Latin America,* DOI: 10.1007/978-3-642-20051-9,

Table A.1 Taxonomic list of the Echinoderms of the Pacific coast of Latin America

	Depth (m)	Habitat	MEX	REV	GUA	SAL	HON	NIC	CRC	COC	PAN	COL	MAL	GAL	ECU	PER	CHL	PAS
Class Crinoidea																		
Order Comatulida																		
Superfamily Antedonacea																		
Family Antedonidae																		
Antedon bifida (Pennant, 1777)																	x	
Anthometra adriani (Bell, 1908)	70–1674																x	
Eumorphometra fraseri John, 1938	410–687																x	
Fariometra parvula (Hartlaub, 1895)		rb	x								x							
Florometra magellanica (Bell, 1882)	20–1017	rb,sb									x					x	x	
Florometra mawsoni A. H. Clark, 1913	38–770																x	
Florometra parvula (Hartlaub, 1895)	589–1969	rb								x								
Florometra serratissima (A. H. Clark, 1907)	12–3234	rb	x								x							
Florometra tanneri (Hartlaub, 1895)	104–207	rb	x								x							
Isometra hordea John, 1938	17–490																x	
Isometra graminea John, 1938																	x	
Isometra vivipara Mortensen, 1917	79–845																x	
Promachocrinus kerguelensis Carpenter, 1880	20–2100																x	
Solanometra antarctica (Carpenter, 1888)	287–1759																x	
Family Zenometridae																		
Psathyrometra bigradata (Hartlaub, 1895)	741–823										x			x				

(continued)

Table A.1 (continued)

	Depth (m)	Habitat	MEX	REV	GUA	SAL	HON	NIC	CRC	COC	PAN	COL	MAL	GAL	ECU	PER	CHL	PAS
Superfamily Notocrinacea																		
Family Notocrinidae																		
Notocrinus virilis Mortensen, 1917																	x	
Notocrinus mortenseni John, 1938																	x	
Superfamily Tropiometracea																		
Family Thalassometridae																		
Thalassometra agassizii (Hartlaub, 1895)	596–1429	rb								x	x			x				
Order Hyocrinida																		
Family Hyocrinidae																		
Calamocrinus diomedae A. Agassiz, 1890	717–1431										x			x				
Hyocrinus foelli (Roux & Pawson, 1999)	3030	rb	x	x														
Ptilocrinus antarcticus (Bather, 1908)	450–500	rb															x	
Class Asteroidea																		
Order Paxillosida																		
Family Astropectinidae																		
Astropecten armatus Gray, 1840	0–160	mb,rb,sb	x			x			x		x	x		x	x	x		
Astropecten armatus erinaceus Gray, 1840	11–60	rb,sb														x		
Astropecten benthophilus Ludwig, 1905	1408									x								
Astropecten brasiliensis peruvianus Verrill, 1870	6–99	mb,sb										x				x		
Astropecten exiguus Ludwig 1905	232–2136										x			x				
Astropecten fragilis Verrill, 1870	1–5	sb									x			x		x		

(continued)

Table A.1 (continued)

	Depth (m)	Habitat	MEX	REV	GUA	SAL	HON	NIC	CRC	COC	PAN	COL	MAL	GAL	ECU	PER	CHL	PAS
Astropecten regalis Gray, 1840	0–204	mb,rb,sb	x						x		x	x						
Astropecten sulcatus Ludwig, 1905	95–121									x		x	x					
Astropecten verrilli de Loriol, 1899	2–488	mb,rb,sb	x			x						x					x	
Astropecten ornatissimus Fisher, 1906	278–286	rb,sb	x															
Bathybiaster loripes Sladen, 1889	80–500	sb															x	
Dipsacaster eximius Fisher, 1905	377–971	mb,sb	x													x		
Dytaster gilberti Fisher, 1905	1573–4335	mb,sb									x	x	x	x		x		
Leptychaster inermis (Ludwig, 1905)	732–1593									x	x			x	x			
Mimastrella cognata (Sladen, 1889)	10–2424	mb,rb															x	
Psilaster charcoti (Koehler, 1906)	20–3248	sb															x	
Psilaster pectinatus (Fisher, 1905)	1866										x							
Psilaster sladeni Ludwig, 1905	1485–1618														x			
Persephonaster armiger Ludwig, 1905										x								
Tethyaster canaliculatus (A. H. Clark, 1916)	23–300	rb	x								x			x		x		
Thrissacanthias penicillatus (Fisher, 1905)	55–1503	sb	x													x		
Family Goniopectinidae																		
Ctenodiscus procurator Sladen, 1889	50–1050	mb															x	
Ctenodiscus crispatus (Retzius, 1805)	10–1946	mb	x								x					x		
Luidia (Platasterias) latiradiata (Gray, 1871)	10–12	mb,sb										x						

(continued)

Table A.1 (continued)

	Depth (m)	Habitat	MEX	REV	GUA	SAL	HON	NIC	CRC	COC	PAN	COL	MAL	GAL	ECU	PER	CHL	PAS
Luidia armata Ludwig, 1905	33–126	sb	x							x	x							
Luidia asthenosoma Fisher, 1906											x							
Luidia bellonae Lütken, 1865	4–55	rb,sb	x											x		x		
Luidia columbia (Gray, 1840)	0–220	mb,sb	x								x	x				x		
Luidia ferruginea Ludwig, 1905	280	mb									x					x		
Luidia foliolata (Grube, 1866)	3–55	mb,sb				x		x			x	x		x				
Luidia latiradiata (Gray, 1871)	1.5–12	mb,sb	x			x		x			x	x						
Luidia magellanica Leipoldt, 1895	0–40	rb,sb														x	x	
Luidia phragma H. L. Clark, 1910	1–386	sb	x								x					x		
Luidia porteri A. H. Clark, 1917	110																x	
Luidia superba A. H. Clark, 1917	3–250	mb,sb,rub	x								x	x		x	x	x		
Luidia tessellata Lütken, 1859	18	sb	x						x		x			x				
Family Porcellanasteridae																		
Eremicaster crassus (Sladen, 1883)	1570–6330	rb	x								x	x	x	x		x		
Eremicaster pacificus (Ludwig, 1905)	1463–5780	mb,sb								x	x	x	x	x	x	x		
Eremicaster vicinus Ludwig, 1907	5204–7200	mb														x		
Porcellanaster ceruleus Wyville-Thomson, 1877	1158–6035	mb								x	x					x		
Styracaster paucispinus Ludwig, 1907	4240–4337	mb														x		
Thoracaster cylindratus Sladen, 1883	2600–5303	mb												x		x		
Family Pseudarchasteridae																		
Pseudarchaster discus Sladen, 1889	117–400	mb,rb,sb															x	
Pseudarchaster pectinifer Ludwig, 1905	1180–3575	sb,rub,mb	x								x			x		x		

(continued)

Table A.1 (continued)

	Depth (m)	Habitat	MEX	REV	GUA	SAL	HON	NIC	CRC	COC	PAN	COL	MAL	GAL	ECU	PER	CHL	PAS
Pseudarchaster pulcher Ludwig, 1905	700–1620	sb,rub	x											x				
Pseudarchaster pusillus Fisher, 1905	98–800	sb,rub	x															
Pseudarchaster verrilli Ludwig, 1905	999										x							
Order Notomyotida																		
Family Benthopectinidae																		
Benthopecten acanthonotus Fisher, 1905	1800–1936	rb	x															
Benthopecten cognatus (Ludwig, 1905)	3058										x							
Benthopecten pectinifer (Ludwig, 1905)	1485–2323	rb	x								x			x				
Benthopecten spinuliger (Ludwig, 1905)	1618–2323									x	x							
Calyptraster tenuissimus Bernasconi, 1966	0–732	sb															x	
Cheiraster (Luidiaster) planeta (Sladen, 1889)	350–500	sb															x	
Cheiraster (Luidiaster) californicus Ziesenhenne, 1942	488–512	mb	x															
Nearchaster aciculosus (Fisher, 1910)	466–1903	mb,sb	x	x														
Pectinaster agassizi Ludwig, 1905	790–2323	mb,rb	x							x	x	x	x			x		
Order Valvatida																		
Family Acanthasteridae																		
Acanthaster planci (Linnaeus, 1758)	0–30	cr,rb	x	x					x	x	x	x	x	x				
Family Asterinidae																		
Asterina fimbriata Perrier, 1875	0–300	mb,rb															x	
Meridiastra modesta (Verrill, 1870)	213	rb,sb	x	x							x			x	x	x		
Patiria chilensis (Lütken, 1859)	2–40	rb,kf														x	x	

(continued)

Table A.1 (continued)

	Depth (m)	Habitat	MEX	REV	GUA	SAL	HON	NIC	CRC	COC	PAN	COL	MAL	GAL	ECU	PER	CHL	PAS
Patiria miniata Verrill, 1913	0–300	cr,rb	x	x														
Patiriella calcarata (Perrier, 1875)	0–20	rb														x		x
Patiriella exigua (Lamarck, 1816)	0–3													x				
Family Asterodiscididae																		
Amphiaster insignis Verrill, 1868	0–128	cr,mb,rb,sb	x			x					x	x	x			x		
Paulia horrida Gray, 1840	33–121	rb	x							x	x			x		x		
Family Asteropseidae																		
Asteropsis carinifera (Lamarck, 1816)	3–60	cr,rb	x						x	x	x	x	x					
Dermasterias imbricata (Grube, 1857)	2–260	rb	x															
Family Ganeriidae																		
Cycethra verrucosa (Philippi, 1857)	0–270	rb,sb														x	x	
Ganeria falklandica Gray, 1847	0–145	rb,sb															x	
Ganeria hahni Perrier, 1891	0–135	rb,sb															x	
Family Goniasteridae																		
Ceramaster grenadensis patagonicus (Sladen, 1889)	75–898													x			x	
Ceramaster patagonicus (Sladen, 1889)	10–898	rb,mb,sb	x													x	x	
Ceramaster leptoceramus (Fisher, 1905)	382–1243	mb,rb	x													x		
Cryptopeltaster lepidonotus Fisher, 1905	188–1244	sb												x			x	
Hippasteria phrygiana (Parelius, 1768)	0–400	rb															x	
Hippasteria falklandica Fisher, 1940	225–1148	rb															x	

(continued)

Table A.1 (continued)

	Depth (m)	Habitat	MEX	REV	GUA	SAL	HON	NIC	CRC	COC	PAN	COL	MAL	GAL	ECU	PER	CHL	PAS
Hippasteria spinosa Verrill, 1909	50–512	rb	x															
Litonotaster tumidus H. L. Clark, 1920	4066–5196	mb														x		
Mediaster elegans Ludwig, 1905	1789									x		x	x					
Mediaster transfuga Ludwig, 1905	840–900	sb	x	x														
Mediaster tenellus Fisher, 1905	580–1192	sb	x															
Nymphaster diomedeae Ludwig, 1905	702–1810	mb,sb	x							x	x			x	x			
Pillsburiaster ernesti (Ludwig, 1905)	2149									x								
Family Mithrodiidae																		
Mithrodia bradleyi Verrill, 1870	0–14	cr,rb	x	x						x	x	x	x	x	x	x	x	
Family Odontasteridae																		
Acodontaster elongatus elongatus (Sladen, 1889)	8–400	rb,sb															x	
Acodontaster elongatus granuliferus (Koehler, 1912)	74–841	rb,sb															x	
Diplodontias singularis (Müller & Troschel, 1843)	0–84	rb,sb														x	x	
Odontaster meridionalis (E.A. Smith, 1876)	0–646	rb,sb															x	
Odontaster penicillatus (Philippi, 1870)	6–400	rb,sb															x	
Family Ophidiasteridae																		
Leiaster coriaceus Peters, 1852											x							
Leiaster glaber Peters, 1852	2–15	cr										x						
Leiaster teres (Verrill, 1871)	1–57	cr,rb	x								x	x	x	x				
Linckia columbiae Gray, 1840	0–100	cr,rb	x							x		x		x		x		

(continued)

Table A.1 (continued)

	Depth (m)	Habitat	MEX	REV	GUA	SAL	HON	NIC	CRC	COC	PAN	COL	MAL	GAL	ECU	PER	CHL	PAS
Linckia guildingi Gray, 1840		rb									x			x	x			
Linckia multifora (Lamarck, 1816)																		x
Narcissia gracilis A.H. Clark, 1916	56–91	rb	x							x				x				
Narcissia gracilis malpeloensis Downey, 1975	20–59	cr, rb										x	x					
Ophidiaster agassizi Perrier, 1881	1–75	rb																x
Ophidiaster ludwigi de Loriol, 1900											x					x		
Pharia pyramidata (Gray, 1840)	0–130	cr,rb,sb	x	x		x			x		x	x	x	x	x	x		
Phataria mionactis Ziesenhenne, 1942															x			
Phataria unifascialis (Gray, 1840)	0–50	cr,rb,sb	x	x		x		x	x	x	x	x	x	x	x	x		
Tamaria obstipa Ziesenhenne, 1942	67–79								x	x								
Tamaria stria Downey, 1975	15–49	rb										x	x					
Family Poraniidae																		
Porania (Porania) antarctica E.A. Smith, 1876	0–900	mb,rb,sb															x	
Poraniopsis echinaster Perrier, 1891	5–450	mb,rb,sb															x	
Poraniopsis inflatus (Fisher, 1906)	48–1094										x			x				
Family Oreasteridae																		
Nidorellia armata (Gray, 1840)	0–183	cr,rb,rub	x			x		x	x	x	x	x	x	x	x	x		
Pentaceraster cumingi (Gray, 1840)	2–92	cr,rb,sb	x						x	x	x	x	x	x		x		
Family Solasteridae																		
Lophaster furcilliger Fisher, 1905	86–4200	mb,rb,rub	x								x			x		x		
Lophaster stellans Sladen, 1889	15–450	rb,sb															x	

(continued)

Table A.1 (continued)

	Depth (m)	Habitat	MEX	REV	GUA	SAL	HON	NIC	CRC	COC	PAN	COL	MAL	GAL	ECU	PER	CHL	PAS
Paralophaster antarcticus Koehler, 1912	88–750	sb															x	
Solaster regularis Sladen, 1889	5–800	rb,sb,mb															x	
Order Velatida																		
Family Korethrasteridae																		
Peribolaster folliculatus Sladen, 1889	10–460	rb															x	
Family Pterasteridae																		
Diplopteraster semireticulatus (Sladen, 1882)	300–1500	rb,sb															x	
Diplopteraster verrucosus (Sladen, 1882)	0–470	sb															x	
Pteraster cf. diaphanus (Ludwig, 1905)	1410									x				x				
Pteraster gibber (Sladen, 1882)	7–460	rb															x	
Pteraster affinis Smith, 1876	80–130	rb															x	
Pteraster affinis lebruni Perrier, 1891	74–341	sb															x	
Hymenaster cremnodes H. L. Clark, 1920	4335	mb														x		
Hymenaster gracilis Ludwig, 1905	2418–3241											x	x	x	x			
Hymenaster pellucidus Thomson, 1873	13–3240	mb	x															
Hymenaster platyacanthus Ludwig, 1905	2487–2877										x			x	x			
Hymenaster quadrispinosus Fisher, 1905	778–3240	mb								x	x					x		
Hymenaster trias H. L. Clark, 1920	5203	mb														x		
Hymenaster violaceus Ludwig, 1905											x							

(continued)

Table A.1 (continued)

	Depth (m)	Habitat	MEX	REV	GUA	SAL	HON	NIC	CRC	COC	PAN	COL	MAL	GAL	ECU	PER	CHL	PAS
Order Spinulosida																		
Family Echinasteridae																		
Echinaster (Echinaster) parvispinus A. H. Clark, 1916	18–28	rb	x															
Echinaster (Othilia) aculeata (Gray, 1840)	3–24	sb						x	x									
Echinaster (Othilia) spinulosus Verrill, 1869	1–55	cr,mb,rb										x						
Echinaster (Othilia) tenuispinus Verrill, 1871	0–18	rb	x															
Echinaster cribella Lütken, 1871																x		
Echinaster cylindricus Meissner, 1892																x		
Echinaster panamensis Leipoldt, 1895											x							
Henricia aspera Fisher, 1906	487–570	rb	x															
Henricia asthenactis Fisher, 1910	91–1250	rb	x															
Henricia clarki Fisher, 1910	226–2001	rb	x	x														
Henricia gracilis (Ludwing, 1905)	267–1244	rb												x				
Henricia leviuscula (Stimpson, 1857)	0–228	rb	x															
Henricia nana (Ludwig, 1905)	57–200	rb	x											x				

(continued)

Table A.1 (continued)

	Depth (m)	Habitat	MEX	REV	GUA	SAL	HON	NIC	CRC	COC	PAN	COL	MAL	GAL	ECU	PER	CHL	PAS
Henricia obesa (Sladen, 1889)	22–940	rb												x		x	x	
Henricia seminudus (A. H. Clark, 1916)	666	rb	x	x														
Henricia studeri Perrier, 1891	0–430	rb,sb,rub														x	x	
Order Forcipulatida																		
Family Asteriidae																		
Anasterias antarctica (Lütken, 1857)	0–350	rb															x	
Anasterias pedicellaris (Köehler, 1923)	0–120	sb															x	
Anasterias spirabilis (Bell, 1881)	34–54																x	
Anasterias varium (Philippi, 1870)	0–350	rb															x	
Astrometis sertulifera (Xantus, 1860)	11–156	cr,rb	x											x		x		
Astrostole platei (Meissner, 1896)	0–20	rb																x
Astrostole paschae (H. L. Clark, 1920)		rb																x
Coronaster marchenus Ziesenhenne, 1942	52–84	cr,rb,rub	x							x				x	x			
Cosmasterias lurida (Philippi, 1858)	0–650	mb,rb,sb															x	
Diplasterias brandti (Bell, 1881)	0–500	mb,rb,sb															x	
Distolasterias robusta (Ludwig, 1905)	3334											x	x	x				

(continued)

Table A.1 (continued)

	Depth (m)	Habitat	MEX	REV	GUA	SAL	HON	NIC	CRC	COC	PAN	COL	MAL	GAL	ECU	PER	CHL	PAS
Evasterias troscheli (Stimpson, 1862)														x				
Leptasterias pusilla Fisher, 1930	37	rb	x															
Meyenaster gelatinosus (Meyen, 1834)	0–22	rb									x					x	x	
Pisaster brevispinus (Stimpson, 1857)	0–102	rb	x															
Pisaster giganteus (Stimpson, 1857)	0–374	rb	x															
Pisaster ochraceus (Brandt, 1835)	0–8	rb	x								x							
Rathbunaster californicus Fisher, 1906		rb	x															
Sclerasterias alexandri (Ludwig, 1905)	61–384									x	x	x	x		x			
Sclerasterias heteropaes Fisher, 1924	18–457	rb	x							x								
Family Coscinasteridae																		
Psalidaster mordax Fisher, 1940	80–600	sb															x	
Family Heliasteridae																		
Heliaster canopus Perrier, 1875	intertidal	rb																x
Heliaster cumingii (Gray, 1840)	0–14	rb						x										
Heliaster helianthus (Lamarck, 1816)	0–20	rb,sb	x													x	x	
Heliaster kubiniji Xantus, 1860	0–20	rb	x											x		x		

(continued)

Table A.1 (continued)

	Depth (m)	Habitat	MEX	REV	GUA	SAL	HON	NIC	CRC	COC	PAN	COL	MAL	GAL	ECU	PER	CHL	PAS
Heliaster microbrachius Xanthus, 1860	0–20	rb	x								x					x		
Heliaster polybrachius H. L. Clark, 1907	0–20	rb	x												x	x		
Heliaster solaris A. H. Clark, 1920		rb							x									
Labidiaster radiosus Lütken, 1871	5–450	rb															x	
Family Pedicellasteridae																		
Ampheraster hyperonchus (H. L. Clark, 1913)	0–520	rb,rub	x													x		
Hydrasterias improvisus (Ludwig, 1905)	1618–2418									x								
Tarsaster cocosanus (Ludwig, 1905)	245									x					x			
Tarsaster galapagensis (Ludwig, 1905)	704													x				
Family Pycnopodiidae																		
Pycnopodia helianthoides (Brandt, 1835)	0–455	rb	x															
Family Stichasteridae																		
Allostichaster capensis (Perrier, 1875)	0–100	rb															x	
Stichaster striatus Müller & Troschel, 1840	0–80	rb,sb,kf														x	x	
Family Zoroasteridae																		

(continued)

Table A.1 (continued)

	Depth (m)	Habitat	MEX	REV	GUA	SAL	HON	NIC	CRC	COC	PAN	COL	MAL	GAL	ECU	PER	CHL	PAS
Cnemidaster nudus (Ludwig, 1905)	1366–2600	mb,rb	x											x				
Cnemidaster wyvillii Sladen, 1889											x							
Myxoderma longispinum (Ludwig, 1905)	980–2418	mb,rb	x								x				x	x		
Myxoderma platyacanthum (H. L. Clark, 1913)	395–650	mb,rb	x															
Myxoderma qawashqari (Moyana & Larrain, 1976)	0–723	sb														x	x	
Myxoderma sacculatum (Fisher, 1905)	1000–1546	mb,rb	x															
Zoroaster magnificus Ludwig, 1905	3056–3667	mb,sb									x			x		x		
Zoroaster ophiurus Fisher, 1905	695–2226	mb,sb														x		
Order Brisingida																		
Family Brisingidae																		
Astrolirus panamensis (Ludwig, 1905)	48–2418									x	x	x	x	x				
Hymenodiscus monacantha H. L. Clark, 1920	4064	mb														x		
Hymenodiscus tenella (Ludwig, 1905)	2418													x				
Family Freyellidae																		
Freyastera benthophila (Sladen, 1889)	4064–4667	mb														x		

(continued)

Table A.1 (continued)

	Depth (m)	Habitat	MEX	REV	GUA	SAL	HON	NIC	CRC	COC	PAN	COL	MAL	GAL	ECU	PER	CHL	PAS
Freyella insignis Ludwig, 1905	3182–3279										x	x	x		x			
Freyella pacifica Ludwig, 1905											x							
Freyella propinqua Ludwig, 1905															x			
Freyellaster scalaris (A. H. Clark, 1916)														x				
Class Ophiuroidea																		
Order Euryalida																		
Family Asteronychidae																		
Asteronyx excavata Lütken & Mortensen, 1899	266–1236	sb	x															
Asteronyx loveni Müller & Troschel, 1842	152–2663	cr,mb,sb	x								x			x				
Asteronyx longifissus Doederlein, 1927	266–1800		x															
Astrodia plana (Lütken & Mortensen, 1899)	717–3058										x	x	x	x				
Family Asteroschematidae																		
Asteroschema rubrum (Lyman, 1882)	350–731																x	
Asteroschema sublaeve Lütken & Mortensen, 1899	1271				x						x							
Family Gorgonocephalidae																		

(continued)

Table A.1 (continued)

	Depth (m)	Habitat	MEX	REV	GUA	SAL	HON	NIC	CRC	COC	PAN	COL	MAL	GAL	ECU	PER	CHL	PAS
Astrocaneum spinosum (Lyman, 1875)	4–183	cr,mb,sb,rb	x						x		x					x		
Astrodendrum galapagensis A. H. Clark, 1916	717													x				
Astrodictyum panamense (Verrill, 1867)	23559	cr,sb	x								x	x	x	x		x		
Astrotoma agassizii Lyman, 1875	2–1180	rb															x	
Gorgonocephalus chilensis (Philippi, 1858)	4–900	rb															x	
Gorgonocephalus diomedeae Lütken & Mortensen, 1899	1271										x							
Order Ophiurida																		
Family Ophiomyxidae																		
Ophiomyxa panamensis Lütken & Mortensen, 1899	60–293	rb,sb	x							x	x			x				
Ophiomyxa vivipara Studer, 1876	15–399	rb															x	
Ophiolycus nutrix (Mortensen, 1936)	70–4538	sb															x	
Family Amphiuridae																		
Amphilepis nuda Tommasi, 1976																x		
Amphilepis patens Lyman, 1879	384–4087	mb									x			x		x		
Family Amphiuridae																		

(continued)

Table A.1 (continued)

	Depth (m)	Habitat	MEX	REV	GUA	SAL	HON	NIC	CRC	COC	PAN	COL	MAL	GAL	ECU	PER	CHL	PAS
Amphicontus minutus Hill, 1940	9–22														x	x		
Amphichondrius granulatus (Lütken & Mortensen, 1899)	280–384	mb, sb	x								x				x			
Amphichondrius laevis Ziesenhenne, 1940	4–15	cr, rb	x		x													
Amphiodia assimilis (Lütken & Mortensen, 1899)	3334										x	x	x					
Amphiodia grisea (Ljungman, 1867)	54										x				x	x		
Amphiodia occidentalis (Lyman, 1860)	0–367	mb, sb	x								x							
Amphiodia oerstedi (Lütken, 1856)	0.2–1	mb,rb				x			x		x	x						
Amphiodia platyspina Nielsen, 1932	350	rb	x								x							
Amphiodia sculptilis Ziesenhenne, 1940	4–15	rb	x					x										
Amphiodia tabogae Nielsen, 1932	0–364	rb	x						x	x	x			x				
Amphiodia urtica (Lyman, 1860)	9–80	rb,sb	x						x									
Amphiodia violacea (Lütken, 1856)									x	x	x							
Amphiodia vicina H. L. Clark, 1940	64	mb							x									
Amphipholis elevata Nielsen, 1932	0–73	sb	x								x							

(continued)

Table A.1 (continued)

	Depth (m)	Habitat	MEX	REV	GUA	SAL	HON	NIC	CRC	COC	PAN	COL	MAL	GAL	ECU	PER	CHL	PAS
Amphipholis granulata (Lütken & Mortensen, 1899)	280–384										x							
Amphipholis laevidisca H. L. Clark, 1909																x	x	
Amphipholis pugetana (Lyman, 1860)	4–1620	sb	x						x							x		
Amphipholis puntarenae (Lütken, 1856)	0–508	mb	x															
Amphipholis squamata (Delle Chiaje, 1828)	0–1200	cr,m,mb,sb,sg,rub,rb						x	x		x	x		x		x		
Amphioplus (Amphioplus) strongyloplax (H. L. Clark, 1911)	4–1408	mb,sb	x															
Amphioplus (Amphioplus) coniortodes H. L. Clark, 1918	1207	mb	x															
Amphioplus daleus (Lyman, 1879)	2690–3219	rb	x								x							
Amphioplus philohelminthius Ziesenhenne, 1940	15–73	mb														x		
Amphioplus magellanica (Mortensen, 1936)	35–174																x	
Amphioplus textilis (Koehler, 1907)																	x	
Amphiura anomala Lyman, 1875	300–1500																	x

(continued)

Table A.1 (continued)

	Depth (m)	Habitat	MEX	REV	GUA	SAL	HON	NIC	CRC	COC	PAN	COL	MAL	GAL	ECU	PER	CHL	PAS
Amphiura arcystata H. L. Clark, 1911	6–849	mb,rb,sb	x							x	x			x				
Amphiura eugeniae Ljungman, 1867	0–582																x	
Amphiura (Amphiura) diomedeae Lütken & Mortensen, 1899	44–3017	sb							x		x			x		x		
Amphiura (Amphiura) magellanica Ljungman, 1867	0–300																x	
Amphiura (Ophionema) hexacantha Nielsen, 1932											x							
Amphiura calbuca Mortensen, 1952	0–300																x	
Amphiura carchara H. L. Clark, 1911	1587	mb,rb,sb	x															
Amphiura gastracantha Lütken & Mortensen, 1899	1207	mb,sb	x															
Amphiura gymnogastra Lütken & Mortensen, 1899	549–2323										x			x				
Amphiura gymnopora Lütken & Mortensen, 1899	333										x							
Amphiura notacantha Lütken & Mortensen, 1899	1236	mb,sb	x															
Amphiura otteri Ljungman, 1872	80	rb	x															

(continued)

Table A.1 (continued)

	Depth (m)	Habitat	MEX	REV	GUA	SAL	HON	NIC	CRC	COC	PAN	COL	MAL	GAL	ECU	PER	CHL	PAS
Amphiura polyacantha Lütken & Mortensen, 1899	1271										x			x				
Amphiura papillata Lütken & Mortensen, 1899	702													x				
Amphiura princeps Koehler, 1907	0–300																x	
Amphiura seminuda Lütken & Mortensen, 1899	9–4096	rb,sb	x															
Amphiura serpentina Lütken & Mortensen 1899	770–1865										x			x				
Amphiura verticillata Ljungman, 1867														x				
Microphiopholis geminata (Le Conte, 1851)	0–82	mb,rb,sb	x			x			x		x	x						
Microphiopholis platydisca (Nielsen, 1932)	0–137	mb,sb	x			x			x		x			x				
Microphiopholis puntarenae (Lütken, 1856)									x					x				
Ophiocnida californica Ziesenhenne, 1940	6–302	mb,rb,sb	x															
Ophiocnida hispida (Le Conte, 1851)	0–794	cr,rb									x					x		

(continued)

Table A.1 (continued)

	Depth (m)	Habitat	MEX	REV	GUA	SAL	HON	NIC	CRC	COC	PAN	COL	MAL	GAL	ECU	PER	CHL	PAS
Ophiophragmus chilensis (Müller & Troschel, 1843)	0–40	rb,sb														x	x	
Ophiophragmus disacanthus Ziesenhenne, 1940														x				
Ophiophragmus marginatus (Lütken, 1856)	0–134	sb	x						x	x	x			x		x		
Ophiophragmus ophiactoides Ziesenhenne, 1940									x		x	x						
Ophiophragmus paucispinus Nielsen, 1932	0–134									x	x			x				
Ophiophragmus stellatus Ziesenhenne, 1940	18–73															x		
Ophiophragmus tabogensis Nielsen, 1932	0–128	sb	x						x		x			x	x			
Ophiostigma tenue Lütken, 1856	1–101	rb	x								x	x						
Triplodia abdita (A. M. Clark, 1970)										x								
Family Hemieuryalidae																		
Amphigyptis perplexa Nielsen, 1932	0–143	mb,rb,sb	x															
Ophiochondrus stelliger Lyman, 1879	73–439																x	
Sigsbeia laevis Ziesenhenne, 1940	55–64										x							

(continued)

Table A.1 (continued)

	Depth (m)	Habitat	MEX	REV	GUA	SAL	HON	NIC	CRC	COC	PAN	COL	MAL	GAL	ECU	PER	CHL	PAS
Sigsbeia lineata Lütken & Mortensen, 1899	43–183	rb								x	x			x				
Family Ophiacanthidae																		
Subfamily Ophiacanthinae																		
Ophiacantha antarctica Koehler, 1900	90–4004	sb															x	
Ophiacantha bathybia H. L. Clark, 1911	1993	rb	x															
Ophiacantha contigua Lütken & Mortensen, 1899	1062–1644										x	x	x					
Ophiacantha cosmica Lyman, 1878	147–4840	mb,rb,sb	x								x					x	x	
Ophiacantha costata Lütken & Mortensen, 1899	733–1271	mb,sb	x								x				x			
Ophiacantha cyrena A. H. Clark, 1916	717													x				
Ophiacantha deruens Koehler, 1899																	x	
Ophiacantha diplasia H. L. Clark, 1911	9–1408	cr,mb,sb	x															
Ophiacantha frigida Koehler, 1909	1667–3914																x	
Ophiacantha hirta Lütken & Mortensen, 1899	1244–1548	mb,sb	x															
Ophiacantha inconspicua Lütken & Mortensen, 1899	702–1485										x			x				

(continued)

Table A.1 (continued)

	Depth (m)	Habitat	MEX	REV	GUA	SAL	HON	NIC	CRC	COC	PAN	COL	MAL	GAL	ECU	PER	CHL	PAS
Ophiacantha marsupialis Lyman, 1878																		x
Ophiacantha moniliformis Lütken & Mortensen, 1899	519–1244	mb,sb	x	x														
Ophiacantha pacifica Lütken & Mortensen, 1899	2877														x			
Ophiacantha paucispina Lütken & Mortensen, 1899	2690										x							
Ophiacantha pentacrinus Lütken, 1869	434–598	sb	x															
Ophiocantha phragma Ziesenhenne, 1940	13–644	rb,sb	x							x	x	x		x				
Ophiacantha quadrispina H. L. Clark, 1917	183–549	rb												x				
Ophiacantha rhachophora H. L. Clark, 1911	115–1552	mb,rb,sb	x															
Ophiacantha rosea Lyman, 1878	32–1538	rb,sb															x	
Ophiacantha savagica Tommasi, 1976	1171–1180														x			
Ophiacantha sentosa Lyman, 1878	2067–5203	mb														x	x	
Ophiacantha similis A. H. Clark, 1916	717													x				
Ophiacantha spinifera Lütken & Mortensen, 1899	999–1865										x							

(continued)

Table A.1 (continued)

	Depth (m)	Habitat	MEX	REV	GUA	SAL	HON	NIC	CRC	COC	PAN	COL	MAL	GAL	ECU	PER	CHL	PAS
Ophiacantha vivipara Ljungman, 1870	0–1097	rb,kf															x	
Ophiolebes mortenseni A. H. Clark, 1916	717													x				
Ophiotreta valenciennesi valenciennesi (Lyman, 1879)	549													x				
Subfamily Ophiohelinae																		
Ophiotholia spathifer (Lyman, 1879)														x				
Subfamily Ophioplinthacinae																		
Ophiomitrella chilensis Mortensen, 1952	25–300																x	
Ophiophthalmus normani (Lyman, 1879)	51–2600	mb,rb,sb	x						x									
Ophiothamnus dupla Tommasi, 1976	1171–1180														x			
Ophiurothamnus laevis (Lütken & Mortensen, 1899)	1008													x				
Subfamily Ophiotominae																		
Ophiotoma paucispina (Lütken & Mortensen, 1899)	1643–4082	mb			x					x					x	x		
Ophiolimna antarctica (Lyman, 1879)	88–2750																x	

(continued)

Table A.1 (continued)

	Depth (m)	Habitat	MEX	REV	GUA	SAL	HON	NIC	CRC	COC	PAN	COL	MAL	GAL	ECU	PER	CHL	PAS
Family Ophiactidae																		
Hemipholis cordifera Bosc, 1802	18–34	cr	x								x							
Hemipholis gracilis Verrill, 1867	34	mb	x			x		x	x		x				x	x		
Histampica duplicata (Lyman, 1875)	125–2870	cr,mb,rb	x								x	x	x	x				
Ophiactis asperula (Philippi, 1858)	0–576	rb,kf														x	x	
Ophiactis kroeyeri Lütken, 1856	0–60	rb,sb,mb,rub				x										x	x	
Ophiactis savignyi (Müller & Troschel, 1842)	0–518	cr,m,rb,sg,rub	x	x		x		x	x	x	x	x	x	x		x		
Ophiactis simplex (Le Conte, 1851)	0–302	cr,rb	x					x	x	x	x	x	x	x		x		
Ophiactis plana Lyman, 1869												x	x	x				
Ophiactis profundi Lütken & Mortensen, 1899	1008													x				
Ophiopholis bakeri McClendon, 1909	9–1006	cr,rb	x															
Family Ophiochitonidae																		
Ophiochiton carinatus Lütken & Mortensen, 1899	589–1355	sb,rub	x								x							
Ophiochiton fastigatus Lyman, 1878	733														x			

(continued)

Table A.1 (continued)

	Depth (m)	Habitat	MEX	REV	GUA	SAL	HON	NIC	CRC	COC	PAN	COL	MAL	GAL	ECU	PER	CHL	PAS
Family Ophiocomidae																		
Ophiocoma aethiops Lütken, 1859	0–30.5	cr,rb,sb	x	x	x	x	x	x	x	x	x	x	x	x	x	x		
Ophiocoma alexandri Lyman, 1860	0–70	cr,rb,sb	x	x	x	x	x	x	x	x	x	x	x	x	x			
Ophiocoma dentata Müller & Troschel, 1842																		x
Ophiocoma erinaceus Müller & Troschel, 1842	8								x									
Ophiocoma longispina H. L. Clak, 1917																		x
Ophiocomella schmitti A. H. Clark, 1939	5–40	cr,rb	x									x		x				
Ophiocomella sexradia (Duncan, 1887)		cr,rb								x	x							
Ophiocomina nigra (Abildgaard, in O.F. Müller, 1789)	236–549	sb												x				
Ophiopsila californica A. H. Clark, 1921	33–201	rb	x															
Family Ophiodermatidae																		
Diopederma danianum (Verrill, 1867)	7–137	cr,mb,rb,sb	x			x			x	x	x							
Ophiocrytus granulosus Nielsen, 1932	0–79	rb,sg	x							x								
Ophioderma appressa (Say, 1825)	0–20	cr,rb							x									
Ophioderma elaps Lütken, 1856														x				

(continued)

Table A.1 (continued)

	Depth (m)	Habitat	MEX	REV	GUA	SAL	HON	NIC	CRC	COC	PAN	COL	MAL	GAL	ECU	PER	CHL	PAS
Ophioderma panamense Lütken, 1859	0–20	cr,rb,sb	x			x		x	x	x	x	x		x	x	x		
Ophioderma pentacantha H. L. Clark, 1917	183				x									x				
Ophioderma teres (Lyman, 1860)	0–54	cr,rb,sb	x			x		x	x		x	x		x		x		
Ophioderma vansyoci Hendler, 1996	15.2–27	rb	x															
Ophioderma variegatum Lütken, 1856	0–110	mb,rb,sb	x	x				x	x	x	x			x				
Ophioderma sodipallaresi Caso, 1986	4–9	rb,sb	x															
Ophiopaepale diplax (Nielsen, 1932)	0–230	mb,sb	x						x		x							
Ophiuroconis bispinosa Ziesenhenne, 1937	4–143	sb	x	x										x				
Family Ophionereididae																		
Ophionereis albomaculata E.A. Smith, 1877	73–155									x	x			x				
Ophionereis annulata (Le Conte, 1851)	0–229	cr,rb,sb	x	x		x			x	x	x	x		x		x		
Ophionereis dubia (Müller & Troschel, 1842)											x							
Ophionereis eurybrachiplax H. L. Clark, 1911		mb								x								

(continued)

Table A.1 (continued)

	Depth (m)	Habitat	MEX	REV	GUA	SAL	HON	NIC	CRC	COC	PAN	COL	MAL	GAL	ECU	PER	CHL	PAS
Ophionereis perplexa Ziesenhenne, 1940	0–73	cr,rb	x									x		x	x			
Ophionereis porrecta Lyman, 1860	0–30	rb												x				
Ophionereis schayeri (Müller & Troschel, 1844)																		x
Family Ophiotrichidae																		
Ophiothrix galapagensis Lütken & Mortensen, 1899	0–549	cr,mb,rb,sb	x											x				
Ophiothela gracilis Nielsen, 1932	7–9	cr,rb	x						x		x							
Ophiothela mirabilis Verrill, 1867	5–20	cr,mb,rb	x			x			x		x	x	x	x		x		
Ophiothrix magnifica Lyman, 1860	0–20	cr,mb,rb,sb												x		x		
Ophiothrix rudis Lyman, 1874	0–1	rb,cr,sb				x												
Ophiothrix spiculata Le Conte, 1851	0–2059	cr,mb,rb,sb	x			x		x	x	x	x	x		x	x	x		
Family Ophiuridae																		
Subfamily Ophiolepidinae																		
Ophiolepis crassa Nielsen, 1932	6–230	mb,rb,sb	x								x							
Ophiolepis grisea H. L. Clark, 1940	7–13	mb			x	x	x	x	x		x							
Ophiolepis pacifica Lütken, 1856	0–18	cr,rb	x						x		x							

(continued)

Table A.1 (continued)

	Depth (m)	Habitat	MEX	REV	GUA	SAL	HON	NIC	CRC	COC	PAN	COL	MAL	GAL	ECU	PER	CHL	PAS
Ophiolepis plateia Ziesenhenne, 1940	1–12	cr,rb,sb,rub	x						x			x						
Ophiolepis variegata Lütken, 1856	1–110	cr,mb,sb	x			x			x		x	x						
Ophiomusium glabrum Lütken & Mortensen, 1899	878–5203	mb,rb,sb	x							x	x	x	x	x	x	x		
Ophiomusium diomedeae Lütken & Mortensen, 1899	702–1485													x				
Ophiomusium lymani Thomson, 1873	51–2906	mb,sb	x							x	x	x	x	x				
Ophiomusium variabile Lütken & Mortensen, 1899	267–902	mb,rb	x	x														
Ophioplocus esmarki Lyman, 1874	0–74	rb,sg	x															
Ophioplocus hancocki Ziesenhenne, 1935					x													
Ophiosphalma jolliense (McClendon, 1909)	17–1230	mb,rb,sb	x		x									x				
Ophiozonella alba (Lütken & Mortensen, 1899)	1408–2487									x				x	x			
Ophiozonella clypeata (Lyman, 1883)														x				
Ophiozonella contigua (Lütken & Mortensen, 1899)	2417–2487													x	x			
Ophiozonella falklandica Mortensen, 1936																	x	

(continued)

Table A.1 (continued)

	Depth (m)	Habitat	MEX	REV	GUA	SAL	HON	NIC	CRC	COC	PAN	COL	MAL	GAL	ECU	PER	CHL	PAS
Subfamily Ophioleucinae																		
Ophiernus adspersus adspersus Lyman, 1883	770–1245	mb,sb	x								x					x		
Ophiernus adspersus annectens Lütken & Mortensen, 1899	770–1158													x				
Ophiernus polyporum Lütken & Mortensen, 1899	1207–1244		x															
Ophiernus seminudus Lütken & Mortensen, 1899	840–4082	mb,sb									x			x		x		
Subfamily Ophiurinae																		
Amphiophiura abcisa (Lütken & Mortensen, 1899)	245–3714	rb,sb								x	x					x		
Amphiophiura irregularis Ziesenhenne, 1940														x				
Amphiophiura obtecta (Lütken & Mortensen, 1899)	2197											x	x	x				
Amphiophiura oligopora (H. L. Clark, 1913)	1152	mb	x															
Amphiophiura paucisquama Ziesenhenne, 1940														x				
Amphiophiura superba (Lütken & Mortensen, 1899)	51–1820	mb	x															

(continued)

Table A.1 (continued)

	Depth (m)	Habitat	MEX	REV	GUA	SAL	HON	NIC	CRC	COC	PAN	COL	MAL	GAL	ECU	PER	CHL	PAS
Amphiophiura vemae Kyte, 1987	3739–4124															x		
Gymnophiura concava Tommasi, 1976																x		
Gymnophiura mollis Lütken & Mortensen, 1899	2417–2487										x			x	x			
Ophiocten amitinum Lyman, 1878	29–920	rb,sb,kf															x	
Ophiocten hastatum Lyman, 1878	1159–2877								x	x	x	x	x	x	x			
Ophiogona doederleini (Koehler, 1901)																	x	
Ophionotus victoriae Bell, 1902	5–752	mb,rb,sb															x	
Ophiomastus bulufonica Tommasi, 1976	1171–1180														x			
Ophiomastus molinae Castillo-Alarcon, 1968	intertidal	rb															x	
Ophiomastus tuberculata Tommasi, 1976	1171–1180														x			
Ophiomisidium leurum Ziesenhenne, 1940														x				
Ophiosteira koehleri A. H. Clark, 1917	733														x			
Ophiophyllum marginatum A. H. Clark, 1916	717													x				

(continued)

Table A.1 (continued)

	Depth (m)	Habitat	MEX	REV	GUA	SAL	HON	NIC	CRC	COC	PAN	COL	MAL	GAL	ECU	PER	CHL	PAS
Ophioplinthus divisa (Lütken & Mortensen, 1899)	2075–2487													x				
Ophioplinthus inflata (Koehler, 1897)	1484													x				
Ophioplinthus inornata (Lyman, 1878)	1485–2418													x			x	
Ophioplinthus nexila (Kyte, 1987)											x							
Ophiotypa simplex Koehler, 1897	3652–3811															x		
Ophiura (Ophiuroglypha) irrorata irrorata (Lyman, 1878)	405-5869	mb,sb	x							x	x	x	x	x	x	x		
Ophiura flagellata (Lyman, 1878)	128–2014	mb	x											x				
Ophiura leptoctenia H. L. Clark, 1911	27–3239	sb,rub	x															
Ophiura luetkenii (Lyman, 1860)	0–1097	mb,rb,sb	x	x														
Ophiura lymani (Ljungman, 1871)	30–800	rb,sb,kf															x	
Ophiura nana (Lütken & Mortensen, 1899)	1650									x								
Ophiura plana (Lütken & Mortensen, 1899)	2070–3241				x						x	x	x	x	x			

(continued)

Table A.1 (continued)

	Depth (m)	Habitat	MEX	REV	GUA	SAL	HON	NIC	CRC	COC	PAN	COL	MAL	GAL	ECU	PER	CHL	PAS
Ophiura stenobrachia H. L. Clark, 1917	3667														x	x		
Theodoria madseni Tommasi, 1976																x		
Class Echinoidea																		
Order Cidaroida																		
Family Cidaridae																		
Austrocidaris canaliculata (A. Agassiz, 1863)	1–845	rb, sb															x	
Austrocidaris spinulosa Mortensen, 1910	13–641																x	
Centrocidaris doederleini (A. Agassiz, 1898)	87–550	sb								x	x			x		x		
Eucidaris thouarsii (Valenciennes, 1846)	0–150	cr,mb,rb,sb	x	x		x		x	x	x	x	x	x		x	x		
Eucidaris thouarsii galapagensis Doederlein, 1887	0–150	rb								x				x				
Hesperocidaris asteriscus H. L. Clark, 1948	2–183	cr,mb,rb,sb	x						x		x	x	x			x		
Hesperocidaris dubia (H. L. Clark, 1907)	55–600								x	x	x				x			
Hesperocidaris houstonianus H. Clark, 1939	45													x				
Hesperocidaris panamensis (A. Agassiz, 1898)	48–274								x	x	x				x			

(continued)

Table A.1 (continued)

	Depth (m)	Habitat	MEX	REV	GUA	SAL	HON	NIC	CRC	COC	PAN	COL	MAL	GAL	ECU	PER	CHL	PAS
Hesperocidaris perplexa (H. L. Clark, 1907)	10–1500	rb,mb,sb	x	x							x	x			x	x		
Family Ctenocidaridae																		
Aporocidaris milleri (A. Agassiz, 1898)	300–3937	mb,rb,sb	x								x	x	x	x				
Family Histocidaridae																		
Histocidaris cobosi (A. Agassiz, 1898)	702													x				
Order Echinothurioida																		
Family Echinothuriidae																		
Araeosoma eurypatum A. Agassiz & H. L. Clark, 1909	1227													x				
Araeosoma leptaleum A. Agassiz & H. L. Clark, 1909	740–1046	rb									x			x				
Tromikosoma hispidum (A. Agassiz, 1898)	1820–3375	rb,sb	x							x	x	x	x	x	x			
Tromikosoma panamense (A. Agassiz, 1898)	2054–3374	mb,rb	x								x	x	x					
Family Kamptosomatidae																		
Kamptosoma asterias (A. Agassiz, 1881)	1020–1278	mb														x		

(continued)

Table A.1 (continued)

	Depth (m)	Habitat	MEX	REV	GUA	SAL	HON	NIC	CRC	COC	PAN	COL	MAL	GAL	ECU	PER	CHL	PAS
Order Diadematoida																		
Family Aspidodiadematidae																		
Plesiodiadema globulosum (A. Agassiz, 1898)	2830–3900	mb										x	x			x		
Plesiodiadema horridum (A. Agassiz, 1898)	1625–3381									x		x	x	x	x			
Family Diadematidae																		
Astropyga pulvinata (Lamarck, 1816)	0–90	cr,mb,rb,sb	x			x		x	x	x	x	x		x	x	x		
Centrostephanus coronatus (Verrill, 1867)	0–125	cr,rb,sb	x						x	x		x	x	x	x	x		
Centrostephanus rodgersii (A. Agassiz, 1863)																	x	
Centrostephanus sylviae Fell, 1975																	x	
Diadema mexicanum A. Agassiz, 1863	0–113	mb,cr,rb,sb	x	x		x		x	x	x	x	x	x	x	x	x		
Diadema paucispinum (A. Agassiz, 1863)																		x
Diadema savignyi Michelin, 1845	1–3	rb, sb																x
Echinothrix calamaris (Pallas, 1774)	0–46	cr,rb								x								
Echinothrix diadema (Linnaeus, 1758)	3–5	cr								x								
Order Pedinoida																		
Family Pedinidae																		

(continued)

Table A.1 (continued)

	Depth (m)	Habitat	MEX	REV	GUA	SAL	HON	NIC	CRC	COC	PAN	COL	MAL	GAL	ECU	PER	CHL	PAS
Caenopedina diomedeae Mortensen, 1939	723–850				x													
Order Salenioida																		
Family Saleniidae																		
Salenocidaris miliaris (A. Agassiz, 1898)	1159–3376									x	x	x	x	x				
Order Arbacioida																		
Family Arbaciidae																		
Arbacia dufresnii (Blainville, 1825)	0–315	rb															x	
Arbacia spatuligera (Valenciennes, 1846)	0–50	sb,rb,mb														x	x	
Arbacia stellata (Blainville, 1825)	0–92	cr,mb,rb, sb	x			x		x	x		x	x	x	x	x	x		
Dialithocidaris gemmifera A. Agassiz, 1898	3193–3279										x	x						
Tetrapygus niger (Molina, 1872)	0–25	rb,sb														x	x	
Order Camarodonta																		
Family Parechinidae																		
Loxechinus albus (Molina, 1782)	0–340	rb														x	x	
Order Temnopleuroida																		
Family Temnopleuridae																		
Pseudechinus magellanicus (Philippi, 1857)	0–820	rb, kf														x	x	

(continued)

Table A.1 (continued)

	Depth (m)	Habitat	MEX	REV	GUA	SAL	HON	NIC	CRC	COC	PAN	COL	MAL	GAL	ECU	PER	CHL	PAS
Family Toxopneustidae																		
Lytechinus anamesus H. L. Clark, 1912	1–207	cr,mb,rb,sb	x															
Lytechinus panamensis Mortensen, 1921											x							
Lytechinus pictus (Verrill, 1867)	0–300	cr,rb,mb,sb	x							x		x	x	x				
Lytechinus semituberculatus (Valenciennes in L. Agassiz, 1846)	0–134	rb,cr,sb												x	x	x		
Tripneustes depressus A. Agassiz, 1863	0–75	cr,rb,sb	x	x				x	x	x	x	x	x	x	x	x		x
Tripneustes gratilla (Linnaeus, 1758)	1–6	rb			x	x												
Toxopneustes roseus (A. Agassiz, 1863)	0–55	cr,rb,sb,mb	x			x		x	x	x	x	x	x	x	x	x		
Family Strongylocentrotidae																		
Allocentrotus fragilis Jackson, 1912	50–1200	mb,rb,sb	x															
Strongylocentrotus franciscanus (A. Agassiz, 1863)	4–84	rb,sb	x															
Strongylocentrotus purpuratus (Stimpson, 1857)	0–161	mb,rb	x															
Order Echinoida																		
Family Echinidae																		
Family Echinometridae																		

(continued)

Table A.1 (continued)

	Depth (m)	Habitat	MEX	REV	GUA	SAL	HON	NIC	CRC	COC	PAN	COL	MAL	GAL	ECU	PER	CHL	PAS
Caenocentrotus gibbosus (L. Agassiz & Desor, 1846)	0–35	rb												x	x	x		
Echinometra insularis H. L. Clark, 1912	0–50	rb,sb	x														x	x
Echinometra mathaei (Blainville, 1825)																x		
Echinometra oblonga (Blainville, 1825)	0–34	cr,rb	x	x						x				x	x			
Echinometra vanbrunti A. Agassiz, 1863	0–106	cr,rb,sb	x	x	x	x	x	x	x	x	x	x	x	x	x	x		
Echinostrephus aciculatus A. Agassiz, 1863																		x
Order Clypeasteroida																		
Family Clypeasteridae																		
Clypeaster elongatus H. L. Clark, 1948	10	sb												x				
Clypeaster europacificus H. L. Clark, 1914	0–402	mb,sb,rb	x	x					x	x	x			x		x		
Clypeaster ochrus H. L. Clark, 1914	0–162	rb,sb	x	x					x	x	x			x				
Clypeaster reticulatus Linnaeus, 1758																		x
Clypeaster rotundus (A. Agassiz, 1863)	0–92	rb,sb,mb	x		x	x	x	x	x	x	x			x				
Clypeaster speciosus Verrill, 1870	0–128	rb,sb	x							x	x			x				
Family Dendrasteridae																		

(continued)

Table A.1 (continued)

	Depth (m)	Habitat	MEX	REV	GUA	SAL	HON	NIC	CRC	COC	PAN	COL	MAL	GAL	ECU	PER	CHL	PAS
Dendraster excentricus (Eschscholtz, 1831)	0–232	mb,rb,sb	x															
Dentraster laevis H. L. Clark, 1948	13–22	sb	x															
Dendraster mexicanus H. L. Clark, 1948	0–30	sb	x															
Dendraster vizcainoensis Grant & Hertlein, 1938	0–30	sb	x															
Family Mellitidae																		
Encope cocosi H. L. Clark, 1948	4–15	sb								x								
Encope ecuadorensis H. L. Clark, 1948	1–3	sb										x						
Encope grandis L. Agassiz, 1841	0–120	sb	x		x	x	x	x				x						
Encope insularis H. L. Clark, 1948	1–10	sb										x						
Encope irregularis H. L. Clark, 1948	0–36.6	sb									x	x						
Encope laevis H. L. Clark, 1948	0–7	sb						x	x		x	x						
Encope michelini L. Agassiz, 1841	3–90	sb							x									
Encope micropora L. Agassiz, 1841	0–82	mb,sb	x			x		x	x	x	x	x		x	x	x		
Encope micropora ecuadorensis H. L. Clark, 1948	1–5	sb										x			x			
Encope micropora insularis H. L. Clark, 1948	1–10	sb										x						

(continued)

Table A.1 (continued)

	Depth (m)	Habitat	MEX	REV	GUA	SAL	HON	NIC	CRC	COC	PAN	COL	MAL	GAL	ECU	PER	CHL	PAS
Encope micropora irregularis H. L. Clark, 1948	1	sb										x						
Encope micropora tetrapora L. Agassiz, 1841	3	sb												x				
Encope pacifica (Verrill, 1867)	0–40	sb														x		
Encope perspectiva L. Agassiz, 1841	8–27	mb,sb	x						x		x			x				
Encope wetmorei A. H. Clark, 1946	1–45	sb	x						x		x	x						
Mellita grantii Mortensen, 1948	0–10	sb	x								x							
Mellita kanakoffi Durham, 1961	0–102.9	sb	x						x		x							
Mellita longifissa Michelin, 1858	0–60	sb	x		x	x	x	x	x		x	x			x	x		
Mellita notabilis H. L. Clark, 1947	0.5–0.9	sb	x			x		x			x							
Mellitella stokesii (L. Agassiz, 1841)	0–10	sb, mb				x			x					x	x	x		
Order Cassiduloida																		
Family Cassidulidae																		
Rhyncholampas pacificus (A. Agassiz, 1863)	6–134	rb,st	x						x		x			x				
Order Holasteroida																		
Family Pourtalesiidae																		
Cystocrepis setigera (A. Agassiz, 1898)	3182														x			

(continued)

Table A.1 (continued)

	Depth (m)	Habitat	MEX	REV	GUA	SAL	HON	NIC	CRC	COC	PAN	COL	MAL	GAL	ECU	PER	CHL	PAS
Pourtalesia tanneri A. Agassiz, 1898														x				
Familly Urechinidae																		
Cystechinus loveni A. Agassiz, 1898					x													
Pilematechinus rathbuni (A. Agassiz, 1898)	3374											x	x					
Urechinus naresianus A. Agassiz, 1879	755–4400										x							
Order Spatangoida																		
Family Aeropsidae																		
Aeropsis fulva (A. Agassiz, 1898)	1455–5200	mb								x	x				x	x		
Family Brissidae																		
Brissopsis columbaris A. Agassiz, 1898	589–3279	mb,sb	x								x			x				
Brissopsis pacifica (A. Agassiz, 1898)	9–237	sb,mb,rb	x						x	x	x			x		x		
Brissus agassizii Doederlein, 1885																		x
Brissus obesus Verrill, 1867	0–240	sb,cr,rb	x						x		x	x	x	x	x			
Brissus latecarinatus (Leske, 1778)														x				
Meoma frangibilis Chesher, 1970	96										x							
Meoma ventricosa grandis Gray, 1851	0–200	sb	x	x					x	x	x			x				
Metalia nobilis Verrill, 1867	0–18	sb	x								x							

(continued)

Table A.1 (continued)

	Depth (m)	Habitat	MEX	REV	GUA	SAL	HON	NIC	CRC	COC	PAN	COL	MAL	GAL	ECU	PER	CHL	PAS
Metalia spatagus (Linnaeus, 1758)			x															
Rhabdobrissus pacificus H. L. Clark, 1940	6–137	sb,mb,sb	x							x	x							
Family Hemiasteridae																		
Hemiaster tenuis (A. Agassiz, 1898)	980–4027										x	x	x			x		
Family Loveniidae																		
Homolampas hastata A. Agassiz, 1898	1785–3376									x	x	x	x					
Lovenia cordiformis A. Agassiz, 1872	0–210	mb,sb	x							x	x			x	x	x		
Family Macropneustidae																		
Argopatagus aculeata (A. Agassiz, 1898)	1952									x								
Family Maretiidae																		
Nacospatangus oblonga (Mortensen, 1950)														x				
Family Palaeotropidae																		
Scrippsechinus fisheri Allison, Durham & Mintz, 1967	270–460															x		
Family Prenasteridae																		
Tripylus excavatus Philippi, 1845	0–130																x	
Family Schizasteridae																		
Abatus cavernosus (Philippi, 1845)	0–760																x	

(continued)

Table A.1 (continued)

	Depth (m)	Habitat	MEX	REV	GUA	SAL	HON	NIC	CRC	COC	PAN	COL	MAL	GAL	ECU	PER	CHL	PAS
Abatus philippii Lovén, 1871	27–804																x	
Aceste ovata A. Agassiz & H. L. Clark, 1907																x		
Agassizia scrobiculata Valenciennes, 1846	0–209	mb,rb,sb	x			x		x	x		x			x	x	x		
Brisaster latifrons (A. Agassiz, 1898)	9–2817	mb,sb	x								x			x				
Brisaster moseleyi (A. Agassiz, 1881)	124–730																x	
Brisaster townsendi (A. Agassiz, 1898)																x		
Moira atropos clotho Michelin, 1855	10–40	mb,sb	x						x		x							
Schizaster rotundatus (Doederlein, 1906)														x				
Tripylaster philippii (Gray, 1851)	0–100	sb															x	
Family Spatangidae																		
Spatangus californicus H. L. Clark, 1917	10–644	mb,rb,sb	x															
Class Holothuroidea																		
Order Dendrochirotida																		
Family Cucumariidae																		
Abyssocucumis abyssorum (Théel, 1886)	3241–4000									x	x	x	x					
Abyssocucumis albatrossi Cherbonnier, 1947	1585–5690															x		

(continued)

Table A.1 (continued)

	Depth (m)	Habitat	MEX	REV	GUA	SAL	HON	NIC	CRC	COC	PAN	COL	MAL	GAL	ECU	PER	CHL	PAS
Cucumaria chilensis Ludwig, 1875	32–162															x		
Cucumaria flamma Solís-Marín & Laguarda-Figueras, 1999	0–15	rb	x					x	x		x				x	x		
Cladodactyla crocea (Lesson, 1830)	0–4300	sb															x	
Heterocucumis godeffroyi (Semper, 1868)	0–379	rb,sb														x	x	
Hemioedema spectabilis (Ludwig, 1882)	14642	sb															x	
Leptopentacta nina Deichmann, 1941											x							
Leptopentacta nova Deichmann, 1941											x							
Leptopentacta panamica Deichmann, 1941											x							
Neocucumis panamensis Heding & Panning, 1954											x							
Neocucumis veleronis (Deichmann, 1941)	0–22	rb	x								x					x		
Pattalus mollis Selenka, 1868	0–45	rb														x	x	
Pentacta panamensis Verrill, 1867											x							
Pseudocnus californicus (Semper, 1868)	0–190	cr,rb,sb	x		x	x	x	x	x		x					x		
Pseudocnus dubiosus (Semper, 1868)	0–300	rb,sb							x									

(continued)

Table A.1 (continued)

	Depth (m)	Habitat	MEX	REV	GUA	SAL	HON	NIC	CRC	COC	PAN	COL	MAL	GAL	ECU	PER	CHL	PAS
Pseudocnus dubiosus leoninus (Semper, 1868)	0–300	rb,sb														x	x	
Pseudocnus perrieri (Eckman, 1927)	0–197	rb,sb															x	
Staurocucumis abyssorum (Théel, 1886)	1587–4088	mb,rb	x												x	x		
Stereoderma laevigata Verrill, 1876																	x	
Thyonella mexicana (Deichmann, 1941)	10–35	sb,rub	x								x							
Trachythyone lechleri (Lampert, 1885)	5–238	rb															x	
Trachythyone peruana (Semper, 1868)	5–42	rb	x													x		
Family Phyllophoridae																		
Allothyone mexicana (Deichmann, 1946)											x					x		
Athyonidium chilensis (Semper, 1868)	0–13	sb,rb,kf														x	x	
Euthyonidium ovulum Deichmann, 1938									x									
Euthyonidium veleronis Deichmann, 1937	0.3					x			x									
Pentamera beebei Deichmann, 1938	73								x									
Pentamera chierchia (Ludwig, 1887)	0–78	cr,sb,rb	x			x			x		x			x		x		
Pentamera chiloensis (Ludwig, 1887)	6–124	sb														x		

(continued)

Table A.1 (continued)

	Depth (m)	Habitat	MEX	REV	GUA	SAL	HON	NIC	CRC	COC	PAN	COL	MAL	GAL	ECU	PER	CHL	PAS
Pentamera zacae Deichmann, 1938	24	mb,rb,sb				x					x					x		
Phyrella aculeata (Ludwig, 1894)											x							
Thyone bidentata Deichmann, 1941	2–30	cr,rb	x			x					x				x			
Thyone neofusus Deichmann, 1941														x				
Thyone parafusus Deichmann, 1941	25–35	mb,sb	x															
Thyone strangeri Deichmann, 1941	0–12	rb	x															
Family Psolidae																		
Lissothuria hancocki (Deichmann, 1941)	1–301	rb	x															
Lissothuria mortenseni Pawson, 1967											x							
Lissothuria ornata Verrill, 1867	0–37	cr,rb	x			x		x	x	x	x							
Lissothuria veleronis (Deichmann, 1941)														x				
Neopsolidium convergens (Herouard, 1901)	15	kf															x	
Psolidium disciformis (Théel, 1886)	8–448	rb															x	
Psolidium dorsipes Ludwig, 1886	11–451	cr,mb,rb,sb	x								x					x	x	
Psolidium ekmani Deichmann, 1941											x							

(continued)

Table A.1 (continued)

	Depth (m)	Habitat	MEX	REV	GUA	SAL	HON	NIC	CRC	COC	PAN	COL	MAL	GAL	ECU	PER	CHL	PAS
Psolidium eubullatum Deichmann, 1941											x							
Psolidium gracile Ludwig, 1894	2323										x							
Psolidium panamense Ludwig, 1894	2323										x							
Psolidium planum Deichmann, 1941	16–110															x		
Psolus antarcticus Philippi, 1857	6–1080	rb															x	
Psolus chitonoides H. L. Clark, 1901	0–247	rb	x															
Psolus digitatus Ludwig, 1894	1271										x							
Psolus diomedeae Ludwig, 1894	13–302	rb,sb	x							x	x			x				
Psolus paradubiosus Carriol & Féral, 1985	10–567	rb															x	
Psolus patagonicus Ekman, 1925	0–110	rb,kf															x	
Psolus squamatus (Koren, 1844)	7–1087	rb	x														x	
Psolus squamatus segregatus Perrier, 1905	7–207	rb															x	
Family Sclerodactylidae																		
Afrocucumis ovulum (Selenka, 1867)	0–7	rb,cr	x			x					x					x		
Apentamera lepra Deichmann, 1941	55–91	sb	x								x							

(continued)

Table A.1 (continued)

	Depth (m)	Habitat	MEX	REV	GUA	SAL	HON	NIC	CRC	COC	PAN	COL	MAL	GAL	ECU	PER	CHL	PAS
Athyone glasselli (Deichmann, 1936)	0–6	sb	x															
Euthyonidiella zacae (Deichmann, 1938)	17–25	rb	x	x										x				
Neopentamera anexigua Deichmann, 1941	8–22	cr	x												x			
Neothyone gibber (Selenka, 1867)	0–50	cr,rb	x			x			x		x			x	x	x		
Neothyone gibbosa Deichmann, 1941	0–14	rb	x			x			x							x		
Neothyone panamensis (Ludwig, 1887)	0–8.6	rb									x					x		
Pachythyone lugubris (Deichmann, 1939)		cr,rb												x				
Pachythyone pseudolugubris Deichmann, 1941	12	cr,rb	x															
Order Dactylochirotida																		
Family Ypsilothuriidae																		
Ypsilothuria bitentaculata (Ludwig, 1893)	225–4082	mb,sb	x							x	x	x	x		x	x	x	
Order Aspidochirotida																		
Family Holothuriidae																		
Actinopyga mauritiana (Quoy & Gaimard, 1833)											x							
Holothuria (Cystipus) casoae Laguarda-Figueras & Solís-Marín, 2009	45–100	sb	x							x								

(continued)

Table A.1 (continued)

	Depth (m)	Habitat	MEX	REV	GUA	SAL	HON	NIC	CRC	COC	PAN	COL	MAL	GAL	ECU	PER	CHL	PAS
Holothuria (Cystipus) inhabilis Selenka, 1867	1–203	cr,rb,sb	x			x				x	x	x		x		x		
Holothuria (Cystipus) rigida (Selenka, 1867)	0–22	mb,sb,rb	x			x			x		x			x				
Holothuria (Halodeima) atra (Jaeger, 1833)	5–26	mb,cr,rb,sb	x						x	x	x			x				
Holothuria (Halodeima) chilensis Semper, 1868																x		
Holothuria (Halodeima) inornata Semper, 1868	0–18	rb	x	x	x	x			x									
Holothuria (Halodeima) kefersteini (Selenka, 1867)	0–27	rb,sb,cr	x	x		x			x	x	x	x	x	x	x	x		
Holothuria (Lessonothuria) pardalis Selenka, 1867	0–306	cr,rb,mb,sb	x	x			x		x	x	x	x	x	x	x			
Holothuria (Mertensiothuria) hilla Lesson, 1830	0.3–52	cr,rb,sb	x	x					x	x	x	x	x	x				
Holothuria (Mertensiothuria) leucospilota (Brandt, 1835)	1–17	rb,sb	x	x						x	x	x	x	x				
Holothuria (Platyperona) difficilis Semper, 1868	0–100	mb,rb,sb,cr	x	x	x	x	x	x	x	x	x	x		x				
Holothuria (Platyperona) parvula (Selenka, 1867)	1–4											x	x					

(continued)

Table A.1 (continued)

	Depth (m)	Habitat	MEX	REV	GUA	SAL	HON	NIC	CRC	COC	PAN	COL	MAL	GAL	ECU	PER	CHL	PAS
Holothuria (Selenkothuria) carere Honey-Escandón & Solís-Marín, 2011	0–6	rb	x											x				
Holothuria (Selenkothuria) lubrica Selenka, 1867	0–55	cr,rb,sb,mb	x	x		x			x		x	x		x		x		
Holothuria (Selenkothuria) portovallartensis Caso, 1954	0–59	cr,rb	x						x		x	x		x	x	x		
Holothuria (Selenkothuria) theeli Deichmann, 1938	0–55	rb	x							x	x			x		x		
Holothuria (Semperothuria) cinerascens (Brandt, 1835)																		x
Holothuria (Semperothuria) imitans Ludwig, 1875	0–22	rb,sb cr	x	x				x	x	x	x	x		x	x	x		
Holothuria (Semperothuria) languens Selenka, 1867	0–91	rb,mb,sb	x						x		x			x				
Holothuria (Stauropora) pluricuriosa Deichmann, 1937	0–130	rb,cr,sb	x	x						x		x	x	x	x			
Holothuria (Theelothuria) paraprinceps Deichmann, 1937	0–64	cr,rb,sb	x							x	x							

(continued)

Table A.1 (continued)

	Depth (m)	Habitat	MEX	REV	GUA	SAL	HON	NIC	CRC	COC	PAN	COL	MAL	GAL	ECU	PER	CHL	PAS
Holothuria (Thymiosycia) arenicola Semper, 1868	1–121	rb,sb,cr	x	x					x	x	x	x		x	x	x		
Holothuria (Thymiosycia) impatiens (Forskaal, 1775)	0–67	cr,rb,sb	x	x		x			x	x	x	x	x	x		x		
Holothuria (Vaneyothuria) zacae Deichmann, 1937	77–250	mb,rb,sb	x								x	x	x	x	x	x		
Labidodemas americanum Deichmann, 1938	0.5–16	rb,cr,sb	x	x		x			x	x	x			x				
Labidodemas maccullochi (Deichmann, 1958)	1–18	rb,sb,cr	x						x		x	x		x				
Family Stichopodidae																		
Apostichopus parvimensis (H. L. Clark, 1913)	0–36	mb,rb,sb	x															
Isostichopus fuscus (Ludwig, 1875)	0–61	cr,rb,mb,sb	x	x	x	x	x	x	x	x	x	x	x	x	x	x		
Parastichopus californicus (Stimpson, 1857)	1–180	mb,rb	x															
Stichopus horrens Selenka, 1867	0–20	cr,rb,sb								x		x	x	x				
Stichopus monotuberculatus (Quoy & Gaimard, 1833)																		x
Family Synallactidae																		

(continued)

Table A.1 (continued)

	Depth (m)	Habitat	MEX	REV	GUA	SAL	HON	NIC	CRC	COC	PAN	COL	MAL	GAL	ECU	PER	CHL	PAS
Bathyplotes moseleyi (Théel, 1886)	24–690	mb,rb														x	x	
Bathyplotes natans (M. Sars, 1868) Östergren, 1896	1644											x	x					
Bathyplotes patagiatus Fisher, 1907	1644											x						
Capheira sulcata Ludwig, 1893	2877–4334	mb													x	x		
Meseres macdonaldi Ludwig, 1894	1644											x	x					
Meseres torvus (Théel, 1886)																	x	
Mesothuria (Mesothuria) multipes Ludwig, 1894	725–4064	mb								x	x			x		x		
Mesothuria (Zygothuria) lactea (Théel, 1886)																x		
Molpadiodemas atlanticus (R. Perrier, 1898)																x		
Molpadiodemas neovillosus O'Loughlin & Ahearn, 2005	2487–3667														x			
Molpadiodemas ustulatus O'Loughlin & Ahearn, 2005																x		
Molpadiodemas villosus (Théel, 1886)	3667														x			
Molpadiodemas violaceus (Théel, 1886)																x		

(continued)

Table A.1 (continued)

	Depth (m)	Habitat	MEX	REV	GUA	SAL	HON	NIC	CRC	COC	PAN	COL	MAL	GAL	ECU	PER	CHL	PAS
Paelopatides confundens Théel, 1886	1571–4070	mb													x	x		
Paelopatides suspecta Ludwig, 1894	2323										x							
Pseudostichopus macdonaldi (Ludwig, 1894)	2149									x								
Pseudostichopus mollis Théel, 1886	100–5203	mb,sb								x	x	x	x	x		x		
Pseudostichopus peripatus (Sluiter, 1901)	1158–3667										x			x	x			
Synallactes aenigma Ludwig, 1894	2418–4334	mb,sb									x	x	x		x	x		
Synallactes alexandri Ludwig, 1894	589–1008										x			x				
Synallactes triplax A. H. Clark, 1920	549													x				
Synallactes virgulasolida Massin & Hendrickx, 2010	1030	mb	x															
Order Elasipodida																		
Family Deimatidae																		
Deima validum pacificum Ludwig, 1894	1618–2487	mb,sb								x				x	x	x		
Oneirophanta mutabilis affinis Ludwig, 1894	3241										x	x	x					
Oneirophanta mutabilis mutabilis Théel, 1879	3241–3670	mb										x	x			x		
Oneirophanta setigera (Ludwig, 1893)	3667–4088														x			

(continued)

Table A.1 (continued)

	Depth (m)	Habitat	MEX	REV	GUA	SAL	HON	NIC	CRC	COC	PAN	COL	MAL	GAL	ECU	PER	CHL	PAS
Family Elpidiidae																		
Achlyonice ecalcarea Théel, 1879	1608–3587	mb,sb	x															
Amperima naresi (Théel, 1882)	2010–7130	mb														x		
Amperima vitjazi Gebruk, 1988																	x	
Elpidia atakama Belyaev, 1971	7720	mb														x		
Elpidia chilensis Belyaev, 1971	2710–4600	mb														x		
Peniagone anamesa (A. H. Clark, 1920)																x		
Peniagone diaphana (Théel, 1882)	4140–4160	mb														x		
Peniagone elongata (Théel, 1879)	4140–4160	mb														x		
Peniagone gracilis (Ludwig, 1894)	2487–4160	mb												x		x		
Peniagone intermedia Ludwig, 1893	2418–3667	mb													x	x		
Peniagone papillata Hansen, 1975											x							
Peniagone vitrea Théel, 1882	1160–4507	mb								x	x	x	x		x	x		
Scotoplanes clarki Hansen, 1975	3570–5107	mb	x													x		
Scotoplanes globosa Théel, 1879	545–6770	mb														x		

(continued)

Table A.1 (continued)

	Depth (m)	Habitat	MEX	REV	GUA	SAL	HON	NIC	CRC	COC	PAN	COL	MAL	GAL	ECU	PER	CHL	PAS
Family Laetmogonidae																		
Laetmogone theeli Ludwig, 1893	2418														x			
Laetmogone violacea Théel, 1879	466–5083																x	
Pannychia moseleyi Théel, 1882	199–2599	mb,sb	x								x			x		x		
Family Pelagothuridae																		
Enypniastes eximia Théel, 1882	4140–4160	pelagic														x		
Pelagothuria natatrix Ludwig, 1893	0–4505	pelagic								x	x	x	x	x	x	x		
Family Psychropotidae																		
Benthodytes incerta Ludwig, 1893	2418													x				
Benthodytes sanguinolenta Théel, 1882	978–2323	mb	x							x	x	x		x		x		
Benthodytes typica Théel, 1882	1158–4700	mb									x			x		x		
Psychronaetes hanseni Pawson, 1983	3852–4189	mb														x		
Psychropotes depressa (Théel, 1882)											x							
Psychropotes longicauda Théel, 1882	3334–4160	mb									x	x	x		x	x		
Psychropotes verrucosa (Ludwig, 1894)	2418–4160	mb									x			x	x	x		
Order Molpadiida																		

(continued)

Table A.1 (continued)

	Depth (m)	Habitat	MEX	REV	GUA	SAL	HON	NIC	CRC	COC	PAN	COL	MAL	GAL	ECU	PER	CHL	PAS
Family Caudinidae																		
Ceraplectana trachyderma H. L. Clark, 1908	5203	mb														x		
Hedingia californica (Ludwig, 1894)	85–2850	sb,mb	x													x		
Paracaudina chilensis chilensis (Müller, 1850)	0–990	mb,sb							x								x	
Family Molpadiidae																		
Molpadia antarctica (Théel, 1886)	80–1218	mb,sb															x	
Molpadia intermedia (Ludwig, 1894)	53–2972	mb	x								x					x		
Molpadia granulata (Ludwig, 1894)	2690–5869	mb									x					x		
Molpadia musculus Risso, 1826	37–6134	mb,sb	x							x	x	x	x	x	x	x	x	
Molpadia spinosa (Ludwig, 1893)	3279										x							
Order Apodida																		
Family Chiridotidae																		
Chiridota aponocrita A. H. Clark, 1920	9–137	cr,rb,sb	x								x							
Chiridota pacifica Heding, 1928										x								
Chiridota pisanii Ludwig, 1886	0–228	mb,sb															x	
Chiridota rigida Semper, 1868	2–9	cr,rb,sb	x															

(continued)

Table A.1 (continued)

	Depth (m)	Habitat	MEX	REV	GUA	SAL	HON	NIC	CRC	COC	PAN	COL	MAL	GAL	ECU	PER	CHL	PAS
Taeniogyrus contortus (Ludwig, 1875)	0–560	mb,sb,kf															x	
Trochodota purpurea (Lesson, 1830)	0–36	rb,sb															x	
Family Myriotrochidae																		
Myriotrochus giganteus A. H. Clark, 1920	3667	mb													x	x		
Myriotrochus (Oligotrochus) bathybius A. H. Clark, 1920	3667	mb														x		
Family Synaptidae																		
Anapta fallax Lampert, 1889	0–350	sb															x	
Epitomapta tabogae Heding, 1928	0–10		x								x							
Euapta godeffroyi (Semper, 1868)	0–79	cr,rb,sb	x						x		x	x		x				
Polyplectana oculata Heding, 1928											x							
Protankyra abyssicola (Théel, 1886)																x		
Protankyra brychia (Verrill, 1885)	3900–4000										x	x	x			x		
Protankyra pacifica (Ludwig, 1894)	870–4990	mb									x	x				x		

Habitat classification: *cr* Coral Reefs, *m* mangroves, *mb* muddy bottom, *rb* rocky bottom, *sb* sandy bottom, *sg* seagrass, *rub* rubble bottom, *kf* kelp forest. Countries abbreviation: *MEX* Mexico, *REV* Revillagigedo Archipelago, *GUA* Guatemala, *SAL* El Salvador, *HON* Honduras, *NIC* Nicaragua, *CRC* Costa Rica, *COC* Cocos Island, *PAN* Panama, *COL* Colombia, *MAL* Malpelo Island, *GAL* Galapagos Archipelago, *ECU* Ecuador, *PER* Perú, *CHL* Chile, *PAS* Eastern Islands

Table A.2 Taxonomic list of the Echiroderms of the Atlantic Ocean and Caribbean Sea of Latin America and the Canary Islands

	Depth (m)	Habitat	MEX	BEL	GUA	HON	NIC	CRC	PAN	COL	VENc	VENa	BRA	URG	ARG	MAV	CUB	HAI	RDO	PRI	CAN
Class Crinoidea																					
Order Comatulida																					
Family Antedonidae																					
Antedon bifida (Pennant, 1777)	0–400	rb,rub									x										x
Antedon nuttingi (A. H. Clark, 1936)	364–429	rb															x				
Caryometra alope A. H. Clark, 1940	475–640	rb															x				
Caryometra atlantidis A. H. Clark, 1940	366–530	mb								x							x				
Caryometra lisa A. H. Clark, 1940	702	rb															x				
Caryometra monilicirra A. H. Clark, 1940																	x				
Caryometra spinosa A. H. Clark, 1940																	x				
Caryometra tenuipes (A. H. Clark, 1908)	386	rb															x			x	
Coccometra guttata A. H. Clark, 1918	288–480	mb,rb,rub	x							x							x				
Coccometra hagenii (Pourtalès, 1867)	14–1046	rb,rub	x														x				
Coccometra nigrolineata A. H. Clark, 1918	40–987	cr,rb,rub	x							x							x			x	
Comatonia cristata (Hartlaub, 1912)	50–396	cr,rb	x																		
Ctenantedon kinziei Meyer, 1972	Sep-49	cr		x					x	x											
Hypalometra defecta (Carpenter, 1888)	46–400	cr,mb,rb	x						x	x	x						x		x		
Isometra vivipara Mortensen, 1917	79–350	mb,sb,rub													x	x					
Leptometra celtica (M'Andrew & Barrett, 1858)																					x
Phrixometra nutrix (Mortensen, 1918)	137–150	mb,rb,sb													x	x					

(continued)

Table A.2 (continued)

	Depth (m)	Habitat	MEX	BEL	GUA	HON	NIC	CRC	PAN	COL	VENc	VENa	BRA	URG	ARG	MAV	CUB	HAI	RDO	PRI	CAN
Poliometra prolixa (Sladen, 1881)	219																		x		
Trichometra cubensis (Pourtalès, 1869)	210–4829	mb,rb	x				x			x								x			
Zenometra columnaris (Carpenter, 1881)	309–1033	mb,rb	x															x			
Family Pentametrocrinidae																					
Pentametrocrinus atlanticus (Perrier, 1883)	860–1674	mb,sb																			x
Superfamily Comasteracea																					
Family Comasteridae																					
Comactinia echinoptera (Müller, 1840)	2–1033	cr,rb,sb,sg,rub	x		x	x	x	x	x	x	x	x	x				x	x	x	x	
Comactinia meridionalis meridionalis (L. Agassiz, 1865)	3–1033	cr,rb,sb	x			x	x		x	x	x	x					x	x	x	x	
Comactinia meridionalis hartlaubi Messing, 1978	58–373	cr,rb								x									x	x	
Comissia venustus (A. H. Clark, 1909)	24–479	cr,mb,rb,rub							x	x	x						x			x	
Davidaster discoideus (Carpenter, 1888)	0–640	cr,br,rub	x	x	x	x	x	x	x	x							x		x	x	
Davidaster rubiginosus (Pourtalès, 1869)	0–334	cr,rb	x	x		x		x	x	x	x		x				x	x	x	x	
Leptonemaster venustus A. H. Clark, 1909	55–777	mb				x												x	x		
Nemaster grandis A. H. Clark, 1909	3–124	cr,rb				x	x		x	x	x								x		
Neocomatella alata (Pourtalès, 1878)	10–560	cr,br	x								x						x		x		
Neocomatella europaea A. H. Clark, 1913	466–960	rb,rub																			x
Neocomatella pulchella (Pourtalès, 1878)	3–695	cr,br	x				x		x	x			x				x				
Family Colobometridae																					
Analcidometra armata (Pourtalès, 1869)	3–155	cr	x	x		x		x	x	x	x						x		x	x	

(continued)

Table A.2 (continued)

	Depth (m)	Habitat	MEX	BEL	GUA	HON	NIC	CRC	PAN	COL	VENc	VENa	BRA	URG	ARG	MAV	CUB	HAI	RDO	PRI	CAN
Analcidometra caribbea (A. H. Clark, 1908)	23–26								x												
Family Charitometridae																					
Crinometra brevipinna brevipinna (Pourtalès, 1867)	69–795	cr,rb	x				x			x	x						x		x	x	
Crinometra brevipinna concinna A. H. Clark, 1909	201–367	rb															x				
Crinometra brevipinna diadema (Hartlaub, 1912)	320–457	rb															x				
Crinometra brevipinna granulifera (Pourtalès, 1878)	185–475	rb															x				
Crinometra brevipinna margaritacea A. H. Clark, 1909	320–457	rb															x				
Crinometra brevipinna pulchra A. H. Clark, 1909	484–567	rb															x				
Family Thalassometridae																					
Horaeometra duplex (Carpenter, 1888)	159–567	rb	x														x			x	
Stylometra spinifera (Carpenter, 1881)	69–658	cr,rb,sb,rub	x	x		x				x	x						x		x	x	
Family Tropiometridae																					
Tropiometra carinata (Lamarck, 1816)	0–84	cr,rb,rub								x	x		x						x		
Family Atelecrinidae																					
Atelecrinus balanoides Carpenter, 1881	512–2890	mb,rb,sb	x							x	x		x				x				
Family Bathycrinidae																					
Bathycrinus gracilis Wyville-Thomson, 1877																					x
Monachocrinus caribbeus (A. H. Clark, 1908)	684–1574	mb,rb	x								x						x				
Family Bourgueticrinidae																					
Conocrinus lofotensis (Sars, 1868)													x								
Democrinus conifer (A. H. Clark, 1909)	155–1750	cr,mb,rb				x	x		x	x			x				x	x			

(continued)

Table A.2 (continued)

	Depth (m)	Habitat	MEX	BEL	GUA	HON	NIC	CRC	PAN	COL	VENc	VENa	BRA	URG	ARG	MAV	CUB	HAI	RDO	PRI	CAN
Democrinus brevis (A. H. Clark, 1909)	374–1574	mb							x		x										
Democrinus rawsoni (Pourtalès, 1874)	70–650	mb,rb	x				x		x								x		x		
Family Septocrinidae																					
Rouxicrinus vestitus Mironov & Pawson, 2010	421–887	mb								x											
Order Isocrinida																					
Family Isocrinidae																					
Endoxocrinus (Diplocrinus) wyvillethomsoni (Jeffreys, 1870)																					x
Endoxocrinus parrae (Gervais, 1835)	154–971	cr,mb,rb	x			x				x							x	x		x	
Endoxocrinus prionodes (H. L. Clark, 1941)	80.7	rb															x				
Isocrinus blakei (Carpenter, 1884)	220–1200	cr,rb,rub	x														x				
Neocrinus decorus Thomson, 1864	154–1219	mb,rb,rub	x			x					x						x		x	x	
Family Isselicrinidae																					
Cenocrinus asterius (Linnaeus, 1767)	170–412	cr,mb,rb	x														x		x		
Order Cyrtocrinida																					
Family Holopodidae																					
Holopus rangii Orbigny, 1837	9–700	rb	x														x		x		
Class Asteroidea																					
Order Paxillosida																					
Family Astropectinidae																					
Astropecten acutiradiatus Tortonese, 1956	35–66	sb									x		x								
Astropecten alligator Perrier, 1881	22–576	cr,mb,sb,rub	x			x	x			x	x		x						x		
Astropecten americanus Verrill, 1880	110–641	mb,sb,rub	x				x			x	x										

(continued)

Table A.2 (continued)

	Depth (m)	Habitat	MEX	BEL	GUA	HON	NIC	CRC	PAN	COL	VENc	VENa	BRA	URG	ARG	MAV	CUB	HAI	RDO	PRI	CAN
Astropecten antillensis Lütken, 1860	3–278	mb,sb,rub	x							x	x		x				x				
Astropecten aranciacus (Linnaeus, 1758)	1–180	sb,rub																			x
Astropecten articulatus Say, 1825	0–256	mb,sb	x	x	x	x	x	x	x	x	x	x	x				x		x	x	
Astropecten brasiliensis Müller & Troschel, 1842	2–60	sb									x		x	x	x						
Astropecten brasiliensis riensis Doederlein, 1917	18–66	sb									x										
Astropecten caribemexicanensis Caso, 1990	51.3	rub	x																		
Astropecten cingulatus Sladen, 1883	16–1350	mb,sb,rub	x				x	x	x	x	x		x	x						x	
Astropecten comptus Verrill, 1915	18–130	sb,rub	x														x				
Astropecten duplicatus Gray, 1840	0–550	mb,sb,sg,rub	x	x		x	x			x	x		x				x		x	x	
Astropecten hermatophilus Sladen, 1883	15–1500																				x
Astropecten irregularis (Pennant, 1777)	10–1000	mb,sb,rub																			x
Astropecten marginatus Gray, 1840	0.5–130	mb,sb,rub	x					x		x	x	x	x						x	x	
Astropecten nitidus Verrill, 1915	11–686	mb,sb,rub	x			x	x			x									x		
Astropecten spiniphorus Madsen, 1950	45												x								
Astropectinides mesactus (Sladen, 1883)	80–165														x						
Bathybiaster loripes Sladen, 1889	80–500	mb,sb											x		x	x					
Blakiaster conicus Perrier, 1881	55–366	sb	x			x											x		x		
Dipsacaster antillensis Halpern, 1968	113–3627	cr								x											

(continued)

Table A.2 (continued)

	Depth (m)	Habitat	MEX	BEL	GUA	HON	NIC	CRC	PAN	COL	VENc	VENa	BRA	URG	ARG	MAV	CUB	HAI	RDO	PRI	CAN
Dytaster grandis nobilis Sladen, 1889	3305–4846														x						
Dytaster insignis Perrier, 1884	2209–3654	sb	x																		
Leptychaster kerguelenensis E. A. Smith, 1876	27–183														x	x					
Lonchotaster tartareus Sladen, 1889	4390																				x
Persephonaster echinulatus H. L. Clark, 1941	196–720	cr,mb,sb	x			x	x		x	x	x						x			x	
Persephonaster leptactis H. L. Clark, 1941	2928–3294																x				
Persephonaster patagiatus (Sladen, 1889)	733–2165	mb,sb	x				x		x	x	x						x	x			
Plutonaster agassizi agassizi Verrill, 1880	70–3110	cr,mb,sb	x							x							x				
Plutonaster bifrons (Wyville-Thomson, 1873)	800–2970												x				x				
Plutonaster efflorescens (Perrier, 1884)	666–2061	mb,sb	x								x									x	
Psilaster andromeda florae (Verrill, 1878)	1409–1629	mb,sb									x										
Psilaster cassiope Sladen, 1889	151–1680	mb,sb	x							x	x		x				x				
Psilaster herwigi (Bernasconi, 1972)	100–800												x		x						
Tethyaster grandis (Verrill, 1899)	24–139	mb,sb	x							x	x								x		
Tethyaster subinermis (Philippi, 1837)	50–1400	mb																			x
Tethyaster vestitus vestitus (Say, 1825)	5–306	mb,sb	x						x	x		x	x							x	
Family Ctenodiscidae																					
Ctenodiscus australis Lütken, 1871	70–4605	sb,rub											x		x	x					

(continued)

Table A.2 (continued)

	Depth (m)	Habitat	MEX	BEL	GUA	HON	NIC	CRC	PAN	COL	VENc	VENa	BRA	URG	ARG	MAV	CUB	HAI	RDO	PRI	CAN
Ctenodiscus crispatus (Retzius, 1805)	1415–1865								x												
Family Goniopectinidae																					
Goniopecten demonstrans Perrier, 1881	300–1250	mb,sb	x	x					x	x	x										
Prionaster elegans Verrill, 1899	208–535	mb,sb	x				x			x			x				x				
Family Luidiidae																					
Luidia alternata alternata (Say, 1825)	0–1500	m,mb,sb	x			x			x	x	x	x	x	x	x		x		x	x	
Luidia bardabensis Perrier, 1881	60–430	mb,rb,sb	x				x			x	x		x						x		
Luidia ciliaris (Philippi, 1837)	1–400	sb																			x
Luidia clathrata (Say, 1825)	0–175	mb,rb,sb,rub	x	x		x	x		x	x	x		x				x	x	x	x	
Luidia heterozona barimae John & A. M. Clark, 1954	16–90	mb,sb	x							x	x	x	x								
Luidia lawrencei Hopkins & Knott, 2010	0–175	mb								x											
Luidia ludwigi scotti Bell, 1917	33–157	mb,rb,sb								x	x		x		x						
Luidia patriae Bernasconi, 1941	100–126	mb,sb													x						
Luidia sagamina aciculata Mortensen, 1933	71–83	sb	x																		
Luidia sarsi Düben & Koren, 1845																					x
Luidia sarsi elegans Perrier, 1875	60–365	mb,sb	x							x											
Luidia senegalensis (Lamarck, 1816)	0–73	mb,sb	x			x	x	x	x	x	x	x	x				x	x	x	x	
Family Porcellanasteridae																					
Eremicaster vicinus Ludwig, 1907	3950–7250	mb,rub													x						
Styracaster horridus Sladen, 1883	3410–5062	sb									x										
Thoracaster cylindratus Sladen, 1883	2540–5990	mb,sb	x								x										x

(continued)

Table A.2 (continued)

	Depth (m)	Habitat	MEX	BEL	GUA	HON	NIC	CRC	PAN	COL	VENc	VENa	BRA	URG	ARG	MAV	CUB	HAI	RDO	PRI	CAN
Family Pseudarchasteridae																					
Pseudarchaster discus Sladen, 1889	140–283														x	x					
Pseudarchaster gracilis gracilis (Sladen, 1889)	170–2940	mb,sb	x		x	x	x			x	x						x	x	x		
Pseudarchaster parelii (Düben & Koven, 1846)	265–869																x				
Order Notomyotida																					
Family Benthopectinidae																					
Benthopecten simplex simplex (Perrier, 1881)	1175–3713	mb,sb,rb	x															x			
Benthopecten spinosus Verrill, 1884	1857–1958	mb															x				
Cheiraster (Barbadosaster) echinulatus (Perrier, 1875)	130–5062	cr,mb,sb,rb	x	x			x			x	x						x		x	x	
Cheiraster (Cheiraster) ludwigi Fisher, 1913	1266	sb,rb	x																		
Cheiraster (Cheiraster) planus Verrill, 1915	226–1339	mb,sb,rb	x			x				x			x								
Cheiraster (Cheiraster) sepitus (Verrill, 1885)	485–5062	mb,sb								x	x		x				x				
Cheiraster (Christopheraster) blakei A. M. Clark, 1981	250–1958	mb,sb,rb	x	x		x				x			x				x				
Cheiraster (Christopheraster) mirabilis (Perrier, 1881)	380–1470	mb,sb,rb	x				x			x							x			x	
Cheiraster (Luidiaster) planeta (Sladen, 1889)	370–500	mb													x	x					
Gaussaster antarcticus (Sladen, 1889)	3305	mb													x						
Pectinaster gracilis Verrill, 1915	576					x															
Order Valvatida																					
Family Asterinidae																					
Allopatiria ocellifera (Gray, 1847)	30–200	mb,rub																			x

(continued)

Table A.2 (continued)

	Depth (m)	Habitat	MEX	BEL	GUA	HON	NIC	CRC	PAN	COL	VENc	VENa	BRA	URG	ARG	MAV	CUB	HAI	RDO	PRI	CAN
Asterina fimbriata Perrier, 1875	0–250	mb,sb,rub													x	x					
Asterina gibbosa (Pennant, 1777)	0–130	rb,rub																			x
Asterina stellifera (Moebius, 1859)	0–50	mb,sb,rub											x	x	x						
Asterinides folium (Lütken, 1860)	0–256	cr,rb,sb,rub	x	x					x	x	x		x							x	
Asterinides hartmeyeri (Doederlein, 1910)	0–1	rb	x														x			x	
Asterinides pompom (A. M. Clark, 1983)	3–6	rb	x						x												
Asterinopsis pilosa (Perrier, 1881)	11–256	mb,sb	x								x										
Tremaster mirabilis Verrill, 1880	150–1060	rb,rub													x	x	x				
Stegnaster wesseli (Perrier, 1875)	0–183	cr,rb,sb,rub	x							x	x									x	
Family Asteropseidae																					
Poraniella echinulata (Perrier, 1881)	3–339	cr,rb		x				x	x								x			x	
Family Chaetasteridae																					
Chaetaster nodosus Perrier, 1875	30–1140	mb,rb,sb									x	x	x				x				x
Family Ganeriidae																					
Cycethra verrucosa (Philippi, 1857)	0–500	sb,rub											x		x	x					
Ganeria falklandica Gray, 1847	0–135	sb,rub													x	x					
Perknaster sladeni (Perrier, 1891)	120–500														x	x					
Vemaster sudatlanticus Bernasconi, 1965	5055–5208														x						
Family Goniasteridae																					
Anthenoides piercei Perrier, 1881	20–844	mb,sb	x	x			x			x	x		x				x		x	x	
Apollonaster yucatanensis Halpern, 1970	1097–1175	mb,sb	x								x										

(continued)

Table A.2 (continued)

	Depth (m)	Habitat	MEX	BEL	GUA	HON	NIC	CRC	PAN	COL	VENc	VENa	BRA	URG	ARG	MAV	CUB	HAI	RDO	PRI	CAN
Astroceramus brachyactis H. L. Clark, 1941	420–1067	mb,sb	x								x						x				
Ceramaster grenadensis (Perrier, 1881)	70–3109	mb,sb	x			x	x				x		x				x				
Ceramaster grenadensis grenadensis (Perrier, 1881)	200–3109	mb																x	x		x
Ceramaster grenadensis patagonicus (Sladen, 1889)	106–192														x	x					
Circeaster americanus (A. H. Clark, 1916)	500–1450	mb,sb					x			x	x						x				
Cladaster rudis Verrill, 1899	150–900	mb,sb	x																		
Diplasiaster productus (A. H. Clark, 1917)	78–567	mb,sb	x																		
Floriaster maya Downey, 1980	933–1024	mb,sb	x																		
Goniaster tessellatus (Lamarck, 1816)	16–155	cr,mb,sb,rb	x							x	x		x								
Hippasteria falklandica Fisher, 1940	251–225														x	x					
Hippasteria phrygiana argentinensis Bernasconi, 1961	108–162														x						
Litonotaster intermedius (Perrier, 1884)	1958–3530	mb,rb,sb	x							x							x		x	x	
Mediaster bairdi (Verrill, 1882)	640–1590	sb,rb	x										x								
Mediaster pedicellaris (Perrier, 1881)	197–580	mb,rb,sb	x										x				x				
Nymphaster arenatus (Perrier, 1881)	60–3000	mb,sb	x			x	x		x	x	x		x				x	x		x	x
Paragonaster grandis H. L. Clark, 1941	257–540	sb	x														x				
Paragonaster subtilis (Perrier, 1881)	1845–4700	rb,sb	x																		
Pawsonaster parvus (Perrier, 1881)	30–600	mb,rb,sb	x							x	x		x								

(continued)

Table A.2 (continued)

	Depth (m)	Habitat	MEX	BEL	GUA	HON	NIC	CRC	PAN	COL	VENc	VENa	BRA	URG	ARG	MAV	CUB	HAI	RDO	PRI	CAN
Peltaster placenta (Müller & Trochel, 1842)	10–1107	mb,sb									x				x		x	x			x
Plinthaster dentatus (Perrier, 1884)	60–2910	mb,rb,sb	x			x	x		x	x	x		x				x				
Rosaster alexandri (Perrier, 1881)	60–2940	cr,mb sb	x				x			x	x		x				x				
Tessellaster notabilis H. L. Clark, 1941	329–575	mb,sb	x														x				
Tosia parva (Perrier, 1881)	100–600												x		x				x		
Family Leilasteridae																					
Leilaster radians (Perrier, 1881)	102–293	sb,rub	x										x							x	
Family Mithrodiidae																					
Mithrodia clavigera (Lamarck, 1816)	0–157	cr,rb,sb	x				x						x				x				
Familia Odontasteridae																					
Acodontaster elongatus granuliferus (Koehler, 1912)	40–840	sb,rub													x	x					
Diplodontias singularis (Müller & Troschel, 1843)	0–84	rb,sb													x						
Odontaster penicillatus (Philippi, 1870)	8–350	mb,sb,rub													x	x					
Odontaster hispidus Verrill, 1880	50–1160	mb,sb	x														x				
Family Ophidiasteridae																					
Copidaster cavernicola Solís-Marín & Laguarda-Figueras, 2010	13–18	mb	x																		
Copidaster lymani A. H. Clark, 1948	0–34	cr,mb,rb	x	x					x								x				
Hacelia attenuata (Gray, 1840)	1–150	sb,rub																			x
Hacelia superba H. L. Clark, 1921	91–200	rb	x																		x
Linckia bouvieri Perrier, 1875	0–380	cr,rb	x														x				
Linckia guildingi Gray, 1840	0–298	cr,sg,rb,sb	x	x	x			x	x	x	x		x				x	x	x	x	

(continued)

Table A.2 (continued)

	Depth (m)	Habitat	MEX	BEL	GUA	HON	NIC	CRC	PAN	COL	VENc	VENa	BRA	URG	ARG	MAV	CUB	HAI	RDO	PRI	CAN
Linckia nodosa Perrier, 1875	0–475	cr,rb,sb	x								x	x	x				x				
Narcissia canariensis (d'Orbigny, 1839)	37–155	rb	x																		x
Narcissia trigonaria Sladen, 1889	37–750	cr,rb,sb	x						x	x	x	x	x								
Ophidiaster alexandri Verrill, 1915	52–585												x				x				
Ophidiaster bayeri A. H. Clark, 1948	intertidal	rub	x																		
Ophidiaster guildingii Gray, 1840	0–330	cr,rb	x	x					x	x							x		x	x	
Ophidiaster ophidianus (Lamarck, 1816)	0–105	rb																			x
Tamaria floridae (Perrier, 1881)	50–600	rb,sb	x	x																	
Tamaria halperni Downey, 1971	180–510	cr,rb,sb	x							x								x			
Family Oreasteridae																					
Oreaster reticulatus (Linnaeus, 1758)	0–1500	cr,m,rb,sb,sg,rub	x	x	x	x	x	x	x	x	x		x				x	x	x	x	x
Family Poraniidae																					
Marginaster pectinatus Perrier, 1881	166–450	cr,mb,rb,rub	x							x			x				x				
Porania (Porania) antarctica magellanica Studer, 1876	18–320	sb,rub													x	x					
Porania (Porania) pulvillus insignis Verrill, 1895	35–680	rb	x																		
Poraniopsis echinaster Perrier, 1891	30–420	sb,rub													x						
Poraniopsis mira (de Loriol, 1904)	0–500												x		x						
Family Solasteridae																					
Laetmaster spectabilis (Perrier, 1881)	1521	rb	x																		
Lophaster verrilli A. H. Clark, 1938	6–1100	mb,rb,sb,rub	x							x			x				x				

(continued)

Table A.2 (continued)

	Depth (m)	Habitat	MEX	BEL	GUA	HON	NIC	CRC	PAN	COL	VENc	VENa	BRA	URG	ARG	MAV	CUB	HAI	RDO	PRI	CAN
Solaster caribbaeus Verrill, 1915	64–869	mb,rb	x														x				
Order Velatida																					
Family Caymanostellidae																					
Caymanostella spinimarginata Belyaev, 1978	3109–3493	mb																x			
Family Korethrasteridae																					
Peribolaster folliculatus Sladen, 1889	81–133														x	x					
Remaster gourdoni Koehler, 1912	10–540														x	x					
Remaster palmatus (Perrier, 1881)	296–585	mb,rb,sb	x	x									x				x				
Family Myxasteridae																					
Pythonaster murrayi Sladen, 1889	3477														x						
Family Pterasteridae																					
Calyptraster coa Sladen, 1882	260–930												x								
Calyptraster personatus (Perrier, 1885)	2151–6560	mb,rb	x							x								x			
Diplopteraster clarki Bernasconi, 1937	82–177														x						
Diplopteraster verrucosus (Sladen, 1882)	74–270														x						
Hymenaster anomalus Sladen, 1882	1984–2606	rb	x																		
Hymenaster pellucidus Thomson, 1873	1784–3294	sb									x						x				
Hymenaster pergamentaceus Sladen, 1882	4846	sb									x				x						
Hymenaster regalis Verrill, 1895	1857																x				
Hymenaster rex Perrier, 1885	1139–2285	rb	x			x															
Pteraster abyssorum (Verrill, 1895)	576–3740	rb	x			x															

(continued)

Table A.2 (continued)

	Depth (m)	Habitat	MEX	BEL	GUA	HON	NIC	CRC	PAN	COL	VENc	VENa	BRA	URG	ARG	MAV	CUB	HAI	RDO	PRI	CAN
Pteraster acicula (Downey, 1970)	196–3713	cr,mb							x	x							x				
Pteraster affinis lebruni Perrier, 1891	74–341														x	x					
Pteraster caribbaeus Perrier, 1881	275–825	mb,sb,rub	x														x				
Pteraster gibber (Sladen, 1882)	27–500														x	x					
Pteraster militarioides militarioides H. L. Clark, 1941	271–466	mb								x							x				
Pteraster militarioides stoibe H. L. Clark, 1941	777	mb															x				
Pteraster militaris (Müller, 1776)	549–594						x														
Pteraster personatus Sladen, 1891	480–1780	mb								x											
Pteraster rugosus H. L. Clark, 1941	91–466	mb,sb								x	x						x				
Pteraster stellifer Sladen, 1882	79–2804														x	x					
Order Spinulosida																					
Family Echinasteridae																					
Echinaster (Echinaster) modestus Perrier, 1881	60–475	mb,rb,sb	x	x			x				x		x				x		x		
Echinaster (Echinaster) sepositus (Retzius, 1783)	2–250	rb,sg,rub																			
Echinaster (Othilia) brasiliensis Müller & Troschel, 1842	0–360	mb,rb,sb,sg				x			x	x			x		x		x				
Echinaster (Othilia) echinophorus (Lamarck, 1816)	0–55	cr,m,mb,rb,sb,rub	x				x		x	x	x		x				x			x	
Echinaster (Othilia) guyanensis A. M. Clark, 1987	13–119	mb,sb			x	x	x		x	x	x	x	x					x	x		
Echinaster (Othilia) sentus (Say, 1825)	0–175	cr,rb,sb,sg,rub	x	x		x				x	x						x	x		x	
Echinaster (Othilia) serpentarius Müller & Troschel, 1842	10–200	cr,mb,sb	x							x											

(continued)

Table A.2 (continued)

	Depth (m)	Habitat	MEX	BEL	GUA	HON	NIC	CRC	PAN	COL	VENc	VENa	BRA	URG	ARG	MAV	CUB	HAI	RDO	PRI	CAN
Echinaster (Othilia) spinulosus Verrill, 1869	1–238	cr,mb sb,rb	x	x		x	x		x	x											
Henricia antillarum (Perrier, 1881)	192–1390	sb,rub	x			x	x				x		x				x				
Henricia downeyae A. M. Clark, 1987	342–1037	sb,rub	x															x			
Henricia obesa (Sladen, 1889)	22–210	sb,rub													x	x					
Henricia sanguinolenta (O. F. Müller, 1776)	0–200																				x
Henricia sexradiata (Perrier, 1881)	29–366	sb,rub	x														x				
Henricia studeri (Perrier, 1891)	74–430	sb,rub													x	x					
Order Forcipulatida																					
Family Asteriidae																					
Anasterias antarctica (Lütken, 1857)	1–183	rub,rb													x	x					
Anasterias minuta Perrier, 1875	0–100														x						
Anasterias pedicellaris (Koehler, 1923)	0–120														x	x					
Anasterias spirabilis (Bell, 1881)	34–54														x						
Asterias forbesi (Desor, 1848)	0–619	rb,rub	x										x				x				
Coscinasterias linearis (Perrier, 1881)	366	rb	x																		
Coscinasterias tenuispina (Lamarck, 1816)	0–165	rb,sb,rub	x										x				x				x
Cosmasterias lurida (Philippi, 1858)	0–650	mb,sb													x	x					
Diplasterias brandti (Bell, 1881)	0–450	sb,mb,rub													x	x					
Lethasterias australis Fisher, 1923	81–155														x	x					
Lysasterias perrieri (Studer, 1885)	0–320														x						
Marthasterias glacialis (Linnaeus, 1758)	0–180	rb																			x

(continued)

Table A.2 (continued)

	Depth (m)	Habitat	MEX	BEL	GUA	HON	NIC	CRC	PAN	COL	VENc	VENa	BRA	URG	ARG	MAV	CUB	HAI	RDO	PRI	CAN
Neosmilaster steineni (Studer, 1885)															x	x					
Perissasterias polyacantha H. L. Clark, 1923	96–760	mb,sb													x						
Psalidaster mordax Fisher, 1940	80–600														x	x					
Sclerasterias contorta (Perrier, 1881)	384–607										x		x								
Sclerasterias tanneri (Verrill, 1895)						x															
Stephanasterias albula (Stimpson, 1853)	33–2300	cr,sg	x							x							x				
Coronaster briareus (Verrill, 1882)	50–700	cr,rb	x							x	x		x				x				
Family Heliasteridae																					
Labidiaster radiosus Lütken, 1871	5–200	sb,rub													x	x					
Family Pedicellasteridae																					
Ampheraster alaminos Downey, 1971	256–3089	sb	x																		
Pedicellaster pourtalesii Perrier, 1881	338–466	mb	x				x										x				
Family Stichasteridae																					
Allostichaster hartti (Rathbun, 1879)	147–380												x								
Smilasterias triremis Sladen, 1889	0–2707														x	x					
Family Zoroasteridae																					
Cnemidaster sigsbeei (Perrier, 1894)	365–735	mb,sb								x	x										
Doraster constellatus Downey, 1970	345–914	mb,rb,sb	x			x			x	x	x		x								
Mammaster sigsbeei (Perrier, 1880)	430–613	mb				x											x				
Zoroaster fulgens Thomson, 1873	220–3000	mb,rb,sb	x						x	x	x		x				x	x			

(continued)

Table A.2 (continued)

	Depth (m)	Habitat	MEX	BEL	GUA	HON	NIC	CRC	PAN	COL	VENc	VENa	BRA	URG	ARG	MAV	CUB	HAI	RDO	PRI	CAN
Order Brisingida																					
Family Brisingidae																					
Brisinga costata Verrill, 1884	630–1903	cr,mb,rb	x								x							x			
Hymenodiscus coronata (G. O. Sars, 1872)	100–2904	mb																			x
Hymenodiscus verticillata (Sladen, 1889)	640	rb															x				
Midgardia xandaros Downey, 1972	366–460	sb	x			x															
Novodinia americana (Verrill, 1880)	408–576	mb								x											
Novodinia antillensis (A. H. Clark, 1934)	366–2700	mb,sb	x							x			x								
Novodinia pandina (Sladen, 1889)	54–990	rb	x														x				
Stegnobrisinga splendens H. L. Clark, 1926	402–933					x					x										
Family Freyellidae																					
Colpaster scutigerula Sladen, 1889	930–2790	mb				x															x
Freyastera mexicana (A. H. Clark, 1939)	2683–5110	rb	x																		
Freyastera tuberculata (Sladen, 1889)	3360–5620																				x
Freyella elegans (Verrill, 1884)	2928–3294																x				
Freyella microspina Verrill, 1894	1848																x				
Class Ophiuroidea																					
Order Euryalida																					
Family Asteronychidae																					
Asteronyx loveni Müller & Troschel, 1842	265–2499	cr,rb,sb	x								x		x				x				
Astrodia tenuispina (Verrill, 1884)	512–935												x				x				

(continued)

Table A.2 (continued)

	Depth (m)	Habitat	MEX	BEL	GUA	HON	NIC	CRC	PAN	COL	VENc	VENa	BRA	URG	ARG	MAV	CUB	HAI	RDO	PRI	CAN
Family Asteroschematidae																					
Asteroschema arenosum Lyman, 1878	677–1555												x				x				
Asteroschema brachiatum Lyman, 1879	108–783	cr	x														x				
Asteroschema elongatum Koehler, 1914	42–708	cr,rb,sb	x														x				
Asteroschema intectum Lyman, 1878	238–475	cr,sb	x														x				
Asteroschema laeve (Lyman, 1875)	42–539	cr	x								x						x		x		
Asteroschema oligactes (Pallas, 1788)	124–521	cr								x							x				
Asteroschema tenue Lyman, 1875	66–180																x				
Ophiocreas lumbricus Lyman, 1869	15–600									x			x								
Ophiocreas oedipus Lyman, 1879	1061–2228	cr	x																		
Ophiocreas spinulosus Lyman, 1883	227–576	cr	x														x				
Family Gorgonocephalidae																					
Asteroporpa annulata Lütken, 1856	15–397	cr,rb,sb,rub	x							x	x		x				x	x	x	x	
Asteroporpa pulchra H. L. Clark, 1915	320–475																x				
Astracme mucronatus (Lyman, 1869)	70–521	cr	x														x				
Astrocaneum herrerai (H. L. Clark, 1918)	0–25	cr	x																		
Astrochele lymani Verrill, 1878	0–300												x								
Astrocyclus caecilia (Lütken, 1856)	20–677	cr,rb,sb	x										x				x				
Astrocnida isidis (Duchassaing, 1850)	20–180	cr								x									x		

(continued)

Table A.2 (continued)

	Depth (m)	Habitat	MEX	BEL	GUA	HON	NIC	CRC	PAN	COL	VENc	VENa	BRA	URG	ARG	MAV	CUB	HAI	RDO	PRI	CAN
Astrogomphus vallatus Lyman, 1869	60–800	cr,rb	x										x				x				
Astropartus mediterraneus (Risso, 1826)	32–265	mb,rb,sb																			x
Astrophyton muricatum (Lamarck, 1816)	0–508	cr,rb,sb,sg,rub	x	x	x	x	x	x	x	x	x	x	x				x	x	x	x	x
Astrotoma agassizii Lyman, 1875	74–1000													x	x	x					
Gorgonocephalus chilensis (Philippi, 1858)	0–500												x	x	x	x					
Schizostella bifurcata A. H. Clark, 1952	12–46	cr		x																x	
Order Ophiurida																					
Family Ophiomyxidae																					
Ophioblenna antillensis Lütken, 1859	1–24	cr,rb	x	x					x								x			x	
Ophiobrachion uncinatus Lyman, 1883	457	mb															x				
Ophiobyrsa perrieri Lyman, 1883	527																x				
Ophiobyrsa serpens Lyman, 1883	51–126	cr,rb	x	x													x				
Astrogeron supinus (Lyman, 1883)	530–1143																x				
Ophioleptoplax brasiliana Tommasi & Abreu, 1974	15–520												x								
Ophiomyxa brevicauda Verrill, 1899	23–360	mb															x				
Ophiomyxa flaccida (Say, 1825)	0–1500	cr,rb,sb,sg	x	x	x	x			x	x	x		x				x	x	x	x	
Ophiomyxa pentagona (Lamarck, 1816)	0–1060	mb,rb,sb,sg,rub																			x
Ophiomyxa stimpsonii (Lyman, 1875)	108–472	mb								x											
Ophiomyxa tumida Lyman, 1883	23–601	cr,mb,rb,sb	x							x			x				x				

(continued)

Table A.2 (continued)

	Depth (m)	Habitat	MEX	BEL	GUA	HON	NIC	CRC	PAN	COL	VENc	VENa	BRA	URG	ARG	MAV	CUB	HAI	RDO	PRI	CAN
Ophiomyxa vivipara Studer, 1876	0–183												x		x	x					
Ophiophrixus quadrispinosus (Koehler, 1914)	265–510																x				
Ophioprium cervicornis (Lyman, 1883)													x								
Ophiosciasma attenuatum Lyman, 1878	130–540												x								
Ophioscolex disacanthus H. L. Clark, 1915	342–549	rb, mb															x			x	
Ophioscolex glacialis Müller & Troschel, 1842	50–2727												x								
Ophioscolex nutrix (Mortensen, 1936)													x		x	x					
Ophioscolex serratus H. L. Clark, 1900	166	sb															x				
Ophiosyzygus disacanthus H. L. Clark, 1911	127–278	cr,mb								x											
Family Amphilepididae																					
Amphilepis norvegica (Ljungman, 1865)	100–2900	mb																			x
Amphilepis sanmatiensis Bernasconi & D'Agostino, 1975	1–145												x		x						
Amphilepis teodorae Tommasi & Abreu, 1974													x								
Family Amphiuridae																					
Amphilimna mirabilis (H. L. Clark, 1941)	200–550	mb								x			x				x				
Amphilimna olivacea (Lyman, 1869)	15–600	mb,rb,sb	x						x	x	x	x	x								
Amphiodia atra (Stimpson, 1852)	1.5–100	sb									x	x	x								
Amphiodia guillermosoberoni Caso, 1979	0.9–3.4	sb	x																		

(continued)

Table A.2 (continued)

	Depth (m)	Habitat	MEX	BEL	GUA	HON	NIC	CRC	PAN	COL	VENc	VENa	BRA	URG	ARG	MAV	CUB	HAI	RDO	PRI	CAN
Amphiodia habilis Albuquerque, Campos-Creasey & Guille, 2001	34												x								
Amphiodia planispina (von Martens, 1867)	0–578	mb,sb,sg,rub							x	x			x	x	x		x			x	
Amphiodia pulchella (Lyman, 1869)	0–370	cr,mb,rb,sb,sg	x	x					x				x	x	x		x			x	
Amphiodia trychna H. L. Clark, 1918	1–160	m,sb,sg,rub	x	x					x		x		x				x			x	
Amphiodia violacea (Lütken, 1856)				x																	
Amphioplus (Amphioplus) abditus (Verrill, 1871)	6–587	mb,sb	x						x		x		x								
Amphioplus (Amphioplus) brasiliensis Tommasi, 1970													x								
Amphioplus (Amphioplus) coniortodes H. L. Clark, 1918	0–1207	cr,mb,rb,sg	x								x						x				
Amphioplus (Amphioplus) sepultus Hendler, 1995	0–82	cr,mb,rb,sg	x														x				
Amphioplus (Amphioplus) thrombodes H. L. Clark, 1918	0.3–0.6	mb,sg															x				
Amphioplus (Unioplus) incisus (Lyman, 1883)	1639								x				x								
Amphioplus albidus (Ljungman, 1867)	1–500	mb,rb,sb											x	x	x						
Amphioplus daleus (Lyman, 1879)													x								
Amphioplus lucyae Tommasi, 1971	5–600	sb											x	x							
Amphioplus mathildae Tommasi & Abreu, 1974	0–120												x								
Amphioplus peregrinator (Koehler, 1912)																x					
Amphioplus tumidus (Lyman, 1878)	70–578	mb								x											

(continued)

Table A.2 (continued)

	Depth (m)	Habitat	MEX	BEL	GUA	HON	NIC	CRC	PAN	COL	VENc	VENa	BRA	URG	ARG	MAV	CUB	HAI	RDO	PRI	CAN
Amphipholis gracillima (Stimpson, 1852)	0–63	mb,sb	x	x													x				
Amphipholis januarii Ljungman, 1867	1–311	cr,mb,rb,sg,rub	x	x									x				x			x	
Amphipholis pachybactera H. L. Clark, 1918											x										
Amphipholis squamata (Delle Chiaje, 1828)	0–1962	cr,m,rb,sb,sg,rub	x	x	x	x			x	x	x		x	x	x		x			x	x
Amphipholis subtilis (Ljungman, 1867)	0–1330												x								
Amphiura (Amphiura) callida Albuquerque, Campos-Creasey & Guille, 2001	50												x								
Amphiura (Amphiura) correcta Koehler, 1907													x								
Amphiura (Amphiura) grandisquama Lyman, 1869																					x
Amphiura (Amphiura) magellanica Ljungman, 1867	0–400													x	x	x					
Amphiura (Amphiura) rosea Tommasi & Oliveira, 1976													x								
Amphiura (Ophionema) intricata Lütken, 1869	1–600												x								
Amphiura algida Koehler, 1911													x								
Amphiura belgicae Koehler, 1900															x	x					
Amphiura chiajei Forbes, 1843	5–1200																				x
Amphiura complanata Ljungman, 1867	0–810												x	x							
Amphiura crassipes Ljungman, 1867	0–60	mb,sb											x	x	x						
Amphiura deichmanni Tommasi, 1965													x								
Amphiura diducta Koehler, 1914																	x				

(continued)

Table A.2 (continued)

	Depth (m)	Habita	MEX	BEL	GUA	HON	NIC	CRC	PAN	COL	VENc	VENa	BRA	URG	ARG	MAV	CUB	HAI	RDO	PRI	CAN
Amphiura eugeniae Ljungman, 1867	0–800	mb												x	x	x					
Amphiura fibulata Koehler, 1913	2–810	cr,rb,st,sg,rub	x	x									x				x				
Amphiura flexuosa Ljungman, 1867	0–50	rub											x	x							
Amphiura iraciae Tommasi & Oliveira, 1976													x								
Amphiura joubini Koehler, 1912	5–3834	rub											x	x	x						
Amphiura kinbergi Ljungman, 1872	3–300												x								
Amphiura latispina Ljungman, 1867	10–50												x		x						
Amphiura muelleri Marktanner-Turneretscher, 1887	134–600												x								
Amphiura otteri Ljungman, 1872	198–3200	mb,sb								x		x	x				x				x
Amphiura palmeri Lyman, 1882	5–479	cr,rb	x								x		x							x	
Amphiura princeps Koehler, 1907	0–107	mb,sb											x	x	x	x					
Amphiura rathbuni Koehler, 1914	29–502	mb								x	x										
Amphiura scabriuscula (Lütken, 1859)	0–26								x											x	
Amphiura semiermis Lyman, 1869	82–1448	mb											x							x	
Amphiura stimpsonii Lütken, 1859	0–2844	cr,rb,st,rub	x	x									x				x	x		x	
Microphiopholis atra (Stimpson, 1852)	1–38	mb								x	x		x								
Microphiopholis gracillima (Stimpson, 1854)	0–26	mb							x		x		x							x	
Microphiopholis subtilis (Ljungman, 1867)													x								

(continued)

Table A.2 (continued)

	Depth (m)	Habitat	MEX	BEL	GUA	HON	NIC	CRC	PAN	COL	VENc	VENa	BRA	URG	ARG	MAV	CUB	HAI	RDO	PRI	CAN
Nudamphiura carvalhoi Tommasi, 1965	15–117												x								
Ophiocnida loveni (Ljungman, 1867)	7–48												x								
Ophiocnida scabra Lyman, 1879	1100												x								
Ophiocnida scabriuscula (Lütken, 1859)	0–70	cr,rb,sg	x										x				x				
Ophionephthys limicola Lütken, 1869	1–12	cr,m,sb,mb		x					x				x				x				
Ophiophragmus chilensis (Müller & Troschel, 1843)	0–112														x	x					
Ophiophragmus cubanus (A. H. Clark, 1917)	1–36	m,mb,sg		x									x				x				
Ophiophragmus filograneus (Lyman, 1875)	0–3			x									x								
Ophiophragmus luetkeni (Ljungman, 1872)	0–50												x								
Ophiophragmus moorei Thomas, 1965	1–2	rb	x																		
Ophiophragmus pulcher H. L. Clark, 1918	0.5–33	cr,m,mb,sg,rub	x	x					x		x		x				x			x	
Ophiophragmus riisei (Lütken, 1859)	1–311	cr,mb,sb	x						x	x	x		x					x			
Ophiophragmus septus (Lütken, 1859)	0.3–116	cr,m,mb,sb	x	x					x	x			x				x			x	
Ophiophragmus wurdemani (Lyman, 1860)	2–11	cr,sb	x								x		x								
Ophiostigma isocanthum (Say, 1825)	0–223	cr,mb,sg,sb,rub	x	x				x	x	x	x		x				x			x	
Ophiostigma siva Hendler, 1995	0.5–99	cr,rb	x	x					x											x	
Family Hemieuryalidae																					
Hemieuryale pustulata Von Martens, 1867	128–148			x					x										x		

(continued)

Table A.2 (continued)

	Depth (m)	Habitat	MEX	BEL	GUA	HON	NIC	CRC	PAN	COL	VENc	VENa	BRA	URG	ARG	MAV	CUB	HAI	RDO	PRI	CAN
Ophiochondrella squamosa (Lyman, 1883)	604																x				
Ophiochondrus convolutus Lyman, 1869	165–720	cr								x							x				
Ophiochondrus crassispinus Lyman, 1883	420																x				
Ophiochondrus gracilis Verrill, 1899	198–468																x				
Ophiochondrus stelliger Lyman, 1879	144–1080														x	x					
Sigsbeia conifera Koehler, 1914	4–450	mb	x	x					x								x				
Sigsbeia murrhina Lyman, 1878	16–706	cr,mb	x	x													x			x	
Family Ophiacanthidae																					
Subfamily Ophiacanthinae																					
Ophiacantha abyssicola G. O. Sars, 1871	35–351	mb,rub																			x
Ophiacantha affinis Koehler, 1914	366–420																x				
Ophiacantha aristata Koehler, 1895	822–1700												x								
Ophiacantha aspera Lyman, 1878	320																x				
Ophiacantha bidentata (Retzius, 1805)	32–4730	cr,mb,rb	x							x											x
Ophiacantha brasiliensis Tommasi & Abreu, 1974	145–380												x								
Ophiacantha cervicornis Lyman, 1883	366–539																x				
Ophiacantha crassidens Verrill, 1885	980–3120	mb	x																		
Ophiacantha densispina Mortensen, 1936																x					
Ophiacantha echinulata Lyman, 1878	304–1958	mb,rb,st								x	x						x				

(continued)

Table A.2 (continued)

	Depth (m)	Habitat	MEX	BEL	GUA	HON	NIC	CRC	PAN	COL	VENc	VENa	BRA	URG	ARG	MAV	CUB	HAI	RDO	PRI	CAN
Ophiacantha ensifera (Verrill, 1899)	198–468																x				
Ophiacantha mesembria H. L. Clark, 1915	252–1125	mb								x							x				
Ophiacantha metallacta H. L. Clark, 1915	295–763	mb	x										x								
Ophiacantha pentacrinus Lütken, 1869	320–1249										x		x				x				
Ophiacantha rosea Lyman, 1878	402–1538	mb,rb	x																		
Ophiacantha scutata Lyman, 1878	521–549																x				
Ophiacantha sertata (Lyman, 1869)	183–618	cr,rb	x																		
Ophiacantha stellata Lyman, 1875	183																x				
Ophiacantha setosa (Retzius, 1805)	40–1480	sb																			x
Ophiacantha vivipara Ljungman, 1870	0–1097												x	x	x	x					
Ophiacanthella troscheli (Lyman, 1878)	185–238	cr,rb,sb	x																		
Ophialcaea nuttingii Verrill, 1899	366																x				
Ophiotreta lineolata (Lyman, 1883)	347–521																x				
Ophiotreta mixta (Lyman, 1878)	338–442																x				
Ophiotreta sertata (Lyman, 1869)	293–567	mb								x							x				
Ophiotreta valenciennesi valenciennesi (Lyman, 1879)	200–1440																				x
Ophiotreta valenciennesi rufescens (Koehler, 1896)	155–1440	mb								x											

(continued)

Table A.2 (continued)

	Depth (m)	Habitat	MEX	BEL	GUA	HON	NIC	CRC	PAN	COL	VENc	VENa	BRA	URG	ARG	MAV	CUB	HAI	RDO	PRI	CAN
Subfamily Ophiohelinae																					
Ophiomyces frutectosus Lyman, 1869	151–410	mb								x											
Ophiomyces mirabilis Lyman, 1869	270–560	rb	x																		
Subfamily Ophioplinthacinae																					
Ophiocamax austera Verrill, 1899	198–1045																x				
Ophiocamax fasciculata Lyman, 1883	208–979	mb,rb,sb				x	x		x	x	x						x		x		
Ophiocamax hystrix Lyman, 1878	171–695	mb								x							x				
Ophiomitra ornata Verrill, 1899	201																x				
Ophiomitra robusta Koehler, 1914	395																x				
Ophiomitra valida Lyman, 1869	131–608	cr								x							x				
Ophiomitrella clavigera (Ljungman, 1865)	250–1500																				x
Ophiomitrella conferta (Koehler, 1922)															x	x					
Ophiomitrella cordifera Koehler, 1909													x								
Ophiomitrella glabra (H. L. Clark, 1901)		cr		x															x		
Ophiomitrella ingrata Koehler, 1908															x						
Ophiomitrella laevipellis (Lyman, 1883)	155–507	cr								x							x				
Ophioplinthaca carduus (Lyman, 1878)	796	mb	x																		
Ophioplinthaca chelys (Wyville-Thomson, 1878)	795–3305										x										x
Ophioplinthaca dipsacos (Lyman, 1878)	777																x				

(continued)

Table A.2 (continued)

	Depth (m)	Habitat	MEX	BEL	GUA	HON	NIC	CRC	PAN	COL	VENc	VENa	BRA	URG	ARG	MAV	CUB	HAI	RDO	PRI	CAN
Ophiothamnus exigua (Lyman, 1878)	439																x				
Subfamily Ophiotominae																					
Ophiolimna littoralis Koehler, 1913	0–63																x				
Ophiopristis hirsuta (Lyman, 1875)	147–830	cr								x											
Ophiotoma gracilis (Koehler, 1914)	490–1236	mb								x											
Family Ophiactidae																					
Hemipholis cordifera Bosc, 1802	0–520	cr,mb,rb,sb,rub	x	x					x				x	x			x			x	
Histampica duplicata (Lyman, 1875)	125–2870	cr,mb								x	x		x								
Histampica rugosa (H. L. Clark, 1941)	1125–1281																x				
Ophiactis abyssicola (M. Sars, 1861)	125–4000	mb,rb																			x
Ophiactis algicola H. L. Clark, 1933	0–24	cr	x	x					x												
Ophiactis asperula (Philippi, 1858)	0–310													x	x	x					
Ophiactis balli (W. Thompson, 1840)	30–1765	rb																			x
Ophiactis brasiliensis Manso, 1988	1.5–163												x								
Ophiactis dispar Verrill, 1899	62								x												
Ophiactis ljungmani Marktanner-Turneretscher, 1887																		x			
Ophiactis loricata Lyman, 1869	198	mb																		x	
Ophiactis lymani Ljungman, 1872	0–600	rb											x								
Ophiactis muelleri Lütken, 1856	14–67	sb							x		x		x				x				

(continued)

Table A.2 (continued)

	Depth (m)	Habitat	MEX	BEL	GUA	HON	NIC	CRC	PAN	COL	VENc	VENa	BRA	URG	ARG	MAV	CUB	HAI	RDO	PRI	CAN
Ophiactis quinqueradia Ljungman, 1871	0–640	cr,mb,rb,sb,sg	x	x					x	x			x				x		x	x	
Ophiactis rubropoda Singletary, 1973	2–32	cr	x																		
Ophiactis savignyi (Müller & Troschel, 1842)	0–1500	cr,m,rb,sb,sg,rub	x	x		x		x	x	x	x		x				x	x	x	x	x
Ophiactis virens (M. Sars, 1857)	0–90	rb																			x
Family Ophiochitonidae																					
Ophiochiton ternispinus Lyman, 1883	377–3550	mb,rb	x							x			x				x				
Ophioplax clarimundae Tommasi, 1970													x								
Ophioplax ljungmani Lyman, 1875	22–471	cr								x							x			x	
Ophioplax pardalis H. L. Clark, 1941	347																x				
Ophioplax spinulifera H. L. Clark, 1941	274–420																x				
Family Ophiocomidae																					
Ophiocoma echinata (Lamarck, 1816)	0–183	cr,m,rb,sb,sg,rub	x	x	x	x	x	x	x	x	x		x				x	x	x	x	
Ophiocoma paucigranulata Devaney, 1974	0–455	cr,rb	x	x				x													
Ophiocoma pumila Lütken, 1859	0–368	cr,rb,sg	x	x		x			x	x	x		x				x	x	x	x	x
Ophiocoma riisei Lütken, 1859	1–2	cr									x							x			
Ophiocoma wendtii Müller & Troschel, 1842	0–384	cr,m,rb,sg,rub	x	x		x		x	x	x			x				x	x	x	x	
Ophiocomella ophiactoides (H. L. Clark, 1901)	0–70	cr,rb,sb,sg,rub	x	x					x	x	x		x							x	
Ophiocomella sexradia (Duncan, 1887)													x								
Ophiocomina nigra (Abildgaard, in O.F. Müller, 1789)	0–400	sb,rub																			x

(continued)

Table A.2 (continued)

	Depth (m)	Habitat	MEX	BEL	GUA	HON	NIC	CRC	PAN	COL	VENc	VENa	BRA	URG	ARG	MAV	CUB	HAI	RDO	PRI	CAN
Ophiopsila aranea Forbes, 1843	0–185	sg,rub																			x
Ophiopsila fulva Lyman, 1878	29–320	cr								x							x				
Ophiopsila guineensis Koehler, 1914	18–110																				x
Ophiopsila hartmeyeri Koehler, 1913	1–183	cr,rb,sb	x					x		x	x	x	x				x				
Ophiopsila maculata (Verrill, 1899)	41–3000												x				x				
Ophiopsila riisei Lütken, 1859	0–366	cr,m,rb,sg	x	x					x	x	x		x				x		x	x	
Ophiopsila vittata H. L. Clark, 1918	4–15	cr,rb	x	x					x												
Family Ophiodermatidae																					
Bathypectinura heros (Lyman, 1879)	276–3150	mb,sb	x			x	x		x	x			x				x				
Ophiorachnella angulata (Lyman, 1883)																	x				
Ophiarachnella petersi (Lyman, 1878)	320–475																x				
Ophioconis forbesi (Heller, 1863)	20–200	rb																			x
Ophioconis miliaria Lyman, 1878	444–759																x				
Ophioderma anitae Hotchkiss, 1982		cr,sb		x																	
Ophioderma appressa (Say, 1825)	0–580	cr,rb,sg,rub	x	x		x		x	x	x	x		x				x	x	x	x	x
Ophioderma besnardi Tommasi, 1970	0–600												x								
Ophioderma brevicaudum Lütken, 1856	0–64	cr,rb,sb,sg	x	x					x	x	x						x	x	x	x	
Ophioderma brevispinum (Say, 1825)	1–223	cr,m,mb,rb,sb,sg,rub	x	x				x	x	x	x		x				x	x	x	x	
Ophioderma cinereum Müller & Troschel, 1842	0–1719	cr,m,sb,sg	x	x		x		x	x	x	x		x				x	x	x	x	

(continued)

Table A.2 (continued)

	Depth (m)	Habitat	MEX	BEL	GUA	HON	NIC	CRC	PAN	COL	VENc	VENa	BRA	URG	ARG	MAV	CUB	HAI	RDO	PRI	CAN
Ophioderma divae Tommasi, 1971													x								
Ophioderma ensiferum Hendler & Miller, 1984	10–30	cr,rb	x	x													x				
Ophioderma guttatum Lütken, 1859	0–30	cr,rb,rub	x	x						x							x				
Ophioderma januarii Lütken, 1856	0–1500	cr	x										x								
Ophioderma longicaudum (Retzius, 1805)	0–200	sg,rub																			x
Ophioderma phoenium H. L. Clark, 1918	1–14	cr,rb,sb	x	x					x								x			x	
Ophioderma rubicundum Lütken, 1856	0–360	cr,sg,rb,sb,rub	x	x					x	x	x						x		x	x	
Ophioderma squamosissimum Lütken, 1856	3–85	cr,rb	x	x													x			x	
Ophiopaepale goesiana Ljungman, 1872	101–436	cr,mb sb								x	x						x				
Ophiurochaeta littoralis (Koehler, 1913)	1–110	cr,sb		x																	
Family Ophionereididae																					
Ophionereis dolabriformis John & A. M. Clark, 1954	10–93	cr,mb								x			x								
Ophionereis olivacea H. L. Clark, 1901	0–500	cr,m,sb	x	x					x	x	x		x				x			x	
Ophionereis reticulata (Say, 1825)	0–1500	cr,m,rb,sb,sg,rub	x	x		x		x	x	x	x		x				x	x	x	x	
Ophionereis sexradia Mortensen, 1936	18–130																				x
Ophionereis squamulosa Koehler, 1913	0.5–110	cr,sb,rb,sg,rub	x	x					x				x				x	x		x	
Ophionereis vittata Hendler, 1995	10–126	cr	x	x					x								x				
Family Ophiotrichidae																					
Ophiothela danae Verrill, 1869													x								

(continued)

Table A.2 (continued)

	Depth (m)	Habitat	MEX	BEL	GUA	HON	NIC	CRC	PAN	COL	VENc	VENa	BRA	URG	ARG	MAV	CUB	HAI	RDO	PRI	CAN
Ophiotrichoides lymani Ludwig, 1882													x								
Ophiothrix (Ophiothrix) ailsae Tommasi, 1970													x								
Ophiothrix (Ophiothrix) trindadensis Tommasi, 1970													x								
Ophiothrix angulata (Say, 1825)	0.5–540	cr,m,mb,rb,sb,sg,rub	x	x					x	x	x		x	x			x	x	x	x	
Ophiothrix angulata violacea (Müller & Troschel, 1842)													x								
Ophiothrix brachyactis H. L. Clark, 1915	1–6	cr,rb,rub	x								x		x				x			x	
Ophiothrix cimar Hendler, 2005	0–10	cr,sg						x	x												
Ophiothrix fragilis (Abildgaard, 1789)	0–1250	rb,sg,rub																			x
Ophiothrix lineata Lyman, 1860	0–74	cr,rb	x	x				x	x	x							x				
Ophiothrix luetkeni Wyville-Thomson, 1873	50–500	rb,sb,rub																			x
Ophiothrix maculata Ljungman, 1872	114–410																				x
Ophiothrix oerstedii Lütken, 1856	0–31	cr,sb,mb,sg,rb,rub	x	x					x	x	x						x	x	x	x	
Ophiothrix pallida Ljungman, 1871	183						x														
Ophiothrix platyactis H. L. Clark, 1939	41	rb	x																		
Ophiothrix rathbuni Ludwig, 1882	8–600												x								
Ophiothrix stri Hendler, 2005	0–10	cr						x	x												
Ophiothrix suensonii Lütken, 1856	0–1000	cr,m,mb,rb,sb	x	x				x	x	x	x		x				x	x	x	x	
Ophiothrix synoecina Schoppe, 1996	0–5	rb								x											

(continued)

Table A.2 (continued)

	Depth (m)	Habitat	MEX	BEL	GUA	HON	NIC	CRC	PAN	COL	VENc	VENa	BRA	URG	ARG	MAV	CUB	HAI	RDO	PRI	CAN
Family Ophiuridae																					
Subfamily Ophiolepidinae																					
Amphipholizona delicata H. L. Clark, 1915	15–600	mb								x			x								
Ophiolepis ailsae Hendler & Turner, 1987	156–353																x				
Ophiolepis elegans Lütkens, 1859	1–329	cr,m,mb,rb,sb,sg,rub	x			x			x	x	x		x				x			x	
Ophiolepis gemma Hendler & Turner, 1987	2–139	cr,rb,sb	x	x													x				
Ophiolepis impressa Lütken, 1859	0–1500	cr,rb,sb,sg,rub	x	x		x		x	x	x	x		x				x		x	x	
Ophiolepis kieri Hendler, 1979	2–8	sg							x				x				x				
Ophiolepis paucispina (Say, 1825)	0–37	cr,m,rb,sb,sg,rub	x	x					x	x			x				x	x		x	x
Ophiolepis pawsoni Hendler, 1988	24.4			x																	
Ophiolipus agassizii Lyman, 1878	146–310	rb	x																		
Ophiomidas dubius (Lyman, 1878)	158–272																x				
Ophiomusium acuferum Lyman, 1875	76–575	cr,mb,rb	x							x			x				x				
Ophiomusium anaelisae Tommasi & Abreu, 1974	180–260												x								
Ophiomusium constrictum Mortensen, 1936															x						
Ophiomusium eburneum Lyman, 1869	35–3477	cr,mb,sb	x							x			x				x				
Ophiomusium leptobrachium H. L. Clark, 1941	1249–1518	sb									x						x				
Ophiomusium lütkeni Lyman, 1878	190–278	rb,sb	x																		
Ophiomusium lymani Thomson, 1873	62–4700	mb,rb,sb,rub	x							x							x				x

(continued)

Table A.2 (continued)

	Depth (m)	Habitat	MEX	BEL	GUA	HON	NIC	CRC	PAN	COL	VENc	VENa	BRA	URG	ARG	MAV	CUB	HAI	RDO	PRI	CAN
Ophiomusium microporum H. L. Clark, 1941	347–530																x				
Ophiomusium moniliforme H. L. Clark, 1941	1125																x				
Ophiomusium planum Lyman, 1879	368–4062	rb,sb	x																		
Ophiomusium sculptum Verrill, 1899	475																x				
Ophiomusium serratum Lyman, 1878	520–3103	rb,sb	x																		x
Ophiomusium stellatum Verrill, 1899	356–521																x				
Ophiomusium testudo Lyman, 1875	60–914	cr,rb,sb	x							x							x				
Ophiomusium validum Ljungman, 1872	108–2732	cr,mb								x							x				x
Ophioplocus januarii (Lütken, 1856)	0–180	rb,sb,rub											x	x	x						
Ophiosphalma armigerum (Lyman, 1878)	262–4024	rb,sb	x										x				x				
Ophiosphalma dyscritum H. L. Clark, 1941	512–713																x				
Ophiosphalma monoplax H. L. Clark, 1915	376–746																x				
Ophiothyreus goesii Ljungman, 1872	144–540	cr					x			x							x				
Ophiozonella clypeata (Lyman, 1883)	158–272																x				
Ophiozonella falklandica Mortensen, 1936															x	x					
Ophiozonella granulifera H. L. Clark, 1941	1006–1098	rb,sb	x														x				
Ophiozonella marmorea Lyman, 1883	484–1385																x				
Ophiozonella molesta (Koehler, 1904)	2115																				x

(continued)

Table A.2 (continued)

	Depth (m)	Habitat	MEX	BEL	GUA	HON	NIC	CRC	PAN	COL	VENc	VENa	BRA	URG	ARG	MAV	CUB	HAI	RDO	PRI	CAN
Ophiozonella nivea (Lyman, 1875)	70–1249	rb,sb	x								x						x				
Ophiozonella tessellata (Lyman, 1879)	320–475																x				
Subfamily Ophioleucinae																					
Ophiostriatus striatus (Mortensen, 1933)	270–3500												x								
Ophiernus adspersus adspersus Lyman, 1883	68–3650	mb,rb	x							x							x				
Ophioleuce depressum (Lyman, 1869)	585																x				
Ophiopyren longispinus Lyman, 1878	329																x				
Ophiostriatus atlanticus (Mortensen, 1933)	296–396	mb								x											
Subfamily Ophiurinae																					
Amphiophiura metabula H. L. Clark, 1915	274–1830	mb,sb	x							x							x				
Amphiophiura oedignatha H. L. Clark, 1915	284–944	mb	x							x							x				
Amphiophiura ornata (Lyman, 1878)													x								
Anthophiura ingolfi Fasmer, 1930	21–1200	cr	x																		
Ophiambix devaneyi Paterson, 1985	146–494	mb								x											
Ophiocten amitinum Lyman, 1878	46–545													x	x	x					
Ophiomastus meridionalis (Lyman, 1879)															x						
Ophiomastus satelitae Tommasi & Abreu, 1974	115–600												x								
Ophiomastus secundus Lyman, 1878	108–2035	mb,sb								x	x						x				
Ophiomisidium pulchellum (Wyville-Thomson, 1878)	11–3061	cr,rb,sb	x										x								x

(continued)

Table A.2 (continued)

	Depth (m)	Habitat	MEX	BEL	GUA	HON	NIC	CRC	PAN	COL	VENc	VENa	BRA	URG	ARG	MAV	CUB	HAI	RDO	PRI	CAN
Ophiomisidium speciosum Koehler, 1914	880–1636	mb	x										x								
Ophiomisidium tommasii Borges, Monteiro & Amaral, 2006	250–808												x								
Ophiopleura inermis (Lyman, 1878)	1786																				x
Ophioplinthus abyssorum (Lyman, 1883)	338–366																x				
Ophioplinthus inornata (Lyman, 1878)	54–3384														x	x					
Ophioplinthus martensi (Studer, 1885)															x						
Ophiophycis mirabilis Koehler, 1901													x								
Ophiura acervata (Lyman, 1869)	20–1900	cr,mb,rb,sb	x							x	x	x	x				x				
Ophiura carinifera (Koehler, 1901)	400													x							
Ophiura carnea Lütken, 1858	40–1260	mb,sb																			x
Ophiura clemens (Koehler, 1904)													x								
Ophiura falcifera (Lyman, 1869)	73–1037	mb								x							x				
Ophiura fallax Cherbonnier, 1959	20–842	mb,sb	x										x								
Ophiura grubei Heller, 1863	10–350	mb,sb,rub																			x
Ophiura (Ophiuroglypha) irrorata concreta (Koehler, 1901)	600–4315	mb,sb																			x
Ophiura (Ophiuroglypha) irrorata irrorata (Lyman, 1878)													x								
Ophiura ljungmani (Lyman, 1878)	46–6398	mb,rb,sb	x										x								x
Ophiura lymani (Ljungman, 1871)	0–463													x	x	x					

(continued)

Table A.2 (continued)

	Depth (m)	Habitat	MEX	BEL	GUA	HON	NIC	CRC	PAN	COL	VENc	VENa	BRA	URG	ARG	MAV	CUB	HAI	RDO	PRI	CAN
Ophiura ophiura (Linnaeus, 1758)	0–300	mb,sb																			x
Ophiura sarsi (Lütken, 1855)	82–827	mb	x																		
Ophiura scomba Paterson, 1985	358–3943	rb,sb	x																		
Ophiura tenera (Lyman, 1883)	158–500	mb								x											
Ophiuraster perissus H.L. Clark, 1939													x								
Class Echinoidea																					
Order Cidaroida																					
Family Cidaridae																					
Austrocidaris canaliculata (A. Agassiz, 1863)	1–424														x	x					
Austrocidaris lorioli (Mortensen, 1903)	160–1081														x						
Austrocidaris spinulosa Mortensen, 1910	124–641														x	x					
Calocidaris micans (Mortensen, 1903)	33–624	mb,rb,sb	x														x				
Cidaris abyssicola (A. Agassiz, 1869)	225–375	mb,rb,sb	x																		
Cidaris blakei (A. Agassiz, 1878)	270–720	mb,rb															x				x
Cidaris cidaris (Linnaeus, 1758)	50–2000	mb															x				x
Cidaris rugosa (H. L. Clark, 1907)	40–540	mb,rb,sb	x	x							x		x								
Eucidaris tribuloides tribuloides (Lamarck, 1816)	0–1500	cr,rb,sb,sg,rub	x	x	x	x	x	x	x	x	x		x				x	x	x	x	
Histocidaris nuttingi Mortensen, 1926	225–740	mb,rb,sb	x														x				
Histocidaris purpurata (Wyville-Thomson, 1872)	300–1800																				x
Histocidaris sharreri (A. Agassiz, 1880)	200 – 740	rb,sb	x														x				

(continued)

Table A.2 (continued)

	Depth (m)	Habitat	MEX	BEL	GUA	HON	NIC	CRC	PAN	COL	VENc	VENa	BRA	URG	ARG	MAV	CUB	HAI	RDO	PRI	CAN
Stereocidaris ingolfiana Mortensen, 1903	300–1745	mb,rb,sb	x														x				
Stylocidaris affinis (Philippi, 1845)	22–1000	cr,mb,rb,sb,rub	x				x		x	x	x		x				x				x
Stylocidaris lineata Mortensen, 1910	20–1717	cr,mb,rb,sb,rub	x			x	x			x	x		x				x		x		
Tretocidaris bartletti (A. Agassiz, 1880)	48–1089	mb,rb,sb	x			x				x	x	x	x						x		
Family Rhabdicidaridae																					
Porocidaris purpurata (Wyville-Thomson, 1872)	640–782	rb,sb	x				x														
Order Echinothurioida																					
Family Echinothuriidae																					
Araeosoma belli Mortensen, 1903	130–1020	mb,sb	x				x			x	x						x		x		
Araeosoma fenestratum (Thomson, 1872)	16–1180	mb,rb,sb	x							x							x		x		x
Calveriosoma hystrix (Wyville-Thomson, 1872)	360–2545	mb,sb	x															x			x
Hygrosoma petersii (A. Agassiz, 1880)	200–3700	mb,sb	x														x		x		
Phormosoma placenta placenta Thomson, 1872	50–3700	mb,sb	x			x	x		x	x	x							x			x
Phormosoma placenta sigsbei A. Agassiz, 1880	200–1800	mb															x	x			
Sperosoma grimaldii Koehler, 1897	300–2300																				x
Order Diadematoida																					
Family Aspidodiadematidae																					
Aspidodiadema jacobyi A. Agassiz, 1880	170–720	cr,mb,rb	x	x			x			x							x		x		
Aspidodiadema tonsum A. Agassiz, 1879	180–925												x								
Plesiodiadema antillarum (A. Agassiz, 1880)	684–3111	mb,rb,sb	x								x		x				x	x			x
Family Diadematidae																					
Astropyga magnifica A. H. Clark, 1934	5–89	cr,rb,sb,sg,rub	x						x	x	x						x		x	x	

(continued)

Table A.2 (continued)

	Depth (m)	Habitat	MEX	BEL	GUA	HON	NIC	CRC	PAN	COL	VENc	VENa	BRA	URG	ARG	MAV	CUB	HAI	RDO	PRI	CAN
Astropyga nuptialis Tommasi, 1958	1–20												x								
Centrostephanus besnardi Bernasconi, 1955													x								
Centrostephanus longispinus (Philippi, 1845)	10–1000	cr,mb,rb,sb,rub	x							x											x
Centrostephanus longispinus rubicingulus H. L. Clark, 1921	33–842	sb,rub									x						x		x		
Diadema antillarum antillarum (Philippi, 1845)	0–800	cr,m,ro,sb,sg,rub	x	x	x	x	x	x	x	x	x		x				x	x	x	x	x
Diadema ascensionis Mortensen, 1909	1–400												x								
Order Pedinoida																					
Family Pedinidae																					
Caenopedina cubensis A. Agassiz, 1869	220–1200	mb,rb,sb	x	x													x				x
Order Salenioida																					
Family Saleniidae																					
Salenia goesiana Lovén, 1874	90–842	rb															x		x		
Salenocidaris profundi (Duncan, 1877)	200–3700	mb,rb,sb	x														x	x			
Salenocidaris varispina A. Agassiz, 1869	250–3015	mb,rb,sb	x								x		x				x	x			x
Order Arbacioida																					
Family Arbaciidae																					
Arbacia dufresnii (Blainville, 1825)	0–340	rb,rub												x	x	x					
Arbacia lixula (Linnaeus, 1758)	0–50	rb,sb											x								x
Arbacia punctulata (Lamarck, 1816)	0–225	cr,rb,sb,sg,rub	x	x					x	x	x						x		x		
Coelopleurus floridanus A. Agassiz, 1871	60–2380	cr,mb,rb,sb	x	x					x	x	x						x		x		x
Habrocidaris scutata (A. Agassiz, 1880)	920–1400																x				
Podocidaris sculpta A. Agassiz, 1869	8–780	cr,mb,ro,sb	x														x				

(continued)

Table A.2 (continued)

	Depth (m)	Habitat	MEX	BEL	GUA	HON	NIC	CRC	PAN	COL	VENc	VENa	BRA	URG	ARG	MAV	CUB	HAI	RDO	PRI	CAN
Order Camarodonta																					
Family Parechinidae																					
Loxechinus albus (Molina, 1782)	30–340	rb,rub													x	x					
Paracentrotus gaimardi (Blainville, 1825)	1–5												x								
Paracentrotus lividus (Lamarck, 1816)	0–80	rb,sg																			x
Psammechinus microtuberculatus (Blainville, 1825)																					x
Order Temnopleuroida																					
Family Temnopleuridae																					
Genocidaris maculata A. Agassiz, 1869	12–500	cr,mb,rb,sb	x							x	x										x
Pseudechinus magellanicus (Philippi, 1857)	1–361	rb,rub												x	x	x					
Trigonocidaris albida A. Agassiz, 1869	70–720	cr,rb,sb	x							x									x		
Family Toxopneustidae																					
Lytechinus callipeplus H. L. Clark, 1912	22–357	rb,sb	x							x											
Lytechinus euerces H. L. Clark, 1912	40–777	mb,rb,sb	x	x														x	x		
Lytechinus variegatus carolinus A. Agassiz, 1863	0–250	rb,sb,rub,sg															x				
Lytechinus variegatus variegatus (Lamarck, 1816)	0–580	cr,mb,rb,sb,sg,rub	x	x	x	x		x	x	x	x		x				x	x	x	x	
Lytechinus williamsi Chesher, 1968	3–256	cr,rb,sb,sg,rub	x	x					x	x							x	x	x		
Pseudoboletia occidentalis H. L. Clark, 1921	69–155										x										
Sphaerechinus granularis (Lamarck, 1816)	3–100	rb																			x
Toxopneustes pileolus (Lamarck, 1816)	0–35	cr,mb,rb,sb	x																		

(continued)

Table A.2 (continued)

	Depth (m)	Habitat	MEX	BEL	GUA	HON	NIC	CRC	PAN	COL	VENc	VENa	BRA	URG	ARG	MAV	CUB	HAI	RDO	PRI	CAN
Tripneustes ventricosus (Lamarck, 1816)	0–842	cr,rb,sb,sg,rub	x	x		x		x	x	x	x		x				x	x	x	x	
Order Echinoida																					
Family Echinidae																					
Echinus esculentus Linneaeus, 1758	10–289	rb	x																		
Echinus tylodes H. L. Clark, 1912	350–760	sb	x																		
Gracilechinus gracilis (A. Agassiz, 1869)	69–457	mb,rb,sb	x								x						x				
Sterechinus agassizii Mortensen, 1910	24–470														x	x					
Sterechinus neumayeri (Meissner, 1900)	13–250															x					
Family Echinometridae																					
Echinometra lucunter lucunter (Linnaeus, 1758)	0–45	cr,rb,sg,rub	x	x	x	x	x	x	x	x	x		x				x	x	x	x	
Echinometra viridis A. Agassiz, 1863	0–40	cr,m,rb,sb,sg,rub	x	x	x	x		x	x	x	x						x	x	x	x	
Order Holectypoida																					
Family Echinoneidae																					
Echinoneus cyclostomus Leske, 1778	0.5–585	cr,mb,rb,sb,sg,rub	x	x		x		x	x	x	x						x	x	x		
Order Clypeasteroida																					
Family Clypeasteridae																					
Clypeaster aloysioi (Brito, 1959)													x								
Clypeaster chesheri Serafy, 1970	18–101	cr,mb,sb	x			x			x	x									x		
Clypeaster cyclopilus H. L. Clark, 1941	69–460	sb									x						x				
Clypeaster euclastus H. L. Clark, 1941	36–530	cr,mb,sb				x				x	x	x					x		x		
Clypeaster lamprus H. L. Clark, 1914	78–450	cr,mb,sb				x	x			x	x		x						x		

(continued)

Table A.2 (continued)

	Depth (m)	Habitat	MEX	BEL	GUA	HON	NIC	CRC	PAN	COL	VENc	VENa	BRA	URG	ARG	MAV	CUB	HAI	RDO	PRI	CAN
Clypeaster luetkeni Mortensen, 1948	9–55	sb								x								x			
Clypeaster oliveirai Krau, 1952	50	sb											x								
Clypeaster prostratus Ravenel, 1848	15–75	sb	x			x				x	x										
Clypeaster ravenelii A. Agassiz, 1869	20–278	mb,sb	x							x			x								
Clypeaster rosaceus (Linnaeus, 1758)	0–285	sb,sg	x	x		x			x	x	x		x				x	x	x	x	
Clypeaster speciosus Verrill, 1870								x													
Clypeaster subdepressus (Gray, 1825)	1–210	mb,rb,sb,sg,rub	x	x					x	x	x		x				x		x		
Family Echinocyamidae																					
Echinocyamus grandiporus Mortensen, 1907	99–2500	mb,sb	x							x			x				x				
Echinocyamus macrostomus Mortensen, 1907	170–2286	sb	x														x				
Echinocyamus pusillus (Müller, 1776)	0–1250	sb,rub																			x
Family Mellitidae																					
Encope aberrans Martens, 1867	13–32	sb	x																		
Encope emarginata (Leske, 1778)	0–60	mb,sb		x	x				x	x	x		x	x	x						
Encope michelini L. Agassiz, 1841	0–240	sb,sg	x								x						x				
Leodia sexiesperforata (Leske, 1778)	0–180	sb,sg,rub	x	x		x		x	x	x	x		x	x			x		x	x	
Mellita quinquiesperforata (Leske, 1778)	0–180	mb,sb,rub	x			x	x	x	x	x			x	x			x		x	x	
Mellita quinquiesperforata latiambulacra (H. L. Clark, 1840)	0–10	sb									x										
Mellita tenuis H. L. Clark, 1940	0–3	sb,rub	x																		
Order Cassiduloida																					
Family Cassidulidae																					

(continued)

Table A.2 (continued)

	Depth (m)	Habitat	MEX	BEL	GUA	HON	NIC	CRC	PAN	COL	VENc	VENa	BRA	URG	ARG	MAV	CUB	HAI	RDO	PRI	CAN
Cassidulus caribaearum Lamarck, 1801	0.5–197	sb	x	x					x										x		
Cassidulus infidus Mortensen, 1948	3.5												x								
Cassidulus mitis Krau, 1954	1–45												x								
Eurhodia relicta Mooi, 1990	69–155										x										
Family Echinolampadidae																					
Conolampas sigsbei (A. Agassiz, 1878)	130–800	mb,rt,sb	x			x	x										x		x		
Echinolampas depressa Gray, 1851	37–366	sb	x								x						x				
Order Holasteroida																					
Family Pourtalesiidae																					
Pourtalesia miranda A. Agassiz, 1869	450–5850																x				
Order Spatangoida																					
Family Asterostomatidae																					
Archaeopneustes hystrix (A. Agassiz, 1880)	22–1610	mb				x	x			x							x				
Family Brissidae																					
Anabrissus damesi (A. Agassiz, 1881)	640												x								
Brissopsis alta Mortensen, 1907	181–329	mb,sb	x																		
Brissopsis atlantica Mortensen, 1907	18–641	mb,sb	x			x	x	x	x	x	x		x				x				
Brissopsis atlantica mediterranea Mortensen, 1913	37–3200	mb								x	x							x			x
Brissopsis elongata elongata Mortensen, 1907	3–270	mb,sb		x		x			x	x	x	x							x	x	
Brissus unicolor (Leske, 1778)	0–250	mb,rb,sb,rub	x	x		x		x	x	x	x						x	x	x	x	x
Linopneustes longispinus (A. Agassiz, 1878)	55–720	mb				x	x			x							x		x		
Meoma ventricosa ventricosa (Lamarck, 1816)	0.3–293	cr,mb,sb,rb,sg,rub	x	x	x	x	x	x	x	x	x		x				x	x	x	x	
Neopneustes micrasteroides (A. Agassiz, 1878)	148–330																x				

(continued)

Table A.2 (continued)

	Depth (m)	Habitat	MEX	BEL	GUA	HON	NIC	CRC	PAN	COL	VENc	VENa	BRA	URG	ARG	MAV	CUB	HAI	RDO	PRI	CAN
Palaeobrissus hilgardi A. Agassiz, 1880	150–1025	mb								x									x		
Paleopneustes cristatus A. Agassiz, 1873	69–805	cr,mb,rb,sb				x				x	x						x		x		
Paleopneustes josephinae Lovén, 1872	350–436																x				
Paleopneustes tholoformis Chesher, 1968	90–525	cr,mb								x									x		
Plagiobrissus grandis (Gmelin, 1788)	1–210	sb,sg	x	x					x	x	x		x				x		x	x	
Plethotaenia angularis Chesher, 1968	567–570	mb,sb	x																		
Plethotaenia spatangoides (A. Agassiz, 1883)	117–619	mb,sb	x				x										x				
Rhabdobrissus costae (Gasco, 1876)	25–200	mb,sb																			x
Rhynobrissus cuneus Cooke, 1957	0–10	sb	x																		
Family Hemiasteridae																					
Holanthus expergitus (Lovén, 1874)	380–4833	mb,sb									x						x				x
Family Loveniidae																					
Echinocardium cordatum (Pennant, 1777)	0–230	sb											x								x
Family Maretiidae																					
Homolampas fragilis (A. Agassiz, 1869)	350–3550	mb,sb	x							x	x		x				x				
Homolampas lovenioides Mortensen, 1948	910												x								
Family Prenasteridae																					
Tripylus excavatus Philippi, 1845	50–113														x	x					
Family Schizasteridae																					
Abatus agassizii (Pfeffer, 1889)	75–970															x					
Abatus cavernosus (Philippi, 1845)	1–676	mb,sb													x	x					
Abatus philippii Lovén, 1871	71–225														x	x					

(continued)

Table A.2 (continued)

	Depth (m)	Habitat	MEX	BEL	GUA	HON	NIC	CRC	PAN	COL	VENc	VENa	BRA	URG	ARG	MAV	CUB	HAI	RDO	PRI	CAN
Aceste bellidifera Thomson, 1877	116–5220	sb									x						x				x
Agassizia excentrica A. Agassiz, 1869	27–900	cr,mb,sb	x	x			x		x	x	x	x	x				x		x	x	
Brisaster fragilis (Düben & Koren, 1846)					x																
Brisaster moseleyi (A. Agassiz, 1881)	401–1100															x					
Hypselaster limicolus (A. Agassiz, 1878)	27–340	mb,sb	x			x				x											
Moira atropos clotho Michelin, 1855	0–445	mb,rb,sb	x	x		x			x	x	x		x				x		x	x	
Ova canaliferus (Lamarck, 1816)	9–100	mb,sb																			x
Tripylaster philippii (Gray, 1851)	13–595														x	x					
Schizaster doederleini (Chesher, 1972)	9–220	mb,sb	x	x		x	x		x			x					x		x		
Schizaster floridiensis (Chesher, 1972)	12–65	sb		x		x			x		x	x	x						x		
Schizaster orbignyanus A. Agassiz, 1880	22–500	mb,sb								x	x	x					x				
Family Spatangidae																					
Spatangus purpureus (Müller, 1776)	15–900	mb,sb,rub																			x
Class Holothuroidea																					
Order Dendrochirotida																					
Family Cucumariidae																					
Aslia lefevrei (Barrois, 1882)	6–25	rb																			x
Cladodactyla crocea (Lesson, 1830)	0–4300	mb,rb,sb,rub													x	x					
surinamensis (Semper, 1868)	0–5	rb,sg							x	x	x		x				x		x	x	
Cucumaria manoelina Tommasi, 1971													x								

(continued)

Table A.2 (continued)

	Depth (m)	Habitat	MEX	BEL	GUA	HON	NIC	CRC	PAN	COL	VENc	VENa	BRA	URG	ARG	MAV	CUB	HAI	RDO	PRI	CAN
Duasmodactyla seguroensis (Deichmann, 1930)	0–50	cr,rb,sb,sg	x								x		x							x	
Euthyonacta solida (Deichmann, 1930)	1.5–183	rb,sb	x								x										
Hemioedema spectabilis (Ludwig, 1882)	2–160	mb,sb													x						
Heterocucumis godeffroyi (Semper, 1868)	0–379	sb,rub													x						
Heterocucumis steineni (Ludwig, 1898)	0–400	rb													x	x					
Leptopentacta deichmannae Domantay, 1958	17–37	mb,sb,rub								x		x									
Neocnus incubans Cherbonnier, 1972	0–5	rb																			x
Ocnus braziliensis (Verrill, 1868)	0–2												x								
Ocnus pygmaeus (Théel, 1886)	0–72	cr,rb,sb,rub	x								x		x							x	
Ocnus suspectus (Ludwig, 1874)	0–1874	rb,sg	x	x					x	x	x						x			x	
Paracolochirus mysticus (Deichmann, 1930)	69–155										x										
Pawsonia saxicola (Brady & Robertson, 1871)	4–130	rb																			x
Pentacta pygmaea (Théel, 1886)	0.6–1.7										x		x								
Pseudocnus dubiosus leoninus (Semper, 1868)	0–340	mb,rb,sb													x	x					
Pseudocnus cornutus (Cherbonnier, 1941)	135–189														x	x					
Pseudocnus perrieri (Ekman, 1927)	0–197	mb,rb,sb												x	x	x					
Pseudrotasfer microincubator Bohn, 2007	271–290														x						
Stereoderma unisemita (Stimpson, 1851)	31.1–40.2												x								
Thyonella gemmata (Pourtalès, 1851)	0–64	cr,rb,sb,mb,sg	x						x								x				

(continued)

Table A.2 (continued)

	Depth (m)	Habitat	MEX	BEL	GUA	HON	NIC	CRC	PAN	COL	VENc	VENa	BRA	URG	ARG	MAV	CUB	HAI	RDO	PRI	CAN
Thyonella pervicax (Théel, 1886)	9–67	rb,sb	x						x				x								
Thyonella sabanillaensis (Deichmann, 1930)	2–30	sb								x	x		x								
Trachythyone crassipeda Cherbonnier, 1961													x								
Trachythyone lechleri (Lampert, 1885)	0–30														x						
Trachythyone parva (Ludwig, 1874)	0–180	mb,rb,sb													x	x					
Trachythyone peruana (Semper, 1868)	90–159														x	x					
Family Phyllophoridae																					
Havelockia inermis (Heller, 1868)	0.5–60								x		x										
Neothyonidium parvum (Ludwig, 1881)	0–3	sg							x		x		x							x	
Pentamera chiloensis (Ludwig, 1887)	0–111	sb,rub													x						
Pentamera pulcherrima Ayres, 1854	0–60	mb								x	x		x								
Phyllophorus (Urodemella) occidentalis (Ludwig, 1875)	0–99	cr,rb,sg	x						x		x		x				x			x	
Stolus cognatus (Lampert, 1885)	0–38	mb,rb,sb,sg	x						x	x	x		x				x				
Thyone cognata (Lampert, 1885)											x							x			
Thyone deichmannae Madsen, 1941	6–366	sg															x				
Thyone fusus (Müller, 1776)	5–615	mb,sb,rub																			x
Thyone montoucheti Tommasi, 1971													x								
Thyone pawsoni Tommasi, 1972	6–51	sb								x	x		x								
Thyone pseudofusus Deichmann, 1930	0.5–46	rb,sb,sg,rub	x	x					x	x	x		x								
Thyone tanyspeira Pawson & Miller, 1988	51–170	mb				x				x											
Family Psolidae																					

(continued)

Table A.2 (continued)

	Depth (m)	Habitat	MEX	BEL	GUA	HON	NIC	CRC	PAN	COL	VENc	VENa	BRA	URG	ARG	MAV	CUB	HAI	RDO	PRI	CAN
Lissothuria antillensis Pawson, 1967	1–17	cr,rb	x	x							x										
Lissothuria braziliensis (Théel, 1886)	0–2										x		x								
Neopsolidium convergens (Herouard, 1901)	0–20	rb													x	x					
Psolidium dorsipes Ludwig, 1886	0–483	rb													x	x					
Psolus antarcticus Philippi, 1857	35–1080	rb													x						
Psolus marcusi Tommasi, 1971	95	rb													x						
Psolus patagonicus Ekman, 1925	0–430	rb											x		x	x					
Psolus squamatus (Koren, 1844)	7–207	rb,rub													x	x					
Psolus tuberculosus Théel, 1886	73–243										x							x			
Psolus victoriae Tommasi, 1971													x								
Family Sclerodactylidae																					
Euthyonidiella dentata Cherbonnier, 1961											x		x								
Euthyonidiella destichada (Deichmann, 1930)	0–7	mb,rb,sb,sg		x					x		x						x			x	
Euthyonidiella trita (Sluiter, 1910)	0–100	rb	x								x									x	
Pseudothyone belli (Ludwig, 1886)	0–37	mb,rb,sb,sg	x	x					x		x		x							x	
Sclerodactyla briareus (Lesueur, 1824)	15										x										
Order Dactylochirotida																					
Family Ypsilothuriidae																					
Echinocucumis asperrima (Théel, 1886)	43–723																x				
Echinocucumis hispida (Barret, 1856)	50–1000																x				

(continued)

Table A.2 (continued)

	Depth (m)	Habitat	MEX	BEL	GUA	HON	NIC	CRC	PAN	COL	VENc	VENa	BRA	URG	ARG	MAV	CUB	HAI	RDO	PRI	CAN
Ypsilothuria talismani talismani Perrier, 1886	270–2684	mb								x							x				
Order Aspidochirotida																					
Family Holothuriidae																					
Actinopyga agassizii (Selenka, 1867)	0–54	cr,rb,sg,rub	x	x	x	x		x	x	x							x	x	x	x	
Holothuria (Cystipus) cubana Ludwig, 1875	0–8	cr,mb,rɔ,sb	x	x					x	x	x						x		x	x	
Holothuria (Cystipus) occidentalis Ludwig, 1875	0–457	mb,sb	x			x				x	x	x					x		x		
Holothuria (Cystipus) pseudofossor Deichmann, 1930	3–370	cr,mb,sb,rub	x	x						x	x		x				x		x		
Holothuria (Halodeima) floridana Pourtalès, 1851	0–10	cr,m,rb,sb,sg,rub	x	x		x			x	x							x	x	x	x	
Holothuria (Halodeima) grisea Selenka, 1867	0–30	cr,mb,rɔ,sb,sg	x	x					x	x	x		x				x	x	x	x	
Holothuria (Halodeima) mexicana Ludwig, 1875	0–183	cr,m,mɔ,rb,sb,sg	x	x	x	x	x	x	x	x	x						x	x	x	x	
Holothuria (Holothuria) dakarensis Panning, 1939	0–15	mb,sb																			x
Holothuria (Holothuria) helleri Marenzeller, 1878	15	sb,rub																			x
Holothuria (Holothuria) mammata Grube, 1840	1–77	rb																			x
Holothuria (Holothuria) tubulosa Gmelin, 1788	0–30	mb,sb,sg																			x
Holothuria (Platyperona) parvula (Selenka, 1867)	0–8	cr,m,mɔ,rb,sb,rub	x							x	x		x						x	x	
Holothuria (Platyperona) rowei Pawson & Gust, 1981				x																	
Holothuria (Platyperona) sanctori Delle Chiaje, 1823	0–30	rb																			x
Holothuria (Panningothuria) forskali Delle Chiaje, 1823	3–193	rb																			x
Holothuria (Roweothuria) arguinensis Koehler & Vaney, 1906	0–30	rb																			x
Holothuria (Roweothuria) polii Delle Chiaje, 1823	0–250	mb,sb,sg																			x

(continued)

Table A.2 (continued)

	Depth (m)	Habitat	MEX	BEL	GUA	HON	NIC	CRC	PAN	COL	VENc	VENa	BRA	URG	ARG	MAV	CUB	HAI	RDO	PRI	CAN
Holothuria (Semperothuria) surinamensis Ludwig, 1875	0–42	cr,m,mb,rb,sb,sg,rub	x	x						x	x		x				x		x	x	
Holothuria (Selenkothuria) glaberrima Selenka, 1867	0–42	mb,rb,sb	x			x			x	x	x						x	x	x	x	
Holothuria (Theelothuria) princeps Selenka, 1867	0–402	cr,mb,rb,sb,sg	x							x	x						x		x	x	
Holothuria (Thymiosycia) arenicola Semper, 1868	0–121	cr,m,mb,sb,sg	x	x	x	x			x	x	x		x				x		x	x	
Holothuria (Thymiosycia) impatiens (Forskaal, 1775)	0–27.4	cr,mb,rb,sb,sg	x	x					x	x	x						x			x	
Holothuria (Thymiosycia) rathbunii Lampert, 1885																				x	
Holothuria (Thymiosycia) thomasi Pawson & Caycedo, 1980	0–30	cr	x	x		x			x	x							x	x	x	x	
Holothuria (Vaneyothuria) lentiginosa von Marenzeller, 1892	20–466	mb,rb,sb	x										x				x				x
Holothuria (Vaneyothuria) lentiginosa enodis Miller & Pawson, 1979	69–466	cr,mb								x											
Family Stichopodidae																					
Astichopus multifidus (Sluiter, 1910)	1–162	cr,rb,sb,sg	x						x	x							x		x	x	
Eostichopus arnesoni Cutress & Miller, 1982	36	sb			x															x	
Eostichopus regalis (Cuvier, 1817)	5–800	mb,sb,sg,rub															x				x
Isostichopus badionotus (Selenka, 1867)	0–274	cr,mb,rb,sb,sg	x	x		x			x	x	x		x				x	x	x	x	
Isostichopus macroparentheses (H. L. Clark, 1922)	0–18	rb	x	x																	
Parastichopus tremulus (Gunnerus, 1767)	20–1960	mb,sb																			x
Family Synallactidae																					
Amphigymnas bahamensis Deichman, 1930	439–900	mb	x							x							x				

(continued)

Table A.2 (continued)

	Depth (m)	Habitat	MEX	BEL	GUA	HON	NIC	CRC	PAN	COL	VENc	VENa	BRA	URG	ARG	MAV	CUB	HAI	RDO	PRI	CAN
Bathyplotes bigelowi Deichmann, 1940	402–585	mb															x				
Bathyplotes natans (M. Sars, 1868) Östergren, 1896	210–1600	cr,mb	x							x											
Bathyplotes pourtalesi (Théel, 1886)	134–2919	mb								x							x				x
Hansenothuria benti Miller & Pawson, 1989	548–903	cr,sb	x																		
Meseres atlanticus (R. Perrier, 1902)											x										
Meseres occultatus (von Marenzeller, 1893)	232–2180	mb								x											
Mesothuria connectens (Perrier, 1898)	1975–2515																				x
Mesothuria gargantua Deichmann, 1930	180–720	mb															x				
Mesothuria intestinalis (Ascanius, 1805)	18–1445	mb	x																		x
Mesothuria (Zygothuria) lactea (Théel, 1886)	484–5100	mb	x							x	x						x				
Mesothuria maroccana Perrier, 1902	914–5062	mb									x						x				
Mesothuria (Mesothuria) rugosa Herouard, 1912	151–3890	mb								x											
Mesothuria (Penichrothuria) verrilli (Théel, 1886)	699–4060	mb								x			x				x				x
Molpadiodemas depressus (Herouard, 1902)	1353–5690												x								
Paelopatides gigantea (Verrill, 1884)	1100–4060																				x
Pseudostichopus occultatus Merenzeller, 1893	232	mb,sb															x				
Pseudostichopus peripatus (Sluiter, 1901)	3411–3459										x										
Scotothuria herringi Hansen, 1978																					x
Synallactes crucifera Perrier, 1898	1000–3476	mb									x						x				

(continued)

Table A.2 (continued)

	Depth (m)	Habitat	MEX	BEL	GUA	HON	NIC	CRC	PAN	COL	VENc	VENa	BRA	URG	ARG	MAV	CUB	HAI	RDO	PRI	CAN
Order Elasipodida																					
Family Deimatidae																					
Deima validum validum Théel, 1879	914–2780	mb									x		x				x				
Orphnurgus asper Théel, 1882	600–1049	mb															x				
Family Elpidiidae																					
Peniagone purpurea (Théel, 1882)	2800–3000																				x
Family Pelagothuridae																					
Enypniastes eximia Théel, 1882	461–689									x											
Family Psychropotidae																					
Benthodytes lingua Perrier, 1896	540–2200	mb															x				x
Benthodytes sanguinolenta Théel, 1882	914–3100	mb	x			x				x											
Benthodytes typica Théel, 1882	315–5046	mb									x						x	x			x
Benthodytes valdiviae Hansen, 1975																					x
Psychropotes depressa (Théel, 1882)	955–4060	mb															x				x
Psychropotes longicauda Théel, 1882	2210–5000																				x
Psychropotes scotiae (Vaney, 1908)																	x				
Psychropotes semperiana Théel, 1882	1143–5600										x										x
Order Molpadiida																					
Family Caudinidae																					
Acaudina molpadioides (Semper, 1867)	3545–3635	mb	x																		
Paracaudina chilensis chilensis (Müller, 1850)	0–990	mb,sb,rub											x	x	x						
Family Molpadiidae																					

(continued)

Table A.2 (continued)

	Depth (m)	Habitat	MEX	BEL	GUA	HON	NIC	CRC	PAN	COL	VENc	VENa	BRA	URG	ARG	MAV	CUB	HAI	RDO	PRI	CAN
Molpadia barbouri Deichman, 1940	440–1529	mb								x							x				
Molpadia blakei (Théel, 1886)	3482–3518	mb								x	x										
Molpadia cubana Deichmann, 1940	24–1464	mb							x	x							x				
Molpadia eltaninae Pawson, 1977	20–855														x	x					
Molpadia musculus Risso, 1826	35–5205	mb	x							x			x		x	x	x				x
Molpadia oolitica (Pourtalès, 1851)	42–1440	mb								x											
Molpadia parva (Théel, 1886)	125–2844	mb	x						x	x							x				
Order Apodida																					
Family Chiridotidae																					
Chiridota marenzelleri R. Perrier, 1904	40–159	mb,rub												x	x						
Chiridota rotifera (Pourtalès, 1851)	0–360	cr,rb,sb,sg	x	x							x		x				x		x	x	
Taeniogyrus contortus (Ludwig, 1875)	82–200	mb,sb													x	x					
Trochodota purpurea (Lesson, 1830)	0–100	sb,rub													x	x					
Family Synaptidae																					
Anapta fallax Lampert, 1889	0–350	sb,rub													x	x					
Epitomapta roseola (Verrill, 1874)	1.5–40	rb,sb	x	x					x												
Eupatinapta acanthia (A. H. Clark, 1899)	10.7								x												
Euapta lappa (Müller, 1850)	0–256	cr,rb,sb,sg	x	x				x	x	x	x						x	x	x	x	x
Leptosynapta brasiliensis Freire & Grohmann, 1989	1–4												x								
Leptosynapta imswe Pawson, 1976	0–2			x																	
Leptosynapta inhaerens (Müller, 1776)	0–173	mb																			x

(continued)

Table A.2 (continued)

	Depth (m)	Habitat	MEX	BEL	GUA	HON	NIC	CRC	PAN	COL	VENc	VENa	BRA	URG	ARG	MAV	CUB	HAI	RDO	PRI	CAN
Leptosynapta multigranula H. L. Clark, 1924	0.5–2.9	sb	x	x																	
Leptosynapta nannoplax Pawson, 1976				x																	
Leptosynapta parvipatina H. L. Clark, 1924	9–12								x												
Leptosynapta roseogradia Pawson, 1976	0.5			x																	
Leptosynapta tenuis (Ayres, 1851)	0.5–167		x								x										
Protankyra benedeni (Ludwig, 1881)	5–100										x		x								
Protankyra brychia (Verrill, 1885)													x								
Protankyra ramiurna Heding, 1928	0–18	mb,rb,sb,sg	x																	x	
Synaptula hydriformis (Lesueur, 1824)	0–10	cr,m,sb,sg	x	x		x	x	x	x	x	x		x				x			x	
Synaptula secreta Ancona-Lopez, 1957	1–10												x								

Habitat classification: *cr* Coral Reefs, *m* mangroves, *mb* muddy bottom, *rb* rocky bottom, *sb* sandy bottom, *sg* seagrass, *rub* rubble bottom. Countries abbreviation: *MEX* Mexico, *BEL* Belize, *GUA* Guatemala, *HON* Honduras, *NIC* Nicaragua, *CRC* Costa Rica, *PAN* Panama, *COL* Colombia, *VENc* Venezuela Caribbean, *VENa* Venezuela Atlantic, *BRA* Brazil, *URG* Uruguay, *ARG* Argentina, *MAV* Malvinas Islands, *CUB* Cuba, *HAI* Haiti, *RDO* Dominican Republic, *PRI* Puerto Ricc, *CAN* Canary Islands

Index

J. J. Alvarado and F. A. Solís-Marín (eds.), *Echinoderm Research and Diversity in Latin America*, DOI: 10.1007/978-3-642-20051-9,

MIX
Papier aus verantwortungsvollen Quellen
Paper from responsible sources
FSC® C105338

If you have any concerns about our products,
you can contact us on
ProductSafety@springernature.com

In case Publisher is established outside the EU,
the EU authorized representative is:
Springer Nature Customer Service Center GmbH
Europaplatz 3, 69115 Heidelberg, Germany

Printed by Libri Plureos GmbH
in Hamburg, Germany